Solid Mechanics and Its Applications

Volume 194

Series Editor

G.M.L. Gladwell
Department of Civil Engineering, University of Waterloo, Canada

For further volumes:
http://www.springer.com/series/6557

Aims and Scope of the Series

The fundamental questions arising in mechanics are: *Why*?, *How*?, and *How much*? The aim of this series is to provide lucid accounts written by authoritative researchers giving vision and insight in answering these questions on the subject of mechanics as it relates to solids.

The scope of the series covers the entire spectrum of solid mechanics. Thus it includes the foundation of mechanics; variational formulations; computational mechanics; statics, kinematics and dynamics of rigid and elastic bodies: vibrations of solids and structures; dynamical systems and chaos; the theories of elasticity, plasticity and viscoelasticity; composite materials; rods, beams, shells and membranes; structural control and stability; soils, rocks and geomechanics; fracture; tribology; experimental mechanics; biomechanics and machine design.

The median level of presentation is the first year graduate student. Some texts are monographs defining the current state of the field; others are accessible to final year undergraduates; but essentially the emphasis is on readability and clarity.

Esben Byskov

Elementary Continuum Mechanics for Everyone

With Applications to Structural Mechanics

Springer

Esben Byskov
Department of Civil Engineering
Aalborg University
Aalborg
Denmark

ISSN 0925-0042
ISBN 978-94-007-5765-3 ISBN 978-94-007-5766-0 (eBook)
DOI 10.1007/978-94-007-5766-0
Springer Dordrecht Heidelberg New York London

Library of Congress Control Number: 2012954065

Printed on acid-free paper

Springer is part of Springer Science+Business Media (www.springer.com)

Preface

Why This Book?

It is my hope that the present book is the final edition of my lecture notes *Elementary Continuum Mechanics for Everyone.* The very first edition was written in connection with my teaching a first course on continuum mechanics at the Technical University of Denmark. I tried to find a text that would suit my intentions, but found that the books on the market either were filled with tensor analysis, including e.g. *curvilinear coordinates* and *Christoffel Symbols*, or they were of the old mechanics traditions, meaning that every new example was treated separately with the result that universally valid principles, such as the *Principle of Virtual Work*,[P.1] never appeared. In my opinion, if teaching continuum mechanics at any level is justified it must contain a strong element of general statements. Then, of course, there is the risk that the treatment becomes mathematically so difficult that it cannot serve as an introduction to the subject. Therefore, this book contains an elementary, but quite general, exposition of the subject, and it is my sincere expectation that most students—*with some effort, of course*—should be able to get a feel for the important concepts of *(generalized) strains*, *(generalized) stresses*, and the *Principle of Virtual Work*. Judging from the experience of many of my predecessors I may have set too high a goal, but still I shall try to reach it.

I feel that a solid understanding of almost any subject may only be obtained through examples, and this is one of the reasons why I have included Parts II–V, which may be viewed as applications of the theory of Part I. Of course, I consider the topics of Parts II–V to be important by themselves, but in the present context it is just as pertinent that they may be based on and illustrate concepts from Part I. Thus, an essential idea behind this book is to present continuum mechanics, not only as a valuable subject in its own right, but as the foundation for all our theories and methods governing the behavior of solids and structures. Although I endorse the use of examples it has been my experience that they tend to obscure the real subject by their sheer number and length. Therefore, I have limited the number of worked examples, except in Part III because there the examples provide useful information about properties of a number of cross-sections. I hope that I have found a reasonable balance in this regard.

[P.1] Later, we shall see that there are more than just one principle of virtual work, but at this point this is not important.

Who Should Read this Book?

The present book is meant as an introduction to continuum mechanics with applications in solid and structural mechanics and, obviously, engineering students are potential readers. It is, however, my hope that engineers who would like to achieve a better understanding of the theories they apply will appreciate this book.

Other Possible Topics

As regards topics not treated here there is a couple that I would have liked very much to include. An important one is *Elastic Fracture Mechanics.* The reason why I have left it out is that in order to do the job right one has to introduce solutions to some fundamental elasticity problems, which can almost only be found by use of complex functional analysis. If I were to introduce the methods of Muskhelishvili (1963) that particular subject would have taken up much more space than I consider feasible in the present context. The possibility of presenting the solutions without proof and then utilizing them as a basis for the theory of linear fracture mechanics may be justified under other circumstances, but not here, where the emphasis is different, namely on full derivations of all[P.2] formulas.

Another subject which might have been covered is perfect plasticity because results from computations by hand based on this assumption have been used over and over during the past sixty years or so. However, with few exceptions, computer analyses that consider the more realistic case of strain hardening have been made possible due to the small cost of personal computers and have therefore made the older methods redundant. In my opinion, today the most important application of perfect plasticity is *Johansen's Yield Line Theory,*[P.3] see e.g. (Johansen 1963) and (Johansen 1972). The beauty of his theory is that it may provide useful results with only a limited effort, even for rather complicated situations. On the other hand, derivation of the Yield Line Theory would take up more space than I consider reasonable in the present context.

My Writing Style

Readers of this book will soon discover that its style is somewhat unusual—probably an older reader will find it appalling, see also the Introduction, but I have my reasons for writing the way I do. I have tried to write in an informal style without sacrificing accuracy. A main idea behind my

[P.2] There is one exception to this, namely the formulas for imperfection sensitivity in Chapter 20. The reason for this is that the formulas by themselves are short, while their derivation is long and very complicated, see e.g. (Budiansky 1974) or (Christensen & Byskov 2010).

[P.3] Johansen himself never acknowledged that his yield line theory actually could be viewed as an example of upper bound solutions of the more general theory of perfect plasticity.

writing is that I attempt not to introduce any concept without first giving a motivation for its usefulness and relevance. I may not have succeeded, but this was my goal.

Many years ago when I was an undergraduate at the Technical University of Denmark I encountered the subject of mechanics in a physics course. It was never explained to us why the subject as a whole was important, and in particular there never was an indication of the reason behind the way the different topics were dealt with—other than the comment: "If you don't understand this, you don't belong here." In my case they may have been right, but at that time I did not feel that a statement like that showed much insight into how the human mind works, and today I think that teachers like that should not have been allowed to stand in front of a class. I intend to avoid such arrogance, but caution that the consequence in some cases is that the presentation will seem unnecessarily lengthy to some readers.

You might say that my ambition has been to explain "why—not just how."

In all probability it is not necessary to mention that English is not my mother tongue, but hopefully the meaning of my efforts is clear in spite of that.

If, on the other hand, the language of this book bears some resemblance to (American) English, my wonderful secretary at Aalborg University, Kirsten Aakjær, who has read all the pages[P.4] of a shorter, previous edition and has suggested many substantial improvements to my writing, deserves to be mentioned. Kirsten Aakjær favors British English, and many times she tried to convert every phrase into that language. Over the years, we have had discussions about the virtues of these two kinds of English, but have never obtained a complete agreement—only a friendly ceasefire.[P.5]

It is also worth mentioning that Kirsten Aakjær in many cases suggested more fundamental improvements to my writing, such as telling me that a particular example had a too abrupt ending.

Being a very stubborn person, I have sometimes chosen to follow my own instincts in spite of the advice by Kirsten Aakjær, so blame me—not her—for the errors.

As you will see later, in order that you may have a quick view of the topics I have put a lot of gray boxes in the margin.

Many gray boxes in margin

[P.4] It must have been a boring job considering the fact that the contents must have been alien to Kirsten Aakjær.

[P.5] In the mid-Seventies my family and I stayed in Massachusetts on a sabbatical leave at Harvard University. My son, Torben, who was five or six years old at that time, once proclaimed: "British English is bad English," which is a statement that I have often quoted since.

To be fair, our year in Massachusetts was the Bicentennial year, and the British were always the bad guys when the kindergarten kids played.

Open Source Programs

I could not have written this book without the use of open source and free programs such as LaTeX 2ε, BibTeX, makeindex, texindy,[P.6] gnuplot, maxima, Octave, xfig, gcc, and g++.

Possible Errors in This Book

I am sure that there must be some errors even in this edition. There is, unfortunately, nobody else but myself to blame: I have pressed the keys on the keyboard, I have moved and clicked the mouse, I have written the text, and I have drawn all figures and produced all the plots.

Esben Byskov
Jægersborg and Aalborg
August 14, 2012

[P.6] Just the thought of using one of the WYSIWIG, sometimes called WYSIAIG for "What You See Is All You Get," text editing programs brings sweat to my face.

Contents

Introduction

What Is Continuum Mechanics?

Quite trivially, continuum mechanics *per se* deals with the description of deformations of three-dimensional *continua*, i.e. models whose properties are independent of scale in that the continuum does not possess a structure.[I.7] Thus, continuum mechanics does not try to model the atomic structure of the involved materials—perhaps not even the crystalline, or spongy, or lumpy structure—but offers a "smeared-out" version of the real world. Also, the desired description depends very much on the needs of the discipline in question. In solid mechanics—*which is the sub-discipline of continuum mechanics that I treat in this book*—material elements that originally were neighbors remain neighbors throughout the lifetime of the material, with the possible exception of a crack forming and separating the neighbors. In fluid mechanics, on the other hand, it is necessary to be able to handle cases where neighboring material elements cease to be neighbors because they have been separated by other material elements that have moved in between them.

Continua—no atomic structure

In this book: Only solid mechanics, no fluid mechanics

Thus, continuum mechanics is many different things depending on the purpose. In the following chapters I give an introductory exposition of the continuum mechanics of solids.

Continuum mechanics is not just one thing

Many readers—in particular older ones—may feel that the emphasis is completely wrong and that subjects which are left out here are much more important than the ones included.

Also, some may think that in order to motivate the students I should have shown lots of results from computer analyses in the form of plots of *strains and stresses*[I.8] and from experiments. I refuse to believe that the students of today are incapable of understanding the importance and usefulness of a subject unless it is presented in terms of cartoons.[I.9] Furthermore, by showing such results in terms of colored plots the students might get the impression that in order to do an analysis you only have to get some commercial computer program, probably a Finite Element Program, and that

You must know the theory behind a commercial computer program before using it

[I.7] This does not preclude continuum theories that entail a *length scale* which is derived from analysis of materials that do have a structure, e.g. wood, fiber reinforced epoxy, and concrete. Such theories, which are more advanced than the theories described in this book, have become quite popular in recent years for analysis of material instabilities, such as "kinkbands" in wood and in fiber reinforced epoxy.

[I.8] These terms will be introduced later, see Part I, Chapter 2.

[I.9] Although I am a cartoon buff myself, I hope that I am correct in this belief.

therefore you do not have to know any theory. Nothing can be more wrong, as any experienced engineer or researcher knows—many mistakes in "the real world" are made by people who know too little about the theoretical background of the method they are using.

The Principle of Virtual Work ties strains and stresses together

Be that as it may, I have made some choices that I feel are proper—knowing that a lot of useful subjects will not be covered. But, and this is my strong opinion, in an introductory course on continuum mechanics it is important that the students get a feel for the nature of strains and stresses and learn the central role played by the *Principle of Virtual Work* rather than being able to juggle *Curvilinear Coordinate Systems*, *Co- and Contravariant Tensors*, *Christoffel Symbols*, *Mohr's Circles*, etc.

Continuum Mechanics: also a foundation for theories for beams, plates, shells etc.

The reader will observe that the same topic is introduced several times. The reason for this is that in a number of cases I considered it easier for the reader to become acquainted with some new concept via an application or an example rather than through general statements. However, I am certain that it should not be too difficult to find the place where the more concise statements or formulas are given.

The most central obligation[I.10] of continuum mechanics of solids is to provide a common basis, including a set of rather strict requirements, for *all* specialized theories, i.e. theories for bars, beams, plates, shells etc.[I.11]

"I Need Continuum Mechanics Like I Need Another Hole in My Head" "What's wrong with P over A?"

Why learn continuum mechanics?

The way the curriculum at the Department of Building Technology and Structural Engineering, Aalborg University, is set up—and to my experience, it is very much like many other curricula—the students have been acquainted with analysis of bars and beams, and possibly plates, as well as structures made up of such elements before they meet the more general continuum mechanics. So, a student might say "I need continuum mechanics like I need another hole in my head now that I have already learned to handle the most important structural elements as well as structures made up of such elements." That some professors share that way of thinking was told to me by one of my very good American friends[I.12] who said that at a meeting at his university[I.13] an older professor exclaimed "What's wrong with P over A?" when the subject of continuum mechanics including strain

[I.10] Again, some may disagree with me on this.

[I.11] I am very well aware that continuum mechanics in itself is a vast subject that has its own justification, but here I am thinking about our particular purpose of giving an introductory course.

[I.12] He shall remain nameless in this connection.

[I.13] The university shall also remain nameless.

and stress tensors was mentioned.

So why bother spending time on learning continuum mechanics? The answer to this contains at least two ingredients. First, it may not be all that obvious to the students that there is a common foundation for the theories for bars, beams, etc. Second, there are a lot of other important cases that have not been covered by the theories already learned, and if the student encounters such theories later it is imperative that he or she knows the foundation for any sound continuum mechanical theory, including the specialized ones. Otherwise, the consequences may be dire.

Continuum mechanics is a foundation for "all" our theories

The Main Emphasis of this Book

General principles

The main technical emphasis of this book is on general principles, most notably the *Principle of Virtual Displacements*[I.14] that govern deformation of structures and structural elements. I consider it just as important to explain why we are interested in a particular subject and even more so not just how, but rather why we establish our theories the way we do.

We may consider the three-dimensional continuum the only "real" model of our world, but at the same time we must acknowledge that in many cases more primitive models may be more feasible in terms of computational effort and clearness of results. On the other hand, establishing equations for specialized continua, such as beams and plates, on an *ad hoc* basis does not appeal to me for several reasons. Probably the most important is that you must be extremely careful—or lucky—to insure that such a theory obeys the general principles. The main stumbling block consists in the problem of ensuring that the *stress* and *strain* quantities are *generalized*, i.e. are work conjugate, see Section 2.4.2 and Chapter 33. If they are not, the fundamental principles, such as the *Principle of Minimum of the Potential Energy* and the upper and lower bounds of the theory of perfect plasticity—a fairly important subject[I.15] which is not dealt with in this book—do not apply. It is my sincere hope that after reading this book—and having done some work yourself—you will appreciate the importance of this.

Ad hoc theories may be dangerous

There is, however, one major disadvantage of this approach, namely that the generalized quantities, usually the generalized stresses, sometimes are difficult to interpret in a physically obvious way, but, as you will see, we succeed as regards most beams, see Section 7.3, and plates, see Section 9.1, but encounter problems with some other beams, see Section 7.4.

Interpretation of generalized stresses

[I.14] For elastic structures I use the *Principle of Minimum of the Potential Energy*, see Chapter 33, more frequently, in part because I think that in some connections this provides a more clear exposition, and in part because some other principles and methods, e.g. the *Rayleigh-Ritz Procedure*, Section 18.3.8, are based on this.

[I.15] Over the years, calculations by hand based on application of the theory of perfect plasticity have provided extremely useful solutions to engineering problems, but, to be fair, the theory makes very strong assumptions as regards the behavior of the materials, and today it is less important because of the advent of computers.

How to Read this Book

Read Part I first

I strongly suggest that you start reading Part I, which deals with the "real" subject of this book.[I.16] Then, if there is some mathematical subject or tool that you need you might try to find it in Part VI, although I cannot guarantee that it is covered there. The Index at the end of the book should be of help in this regard.

You may not appreciate the book until you have been through Parts II–VI,[I.17] so I recommend that you do not stop after Part I.

Finally, I urge you to read the footnotes.[I.18]

Expected Prerequisites

A reader with a working knowledge of calculus, linear algebra, and a handle on the concepts of rational mechanics, i.e. of forces and moments etc., should be able to read and understand this book. However, much of its contents may seem abstract unless the reader also has some previous experience with strength of materials and basic structural mechanics.

What this Book Is About—and what it Is Not

Part I, Continuum Mechanics

This book is about continuum mechanics and its central position in all our theories and methods for analysis of solids and structures. Although Part I is called *Continuum Mechanics* this book is not intended as a textbook covering all available theories of continuum mechanics, just the most essential topics of continuum mechanics—at least the ones that I have found the most important.

For some reason, it seems to be the case that often the *Principle of Virtual Work* does not get the attention it deserves, both in its own right and as a solid foundation for theories concerned with *Specialized Continua*—for my definition of this term, see page 27—such as beams, plates and shells.

What Are these Other Parts About?

The justification for the Parts II–VI is two-fold in that, on the one hand they provide theories and results that are useful by themselves, and on the other hand they serve as examples of application of the general principles and fundamental expressions of Continuum Mechanics, Part I. It is my intention

[I.16] Personally, I don't like the way some books on mechanics are organized, namely beginning with a long part of math which one may not see the need for until much later. I prefer to get on with the "real" subject of mechanics and look up the math I don't know by heart.

[I.17] Actually, you may never appreciate it.

[I.18] In more than one case, a footnote contains important information, either of a technical nature or expressing my personal view on a particular subject.

that, together, Parts I–VI form an entirety in that the same notation is used and, even more important, because the same fundamental approach lies beneath every chapter.

Part II, Specialized Continua

Theories for some specialized continua,[I.19] namely beams and plates, are developed in Part II, but it is not meant as a book about beams and plates and other specialized continua.[I.20] This part furnishes examples of theories for—very important—specialized continua developed from the general principles mentioned above.

Part III, Beams with Cross Sections and Plates with Thickness

In Part II beams are viewed as one-dimensional bodies and plates as two-dimensional bodies, both with some cross-sectional properties that are assumed known in some way. In Part III approximate continuum mechanics solutions are utilized to determine axial, shear and bending stiffness of important types of cross-sections. In addition to this, unconstrained torsion is treated and the torsional stiffness of a number of cross-sections is found. Finally, in order to show that finite elements need not be based on energy or virtual work a finite element procedure, which takes the value of the stress function for torsion as the unknowns, is developed.

Part IV, Buckling

The subject of buckling, including initial postbuckling and imperfection sensitivity, of linear elastic structures is covered to some extent in Part IV, but this book is not a book about stability and instability. Elastic-plastic buckling is studied via the so-called Shanley model column which provides a foundation for understanding this difficult subject. This part as a whole introduces fundamental concepts and, based on the general principles of continuum mechanics, it establishes some very important theories and methods relevant for the study of stability of structures.

Part V, Introduction to the Finite Element Method

In Part V, this useful tool is introduced, but this is not a finite element book. However, the method is introduced on the basis of the general principles of continuum mechanics. Some causes for errors in usual finite element analyses are investigated, and ways to deal with such problems are given.

[I.19] This is probably not a common terminology, but I consider it clear and useful.
[I.20] Other specialized continua, such as shells are not treated.

Part VI, Mathematical Preliminaries

Part VI, contains some mathematical subjects, e.g. *Variational Principles*, and a number of fundamental relations and expressions of continuum mechanics formulated in terms of a compact notation, the so-called *Budiansky-Hutchinson Notation*, but this book is not about mathematics. This part merely serves to introduce the notation and some fundamental expressions that are employed in the rest of the book.

Some Comments on Notation

For any particular purpose, a certain notation may be superior to others, yet cumbersome for most other purposes. Here, we shall employ two notations which I have found to be convenient in the present connection.

Notation depends on purpose

First, we encounter a notation that entails indices.[I.21] This notation is very useful for deriving the equations of deformation and stress of a three-dimensional continuum or a plate, i.e. to establish equations for a three-dimensional continuum or a two-dimensional, specialized one. It is, however, rather awkward in connection with general statements, such as the *Principle of Virtual Work*, where we shall employ the so-called *Budiansky-Hutchinson Notation*. Both notations will be introduced in detail later, see Part VI.

Some Comments on Length

To ease reading formulas are sometimes repeated

Some sections may seem too long because I cite formulas from other sections verbatim instead of just giving a reference—which, by the way, is done in most cases. This might seem in conflict with my wish to utilize compact notations, but the reason is that in some cases[I.22] it proves to be very disturbing to have to flip from one page to another and then back to the original one in order to see the referenced formulas. Back when I was a graduate student I read Muskhelishvili's book *Some Basic Problems of the Mathematical Theory of Elasticity* (Muskhelishvili 1963), which is a book of approximately 700 pages, in a fairly short time,[I.23] and in that book it is rather the rule than the exception that earlier formulas are cited verbatim.

[I.21] We shall only use *subscripts* and avoid *superscripts*, except as labels, because the description will be in *Cartesian Coordinates* where the distinction between co- and contravariant components of tensors disappears.

[I.22] Again, I have made choices that some readers may not like.

[I.23] I am a slow reader, but it only took me about a month to read the first 550 pages.

Part I

Continuum Mechanics

Chapter 1

The Purpose of Continuum Mechanics

Introduction

Continuum mechanics is a vast subject in itself, see e.g. (Washizu 1982), (Malvern 1969), (Flügge 1972), (Truesdell 1965), (Green & Zerna 1960), or (Sokolnikoff 1956). Here, I intend to cover only the most basic elements of continuum mechanics of solids and structures and refer the interested reader to the above-mentioned books.

In this book only solids and structures

In the present connection, the main purpose of continuum mechanics is to provide a sound foundation for solid and structural mechanics, which in turn are used to provide information about the displacements, deformations, stability, vibrations, and internal forces in structures and structural elements subjected to prescribed loads, prescribed displacements, or combinations of the two. In order to do that, the description must entail several topics such as loads, kinematics (geometric description), statics, dynamics, and constitutive laws.[1.1]

Kinematic, static, and constitutive equations must always be satisfied.

First, we focus on the kinematic and static description and use less space on constitutive relations. However, in order to find the solution to any static problem we need the full set of equations, namely the kinematic, the static, and the constitutive equations. Otherwise, it is not possible to set up the governing boundary value problem.[1.2] In most cases, we shall formulate the ensuing boundary value problem either in terms of a variational principle, e.g. some principle of virtual work, or sometimes as a set of differential field equations with appropriate boundary conditions.

Although dynamic effects are extremely important in many applications,

[1.1] Constitutive laws, which are always *models*—not laws, express the physical characteristics of the material. In connection with the subjects covered in this book they express the connection between the *strains*, i.e. deformation of the material, and the *stresses*, i.e. internal forces, in a body of a given material—in some cases, the constitutive model entails the rate of these quantities. The terms strain and stress will be defined in later chapters, see in particular Section 2.2 and Section 2.3.

[1.2] The term boundary value problem should be familiar to you from some math course. Otherwise, go look it up.

E. Byskov, *Elementary Continuum Mechanics for Everyone*,
Solid Mechanics and Its Applications 194, DOI: 10.1007/978-94-007-5766-0_1,

for example wave propagation, I do not include them here. The reason is that I wish to introduce the most basic concepts and therefore must leave many important subjects untouched.

In spite of the fact that later we focus on kinematically[1.3] linear theories we do *not* begin with establishing an infinitesimal, i.e. linear, theory but with a theory which entails large displacements and deformations. Among the reasons for this choice is the fact that—to my experience—a linear theory is rather difficult to understand if it is derived without reference to a more general, nonlinear one. In another context I used to ask the students:

Kinematically linear theories seem suspect

"If a person came in from the street and tried to sell you a theory, which goes as follows:

1. Assume that we may disregard all displacements.
2. Based on this assumption, compute the internal forces, i.e. the stresses,[1.4] caused by the external loads.
3. With these stresses in hand, compute the internal deformations, i.e. the strains.[1.5]
4. Integrate the strains to get the displacements of the body—*which we assumed to vanish.*

would you buy it?"[1.6]

There are also some other aspects in favor of deriving a full nonlinear theory. For example, all consistent continuum theories must fulfill the so-called *Rigid-Body Criterion*, which states that the strain measure must vanish for all rigid-body motions. In the theory which is derived in Section 2.2 this requirement is fulfilled *ab initio*, and is therefore not of any concern. In *any* theory which assumes the displacements—translatory and/or rotatory—to be limited, the *Rigid Body Criterion* must be interpreted as "... all rigid body motions *taking the imposed limitations into consideration.*" This is a point that I shall come back to in Part II in connection with theories for specialized continua (bars, beams, plates etc.) as well as in the chapter on the infinitesimal theory, see Chapter 4.

Rigid-Body Criterion: no strains from rigid-body displacements.

My point is that it is usually much easier to understand a linear theory if it is derived as a special case of a general one. In most cases, however, the full nonlinear theory results in mathematical problems that are much more difficult to handle—a linear problem is usually much easier to solve than the corresponding nonlinear one.

[1.3] Sometimes the term *geometrically linear* is used instead of *kinematically linear.* Loosely speaking kinematics is concerned with movement, while geometry is concerned with configuration. Therefore, the term *kinematically (non)linear* should be preferred in our connection.

[1.4] The term stress is defined later, see Section 2.3.

[1.5] The term strain is defined later, see Section 2.2.

[1.6] In one of his books the British author P.G. Wodehouse writes about a person who was so dumb that he bought a gold bullion from a man at the door.

Chapter 2

Large Displacements and Large Strains

2.1 Introduction

Here, we focus on the kinematic and static descriptions which are independent of the constitutive equations, which connect the strains (deformations in the body) with the stresses (internal forces in the body). For the sake of preventing misunderstandings regarding the issue of this I emphasize that, in order to find the solution to any static problem in solid or structural mechanics, we need the full set of equations, namely the kinematic, the static, and the constitutive equations. Otherwise, it is not possible to set up the governing boundary-value problem. In most cases, we shall formulate boundary-value problems in terms of a variational principle, e.g. some *Principle of Virtual Work*, or sometimes as a set of differential equations with associated boundary conditions.

Lagrange Strains, Piola-Kirchhoff Stresses

Principle of Virtual Work

In the first part of this chapter we introduce three-dimensional continuum mechanics using *Lagrange Strains* and *Piola-Kirchhoff Stresses* as measures of internal deformations and internal forces, respectively. The *Principle of Virtual Work* is then derived. When dealing with specialized theories, e.g. theories for beams, plates, or shells the postulate of a Principle of Virtual Work—in particular the *Principle of Virtual Displacements*—together with a definition of the "generalized" strains, will serve as the basis, while the definition of the associated "generalized" stresses will follow from these postulates.[2.1]

As mentioned above, both the kinematics and the statics of a body entail establishing field equations and boundary conditions. For three-dimensional bodies the kinematic boundary conditions rarely present problems, while the static boundary conditions turn out to be somewhat more involved, see Section 2.3.2.

[2.1] By the term "generalized strains and stresses" we mean strains and stresses that are work conjugate in a principle of virtual work.

If this sounds cryptic, don't worry. You will realize that it is all quite straightforward, see Part II and Chapter 33.

E. Byskov, *Elementary Continuum Mechanics for Everyone*,
Solid Mechanics and Its Applications 194, DOI: 10.1007/978-94-007-5766-0_2,

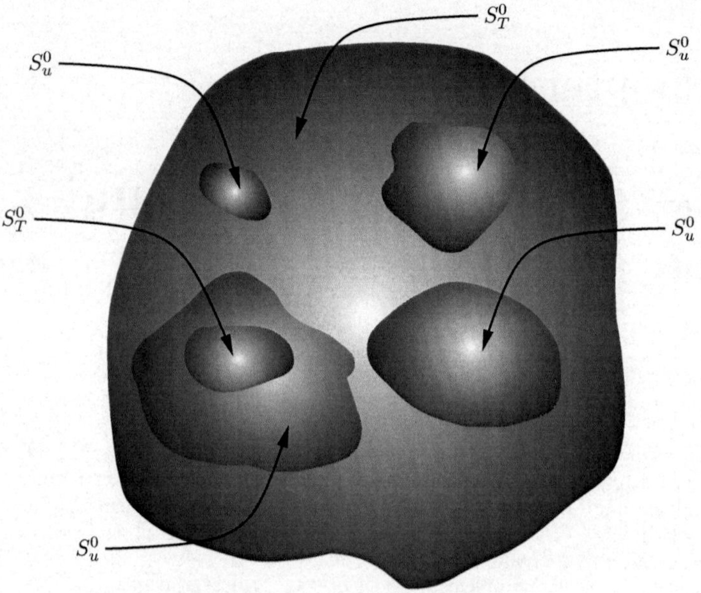

Fig. 2.1: Three-dimensional body.

Continuum "Potatoes" instead of real structures

To set the stage, consider the deformation and equilibrium of a three-dimensional body, see Fig. 2.1. In introductions to continuum mechanics such "potatoes" are often used instead of bodies of more regular configurations. The reason is that the author does not want the reader to put too much emphasis on the shape of the body because the reader might object to the practical relevance of the shape of the body which the author chooses. I shall stick with this fairly common habit.

Kinematic boundary S_u^0

Static boundary S_T^0

The body may be subjected to prescribed displacements on the boundary and, occasionally, in the interior as well as prescribed forces on the boundary and in the interior. The objective of continuum mechanics then is to set up equations that determine the deformed configuration of the body. For later purposes here we provide a classification of the different parts of the body. It consists of the *interior* V^0 and the surface S^0 where upper index ^{0} indicates that the quantity is associated with the configuration before any deformation occurs. Further, the surface is divided into two parts, namely the so-called *Kinematic Boundary* S_u^0, where the displacements are prescribed, and the so-called *Static Boundary* S_T^0, where the surface loads, the surface tractions are prescribed. It is well worth mentioning that S_u^0 and S_T^0 may occupy the same area of the surface, but in that case only some of the displacement components and some unrelated stress component(s) are prescribed over the same area.

2.2 Kinematics and Deformation

We assume that we know the configurations of a structure in two states, namely the initial, undeformed—or *virgin*—state and the deformed state.[2.2] Any strain measure serves the purpose of telling how much the material at a point of the structure suffers because of the deformation. Physically, the strains at a point in the body are measures of the intensity of the deformation at that point. There is no "best" way to define the strain measure, in particular in kinematically nonlinear problems[2.3] there exists a wealth of useful strain definitions. Here, I shall consider only the strains that are known as *Lagrange Strains* because usually they are the most convenient for our later purposes, in particular see Part IV.

Virgin state

No "best" strain definition

Lagrange strains

Mathematically speaking, a strain measure describes the deformation of the immediate neighborhood of a point. Once we know the strain at all points of the structure, we can compute the shape of the deformed structure. We may also check to see whether the magnitude of the strains have exceeded some maximum criterion, which then would tell us if the material has ruptured.

The kinematic description entails two sets of equations, namely the *Kinematic Field Equations* and *Kinematic Boundary Conditions* which are derived below.

Kinematic field equations
Kinematic boundary conditions

2.2.1 Kinematics and Strain

There are a number of ways in which we can define the Lagrange Strain. Here, we shall employ the one that seems to be the most satisfactory, namely one that considers deformation of an infinitesimal sphere, which in the initial, undeformed, state has the center P^0 and the radius ds^0—recall that upper index ^{0} indicates the *undeformed* state, see Fig. 2.2. After deformation, P^0 has moved to the position P, and the sphere has changed shape. Once we know how the infinitesimal sphere is deformed, we possess sufficient *local* information about the deformed state. This amounts to, for all directions, computing the change in length and direction of an infinitesimal vector from P^0 to a neighboring point Q^0.

Initial state
= Virgin state

[2.2] Usually we do *not* know the deformed configuration in advance, but this is a necessary assumption here. Later we shall see that we do get the tools to compute the deformed state once the undeformed geometry, the constitutive laws and the loads are known.

[2.3] Kinematic *linearity* implies that all deformations are *infinitesimal*, and thus all nonlinearity has been excluded from the description of the problem with the result that the reasonable choices are much more limited.

Cartesian coordinate system

For the description we employ a three-dimensional *Cartesian* coordinate system, i.e. one that has three orthogonal axes x_1, x_2 and x_3 with base

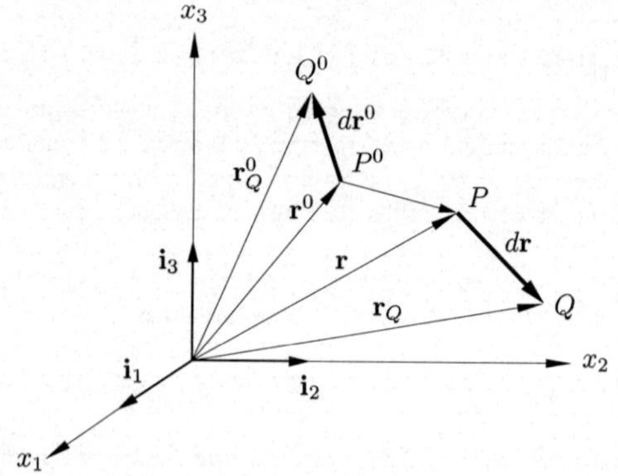

Fig. 2.2: Geometry—Kinematics.

vectors $\mathbf{i}_j$, $j = [1, 2, 3]$, of unit length, i.e.[2.4]

Base vectors $\mathbf{i}_j$

$$|\mathbf{i}_j| = 1 \tag{2.1}$$

Summation convention

Dummy index = Repeated index

In the following—as in most of this book—we utilize the *Summation Convention*, see Part VI, which states that a repeated *lowercase* index indicates summation over the range of that index. The repeated index, the *Dummy Index*, must appear exactly twice in each product in order that the operation is defined. Thus, for a three-dimensional coordinate system we may write the following expression

$$dx_i dx_i = dx_1 dx_1 + dx_2 dx_2 + dx_3 dx_3 \tag{2.2}$$

In the undeformed configuration the position vectors of point P^0 and the neighboring point Q^0 are $\mathbf{r}^0$ and $\mathbf{r}^0_Q$, respectively, see Fig. 2.2. The distance between P^0 and Q^0 is ds^0 and thus

$$\left(ds^0\right)^2 = d\mathbf{r}^0 \cdot d\mathbf{r}^0 = dx_i\, dx_i = \delta_{ij}\, dx_i\, dx_j \tag{2.3}$$

Kronecker delta δ_{ij}

where, δ_{ij} denotes the *Kronecker delta* δ_{ij} which is defined by (2.4) below and by (31.9) in Chapter 31. Here, we have used the Kronecker delta to rewrite the formula for the distance between points P^0 and Q^0 in a form,

[2.4] Vectors are indicated by boldface letters.

which is often useful, see e.g. (2.10).

$$\delta_{ij} \equiv \begin{cases} 1 & \text{for} \quad j = i \\ 0 & \text{for} \quad j \neq i \end{cases} \tag{2.4}$$

The Kronecker delta δ_{ij}

Because of (2.1) and the fact that the base vectors are orthogonal their *scalar product*, also denoted the *inner product*, is given by the *Kronecker delta*

$$\delta_{ij} = \mathbf{i}_i \cdot \mathbf{i}_j \tag{2.5}$$

where $\cdot$ indicates a scalar product.

After deformation the material points P^0 and Q^0 are moved to the positions P and Q with position vectors $\mathbf{r}$ and $\mathbf{r}_Q$, respectively. The length of the infinitesimal line element has changed from ds^0 to ds. Then

$$(ds)^2 = d\mathbf{r} \cdot d\mathbf{r} = (\mathbf{r}_{,i} dx_i) \cdot (\mathbf{r}_{,j} dx_j) = (\mathbf{r}_{,i} \cdot \mathbf{r}_{,j})\, dx_i dx_j \tag{2.6}$$

where the *Comma Notation* indicates partial derivatives

$$(\,)_{,j} \equiv \frac{\partial}{\partial x_j}(\,) \tag{2.7}$$

Comma notation $(\,)_{,j}$

For later purposes introduce the quantities $\mathbf{g}_i$ and g_{ij} by

$$\mathbf{g}_i \equiv \mathbf{r}_{,i} \quad \text{and} \quad g_{ij} \equiv \mathbf{g}_i \cdot \mathbf{g}_j \tag{2.8}$$

where the geometric interpretation of $\mathbf{g}_i$ and g_{ij} will be clear subsequently. For now it suffices to think of them as convenient shorthand notations.

By (2.8a) and (2.8b) Eq. (2.6) may be written

$$(ds)^2 = \mathbf{g}_i \cdot \mathbf{g}_j\, dx_i dx_j = g_{ij} dx_i dx_j \tag{2.9}$$

The change in the square of the length of ds^0 provides as much information as the change in length itself and is more easily applied in the following. Therefore, we compute

$$(ds)^2 - (ds^0)^2 = (g_{ij} - \delta_{ij})\, dx_i dx_j \tag{2.10}$$

Change in square of length of ds^0

2.2.2 Kinematic Field Equations—Lagrange Strain

Now, we define the *Lagrange Strain Measure* γ_{ij} by

$$\gamma_{ij} \equiv \tfrac{1}{2}(g_{ij} - \delta_{ij}) \tag{2.11}$$

Lagrange strain γ_{ij}

which may be introduced into (2.10) to give

$$(ds)^2 - (ds^0)^2 = 2\gamma_{ij} dx_i dx_j \tag{2.12}$$

Since both g_{ij} and δ_{ij} are symmetric it is obvious from its definition that

The Lagrange strain γ_{ij} is symmetric

γ_{ij} is symmetric in its indices

$$\gamma_{ij} = \gamma_{ji} \tag{2.13}$$

Lagrange strain γ_{ij} Infinitesimal strains ε_{ij} or e_{ij}

It is a fact that there exist more different notations for strains than anyone could wish for, which occasionally causes some confusion.[2.5] For instance, sometimes we employ the notation ε_{ij} for nonlinear strains instead of γ_{ij}, see Part II. Here, however, we retain γ_{ij} because we wish to emphasize that the Lagrange Strains are nonlinear. It is unfortunate that the infinitesimal strains, which often are denoted e_{ij}, see (2.36), also are designated ε_{ij}. Since these notations, confusing as they are, all are very common, I have decided to use them and in each case try to be careful to note when ε_{ij} indicates a nonlinear or a linear strain, respectively.

It may be worthwhile mentioning that to lowest order there is no difference between the infinitesimal strain measure and the Lagrange Strain, see Chapter 4, Infinitesimal Theory.

In the following we assume that the displacement field, given by the vector field $\mathbf{u}(\mathbf{r}^0)$ or equivalently by its components $u_i(x_k)$, is known[2.6] and establish expressions for the strains $\gamma_{ij}(x_k)$ in terms of the displacement gradients $u_{i,j}(x_k)$.

From the geometry of the undeformed and the deformed configurations we may get

$$\mathbf{r} = \mathbf{r}^0 + \mathbf{u} = (x_j + u_j)\,\mathbf{i}_j \tag{2.14}$$

which by differentiation with respect to x_m furnishes

$$\mathbf{r}_{,m} = (\delta_{jm} + u_{j,m})\,\mathbf{i}_j \tag{2.15}$$

Because of (2.8a) we may get

$$\mathbf{g}_m = (\delta_{jm} + u_{j,m})\,\mathbf{i}_j \tag{2.16}$$

From Fig. 2.3 it is observed that by the deformation the base vector $\mathbf{i}_m$, where $\mathbf{i}_m$ indicates any of the base vectors, is displaced and deformed into the vector $\mathbf{j}_m$, where

$$\mathbf{j}_m = (\mathbf{u} + \mathbf{u}_{,m}\,|\mathbf{i}_j|) - \mathbf{u} + \mathbf{i}_m = (\delta_{jm} + u_{j,m})\,\mathbf{i}_j \tag{2.17}$$

where it has been exploited that the length $|\mathbf{i}_j|$ of the base vectors is unity, see (2.1).

Comparison between (2.16) and (2.17) shows that $\mathbf{j}_m = \mathbf{g}_m$, which means that $\mathbf{g}_m$ is the base vector $\mathbf{i}_m$ after deformation and is therefore called the

[2.5] I only expose you to three of them, namely γ_{ij}, ε_{ij} and e_{ij}, see below.
[2.6] See footnote on page 35.

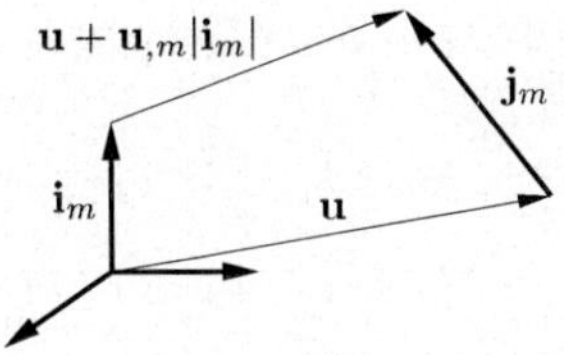

Fig. 2.3: Connection between $\mathbf{i}_m$ and $\mathbf{g}_m$.

deformed base vector. Note that, in general, $\mathbf{g}_m$ is not a unit vector.

Deformed base vector $\mathbf{g}_m$

From (2.8b) and (2.17) we may get

$$\begin{aligned} g_{mn} &= \mathbf{g}_m \cdot \mathbf{g}_n = (\delta_{jm} + u_{j,m})(\delta_{kn} + u_{k,n})\,\mathbf{i}_j \cdot \mathbf{i}_k \\ &= (\delta_{jm}\delta_{kn} + \delta_{jm}u_{k,n} + \delta_{kn}u_{j,m} + u_{j,m}u_{k,n})\,\delta_{jk} \\ \Rightarrow \quad g_{mn} &= \delta_{mn} + u_{m,n} + u_{n,m} + u_{k,m}u_{k,n} \end{aligned} \tag{2.18}$$

and thus the components of the Lagrange Strain are given by

Lagrange strain γ_{mn}

$$\gamma_{mn} = \tfrac{1}{2}(u_{m,n} + u_{n,m}) + \tfrac{1}{2}u_{k,m}u_{k,n} \tag{2.19}$$

2.2.2.1 "Fiber" Elongation

The relative elongation of a "fiber" of the material may be given by the quantity γ, where

Fiber elongation γ = Change of length

$$\gamma \equiv \frac{(ds)^2 - (ds^0)^2}{2(ds^0)^2} \tag{2.20}$$

where the reason for the factor 2 in the denominator may be seen from the expansion of (2.20) and from the definition of the Lagrange strain

Fiber elongation γ for infinitesimal deformation

$$\gamma = \frac{(ds + ds^0)(ds - ds^0)}{2ds^0 ds^0} \approx \frac{ds - ds^0}{ds^0} \quad \text{for} \quad ds \approx ds^0 \tag{2.21}$$

which shows that γ indeed is equal to the relative change in length for small deformations.

You may, of course, ask: why not use the relative change in length, i.e. $(ds - ds^0)/ds^0$, instead of the quantity γ defined by (2.20)? The reason is that our strain measure is the Lagrange strain, which is associated with the difference between the square of the length of the line element after and before deformation, and we wish to express the elongation in terms of the Lagrange strain. Utilize (2.12) to get

$$\gamma = \frac{2\gamma_{ij}dx_i dx_j}{2ds^0 ds^0} \tag{2.22}$$

and realize that the unit vector $\overset{0}{\mathbf{n}}$ in the direction of the "fiber" before deformation has the components $\overset{0}{n}_j$

$$\overset{0}{n}_j = \frac{dx_j}{ds^0} \tag{2.23}$$

to get

$$\gamma = \gamma_{ij}\overset{0}{n}_i\overset{0}{n}_j \tag{2.24}$$

Thus, when we have determined the Lagrange strain γ_{ij} we may determine the change in length of an arbitrary unit vector.

If the vector n_j is directed along one of the coordinate axes, say number I, then (2.24) provides

$$\gamma = \gamma_{II} \quad \text{(no sum over capital indices)} \tag{2.25}$$

An obvious question to ask is: in which direction do we find the maximum (or minimum) elongation? We shall not pursue this question here, but refer to Section 4.2.6, where this subject is covered for the kinematically linear case.

2.2.2.2 *Change of Angle*

Change of Angle

The change of angle between two initially orthogonal directions is another important measure of deformation. In the undeformed configuration introduce two orthogonal unit vectors $\overset{0}{\mathbf{n}}^{(1)}$ and $\overset{0}{\mathbf{n}}^{(2)}$ with components $\overset{0}{n}{}_j^{(1)}$ and $\overset{0}{n}{}_j^{(2)}$. Then,

$$\overset{0}{n}{}_j^{(1)}\overset{0}{n}{}_j^{(2)} = 0\,, \quad \overset{0}{n}{}_j^{(1)}\overset{0}{n}{}_j^{(1)} = 1 \quad \text{and} \quad \overset{0}{n}{}_j^{(2)}\overset{0}{n}{}_j^{(2)} = 1 \tag{2.26}$$

Let us denote these vectors after deformation by $\mathbf{m}^{(1)}$ and $\mathbf{m}^{(2)}$ with components $m_j^{(1)}$ and $m_j^{(2)}$, respectively. Then

$$m_j^{(1)}m_j^{(2)} = |\mathbf{m}^{(1)}||\mathbf{m}^{(2)}|\cos(\psi^{(12)}) \tag{2.27}$$

where $\psi^{(12)}$ denotes the angle between $\mathbf{m}^{(1)}$ and $\mathbf{m}^{(2)}$.

The components $m_j^{(1)}$ and $m_j^{(2)}$ may be found to be

$$m_j^{(1)} = \overset{0}{n}{}_j^{(1)} + \overset{0}{n}{}_i^{(1)}u_{j,i} \quad \text{and} \quad m_j^{(2)} = \overset{0}{n}{}_j^{(2)} + \overset{0}{n}{}_i^{(2)}u_{j,i} \tag{2.28}$$

Another way of expressing (2.27) therefore is

$$\begin{aligned} m_j^{(1)}m_j^{(2)} &= (\overset{0}{n}{}_j^{(1)} + \overset{0}{n}{}_i^{(1)}u_{j,i})(\overset{0}{n}{}_j^{(2)} + \overset{0}{n}{}_k^{(2)}u_{j,k}) \\ &= \overset{0}{n}{}_j^{(1)}\overset{0}{n}{}_j^{(2)} + \overset{0}{n}{}_j^{(1)}\overset{0}{n}{}_k^{(2)}u_{j,k} + \overset{0}{n}{}_i^{(1)}\overset{0}{n}{}_j^{(2)}u_{j,i} + \overset{0}{n}{}_i^{(1)}\overset{0}{n}{}_k^{(2)}u_{j,i}u_{j,k} \\ &= 0 + \overset{0}{n}{}_j^{(1)}\overset{0}{n}{}_i^{(2)}(u_{j,i} + u_{i,j} + u_{k,j}u_{k,i}) \end{aligned} \tag{2.29}$$

where we have changed dummy indices in several places. By use of (2.19) we may introduce the Lagrange strain γ_{ij} and get

$$m_j^{(1)} m_j^{(2)} = 2\overset{0}{n}{}_i^{(1)} \overset{0}{n}{}_j^{(2)} \gamma_{ij} \tag{2.30}$$

In order to find an expression for $\cos(\psi^{(12)})$ we need expressions for the length of $\mathbf{m}^{(1)}$ and $\mathbf{m}^{(2)}$, see (2.27). First, let us determine $|\mathbf{m}^{(1)}|^2$

$$\begin{aligned} |\mathbf{m}^{(1)}|^2 &= m_j^{(1)} m_j^{(1)} \\ &= (\overset{0}{n}{}_j^{(1)} + \overset{0}{n}{}_i^{(1)} u_{j,i})(\overset{0}{n}{}_j^{(1)} + \overset{0}{n}{}_k^{(1)} u_{j,k}) \\ &= \overset{0}{n}{}_j^{(1)} \overset{0}{n}{}_j^{(1)} + \overset{0}{n}{}_j^{(1)} \overset{0}{n}{}_k^{(1)} u_{j,k} + \overset{0}{n}{}_i^{(1)} \overset{0}{n}{}_j^{(1)} u_{j,i} + \overset{0}{n}{}_i^{(1)} \overset{0}{n}{}_k^{(1)} u_{j,i} u_{j,k} \\ &= 1 + \overset{0}{n}{}_j^{(1)} \overset{0}{n}{}_i^{(1)} (u_{j,i} + u_{i,j} + u_{k,j} u_{k,i}) \\ &= 1 + 2\overset{0}{n}{}_i^{(1)} \overset{0}{n}{}_j^{(1)} \gamma_{ij} \end{aligned} \tag{2.31}$$

By substituting $^{(2)}$ for $^{(1)}$ in (2.31) the expression for $|\mathbf{m}^{(2)}|^2$ is found to be

$$|\mathbf{m}^{(2)}|^2 = 1 + 2\overset{0}{n}{}_i^{(2)} \overset{0}{n}{}_j^{(2)} \gamma_{ij} \tag{2.32}$$

We are now able to establish an expression for the angle $\psi^{(12)}$, but we are probably more interested in the *change* of angle between $\mathbf{m}^{(1)}$ and $\mathbf{m}^{(2)}$. Let $\varphi^{(12)}$ denote this change. Note that

Change of angle $\varphi^{(12)}$

$$\varphi^{(12)} = \tfrac{1}{2}\pi - \psi^{(12)} \tag{2.33}$$

Then,

$$\sin(\varphi^{(12)}) = \cos(\psi^{(12)}) \tag{2.34}$$

i.e.

$$\sin(\varphi^{(12)}) = \frac{2\gamma_{ij} \overset{0}{n}{}_i^{(1)} \overset{0}{n}{}_j^{(2)}}{\sqrt{(1 + 2\overset{0}{n}{}_i^{(1)} \overset{0}{n}{}_j^{(1)} \gamma_{ij})(1 + 2\overset{0}{n}{}_m^{(2)} \overset{0}{n}{}_n^{(2)} \gamma_{mn})}} \tag{2.35}$$

Thus, when we have determined the Lagrange strain γ_{ij} we may determine the change of angle between two arbitrary orthogonal unit vectors.

2.2.3 Infinitesimal Strains and Infinitesimal Rotations

The *infinitesimal strain* tensor[2.7] e_{mn} is defined by

Infinitesimal strain e_{mn}

$$e_{mn} \equiv \tfrac{1}{2}(u_{m,n} + u_{n,m}) = e_{nm} \tag{2.36}$$

which shows that e_{mn} is symmetric.

[2.7] Do not put too much emphasis on the term *tensor*—in our case it is merely a fancy word used to impress some.

Infinitesimal rotation ω_{mn}

The *infinitesimal rotation* tensor ω_{mn}[2.8] is defined by

$$\omega_{mn} \equiv \tfrac{1}{2}\left(u_{m,n} - u_{n,m}\right) = -\omega_{nm} \tag{2.37}$$

which shows that ω_{mn} is antisymmetric in its indices.

From (2.36) and (2.37)

$$u_{m,n} = e_{mn} + \omega_{mn} \tag{2.38}$$

The *Lagrange Strain* γ_{mn} may then be expressed in terms of e_{mn} and ω_{mn}

$$\gamma_{mn} = e_{mn} + \tfrac{1}{2}\left(e_{km} + \omega_{km}\right)\left(e_{kn} + \omega_{kn}\right) \tag{2.39}$$

or

$$\gamma_{mn} = e_{mn} + \tfrac{1}{2}e_{km}e_{kn} + \tfrac{1}{2}\left(e_{km}\omega_{kn} + e_{kn}\omega_{km}\right) + \tfrac{1}{2}\omega_{km}\omega_{kn} \tag{2.40}$$

For small strains, i.e. $|e_{mn}| \ll 1$

$$\gamma_{mn} \approx e_{mn} + \tfrac{1}{2}\left(e_{km}\omega_{kn} + e_{kn}\omega_{km}\right) + \tfrac{1}{2}\omega_{km}\omega_{kn} \tag{2.41}$$

and for small strains and moderately small rotations, i.e. $|\omega_{mn}| \ll 1$, but $|\omega_{mn}| > |e_{mn}|$

$$\gamma_{mn} \approx e_{mn} + \tfrac{1}{2}\omega_{km}\omega_{kn} \tag{2.42}$$

The approximate strain measure given by (2.42) forms the basis of *Kinematically Moderately Nonlinear Theories*, which assume that the strains are infinitesimal and that the rotations are small, but finite. Most kinematically nonlinear analyses of beam, plate and shell structures utilize a theory of this type.

2.2.4 Compatibility Equations

The compatibility equations ensure that a strain field can be integrated to provide a displacement field

It is a mathematical fact that not all strain fields are compatible in the sense that there is no guarantee that a given strain field γ_{ij} with 6 independent terms can be integrated to a displacement field u_m of 3 independent terms. In order that this is the case the so-called *Compatibility Equations* must be satisfied. Even in the linear (infinitesimal) case these equations have a rather complicated structure, and in the kinematically nonlinear cases it is, of course, even worse. But, fortunately we do not always need them for our purposes because in our applications we often determine the displacement field first and derive the strains from the displacements, and thus the strains satisfy the compatibility equations automatically. We shall therefore neither derive, nor cite the compatibility equations here, but refer the interested

[2.8] For the sake of completeness I emphasize that, as its name indicates, only for small rotations is ω_{mn} a valid measure of rotation.

reader to the book by Malvern (1969).

We shall, however, derive the compatibility equations in the case of kinematic linearity, see Chapter 4.2, Infinitesimal Theory, in particular Section 4.2.2. One reason for this is that in formulation of theories based on stress functions, such as the Airy Stress Function, see Section 9.2.5, the compatibility conditions are not satisfied a priori.

2.2.5 Kinematic Boundary Conditions

For a three-dimensional body the *Kinematic Boundary Conditions* usually are quite obvious, and we shall defer discussion of this matter to the theories for specialized continua, see Part II.

Kinematic boundary conditions

2.3 Equilibrium Equations

Like the kinematic equations, the equilibrium equations fall into two parts, namely the *Static Field Equations* and the *Static Boundary Conditions*. The static boundary equations usually do not present difficulties for the three-dimensional continuum, but for the specialized continua, see Part II, the situation often is quite different.

Static field equations and static boundary conditions

2.3.1 Static Field Equations

We may derive the continuum equilibrium equations from the equilibrium of a deformed, infinitesimal sphere and thus apply a procedure which is analogous to the one from Section 2.2.1. Here, however, I prefer to go about the task in another way and establish the equilibrium equations of a (deformed) infinitesimal parallelepiped, see Fig. 2.4. In the undeformed state the parallelepiped, whose volume is dV^0, is spanned by the vectors $\mathbf{i}_1 dx_1$, $\mathbf{i}_2 dx_2$, and $\mathbf{i}_3 dx_3$. From Section 2.2.1 we know that these vectors deform into the vectors $\mathbf{g}_1 dx_1$, $\mathbf{g}_2 dx_2$, and $\mathbf{g}_3 dx_3$, which span a deformed parallelepiped with the volume dV, where

Equilibrium of an infinitesimal parallelepiped

$$dV = (\mathbf{g}_1 \times \mathbf{g}_2 \cdot \mathbf{g}_3)\, dV^0 \tag{2.43}$$

where we assume that $dV > 0$ because otherwise the cube would have collapsed.

Since the loads on the structure act on the deformed structure, it is the equilibrium of the deformed parallelepiped which must be analyzed. In spite of this fact, we shall measure all forces in terms of *undeformed* areas, simply because it proves to be more convenient.

The force acting on the side, which before deformation had the normal $-\mathbf{i}_1$, is $-\mathbf{F}_1$, and the force on the other side of the parallelepiped is then $\mathbf{F}_1 + d\mathbf{F}_1$.

In terms of the undeformed area.

Measure all forces in terms of *undeformed* areas

$$-\mathbf{F}_1 = -\mathbf{t}_1 dx_2 dx_3 \tag{2.44}$$

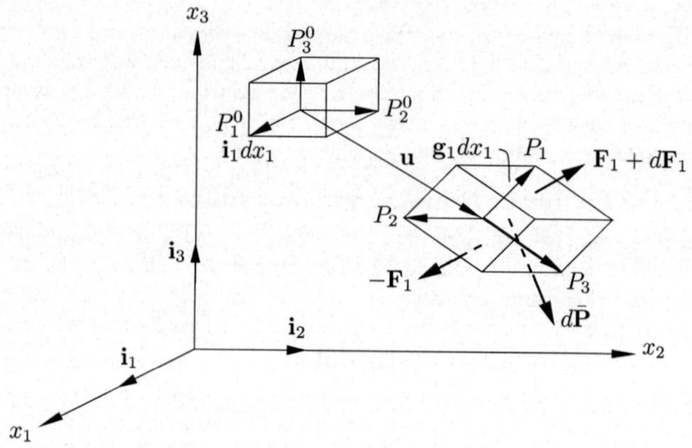

Fig. 2.4: Statics of an infinitesimal element. Not all forces acting on the element are shown.

and on the other side of the parallelepiped

$$\mathbf{F}_1 + d\mathbf{F}_1 = \mathbf{t}_1 dx_2 dx_3 + \left(\mathbf{t}_1 dx_2 dx_3\right)_{,1} dx_1 \tag{2.45}$$

with analogous relations for the other two directions. The total load, or total body force, acting within the volume dV is $d\bar{\mathbf{P}}$, which is measured in terms of the *undeformed* volume dV^0

Total body force in dV^0

$$d\bar{\mathbf{P}} = \bar{\mathbf{q}} dx_1 dx_2 dx_3 \tag{2.46}$$

where $\bar{\mathbf{q}}$ is the body force acting within the element per unit volume.

Force equilibrium of dV requires that

Force equilibrium of dV

$$d\mathbf{F}_1 + d\mathbf{F}_2 + d\mathbf{F}_3 + d\bar{\mathbf{P}} = \mathbf{0} \tag{2.47}$$

which, in light of (2.44), (2.45), and (2.46), provides

Force equilibrium of dV

$$\mathbf{t}_{i,i} + \bar{\mathbf{q}} = \mathbf{0} \tag{2.48}$$

We wish to express (2.48) in component form, and to this end we resolve the vectors $\mathbf{t}_i$ with respect to the *deformed* base vectors $\mathbf{g}_j$ although the loads are measured in terms of the *undeformed* area. This is, of course, not the only possible choice—and not an obvious one either—but it will prove to be convenient. Thus,

$$\mathbf{t}_i = t_{ij} \mathbf{g}_j \tag{2.49}$$

where t_{ij} is the (second)[2.9] *Piola-Kirchhoff Stress Tensor.* For a good reason it is often referred to as a *pseudo* stress because it is measured as the force per unit *undeformed* area resolved in terms of the *deformed* base vectors $\mathbf{g}_j$, which, as mentioned above, in general are not unit vectors.

The Piola-Kirchhoff *pseudo* stress t_{ij} is measured on the *undeformed* area, but resolved in terms of the *deformed* base vectors

When higher order terms in dx_i are neglected, moment equilibrium of dV requires that[2.10]

$$\mathbf{g}_1 \times \mathbf{F}_1 dx_1 + \mathbf{g}_2 \times \mathbf{F}_2 dx_2 + \mathbf{g}_3 \times \mathbf{F}_3 dx_3 = \mathbf{0} \tag{2.50}$$

or because of (2.44)

$$(\mathbf{g}_i \times \mathbf{t}_i)\, dx_1 dx_2 dx_3 = \mathbf{0} \tag{2.51}$$

and thus

$$\mathbf{g}_i \times \mathbf{t}_i = \mathbf{0} \quad \Rightarrow \quad t_{ij}\mathbf{g}_i \times \mathbf{g}_j = \mathbf{0} \tag{2.52}$$

which, written out, gives the following three equations

$$(t_{12} - t_{21})\,\mathbf{g}_1 \times \mathbf{g}_2 + (t_{23} - t_{32})\,\mathbf{g}_2 \times \mathbf{g}_3 + (t_{31} - t_{13})\,\mathbf{g}_3 \times \mathbf{g}_1 = \mathbf{0} \tag{2.53}$$

which do not seem to provide much information about the properties of the stress tensor t_{ij}. However, barring deformations that annihilate the initial element[2.11] the vectors $\mathbf{g}_1 \times \mathbf{g}_2$, $\mathbf{g}_2 \times \mathbf{g}_3$, and $\mathbf{g}_3 \times \mathbf{g}_1$ are linearly independent and therefore

The Piola-Kirchhoff stress tensor t_{ij} is symmetric

$$t_{ij} = t_{ji} \tag{2.54}$$

meaning that the Piola-Kirchhoff stress tensor is symmetric.[2.12]

We wish to express the equilibrium equation (2.48) in component form and start with resolving the body force $\bar{\mathbf{q}}$ in terms of the *undeformed* base vectors $\mathbf{i}_j$

$$\bar{\mathbf{q}} = \bar{q}_j \mathbf{i}_j \tag{2.55}$$

[2.9] There is, as you may have guessed, a first Piola-Kirchhoff stress tensor, see e.g. (Malvern 1969). For our purposes it is not the one that we want.

[2.10] We exclude possible distributed moment loads $\bar{c}_j$ because they do not appear directly in most structural problems, except in connection with e.g. magneto-elasticity, which rightfully may be considered a very specialized field that we do not intend to discuss.

On the other hand, since about 1980 there has been great interest in continuum theories that involve *couple stresses* μ_{ij}. The reason is that in the description of materials with (micro)structure there is a need for a length scale in order to handle problems such as *strain localization*, sometimes in the form of *kinkbands* in wood, concrete and fiber reinforced epoxy.

[2.11] Collapse of the body would mean that $dV \not> 0$, which on physical grounds is not an acceptable idea.

[2.12] In all fairness, this is not the only useful stress tensor, which exhibits this nice property.

Again, this may not seem like a natural choice because the stress tensor t_{ij} is resolved in terms of $\mathbf{g}_j$—not $\mathbf{i}_j$—but in many cases the load keeps its direction throughout the deformation history. Should this not be the case, then it is a fairly easy task to take this into account and redo the following derivations. Now, (2.48), (2.49) and (2.55) give

$$\left(t_{ij}\mathbf{g}_j\right)_{,i} + \bar{q}_j\mathbf{i}_j = \mathbf{0} \tag{2.56}$$

Recall (2.16) and get

$$\left(t_{ij}(\delta_{mj} + u_{m,j})\mathbf{i}_m\right)_{,i} + \bar{q}_m\mathbf{i}_m = \mathbf{0} \tag{2.57}$$

which after some trivial manipulations provides

$$\left(t_{im,i} + \left(t_{ij}u_{m,j}\right)_{,i} + \bar{q}_m\right)\mathbf{i}_m = \mathbf{0} \tag{2.58}$$

Take the inner product with $\mathbf{i}_k$ on both sides and note the expression for the scalar product of two base vectors in terms of the Kronecker delta (2.5) ($\delta_{ij} = \mathbf{i}_i \cdot \mathbf{i}_j$) to get[2.13]

Internal equilibrium. Static field equations

$$t_{ik,i} + \left(t_{ij}u_{k,j}\right)_{,i} + \bar{q}_k = 0\,, \quad k \in [1,2,3] \tag{2.59}$$

These equations connect the components of the Piola-Kirchhoff stress with the body forces and the displacement gradients and thus they express internal equilibrium, i.e. (2.59) are the static field equations. In a way the term "static field equations" is somewhat misleading because (2.59) entails not only static quantities, but also kinematic ones. However, this is the usual nomenclature, and in the case of infinitesimal displacements and infinitesimal displacement gradients the second term vanishes and then the name is clearly justified, see Chapter 4, in particular (4.75).

2.3.2 Properties of the Stress Vector—Static Boundary Conditions

Static boundary conditions

In order to establish the static boundary conditions we consider an infinitesimal tetrahedron, see Fig. 2.5. Three of its faces are parallel to the coordinate planes, while the fourth is inclined at an angle given by its unit normal vector $\mathbf{n}^0$. It is our intention to establish the equilibrium equations of the tetrahedron *after* displacement and deformation, see Fig. 2.5. The total force on the face, which in the undeformed configuration had the normal $-\mathbf{i}_1$, is $-d\mathbf{F}_1$ with self-evident analogies for the two other faces. The force on the inclined face $P_1P_2P_3$ with the normal $\mathbf{n}^0$ is $d\mathbf{F}_n$. Equilibrium of the deformed tetrahedron requires that

$$-d\mathbf{F}_1 - d\mathbf{F}_2 - d\mathbf{F}_3 + d\mathbf{F}_n + d\bar{\mathbf{P}} = \mathbf{0} \tag{2.60}$$

where $d\bar{\mathbf{P}}$ is the total body force.

[2.13] We may also find (2.59) by appealing to the linear independence of the base vectors $\mathbf{i}_m$.

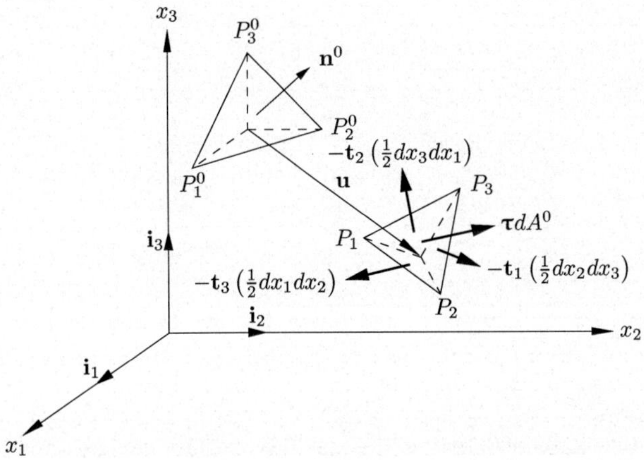

Fig. 2.5: Statics of an infinitesimal tetrahedron.

As before, we measure the intensity of all forces on the *undeformed* configuration. The forces $-d\mathbf{F}_i, i = [1, 2, 3]$, are simply

$$-d\mathbf{F}_1 = -\mathbf{t}_1\left(\tfrac{1}{2}dx_2dx_3\right) \tag{2.61}$$

etc., by analogy with (2.44). The force intensity of the inclined face is $\boldsymbol{\tau}$, and thus

$$d\mathbf{F}_n = +\boldsymbol{\tau}dA^0 \tag{2.62}$$

The following expressions, which are associated with the *undeformed* configuration, prove to be useful

$$\begin{aligned} dA^0\mathbf{n}^0 \cdot \mathbf{i}_1 &= \tfrac{1}{2}dx_2dx_3 \\ dA^0\mathbf{n}^0 \cdot \mathbf{i}_2 &= \tfrac{1}{2}dx_3dx_1 \\ dA^0\mathbf{n}^0 \cdot \mathbf{i}_3 &= \tfrac{1}{2}dx_1dx_2 \end{aligned} \tag{2.63}$$

As in Section 2.3.1, let $\bar{\mathbf{q}}$ denote the body force intensity, and recall that the volume of the undeformed tetrahedron is $\frac{1}{6}dx_1dx_2dx_3$, and rewrite (2.60) and get

$$\begin{aligned} \mathbf{0} = &-\mathbf{t}_1\tfrac{1}{2}dx_2dx_3 - \mathbf{t}_2\tfrac{1}{2}dx_3dx_1 - \mathbf{t}_3\tfrac{1}{2}dx_1dx_2 \\ &+ \boldsymbol{\tau}dA^0 + \bar{\mathbf{q}}\tfrac{1}{6}dx_1dx_2dx_3 \end{aligned} \tag{2.64}$$

or

$$\mathbf{0} = -\mathbf{t}_j \mathbf{n}^0 \cdot \mathbf{i}_j dA^0 + \boldsymbol{\tau} dA^0 + \bar{\mathbf{q}}\tfrac{1}{3} dA^0 \mathbf{n}^0 \cdot \mathbf{i}_1 dx_1 \tag{2.65}$$

The load term is clearly of order 3 in the differentials dx_j because dA^0 is of order 2, while the other terms are of order 2. Therefore, the last term vanishes in the limit and may be omitted. Then, if we resolve $\boldsymbol{\tau}$ in terms of the *undeformed* base vectors $\mathbf{i}_j$

$$\boldsymbol{\tau} = \tau_j \mathbf{i}_j \tag{2.66}$$

and recall (2.49), we may get

$$-t_{jk}\mathbf{g}_k \left(\mathbf{n}^0 \cdot \mathbf{i}_j\right) + \tau_j \mathbf{i}_j = \mathbf{0} \tag{2.67}$$

In the same way as leads to (2.59) we take the inner product with $\mathbf{i}_m$ on both sides and get

$$-t_{jk}\left(\mathbf{g}_k \cdot \mathbf{i}_m\right)\left(\mathbf{n}^0 \cdot \mathbf{i}_j\right) + \tau_j \delta_{jm} = 0 \tag{2.68}$$

and arrive at

$$\tau_m = \left(\mathbf{g}_k \cdot \mathbf{i}_m\right)\left(\mathbf{n}^0 \cdot \mathbf{i}_j\right) t_{jk} \tag{2.69}$$

Utilize (2.16) and resolve $\mathbf{n}^0$ in terms of $\mathbf{i}_j$ and get

$$\tau_m = \left(\left(\delta_{nk} + u_{n,k}\right)\mathbf{i}_n \cdot \mathbf{i}_m\right) n_j^0 t_{jk} \tag{2.70}$$

which with (2.4) yields the following expression for the *surface tractions* τ_m

Surface tractions τ_m

$$\tau_m = \left(\delta_{mk} + u_{m,k}\right) n_j^0 t_{jk} \tag{2.71}$$

This equation expresses the stress τ_m, the *Surface traction*, on any surface with the normal n_j^0 in the *undeformed* geometry in terms of the components of the Piola-Kirchhoff stress tensor t_{jk} and the displacement gradients $u_{m,k}$. As is the case for the static field equations (2.59) the displacement gradients enter the static boundary conditions (2.71), which makes the relations nonlinear. On the static boundary S_T^0 the surface tractions are prescribed, i.e. $\tau_m = \bar{\tau}_m$, and the static boundary conditions become

Static boundary conditions

$$\left(\delta_{mk} + u_{m,k}\right) n_j^0 t_{jk} = \bar{\tau}_m \,, \quad x_n \in S_T^0 \tag{2.72}$$

In the kinematically linear case, see Chapter 4, the static equations are linear and we shall study them in more detail.

2.4 Principle of Virtual Displacements

The *Principle of Virtual Work* takes different forms depending on the purpose. In the following we concentrate on the particular version, which is known as the *Principle of Virtual Displacements* because this is the most convenient one for kinematically nonlinear problems.

Principle of Virtual Work. Special case: Principle of Virtual Displacements

As hinted at in the Introduction of the present chapter, the principle of virtual work plays a central role in continuum mechanics. This may, however, not become obvious until Parts II and IV–VI where we exploit the principle of virtual work over and over for various purposes.

Actually, if we stopped after having derived the principle of virtual displacements, the whole idea of establishing the principle could seem like an exercise in futility because we start out with three equilibrium equations and end up with only one equation instead, suggesting that information has been lost. This is, however, not the case, as we shall see below. Also, in the manipulations below the direction we are headed may be unclear until the final stage, so the reader must trust that something positive and useful eventually results.

For convenience repeat (2.48), which expresses the three equilibrium equations

Equilibrium equations

$$\mathbf{t}_{i,i} + \bar{\mathbf{q}} = \mathbf{0} \tag{2.73}$$

Since this statement holds everywhere in the body we may multiply by an arbitrary, smooth vector field[2.14] $\boldsymbol{\alpha}$ to get

$$(\mathbf{t}_{i,i} + \bar{\mathbf{q}}) \cdot \boldsymbol{\alpha} = 0 \;\forall\; \boldsymbol{\alpha} \tag{2.74}$$

Already here, we have transformed the three equilibrium equations into one scalar equation which, on the other hand, does not mean that we have lost information, because $\boldsymbol{\alpha}$ is arbitrary.

Integrate (2.74) over the (undeformed) volume V^0 with the result

$$\int_{V^0} (\mathbf{t}_{i,i} + \bar{\mathbf{q}}) \cdot \boldsymbol{\alpha} dV^0 = 0 \;\forall\; \boldsymbol{\alpha} \tag{2.75}$$

which again possesses as much information as (2.73) because of the arbitrariness of $\boldsymbol{\alpha}$. For reasons that become clear later rewrite (2.75)

$$\int_{V^0} \Big((\mathbf{t}_i \cdot \boldsymbol{\alpha})_{,i} - \mathbf{t}_i \cdot \boldsymbol{\alpha}_{,i} + \bar{\mathbf{q}} \cdot \boldsymbol{\alpha} \Big)\, dV^0 = 0 \;\forall\; \boldsymbol{\alpha} \tag{2.76}$$

By application of the *Divergence Theorem* the first term is transformed

[2.14] The field must not contain singularities, but we shall not state this explicitly in the following and by $\forall\; \boldsymbol{\alpha}$ imply "$\forall$ non-singular $\boldsymbol{\alpha}$."

into a surface integral instead of a volume integral[2.15]

$$\int_{S^0} \mathbf{t}_i \cdot \boldsymbol{\alpha} n_i^0 dS^0 + \int_{V^0} \bar{\mathbf{q}} \cdot \boldsymbol{\alpha} dV^0 = \int_{V^0} \mathbf{t}_i \cdot \boldsymbol{\alpha}_{,i} dV^0 \; \forall \, \boldsymbol{\alpha} \tag{2.77}$$

In principle, (2.77) could be useful as it stands, but we shall offer an interpretation of the vector field $\boldsymbol{\alpha}$. First, let us add a small *variation*,[2.16] $\epsilon\delta\mathbf{r}$ to the position vector $\mathbf{r}$ and let $\mathbf{r}_{\text{tot}}$ denote the total value

Variations $\epsilon\delta\mathbf{r}$

$$\mathbf{r}_{\text{tot}} = \mathbf{r} + \epsilon\delta\mathbf{r} \tag{2.78}$$

Shape $\delta\mathbf{r}$
Amplitude ϵ

where $\delta\mathbf{r}$ is the *shape* and ϵ is the *amplitude*[2.17] of the *variation* of $\mathbf{r}$. For reasons that, hopefully, will be clear later, we shall only concern ourselves with values of ϵ which observe

$$|\epsilon| \ll 1 \tag{2.79}$$

As an illustration, Fig. 2.6 shows some function, or a field, $f(x)$ and its

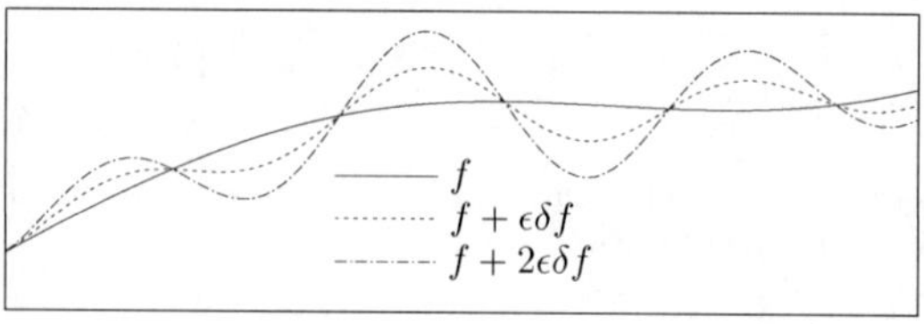

Fig. 2.6: A function and variations.

variations $f + \epsilon\delta f$ for two different values of ϵ. In order to make the drawing clear the magnitude of ϵ has been exaggerated. In this particular case the conditions enforced on the variation are that its value and first derivative at the left-hand end of the interval vanish, but many other conditions may be imposed.

Then, choose

$$\boldsymbol{\alpha} = \delta\mathbf{r} \;\Rightarrow\; \boldsymbol{\alpha}_{,i} = \delta\mathbf{r}_{,i} = \delta\mathbf{g}_i \tag{2.80}$$

[2.15] I expect you to know the divergence theorem—if you don't, go pick up your good old math books.

[2.16] I have chosen to attack the problem of variations directly here, but otherwise refer to Chapter 32 where the subject is dealt with in more detail. The derivation becomes a little more pedestrian this way, but I hope that it makes for easier reading.

[2.17] The "American" epsilon ϵ used here to denote a small quantity must not be confused with the other epsilon ε, which—with or without a number of subscripts—is employed as the symbol for strains.

When we recall (2.49)

$$\mathbf{t}_i = t_{ij}\mathbf{g}_j \tag{2.81}$$

and utilize (2.80) we may rewrite the right-hand side of (2.77)

$$\int_{S^0} t_{ij}\mathbf{g}_j \cdot \delta\mathbf{r} n_i^0 dS^0 + \int_{V^0} \bar{q}_i\mathbf{i}_i \cdot \delta\mathbf{r} dV^0 = \int_{V^0} t_{ij}\mathbf{g}_j \cdot (\delta\mathbf{r})_{,i} dV^0 \;\forall\; \delta\mathbf{r} \tag{2.82}$$

where we have resolved $\mathbf{t}_i$ in terms of the *deformed* base vectors $\mathbf{g}_j$ and the load $\bar{\mathbf{q}}$ in terms of the *undeformed* base vectors $\mathbf{i}_j$. Further, resolve $\mathbf{r}$ and $\delta\mathbf{r}$ in terms of the *undeformed* base vectors $\mathbf{i}_j$ and note that $(\delta\mathbf{r})_{,i} = \delta\mathbf{g}_i$ to get

$$\begin{aligned}&\int_{S^0} t_{ij}\mathbf{g}_j \cdot (\delta r_m \mathbf{i}_m)\, n_i^0 dS^0 + \int_{V^0} \bar{q}_i\mathbf{i}_i \cdot (\delta r_j \mathbf{i}_j)\, dV^0 \\ &= \int_{V^0} t_{ij}\mathbf{g}_j \cdot \delta\mathbf{g}_i dV^0 \;\forall\; \delta\mathbf{r}\end{aligned} \tag{2.83}$$

Before we proceed we need another expression for $\mathbf{g}_j \cdot \delta\mathbf{g}_i$. In order to get this, we compute

$$\begin{aligned}(\mathbf{g}_i \cdot \mathbf{g}_j)_{\text{tot}} &= (\mathbf{g}_i + \epsilon\delta\mathbf{g}_i) \cdot (\mathbf{g}_j + \epsilon\delta\mathbf{g}_j) \\ &= \mathbf{g}_i \cdot \mathbf{g}_j + \epsilon\,(\mathbf{g}_i \cdot \delta\mathbf{g}_j + \mathbf{g}_j \cdot \delta\mathbf{g}_i) + O(\epsilon^2)\end{aligned} \tag{2.84}$$

But, by (2.8b)

$$(\mathbf{g}_i \cdot \mathbf{g}_j)_{\text{tot}} = (g_{ij})_{\text{tot}} = g_{ij} + \epsilon\delta g_{ij} + O(\epsilon^2) \tag{2.85}$$

and therefore, under the assumption that ϵ is small, see (2.79)

$$\delta g_{ij} = \delta\,(\mathbf{g}_i \cdot \mathbf{g}_j) = \mathbf{g}_i \cdot \delta\mathbf{g}_j + \mathbf{g}_j \cdot \delta\mathbf{g}_i \tag{2.86}$$

Recall (2.14)

$$\mathbf{r} = \mathbf{r}^0 + \mathbf{u} = (x_j + u_j)\,\mathbf{i}_j \tag{2.87}$$

and the fact that $\mathbf{r}^0$ and x_j are given once the virgin state of the body is given. Then, all variations of $\mathbf{r}^0$ and x_j vanish with the result that (2.83) becomes

$$\int_{S^0} t_{ij}\mathbf{g}_j \cdot \mathbf{i}_m \delta u_m n_i^0 dS^0 + \int_{V^0} \bar{q}_j \delta u_j dV^0 = \int_{V^0} t_{ij}\tfrac{1}{2}\delta g_{ij} dV^0 \quad \forall\; \delta u_j \tag{2.88}$$

In order to get this, we have exploited that the *Kronecker delta* δ_{ij} is equal to $\mathbf{i}_i \cdot \mathbf{i}_j$, see (2.5), that t_{ij} and g_{ij} are symmetric, and that

$$\begin{aligned}t_{ij}\mathbf{g}_i \cdot \delta\mathbf{g}_j &= \tfrac{1}{2}\,(t_{ij}\mathbf{g}_i \cdot \delta\mathbf{g}_j + t_{ji}\mathbf{g}_j \cdot \delta\mathbf{g}_i) \\ &= \tfrac{1}{2}t_{ij}\,(\mathbf{g}_i \cdot \delta\mathbf{g}_j + \mathbf{g}_j \cdot \delta\mathbf{g}_i) \\ &= \tfrac{1}{2}t_{ij}\delta\,(\mathbf{g}_i \cdot \mathbf{g}_j) \\ &= \tfrac{1}{2}t_{ij}\delta g_{ij}\end{aligned} \tag{2.89}$$

As a consequence of the definition (2.11) of *Lagrange Strains*

$$\gamma_{ij} \equiv \tfrac{1}{2}\left(g_{ij} - \delta_{ij}\right) \tag{2.90}$$

we get

$$\delta\gamma_{ij} = \tfrac{1}{2}\delta\left(g_{ij} - \delta_{ij}\right) = \tfrac{1}{2}\delta g_{ij} \tag{2.91}$$

because the *Kronecker delta* is constant. By (2.19) this gives

$$\delta\gamma_{ij} = \tfrac{1}{2}\left(\delta u_{i,j} + \delta u_{j,i}\right) + \tfrac{1}{2}\left(u_{k,i}\delta u_{k,j} + \delta u_{k,i} u_{k,j}\right) \tag{2.92}$$

From (2.69)

$$\tau_m = \left(\mathbf{g}_k \cdot \mathbf{i}_m\right)\left(\mathbf{n}^0 \cdot \mathbf{i}_j\right) t_{jk} \tag{2.93}$$

it follows that

$$t_{ij}\mathbf{g}_j \cdot \mathbf{i}_m \delta u_m n_i^0 = \tau_m \delta u_m \tag{2.94}$$

and finally we arrive at the *Principle of Virtual Displacements*

Principle of Virtual Displacements

$$\int_{V^0} t_{ij}\delta\gamma_{ij} dV^0 = \int_{S^0} \tau_i \delta u_i dS^0 + \int_{V^0} \bar{q}_j \delta u_j dV^0 \;\forall\; \delta u_j \tag{2.95}$$

where we have interchanged the right-hand and left-hand sides.

Principle of virtual work: *Not* real work

When we investigate (2.95) we may see that the left-hand side is equal to the *virtual* work done by the *Piola-Kirchhoff Stresses* t_{ij} together with the *variation* $\delta\gamma_{ij}$ of the *Lagrange Strains* γ_{ij}, while the right-hand side expresses the *virtual* work done by the applied body force $\bar{q}$ and the surface tractions, i.e. applied loads $\bar{\tau}_i$ on S_T^0 and reactions τ_i on S_u^0 together with their associated displacement variations.[2.18]

In many cases we wish to[2.19]—and are able to—fulfill homogeneous kinematic boundary conditions on δu_i

$$\delta u_i = 0\;,\; x_j \in S_u^0 \tag{2.96}$$

and then the principle is

Principle of Virtual Displacements for $\delta u_i = 0,\ x_j \in S_u^0$

$$\int_{V^0} t_{ij}\delta\gamma_{ij} dV^0 = \int_{S_T^0} \bar{\tau}_i \delta u_i dS^0 + \int_{V^0} \bar{q}_j \delta u_j dV^0 \qquad \forall\; \delta u_j = 0,\ x_j \in S_u^0 \tag{2.97}$$

Actually, in most applications it is (2.97) rather than (2.95) we employ,

[2.18] The reason why I emphasize the word *virtual* here is that it is extremely important to note that the principle of virtual work does *not* entail *real* work.

[2.19] The reasons for this are probably not obvious at this point, so the reader will have to trust me on this.

but, since (2.97) is a special case of (2.95), you might want to focus on the latter only.

It is important to note that the assumptions behind the derivation of the *Principle of Virtual Displacements*, both (2.95) and (2.97), are[2.20]

- the loads and the stresses are in equilibrium and satisfy (2.59). Therefore, *no kinematic or constitutive* relations need apply to the stress field,
- the strain variations are derived from the displacements according to the strain-displacement relation, here given by (2.92), and that the displacement variation satisfies the *kinematic* boundary conditions. Thus, *no static or constitutive* conditions need apply to the variations of the strain-displacement field.

Kinematically admissible displacement field

A (variation of a) strain-displacement field which satisfies the strain-displacement relation and observes the appropriate[2.21] continuity and boundary conditions is called a *kinematically admissible displacement field.* Only such fields may be utilized in the *Principle of Virtual Displacements.*

2.4.1 The Budiansky-Hutchinson Dot Notation

Budiansky-Hutchinson Dot Notation

It is the property of (virtual) work inherent in the *Principle of Virtual Displacements*—and in other forms of the *Principle of Virtual Work*[2.22]—that makes the principle such a strong tool and foundation. I shall come back to this in Part II where we shall see how the principle serves as a useful and convenient basis for deriving theories for specialized continua such as beams and plates. In this connection, the so-called *Budiansky-Hutchinson (Dot) Notation*, see Chapter 33, proves to be a very convenient tool. When we utilize this, the principle may be expressed in the short form (33.14)

Principle of Virtual Displacements

$$\boldsymbol{\sigma} \cdot \delta \boldsymbol{\varepsilon} = \overline{\boldsymbol{T}} \cdot \delta \boldsymbol{u} \tag{2.98}$$

which is valid when

- the stress field $\boldsymbol{\sigma}$ is in equilibrium with the applied loads $\overline{\boldsymbol{T}}$,

[2.20] I emphasize the statements below so much because the experience of a long life as a teacher has proved to me that almost no student remembers the assumptions two days after I have gone through the derivations and told the class that they are very important.

[2.21] The meaning of the term "appropriate" depends on the actual version of the *Principle of Virtual Displacements*, see for example the differences between the restrictions on δu_j in (2.95) and in (2.97).

Also, in the above derivations we have assumed that the variation of the displacement field is continuous everywhere inside the body, i.e. continuous where the real displacement field is continuous, but sometimes, in particular in connection with formulation of *Finite Element Equations*, it proves convenient to abandon this requirement and add the appropriate terms to the principle. Since the presentation here is introductory, I have decided to avoid such complications.

[2.22] There exist other kinds of *Principle of Virtual Work* such as the *Principle of Virtual Forces* which is sometimes used in the linear case. In the nonlinear cases its formulation presents so great problems that it is rarely applied.

- the displacements $\boldsymbol{u}$ are "sufficiently smooth"[2.23] and satisfy the kinematic boundary conditions $\boldsymbol{u} = \bar{\boldsymbol{u}}$ on the kinematic boundary S_u,
- the displacement variations $\delta\boldsymbol{u}$ are "sufficiently smooth" and satisfy the homogeneous kinematic boundary conditions $\delta\boldsymbol{u} = \delta\bar{\boldsymbol{u}}$, i.e. vanish on the kinematic boundary S_u,
- the strain variations $\delta\boldsymbol{\varepsilon}$ are given by the strain-displacement relation, see e.g. (2.92), which is valid for the general three-dimensional case, or (33.18)

Strain variation $\delta\varepsilon$

$$\delta\boldsymbol{\varepsilon} = \mathbf{l}_1(\delta\boldsymbol{u}) + \mathbf{l}_{11}(\boldsymbol{u}, \delta\boldsymbol{u}) \tag{2.99}$$

which is written by use of the *Budiansky-Hutchinson Notation* and covers (2.92) as well as many other relevant strain-displacement relations.

Clearly, (2.98) does not display any information about the dimension of the body—it could just as well be a one-dimensional body such as a beam instead of the three-dimensional body treated here. Of course, in that case the different fields must be reinterpreted accordingly. In Part II we shall see how this is done and at this point merely note that the strength of the short notation employed in (2.98) is that it covers all sorts of bodies. Part VI, Section 33.8, contains a summary of interpretations of (2.98) as well as other relevant formulas for a number of different structures.

2.4.2 Generalized Strains and Stresses

Generalized strains and stresses

Work conjugate quantities

The strain and stress measures $\boldsymbol{\varepsilon}$ and $\boldsymbol{\sigma}$ are called *Generalized* in the sense that they do work together—are *Work Conjugate*. In Part II we derive continuum theories for a number of specialized continua and our guideline in this connection will be the requirement that the strain and stress measures are each other's work conjugate —are generalized quantities. At this point it is probably not self-evident why it is important that the strain and stress measures possess this property, and I refer the reader to Part II. It should be mentioned that the validity of minimum principles such as the *Principle of Minimum Potential Energy* and the upper- and lower-bound theorems of the theory of plasticity hinges on the fact that the participating strain and stress measures are generalized.[2.24]

[2.23] By this term I mean so smooth that the continuity of the body is not violated. This requirement is particular to the different types of structures. As an example, for the three-dimensional body the displacements themselves must be differentiable, while for a plate also the first derivatives of the transverse displacement component must also be differentiable.

[2.24] If the stresses and strains are not generalized you may find upper bound solutions to problems of perfect plasticity which appear to be *lower* than the equivalent lower bound solutions which, obviously, is wrong. Actually, this happened to one of my former colleagues when he worked on his Master's Thesis. The error disappeared when he used appropriate stress and strain measures.

2.5 Principle of Virtual Forces

There is a duality between kinematic and static quantities, which sometimes may be exploited.[2.25] By duality I refer to the fact that kinematic and static quantities appear in pairs, e.g. as generalized strains and their work conjugate generalized stresses, see Sections 2.4.2 and 33.2, and that there exist dual principles such as the *Principle of Virtual Displacements* and the *Principle of Virtual Forces.*

While we were able to establish the *Principle of Virtual Displacements* quite easily, see above, there seems to be no universally accepted *Principle of Virtual Forces* for the present case of large displacements and large strains. I shall therefore postpone derivation of this principle to the chapter on infinitesimal displacements and infinitesimal strains.

For large displacements: Problems with Principle of Virtual Forces

2.6 Constitutive Relations

The purpose of *Constitutive Relations* is to connect the strain and stress measures. This connection may be as simple as a linear relation between the stresses t_{ij} and the strains γ_{kl}, the so-called *Generalized Hooke's Law*,[2.26] or it may be more complicated and for instance involve information about the loading history of the material point in question, e.g. whether the point is subjected to further loading or it is experiencing unloading. Such materials display *Plastic* properties.

Generalized Hooke's "Law"

2.6.1 Hyperelastic Materials

The term *elastic* implies that all deformation is reversible, e.g. no matter how hard we pull on a bar of an elastic material it will always recover its original shape when the loading is removed. *Hyperelastic* materials are defined by the assumption of the existence of a *Strain Energy Function* $W(\gamma_{ij})$, also called the *Strain Energy Density*, with the property that

Hyperelasticity Strain energy function $W(\gamma_{ij})$ = Strain energy density $W(\gamma_{ij})$

$$t_{mn} = \frac{\partial W(\gamma_{ij})}{\partial \gamma_{mn}} \tag{2.100}$$

which presupposes that the stress is independent of the strain history. At this point we do not intend to proceed investigating the general case but limit ourselves to linear hyperelasticity. Omitting inconsequential constants, for linear hyperelasticity the form of $W(\gamma_{ij})$ must be

Linear hyperelasticity

$$W(\gamma_{ij}) = \tfrac{1}{2} E_{ijkl} \gamma_{ij} \gamma_{kl} \tag{2.101}$$

[2.25] Earlier this duality was often used to compute the displacements of beams by determining the bending moments in a so-called "conjugate beam," which was loaded by the curvature of the real beam. Such methods do not seem to be used anymore.

[2.26] It is important to note that *Hooke's Law* is a *material model* and not a law. No material obeys "Hooke's Law," but the relation (2.101) is a simple and useful material model which, by the way, is *the* most commonly used material model.

because differentiation of W given by (2.101) provides

$$t_{ij} = E_{ijkl}\gamma_{kl} \tag{2.102}$$

When we define linear hyperelasticity according to (2.101) it is obvious that E_{ijkl} may be assumed to possess *group symmetry* in the sense that

$$E_{ijkl} = E_{klij} \tag{2.103}$$

since any antisymmetric part of E_{ijkl} vanishes from the product on the right-hand side of (2.101). The linear (hyper)elastic model entails two more symmetry properties of E_{ijkl}, which we derive below. First, because t_{ij} is symmetric in its indices we must have

$$E_{ijkl} = E_{jikl} \tag{2.104}$$

and, secondly, since γ_{kl} is symmetric in its indices, without loss of generality, we may take

$$E_{ijkl} = E_{ijlk} \tag{2.105}$$

Isotropy

These relations hold for all materials whether they are *Isotropic*, i.e. their properties are independent of direction,[2.27] or they are *Anisotropic*,[2.28] which means that their properties depend on direction. Because of the symmetry relations (2.104) and (2.105) the original 81 constants in E_{ijkl} are reduced to only 36, and the group symmetry further reduces the number to 21 independent constants in the general, anisotropic case. If the material is special, e.g. isotropic, then the number of different constants is reduced further, but except for this very short introduction to the subject we defer discussion of constitutive relations, i.e. material models, to the section on the kinematically linear theory, Section 5.

Anisotropy

2.6.2 Plastic Materials

In Chapter 5 we discuss plastic material models and do not pursue the subject here.

2.7 Potential Energy

Principle of Virtual Displacements No information about material

The *Principle of Virtual Displacements*, see (2.97) or (2.98), does not entail any information about the material. This is not surprising since the principle was derived as an auxiliary way of expressing equilibrium. There exists a very important principle, which is valid for (hyper)elastic solids and structures and thus exploits the constitutive information. This principle is called the *Principle of Stationary Potential Energy.*

As usual in continuum mechanics there are several ways to arrive at

[2.27] To a good approximation, steel is such a material.
[2.28] Wood is such a material.

a result. Here, we postulate that the *Potential Energy*[2.29] of the three-dimensional *Hyperelastic* body is

$$\mathbf{\Pi}_{\mathrm{P}}(u_i) = \int_{V^0} W(\gamma_{ij})dV^0 - \int_{V^0} \bar{q}_i u_i dV^0 - \int_{S_T^0} \bar{\tau}_i u_i dS^0 \tag{2.106}$$

Potential energy, hyperelasticity

and investigate its properties below. Observe that for a given structure with a given load $\mathbf{\Pi}_{\mathrm{P}}$ does not depend on the stresses, but only on kinematic quantities, namely the displacements u_m and the strains $\gamma_{ij}(u_m)$, which according to (2.19) are given by the displacements

$$\gamma_{mn} = \tfrac{1}{2}\left(u_{m,n} + u_{n,m}\right) + \tfrac{1}{2} u_{k,m} u_{k,n} \tag{2.107}$$

where the displacement field u_m obviously must satisfy the condition that its first derivatives are defined in V^0. This, however, is not sufficient because we shall appeal to the *Principle of Virtual Displacements*, which presupposes that the displacements satisfy the kinematic boundary conditions $u_m = \bar{u}_m$ on S_u^0.

Furthermore, the variation $\delta\gamma_{mn}$ of the strain must be derived from the displacements u_m according to (2.92)

Strain variation $\delta\gamma_{mn}$ derived from displacement u_m and displacement variation δu_m

$$\delta\gamma_{mn} = \tfrac{1}{2}\left(\delta u_{m,n} + \delta u_{n,m}\right) + \tfrac{1}{2}\left(\delta u_{k,m} u_{k,n} + u_{k,m} \delta u_{k,n}\right) \tag{2.108}$$

According to Chapter 33, (33.20), the first variation of $\mathbf{\Pi}_{\mathrm{P}}$ is

$$\delta\mathbf{\Pi}_{\mathrm{P}}(u_i) = \int_{V^0} \frac{\partial W(\gamma_{ij})}{\partial \gamma_{kl}} \delta\gamma_{kl} dV^0 - \int_{V^0} \bar{q}_i \delta u_i dV^0 - \int_{S_T^0} \bar{\tau}_i \delta u_i dS^0 \tag{2.109}$$

which, when we utilize (2.100), gives

$$\delta\mathbf{\Pi}_{\mathrm{P}}(u_i) = \int_{V^0} t_{kl} \delta\gamma_{kl} dV^0 - \int_{V^0} \bar{q}_i \delta u_i dV^0 - \int_{S_T^0} \bar{\tau}_i \delta u_i dS^0 \tag{2.110}$$

When we require that $\delta\mathbf{\Pi}_{\mathrm{P}}(u_i)$ vanishes we arrive at the *Principle of Virtual Displacements* (2.97)

$$\begin{gathered}\int_{V^0} t_{ij} \delta\gamma_{ij} dV^0 = \int_{S_T^0} \bar{\tau}_i \delta u_i dS^0 + \int_{V^0} \bar{q}_j \delta u_j dV^0 \\ \forall \ \left(\delta u_j = 0 \,, x_j \in S_u^0\right)\end{gathered} \tag{2.111}$$

Note that in the derivation of (2.111) we have assumed hyperelasticity to be valid, while (2.97) is only based on equilibrium and is therefore valid for *all* material models.

[2.29] Potentials are discussed in some detail in Chapter 32 and in Chapter 33.

2.7.1 Linear Elasticity

For linear (hyper)elasticity, i.e. for *Hooke's "Law"* the expression for the potential energy becomes

Potential energy, linear hyperelasticity

$$\mathbf{\Pi}_{\mathrm{P}}(u_i) = \tfrac{1}{2}\int_{V^0} E_{ijkm}\gamma_{ij}\gamma_{km}dV^0 - \int_{V^0}\bar{q}_i u_i dV^0 - \int_{S_T^0}\bar{\tau}_i u_i dS^0 \quad (2.112)$$

This is an expression which is very often used in various connections, e.g. as a foundation for study of elastic buckling and other nonlinear problems.

2.8 Complementary Energy

We shall not attempt to establish a *Complementary Energy* for the present case of large displacements and large strains, see the comment in Section 2.5, and once more refer to the chapter on infinitesimal displacements and infinitesimal strains.

2.9 Static Equations by the Principle of Virtual Displacements

Derivation of equilibrium equations for generalized strains and stresses by the principle of virtual displacements

Deriving static equations in the spirit of Section 2.3 is not always a straightforward task, in particular if the type of continuum is a specialized one such as a plate or shell. Fortunately, there exists another way of getting the equilibrium equations, namely via the principle of virtual displacements. This approach is discussed below and utilized in Part II.

We may derive the equilibrium equations by use of the principle of virtual displacements observing the kinematic relations, i.e. the strain-displacement relation, the compatibility conditions, the kinematic (dis)continuity conditions, and the kinematic boundary conditions, rather than attempt to establish them in a more direct way. One reason for doing this is that in this way we insure that the stresses and strains are *generalized* in the sense that these quantities work together in producing the correct internal virtual work.

Difficult static plate boundary condition easily established by use of the principle of virtual displacements

Earlier, particularly in the twentieth century, when this approach was not *en vogue*—not known then, to be exact—there was much discussion about how to derive the static boundary conditions, in particular boundary conditions involving the shear force and the torsional moment in plates, see (9.25c). Another reason for the newer approach is that in specialized, kinematically nonlinear continuum theories, e.g. nonlinear theories for beams, plates, shells, etc., it is often extremely difficult to choose the static quantities in a meaningful way. If, however, they are defined through the *Principle of Virtual Displacements* they will always have a sound interpretation, albeit sometimes not very evident.

At a more philosophical level I mention that nobody has been able to measure continuum mechanical stresses, while sometimes it is possible to measure strains quite accurately. The reason for this is that the stress components are obtained as the limit of force per area, whereas the the strain components are given as the limit of changes in distance and direction, which is much easier to measure.

Stresses are figments of our imagination—strains are more real

Chapter 3

Kinematically Moderately Nonlinear Theory

In Chapter 2 a full nonlinear description of kinematics and statics was given. Since the derivation of the kinematics of the kinematically moderately nonlinear case follows exactly the same line, we will not redo that part of the foundation but refer the reader to Section 2.2. Moreover, in the three-dimensional case the savings by using this reduced theory are negligible, so we do not travel this path further, but refer to the formulas (2.36)–(2.41). And, finally, the equilibrium equations become *more* complicated in this case. Thus, instead of (2.59)

$$t_{ik,i} + \left(t_{ij}u_{k,j}\right)_{,i} + \bar{q}_k = 0 \tag{3.1}$$

by use of (2.38) we get

$$t_{ik,i} + \left(t_{ij}\left(e_{kj} + \omega_{kj}\right)\right)_{,i} + \bar{q}_k = 0 \tag{3.2}$$

The assumption that the strains are small and that the rotations are moderate $|\omega_{mn}| \ll 1$, but $|\omega_{mn}| > |e_{mn}|$, see page 42, entails that

$$t_{ik,i} + \left(t_{ij}\omega_{kj}\right)_{,i} + \bar{q}_k = 0 \tag{3.3}$$

which may be expressed in terms of the displacements instead of the rotation, see (2.37)

$$t_{ik,i} + \left(t_{ij}\left(u_{k,j} - u_{j,k}\right)\right)_{,i} + \bar{q}_k = 0 \tag{3.4}$$

which looks more complicated than (3.1) and seems to offer no advantages over the latter.

On the other hand, for one- and two-dimensional bodies such as beams and plates the kinematically moderately nonlinear theories offer great advantages in terms of computational ease, and we shall return to them in Part II.

E. Byskov, *Elementary Continuum Mechanics for Everyone*,
Solid Mechanics and Its Applications 194, DOI: 10.1007/978-94-007-5766-0_3,

Chapter 4

Infinitesimal Theory

4.1 Introduction

Probably some 99% of all structural analyses are based on the assumption of kinematic linearity. Therefore, we shall study this special case in much detail, but caution that you must understand the limitations of the linear theories in order to use them judiciously. If you do not know when kinematic nonlinearities are important then you may—and most likely will—commit serious errors when you apply a linear theory instead of a nonlinear one. Choosing a theory of a more limited range of validity is central to a good structural engineer and at the same time a task that sometimes is very difficult.

Most computations are kinematically linear

4.2 Kinematics and Deformation

Here, we exploit the kinematic relations derived in Section 2.2, but emphasize that we might go about the task in another way and establish the formulas by assuming linearity *ab initio*. However, because of the reasons given in Chapter 1, we utilize the results from the general case.

Linear theory by linearization of nonlinear theory

4.2.1 Kinematics and Strain

Recall the formula (2.19) for the Lagrange Strain γ_{mn}

$$\gamma_{mn} = \tfrac{1}{2}\left(u_{m,n} + u_{n,m}\right) + \tfrac{1}{2}u_{k,m}u_{k,n} \tag{4.1}$$

Lagrange strain γ_{mn}

which expresses the Lagrange Strain in terms of the displacement gradients $u_{m,n}$. The *Infinitesimal Strain Tensor* e_{mn} is given by (2.36)

$$e_{mn} \equiv \tfrac{1}{2}\left(u_{m,n} + u_{n,m}\right) \tag{4.2}$$

Infinitesimal strain e_{mn}

and the *Infinitesimal Rotation Tensor* ω_{mn} is defined by (2.37)

$$\omega_{mn} \equiv \tfrac{1}{2}\left(u_{m,n} - u_{n,m}\right) = -\omega_{nm} \tag{4.3}$$

Infinitesimal rotation ω_{mn}

Finally, the expression (2.39)

$$\gamma_{mn} = e_{mn} + \tfrac{1}{2}\left(e_{km} + \omega_{km}\right)\left(e_{kn} + \omega_{kn}\right) \tag{4.4}$$

provides insight into the approximations inherent in the linearization of the

E. Byskov, *Elementary Continuum Mechanics for Everyone*,
Solid Mechanics and Its Applications 194, DOI: 10.1007/978-94-007-5766-0_4,

strain measure. When we define the infinitesimal strain tensor ε_{mn} as[4.1]

Infinitesimal strain $\varepsilon_{mn} = e_{mn}$

$$\varepsilon_{mn} \equiv e_{mn} = \tfrac{1}{2}\left(u_{m,n} + u_{n,m}\right) \tag{4.5}$$

we may see that not only the strains, but also the rotations are assumed to be small, i.e. $|\varepsilon_{mn}| \ll 1$ and $|\omega_{mn}| \ll 1$, respectively. If we compare (4.5) with (4.1) we realize that the assumption of linearity also implies that the displacement gradients are small, i.e. $|u_{m,n}| \ll 1$.

4.2.2 Strain Compatibility Equations

Compatibility equations insure integrability of strains

In Section 2.2.4 I mentioned that not all strain fields are compatible and therefore there is no guarantee that a given strain field can be integrated to a displacement field. The so-called *Strain Compatibility Equations* must be satisfied in order for this to be the case. In principle, the derivation goes as follows, see Fig. 4.1:

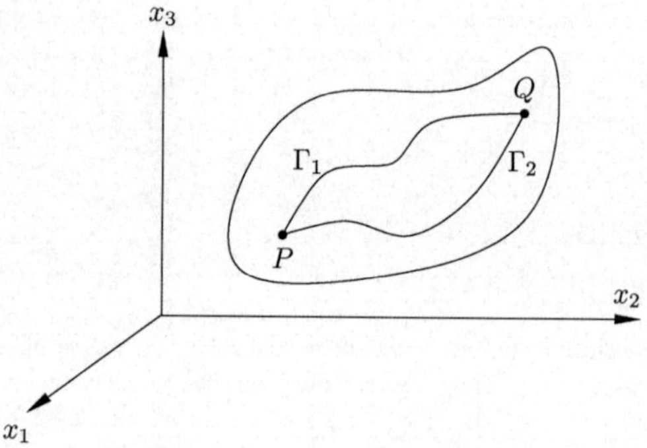

Fig. 4.1: Two integration paths.

1. Assume that the displacements $u_j(x_k^P)$ at point P are known.
2. Find the displacements $u_j(x_k^Q)$ at another point Q by integrating the strains ε_{ij} and the rotations ω_{ij} or, equivalently, the displacement gradients $u_{i,j}$.
3. Demand that the integral is path independent. This is an obvious requirement because the displacements of a point must not depend on the integration path one has traveled to get to that point. In Fig. 4.1 two such paths, Γ_1 and Γ_2 are shown.

[4.1] It may be confusing that the same quantity is denoted ε_{mn} as well as e_{mn}. The reason that both notations are kept here is that they both are very common in the international literature. To make the situation even more complex the permutation symbol is denoted e_{ijk} and e_{ij} in the two- and three-dimensional case, respectively, see (31.19) and (31.12).

Other ways of obtaining the compatibility equations

There is another way of getting the compatibility equations where different expressions for the strains are differentiated and combined to get the required result. Personally, I have never liked that procedure because it does not address the fundamental issue of integrability, i.e. path independence, directly, but see e.g. (Malvern 1969).

Among the procedures that are based on requiring path independence Sokolnikoff (1956) and Pearson (1959) are worth mentioning. Here, however, the derivation below lies closer to the ones found in the books by Fung (1965), by Fung & Tong (2001) and by Reismann & Pawlik (1980).[4.2]

As you will see, not all of the manipulations below are straightforward—actually a number of them seem strange, but I intend to explain them as far as is possible for me.

In order to find the displacements u_j^Q at point Q simply write

$$u_j^Q = u_j^P + \int_P^Q du_j \quad \text{with} \quad u_j^Q \equiv u_j(x_i^Q) \quad \text{and} \quad u_j^P \equiv u_j(x_i^P) \tag{4.6}$$

or

$$u_j^Q = u_j^P + \int_P^Q u_{j,k} dx_k = u_j^P + \int_P^Q \varepsilon_{jk} dx_k + \int_P^Q \omega_{jk} dx_k \tag{4.7}$$

where we have exploited the fact that (4.5) and (4.3) combined provide

$$\varepsilon_{jk} + \omega_{jk} = \tfrac{1}{2}(u_{j,k} + u_{k,j}) + \tfrac{1}{2}(u_{j,k} - u_{k,j}) = u_{j,k} \tag{4.8}$$

The question of path independence is now equivalent to establishing the conditions for path independence of the sum of the two integrals in (4.7).

If we can get rid of the rotation term in (4.7)—and hopefully write it in terms of the strain—it seems likely that we can get the relation between the strains or strain derivatives that is necessary for integrability. Therefore, we try to rewrite in terms of the displacement gradients in the hope that we may end up with an expression involving only the strains, possibly their derivatives. First, however, write

$$\omega_{jk} dx_k = d(\omega_{jk} x_k) - x_k \omega_{jk,l} dx_l \tag{4.9}$$

With this in hand, get

$$u_j^Q = u_j^P + \left[\omega_{jk} x_k\right]_P^Q + \int_P^Q \left(\varepsilon_{jl} - x_k \omega_{jk,l}\right) dx_l \tag{4.10}$$

We wish to write the term involving the derivative of the rotation in terms of the derivatives of the strain.

[4.2] Originally, my inspiration was notes found on the Internet and written by L. Schmerr of *Iowa State University* for the course *EM 424*. Dr. Schmerr was so kind as pointing me in the direction of Fung (1965) and Reismann & Pawlik (1980).

Therefore, note that

$$\omega_{jk,l} = \tfrac{1}{2}(u_{j,kl} - u_{k,jl}) = \tfrac{1}{2}(u_{j,kl} - u_{k,jl}) + \tfrac{1}{2}(u_{l,jk} - u_{l,kj}) \tag{4.11}$$

where the last term obviously vanishes because of symmetry in indices k and j. Rearranging of terms provides

$$\begin{aligned}\omega_{jk,l} &= \tfrac{1}{2}(u_{j,lk} - u_{l,jk}) - \tfrac{1}{2}(u_{k,lj} - u_{l,kj}) \\ &= \tfrac{1}{2}(u_{j,l} - u_{l,j})_k - \tfrac{1}{2}(u_{k,l} - u_{l,k})_j \\ &= \varepsilon_{jl,k} - \varepsilon_{kl,j}\end{aligned} \tag{4.12}$$

Thus, (4.6) or (4.7) may be written

$$u_j^Q = u_j^P + \big[\omega_{jk} x_k\big]_P^Q + \int_P^Q \Big(\varepsilon_{jl} - x_k\big(\varepsilon_{jk,l} - \varepsilon_{kl,j}\big)\Big) dx_l \tag{4.13}$$

where the bracketed term $[\omega_{jk} x_k]_P^Q$ contains two terms that are associated with the end-points, not the integration path. Therefore, we may concentrate on the integral, and inspection of (4.13) may show us that it is of the form

$$\int_P^Q \Phi_{jl} dx_l \,, \text{ where } \Phi_{jl} \equiv \varepsilon_{jl} - x_k\big(\varepsilon_{jl,k} - \varepsilon_{kl,j}\big) \tag{4.14}$$

The necessary and sufficient condition for path independence of the integral in (4.14) is that the integrand is an exact differential, see e.g. (Kreyszig 1993)

$$\Phi_{jl,m} = \Phi_{jm,l} \quad \text{or} \quad 0 = \Phi_{jl,m} - \Phi_{jm,l} \tag{4.15}$$

Therefore, compute

$$\begin{aligned}\Phi_{jl,m} &= \varepsilon_{jl,m} - \delta_{km}\big(\varepsilon_{jl,k} - \varepsilon_{kl,j}\big) - x_k\big(\varepsilon_{jl,km} - \varepsilon_{kl,jm}\big) \\ &= \varepsilon_{jl,m} - \varepsilon_{jl,m} - \varepsilon_{ml,j} - x_k\big(\varepsilon_{jl,km} - \varepsilon_{kl,jm}\big) \\ \Phi_{jm,l} &= \varepsilon_{jm,l} - \delta_{kl}\big(\varepsilon_{jm,k} - \varepsilon_{km,j}\big) - x_k\big(\varepsilon_{jm,kl} - \varepsilon_{km,jl}\big) \\ &= \varepsilon_{jm,l} - \varepsilon_{jm,l} - \varepsilon_{lm,j} - x_k\big(\varepsilon_{jm,kl} - \varepsilon_{km,jl}\big)\end{aligned} \tag{4.16}$$

When we exploit the symmetry in the indices of ε_{ij} the first two terms on the right-hand sides vanish and, utilizing the same symmetry, the third terms in $\Phi_{jl,m}$ and $\Phi_{jm,l}$ cancel each other when inserted into (4.15b). Thus, the requirement for integrability becomes

$$0 = x_k\big(\varepsilon_{jl,km} - \varepsilon_{kl,jm} - \varepsilon_{jm,kl} + \varepsilon_{km,jl}\big) \tag{4.17}$$

and, since this must hold independent of the value of x_k,the following 81 equations ensue

$$0 = \varepsilon_{jl,km} - \varepsilon_{kl,jm} - \varepsilon_{jm,kl} + \varepsilon_{km,jl} \tag{4.18}$$

Compatibility equations

but some of these are just identities and therefore do not provide useful information, and others are mere repetitions since many of the expressions exhibit symmetry, and in the end there are only six seemingly independent equations

$$\begin{aligned}
\varepsilon_{11,23} &= \left(-\varepsilon_{23,1} + \varepsilon_{31,2} + \varepsilon_{12,3}\right)_{,1} \\
\varepsilon_{22,31} &= \left(-\varepsilon_{31,2} + \varepsilon_{12,3} + \varepsilon_{23,1}\right)_{,2} \\
\varepsilon_{33,12} &= \left(-\varepsilon_{12,3} + \varepsilon_{23,1} + \varepsilon_{31,2}\right)_{,3} \\
2\varepsilon_{12,12} &= \varepsilon_{11,22} + \varepsilon_{22,11} \\
2\varepsilon_{23,23} &= \varepsilon_{22,33} + \varepsilon_{33,22} \\
2\varepsilon_{31,31} &= \varepsilon_{33,11} + \varepsilon_{11,33}
\end{aligned} \tag{4.19}$$

The 6 (independent?) compatibility equations

Most authors claim that in reality there are only three independent relations, while others insist that there are four. I do not intend to investigate this problem, and conclude with the comment that if a given strain field satisfies all six equations, then it is compatible and may therefore be integrated and thus provide a displacement field. In that case, it is less important if the six relations are not all independent. Furthermore, in many cases we determine the displacement field without integrating the strains and, in that case, the strains are obviously compatible.

Number of independent compatibility equations between 3 and 6

Although it does not appear explicitly in the above derivations the conditions for strain compatibility presuppose that the body—the domain—is simply connected, i.e. that it does not contain holes. In the case of a multiply connected region it may be converted into a simply connected one by application of the trick used in Section 13.6.3.2, see especially Fig. 13.6. However, since most of this book is concerned with methods that determine the displacements without first postulating a strain field we shall leave the question of compatibility here, but refer to the book by Fung (1965).

4.2.3 Kinematic Boundary Conditions

Especially for the present case with infinitesimal displacements the kinematic boundary conditions for a three-dimensional body are usually quite self-evident. Therefore, we do not go into any detail at this point but refer to Part II, where the problem is much more prominent.

Kinematic boundary conditions

4.2.4 Interpretation of Strain Components

In order to interpret the components of the strain tensor ε_{mn} it is customary to consider drawings of "infinitesimal" squares, which is all right, but not

Interpretation of strains

the most revealing way to do things, especially since we have established the strain-displacement relations by vector algebra, see Section 2.2. Therefore, we intend to exploit the definitions and other relations given in Section 2.2, in particular (2.8) and (2.11)

$$\mathbf{g}_i \equiv \mathbf{r}_{,i}\,, \quad g_{ij} \equiv \mathbf{g}_i \cdot \mathbf{g}_j \quad \text{and} \quad \gamma_{ij} \equiv \tfrac{1}{2}\left(g_{ij} - \delta_{ij}\right) \tag{4.20}$$

On the other hand, it is pertinent to realize that in the present context we are dealing with a linearized theory.

There are two basically different components of the strain tensor, namely the ones where both indices are the same, and the ones where the indices are different.

4.2.4.1 Both Indices Equal

No summation over upper-case indices

Recall that the summation convention only applies to repeated *lower-case* indices and note that

$$\gamma_{JJ} \equiv \tfrac{1}{2}\left(g_{JJ} - \delta_{JJ}\right) = \tfrac{1}{2}\left(g_{JJ} - 1\right) \tag{4.21}$$

and thus

$$\begin{aligned} \gamma_{JJ} &= \tfrac{1}{2}\left(\mathbf{g}_J \cdot \mathbf{g}_J - 1\right) = \tfrac{1}{2}\left(|\mathbf{g}_J|^2 - |\mathbf{i}_J|^2\right) \\ &= \tfrac{1}{2}\left(|\mathbf{g}_J| + |\mathbf{i}_J|\right)\left(|\mathbf{g}_J| - |\mathbf{i}_J|\right) \end{aligned} \tag{4.22}$$

and

$$\varepsilon_{JJ} \approx \left(|\mathbf{g}_J| - |\mathbf{i}_J|\right) \tag{4.23}$$

because

$$|\mathbf{g}_J| + |\mathbf{i}_J| \approx 2 \tag{4.24}$$

Equal indices $\sim$ change of length

Thus, (4.23) shows that the components of the strain tensor, which have both indices equal, express the change in length of the line element along the axis in question *for the kinematically linear theory.* When you stop and think about it, it will not come as a surprise since the *Lagrange Strain Measure* γ_{ij} is based on the change of length of arbitrary radii of a unit sphere. If we consider the definition of the infinitesimal strain tensor ε_{ij}, (4.5)

$$\varepsilon_{mn} \equiv e_{mn} = \tfrac{1}{2}\left(u_{m,n} + u_{n,m}\right) \tag{4.25}$$

we may immediately see that

$\varepsilon_{JJ} \sim$ change in length of $\mathbf{i}_J$

$$\varepsilon_{JJ} = e_{JJ} = u_{J,J} \tag{4.26}$$

which clearly expresses a change of length.

4.2.4.2 Different Indices

Here,

$$\gamma_{JK} \equiv \tfrac{1}{2}\left(g_{JK} - \delta_{JK}\right) = \tfrac{1}{2} g_{JK} \tag{4.27}$$

i.e.

$$2\gamma_{JK} = \mathbf{g}_J \cdot \mathbf{g}_K = |\mathbf{g}_J||\mathbf{g}_K| \cos(\psi_{JK}) \tag{4.28}$$

where ψ_{JK} is the angle between $\mathbf{g}_J$ and $\mathbf{g}_K$. Introduce the *decrease* of the angle between $\mathbf{i}_J$ and $\mathbf{i}_K$, i.e. $\varphi_{JK} \equiv \frac{\pi}{2} - \psi_{JK}$

$$\begin{aligned} 2\gamma_{JK} &= |\mathbf{g}_J||\mathbf{g}_K| \cos(\tfrac{\pi}{2} - \varphi_{JK}) = |\mathbf{g}_J||\mathbf{g}_K| \sin(\varphi_{JK}) \\ &\approx |1 + \gamma_{JJ}||1 + \gamma_{KK}| \varphi_{JK} \approx \varphi_{JK} \end{aligned} \tag{4.29}$$

For $J \neq K$: $\varepsilon_{JK} \sim \frac{1}{2}\times$ decrease in angle between $\mathbf{i}_J$ and $\mathbf{i}_K$

which means that

$$\varepsilon_{JK} = \tfrac{1}{2}\varphi_{JK} \tag{4.30}$$

Thus, the component ε_{JK} for $J \neq K$ expresses half the decrease in angle between the deformed base vectors $\mathbf{g}_J$ and $\mathbf{g}_K$ in relation to the angle between the undeformed base vectors $\mathbf{i}_J$ and $\mathbf{i}_K$, which always are at right angles in a Cartesian coordinate system.

You may, of course, make a drawing of $u_{J,K}$ and $u_{K,J}$ and in that way visualize the change in angle. If you do it right, you are bound to come up with the same answer as the one given by (4.30). I shall, however, leave this task to your own initiative.

4.2.5 Transformation of Strain

Transformation of strain

In a number of situations we would like to know how a given strain state is expressed in two different (Cartesian) coordinate systems. This is a subject that, obviously, is closely connected with Section 4.2.6 below. Let the first coordinate system be given by its base vectors $\mathbf{i}_j$ and denote the coordinates in this system by x_k. Similarly, the base vectors of the second coordinate systems are $\mathbf{i}^\star_j$ and the coordinates given by $x^\star_k$. All quantities that are expressed in terms of the second coordinate system are supplied with a star: *.

Note that the indices in both coordinate systems are denoted $_i$, $_j$, etc. and that there is no (compelling) reason to distinguish the indices by e.g. introducing $_{i^\star}$, $_{j^\star}$ etc. for the second coordinate system. Of course, not doing that means that we must be a little careful, but recall that the indices merely serve as counters.

We wish to express the same physical quantities in the two coordinate systems, beginning with the expressions for the strain tensor

$$\varepsilon_{ij} = \tfrac{1}{2}\left(u_{i,j} + u_{j,i}\right) \text{ and } \varepsilon^\star_{ij} = \tfrac{1}{2}\left(\frac{\partial u^\star_i}{\partial x^\star_j} + \frac{\partial u^\star_j}{\partial x^\star_i}\right) \tag{4.31}$$

where we have been forced to use the long notation for the partial derivatives in the second coordinate system because the comma notation has been reserved for derivatives in the first.

4.2.5.1 Transformation of Coordinates

The position vector $\mathbf{r}$ of a point, say P, is

Transformation of coordinates

$$\mathbf{r} = x_i \mathbf{i}_i = x_i^\star \mathbf{i}_i^\star \tag{4.32}$$

From this we get

$$x_i \mathbf{i}_i \cdot \mathbf{i}_j = x_i^\star \mathbf{i}_i^\star \cdot \mathbf{i}_j \;\text{ or }\; x_j = \mathbf{i}_i^\star \cdot \mathbf{i}_j x_i^\star \;\text{ because }\; \mathbf{i}_i \cdot \mathbf{i}_j = \delta_{ij} \tag{4.33}$$

where δ_{ij} is the *Kronecker delta*, see (2.5). Similarly,

$$x_i^\star \mathbf{i}_i^\star \cdot \mathbf{i}_j^\star = x_i \mathbf{i}_i \cdot \mathbf{i}_j^\star \;\text{ or }\; x_j^\star = \mathbf{i}_i \cdot \mathbf{i}_j^\star x_i \;\text{ because }\; \mathbf{i}_i^\star \cdot \mathbf{i}_j^\star = \delta_{ij} \tag{4.34}$$

The products $\mathbf{i}_i^\star \cdot \mathbf{i}_j$ and $\mathbf{i}_i \cdot \mathbf{i}_j^\star$ designate the *directional cosines*. For brevity we introduce

$$\alpha_{ij} \equiv \mathbf{i}_i^\star \cdot \mathbf{i}_j \;\text{ and }\; \alpha_{ij}^\star \equiv \mathbf{i}_i \cdot \mathbf{i}_j^\star \tag{4.35}$$

and thus we may express the transformations as

$$x_j = \alpha_{ji}^\star x_i^\star \;\text{ and }\; x_j^\star = \alpha_{ji} x_i \tag{4.36}$$

Because the scalar product obeys the commutative law the following relation holds

$$\alpha_{ij} = \alpha_{ji}^\star \tag{4.37}$$

From (4.36a) and (4.36b)

$$x_j = \alpha_{ji}^\star x_i^\star = \alpha_{ji}^\star \alpha_{ik} x_k \tag{4.38}$$

which allows us to conclude that

$$\alpha_{ji}^\star \alpha_{ik} = \delta_{jk} \;\Rightarrow\; \alpha_{ji}^\star = (\alpha_{ji})^{-1} = (\alpha_{ij}^\star)^{-1} = (\alpha_{ij}^\star)^T \tag{4.39}$$

where superscript $^{-1}$ denotes the inverse and superscript ^{T} the transpose.

4.2.5.2 Transformation of Components of Displacement and Components of Strain

We may express the displacement vector $\mathbf{u}$ in both coordinate systems

$$\mathbf{u} = u_i \mathbf{i}_i = u_i^\star \mathbf{i}_i^\star \tag{4.40}$$

By manipulations analogous to the ones that led to (4.36) we may get expressions for the displacements in the two coordinate systems

Transformation of components of displacement

$$u_j = \alpha^\star_{ji} u^\star_i \quad \text{and} \quad u^\star_j = \alpha_{ji} u_i \tag{4.41}$$

and for the displacement gradients $u_{j,k}$

$$\begin{aligned} u_{j,k} &= \frac{\partial u_j}{\partial x_k} = \frac{\partial(\alpha^\star_{jm} u^\star_m)}{\partial x_k} = \alpha^\star_{jm} \frac{\partial u^\star_m}{\partial x^\star_n} \frac{\partial x^\star_n}{\partial x_k} \\ &= \alpha^\star_{jm} \frac{\partial u^\star_m}{\partial x^\star_n} \frac{\partial(\alpha_{np} x_p)}{\partial x_k} = \alpha^\star_{jm} \frac{\partial u^\star_m}{\partial x^\star_n} \alpha_{np} \delta_{pk} \end{aligned} \tag{4.42}$$

$$\text{i.e} \quad u_{j,k} = \alpha^\star_{jm} \alpha^\star_{kn} \frac{\partial u^\star_m}{\partial x^\star_n} \tag{4.43}$$

which makes it possible to express the transformation between ε_{ij} and $\varepsilon^\star_{km}$

$$\begin{aligned} \varepsilon_{ij} &= \tfrac{1}{2} \left(\alpha^\star_{ik} \alpha^\star_{jm} \frac{\partial u^\star_k}{\partial x^\star_m} + \alpha^\star_{jk} \alpha^\star_{im} \frac{\partial u^\star_k}{\partial x^\star_m} \right) \\ &= \tfrac{1}{2} \left(\alpha^\star_{ik} \alpha^\star_{jm} \frac{\partial u^\star_k}{\partial x^\star_m} + \alpha^\star_{jm} \alpha^\star_{ik} \frac{\partial u^\star_m}{\partial x^\star_k} \right) \\ &= \alpha^\star_{ik} \alpha^\star_{jm} \tfrac{1}{2} \left(\frac{\partial u^\star_k}{\partial x^\star_m} + \frac{\partial u^\star_m}{\partial x^\star_k} \right) \end{aligned} \tag{4.44}$$

Transformation of components of strain

$$\text{i.e} \quad \varepsilon_{ij} = \alpha^\star_{ik} \alpha^\star_{jm} \varepsilon^\star_{km} \tag{4.45}$$

Above, we have exploited the fact that indices $_k$ and $_m$ are dummy indices and therefore may be exchanged. By going through similar operations but taking the expression for $\varepsilon^\star_{km}$ as our point of departure, we may get the inverse relationship

Transformation of components of strain

$$\varepsilon^\star_{ij} = \alpha_{ik} \alpha_{jm} \varepsilon_{km} \tag{4.46}$$

Relations analogous to (4.45) and (4.46) are, of course, also needed if one—or both—of the coordinate systems are curvilinear. However, since we have limited our presentation to Cartesian coordinate systems we do not cite such formulas, but refer the reader to more advanced books, e.g. (Malvern 1969).

4.2.6 Principal Strains

It is a natural thought that the maximum or minimum strain may determine if a material breaks or stays in one piece.[4.3] Therefore—and for other reasons, which include formulation of constitutive equations[4.4] —we investigate

[4.3] At least for *isotropic* materials, see Sections 2.6 and 5, it is relevant, while for anisotropic materials this assumption does not hold. For instance, an anisotropic material may be able to absorb large deformations in one direction and, though the strain in that direction may be larger than in another, it may very well be the second, smaller strain that causes the material to fracture.

[4.4] Again, primarily for isotropic materials.

the magnitude of the strain at a given point as it varies with direction.Our intention is therefore to find relations that determine the strain extrema, the *Principal Strains*. Thus, we wish to compute the strain at a point P in a direction given by the unit vector $\mathbf{n}$ with components n_j. For this purpose we shall exploit (4.46). First, introduce a new Cartesian coordinate system given by its base vectors $\mathbf{i}_i^\star$, where

Principal strains

$$\mathbf{i}_1^\star = n_j \mathbf{i}_j \tag{4.47}$$

where the direction of the two other axes is unimportant, except that they must be mutually orthogonal and at right angles to $\mathbf{i}_i^\star$, i.e. $\mathbf{i}_1^\star \cdot \mathbf{i}_2^\star = \mathbf{i}_1^\star \cdot \mathbf{i}_3^\star = \mathbf{i}_2^\star \cdot \mathbf{i}_3^\star = 0$. Thus, in this case, $\mathbf{i}_1^\star$ is the same as the unit vector $\mathbf{n}$. Then, the strain $\varepsilon_{11}^\star$ describes the change in length of $\mathbf{n}$, see Section 4.2.4. Now, (4.46) provides

$$\varepsilon_{11}^\star = \alpha_{1k}\alpha_{1m}\varepsilon_{km} \tag{4.48}$$

and (4.35a) gives

$$\alpha_{1k} = \mathbf{i}_1^\star \cdot \mathbf{i}_k \tag{4.49}$$

and thus

$$\alpha_{1k} = n_j \mathbf{i}_j \cdot \mathbf{i}_k = n_j \delta_{jk} = n_k \tag{4.50}$$

The strain tensor is ε_{mn} and its projection, or component, $\varepsilon^{(n)}$ in the direction of n_j then is

$$\varepsilon_{11}^\star = \varepsilon^{(n)} = n_m n_n \varepsilon_{mn} \tag{4.51}$$

where $^{(n)}$ indicates the dependence on the unit vector $\mathbf{n}$.

Extrema of strain

In order to determine the extrema—or at least stationary values—of $\varepsilon^{(n)}$ we may not indiscriminately differentiate (4.51) with respect to n_k because we must fulfill the auxiliary condition that n_k is a unit vector

$$n_k n_k = 1 \tag{4.52}$$

Lagrange Multiplier

There are several ways to handle the problem of satisfying this auxiliary condition. Here, I choose the method of *Lagrange Multipliers*, see Chapter 32 and Chapter 33, because it is simpler and usually supplies us with more information than more "direct methods." In (4.51) $\varepsilon^{(n)}$ is a functional of n_k alone since ε_{jm} are constants—which, of course, usually vary from one point in the structure to the next. Thus, our task is to make $\varepsilon^{(n)}$ stationary with respect to n_k, subject to the auxiliary condition (4.52), which we rewrite

$$\mathbf{\Psi}(n_k) = 0 \text{ where } \mathbf{\Psi}(n_k) \equiv 1 - n_k n_k \tag{4.53}$$

For conformity with Chapter 32 we introduce the notation

$$\mathbf{\Pi}(n_k) = n_j n_m \varepsilon_{jm} \, , \text{ i.e. } \mathbf{\Pi}(n_k) = \varepsilon^{(n)} \tag{4.54}$$

From (4.53) and (4.54) we construct the *Modified Potential*, or *Modified Functional*, $\mathbf{\Pi}_\mathrm{M}$

$$\mathbf{\Pi}_\mathrm{M}(n_k, \eta) \equiv \mathbf{\Pi}(n_k) + \eta \mathbf{\Psi}(n_k) \tag{4.55}$$

Modified potential $\mathbf{\Pi}_\mathrm{M}$

where η is the *Lagrange Multiplier*, see Chapter 32 and Chapter 33. More explicitly, (4.55) is

Lagrange Multiplier η

$$\mathbf{\Pi}_\mathrm{M}(n_k, \eta) = n_j n_m \varepsilon_{jm} + \eta(1 - n_k n_k) \tag{4.56}$$

In (4.55) and (4.56) there are no auxiliary conditions to fulfill, and we may therefore take variations with respect to n_l and η and require that the variation vanishes

$$\delta \mathbf{\Pi}_\mathrm{M}(n_k, \eta) = \frac{\partial \mathbf{\Pi}_\mathrm{M}}{\partial n_l} \delta n_l + \frac{\partial \mathbf{\Pi}_\mathrm{M}}{\partial \eta} \delta \eta = 0 \; \forall \; (\delta n_l, \delta \eta) \tag{4.57}$$

i.e.

$$0 = \left(\varepsilon_{jm} n_j \frac{\partial n_m}{\partial n_l} + \varepsilon_{jm} \frac{\partial n_j}{\partial n_l} n_m - \eta 2 n_k \frac{\partial n_k}{\partial n_l} \right) \delta n_l + (1 - n_k n_k) \, \delta \eta \tag{4.58}$$

Since the partial derivative of n_k with respect to n_l is the *Kronecker delta*

$$\frac{\partial n_k}{\partial n_l} = \delta_{kl} \tag{4.59}$$

and because ε_{jm} is symmetric (4.58) may provide

$$0 = 2 \left(\varepsilon_{jl} n_j - \eta n_l \right) \delta n_l + (1 - n_k n_k) \, \delta \eta \tag{4.60}$$

which, by use of the *Kronecker delta*, gives

$$0 = 2 \left(\varepsilon_{jl} - \eta \delta_{jl} \right) n_j \delta n_l + (1 - n_k n_k) \, \delta \eta \tag{4.61}$$

As δn_k and $\delta \eta$ are arbitrary their coefficients must vanish, and thus (4.61) furnishes both (4.62)

$$\left(\varepsilon_{jl} - \eta \delta_{jl} \right) n_j = 0 \text{ and } 1 - n_k n_k = 0 \tag{4.62}$$

The second of these equations is the auxiliary condition (4.52) associated with the original functional $\mathbf{\Pi}$, and the first is a *Linear Eigenvalue Problem* in η. From linear algebra it is known that non-trivial solutions to (4.62) require that the determinant of $(\varepsilon_{jl} - \eta\delta_{jl})$ vanishes

$$|\varepsilon_{jl} - \eta\delta_{jl}| = 0 \quad \text{or} \quad \begin{vmatrix} (\varepsilon_{11} - \eta) & \varepsilon_{12} & \varepsilon_{13} \\ \varepsilon_{12} & (\varepsilon_{22} - \eta) & \varepsilon_{23} \\ \varepsilon_{13} & \varepsilon_{23} & (\varepsilon_{33} - \eta) \end{vmatrix} = 0 \tag{4.63}$$

where we have exploited the symmetry of ε_{jl}.

After a number of elementary, but tedious, manipulations (4.63b) may provide the following equation for η

Strain invariants J_1, J_2, J_3

$$\eta^3 - J_1\eta^2 + J_2\eta - J_3 = 0 \tag{4.64}$$

where the *Strain Invariants* J_1, J_2, and J_3 are given by

J_1: Trace of ε_{ij}

$$J_1 = \varepsilon_{11} + \varepsilon_{22} + \varepsilon_{33} = \varepsilon_{jj} \tag{4.65}$$

J_2: Quadratic invariant of ε_{ij}

$$J_2 = \begin{vmatrix} \varepsilon_{11} & \varepsilon_{12} \\ \varepsilon_{21} & \varepsilon_{22} \end{vmatrix} + \begin{vmatrix} \varepsilon_{22} & \varepsilon_{23} \\ \varepsilon_{32} & \varepsilon_{33} \end{vmatrix} + \begin{vmatrix} \varepsilon_{33} & \varepsilon_{31} \\ \varepsilon_{13} & \varepsilon_{11} \end{vmatrix} \tag{4.66}$$

and

J_3: Determinant of ε_{ij}

$$J_3 = \begin{vmatrix} \varepsilon_{11} & \varepsilon_{12} & \varepsilon_{13} \\ \varepsilon_{21} & \varepsilon_{22} & \varepsilon_{23} \\ \varepsilon_{31} & \varepsilon_{32} & \varepsilon_{33} \end{vmatrix} \tag{4.67}$$

The first invariant J_1 is the *trace* of the strain matrix given by ε_{ij}, the second invariant J_2 is sometimes called the quadratic invariant of the strain matrix, while the third invariant J_3 is the determinant of the strain matrix.

There are several important observations to be made at this point. First, J_1, J_2, and J_3 really do deserve the label invariant (with respect to a change to another Cartesian coordinate system) because they express the condition for stationary values of $\varepsilon_j^{(n)}$, which cannot depend on the choice of coordinate system. Second, since $(\varepsilon_{jl} + \eta\delta_{jl})$ is symmetric, the three eigenvalues of (4.63a), or equivalently (4.63b), are known from linear algebra to be real-valued, i.e. they do not possess an imaginary part, and their eigenvectors are orthogonal.[4.5]

J_1, J_2 and J_3 have physical meanings—for isotropic materials

We should not consider the strain invariants only as a set of exotic quantities that only come about because of some coincidences, see footnote on page 71. As mentioned in that footnote, it is such that many constitutive equations, especially for isotropic materials, are formulated in terms of—or at least by use of—the strain invariants and their static counterparts, the

[4.5] If two of the eigenvalues are equal, then we may choose their associated eigenvectors such that they are orthogonal with respect to each other. A similar statement holds when all three eigenvalues are equal.

stress invariants I_1, I_2 and I_3, see Section 4.3. At this point I shall not expand on the subject of constitutive equations and convenient ways of formulating them. The emphasis on the strain invariants is to some degree a relic from an older age when all computations were done by hand and the number of unknowns should be limited as much as possible. Instead of formulating everything in terms of *six*[4.6] strain components ε_{11}, ε_{12}, ε_{13}, ε_{22}, ε_{23}, and ε_{33} it was considered better to utilize the *three* strain invariants J_1, J_2, and J_3 as much as possible. To some extent, this trend may be considered old-fashioned and, since the strain invariants are relevant only for isotropic materials, in most cases we may as well employ formulations in terms of the strains themselves.

4.3 Equilibrium Equations

Below we derive the equilibrium equations by application of the *Principle of Virtual Displacements* and later give an interpretation of the the stress measures. Recall the Principle of Virtual Displacements (2.98)

Principle of Virtual Displacements

$$\boldsymbol{\sigma} \cdot \delta\boldsymbol{\varepsilon} = \overline{\boldsymbol{T}} \cdot \delta\boldsymbol{u} \tag{4.68}$$

Because of the fact that the infinitesimal theory assumes $|u_{i,j}| \ll 1$ there is no reason to distinguish between the deformed volume and the undeformed volume in the following. Therefore, we omit superscript ^{0} on all quantities.

Stress tensor σ_{mn}

σ_{mn} and $\delta\varepsilon_{mn}$ are work conjugate

First, introduce a tensor[4.7] σ_{mn}, which we later shall interpret as a stress tensor, but at present is defined by the requirement that it must give the proper virtual work together with the variation of infinitesimal strain tensor ε_{mn}, i.e. the internal virtual $\boldsymbol{\sigma} \cdot \delta\boldsymbol{\varepsilon}$, see Chapter 33, must be

Internal virtual work

$$\boldsymbol{\sigma} \cdot \delta\boldsymbol{\varepsilon} = \int_V \sigma_{mn} \delta\varepsilon_{mn} dV \tag{4.69}$$

where

$$\delta\varepsilon_{mn} = \tfrac{1}{2}\left(\delta u_{m,n} + \delta u_{n,m}\right) \tag{4.70}$$

At this point we do not know "how σ_{mn} looks," but we may already at this point realize that without loss of generality σ_{mn} may be assumed to be symmetric because of the symmetry of ε_{mn}, see Section 31.3.2.2.

Next, we demand that the internal virtual work is balanced by the external virtual work $\overline{\boldsymbol{T}} \cdot \delta\boldsymbol{u}$, see Chapter 33, produced by the body forces $\bar{q}_m$ and the applied boundary tractions $\bar{\tau}_m$ working through the variation of the displacements u_m, see (2.97) or (2.98). In the present notation $\overline{\boldsymbol{T}} \cdot \delta\boldsymbol{u}$ is

[4.6] Recall that ε_{mn} is symmetric.

[4.7] Indeed σ_{mn} designates generalized stresses, but at this point we do not know their properties except for the fact that σ_{mn} produces the correct virtual work through $\delta\varepsilon_{mn}$, see (4.69).

External virtual work

$$\overline{\boldsymbol{T}} \cdot \delta \boldsymbol{u} = \int_{S_T} \bar{\tau}_i \delta u_i dS + \int_V \bar{q}_j \delta u_j dV \,, \text{ with } \delta u_j = 0 \,,\ x_j \in S_u \tag{4.71}$$

Thus the *Principle of Virtual Displacements* is

Principle of Virtual Displacements

$$\int_V \sigma_{ij} \delta \varepsilon_{ij} dV = \int_{S_T} \bar{\tau}_i \delta u_i dS + \int_V \bar{q}_j \delta u_j dV \;\forall\; \delta u_j = 0 \,,\ x_j \in S_u \tag{4.72}$$

In the following we employ the principle of virtual displacements to derive the equilibrium equations, while in Section 2.4 we started with the equilibrium equations and established that principle. Therefore, below we utilize the *Divergence Theorem* in the opposite direction

$$\begin{aligned}
\boldsymbol{\sigma} \cdot \delta \boldsymbol{\varepsilon} &= \int_V \sigma_{mn} \delta \varepsilon_{mn} dV \\
&= \tfrac{1}{2} \int_V \sigma_{mn} \left(\delta u_{m,n} + \delta u_{n,m} \right) dV = \int_V \sigma_{mn} \delta u_{m,n} dV \\
&= \int_V \left(\sigma_{mn} \delta u_m \right)_{,n} dV - \int_V \sigma_{mn,n} \delta u_m dV \\
&= \int_{S_T} \sigma_{mn} n_n \delta u_m dS - \int_V \sigma_{mn,n} \delta u_m dV
\end{aligned} \tag{4.73}$$

where we have exploited the symmetry of σ_{mn}. Then, by equating the internal to the external virtual work and after rearrangement of terms we get

$$\int_V \left(\sigma_{mn,n} + \bar{q}_m \right) \delta u_m dV + \int_{S_T} \left(\bar{\tau}_m - \sigma_{mn} n_n \right) \delta u_m dS = 0 \tag{4.74}$$

In order for this to be true for all kinematically admissible δu_m both integrands must vanish identically with the result that we get the expression for the static field equations

Static field equations

$$\sigma_{mn,n} + \bar{q}_m = 0 \,,\ x_j \in V \tag{4.75}$$

and for the static boundary conditions

Static boundary conditions

$$\sigma_{mn} n_n = \bar{\tau}_m \,,\ x_j \in S_T \tag{4.76}$$

These are the equilibrium equations that govern the stress tensor σ_{mn}.

4.3.1 Interpretation of Stress Components

Interpretation of stresses

In order to interpret the components of σ_{mn} consider equilibrium of an infinitesimal rectangular parallelepiped, see Fig. 4.2.

4.3.1.1 Internal Equilibrium

We introduce the stresses $\tilde{\sigma}_{mn}$ as the force per unit area in the direction of x_n acting on the face with unit normal $\mathbf{n}_m$, which points in the positive direction of the x_m-axis. At this point we do not know whether these stresses are equal to the generalized stresses σ_{mn} although we probably have a strong suspicion that they are. And, indeed it is our intention to prove that $\tilde{\sigma}_{mn} = \sigma_{mn}$.

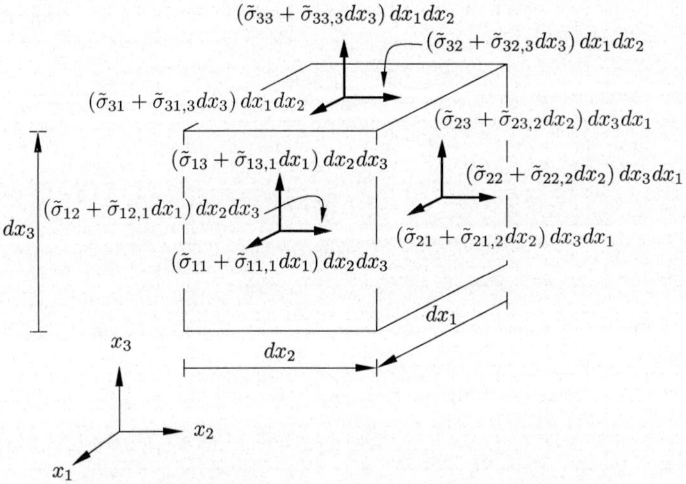

Fig. 4.2: Stress resultants on the positive faces of an infinitesimal parallelepiped.
The body forces and the stress resultants on the negative faces are not shown.

Let us establish equilibrium in the x_1-direction first. In order to do so, note that if the forces acting on the face with a normal in the *negative* x_1-direction are $\tilde{\sigma}_{1j}dx_2dx_3$, and that they must be $(\tilde{\sigma}_{1j} + \tilde{\sigma}_{1j,1}dx_1)\, dx_2dx_3$ at the opposite face with analogous expressions for forces on the other faces.

Thus, equilibrium in the x_1-direction requires

$$\begin{aligned} 0 = &+ (\tilde{\sigma}_{11} + \tilde{\sigma}_{11,1}dx_1)\, dx_2dx_3 - \tilde{\sigma}_{11}dx_2dx_3 \\ &+ (\tilde{\sigma}_{21} + \tilde{\sigma}_{21,2}dx_2)\, dx_3dx_1 - \tilde{\sigma}_{21}dx_3dx_1 \\ &+ (\tilde{\sigma}_{31} + \tilde{\sigma}_{31,3}dx_3)\, dx_1dx_2 - \tilde{\sigma}_{31}dx_1dx_2 \\ &+ \bar{q}_1 dx_1dx_2dx_3 \end{aligned} \qquad (4.77)$$

Equilibrium in the x_1-direction

where $\bar{q}_1$ is the component in the x_1-direction of the body force.

Since $dx_1dx_2dx_3$ by implication does not vanish, we get

$$\tilde{\sigma}_{n1,n} + \bar{q}_1 = 0 \tag{4.78}$$

which is the same as (4.75) with index $m = 1$ when the symmetry of σ_{mn} in (4.75) is exploited. By analogy we may arrive at equations expressing equilibrium in the two other directions and realize that we recover (4.75). Now we know much about "how $\tilde{\sigma}_{ij}$ looks." However, we still need to investigate moment equilibrium in order to confirm the symmetry property of $\tilde{\sigma}_{mn}$, and we must also interpret the static boundary conditions (4.76) in order to get the full interpretation of the stresses, see below.

As regards moment equilibrium we compute the moment about the center of the parallelepiped and commence by writing the moment about the x_1-direction

Moment equilibrium about the x_1-direction

$$\begin{aligned} 0 = & -\tilde{\sigma}_{32}dx_1dx_2\tfrac{1}{2}dx_3 - \big((\tilde{\sigma}_{32} + \tilde{\sigma}_{32,3}dx_3)\,dx_1dx_2\big)\tfrac{1}{2}dx_3 \\ & + \tilde{\sigma}_{23}dx_3dx_1\tfrac{1}{2}dx_2 + \big((\tilde{\sigma}_{23} + \tilde{\sigma}_{23,2}dx_2)\,dx_3dx_1\big)\tfrac{1}{2}dx_2 \\ \text{or} \quad 0 = & -\tilde{\sigma}_{32} - \tfrac{1}{2}\tilde{\sigma}_{32,3}dx_3 + \tilde{\sigma}_{23} + \tfrac{1}{2}\tilde{\sigma}_{23,2}dx_2 \end{aligned} \tag{4.79}$$

Because $\tilde{\sigma}_{mn}$ and $\tilde{\sigma}_{mn,k}$ are finite while dx_k is infinitely small, (4.79b) implies

$$\tilde{\sigma}_{32} = \tilde{\sigma}_{23} \tag{4.80}$$

If we repeat the above procedure for the moment equilibrium about the two other directions we realize that we have proved the symmetry property of $\tilde{\sigma}_{mn}$

Symmetry of $\tilde{\sigma}_{mn}$

$$\tilde{\sigma}_{mn} = \tilde{\sigma}_{nm} \tag{4.81}$$

4.3.1.2 Static Boundary Conditions

In order to establish the static boundary conditions consider Fig. 4.3, which is spanned by dx_1, dx_2 and dx_3 in the directions of the respective axes. The total body force on the tetrahedron is $d\bar{P}$, and $-d\mathbf{F}_j, j \in [1,3]$ is the total load acting on the face with the normal in the *negative* x_j-direction. Finally, $d\mathbf{F}_n$ is the total load on the inclined face. Equilibrium of the tetrahedron requires

Equilibrium of tetrahedron

$$\mathbf{0} = -d\mathbf{F}_1 - d\mathbf{F}_2 - d\mathbf{F}_3 + d\mathbf{F}_n + d\bar{\mathbf{P}} \tag{4.82}$$

The intensity of the force vector $-d\mathbf{F}_1$ is $\tilde{\mathbf{t}}_1$ etc., the intensity of $d\mathbf{F}_n$ is $\tilde{\boldsymbol{\tau}}$, and the intensity of $d\bar{\mathbf{P}}$ is $\bar{\mathbf{q}}$, and thus (4.82) requires

$$\mathbf{0} = -\tilde{\mathbf{t}}_1\tfrac{1}{2}dx_2dx_3 - \tilde{\mathbf{t}}_2\tfrac{1}{2}dx_3dx_1 - \tilde{\mathbf{t}}_3\tfrac{1}{2}dx_1dx_2 + \tilde{\boldsymbol{\tau}}dA + \bar{\mathbf{q}}dV \tag{4.83}$$

where dA is the area of the inclined face, whose normal is $\mathbf{n}$, and dV is the volume of the tetrahedron. Recall that $\tilde{\sigma}_{ij}$ are the force intensities resolved in terms of the base vectors $\mathbf{i}_m, m \in [1,3]$

$$\tilde{\mathbf{t}}_i = \tilde{\sigma}_{ij}\mathbf{i}_j \tag{4.84}$$

Further, resolve $\widetilde{\boldsymbol{\tau}}$ in terms of the base vectors $\mathbf{i}_j, j \in [1,3]$ to get

$$\widetilde{\boldsymbol{\tau}} = \tilde{\tau}_j\mathbf{i}_j \tag{4.85}$$

and get

$$\begin{aligned}\mathbf{0} = &-\tilde{\sigma}_{1j}\mathbf{i}_j\tfrac{1}{2}dx_2dx_3 - \tilde{\sigma}_{2j}\mathbf{i}_j\tfrac{1}{2}dx_3dx_1 - \tilde{\sigma}_{3j}\mathbf{i}_j\tfrac{1}{2}dx_1dx_2\\ &+\tilde{\tau}_j\mathbf{i}_jdA + \bar{\mathbf{q}}dV\end{aligned} \tag{4.86}$$

Obviously, the last term is of order 3 in the differentials, while the others are of order 2, and therefore it may be omitted in the following.

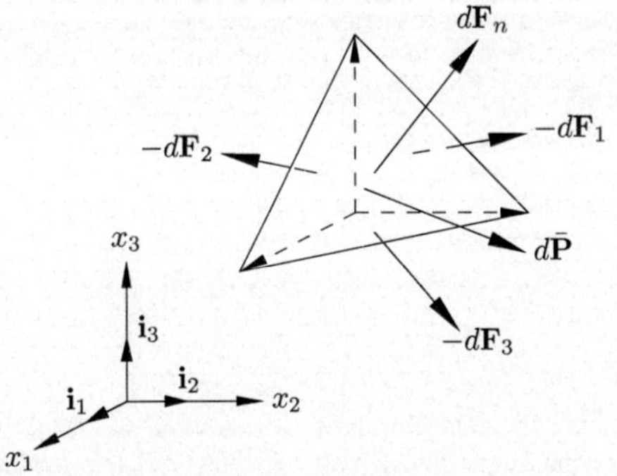

Fig. 4.3: Stresses and loads on an infinitesimal tetrahedron.

The areas of the faces that are parallel with the coordinate planes may be computed as

$$\begin{aligned}dA_1 &= \tfrac{1}{2}dx_2dx_3 = \mathbf{n}\cdot\mathbf{i}_1dA\\ dA_2 &= \tfrac{1}{2}dx_3dx_1 = \mathbf{n}\cdot\mathbf{i}_2dA\\ dA_3 &= \tfrac{1}{2}dx_1dx_2 = \mathbf{n}\cdot\mathbf{i}_3dA\end{aligned} \tag{4.87}$$

and thus (4.86) may yield

$$\begin{aligned}\mathbf{0} = &-\tilde{\sigma}_{1j}\mathbf{i}_jdA\,(\mathbf{n}\cdot\mathbf{i}_1) - \tilde{\sigma}_{2j}\mathbf{i}_jdA\,(\mathbf{n}\cdot\mathbf{i}_2) - \tilde{\sigma}_{3j}\mathbf{i}_jdA\,(\mathbf{n}\cdot\mathbf{i}_3)\\ &+\tilde{\tau}_j\mathbf{i}_jdA\end{aligned} \tag{4.88}$$

Resolve $\mathbf{n}$ is resolved in terms of $\mathbf{i}_k$

$$\mathbf{n} = n_k\mathbf{i}_k \;\Rightarrow\; \mathbf{n}\cdot\mathbf{i}_m = n_k\mathbf{i}_k\cdot\mathbf{i}_m = n_k\delta_{km} = n_m \tag{4.89}$$

and (4.88) becomes

$$\mathbf{0} = -\tilde{\sigma}_{ij}\mathbf{i}_j n_i + \tilde{\tau}_j\mathbf{i}_j \tag{4.90}$$

Since this must hold for all $\mathbf{i}_j$, which are linearly independent, we get the final result

Projection of stress

$$\tau_j = \tilde{\sigma}_{ij}n_i \tag{4.91}$$

This expression, which holds for all inclined surfaces—not only on the static boundary S_T—is similar to (4.76).

$\sigma_{mn} = \tilde{\sigma}_{mn}$

Now that we have seen that the equilibrium equations derived from the principle of virtual displacements and by direct formulation of equilibrium are identical, we may finally interpret σ_{mn} as the stress components $\tilde{\sigma}_{mn}$, which are the forces per unit area in the direction of x_n acting on the face with unit normal $\mathbf{n}_m$.

Actually, we might just as well have acknowledged the fact that in the infinitesimal theory $|u_{i,j}| \ll 1$ and therefore (2.59)

$$t_{ik,i} + \left(t_{ij}u_{k,j}\right)_{,i} + \bar{q}_k = 0 \tag{4.92}$$

would become

$$t_{ik,i} + \bar{q}_k = 0 \tag{4.93}$$

which, except for notational differences, is the same as (4.75). Since the infinitesimal theory does not distinguish between undeformed and deformed configuration[4.8] the components of the stress σ_{ij} must therefore be the force per unit undeformed area resolved in terms of the undeformed base vectors $\mathbf{i}_j$, just as the Piola-Kirchhoff stresses are the forces per unit *undeformed* area resolved in terms of the *deformed* base vectors $\mathbf{g}_j$, see Section 2.3.1.

If we follow this line of reasoning, we may immediately write the equivalent of (2.71)

$$\tau_m = \left(\delta_{mk} + u_{m,k}\right)n_j^0 t_{jk} \tag{4.94}$$

[4.8] Strictly speaking, this is not completely true in that the strains and the displacements are expressions of the difference between undeformed and deformed configurations. For example, while $1 + \varepsilon_{11} \approx 1$ is true because the theory assumes that $|\varepsilon_{11}|$ is small, it does not necessarily imply that $(1 + \varepsilon_{11}) - 1 \approx 0$. In most other instances it is neither possible nor desirable to try to differentiate between the two configurations when we employ the infinitesimal theory. This is a fundamental difficulty inherent in the infinitesimal theory, as mentioned in Chapter 1, and the most important reason why I chose to begin with a theory that does not have such paradoxes built in.

for the infinitesimal theory, namely

$$\tau_m = n_j \sigma_{jm} \tag{4.95}$$

which gives the relation between the stress tensor σ_{jm} and the stress on a plane with the unit normal n_j. On the static boundary S_T^0, where the boundary stress resultant $\bar{\tau}_m$ is given, (4.95) becomes

Static boundary conditions

$$\bar{\tau}_m = n_j \sigma_{jm} \tag{4.96}$$

Thus, the static boundary condition (4.76) may be interpreted in a straightforward manner through (4.96).

4.3.2 Transformation of Stress

Transformation of stress

Closely connected with the topic of static boundary conditions, see Section 4.3, and with the question of principal stresses, see Section 4.3.3 below, is the problem of transformation of the stress tensor between two (Cartesian) coordinate systems.

There are several ways to establish the relation between the stress tensor in one coordinate system and another. It seems an obvious idea to express the same stress state in the two coordinate systems in terms of the equilibrium equations in both systems. However, in Section 4.3 we defined the stresses by use of the principle of virtual displacements and the definition of the strain tensor. Thus, the consistent—and therefore the most satisfactory—way to derive the rule for transformation of the stress tensor is through application of the same procedure. Also, it proves to be the easiest way. In Section 4.2.5 we introduced two (Cartesian) coordinate systems, one with the base vectors $\mathbf{i}_i$, the other with the base vectors $\mathbf{i}_i^\star$, and supplied quantities expressed in the second coordinate system with a star * as superscript.

Recall that the internal virtual work is, see e.g. (4.69)

Internal virtual work

$$\boldsymbol{\sigma} \cdot \delta\boldsymbol{\varepsilon} = \int_V \sigma_{ij} \delta\varepsilon_{ij} dV \tag{4.97}$$

which in the starred coordinate system is

$$\boldsymbol{\sigma}^\star \cdot \delta\boldsymbol{\varepsilon}^\star = \int_V \sigma_{km}^\star \delta\varepsilon_{km}^\star dV \tag{4.98}$$

where the change of dummy indices proves to be convenient later, but otherwise is inconsequential.

The rule for transformation of strain is, see (4.45)

Transformation of strain

$$\varepsilon_{ij} = \alpha_{ik}^\star \alpha_{jm}^\star \varepsilon_{km}^\star \tag{4.99}$$

Because the strain measures are linear in the displacements an equivalent relation holds for transformation of strain variations.

Then, for the strain variations

$$\delta\varepsilon_{ij} = \alpha^\star_{ik}\alpha^\star_{jm}\delta\varepsilon^\star_{km} \tag{4.100}$$

Insert (4.100) into (4.97)

Internal virtual work in starred coordinate system

$$\boldsymbol{\sigma} \cdot \delta\boldsymbol{\varepsilon} = \int_V \sigma_{ij}\alpha^\star_{ik}\alpha^\star_{jm}\delta\varepsilon^\star_{km}dV \tag{4.101}$$

Since this relation and (4.98) hold for any $\delta\varepsilon^\star_{km}$ and by use of (4.37) we may conclude that

Transformation of stress

$$\sigma^\star_{km} = \alpha^\star_{ik}\alpha^\star_{jm}\sigma_{ij} \;\text{ or }\; \sigma^\star_{km} = \alpha_{ki}\alpha_{mj}\sigma_{ij} \tag{4.102}$$

In order to arrive at the inverse relations we may premultiply by $\alpha_{nk}\alpha_{pm}$ in (4.102a). After utilizing (4.39) and some—as a matter of fact, quite a lot—of index manipulations (4.102a) may provide (4.103a), and in a similar way (4.102b) may give (4.103b)

Transformation of stress

$$\sigma_{km} = \alpha_{ik}\alpha_{jm}\sigma^\star_{ij} \;\text{ or }\; \sigma_{km} = \alpha^\star_{ki}\alpha^\star_{mj}\sigma^\star_{ij} \tag{4.103}$$

As is the case for the rules for transformation of strain, for some applications there is a need for the equivalent of (4.103) in curvilinear coordinates, but, again, we refer to more advanced presentations, such as (Malvern 1969).

4.3.3 Principal Stresses

Principal stresses—relevant for isotropic materials

The question of finding the extrema of the normal stress at a point—depending on the orientation of the plane—is closely connected to the previous section. The reason why we wish to find stress extrema, also called *Principal Stresses* is that many criteria for predicting stress failure entail the normal stress. This is certainly the case for *isotropic* materials, see Sections 2.6 and 5, but for *anisotropic* materials the case is often different. For example, a material such as wood is much stronger in one direction—the fiber direction—than the others, i.e. is able to sustain larger stresses in that direction, which means that a smaller stress transverse to the fibers may fracture the specimen, see also the discussion in Section 4.2.6.

Let the orientation of the plane through point P be given by its unit normal n_j. According to (4.95) the stress $\tau^{(n)}_m$ is given by

$$\tau^{(n)}_m = n_j\sigma_{jm} \tag{4.104}$$

where $^{(n)}$ indicates that the stress depends on the unit normal. The component in the direction of the normal is

$$\sigma^{(n)} = n_j n_m \sigma_{jm} \tag{4.105}$$

Extrema of stress

In order to determine the extrema—or at least stationary values—of $\sigma^{(n)}$ we face the same problem as in Section 4.2.6, where (4.51) was subjected to the condition (4.51), in that (4.105) must fulfill the same auxiliary condition, namely that n_k is a unit vector

$$n_k n_k = 1 \tag{4.106}$$

The two expressions (4.105) and (4.106) are completely analogous to (4.51) and (4.52), respectively. Therefore, and since σ_{mn}, like ε_{mn} is symmetric in its indices, we may exploit the results from Section 4.2.6. The outcome is that the governing eigenvalue problem here is

$$|\sigma_{jl} - \eta\delta_{jl}| = 0 \quad \text{or} \quad \begin{vmatrix} (\sigma_{11} - \eta) & \sigma_{12} & \sigma_{13} \\ \sigma_{12} & (\sigma_{22} - \eta) & \sigma_{23} \\ \sigma_{13} & \sigma_{23} & (\sigma_{33} - \eta) \end{vmatrix} = 0 \tag{4.107}$$

where we have exploited the symmetry of σ_{jl}.

Again, after a number of elementary, but tedious, manipulations (4.107b) provides the following equation for η

Stress invariants I_1, I_2, I_3

$$\eta^3 - I_1\eta^2 + I_2\eta - I_3 = 0 \tag{4.108}$$

where the *Stress Invariants* I_1, I_2, and I_3 are given by

I_1: Trace of σ_{ij}

$$I_1 = \sigma_{11} + \sigma_{22} + \sigma_{33} = \sigma_{jj} \tag{4.109}$$

I_2: Quadratic invariant of σ_{ij}

$$I_2 = \begin{vmatrix} \sigma_{11} & \sigma_{12} \\ \sigma_{21} & \sigma_{22} \end{vmatrix} + \begin{vmatrix} \sigma_{22} & \sigma_{23} \\ \sigma_{32} & \sigma_{33} \end{vmatrix} + \begin{vmatrix} \sigma_{33} & \sigma_{31} \\ \sigma_{13} & \sigma_{11} \end{vmatrix} \tag{4.110}$$

and

I_3: Determinant of σ_{ij}

$$I_3 = \begin{vmatrix} \sigma_{11} & \sigma_{12} & \sigma_{13} \\ \sigma_{21} & \sigma_{22} & \sigma_{23} \\ \sigma_{31} & \sigma_{32} & \sigma_{33} \end{vmatrix} \tag{4.111}$$

The first invariant I_1 is the *trace* of the stress matrix given by σ_{ij}, the second invariant I_2 is sometimes called the quadratic invariant of the stress matrix, while the third invariant I_3 is the determinant of the stress matrix.

For a discussion of the properties of I_1, I_2, and I_3 and their importance in connection with formulation of constitutive equations for isotropic materials, see Section 4.2.6.

4.4 Potential Energy

In Section 2.7 we postulated the expression (2.106) for the potential energy associated with large displacements and large strains. In the present case where the displacements and strains are infinitesimal the equivalent formula is

$$\mathbf{\Pi}_\mathrm{P}(u_i) = \int_{V^0} W(\varepsilon_{ij}) dV^0 - \int_{V^0} \bar{q}_i u_i dV^0 - \int_{S_T^0} \bar{\tau}_i u_i dS^0 \tag{4.112}$$

because the strains are now ε_{ij} instead of γ_{ij}.

4.4.1 Linear Elasticity

For linear (hyper)elasticity, i.e. for *Hooke's "Law"* the expression for the potential energy becomes

Potential energy, linear hyperelasticity

$$\Pi_P(u_i) = \tfrac{1}{2}\int_{V^0} E_{ijkm}\varepsilon_{ij}\varepsilon_{km}dV^0 - \int_{V^0}\bar{q}_i u_i dV^0 - \int_{S_T^0}\bar{\tau}_i u_i dS^0 \quad (4.113)$$

In connection with establishing finite elements this expression is probably the one that is utilized more frequently than all other formulas, see e.g. (23.71).

4.5 Principle of Virtual Forces

Principle of Virtual Forces

Below, we derive the *Principle of Virtual Forces* which is the dual principle of the *Principle of Virtual Displacements*, see Section 2.4. While we established the latter principle from the equilibrium equations, see Section 2.4, we use the kinematic relations, i.e. the strain-displacement relation and the kinematic boundary conditions, as our base here. We limit ourselves to infinitesimal displacements and note the definition of the strains ε_{ij}, see (4.5)

Strain definition

$$\varepsilon_{ij} \equiv \tfrac{1}{2}(u_{i,j} + u_{j,i}) \; , \; x_k \in V \quad (4.114)$$

where V denotes the interior of the body. The kinematic boundary conditions are of the form

Kinematic boundary conditions

$$u_i = \bar{u}_i \; , \; x_k \in S_u \quad (4.115)$$

where S_u is the kinematic boundary, i.e. that part of the boundary where kinematic boundary conditions are prescribed.

The derivations are very similar to the ones of Section 2.4. First, we rewrite the strain-displacement relation such that we move all terms to the right-hand side of the equation. Then, we multiply this by the tensor field α_{ij}, which is supposed to be as smooth as we need.[4.9] Finally, we integrate over the volume and note that the result still equals zero

$$\begin{aligned} 0 &= \int_V \alpha_{ij}\left(\varepsilon_{ij} - \tfrac{1}{2}(u_{i,j} + u_{j,i})\right)dV \\ \Rightarrow 0 &= \int_V \alpha_{ij}\varepsilon_{ij}dV - \int_V \alpha_{ij}u_{i,j}dV \end{aligned} \quad (4.116)$$

where we have exploited the fact that the symmetry of ε_{ij} implies that, without loss of generality, we may take α_{ij} to be symmetric, too.

By a simple reformulation we may get

$$0 = \int_V \alpha_{ij}\varepsilon_{ij}dV - \int_V \left((\alpha_{ij}u_i)_{,j} - \alpha_{ij,j}u_i\right)dV \quad (4.117)$$

[4.9] Note that the meaning of α_{ij} is not the same as in Subsection 4.3.2.

and after application of the divergence theorem

$$0 = \int_V \alpha_{ij}\varepsilon_{ij}dV + \int_V \alpha_{ij,j}u_i dV - \int_S \alpha_{ij}n_j u_i dS \qquad (4.118)$$

where n_j is the unit normal to S.

Let us choose to interpret α_{ij} as the variation $\delta\sigma_{ij}$ of the stress

$$\alpha_{ij} = \delta\sigma_{ij} \qquad (4.119)$$

The static field equations, see (4.75)

$$\sigma_{ij,j} + \bar{q}_i = 0\,,\ x_j \in V \qquad (4.120)$$

Static field equations

imply that—note that $\bar{q}_i$ is prescribed and may therefor not be varied

$$\delta\sigma_{ij,j} = 0\,,\ x_j \in V \qquad (4.121)$$

Recall the static boundary conditions (4.76)

$$T_i \equiv \sigma_{ij}n_j = \bar{T}_i\,,\ x_j \in S_T \qquad (4.122)$$

Static boundary conditions

which furnishes

$$\delta\sigma_{ij}n_j = 0\,,\ x_j \in S_T \qquad (4.123)$$

When we require that $\delta\sigma_{ij}$ satisfies (4.121) and (4.123) we get the *Principle of Virtual Forces*

$$\int_V \delta\sigma_{ij}\varepsilon_{ij}dV = \int_{S_u} \delta\sigma_{ij}n_j\bar{u}_i dS \qquad (4.124)$$

Principle of Virtual Forces

As was the case for the *Principle of Virtual Displacements* the *Principle of Virtual Forces* may be expressed using the *Budiansky-Hutchinson Dot Notation*, see Section 33.2

$$\boldsymbol{\varepsilon} \cdot \delta\boldsymbol{\sigma} = \bar{\boldsymbol{u}} \cdot \delta\boldsymbol{T} \qquad (4.125)$$

Principle of Virtual Forces

4.6 Complementary Strain Energy Function

We shall only be concerned with linearly elastic materials, and thus the constitutive model is the so-called *Hooke's Law*, which may be given by the strain energy function $W(\varepsilon_{ij})$. For finite strains γ_{ij} we saw that the strain energy could be expressed as, see (2.102)

$$W(\gamma_{ij}) = \tfrac{1}{2}E_{ijkl}\gamma_{ij}\gamma_{kl} \qquad (4.126)$$

Strain energy function $W(\gamma_{ij})$

Quite simply, for the case of infinitesimal strains ε_{ij} the expression for

Strain energy function $W(\varepsilon_{ij})$

the strain energy is

$$W(\varepsilon_{ij}) = \tfrac{1}{2}E_{ijkl}\varepsilon_{ij}\varepsilon_{kl} \tag{4.127}$$

In the present connection we need the *Complementary Strain Energy Function* $W_C(\sigma_{ij})$, which by differentiation with respect to the stress σ_{kl} furnishes the strain ε_{kl} in the same way that differentiation of $W(\varepsilon_{ij})$ with respect to the strain ε_{kl} provides the stress σ_{kl}. Again, we limit ourselves to the case of linear elasticity

Complementary strain energy function $W_C(\varepsilon_{ij})$

$$W_C(\sigma_{ij}) = \tfrac{1}{2}C_{ijkl}\sigma_{ij}\sigma_{kl} \tag{4.128}$$

where the tensor C_{ijkl} is the inverse of the elasticity tensor E_{ijkl} and is an expression of the *flexibility* of the material in the same sense that E_{ijkl} denotes the *stiffness* of the material. Differentiation of $W_C(\sigma_{ij})$ in (4.128) provides

$$\varepsilon_{mn} = C_{mnkl}\sigma_{kl} \tag{4.129}$$

In a similar way as was done in Section 2.6 we may show that C_{ijkl} possesses the same symmetries as E_{ijkl}

$$C_{ijkl} = C_{jikl} = C_{klij} = C_{ijlk} \tag{4.130}$$

4.7 Complementary Energy

Complementary energy Π_C

In general, the *Principle of Stationarity of the Potential Energy* Π_P, see (2.106), receives much more attention than the *Principle of Stationarity of the Complementary Energy* Π_C which we shall define below. This is in part because the latter often proves to be more difficult to apply to specific structures. For instance, it is usually easier to derive finite elements based on Π_P than on Π_C. There is, however, a very useful kind of finite elements that are derived from a modified version of Π_C, namely the *Stress Hybrid Finite Elements*, which take a certain variant Π_{CM} of Π_C as their starting point, see Section 27.3. One reason for the interest in finite elements based on other functionals than Π_P is the major disadvantage of finite elements based on Π_P, namely that they model a too stiff behavior, see Section 33.4.3, while the stress hybrid finite elements often are more flexible. It is, however, necessary to mention that the vast majority of all finite element codes use elements that are based on Π_P.

We define the *Complementary Energy* $\Pi_C(\sigma_{ij})$ as

Complementary energy Π_C

$$\Pi_C(\sigma_{ij}) \equiv \tfrac{1}{2}\int_V C_{ijkl}\sigma_{ij}\sigma_{kl}dV - \int_{S_u} \tau_i\bar{u}_i dS \tag{4.131}$$

where we require that σ_{ij} and τ_i satisfy all static conditions, namely internal equilibrium (4.120) and static boundaryconditions (4.122), which serve as the auxiliary conditions on $\boldsymbol{\Pi}_\mathrm{C}(\sigma_{ij})$. Note the duality between the complementary energy and the potential energy, see e.g. (33.25), which for kinematic linearity and linear elasticity becomes

$$\boldsymbol{\Pi}_\mathrm{P}(u_i) = \tfrac{1}{2}\int_V E_{ijkl}\varepsilon_{ij}\varepsilon_{kl}dV - \int_V \bar{q}_i u_i dV - \int_{S_T} \bar{\tau}_i u_i dS \qquad (4.132)$$

Potential energy $\boldsymbol{\Pi}_\mathrm{P}$

You may wonder how $\boldsymbol{\Pi}_\mathrm{C}(\sigma_{ij})$ accounts for the applied loads and may yield any useful results. The reason is that σ_{ij} satisfies all static field equations and static boundary conditions and thus the applied loads enter through these requirements.

As a consequence of (4.122) note that across any internal surface we must demand that

$$\left(\sigma_{ij}n_j\right)^+ = \left(\sigma_{ij}n_j\right)^- \qquad (4.133)$$

In (4.133) we take $n_j^+ = n_j^-$, cf. (27.2c), where n_j^+ and n_j^- belong to two different (sub-)bodies.[4.10]

Quite often the problem at hand is such that the prescribed displacements all vanish with the result that the expression for $\boldsymbol{\Pi}_\mathrm{C}(\sigma_{ij})$ simplifies accordingly

$$\boldsymbol{\Pi}_\mathrm{C}(\sigma_{ij}) \equiv \tfrac{1}{2}\int_V C_{ijkl}\sigma_{ij}\sigma_{kl}dV \ \text{ for } \bar{u}_i = 0 \qquad (4.134)$$

Complementary energy $\boldsymbol{\Pi}_\mathrm{C}$ for $\bar{u}_i = 0$

When we require that the variation of $\boldsymbol{\Pi}_\mathrm{C}(\sigma_{ij})$ vanishes we may get the principle of virtual forces with the strains expressed in terms of the stresses and the constitutive relation

$$\delta\boldsymbol{\Pi}_\mathrm{C}(\sigma_{ij}) = 0 \,\forall\, \delta\sigma_{ij} \;\Rightarrow\; \int_V C_{ijkl}\sigma_{ij}\delta\sigma_{kl}dV = \int_{S_u} \delta\sigma_{ij}n_j\bar{u}_i dS \qquad (4.135)$$

cf. (4.123). If we proceed by converting the surface integral we may get the strain-displacement relation (4.114)

$$\begin{aligned}
0 &= \int_V C_{ijkl}\sigma_{ij}\delta\sigma_{kl}dV - \int_V \left(\delta\sigma_{ij}u_i\right)_{,j}dV \\
\Rightarrow 0 &= \int_V C_{ijkl}\sigma_{kl}\delta\sigma_{ij}dV - \int_V \delta\sigma_{ij,j}u_i dV - \int_V \delta\sigma_{ij}u_{i,j}dV \qquad (4.136)\\
\Rightarrow 0 &= \int_V \left(C_{ijkl}\sigma_{kl} - \tfrac{1}{2}(u_{i,j}+u_{j,i})\right)\delta\sigma_{ij}dV
\end{aligned}$$

where we have exploited the fact that $\delta\sigma_{ij}$ satisfies the homogeneous equilibrium equations with the result that $\delta\sigma_{ij,j}$ vanishes, and where we have

[4.10] Later, when we construct a modified complementary energy we abandon the condition (4.133).

insisted that the displacement gradients enter in a symmetric fashion because of the symmetry of $C_{ijkl}\sigma_{kl}$. Use of the constitutive relation (4.129) provides the strain-displacement relation (4.114).

Chapter 5

Constitutive Relations

Sections 5.2–5.2.1 (re)introduce[5.1] the theory of (linear) elasticity and Example Ex 5-3 provides expressions for important special strain and stress states in elastic bodies. The topic of plasticity is introduced briefly in Section 5.4.

This chapter is mainly concerned with constitutive relations for the infinitesimal theory, i.e. the mathematically linear theory. However, most of the formulas may be made valid for kinematically nonlinear cases provided sufficient care is taken in the selection of strain and stress measures.

In connection with the formulation of constitutive relations it sometimes proves convenient to rearrange the strain and stress measures as discussed in Section 5.1 below. This is why that section is in this chapter rather than elsewhere, although the rearrangement in itself does not entail any reference to constitutive relations. Also, the special states discussed in Section Ex 5-3 are not necessarily limited to linear elasticity, but the derivations and formulas become more involved for other constitutive relations.

5.1 Rearrangement of Strain and Stress Components

Rearrangement of Strain and Stress Components

For the discussions below—as well as for many practical purposes—it proves convenient to rearrange the strain and stress tensors. There are two main reasons for doing this. The first is that ε_{ij} and σ_{ij} are both symmetric and therefore contain more terms than are independent. The second has to do with the fact that both quantities are second order tensors and therefore even the simplest constitutive relation, see (2.101), (2.102) or (5.8) and (5.9) below, entails fourth order tensors, e.g. E_{ijkl}. Let it be noted that although the expressions in terms of the second order tensors seem more involved than the ones introduced below, which are given in terms of strains and stresses with only one index, a number of manipulations are much easier to perform using the former quantities because they exhibit a larger degree of symmetry and regularity of many of the equations. Derivations in connection with the

[5.1] The reason for "(re)" is that in connection with introduction of potential energy we already touched the subject of elasticity, see Sections 2.7 and 4.4.

E. Byskov, *Elementary Continuum Mechanics for Everyone*,
Solid Mechanics and Its Applications 194, DOI: 10.1007/978-94-007-5766-0_5,

Finite Element Method, see Part V, are often done by use of matrix algebra which does not work for fourth order tensors, but fits well with vectors and matrices, i.e. quantities with one or two indices, respectively, see below.

The most common way to rearrange σ_{ij} is[5.2]

Rearranged stress components σ_j

$$\{\sigma\} \equiv \begin{bmatrix} \sigma_1 \\ \sigma_2 \\ \sigma_3 \\ \sigma_4 \\ \sigma_5 \\ \sigma_6 \end{bmatrix} \equiv \begin{bmatrix} \sigma_{11} \\ \sigma_{22} \\ \sigma_{33} \\ \sigma_{23} \\ \sigma_{31} \\ \sigma_{12} \end{bmatrix} \tag{5.1}$$

Generalized stresses σ_j. Generalized strains ε_j

If we insist—which we do most emphatically—that the new stresses are generalized, i.e. that there exist strains which work with the stresses to produce internal virtual work, we are forced to introduce the new strains by[5.3] the vector $\{\varepsilon\}$

Rearranged, work conjugate strain components σ_j

$$\{\varepsilon\} \equiv \begin{bmatrix} \varepsilon_1 \\ \varepsilon_2 \\ \varepsilon_3 \\ \varepsilon_4 \\ \varepsilon_5 \\ \varepsilon_6 \end{bmatrix} \equiv \begin{bmatrix} \varepsilon_{11} \\ \varepsilon_{22} \\ \varepsilon_{33} \\ 2\varepsilon_{23} \\ 2\varepsilon_{31} \\ 2\varepsilon_{12} \end{bmatrix} \tag{5.2}$$

because then the internal virtual work, which originally was expressed

Internal virtual work

$$\boldsymbol{\sigma} \cdot \delta\varepsilon = \int_V \sigma_{ij}\delta\varepsilon_{ij} dV, \quad \text{sum over } i, j = 1, 2, 3 \tag{5.3}$$

may now be written

Internal virtual work

$$\boldsymbol{\sigma} \cdot \delta\varepsilon = \int_V \sigma_i \delta\varepsilon_i dV, \quad \text{sum over } i = 1, 2, \ldots, 6 \tag{5.4}$$

or in a common vector notation, see Chapter 30

Internal virtual work

$$\boldsymbol{\sigma} \cdot \delta\varepsilon = \int_V \{\sigma\}^T \delta\{\varepsilon\} dV \tag{5.5}$$

Note that $\varepsilon_4 = 2\varepsilon_{23}$ etc. fits very well with (4.30)

$$2\varepsilon_{JK} = \varphi_{JK} \tag{5.6}$$

[5.2] Contrary to the habit of almost the rest of this book we begin with the *static*, not the *kinematic* quantities. The reason for doing this should be clear very soon, hopefully.

[5.3] The factor 2 on the the right-hand side of the last three rows of (5.2) is the important feature.

which showed that $2\varepsilon_{JK}$ denotes the decrease in angle between the base vectors $\mathbf{i}_J$ and $\mathbf{i}_K$ caused by deformation of the solid.

There are, of course, other possibilities for the definition of the new strain and stress components. We could have moved the factor 2 from the shear strains to the shear stresses, or we could have exhibited a democratic mind and have supplied them both with the factor $\sqrt{2}$, and still have preserved the property of generalized quantities. While the first of these other possibilities does not seem to have been used, the second, which indeed has its virtues, has received some attention recently. I shall, however, stick to the conventional rearrangement.

5.2 Linear Elasticity

When we take notice of the notational differences (2.100)–(2.102) become

$$\sigma_{mn} = \frac{\partial W(\varepsilon_{ij})}{\partial \varepsilon_{mn}} \tag{5.7}$$

$$W(\varepsilon_{ij}) = \tfrac{1}{2} E_{ijkl}\varepsilon_{ij}\varepsilon_{kl} \tag{5.8}$$

Linear elasticity

and

$$\sigma_{ij} = E_{ijkl}\varepsilon_{kl} \tag{5.9}$$

Linear elasticity = Hooke's "law"

where the meaning of E_{ijkl} is changed slightly in that it connects the *linear* measures σ_{ij} and ε_{kl} instead of the *nonlinear* measures t_{ij} and γ_{kl}. However, the symmetry properties given in Section 2.6 still hold

$$E_{ijkl} = E_{klij}, \quad E_{ijkl} = E_{jikl} \quad \text{and} \quad E_{ijkl} = E_{ijlk} \tag{5.10}$$

We note again that because of these symmetry properties the number of independent constants is 21 for the most general case. In many—actually most—applications the material is assumed to display some kind of symmetry. In the following we shall discuss materials of varying degrees of symmetry, but note that the exposition will not be a very thorough one. First, however, let us rewrite (5.8) and (5.9) by use of (5.1) and (5.2)

$$W(\varepsilon_i) = \tfrac{1}{2} E_{ij}\varepsilon_i\varepsilon_j \tag{5.11}$$

Linear elasticity

and

$$\sigma_i = E_{ij}\varepsilon_j \tag{5.12}$$

Linear elasticity = Hooke's "law"

where we, as below, imply summation from 1 to 6 over lower-case roman subscripts. From (5.11) it is clear that we may take E_{ij} to be symmetric

$$E_{ij} = E_{ji} \tag{5.13}$$

E_{ij} is symmetric

and thus there are still at the most 21 different material constants.

Isotropy

For *isotropic* materials[5.4] the properties are independent of direction, and the number of elastic constants is reduced to 2, as we shall see later. This material model, by the way, is probably used more than all other models combined. The reasons are that, to a good approximation, many materials, such as metals and concrete,[5.5] may be assumed to obey this "law."

Orthotropy

Another very important class of materials exhibits *orthotropy* which means that there are three orthogonal planes of symmetry. Rolled steel plates are examples of such materials, and in many cases wood may be assumed to behave in this way, but see a comment on this below. Modern materials such as fiber reinforced plastics and fiber reinforced concrete are often made so that they exhibit orthotropy.[5.6] If the coordinate planes are oriented parallel to the planes of symmetry there are only nine independent constants of orthotropic materials.

Other types of material symmetry

There are other types of material symmetry such as transverse isotropy combined with axial symmetry, and a trunk of wood falls into this category. To a good approximation its properties may be taken to be independent of the height above ground of the cross-section and therefore this direction is at right angles to the cross-section of the tree. In the cross-section, the properties vary with the radius, but may be assumed to be independent of the angle. On the other hand, often boards are cut in the circumferential or in the tangential direction and may therefore be assumed to exhibit orthotropy instead.

In most presentations of linear elasticity, relations such as (5.12) are taken as the basis for the discussion of the various cases of symmetry, and more and more specialized versions of E_{ij} are derived through manipulations that acknowledge the degree of symmetry. This is clearly the most satisfactory way to do it, but, since these derivations are somewhat involved, see e.g. (Malvern 1969), I shall pursue another path in that we simply postulate some of the most common versions of E_{ij}.

5.2.1 Isotropic Linear Elasticity

In an *isotropic* material all properties are independent of direction. This means that

Isotropy

$$\begin{aligned} &E_{11} = E_{22} = E_{33}\ , \quad E_{44} = E_{55} = E_{66} \\ &E_{12} = E_{13} = E_{23}\ , \quad E_{15} = E_{16} = E_{24} = E_{26} = E_{34} = E_{35} \\ &E_{45} = E_{46} = E_{56}\ , \quad E_{14} = E_{25} = E_{36} \end{aligned} \tag{5.14}$$

5.4 By this we really refer to isotropic *material models*.

5.5 The fact that the material properties of concrete vary spatially does not make the assumption of isotropy invalid. Isotropy focuses on the properties at a point—not the variation from one point to its neighbors.

5.6 It must be acknowledged that in order to optimize either strength or stiffness the fibers in a reinforced plastic are sometimes arranged in layers at "funny" angles.

This brings down the number of different material constants to 6. However, by going through the derivations hinted to above the number is decreased to only 2, namely E, which is called the *Modulus of Elasticity* or *Young's Modulus*, and *Poisson's Ratio* ν. A third constant, which is used frequently, is the *Shear Modulus* G

Modulus of elasticity E. Poisson's Ratio ν

$$G \equiv \frac{E}{2(1+\nu)} \tag{5.15}$$

Shear modulus G

The components of E_{ij} may now be written

$$E_{ij} = \frac{E}{(1+\nu)(1-2\nu)} \begin{bmatrix} \widehat{E}_{11} & \widehat{E}_{12} & \widehat{E}_{12} & 0 & 0 & 0 \\ \widehat{E}_{12} & \widehat{E}_{11} & \widehat{E}_{12} & 0 & 0 & 0 \\ \widehat{E}_{12} & \widehat{E}_{12} & \widehat{E}_{11} & 0 & 0 & 0 \\ 0 & 0 & 0 & \widehat{E}_{44} & 0 & 0 \\ 0 & 0 & 0 & 0 & \widehat{E}_{44} & 0 \\ 0 & 0 & 0 & 0 & 0 & \widehat{E}_{44} \end{bmatrix} \tag{5.16}$$

where

$$\widehat{E}_{11} \equiv (1-\nu)\,, \quad \widehat{E}_{12} \equiv \nu \quad \text{and} \quad \widehat{E}_{44} \equiv \tfrac{1}{2}(1-2\nu) \tag{5.17}$$

Sometimes, two other constants are introduced instead, namely the *Lamé Constants* μ and λ, where

$$\mu = G = \frac{E}{2(1+\nu)} \quad \text{and} \quad \lambda = \frac{\nu E}{(1+\nu)(1-2\nu)} \tag{5.18}$$

Lamé constants μ and λ

For the sake of convenience I provide the inverse relations

$$\nu = \frac{\lambda}{2(\lambda+\mu)} \quad \text{and} \quad E = \frac{\mu(3\lambda+2\mu)}{\lambda+\mu} \tag{5.19}$$

See also Table 5.1, p. 99 which contains more relations.

For some purposes it is useful to express isotropic elasticity as

$$\sigma_{ij} = 2\mu\varepsilon_{ij} + \lambda\varepsilon_{kk}\delta_{ij} \tag{5.20}$$

Stress-strain relation

or

$$\sigma_{ij} = \frac{E}{(1+\nu)}\varepsilon_{ij} + \frac{\nu E}{(1+\nu)(1-2\nu)}\varepsilon_{kk}\delta_{ij}, \quad i = 1,2,3 \tag{5.21}$$

Stress-strain relation

where you may prove that (5.20) and (5.21) express the same relation as (5.9) with (5.16).

The inverse relationship is also important. In order to derive this, we write (5.20) for $i = j = m$ and note that $\delta_{mm} = 3$

$$\sigma_{mm} = 2\mu\varepsilon_{mm} + \lambda\varepsilon_{kk}\delta_{mm} = 2\mu\varepsilon_{mm} + 3\lambda\varepsilon_{kk} = (3\lambda + 2\mu)\varepsilon_{kk} \tag{5.22}$$

and solve for ε_{kk}

$$\varepsilon_{kk} = \frac{1}{3\lambda + 2\mu}\sigma_{mm} \tag{5.23}$$

and finally we insert this result into (5.20) to get

Strain-stress relation

$$\varepsilon_{ij} = \frac{1}{2\mu}\left(\sigma_{ij} - \frac{\lambda}{3\lambda + 2\mu}\sigma_{kk}\delta_{ij}\right) \tag{5.24}$$

or

Strain-stress relation

$$\varepsilon_{ij} = \frac{1+\nu}{E}\sigma_{ij} - \frac{\nu}{E}\sigma_{kk}\delta_{ij}, \quad i = 1, 2, 3 \tag{5.25}$$

where exploitation of (5.19) facilitates the derivation. This may also be expressed equivalently to (5.12)

Linear elasticity = Hooke's "law"

$$\varepsilon_i = C_{ij}\sigma_j \tag{5.26}$$

where

$$C_{ij} = \frac{1}{E}\begin{bmatrix} 1 & -\nu & -\nu & 0 & 0 & 0 \\ -\nu & 1 & -\nu & 0 & 0 & 0 \\ -\nu & -\nu & 1 & 0 & 0 & 0 \\ 0 & 0 & 0 & 2(1+\nu) & 0 & 0 \\ 0 & 0 & 0 & 0 & 2(1+\nu) & 0 \\ 0 & 0 & 0 & 0 & 0 & 2(1+\nu) \end{bmatrix} \tag{5.27}$$

where $C_{ij} = E_{ij}^{-1}$, as always. That this is the case may be seen by combining (5.9) and (5.26).

5.2.1.1 *The Value of Poisson's Ratio*

The value of ν

Some comments as regards the value of ν may be needed. If we consider a state given by $\sigma_{11} \neq 0$ and all other $\sigma_{ij} = 0$, then $\varepsilon_{11} = \sigma_{11}/E$, $\varepsilon_{22} = \varepsilon_{33} = -\sigma_{11}\nu/E$ and all other $\varepsilon_{ij} = 0$. If $\sigma_{11} > 0$ the material has experienced an elongation of σ_{11}/E in the x_1-direction and a contraction (provided $\nu > 0$) of $\sigma_{11}\nu/E$ in the x_2- and x_3-directions.

By physical reasoning we must insist that the strain energy function $W(\varepsilon_{pq}) = \frac{1}{2}E_{jkmn}\varepsilon_{jk}\varepsilon_{mn}$ or, in another notation, $W(\varepsilon_m) = \frac{1}{2}E_{jk}\varepsilon_j\varepsilon_k$, see (5.16), is positive semi-definite.

If this was not the case we could extract energy from the material, thus we demand that

$$W(\varepsilon_m) = \tfrac{1}{2}E_{jk}\varepsilon_j\varepsilon_k \begin{cases} > 0 \text{ for } \varepsilon_m \not\equiv 0 \\ = 0 \text{ for } \varepsilon_m \equiv 0 \end{cases} \qquad (5.28)$$

$W(\varepsilon_m)$ positive semi-definite

This is equivalent to the following requirement for the determinant $|E_{jk}|$ of $E_{jk} \geq 0$. As you can see, it is much easier to use the version with only one index on the strain than the one with two indices

$$|E_{jk}| = \frac{E^6}{8(1+\nu)^5(1-2\nu)} \geq 0 \;\Rightarrow\; (1+\nu)(1-2\nu) \geq 0 \qquad (5.29)$$

Demand that $|E_{jk}| \geq 0$

We may now conclude that the only physically acceptable values of ν obey

$$-1 \leq \nu \leq \tfrac{1}{2} \qquad (5.30)$$

$-1 \leq \nu \leq \frac{1}{2}$

There is another, maybe a little less direct but more instructive, way of getting the limits of ν, which I shall describe below. But, first we need some more formulas. Introduce the *Mean Strain* ε and the *Strain Deviator* ε'_{jk}

$$\varepsilon \equiv \tfrac{1}{3}\varepsilon_{jj} \quad \text{and} \quad \varepsilon'_{jk} \equiv \varepsilon_{jk} - \tfrac{1}{3}\varepsilon_{mm}\delta_{jk} = \varepsilon_{jk} - \varepsilon\delta_{jk} \qquad (5.31)$$

Mean strain e
Strain deviator ε'_{jk}

and, similarly, the *Mean Stress* σ, and the *Stress Deviator* σ'_{jk} by

$$\sigma \equiv \tfrac{1}{3}\sigma_{jj} \quad \text{and} \quad \sigma'_{jk} \equiv \sigma_{jk} - \tfrac{1}{3}\sigma_{mm}\delta_{jk} = \sigma_{jk} - \sigma\delta_{jk} \qquad (5.32)$$

Mean Stress σ
Stress deviator σ'_{jk}

For the two deviators it is clearly such that they vanish when their indices are equal and summation is implied.[5.7]

$$\begin{aligned} \varepsilon'_{jj} &= \varepsilon_{jj} - \tfrac{1}{3}\delta_{jj}\varepsilon_{kk} = \varepsilon_{jj} - \tfrac{1}{3}3\varepsilon_{kk} = 0 \\ \sigma'_{jj} &= \sigma_{jj} - \tfrac{1}{3}\delta_{jj}\sigma_{kk} = \sigma_{jj} - \tfrac{1}{3}3\sigma_{kk} = 0 \end{aligned} \qquad (5.33)$$

$\varepsilon'_{jj} = 0$
$\sigma'_{jj} = 0$

We need one more material constant, namely the *Bulk Modulus* B

$$B = \frac{E}{3(1-\nu)} = \lambda + \tfrac{2}{3}\mu \qquad (5.34)$$

Bulk modulus B

where the reason for the name bulk modulus[5.8] ought to be clear from the following Example Ex 5-1.

[5.7] This is basically the definition of the deviators.
[5.8] In many textbooks the bulk modulus is denoted K.

Ex 5-1 Expression for the Bulk Modulus

Consider a case of hydrostatic pressure p, then[5.9]

Hydrostatic pressure p

$$\sigma_{jk} = \begin{cases} -p & \text{for } j = k \\ 0 & \text{for } j \neq k \end{cases} \tag{Ex. 5-1.1}$$

resulting in the following strains

$$\varepsilon_{jk} = \begin{cases} -(1-2\nu)\dfrac{p}{E} & \text{for } j = k \\ 0 & \text{for } j \neq k \end{cases} \tag{Ex. 5-1.2}$$

Let V_0 denote a volume before application of the hydrostatic stress, then to lowest order, that is for infinitesimal strains, the deformed volume V is

Deformed volume V

$$V = (1+\varepsilon_{jj})V_0 = \big(1 - 3(1-2\nu)\big)\frac{p}{E}V_0 \tag{Ex. 5-1.3}$$

and thus, the relative change in volume is

Relative volume change

$$\frac{V - V_0}{V_0} = \varepsilon_{jj} = -3(1-2\nu)\frac{p}{E} = -\frac{p}{B} \tag{Ex. 5-1.4}$$

which justifies the term *Bulk Modulus* for B, which was defined in (5.34). We may now see that

$$\sigma_{jj} = 3B\varepsilon_{kk} \tag{Ex. 5-1.5}$$

The stress-strain relation (5.20), which was given in terms of the Lamé constants λ and μ may now be rewritten in terms of the bulk modulus B and the shear modulus G, where $G = \mu$, see (5.18a)

Stress-strain relation

$$\sigma_{jk} = 2G\varepsilon'_{jk} + B\delta_{jk}\sigma_{mm} \tag{5.35}$$

With this relation in hand we may write the strain energy density $W(\varepsilon_{jk})$ in the following compact form

Strain energy $W(\varepsilon_{jk}$

$$W(\varepsilon_{jk}) = \tfrac{1}{2}\sigma_{jk}(\varepsilon_{pq})\varepsilon_{jk} = \tfrac{1}{2}\big(2G\varepsilon'_{jk} + B\delta_{jk}\varepsilon_{mm}\big)\varepsilon_{jk} \tag{5.36}$$

where it is indicated that the stresses must be given in terms of the strains. This may be written in terms of the strain deviator ε'_{pq}

Strain energy $W(\varepsilon_{jk}$

$$\begin{aligned} W(\varepsilon_{jk}) &= \tfrac{1}{2}\big(2G\varepsilon'_{jk}\varepsilon'_{jk} + 2G\varepsilon'_{jk}\tfrac{1}{3}\delta_{jk}\varepsilon_{mm} + B\varepsilon_{mm}\varepsilon_{nn}\big) \\ &= \tfrac{1}{2}\big(2G\varepsilon'_{jk}\varepsilon'_{jk} + \tfrac{2}{3}G\varepsilon'_{kk}\varepsilon_{mm} + B\varepsilon_{mm}\varepsilon_{nn}\big) \\ &= \tfrac{1}{2}\big(2G\varepsilon'_{jk}\varepsilon'_{jk} + B\varepsilon_{mm}\varepsilon_{nn}\big) \end{aligned} \tag{5.37}$$

or, exploiting (5.33a)

Strain energy $W(\varepsilon_{jk}$

$$W(\varepsilon_{jk}) = \tfrac{1}{2}\big(2G\varepsilon'_{jk}\varepsilon'_{jk} + B\varepsilon_{mm}\varepsilon_{nn}\big) \tag{5.38}$$

Since this expression is quadratic in the strain deviator ε'_{pq} and the mean strain ε_{pp} both the shear modulus G and the bulk modulus B must

[5.9] Here, we switch back to using two indices on stresses and strains.

be greater than or equal to zero in order to insure that the strain energy is positive semi-definite. Thus we have once again established (5.30). These requirements regarding the bulk modulus and the shear modulus are also known as the *Thermodynamic Restriction.*

Thermodynamic Restriction

Ex 5-2 Is Our Expression for the Strain Energy Valid?

For infinitesimal strains (2.100) becomes

$$\sigma_{pq} = \frac{\partial W(\varepsilon_{ij})}{\partial \varepsilon_{pq}} \tag{Ex. 5-2.1}$$

which by (5.38) provides

$$\sigma_{pq} = 2G\varepsilon'_{jk}\frac{\partial \varepsilon'_{jk}}{\partial \varepsilon_{pq}} + B\varepsilon_{mm}\frac{\partial \varepsilon_{nn}}{\partial \varepsilon_{pq}} \tag{Ex. 5-2.2}$$

The last factor in the first term deserves special attention

$$\frac{\partial \varepsilon'_{jk}}{\partial \varepsilon_{pq}} = \frac{\partial \varepsilon_{jk}}{\partial \varepsilon_{pq}} - \tfrac{1}{3}\delta_{jk}\frac{\varepsilon_{mm}}{\partial \varepsilon_{pq}} \tag{Ex. 5-2.3}$$

where you may note that the first term on the right-hand side may be written[5.10]

$$\frac{\partial \varepsilon_{jk}}{\partial \varepsilon_{pq}} = \tfrac{1}{2}(\delta_{jp}\delta_{kq} + \delta_{jq}\delta_{kp}) \tag{Ex. 5-2.4}$$

Utilizing (*Ex. 5-2.2*)–(*Ex. 5-2.4*) we may find

$$\begin{aligned}
\sigma_{pq} &= 2G\varepsilon'_{jk}\big(\tfrac{1}{2}(\delta_{jp}\delta_{kq} + \delta_{jq}\delta_{kp}) - \tfrac{1}{3}\delta_{jk}\delta_{pq}\big) \\
&\quad + B\varepsilon_{mm}\delta_{pq} \\
&= 2G\big(\tfrac{1}{2}(\varepsilon'_{pq} + \varepsilon'_{qp}) - \tfrac{1}{3}\varepsilon'_{jk}\delta_{jk}\delta_{pq}\big) + B\varepsilon_{mm}\delta_{pq} \\
&= 2\mu\big(\varepsilon'_{pq} - \tfrac{1}{3}\varepsilon'_{kk}\delta_{pq}\big) + \big(\lambda + \tfrac{2}{3}\mu\big)\,\varepsilon_{mm}\delta_{pq} \\
&= 2\mu\big(\varepsilon_{pq} - \tfrac{1}{3}\varepsilon_{kk}\delta_{pq}\big) + \big(\lambda + \tfrac{2}{3}\mu\big)\,\varepsilon_{mm}\delta_{pq}
\end{aligned} \tag{Ex. 5-2.5}$$

where (5.33a) has been exploited.

Thus

$$\sigma_{pq} = 2\mu\varepsilon_{pq} + \lambda\varepsilon_{mm}\delta_{pq} \tag{Ex. 5-2.6}$$

which, except for change of free and dummy indices, is the same expression for the stress tensor as (5.20) which proves the validity of (5.38).[5.11]

[5.10] The simplest way to see this is simply to write a table of the result of the left-hand and right-hand sides of (*Ex. 5-2.3*)—the work is not that great.

[5.11] This should not come as a surprise because we established the strain energy density $W(\varepsilon_{jk})$ on the basis of (5.20). Therefore, the above manipulations may be viewed as an exercise in handling contraction using the Kronecker delta δ_{jk} and related issues.

Relation Between Elastic Constants

Already you may have observed that there are many elastic constants which have been used in the present connection. As mentioned earlier, in isotropic linear elasticity there are only two independent material constants, such as Young's Modulus E and Poisson's Ratio ν. So, why do we also use the Lamé Constants μ and λ? The reason is that some expressions become much simpler by use of one set of constants than another—and which ones to choose depends on the purpose. Therefore, we may need to know the relation between all these constants, and the following table, Table 5.1, p. 99, may help.

The table is not complete, but ought to contain the most commonly used relations.

The Value of Poisson's Ratio for Artificial "Materials"

In general, we expect that pulling a rod makes it thinner, meaning that $\nu > 0$, but artificial materials such as the one shown in Fig. 5.1, see (Lakes 1987),

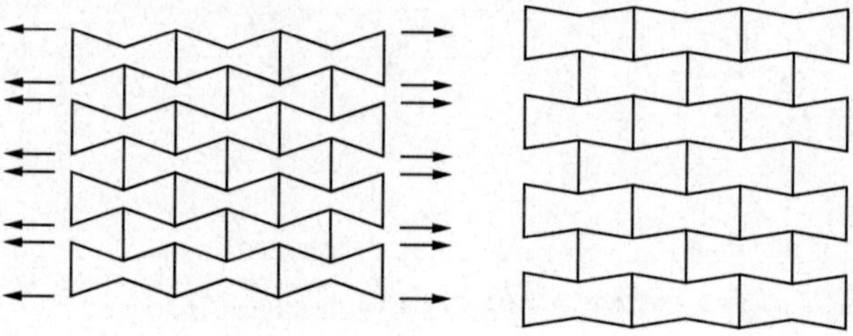

Fig. 5.1: A "material" with negative Poisson's ratio.

expands in the direction transverse to the applied force.

The expression for the shear modulus G given in (5.15) indicates that we must insist that $\nu \geq -1$ because otherwise a block of material which we subject to shear would pull in the same direction, again implying that we could harvest energy. Although $\nu = -1$ means that the shear stiffness is infinite, which seems to be unacceptable, it may be a practical approximation in some cases.

The Value of Poisson's Ratio for "Real" Materials

As mentioned above, most materials contract transverse to a stretching and thus ν is usually positive. As a fairly safe bet you may assume that

ν for "real" materials

$$0 \leq \nu \leq \tfrac{1}{2} \text{ for most "real" materials} \tag{5.39}$$

with the comment that for many engineering materials $\frac{1}{4} \lesssim \nu \lesssim \frac{1}{3}$.

Symbol	Also equal to				
λ	$\dfrac{2\mu\nu}{1-2\nu}$	$\dfrac{\mu(E-2\mu)}{3\mu-E}$	$B-\frac{2}{3}\mu$	$\dfrac{E\nu}{(1+\nu)(1-2\nu)}$	$\dfrac{3B\nu}{1+\nu}$
μ	G	$\dfrac{\lambda(1-2\nu)}{2\nu}$	$\frac{3}{2}(B-\lambda)$	$\dfrac{E}{2(1+\nu)}$	$\dfrac{3B(1-2\nu)}{2(1+\nu)}$
ν	$\dfrac{\lambda}{2(\lambda+\mu)}$	$\dfrac{\lambda}{3B-\lambda}$	$\dfrac{E}{2B}-1$	$\dfrac{3B-2\mu}{2(3B+\mu)}$	$\dfrac{3B-E}{6B}$
E	$\dfrac{\mu(3\lambda+2\mu)}{\lambda+\mu}$	$\dfrac{\lambda(1+\nu)(1-2\nu)}{\nu}$	$2\mu(1+\nu)$	$\dfrac{9B\mu}{3B+\mu}$	$3B(1-2\nu)$
B	K	$\lambda+\frac{2}{3}\mu$	$\dfrac{\lambda(1+\nu)}{3\nu}$	$\dfrac{2\mu(1+\nu)}{3(1-2\nu)}$	$\dfrac{E}{3(1-2\nu)}$

Table 5.1: Relation between elastic constants.

Ex 5-3 Special Two-Dimensional Strain and Stress States in Elastic Bodies

There are a number of special strain and stress states that are often encountered or assumed. Among the most prominent are the cases of *Plane Strain* and *Plane Stress*, which we treat briefly below, but mention that there exist several other important special states such as states that entail axial symmetry, for example.

Ex 5-3.1 Plane Strain

Plane strain: Thick plates

In a thick plate loaded in its own plane the strain state may be assumed to be constant throughout the thickness, and the strain transverse to the plane of the plate may be taken to vanish. Let $x_\alpha, \alpha \in [1,2]$ be the coordinates in the plane of the plate[5.12] and let x_3 be the coordinate at right angles to x_α. Then,

$$u_\alpha(x_j) = u_\alpha(x_\beta) \text{ and } u_3(x_j) \equiv 0 \qquad (Ex.\ 5\text{-}3.1)$$

Recall the strain-displacement relation (4.5)

$$\varepsilon_{mn} \equiv e_{mn} = \tfrac{1}{2}\left(u_{m,n} + u_{n,m}\right) \qquad (Ex.\ 5\text{-}3.2)$$

The assumption (*Ex. 5-3.1*b) implies that

$$u_{i,3} \equiv 0 \qquad (Ex.\ 5\text{-}3.3)$$

and therefore

$$\varepsilon_{33} = u_{3,3} \equiv 0 \qquad (Ex.\ 5\text{-}3.4)$$

The only non-vanishing strains then are

$$\varepsilon_{\alpha\beta} = \tfrac{1}{2}\left(u_{\alpha,\beta} + u_{\beta,\alpha}\right) \qquad (Ex.\ 5\text{-}3.5)$$

which, obviously, reflects on the constitutive equations given by (5.16) or (5.20). The latter becomes

$$\begin{aligned} \sigma_{\alpha\beta} &= 2\mu\varepsilon_{\alpha\beta} + \lambda\varepsilon_{\gamma\gamma}\delta_{\alpha\beta} \\ \sigma_{33} &= \lambda\varepsilon_{\gamma\gamma} \\ \sigma_{3\alpha} &= \sigma_{\alpha 3} \equiv 0 \end{aligned} \qquad (Ex.\ 5\text{-}3.6)$$

or, in terms of Young's modulus E and Poisson's ratio ν

$$\begin{aligned} \sigma_{\alpha\beta} &= \frac{E}{1+\nu}\varepsilon_{\alpha\beta} + \frac{\nu E}{(1+\nu)(1-2\nu)}\varepsilon_{\gamma\gamma}\delta_{\alpha\beta} \\ \sigma_{33} &= \frac{\nu E}{(1+\nu)(1-2\nu)}\varepsilon_{\gamma\gamma} \\ \sigma_{3\alpha} &= \sigma_{\alpha 3} \equiv 0 \end{aligned} \qquad (Ex.\ 5\text{-}3.7)$$

We may also be interested in expressing the non-vanishing strains in terms of the stresses. Realize that $\delta_{\omega\omega} = 2$ and utilize (*Ex. 5-3.6*a) to get

$$\sigma_{\omega\omega} = 2\mu\varepsilon_{\omega\omega} + 2\lambda\varepsilon_{\gamma\gamma} = 2(\lambda+\mu)\varepsilon_{\gamma\gamma} \qquad (Ex.\ 5\text{-}3.8)$$

[5.12] As regards the notation using Greek lower-case indices to cover the values 1 and 2, see e.g. Chapter 31, in particular Section 31.1.

Solve for $\varepsilon_{\gamma\gamma}$

$$\varepsilon_{\gamma\gamma} = \frac{1}{2(\lambda+\mu)}\sigma_{\gamma\gamma} \tag{Ex. 5-3.9}$$

Then, this expression is inserted into (*Ex. 5-3.6*a), which is solved for $\varepsilon_{\alpha\beta}$, with the result that

$$\begin{aligned} \varepsilon_{\alpha\beta} &= \frac{1}{2\mu}\left(\sigma_{\alpha\beta} - \frac{\lambda}{2(\lambda+\mu)}\sigma_{\gamma\gamma}\delta_{\alpha\beta}\right) \\ \varepsilon_{3j} &\equiv 0 \end{aligned} \tag{Ex. 5-3.10}$$

or, in terms of E and ν

$$\begin{aligned} \varepsilon_{\alpha\beta} &= \frac{1+\nu}{E}\left(\sigma_{\alpha\beta} - \nu\sigma_{\gamma\gamma}\delta_{\alpha\beta}\right) \\ \varepsilon_{3j} &\equiv 0 \end{aligned} \tag{Ex. 5-3.11}$$

Ex 5-3.2 Plane Stress

Plane stress: Thin plates

Thin plates that are loaded in their own plane may be assumed to be in a state of *Plane Stress*, which entails that the following stress components vanish

$$\sigma_{3j} \equiv 0 \tag{Ex. 5-3.12}$$

This state is the counterpart of the case of plane strain, and the derivations below follow the same patterns as in Ex 5-3.1. We employ the same notation as in Ex 5-3.1, in particular as regards the Greek indices. From (5.24) and (*Ex. 5-3.12*) we immediately get

$$\varepsilon_{\alpha\beta} = \frac{1}{2\mu}\left(\sigma_{\alpha\beta} - \frac{\lambda}{3\lambda+2\mu}\sigma_{\gamma\gamma}\delta_{\alpha\beta}\right) \tag{Ex. 5-3.12}$$

$$\varepsilon_{33} = -\frac{1}{2\mu}\frac{\lambda}{3\lambda+2\mu}\sigma_{\alpha\alpha} \tag{Ex. 5-3.12}$$

or, in terms of Young's modulus E and Poisson's ratio ν

$$\varepsilon_{\alpha\beta} = \frac{1+\nu}{E}\sigma_{\alpha\beta} - \frac{\nu}{E}\sigma_{\gamma\gamma}\delta_{\alpha\beta} \tag{Ex. 5-3.12}$$

$$\varepsilon_{33} = -\frac{\nu}{E}\sigma_{\alpha\alpha} \tag{Ex. 5-3.12}$$

We wish to solve this for $\sigma_{\alpha\beta}$ and use (*Ex. 5-3.12*) as the basis. For equal indices (*Ex. 5-3.12*) gives

$$\varepsilon_{\omega\omega} = \frac{1}{2\mu}\left(\sigma_{\omega\omega} - \frac{2\lambda}{3\lambda+2\mu}\sigma_{\gamma\gamma}\right) = \frac{1}{2\mu}\frac{\lambda+2\mu}{3\lambda+2\mu}\sigma_{\gamma\gamma} \tag{Ex. 5-3.13}$$

and thus

$$\sigma_{\gamma\gamma} = 2\mu\frac{3\lambda+2\mu}{\lambda+2\mu}\varepsilon_{\gamma\gamma} \tag{Ex. 5-3.14}$$

In terms of E and ν this becomes

$$\sigma_{\gamma\gamma} = \frac{E}{1-\nu}\varepsilon_{\gamma\gamma} \qquad (Ex.\ 5\text{-}3.15)$$

Now we insert (*Ex. 5-3.15*) into (*Ex. 5-3.12*) with the result

$$\varepsilon_{\alpha\beta} = \frac{1}{2\mu}\left(\sigma_{\alpha\beta} - 2\mu\frac{\lambda}{\lambda+2\mu}\varepsilon_{\gamma\gamma}\delta_{\alpha\beta}\right) \qquad (Ex.\ 5\text{-}3.16)$$

and solve this for $\sigma_{\alpha\beta}$. The expressions for the stresses then are

$$\begin{aligned}\sigma_{\alpha\beta} &= 2\mu\left(\varepsilon_{\alpha\beta} + \frac{\lambda}{\lambda+2\mu}\varepsilon_{\gamma\gamma}\delta_{\alpha\beta}\right)\\ \sigma_{3j} &\equiv 0\end{aligned} \qquad (Ex.\ 5\text{-}3.17)$$

or in terms of E and ν

$$\begin{aligned}\sigma_{\alpha\beta} &= \frac{E}{1+\nu}\left(\varepsilon_{\alpha\beta} + \frac{\nu}{1-\nu}\varepsilon_{\gamma\gamma}\delta_{\alpha\beta}\right)\\ \sigma_{3j} &\equiv 0\end{aligned} \qquad (Ex.\ 5\text{-}3.18)$$

Ex 5-3.3 Principle of Virtual Displacements

Membrane force $N_{\alpha\beta}$ work conjugate on strain $\varepsilon_{\alpha\beta}$

Whether we are talking about *Plane Strain* or *Plane Stress*—or the more generalized types mentioned below—the internal virtual work of the *Principle of Virtual Displacements* can be written

$$\boldsymbol{\sigma} \cdot \delta\boldsymbol{\varepsilon} = \int_V \sigma_{\alpha\beta}\delta\varepsilon_{\alpha\beta}dV = \int_A N_{\alpha\beta}\delta\varepsilon_{\alpha\beta}dA \qquad (Ex.\ 5\text{-}3.19)$$

with

$$N_{\alpha\beta} \equiv \int_t \sigma_{\alpha\beta}dx_3 \qquad (Ex.\ 5\text{-}3.20)$$

where t denotes the thickness of the body and $N_{\alpha\beta}$ denotes the membrane forces, which are work conjugate on the membrane strains $\varepsilon_{\alpha\beta}$.

More generalized plane states

The above plane states are the ones that are most often employed in analyses of plates loaded in their own plane, but other plane states, such as *Generalized Plane Strain* and *Generalized Plane Stress*, the latter, also known as *Modified Plane Stress*, may sometimes be encountered. These states are characterized by less strict assumptions than the above. We shall, however, not deal with these states in this book and leave their treatment to more advanced texts.

5.3 Nonlinear Constitutive Models

Nonlinear elasticity

There are several important types of nonlinear material behavior, such as nonlinear elasticity, and plasticity. The former is characterized by the property that, even though the stress-strain relation is nonlinear, when the load on a body is removed after any kind of deformation then the body returns to its original shape, and all stresses and strains vanish. For a body loaded into the plastic regime at least part of the strains are non-vanishing after the load has been removed and thus there are residual strains and stresses after

Plasticity

unloading. This means that for a material exhibiting plasticity, loading and unloading take different paths after plasticity has occurred, see Figs. 5.2 and 5.3.

5.4 Plasticity

Plasticity

In this book I shall only give a very summary description of plasticity and do not intend to derive any of the important theories or relations that are used.[5.13] If you wish to study classical theory of plasticity you might consider reading (Prager & Hodge, Jr. 1968) or (Kachanov 1974).

5.4.1 One-Dimensional Case

The simplest case of strain and stress, namely the one-dimensional state in a straight bar, is shown in Fig. 5.1. Even when we assume a linear relationship

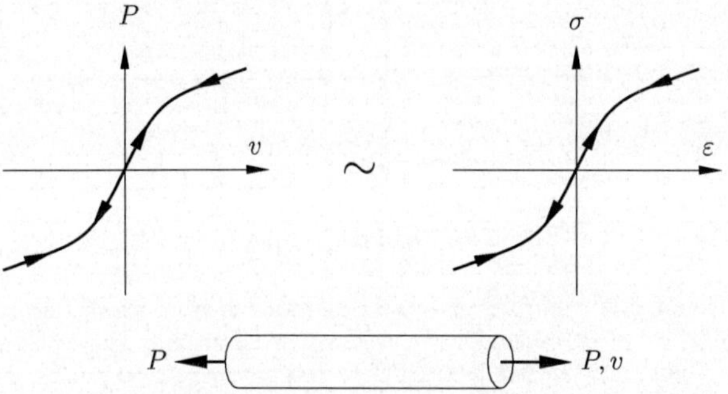

Fig. 5.1: One-dimensional state: Bar with end load and end displacement. Nonlinear elastic model.

between the end displacement v and the strain ε we may see a nonlinear response caused by material nonlinearities.

In general, a nonlinear constitutive model complicates the governing equations of any structural problem, and therefore engineers tend to apply linear elasticity or rigid, perfect plasticity, see below, when they feel confident that it can be justified. Especially in connection with analysis of reinforced concrete walls and deep beams application of the popular rigid,

[5.13] There are at least three very good reasons for not doing this; one has to do with the fact that plasticity is such a large subject in itself that it would change the emphasis of this book completely; another reason is that, while some kinds of theory of plasticity are well established, others entail so many important unanswered questions that I deemed it unfeasible to try to cover the subject in detail; the third reason is that I do not consider myself an expert on the subject.

perfectly plastic model seems extremely problematic.[5.14]

5.4.1.1 Rigid, Perfect Plasticity

Rigid, perfect plasticity

There is, however, one prominent type of nonlinear material model, which engineers like, namely *rigid, perfect plasticity*,[5.15] see Fig. 5.2. The qualifier *rigid* indicates that no strains are assumed to develop until plasticity has been encountered. The term *perfect* means that the stress cannot exceed a certain limit, the absolute value of the yield stress, which may be different in tension and compression, but is often taken to be the same, except for the sign, and then $\sigma_Y^- = -\sigma_Y$, see Fig. 5.2. The mere fact that the stress,

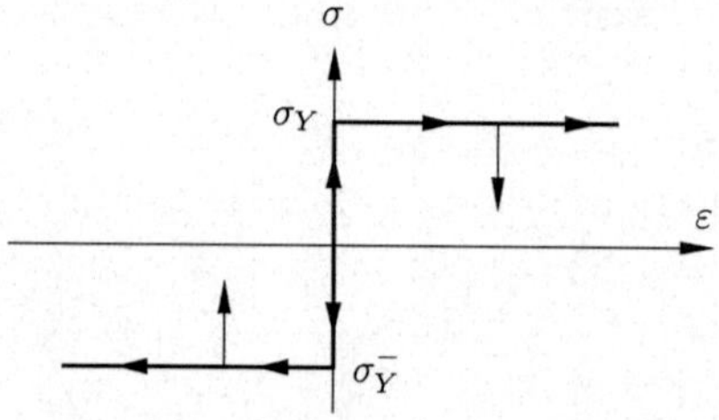

Fig. 5.2: Rigid perfect plasticity.

when the material deforms, is always equal to plus or minus the yield stress σ_Y makes life a lot easier for the engineer which is why this constitutive model is popular. In particular before the advent of computers, perfect plasticity dominated the picture completely when speaking of problems with material nonlinearity. Today, this model has lost some of its popularity, in part because of a wish to account for strains before the yield stress and to include strain hardening, see Subsection 5.4.1.5.

5.4.1.2 Steel

Steel

Choosing a relevant constitutive law, even for a particular material, may depend on several circumstances, e.g. the magnitude of the strains that are expected to develop for the problem, whether one insists on doing the analysis by hand or on a computer. As an example of this, consider the stress-strain curve for mild steel, which is sketched in Fig. 5.3. Note that the unloading path is parallel to the initial slope of the stress-strain curve which, for many materials, is observed experimentally. In the present context you should not pay too much attention to the small mound which precedes the

[5.14] As a matter of fact, sometimes they close one eye and apply it anyway. Maybe they use Garrison Keillor's statement "There comes a point where you have to stand up to reality and deny it," see (Keillor 1987), p. 40 of the English edition, as an excuse.

[5.15] Sometimes it is called *rigid, ideal plasticity*.

horizontal part of the stress-strain curve. If you anticipate most strains in a

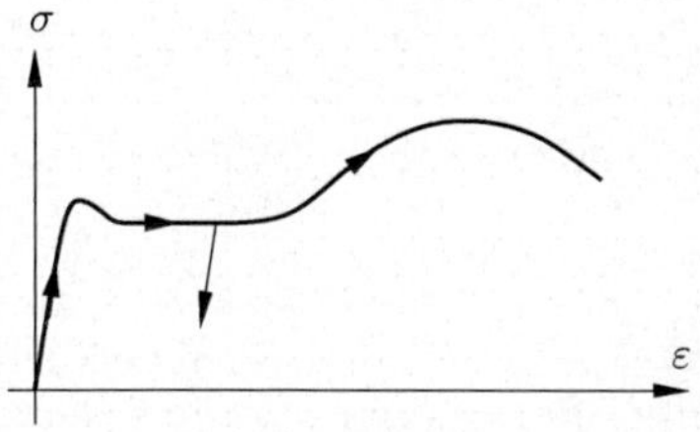

Fig. 5.3: Stress-strain curve for mild steel. Only positive strains and stresses are shown.

body to be in the interval which corresponds to the plateau, then you would make good use of the perfect plasticity model. On the other hand, if you are confident that the strains are fairly small you would apply the linearly elastic model. Finally, if most of the important strains are larger than the strains associated with the plateau then you might apply a nonlinear model like the one in Fig. 5.1.

Choice of constitutive model depends on purpose

Other materials behave quite differently, and as examples of two of the other most common building materials[5.16] we describe the behavior of concrete and wood in some detail, see Figs. 5.4 and 5.5.

5.4.1.3 Concrete

Ordinary concrete is very week in tension compared with its compression strength which is on the order of ten times higher. Its tension strength is

Concrete: strong in compression—weak in tension

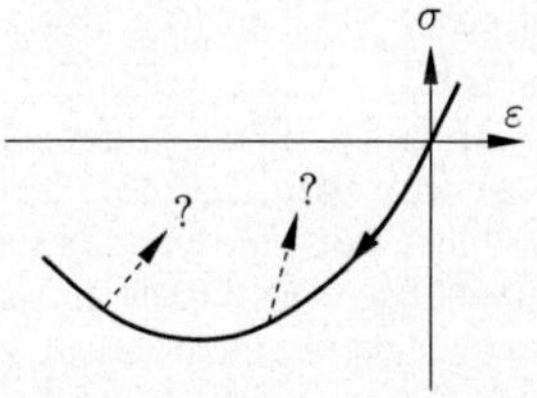

Fig. 5.4: Stress-strain curve for concrete.

therefore often neglected in structural computations. In compression con-

[5.16] I do know that soils like sand and clay are extremely common building materials in that the foundation of many buildings consists of either one or the other of these materials. Still, I have chosen not to discuss these materials because they behave even more differently from steel than concrete and wood.

crete behaves very nonlinearly, even for relatively small absolute values of the strain. Thus, the concept of a yield stress is not an obvious one for concrete.

On the other hand, many analyses of *reinforced* concrete structures have been performed successfully under the assumption of perfect plasticity. The reason why these computations agree well with experimental results hinges on the fact that the steel reinforcement bars do have a pronounced yield stress, see the plateau in Fig. 5.3, and, since the load-carrying capacity for many reinforced concrete structures is determined by the characteristics of the steel rather than the concrete itself, which should just be "strong enough," application of perfect plasticity may provide good results. This is certainly often the case for beams and plates in bending, see e.g. (Johansen 1963) and (Johansen 1972). For reinforced concrete plates loaded in their own plane the load-carrying capacity is to a much greater extent governed by the properties of the concrete, and its complicated behavior should be modeled accordingly, see also page 103 and the footnote 5.14.

There exist strong indications that the behavior of concrete after the "peak" stress, i.e. the minimum stress, cannot be described by a single curve and that factors such as the length of the test specimen enter the problem. Several analyses confirm this and the trend seems to be that the drop in absolute value of the stress becomes steeper the longer the test specimen. Thus, only the stress-strain curve in tension and down to the minimum stress may be considered a material characteristic.

5.4.1.4 Wood

Wood is strong in tension, weaker in compression

Another intriguing material is wood,[5.17] which is typically three times as strong in tension as in compression, see Fig. 5.5. In tension the behavior is fairly linear up to the ultimate stress, at which point the fibers break. The behavior in compression is much more complicated and only in an overall way understood today, whereas no fully satisfactory analysis has been conducted yet.[5.18] Around the little valley on the stress-strain curve the wood fibers start to buckle with the result that a so-called kinkband is formed. When the absolute value of the axial strain on the specimen is increased then the kinkband becomes thicker, but after the valley the load stays the same for very large compressive strains.

5.4.1.5 Strain Hardening

For materials such as steel and wood mentioned above the stress may exceed the initial yield stress σ_Y, and models like the ones in Fig. 5.6 and in Fig. 5.7

[5.17] Here we are talking about "clear wood," which is wood without defects such as knots. The behavior of structural wood is somewhat different.

[5.18] This was written in 2012, and since there is an intense effort taking place at several research laboratories and universities the final answer may have been found when you read this.

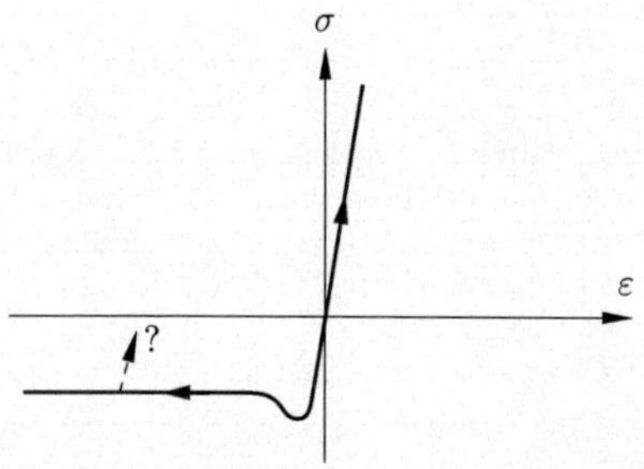

Fig. 5.5: Stress-strain curve for wood.

are sometimes applied.[5.19] The rise in stress above σ_Y is called *strain*

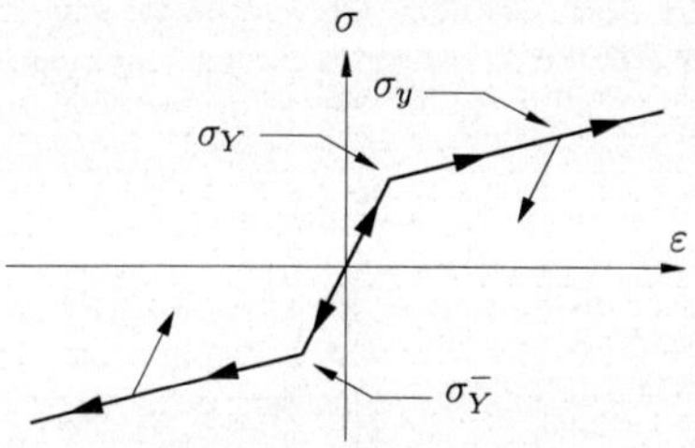

Fig. 5.6: Material exhibiting linear strain hardening.

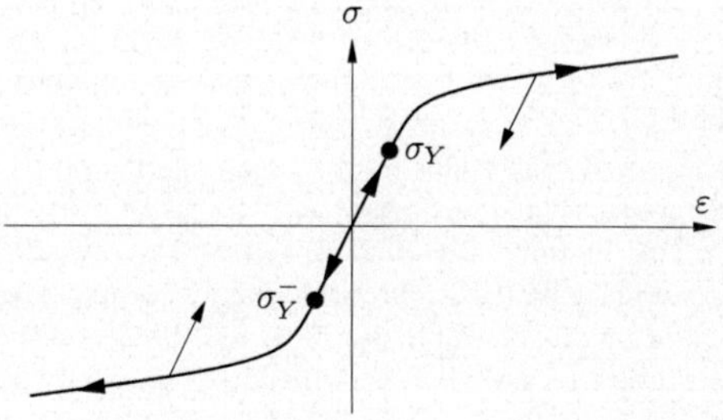

Fig. 5.7: Material exhibiting nonlinear strain hardening.

[5.19] For concrete, the idea of a definite yield stress is not a very obvious one as can be seen from the sketch in Fig. 5.4.

Strain hardening not an obvious term

hardening, which—to my opinion—is a misleading term in this connection because the material actually gets more *flexible*, not *stiffer*, after yielding. This is, however, the commonly used term and therefore I employ it too.[5.20]

Strain softening

The term *strain softening* has been reserved for that part of a stress-strain curve which displays a negative slope which for instance is the case for steel in tension after maximum stress and for concrete in compression after minimum stress, see Figs. 5.3 and 5.4.

The model with linear strain hardening is quite popular in some circles, but it seems fair to warn readers that the sharp corner may cause numerical difficulties. Of course, it is a simpler model than one with a continuous slope over the entire range, but today, when almost all nonlinear computations are done on a computer, the analytical expression for the stress-strain may be fairly complicated and still the computations may go smoothly.

5.4.1.6 Unloading—Reloading

Unloading–reloading

In the preceding sections unloading from a plastic state was not discussed in great detail. This is not because unloading is unimportant, but due to the fact that when unloading enters the picture modeling of the constitutive

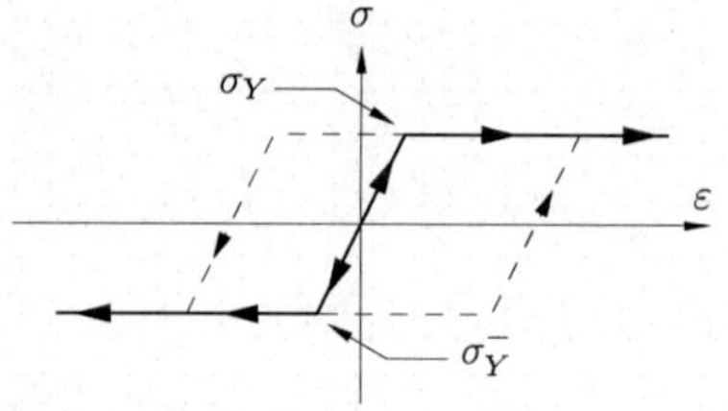

Fig. 5.8: Linear elastic-perfect plastic material model.

behavior becomes even more difficult, and sometimes dubious—even in the one-dimensional case.

Without reference to particular materials we shall consider some simple material models below. The first is the linear elastic-perfectly plastic model, see Fig. 5.8. In this model, as well as almost all other models, unloading is assumed to take place at the same slope as the original elastic slope, the initial stiffness. For metals there is experimental evidence that this is indeed a very good assumption, and, since it is a simple one, it is employed very often.

Bauschinger effect

The next model is the linear elastic-linear strain hardening model, which exhibits a *Bauschinger Effect* meaning that the compressive yield stress is

[5.20] The term strain hardening makes sense in its original connection, namely work hardening of metals which undergo plastic strains in order to make better, harder materials for tools.

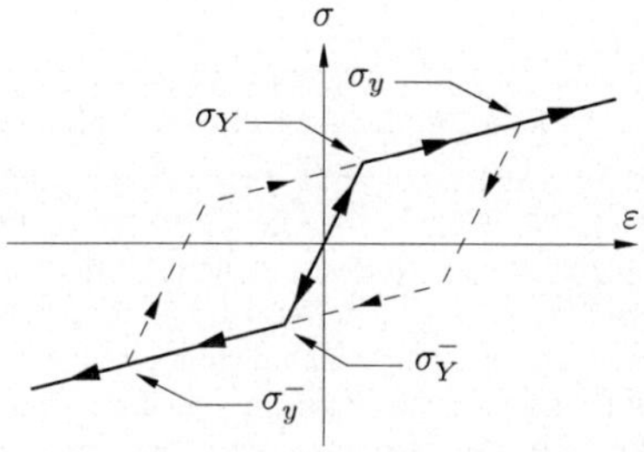

Fig. 5.9: Linear elastic linear strain hardening plastic material model exhibiting *Bauschinger Effect.*

changed when the material yields in tension, and *vice versa*, as shown in Fig. 5.9. The process in tension is such that the *initial yield stress* σ_Y is increased to a *subsequent yield stress* σ_y, and at the same time the negative yield stress is changed by the same amount from σ_Y^- to σ_y^-. In metals there is experimental evidence for the *Bauschinger effect*, and this fairly simple model is quite popular.

Initial yield stress σ_Y and subsequent yield stress σ_y

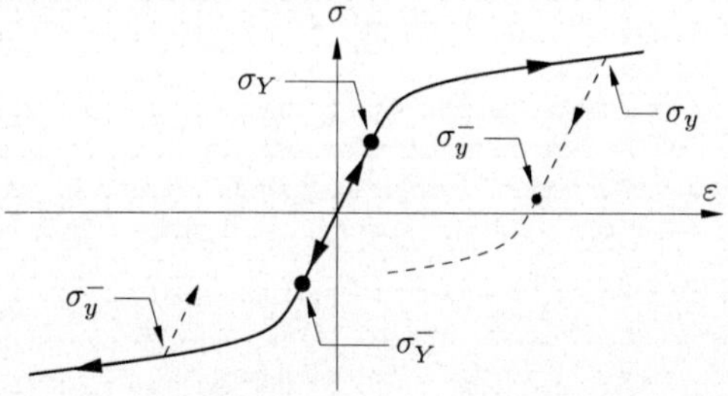

Fig. 5.10: Linear elastic-perfectly general strain hardening plastic material model.

Common to all these models is the assumption that unloading is elastic and that reloading also takes place at the initial slope, the *Initial Young's Modulus* E or E_0, until the (subsequent) yield stress is reached. Then, further loading is plastic with a lower stiffness of the material, the *Tangent Modulus* E_T.

The above models all fall into the category of the *Incremental Theory of Plasticity*[5.21] since the associated behavior must be formulated as relationships involving strain increments and stress increments.

Incremental Theory of Plasticity

Clearly, for a structure of a plastic or an elastic-plastic material the

[5.21] The fact that in most cases you would include an initial elastic behavior is not recognized by this name.

stress-strain relation is non-unique in that the stress cannot be determined just from the value of the strain, but depends on the loading history. This fact makes analysis of (elastic-)plastic structures complicated and therefore other models have been suggested. The most prominent of these is the *Deformation Theory of Plasticity*, also called *Total Theory of Plasticity*, which assumes that the stress indeed can be computed once the strain is known. Today, however, such theories are not considered valid because of lack of experimental support, and the incremental theories are preferred. On the other hand, they are used in cases of no unloading and when the loading is close to proportional.

Deformation, or total, theory of plasticity

5.4.2 Multi-Axial Plastic States

Except for the structural members of trusses and other such simple examples one-dimensional states are almost never encountered. Therefore, there is great interest in plasticity under multi-axial conditions. Both experimentally and theoretically this area is extremely difficult and, at least as far as the experiments are concerned, very expensive. The usual procedure for deriving multi-axial plasticity relations is to generalize the one-dimensional relations and postulate that plasticity occurs when a particular combination of the stress components, or the largest shear stress, reaches a certain value. Without much doubt the two most common generalizations are *von Mises' "Law"* and *Tresca's "Law,"* which is sometimes referred to as *Guest's "Law."* The objective of the generalization is to construct what is known as a *Yield Surface* which, in the general case, is a hypersurface in a multi-dimensional space that is spanned by the components of the stress tensor σ_{jk}.[5.22]

Yield surface

5.4.2.1 von Mises' "Law"

Before we state this relation we need some more stress quantities. The first is the stress deviator σ'_{jk}, which is the deviatoric part of the stress tensor σ_{jk}, see also Subsection 5.2.1.1, p. 94

Stress deviator σ'_{jk}
Mean stress σ

$$\sigma'_{jk} \equiv \sigma_{jk} - \sigma\delta_{jk} \quad \text{where} \quad \sigma \equiv \tfrac{1}{3}\sigma_{jj} = \tfrac{1}{3}I_1 \tag{5.44}$$

Here, σ is the mean, or hydrostatic, part of the stress tensor and I_1 is the first stress invariant, see (4.109). Thus, the stress deviator is equal to the stress tensor minus the hydrostatic, or mean, stress.

We need one more stress quantity, namely the so-called *Effective Stress* σ_e, where

Effective stress σ_e

$$\sigma_e^2 \equiv \tfrac{3}{2}\sigma'_{ij}\sigma'_{ij} \tag{5.45}$$

If we utilize (5.44) we may express this in terms of the components of

[5.22] This may sound fancy, but is not that difficult once you get familiar with the lingo of plasticity theory. That, however, is not going to happen just from reading this book.

stress

$$\sigma_e^2 = \frac{1}{2}\left((\sigma_{11}-\sigma_{22})^2 + (\sigma_{22}-\sigma_{33})^2 + (\sigma_{33}-\sigma_{11})^2\right) \\ +3(\sigma_{12}^2 + \sigma_{23}^2 + \sigma_{31}^2) \qquad (5.46)$$

Effective stress σ_e

Now, *von Mises' "Law"* states that plasticity occurs when

$$\sigma_e = \sigma_Y \Rightarrow \text{plasticity} \qquad (5.47)$$

von Mises' "Law"

where σ_Y is the initial yield stress determined by a one-dimensional test. What happens when the strains are increased beyond the level that caused plasticity to develop is open for discussion, as we shall see subsequently.

To be specific, consider a two-dimensional case, see the sketch Fig. 5.11(a). As mentioned in Sections 4.3 and 4.3.3 this kind of theory presupposes

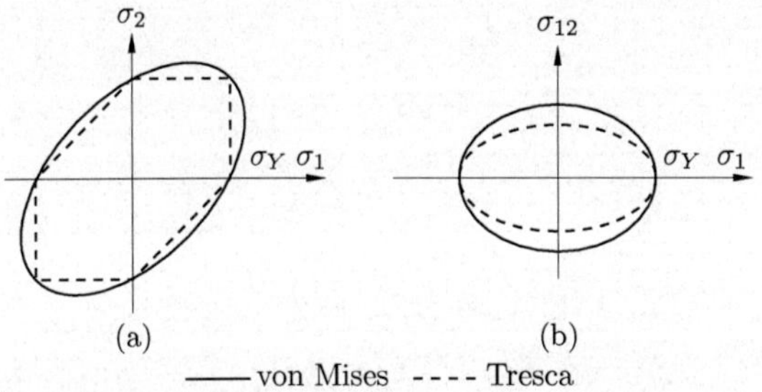

Fig. 5.11: Tresca's and von Mises' "Laws."

isotropy and is therefore quite limited in scope.

5.4.2.2 Tresca's "Law"

The other popular relation, *Tresca's "Law"* states that plasticity occurs when the maximum absolute shear stress reaches a limit. In terms of the principal stresses, see Section 4.3.3, this relation is

$$\sigma_1 - \sigma_3 = \sigma_Y \Rightarrow \text{plasticity} \qquad (5.48)$$

Tresca's "Law"

where σ_1 is the maximum and σ_3 the minimum principal stress, respectively.

In Fig. 5.11 both *von Mises'* and *Tresca's "Laws"* are shown. For a case with $\sigma_{11} \neq 0$, $\sigma_{12} \neq 0$ and all other $\sigma_{ij} = 0$ the two constitutive relations are compared in Fig. 5.11(b).

Depending on the purpose and the method of analysis one or the other of these relations is used. In general, it seems that most people find *von Mises' "Law"* the more appealing because of its smoothness.

5.4.2.3 *Unloading—Reloading*

Unloading and reloading

The question of unloading and reloading becomes very complicated in the multi-axial case. Experiments on metals indicate that the yield surface develops "corners" when plasticity occurs, see Fig. 5.12. Although these corners do not seem to be sharp some theories model them as infinitely sharp and, apparently, with good results. Others neglect the corners and assume that the yield surface stays smooth throughout the loading history. The two most popular theories entail *Kinematic Hardening*, see Fig 5.13 or

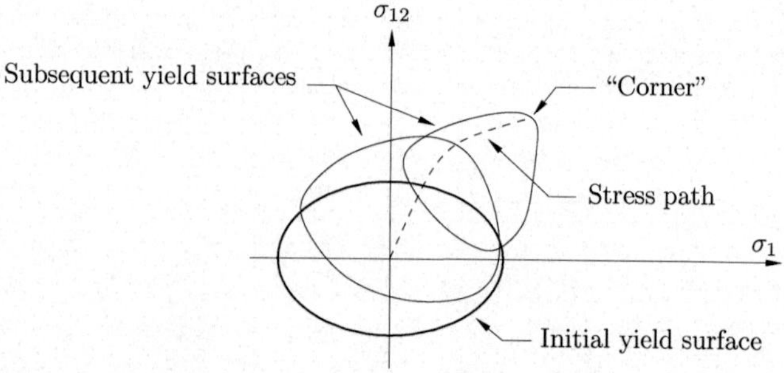

Fig. 5.12: Yield surface developing corners.

Isotropic Hardening, see Fig. 5.14, respectively. None of these hardening theories are satisfactory, as we shall see below.

5.4.2.3.1 *Kinematic Hardening*

Kinematic hardening: Bauschinger effect

This model implies an assumption that the yield surface is unaltered in shape but is translated when plasticity develops, see Fig 5.13. Thus, the

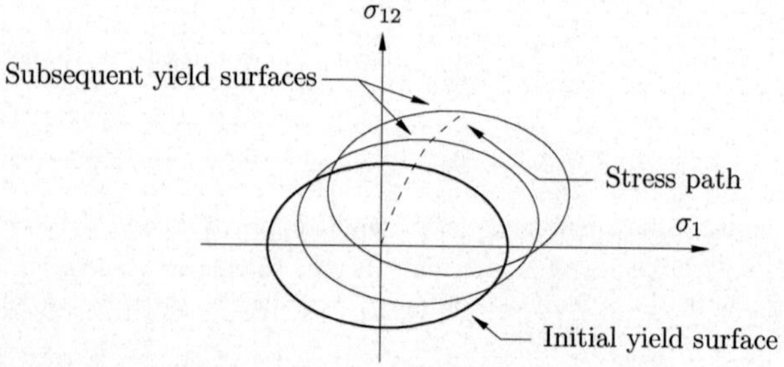

Fig. 5.13: Yield surface with kinematic hardening.

Bauschinger Effect is modeled, while corners are not assumed to appear.

5.4.2.3.2 *Isotropic Hardening*

Isotropic hardening: *Negative* Bauschinger effect

According to this model the yield surface does not move, but expands isotropically, see Fig. 5.14. A major problem with this model is the fact

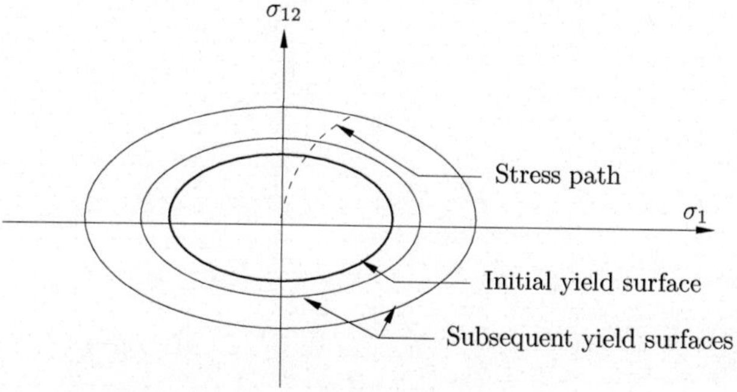

Fig. 5.14: Yield surface with isotropic hardening.

that it predicts a "negative" Bauschinger effect, which disagrees with all experiments. On the other hand, if only minor unloading takes place, then this deficiency is not noticed and therefore not important, and the computational advantages of this simple theory makes it an appealing alternative to the more realistic ones. It may be mentioned that combinations of the two models are also sometimes used.

Part II

Specialized Continua

Chapter 6

The Idea of Specialized Continua

Introduction

In this Part we develop theories for various types of structural elements, such as beams and plates. We may derive the theories in several different ways. The first two mentioned here are often used, while the third—which I find much more safe and satisfactory—receives less attention.

- **Always a New Topic**
  One possibility consists in regarding each type as a completely new topic and establish the formulas without regard to general continuum mechanics—and to some extent also knowledge about other kinds of structural elements. This is often done in courses on mechanics simply because the students have not been taught continuum mechanics and because of tradition.
- **Simplified Three-Dimensional Body**
  Another way is to consider a structural element as a three-dimensional body subject to some constraints, be they of a kinematic or static nature. This procedure works well in many cases, but it is often difficult to see whether the set of formulas derived in this way are internally consistent, e.g. whether a *principle of virtual work* applies. In Part III we shall use this method for a particular purpose, namely to determine cross-sectional properties of beams, not to establish a theory for beams as such.
- **Specialized Continuum**
  The one used here differs from the first ones in that here we insist that general principles form the basis for the theories. In particular, we shall demand that the *principle of virtual work*,[6.1] is valid for our theory.
  First, we define the so-called *generalized* kinematic quantities, i.e. *gen-*

[6.1] Either the *principle of virtual forces* or, almost always, the *principle of virtual displacements*.

E. Byskov, *Elementary Continuum Mechanics for Everyone*,
Solid Mechanics and Its Applications 194, DOI: 10.1007/978-94-007-5766-0_6,

eralized displacements, which e.g. may include rotations,[6.2] and postulate the relations between the *generalized* displacements and the *generalized* strains, which may also entail change in curvatures.[6.3]
How we select the relations between generalized displacements and generalized strains is the crux of the procedure.[6.4] In many cases, this does not cause problems, but under some circumstances, for instance for beams with strong initial curvatures, the choice may not be that obvious.
When we have set up these relations we stick them into the *principle of virtual displacements* and, after a number of manipulations, we may get the static equations, i.e. the equilibrium equations that connect the *generalized* stresses to the *generalized* loads.[6.5]
This means that the static quantities defined in this way are *generalized* in the sense that they are the work conjugate of the proper kinematic quantities, i.e. that the applied *generalized* loads produce virtual work together with the appropriate *generalized* displacements and that the *generalized* stresses and the *generalized* strains do the same.
In this way we make sure that our theory is valid, for instance for the theorems concerning extrema. Only when the static quantities are the work conjugate of the kinematic ones does the *Principle of Minimum Value of the Potential Energy* for elastic structures apply, and the same is the case of the upper and lower bounds of the load-carrying capacity of structures made of a material obeying perfect plasticity. Otherwise you may find that upper and lower bound solutions change place, which must never happen.

I shall try to explain my reason for saying that the third procedure is concerned with specialized continua. While Euclidean geometry describes the three-dimensional world, Riemannian geometry is concerned with other types of "spaces," such as the two-dimensional surface of a sphere. In order for a theory to be valid for the description of such specialized spaces it must include certain subjects, and among the most central ones are a measure of length etc. through a metric. In much the same way a beam or plate may be considered to be a specialized continuum when we insist that it obeys some fundamental principles, most importantly the principle of virtual work.

[6.2] In order to describe beam and plate bending some measure of rotation is needed.
[6.3] Again, bending of beams and plates necessitates such generalized strains.
[6.4] The choice of strain-displacement relations determines whether we get a good theory.
[6.5] In the case of kinematic nonlinearity the static equations may include displacements, as we have seen in Chapter 2.

Chapter 7

Plane, Straight Beams

In this chapter we derive the kinematic and static equations for plane, straight beams. The reason why we do not treat beams in three-dimensional space is that this is not a book on beam theory—but rather a book on continuum mechanics which also shows how to derive theories for specialized continua such as beams and plates. Basically, the principles for the derivations remain the same independent of the dimension of the space.[7.1]

We shall first consider straight beams. There are at least two reasons for this, namely that straight beams are simpler than curved ones and that most beams which are used as structural members are not curved.

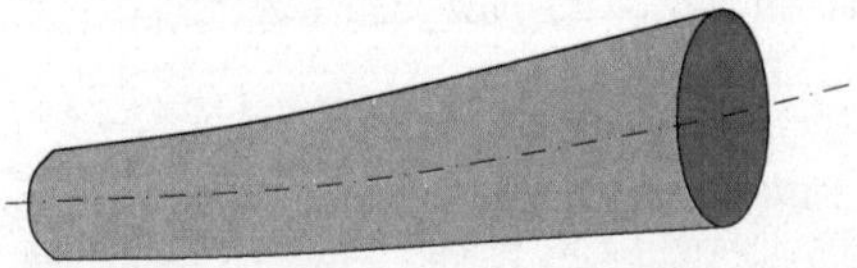

Fig. 7.1: A curved three-dimensional beam.

As mentioned in Chapter 6, there are (at least) three different ways to define the term *beam*. We shall only consider one-dimensional beams, but begin by mentioning that we could take the three-dimensional continuum as our point of departure and define a beam as a three-dimensional body characterized by the fact that one of its dimensions, the length, is much larger than the other two, the cross-sectional dimensions, see Fig. 7.1. This, of course, is a realistic way of defining the term beam.

Beams as three-dimensional bodies

However, it is the third, more abstract definition, which we shall employ here. According to this definition, a beam is a one-dimensional "body" which we supply with certain properties. This, in itself, is not very helpful. But, it proves to be a much easier way to arrive at our destination because the most fundamental decisions we must make are only concerned with the degree

Here: Beams as one-dimensional bodies

[7.1] In all fairness it must be mentioned that in deriving beam theories in three-dimensional space there are many more issues to deal with, making the process more difficult.

E. Byskov, *Elementary Continuum Mechanics for Everyone*,
Solid Mechanics and Its Applications 194, DOI: 10.1007/978-94-007-5766-0_7,

Fig. 7.2: A curved one-dimensional beam.

of nonlinearity and the kind of generalized strain measures we consider necessary to include. By application of the principle of virtual displacements we then get the equilibrium equations which define the generalized stresses.

The last choice, which often proves to be a difficult one, is concerned with the constitutive relation connecting the generalized stresses and generalized strains, and here we may have to apply a (simplified) three-dimensional analysis of a short section of the beam, see Part III. This is usually much easier than to perform a three-dimensional analysis of the entire beam. Although I strongly recommend the above procedure I must admit that there is a very important drawback to this method in that there is no surefire way to choose the strain measures correctly—rather in the most convenient way. But, if we make our decision based on experiences from the three-dimensional, real world we may get something useful. I postpone further discussion of this subject to the particular cases below.

7.1 Beam Deformation Modes

In order to establish a sound foundation for picking good strain measures for a one-dimensional beam first we discuss the most fundamental deformation modes of a *straight* beam, namely bar and beam deformation, respectively.

7.1.1 Axial Deformation

Bar deformation: Change of length

The first type of deformation we consider is *bar deformation*, i.e. extension or shortening of the beam, see Fig. 7.3.

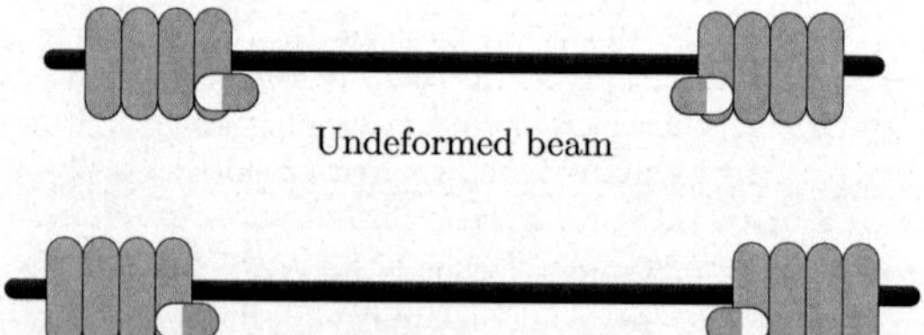

Fig. 7.3: The one-dimensional body, the ruler, in undeformed and stretched state.

To be quite honest, it takes more than usual human strength to extend even a plastic ruler as much as shown in the figure.

In this case the only strain in the bar is an axial one, e.g. γ_{11}, see (2.11) and (2.20)–(2.25) for kinematic nonlinearity, and for kinematic linearity, the axial strain may be ε_{11}, see (4.5).

7.1.2 Shear Deformation

The second basic kind of beam deformation, namely *shear deformation*, see Fig. 7.4, is not as easy to realize as the first ones—in a beam it is actually impossible to impose this deformation mode without introducing some bending.

Shear deformation: Not easy to realize

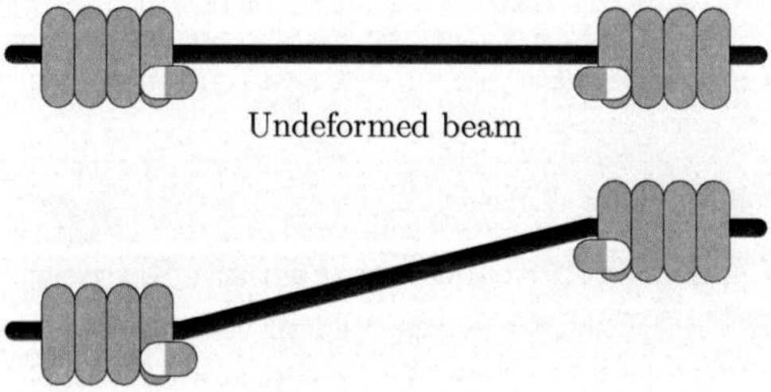

Fig. 7.4: The one-dimensional body, the ruler, in undeformed and after shear deformations.

In Fig. 7.4 there must be an ingredient of bending at the hands. Over the rest of the beam between the hands shearing is the only deformation mode. The shear strain, which we shall call φ, is closely connected to the one found in three-dimensional theory, e.g. γ_{12} for kinematic nonlinearity or ε_{12} for kinematic linearity, see (2.34) and (4.30), respectively.

7.1.3 Bending Deformation

The third, very basic type of deformation bodies is *beam bending*, which we intend to demonstrate by the plastic ruler mentioned above. Imagine that you stretch out your arms in front of you and try to bend the ruler by twisting your arms, see Fig. 7.5. In this case it does not take superhuman strength to achieve a deformation like the one shown in the figure.

Bending deformation: Change of curvature

Unlike in the previous cases, here we cannot appeal directly to the three-dimensional theory and find a relevant kinematic measure of bending,[7.2]

[7.2] The first idea, which comes to mind, but proves to be wrong for *curved* beams, is the change in geometric curvature. The correct one is the change in angle per length of the undeformed beam, see (8.25) and (8.27), but in the present case of a straight beam change in geometric curvature is a valid measure of bending.

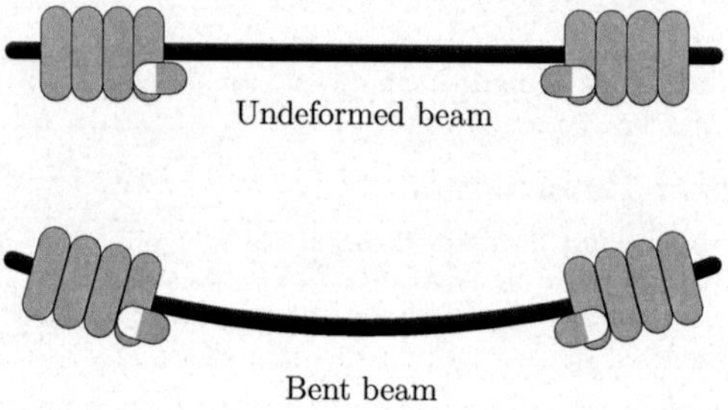

Fig. 7.5: The one-dimensional body, the ruler, in undeformed and bent state.

Curvature strain κ

which we denote κ and call *curvature strain* or *bending strain.*

7.1.4 The Three Fundamental Beam Strains

To sum up, in Fig. 7.6 the three most fundamental beam strains are visualized for the case of an initially straight beam.

Fig. 7.6: The three fundamental beam strains.

7.1.5 Choice of Deformation Modes

Axial strain ε

Curvature strain κ

Bernoulli-Euler Beams

Before we may set up a theory for the *plane* beam shown in Fig. 7.2 we must decide which strain measures we consider the most imperative to include in our theory. In order to describe the elongation of the beam we need a measure of the *axial strain* ε, which, as mentioned above, is the counterpart of e.g. γ_{11} of the three-dimensional continuum theory. The other important strain measure is the *curvature strain* κ, which, as mentioned above, represents the change in curvature of the beam, i.e. it is an expression of how much the beam has been bent. The curvature strain does not enter the three-dimensional continuum theories presented in this book and must therefore be considered separately. These are considered the two most fundamental strain measures of a beam because they appear in deformation of beams, whether they are long or short. Beams that are assumed to entail only axial and bending deformation are called *Bernoulli-Euler Beams.*

Experience shows that the third (above, the second) strain, namely the shear strain φ, which is the equivalent of e.g. γ_{12}, see (2.19), of the three-dimensional theory is only important for short beams.

Shear strain φ
Only important for short beams

Analyses of three-dimensional bodies and experiments show that when the length of a homogeneous beam made of an isotropic, linear elastic material is less than about four to five times the depth of the beam, this strain becomes significant and must be included in the analysis. Beams that are analyzed according to a theory which involves all of the above three strains are called *Timoshenko Beams.*

Timoshenko Beams

Some other types of beam theories take account of more complicated strain patterns, but I do not intend to cover such theories here.

Kinematically, there are three cases of interest, namely the fully nonlinear, the kinematically moderately nonlinear, and the linear case, respectively.

7.2 Fully Nonlinear Beam Theory

Except for the famous *Elastica*, see Example Ex 8-2, I do not intend to derive a fully nonlinear beam theory, but limit myself to considering kinematic relations of various degrees of kinematic nonlinearity, see below. On the other hand, here, we derive some kinematically full nonlinear relations because we exploit the results both in the section on the kinematically moderately nonlinear straight beams, Section 7.3, and in the section on kinematically linear straight beams, Section 7.5.

7.2.1 Kinematics

The only kinematic relations we wish to study here are associated with the axial and bending deformation of the beam.

Kinematics:
Axial and bending deformation

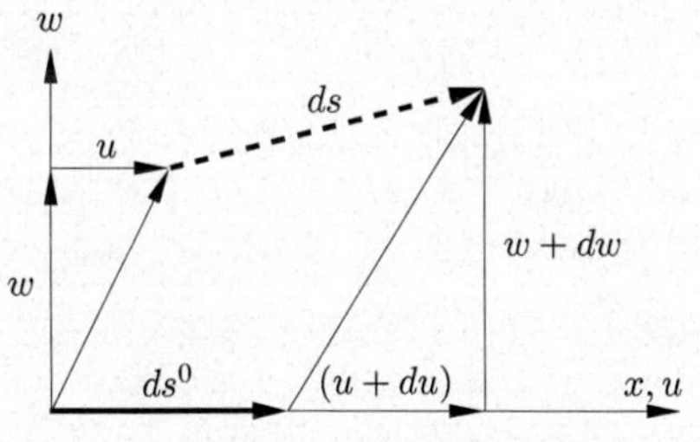

Fig. 7.7: Straight beam element.

Figure 7.7 shows a beam element whose length in the undeformed configuration is ds^0 and in the deformed configuration ds. The axial component of the displacement vector $\mathbf{u}$ is u and the transverse component is w. Then,

Theorem of Pythagoras

by application of the *Theorem of Pythagoras* we get

$$(ds)^2 = \left(ds^0 + (u+du) - u\right)^2 + \left((w+dw) - w\right)^2 \tag{7.1}$$

When we utilize the undeformed abscissa x as the axial coordinate and introduce the notation

$$(\,)' \equiv \frac{d(\,)}{dx} \tag{7.2}$$

we may get

$$(ds)^2 = \left((1+u')^2 + (w')^2\right)\left(ds^0\right)^2 \tag{7.3}$$

By analogy with the expression (2.12) relating the Lagrange strain to the change in the length of the line element we define the *Lagrange axial strain* γ by

$$(ds)^2 - \left(ds^0\right)^2 = 2\gamma\left(ds^0\right)^2 \tag{7.4}$$

and thus

$$\gamma = u' + \tfrac{1}{2}(u')^2 + \tfrac{1}{2}(w')^2 \tag{7.5}$$

As a measure of the bending deformations of the beam it is tempting to take the geometric curvature k which from differential geometry is known to be

Geometric curvature k: Not a good curvature strain measure

$$k \equiv \frac{d\omega}{ds} = \frac{\dfrac{du}{dt}\dfrac{d^2w}{dt^2} - \dfrac{dw}{dt}\dfrac{d^2u}{dt^2}}{\left(\left(\dfrac{du}{dt}\right)^2 + \left(\dfrac{dw}{dt}\right)^2\right)^{\frac{3}{2}}} \tag{7.6}$$

where t is a parameter that increases monotonically with s. If we make the particular choice $t = x$, then

Geometric curvature k: Not a good curvature strain measure

$$k = \frac{u'w'' - w'u''}{\left((u')^2 + (w')^2\right)^{\frac{3}{2}}} \tag{7.7}$$

For straight beams the angle before deformation is $\omega^0 = 0$ and after deformation it is ω, where

$$\omega = \sin^{-1}\left(\frac{dw}{ds}\right) = \sin^{-1}\left(\frac{w'}{\sqrt{1+2u'+(u')^2+(w')^2}}\right) \tag{7.8}$$

experience, however,[7.3] shows that a better curvature strain is defined as the derivative of the change in angle of the beam axis.[7.4] Then, the curvature strain κ is defined as

$$\kappa \equiv \frac{d\omega}{dx} \tag{7.9}$$

Correct curvature strain κ

At this point we leave the fully nonlinear straight beams in favor of the fully nonlinear curved beams in Section 8.1. We shall, however, utilize the above formulas as a basis for the next sections which cover kinematically moderately nonlinear and linear straight beams.

7.3 Kinematically Moderately Nonlinear Straight Bernoulli-Euler Beams

By the term kinematically moderately nonlinear we imply the same restriction on the displacement field as in Chapter 3, namely that the strains—here the axial and curvature strains—are small, but that the rotations are moderately small, see below and (2.42).

Kinematically moderately nonlinear theory

Here, we intend to establish a theory which excludes the shear strain, i.e. we are dealing with a Bernoulli-Euler beam theory.

Bernoulli-Euler beam: Shear strain φ assumed $\equiv 0$

7.3.1 Kinematics

Here, we assume that the beam is slender and omit the shear strain, as mentioned above. Thus, we only take the *axial strain* ε and the *curvature strain* κ into account. Provided that the beam axis does not experience significant stretching we may assume that

Only strains: axial ε curvature κ

$$|u'| \ll 1 \tag{7.10}$$

Little stretching

Further, since we assume that the rotations are moderate, we take

$$(w')^2 \ll 1 \tag{7.11}$$

Moderate rotations

Therefore, the expression (7.5) for the axial strain, which we denote the *moderately nonlinear axial strain* ε, becomes

$$\varepsilon = u' + \tfrac{1}{2}(w')^2 \tag{7.12}$$

Axial strain: $\varepsilon = u' + \frac{1}{2}(w')^2$

This is our axial strain measure in the kinematically moderately nonlinear theory. It is important that we make the same assumptions regarding the order of $|u'|$ and $(w')^2$ in the expressions for both the axial and the curvature strain. Otherwise, we derive a theory that is internally inconsis-

Consistency in strain measures is important

[7.3] The experience comes from observations of curved beams with finite displacements and finite strains, see the discussion Section 8.1.3.2.

[7.4] Actually, it is the derivative of the change in angle of the beam *cross-section*, but for beams, which do not experience shear strains, there is no distinction.

tent with respect to order of terms, which means that basically it is useless. Thus, in view of (7.10) and (7.11) the expression for the rotation ω (7.8) simplifies to

Rotation of beam axis: $\omega = w'$

$$\omega = w' \tag{7.13}$$

Then, the curvature strain κ is given by

Curvature strain: $\kappa = w''$

$$\kappa = w'' \tag{7.14}$$

In this theory, (7.6) yields the same result as (7.14) because of (7.11), but it is for a wrong reason because (7.6) is not correct. Note that one of the strain measures, namely the curvature strain κ, is given in terms of a second derivative of the displacements, not just the gradients, i.e. the first derivatives. This is a common feature of many theories of specialized continua, such as beams, plates, and shells. Further, note that the only nonlinear term in the strains is $\frac{1}{2}(w')^2$ in (7.12).

For the sake of completeness we give an interpretation of the generalized displacements $\boldsymbol{u}$ and the operators $\mathbf{l}_1$, $\mathbf{l}_2$, and $\mathbf{l}_{11}$, which enter the expression for the generalized strains $\boldsymbol{\varepsilon}$, see Chapter 33

$$\boldsymbol{u} \sim \begin{bmatrix} u \\ w \end{bmatrix} \quad \text{and} \quad \boldsymbol{\varepsilon} \sim \begin{bmatrix} \varepsilon \\ \kappa \end{bmatrix} = \mathbf{l}_1(\boldsymbol{u}) + \tfrac{1}{2}\mathbf{l}_2(\boldsymbol{u}) \tag{7.15}$$

with

$$\mathbf{l}_1(\boldsymbol{u}) \sim \begin{bmatrix} u' \\ w'' \end{bmatrix}, \ \mathbf{l}_2(\boldsymbol{u}) \sim \begin{bmatrix} (w')^2 \\ 0 \end{bmatrix} \ \text{and} \ \mathbf{l}_{11}(\boldsymbol{u}^a, \boldsymbol{u}^b) \sim \begin{bmatrix} w'_a w'_b \\ 0 \end{bmatrix} \tag{7.16}$$

where (subscripts) $_a$ and $_b$ denote two different displacement fields.

Ex 7-1 Rigid Rotation of a Beam

Let us consider a very simple example of a beam "deformation," see Fig. Ex. 7-1.1. The beam is subjected to a pure rotation θ about the origin, as shown in the figure, and should therefore not experience any strains. The displacements are exactly given by

$$\begin{bmatrix} u \\ w \end{bmatrix} = \begin{bmatrix} x\cos(\theta) - x \\ x\sin(\theta) \end{bmatrix}$$
$$\Rightarrow \begin{bmatrix} u' \\ w' \end{bmatrix} = \begin{bmatrix} \cos(\theta) - 1 \\ \sin(\theta) \end{bmatrix} \tag{Ex. 7-1.1}$$

Thus, the fully nonlinear axial strain measure given by (7.5) provides

$$\gamma = (\cos(\theta) - 1) + \tfrac{1}{2}\left(\cos(\theta) - 1\right)^2 + \tfrac{1}{2}\sin^2(\theta) = 0 \tag{Ex. 7-1.2}$$

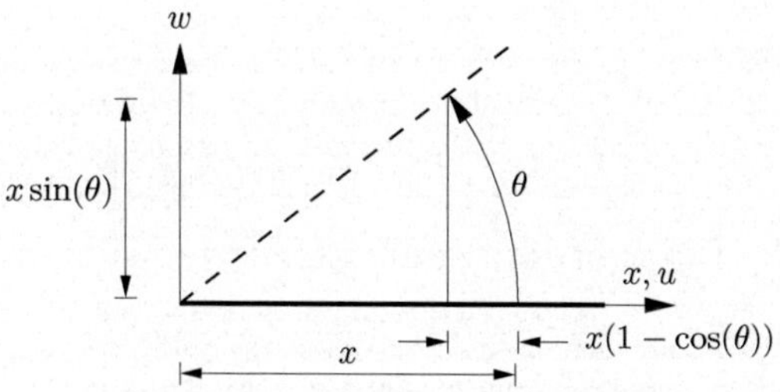

Fig. Ex. 7-1.1: *A rigid rotated beam.*

while the moderately nonlinear axial strain measure, see (7.12), predicts

$$\begin{aligned}\varepsilon &= (\cos(\theta) - 1) + \tfrac{1}{2}\sin^2(\theta)\\ &= \left(1 - \tfrac{1}{2}\theta^2 - 1 + O(\theta^4)\right) + \left(\tfrac{1}{2}\theta^2 + O(\theta^4)\right)\\ &= O(\theta^4) \not\equiv 0\end{aligned} \qquad \textit{(Ex. 7-1.3)}$$

The question is, of course, does this mean that the strain measure ε given by (7.12) is invalid? The answer to this is that it is indeed a good strain measure *provided* that you only employ it for cases which comply with the requirement that θ is "small enough." This example indicates that if $\theta^4 \ll 1$ then the strain measure is sufficiently accurate. We must, however, not draw too wide conclusions from just one very simple example.

Is ε by (7.12) valid?

Note that here the curvature strain vanishes because $w'' = 0$ along the length of the beam.

The two assumptions (7.10) and (7.11) imply that, for the theory to be consistent with respect to order of terms, the rotations ω must be small in the sense that

Consistency is very important

$$\omega^2 \ll 1 \qquad (7.17)$$

Unfortunately, we may not conclude that we have employed a valid theory[7.5] if a computation based on the kinematically moderately nonlinear theory gives results that obey (7.17). The reason is that for the particular purpose omission of terms of order ω^2 erroneously might cause the predictions to comply with (7.17). Basically, if you linearize a problem you cannot catch nonlinear effects, i.e. if you truncate a series expansion after degree n,

[7.5] In this case: a theory which is consistent with respect to order of terms.

Killing all nonlinear effects ⇒ no nonlinear effect are retained

then you may not expect to describe effects of order $n+1$ and higher. Or, in plain English, if you kill all nonlinear effects you will not be able to see any nonlinearity. Therefore, as is the case in many applications of structural mechanics, there is no substitute for engineering experience.

7.3.2 Equilibrium Equations

Generalized stresses given by Principle of Virtual Work

Now that we have chosen the strain measures, the generalized strains, we have lost our freedom to select the stress measures, the generalized stresses, but must observe the principle of virtual displacements.

Let N and M denote the *generalized stresses* that are the work conjugate of ε and κ, respectively. Further, let the beam end points be located at $x = a$ and $x = b$. Then, the internal virtual work $\boldsymbol{\sigma} \cdot \delta\boldsymbol{\varepsilon}$ is

Internal virtual work

$$\boldsymbol{\sigma} \cdot \delta\boldsymbol{\varepsilon} = \int_a^b (N\delta\varepsilon + M\delta\kappa)dx \tag{7.18}$$

In order to proceed we need an expressions for $\delta\varepsilon$ and $\delta\kappa$. We may either get them through use of (7.16) or directly from (7.12) and (7.14)

Variation of strains

$$\delta\varepsilon = \delta u' + w'\delta w' \quad \text{and} \quad \delta\kappa = \delta w'' \tag{7.19}$$

and thus (7.18) becomes

Internal virtual work

$$\begin{aligned} \boldsymbol{\sigma} \cdot \delta\boldsymbol{\varepsilon} &= \int_a^b \big(N(\delta u' + w'\delta w') + M\delta w''\big)dx \\ &= [N\delta u]_a^b + [(-M' + Nw')\delta w]_a^b + [M\delta w']_a^b \\ &\quad - \int_a^b N'\delta u dx + \int_a^b (M'' - (Nw')')\delta w dx \end{aligned} \tag{7.20}$$

Let the beam be subjected to the distributed loads $\bar{p}_u(x)$ and $\bar{p}_w(x)$ $x \in]\,a, b\,[$ and at the ends to the loads $P_u(a)$, $P_w(a)$, $C(a)$, $P_u(b)$, $P_w(b)$, $C(b)$, see Fig. 7.2. The notation should be obvious, while the convention and

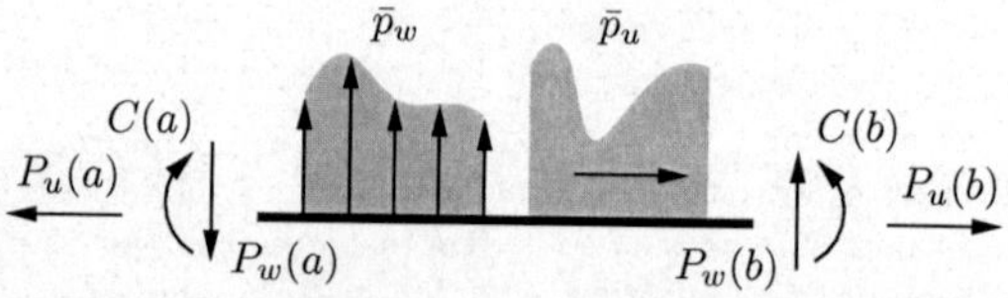

Fig. 7.2: Undeformed beam with loads and end forces.

consistency for the direction of the loads at $x = a$ may seem awkward, but is chosen for convenience. The loads at the ends may either be prescribed or they may be reactions. The external virtual work $\overline{\boldsymbol{T}} \cdot \delta \boldsymbol{u}$ is thus

$$\begin{aligned}\overline{\boldsymbol{T}} \cdot \delta \boldsymbol{u} = & \int_a^b (\bar{p}_u \delta u + \bar{p}_w \delta w) dx \\ & - P_u(a)\delta u(a) - P_w(a)\delta w(a) - C(a)\delta w'(a) \\ & + P_u(b)\delta u(b) + P_w(b)\delta w(b) + C(b)\delta w'(b)\end{aligned} \tag{7.21}$$

External virtual work

where we have exploited (7.13).

We proceed much in the same way as in Section 4.3 where we derived the equilibrium equations for the case of a kinematically linear theory for three-dimensional bodies, the important difference being the kinematically nonlinear terms.

By equating the internal with the external virtual work we arrive at the following variational equation (7.22) which at the first glance may look intimidating. However, interpreted in the right way and inspected closely, it is just a collection of equations which are rather simple by themselves, as we shall see below

$$\begin{aligned}0 = & \int_a^b (N' + \bar{p}_u)\delta u dx - \int_a^b (M'' - (Nw')' - \bar{p}_w)\delta w dx \\ & -(P_u(a) - N(a))\delta u(a) \\ & -(P_w(a) + (M'(a) - N(a)w'(a))\delta w(a) \\ & -(C(a) - M(a))\delta w'(a) \\ & +(P_u(b) - N(b))\delta u(b) \\ & +(P_w(b) + (M'(b) - N(b)w'(b))\delta w(b) \\ & +(C(b) - M(b))\delta w'(b) \\ & \forall(\delta u, \delta w, \delta u(a), \delta w(a), \delta w'(a), \delta u(b), \delta w(b), \delta w'(b))\end{aligned} \tag{7.22}$$

where "$\forall \ldots$" must be interpreted as "$\forall$ kinematically admissible ...".

Note that $\delta w'$ must be derived from δw *in the field*, but that at the ends it is permissible to vary e.g. $\delta w'(a)$ independently of $\delta w(a)$.[7.6] The variations vanish at the supports, but at all other points they are arbitrary. Therefore, at the latter points, i.e. unsupported ends and the entire field, the coefficients to the variations must equal zero in order for the the principle of virtual displacements to be fulfilled. Since the variations are arbitrary and non-vanishing except at the supports all their coefficients must vanish in

[7.6] Both $\delta w(a)$ and $\delta w'(a)$ are associated with a point. We may therefore choose a variation for which $\delta w(a) = 0$ and $\delta w'(a) \neq 0$ with $\delta w = 0$ except for an infinitesimally small neighborhood at the left-hand end. Then, we may conclude that we may take $\delta w'(a)$ to be independent of $\delta w(a)$.

order for (7.22) to hold. Therefore, (7.22) provides the static field equations

Static field equations

$$\left.\begin{aligned} N' + \bar{p}_u &= 0 \\ M'' - (Nw')' - \bar{p}_w &= 0 \end{aligned}\right\} x \in]a, b[\tag{7.23}$$

and the *possible* static boundary conditions

Possible static boundary conditions

$$\begin{aligned} N(a) &= \bar{P}_u(a)\,, & N(b) &= \bar{P}_u(b) \\ -M'(a) + N(a)w'(a) &= \bar{P}_w(a)\,, & -M'(b) + N(b)w'(b) &= \bar{P}_w(b) \\ M(a) &= \bar{C}(a)\,, & M(b) &= \bar{C}(b) \end{aligned} \tag{7.24}$$

where the overbar indicates that the static boundary conditions only apply where the end loads are prescribed, i.e. at the static boundary—which is no big surprise.

Many, probably most, authors find it convenient to introduce a static quantity V,[7.7] which is *not* a generalized stress because it has no kinematic counterpart to do internal work with

The static quantity V is not generalized

$$V \equiv -(M' - Nw') \tag{7.25}$$

The primary reason for the introduction of V is that it may be interpreted as the *shear force* in the beam in the same spirit as N is interpreted as the *axial force*, and M is the *bending moment*. After introduction of (7.25) the static equations (7.23)–(7.24) become

Static field equations including V

$$\left.\begin{aligned} N' + \bar{p}_u &= 0 \\ V' + \bar{p}_w &= 0 \\ M' - Nw' + V &= 0 \end{aligned}\right\} x \in]a, b[\tag{7.26}$$

and the *possible* static boundary conditions

Possible static boundary conditions including V

$$\begin{aligned} &N(a) = \bar{P}_u(a)\,,\ N(b) = \bar{P}_u(b) \\ &V(a) = \bar{P}_w(a)\,,\ V(b) = \bar{P}_w(b) \\ &M(a) = \bar{C}(a)\,,\ M(b) = \bar{C}(b) \end{aligned} \tag{7.27}$$

Since we are dealing with a kinematically nonlinear theory it is not surprising that the displacements enter the equilibrium equations, see (7.23)–(7.27). Notice, however, that no displacement components enter the equilibrium equations (7.23a), (7.24a–b), (7.26a) and (7.27a–b) for N which means that sometimes N can be determined separately. Also, note that none of the static equilibrium equations contain u and thus w is the only displacement component appearing in the equilibrium equations.

[7.7] In literature in the German tradition this quantity is usually called Q for *Querkraft*.

7.3.3 Interpretation of the Static Quantities

Interpretation of static quantities

We define the static quantities $\widetilde{N}$, $\widetilde{V}$ and $\widetilde{M}$ as shown in Fig. 7.3, i.e. such

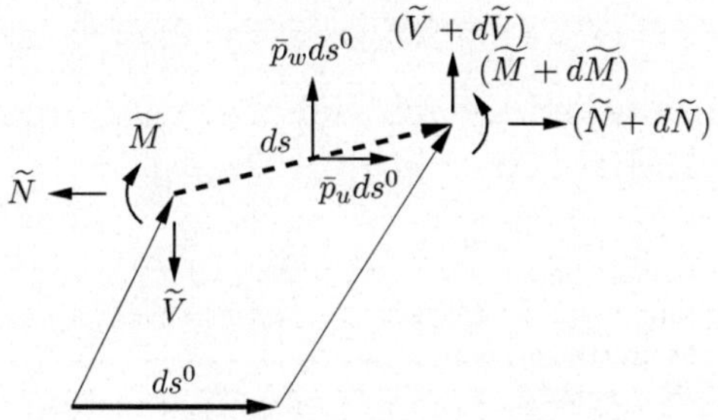

Fig. 7.3: Beam element with loads and end forces.

that $\widetilde{N}$ and $\widetilde{V}$ are parallel with and perpendicular to the undeformed beam axis, respectively. Then, we may immediately write two of the equilibrium equations

$$\left.\begin{aligned} \widetilde{N}' + \bar{p}_u &= 0 \\ \widetilde{V}' + \bar{p}_w &= 0 \end{aligned}\right\} x \in]a, b[\tag{7.28}$$

while the third requires a little care. Moment equilibrium about the left-hand end of the beam element furnishes

$$0 = (\widetilde{M} + d\widetilde{M}) - \widetilde{M} - (\widetilde{N} + d\widetilde{N})ds\sin(\omega) + (\widetilde{V} + d\widetilde{V})ds\cos(\omega) \tag{7.29}$$

where ω, as before, is the rotation of the beam axis.

Without approximations this may be rewritten

$$\begin{aligned} 0 = \widetilde{M}'dx &-(\widetilde{N} + \widetilde{N}'dx)\sqrt{1+2\gamma}\,dx\sin(\omega) \\ &+(\widetilde{V} + \widetilde{V}'dx)\sqrt{1+2\gamma}\,dx\cos(\omega) \end{aligned} \tag{7.30}$$

where the relation (7.4) has been utilized.

Because of the restriction on the magnitude of ω we may exploit (7.17) and introduce ε, which is small compared to 1, instead of γ to get

$$0 = \widetilde{M}' - \widetilde{N}\sqrt{1+2\varepsilon}\,w' + \widetilde{V}\sqrt{1+2\varepsilon}\,(1 + O((w')^2)) \tag{7.31}$$

where we have realized that both $\widetilde{N}'$ and $\widetilde{V}'$ are finite while dx is infinitesimal. The assumptions (7.10) and (7.11) result in

$$0 = \widetilde{M}' - \widetilde{N}w' + \widetilde{V} \tag{7.32}$$

Thus, the equilibrium equations (7.26) are completely equivalent to (7.28) in

Conclusion: $\widetilde{N} = N$, $\widetilde{V} = V$ and $\widetilde{M} = M$

conjunction with (7.29) and we may therefore interpret N as theaxial force $\widetilde{N}$, V as the shear force $\widetilde{V}$, which are directed along the undeformed beam axis and perpendicular to it, respectively, and M as bending moment $\widetilde{M}$.

7.4 Kinematically Moderately Nonlinear Straight Timoshenko Beams

Kinematically moderately nonlinear Timoshenko beams

The so-called *Timoshenko beam theory* developed in the following is applicable to relatively short beams where the shear strains may not be neglected. As a rule of thumb, for an elastic beam of an isotropic material with a rectangular cross-section the effect of shear strains becomes important when the length of the beam is less than about four to five times its

depth. The most important feature of Timoshenko beam theory compared with Bernoulli-Euler beam theory is the inclusion of the shear strain. For a visualization of the shear strain, see Section 7.6.

Timoshenko beam theory includes shear strain

Rotation ω of cross-section uncoupled from rotation w' of beam axis

Regarding the degree of nonlinearity we shall here make assumptions similar to the ones in Section 7.3, i.e. that the strains—here the axial, shear, and curvature strains—are small, while that the rotations are moderately small. But here, the rotation ω of the beam axis is uncoupled from the first derivative w' of the transverse displacement component, and we assume that w' is of the same order as ω.

7.4.1 Kinematics

We shall employ the same generalized strains later in the kinematically linear case, see Section 7.6, namely the axial strain ε, the shear strain φ and the curvature strain κ. The only difference is that the axial strain is nonlinear in the same way as in the case of the moderately kinematically nonlinear Bernoulli-Euler beams, see (7.12).

$$|u'| \ll 1\,, \quad (w')^2 \ll 1 \text{ and } \quad \omega^2 \ll 1 \tag{7.33}$$

7.4.1.1 *Axial Strain*

Axial strain ε

$$\varepsilon = u' + \tfrac{1}{2}(w')^2 \tag{7.34}$$

7.4.1.2 *Shear Strain*

Shear strain $\varphi = w' - \omega$

$$\varphi = w' - \omega \tag{7.35}$$

7.4.1.3 *Curvature Strain*

Curvature strain κ

$$\kappa \equiv \frac{d\omega}{dx} = \omega' \tag{7.36}$$

Here, the interpretation of the operators $\mathbf{l}_1$, $\mathbf{l}_2$, and $\mathbf{l}_{11}$, which enter the expression for the generalized strains ε, see Chapter 33, are

$$\boldsymbol{u} \sim \begin{bmatrix} u \\ w \\ \omega \end{bmatrix} \quad \text{and} \quad \boldsymbol{\varepsilon} \sim \begin{bmatrix} \varepsilon \\ \varphi \\ \kappa \end{bmatrix} = \mathrm{l}_1(\boldsymbol{u}) + \frac{1}{2}\mathrm{l}_2(\boldsymbol{u}) \tag{7.37}$$

with

$$\mathrm{l}_1(\boldsymbol{u}) \sim \begin{bmatrix} u' \\ (w' - \omega) \\ \omega' \end{bmatrix}, \ \mathrm{l}_2(\boldsymbol{u}) \sim \begin{bmatrix} (w')^2 \\ 0 \\ 0 \end{bmatrix}, \ \mathrm{l}_{11}(\boldsymbol{u}^a, \boldsymbol{u}^b) \sim \begin{bmatrix} w'_a w'_b \\ 0 \\ 0 \end{bmatrix} \tag{7.38}$$

where (subscripts) $_a$ and $_b$ denote two different displacement fields.

7.4.2 Equilibrium Equations

As usual, our choice of strain measures, the generalized strains, dictates our stress measures, the generalized stresses, because the two sets of must be each other's work conjugate in the principle of virtual displacements. Since we work with three generalized strains we must have three generalized stresses which we shall denote $\widehat{N}$, $\widehat{V}$ and $\widehat{M}$. They are the work conjugate of ε, φ and κ, respectively.

Generalized strains ε, φ and κ
Generalized stresses $\widehat{N}$, $\widehat{V}$ and $\widehat{M}$

When the beam end points are located at $x = a$ and $x = b$ the internal virtual work $\boldsymbol{\sigma} \bullet \delta\boldsymbol{\varepsilon}$ is

$$\boldsymbol{\sigma} \bullet \delta\boldsymbol{\varepsilon} = \int_a^b (\widehat{N}\delta\varepsilon + \widehat{V}\delta\varphi + \widehat{M}\delta\kappa)dx \tag{7.39}$$

Internal virtual work

In order to proceed we need an expressions for $\delta\varepsilon$, $\delta\varphi$ and $\delta\kappa$. The strain definitions (7.34), (7.35) and (7.36) provide

$$\delta\varepsilon = \delta u' + w'\delta w' \ , \quad \delta\varphi = (\delta w' - \delta\omega) \ \text{ and } \ \delta\kappa = \delta\omega' \tag{7.40}$$

and thus (7.39) becomes

$$\begin{aligned} \boldsymbol{\sigma} \bullet \delta\boldsymbol{\varepsilon} &= \int_a^b \big(\widehat{N}(\delta u' + w'\delta w') + \widehat{V}(\delta w' - \delta\omega) + \widehat{M}\delta\omega'\big)dx \\ &= [\widehat{N}\delta u]_a^b + [(\widehat{V} + \widehat{N}w')\delta w]_a^b + [\widehat{M}\delta\omega]_a^b \\ &\quad - \int_a^b \widehat{N}'\delta u dx - \int_a^b \big(\widehat{V}' + (\widehat{N}w')'\big)\delta w dx \\ &\quad - \int_a^b (\widehat{M}' + \widehat{V})\delta\omega dx \end{aligned} \tag{7.41}$$

Internal virtual work

Provided that there are no distributed moment loads, i.e. that the loads

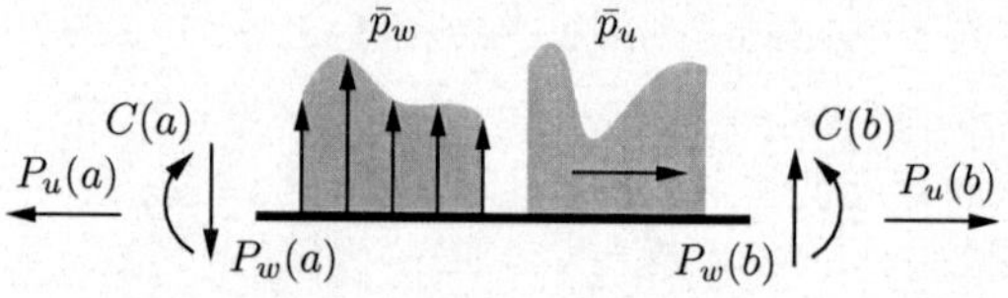

Fig. 7.4: Undeformed beam with loads and end forces.

are the ones shown in Fig. 7.2, which for convenience is repeated here as Fig. 7.4, the external virtual work $\overline{\boldsymbol{T}} \cdot \delta \boldsymbol{u}$ is

External virtual work

$$\begin{aligned} \overline{\boldsymbol{T}} \cdot \delta \boldsymbol{u} = & \int_a^b (\bar{p}_u \delta u + \bar{p}_w \delta w) dx \\ & - P_u(a)\delta u(a) - P_w(a)\delta w(a) - C(a)\delta w'(a) \\ & + P_u(b)\delta u(b) + P_w(b)\delta w(b) + C(b)\delta w'(b) \end{aligned} \tag{7.42}$$

which obviously is the same as (7.21).

By equating the internal and external virtual work, given by (7.41) and (7.42), respectively, we can immediately conclude that

$$\widehat{M}' + \widehat{V} = 0 \,, \; x \in]a, b[\tag{7.43}$$

which indicates that $\widehat{V}$ is not the same as the static quantity V defined by (7.25), and thus we do not recover the equilibrium equations (7.26) and (7.27) or (7.23) and (7.24). Again, by equating the internal and external virtual work we may get the other two static field equations

$$\widehat{V}' + (\widehat{N} w')' + \bar{p}_w = 0 \,, \; x \in]a, b[\tag{7.44}$$

which also does not agree with (7.26b), while

$$\widehat{N}' + \bar{p}_u = 0 \,, \; x \in]a, b[\tag{7.45}$$

is the same as (7.26a).

Combining (7.43) with (7.44) we may get

$$\widehat{M}'' - (\widehat{N} w')' - \bar{p}_w = 0 \,, \; x \in]a, b[\tag{7.46}$$

and we have recovered (7.23). In order to compare with the static field equations (7.62) for the kinematically moderately nonlinear Bernoulli-Euler case, we recapitulate the three static field equations (7.26) but rewrite the third in order to compare with (7.46)

Static field equations for moderately nonlinear *Bernoulli-Euler* beam

$$\left. \begin{aligned} & N' + \bar{p}_u = 0 \\ & V' + \bar{p}_w = 0 \\ & M'' - (N w')' - \bar{p}_w = 0 \end{aligned} \right\} \; x \in]a, b[\;\; \text{(Bernoulli-Euler)} \tag{7.47}$$

and collect the equilibrium equations for the present case, i.e. kinematically moderately nonlinear Timoshenko beam in the same way

$$\left.\begin{aligned} &\widehat{N}' + \bar{p}_u = 0 \\ &\widehat{V}' + (\widehat{N}w')' + \bar{p}_w = 0 \\ &\widehat{M}'' - (\widehat{N}w')' - \bar{p}_w = 0 \end{aligned}\right\} \quad x \in]a, b[\quad \text{(Timoshenko)} \tag{7.48}$$

Static field equations for moderately nonlinear *Timoshenko* beam

As regards the *possible* static boundary conditions, here we get

$$\begin{aligned} &\widehat{N}(a) = \bar{P}_u(a)\,, && \widehat{N}(b) = \bar{P}_u(b) \\ &\widehat{V}(a) + \widehat{N}(a)w'(a) = \bar{P}_w(a)\,, && \widehat{V}(b) + \widehat{N}(b)w'(b) = \bar{P}_w(b) \\ &\widehat{M}(a) = \bar{C}(a)\,, && \widehat{M}(b) = \bar{C}(b) \end{aligned} \tag{7.49}$$

Possible static boundary conditions for *Timoshenko* beams

and also here the discrepancy between these boundary conditions and (7.27) is obvious. On the other hand, the boundary conditions for the Timoshenko beams appear to be consistent with its differential equations.

The question is then to interpret $\widehat{V}$ along with the other two static quantities $\widehat{N}$ and $\widehat{M}$, but first note that, whether we succeed in this respect, the theory developed above is a consistent one in that its stress and strain measures are each other's work conjugate and therefore it is satisfactory from a theoretical standpoint. Therefore, we may not necessarily need to

Consistent theory, but problems with interpretations

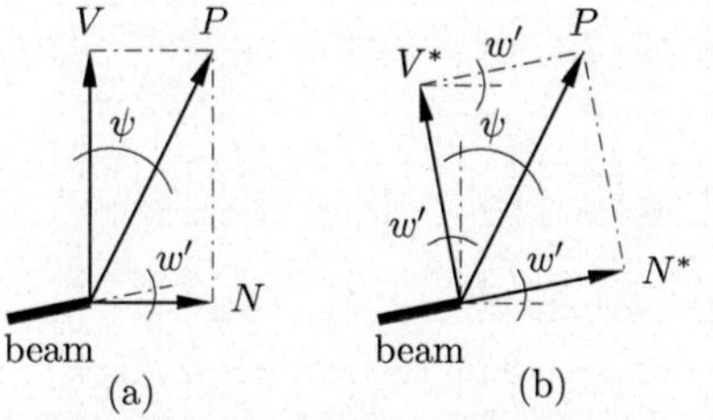

Fig. 7.5: End load and projections.
Bending moments M and M^* are not shown.

interpret the static quantities in the field $x \in]a, b[$, except in order to be able to choose meaningful constitutive relations. Without mentioning it, we have encountered a similar issue as regards the stresses and strains of the theory for three-dimensional bodies developed in Chapter 2, in particular Section 2.6.1. There, we assumed a linear relationship between the second Piola-Kirchhoff pseudo stresses and the Lagrange strains and, as is indicated by "pseudo," that stress does not measure force per present area. However, Koiter (2008) has proved that the possible error in assuming this linear relationship is small, namely of the order of the absolute value of the largest strain compared to 1.[7.8]

In order to try to interpret $\widehat{N}$, $\widehat{V}$ and $\widehat{M}$ consider Fig. 7.5. Let a cross-section of the beam be subjected to the load P and resolve it in two different ways, as sketched in Fig. 7.5(a) and (b). Then,

$$\begin{aligned} N &= P\sin(\psi)\,, & V &= P\cos(\psi) \\ N^* &= P\sin(\psi + w')\,, & V^* &= P\cos(\psi + w') \end{aligned} \tag{7.50}$$

where we have utilized the fact that we assume that the rotation of the beam axis is small, i.e. $(w')^2 \ll 1$, in accordance with our basic assumption (7.33b). Then,

$$\begin{aligned} N^* &= P\big(\sin(\psi)\cos(w') + \cos(\psi)\sin(w')\big) \approx P\big(\sin(\psi) + \cos(\psi)w'\big) \\ V^* &= P\big(\cos(\psi)\cos(w') - \sin(\psi)\sin(w')\big) \approx P\big(\cos(\psi) - \sin(\psi)w'\big) \end{aligned} \tag{7.51}$$

i.e.

$$\begin{bmatrix} N^* \\ V^* \end{bmatrix} \approx \begin{bmatrix} 1 & w' \\ -w' & 1 \end{bmatrix} \begin{bmatrix} N \\ V \end{bmatrix} \tag{7.52}$$

with the inverse relation

$$\begin{bmatrix} N \\ V \end{bmatrix} \approx \frac{1}{1 + (w')^2} \begin{bmatrix} 1 & -w' \\ w' & 1 \end{bmatrix} \begin{bmatrix} N^* \\ V^* \end{bmatrix} \approx \begin{bmatrix} 1 & -w' \\ w' & 1 \end{bmatrix} \begin{bmatrix} N^* \\ V^* \end{bmatrix} \tag{7.53}$$

where we have exploited the condition (7.33b).

The first resolution of the forces is the one that is valid for the moderately nonlinear Bernoulli-Euler beam, while the second is the more intuitive one where the axial force is directed along the deformed beam axis and the shear force is perpendicular to it. The latter is the one which applies to the Elastica, see Example Ex 8-2. One might think that a combination of these resolutions, for example considering N together with V^* could furnish the equilibrium equations (7.48). This is, however, not possible[7.9] and we must live with the fact that we cannot interpret $\widehat{N}$, $\widehat{V}$ and $\widehat{M}$. Fortunately, as we shall see later in Example Ex 18-2, which is concerned with determining the buckling load of a linearly elastic pinned column using the above theory, the theory can be applied with success.

[7.8] Note that this does not preclude finite displacements and rotations.
[7.9] At least some very qualified people and I have not been able to do so.

7.5 Kinematically Linear Straight Bernoulli-Euler Beams

A kinematically linear theory for straight beams is easily derived from the theory given in Section 7.3. In addition to the assumption (7.10)

Kinematically Linear Bernoulli-Euler Beams

$$|u'| \ll 1 \tag{7.54}$$

we impose the restriction

$$|w'| \ll 1 \tag{7.55}$$

in contrast to (7.11)

$$(w')^2 \ll 1 \tag{7.56}$$

Then, all kinematic nonlinearity vanishes with the result that the generalized strains become

Generalized strains ε and κ

$$\varepsilon = u' \quad \text{and} \quad \kappa = w'' \tag{7.57}$$

while the kinematically moderately nonlinear theory has the generalized strains given as

$$\varepsilon = u' + \tfrac{1}{2}(w')^2 \quad \text{and} \quad \kappa = w'' \tag{7.58}$$

By going through the derivations of Section 7.3.2 and omitting all kinematically nonlinear terms we get the following static field equations

Static field equations

$$\left.\begin{aligned} N' + \bar{p}_u &= 0 \\ M'' - \bar{p}_w &= 0 \end{aligned}\right\} x \in]\, a, b\,[\tag{7.59}$$

and the *possible* static boundary conditions

Possible static boundary conditions

$$\begin{aligned} N(a) &= \bar{P}_u(a)\,, & N(b) &= \bar{P}_u(b) \\ -M'(a) &= \bar{P}_w(a)\,, & -M'(b) &= \bar{P}_w(b) \\ M(a) &= \bar{C}(a)\,, & M(b) &= \bar{C}(b) \end{aligned} \tag{7.60}$$

Just to be sure, we note that, as in the case of moderate kinematic nonlinearity, N and M are the generalized stresses.

Introduce the shear force V by (recall that in these theories V is *not* a generalized stress)

$$V \equiv -M' \tag{7.61}$$

then, the static field equations may be written in the following fashion which is similar to and may be derived from (7.27).

The result is

Static field equations including V

$$\left.\begin{aligned} N' + \bar{p}_u &= 0 \\ V' + \bar{p}_w &= 0 \\ M' + V &= 0 \end{aligned}\right\} \; x \in]\, a, b\, [\tag{7.62}$$

and the *possible* static boundary conditions become

Possible static boundary conditions including V

$$\begin{aligned} N(a) &= \bar{P}_u(a)\,, \quad N(b) = \bar{P}_u(b) \\ V(a) &= \bar{P}_w(a)\,, \quad V(b) = \bar{P}_w(b) \\ M(a) &= \bar{C}(a)\,, \quad M(b) = \bar{C}(b) \end{aligned} \tag{7.63}$$

For the interpretation of the static quantities the reader is referred to Section 7.3. In the present, kinematically linear case, the generalized stresses, i.e. the axial force and the bending moment, act on the *undeformed* beam, which is a fact that must be taken into consideration in the interpretation of the generalized stresses, in particular in Fig. 7.3.

7.6 Kinematically Linear Straight Timoshenko Beams

We shall here make the same assumptions as in Section 7.5 regarding the degree of nonlinearity, i.e. that the strains—here the axial, shear, and curvature strains—as well as the rotations are small.

Kinematically Linear Timoshenko Beams

As mentioned in Section 7.4, *Timoshenko beam theories* are applicable to relatively short beams where the shear strains may not be neglected. Recall that as a rule of thumb, for an elastic beam with a rectangular cross-section the effect of shear strains becomes important when the length of the beam is less than about four to five times its depth. If the beam is made of, say an orthotropic material with the stiff direction along the beam axis, then shear strains may always play an important role. A similar observation holds for sandwich beams.

The derivations below follow the same pattern as that of Section 7.4 and of Section 7.5. As mentioned in Section 7.4 we did not visualize the shear strain there, but chose to defer it to the present case. In part the reason for this is that the theory developed in Section 7.4 was more complicated than I like—and probably not completely satisfactory. As before, the first consequence of considering shear strains is that we need three generalized displacements to describe deformations of the beam, namely the axial displacement component u, the transverse displacement component w and, in addition, to these the rotation ω of the beam "cross-section."[7.10]

[7.10] The reason for the quotation marks is that we are dealing with a one-dimensional beam theory, i.e. the beam theory does not, by itself, entail a thickness, but the meaning should be clear from Fig. 7.6.

7.6.1 Kinematics

Here, we must define the three generalized strains, which are strains sketched in Fig. 7.6, namely the axial strain ε, the shear strain φ, and the curvature strain κ, to the same degree of nonlinearity as in Section 7.3.1, namely

Timoshenko beam: Shear strain $\varphi \not\equiv 0$

$$|u'| \ll 1\,, \quad |w'| \ll 1 \text{ and } |\omega| \ll 1 \tag{7.64}$$

7.6.1.1 Axial Strain

Once more (7.57) defines the axial strain

$$\varepsilon = u' \tag{7.65}$$

Axial strain ε

7.6.1.2 Shear Strain

We take the shear strain φ to be the same as for the moderately nonlinear version of the Timoshenko beam theory, see (7.35).

$$\varphi = w' - \omega \tag{7.66}$$

Shear strain $\varphi = w' - \omega$

7.6.1.3 Curvature Strain

Because we have abandoned the assumption of kinematic coupling between the transverse displacement component w and the rotation ω the curvature strain κ must not be defined in accordance with (7.57) but as in (7.9)

$$\kappa \equiv \frac{d\omega}{dx} = \omega' \tag{7.67}$$

Curvature strain $\kappa \neq w''$

which, by the way, holds regardless of the degree of nonlinearity.

7.6.1.4 Interpretation of Shear Strain

While the physical interpretation of the two strains ε and κ probably is clear, both from Fig 7.6 and from our intuition, the shear strain φ deserves special attention. Consider a two-dimensional element of the beam, see Fig. 7.6,[7.11] and subject it to the two simple displacement fields shown. But, first we need to define two directions. One is the direction of the beam axis, which is indicated by the dash-dot line, and the other is the normal to the cross-section of the beam and is given by the arrow. Note that, here, the term cross-section should be taken literally because we are dealing with a two-dimensional beam.

[7.11] The deformation patterns shown in the figure are not possible for an entire beam cross-section because some ingredient of S-shape is unavoidable. So, the sketches must be taken with a grain of salt.

You may, however, ask if such an S-shape does not violate the idea associated with extension and pure bending that the cross-sections remain plane. The answer is that these different kinds of displacements do not interact. A similar conclusion holds for torsion of beams where most cross-sections experience warping.

Case (a)

Recall that all displacements are infinitesimal.

Then, the displacements may be determined as

$$u_1^{(a)} = 0 \text{ and } u_2^{(a)} = w' x_1 \tag{7.68}$$

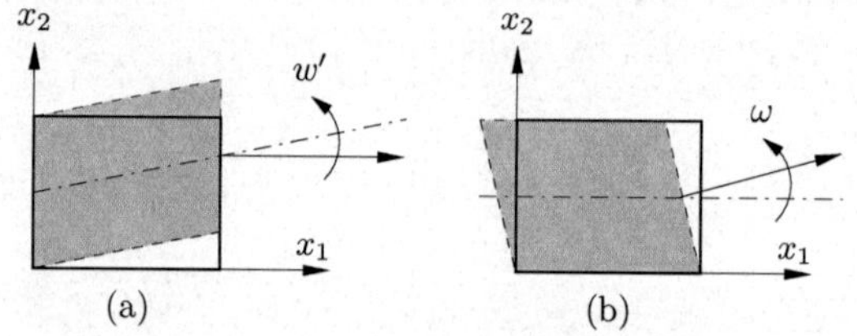

Fig. 7.6: Infinitesimal element of Timoshenko beam. Two fundamental cases of shear strain. In order to avoid clogging of the figure, the x_1-axis is not coinciding with the beam axis.

The result is that the strains (4.5) become

$$\varepsilon_{11}^{(a)} = 0\,, \;\; \varepsilon_{12}^{(a)} = \varepsilon_{21}^{(a)} = \tfrac{1}{2} w' \text{ and } \varepsilon_{22}^{(a)} = 0 \tag{7.69}$$

and thus the beam shear strain $\varphi^{(a)}$ is

$$\varphi^{(a)} = 2\varepsilon_{12}^{(a)} = w' \tag{7.70}$$

Case (b)

In this case the displacements are

$$u_1^{(b)} = -\omega x_2 \text{ and } u_2^{(b)} = 0 \tag{7.71}$$

and

$$\varepsilon_{11}^{(b)} = 0\,, \;\; \varepsilon_{12}^{(b)} = \varepsilon_{21}^{(b)} = -\tfrac{1}{2}\omega \text{ and } \varepsilon_{22}^{(b)} = 0 \tag{7.72}$$

and thus

$$\varphi^{(b)} = 2\varepsilon_{12}^{(b)} = -\omega \tag{7.73}$$

Combination of Cases (a) and (b)

Any strain situation can be composed of cases (a) and (b) and therefore we may take the beam shear strain φ to be

Shear strain
$\varphi = w' - \omega$

$$\varphi = w' - \omega \tag{7.74}$$

This is the strain utilized in Timoshenko beam theories.

7.6.1.5 Interpretation of Operators

Here, we only need to interpret the linear operator $\mathbf{l}_1$ which enters the expression for the generalized strains $\boldsymbol{\varepsilon}$, see Chapter 33. The generalized displacements are

$$\boldsymbol{u} \sim \begin{bmatrix} u \\ w \\ \omega \end{bmatrix} \tag{7.75}$$

and the interpretation of the linear operator $\mathbf{l}_1$ is

$$\mathbf{l}_1(\boldsymbol{u}) \sim \begin{bmatrix} u' \\ (w' - \omega) \\ \omega' \end{bmatrix} \tag{7.76}$$

while the other operators $\mathbf{l}_2$, and $\mathbf{l}_{11}$ do not enter the description due to its linearity.

7.6.2 Generalized Stresses

As mentioned above, our choice of strain measures, the generalized strains, dictates our stress measures, the generalized stresses, because these two sets must be each other's work conjugate in the *Principle of Virtual Displacements.* Since we work with three generalized strains we must have three generalized stresses which, as before, we shall denote N, V and M. They are the work conjugate of ε, φ and κ, respectively.

7.6.3 Equilibrium Equations

When the beam end points are located at $x = a$ and $x = b$ the internal virtual work $\boldsymbol{\sigma} \cdot \delta\boldsymbol{\varepsilon}$ is

$$\boldsymbol{\sigma} \cdot \delta\boldsymbol{\varepsilon} = \int_a^b (N\delta\varepsilon + V\delta\varphi + M\delta\kappa)dx \tag{7.77}$$

Internal virtual work

In order to proceed we need an expression for $\delta\varepsilon$, $\delta\varphi$ and $\delta\kappa$. The strain definitions (7.65), (7.74) and (7.67) provide

$$\delta\varepsilon = \delta u' \,, \quad \delta\varphi = (\delta w' - \delta\omega) \;\text{ and }\; \delta\kappa = \delta\omega' \tag{7.78}$$

Generalized strains ε, φ and κ

and thus (7.77) becomes

$$\begin{aligned} \boldsymbol{\sigma} \cdot \delta\boldsymbol{\varepsilon} &= \int_a^b (N\delta u' + V(\delta w' - \delta\omega) + M\delta\omega')dx \\ &= [N\delta u]_a^b + [V\delta w]_a^b + [M\delta\omega]_a^b \\ &\quad - \int_a^b N'\delta u dx - \int_a^b V'\delta w dx - \int_a^b (M' + V)\delta\omega dx \end{aligned} \tag{7.79}$$

Internal virtual work

Provided that there are no distributed moment loads, i.e. that the loads are the ones shown in Fig. 7.2, which for convenience is repeated here

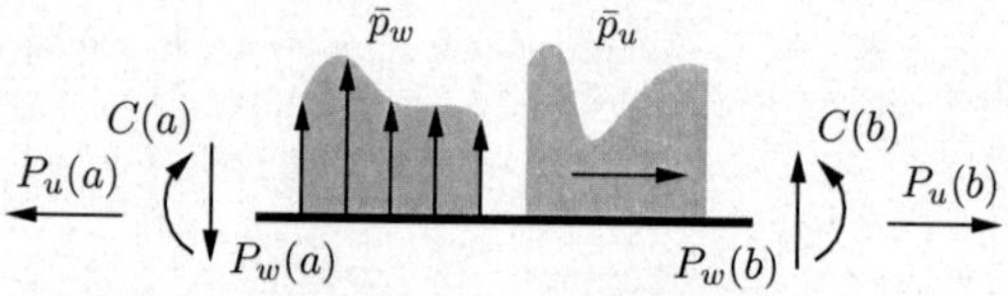

Fig. 7.7: Undeformed beam with loads and end forces.

as Fig. 7.7, the external virtual work $\overline{\boldsymbol{T}} \cdot \delta \boldsymbol{u}$ is

External virtual work

$$\begin{aligned}\overline{\boldsymbol{T}} \cdot \delta \boldsymbol{u} = & \int_a^b (\bar{p}_u \delta u + \bar{p}_w \delta w) dx \\ & - P_u(a)\delta u(a) - P_w(a)\delta w(a) - C(a)\delta w'(a) \\ & + P_u(b)\delta u(b) + P_w(b)\delta w(b) + C(b)\delta w'(b)\end{aligned} \tag{7.80}$$

which obviously is the same as (7.21).

By equating the internal and external virtual work we can immediately conclude that

First static field equation

$$M' + V = 0\,,\ x \in]a, b[\tag{7.81}$$

which indicates that V is the same as the static quantity V introduced by (7.61). Furthermore, from (7.79) and (7.80) we may also get the other two static field equations. The first is

Second static field equation

$$V' + \bar{p}_w = 0\,,\ x \in]a, b[\tag{7.82}$$

which agrees with (7.62b), and the second is

Third static field equation

$$N' + \bar{p}_u = 0\,,\ x \in]a, b[\tag{7.83}$$

which is the same as (7.62a). Thus we recover the equilibrium equations (7.62).

As regards the *possible* static boundary conditions we get

Possible static boundary conditions

$$\begin{aligned} N(a) &= \bar{P}_u(a)\,, \quad N(b) = \bar{P}_u(b) \\ V(a) &= \bar{P}_w(a)\,, \quad V(b) = \bar{P}_w(b) \\ M(a) &= \bar{C}(a)\,, \quad M(b) = \bar{C}(b)\end{aligned} \tag{7.84}$$

which are the same as (7.63).

Kinematic *linearity*: Same static equations for B-E and Timoshenko

We may now realize that equilibrium equations of the linear Bernoulli-Euler theory and the linear Timoshenko theory are the same, although the generalized strains are different,[7.12] but recall that in the Bernoulli-Euler theory the shear force V is not a generalized quantity.

[7.12] One could have hoped that the same held for moderate kinematic nonlinearity.

7.7 Plane, Straight Elastic Bernoulli-Euler Beams

In applications of beam theories, be they kinematically nonlinear or linear, there are two constitutive models which dominate the picture completely. One is linear (hyper)elasticity, the other is perfect plasticity. I do not address the latter issue here, but devote much of Part III to determination of cross-sectional properties of linearly elastic beams. Whether the Bernoulli-Euler beam is kinematically nonlinear or linear the elastic constitutive model, Hooke's "law,"[7.13] is taken to be

$$\begin{bmatrix} N \\ M \end{bmatrix} = \begin{bmatrix} D_{11} & 0 \\ 0 & D_{22} \end{bmatrix} \begin{bmatrix} \varepsilon \\ \kappa \end{bmatrix} \tag{7.85}$$

Hooke's "law"

where D_{11} signifies the axial stiffness and D_{22} the bending stiffness, respectively, and the cross-sectional properties are referred to an axis about which coupling between axial and bending actions does not occur. For a homogeneous, isotropic cross-section (7.85) is usually written

$$\begin{bmatrix} N \\ M \end{bmatrix} = \begin{bmatrix} EA & 0 \\ 0 & EI \end{bmatrix} \begin{bmatrix} \varepsilon \\ \kappa \end{bmatrix} \tag{7.86}$$

Hooke's "law"

where E is Young's Modulus, A is the cross-sectional area, I is the *moment of inertia* of the cross-section, and EA and EI denote the *axial stiffness* and the *bending stiffness* of the beam, respectively. There is an important observation to make in connection with the structure of (7.86), namely that the constitutive matrix is a diagonal one and, therefore all coupling between axial and transverse displacement components must come either from the boundaries or from a kinematically nonlinear effect, which is described by the term $\frac{1}{2}(w')^2$ of the axial strain measure, see (7.12), and in the differential equation (7.23b) is given by the term $-(Nw')'$.

Axial stiffness EA
Bending stiffness EI

When we introduce the relation between M and κ, see (7.86), in the static field equation of the kinematically moderately nonlinear theory (7.23) we obtain

$$N' + \bar{p}_u = 0 \quad \text{and} \quad (EI\kappa)'' - (Nw')' - \bar{p}_w = 0 \tag{7.87}$$

and when the curvature-displacement relation (7.14)

$$\kappa = w'' \tag{7.88}$$

is exploited the result is

$$\left.\begin{aligned} & N' + \bar{p}_u = 0 \\ & (EIw'')'' - (Nw')' - \bar{p}_w = 0 \end{aligned}\right\} x \in]a, b[\tag{7.89}$$

Equilibrium equations in terms of N and w

[7.13] Remember, it is a *model*—not a *law*, but this is the traditional name.

Sometimes N can be determined independent of M and w. This true for statically determinate structures

These relations deserve some comments. First, using A and I in the constitutive relation may be justified through application of the continuum mechanical theories developed in Part III with the comment that the moment-curvature relation is less obvious than the axial force-axial strain relation. Second, it may not seem like a good idea to utilize only the constitutive relation that expresses the bending moment M in terms of the curvature strain κ and let the axial force N remain untouched. The reason why this is sometimes useful is that in many cases N may be determined without recourse to the constitutive relation because it is, *statically determinate* i.e. it may be found by use of the static equations, only. In some other cases, N may not be statically determinate, but may be found through an iterative process which entails that N is assumed known before each step and corrected afterwards. This is often the case when you do a computation by hand, while in more general purpose procedures such as a Finite Element program the constitutive relation connecting N and ε is always invoked. Third, it is important to bear in mind that the constitutive relation for the bending moment is the second row of (7.86), i.e. $M = EI\kappa$, and *not* $M = EIw''$ because in order to get the latter expression the *kinematic* relation $\kappa = w''$, which only holds for kinematically linear or moderately nonlinear Bernoulli-Euler beams, has been exploited.[7.14]

Ex 7-2 When Is the Linear Theory Valid?

It would be comforting to have a strict criterion for the validity of the kinematically linear beam theory. The conditions given by (7.54)

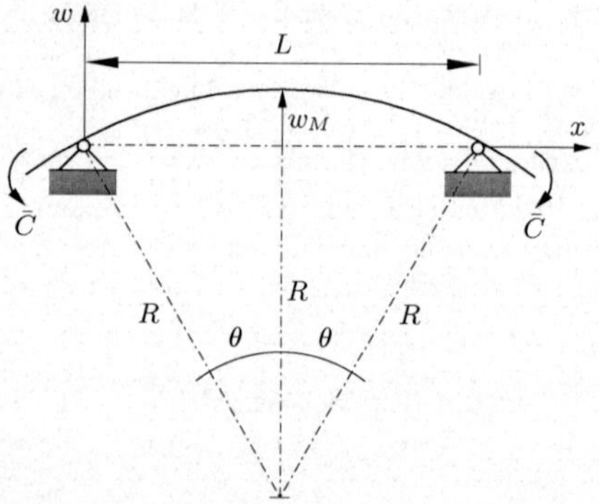

Fig. Ex. 7-2.1: *Beam loaded by couples at the ends.*

and (7.55) are useful but not very precise. We may concentrate on

[7.14] The reason why I stress the third point is that too many people get these topics mixed up.

the approximations inherent in the choice (7.57) of curvature strain. For this purpose, consider the beam shown in Fig. Ex. 7-2.1. The beam is allowed to slide on the rollers, thus keeping their distance at L during the loading history which is important for the full nonlinear theory, while the kinematically linear theory does not recognize kinematic shortening which implies that the curve is longer than the chord.

Ex 7-2.1 Full Nonlinear Analysis

This is one of the very few cases where the full nonlinear analysis is as easy as the kinematically linear one. It is clear that, independent of the theory, the bending moment M is constant and equal to $-\bar{C}$. Then,

$$R = \frac{1}{\kappa} = \frac{EI}{\bar{C}} \qquad (Ex.\ 7\text{-}2.1)$$

Furthermore,

$$\begin{aligned} w_M^{\mathrm{NL}} &= R(1 - \cos(\theta) = R\bar{C}\left(1 - \sqrt{1 - \left(\frac{L}{2R}\right)^2}\right) \\ &= \frac{EI}{\bar{C}}\left(1 - \sqrt{1 - \left(\frac{\bar{C}L}{2EI}\right)^2}\right) \end{aligned} \qquad (Ex.\ 7\text{-}2.2)$$

This may look somewhat suspicious because it seems as if the displacement is inversely proportional to the applied load. A Taylor expansion about $\bar{C} = 0$ provides

$$w_M^{\mathrm{NL}} \approx \tfrac{1}{8}\left(\frac{\bar{C}L}{EI}\right) + \frac{1}{128}\left(\frac{\bar{C}L}{EI}\right)^3 + \cdots \qquad (Ex.\ 7\text{-}2.3)$$

which is comforting to know.

For convenience let us non-dimensionalize in the following fashion

$$\widetilde{w} \equiv \frac{w}{L}, \quad \widetilde{C} \equiv \frac{\bar{C}L^2}{EI} \qquad (Ex.\ 7\text{-}2.4)$$

to get

$$\widetilde{w}_M^{\mathrm{NL}} = \frac{1}{\widetilde{C}}\left(1 - \sqrt{1 - \tfrac{1}{4}\widetilde{C}^2}\right) \approx \tfrac{1}{8}\widetilde{C} + \tfrac{1}{128}\widetilde{C}^3 + \cdots \qquad (Ex.\ 7\text{-}2.5)$$

Note also that

$$\theta^{NL} = \sin^{-1}(\tfrac{1}{2}\widetilde{C}) \qquad (Ex.\ 7\text{-}2.6)$$

Ex 7-2.2 Linear Analysis

Using (7.89b) with $N = 0$ and $\bar{p}_w = 0$ we may find the solution

$$\widetilde{w}^L = \tfrac{1}{2}\widetilde{C}\xi(1 - \xi) \text{ where } \xi \equiv \frac{x}{L} \qquad (Ex.\ 7\text{-}2.7)$$

and thus

$$\widetilde{w}_M^{\mathrm{L}} = \tfrac{1}{8}\widetilde{C} \qquad (Ex.\ 7\text{-}2.8)$$

and

$$\theta^{\mathrm{L}} = \left.\frac{dw}{dx}\right|_{x=0} = \tfrac{1}{2}\widetilde{C} \qquad (Ex.\ 7\text{-}2.9)$$

We may see that the linear approximations are equal to the first term of the Taylor expansion of the full nonlinear solution, as expected, and

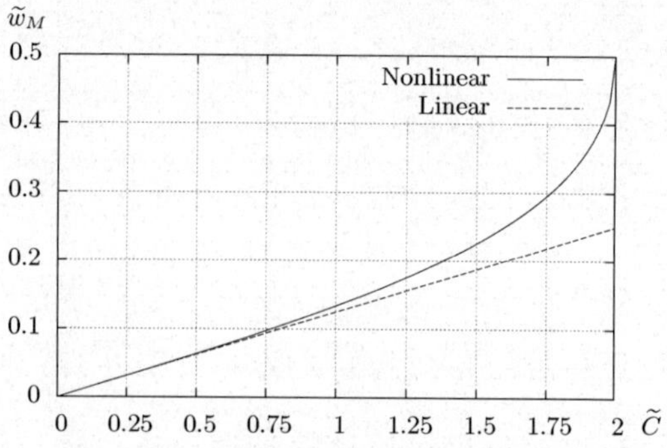

Fig. Ex. 7-2.2: *Beam with couples at the ends. Maximum normalized displacement.*

seen from Figs. Ex. 7-2.2 and Ex. 7-2.3. Clearly, Fig. Ex. 7-2.2 indicates that the linear theory is valid for quite large loads, i.e. loads which cause a maximum displacement of about one tenth of the length of the beam which is more than is acceptable in most problems of structural engineering. The same load is associated with a rotation of about 40° at the supports, see Fig. Ex. 7-2.3, which is much more than we usually allow. Are we therefore justified in claiming that the kinematically linear theory is valid even for very large loads? The answer is

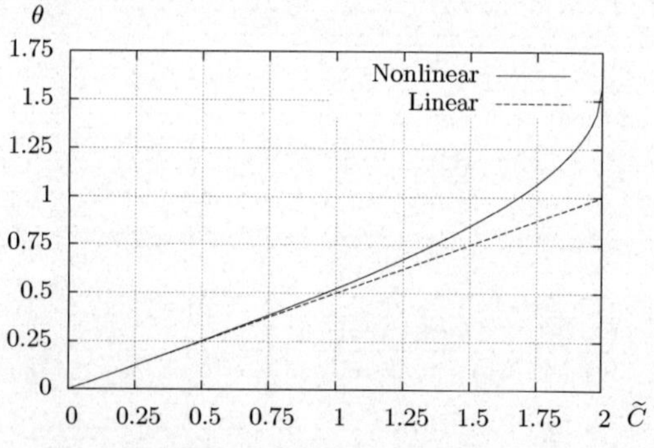

Fig. Ex. 7-2.3: *Beam with couples at the ends. Rotation at support.*

simply: no. The reason is that even a small axial load would increase the displacements considerably if computed by the nonlinear theory, while the results from the linear theory would remain the same. An indication of this may be inferred from Example Ex 8-2, the *Elastica*. We are therefore not much closer to an estimate of the validity of the kinematically beam theory. On the other hand, the condition (7.55), which often is interpreted as $|w'| \lesssim 0.1\,\text{rad} \approx 5^\circ$ still seems to be a good estimate *provided* that no axial forces cause a strong increase in kinematic nonlinearity.

Ex 7-2.3 Moderately Nonlinear Analysis

You might ask why I did not apply the kinematically moderately nonlinear theory to the above problem. The reason for not doing it is that (7.26c) clearly does not predict any kinematic nonlinearity unless there is an axial force.

In Example Ex 8-2 we may see that the kinematically moderately nonlinear theory does not perform well in postbuckling of the *Elastica*, see Fig. Ex. 8-2.2. On the other hand, in most cases, kinematically nonlinear analyses of frames are based, not on the full nonlinear theory, but on the kinematically moderately nonlinear theory and often with good results. There are several reasons for this. Among these is that both nonlinear theories predict the same classical buckling load for most structures. Another reason is that in postbuckling there are other effects that govern than the ones related to the strain measures. A third reason is that in the presence of geometric imperfections the displacements at maximum load are often so small that the difference between the postbuckling predictions according to the full and the moderately nonlinear theories become unimportant. Finally, when plasticity enters then the displacements at maximum load are usually even smaller than in the elastic cases.

We shall return to some of the above issues in Part IV, *Buckling*.

For our next example, the *Euler Column*, as for the *Elastica*, it would be meaningless to try to apply the kinematically linear theory because the only load is an *axial force* and we are interested in seeing whether such a load might cause the column to experience transverse displacements.

Ex 7-3 The Euler Column

Probably the most prominent example of a nonlinear beam is the "Euler Column" with pinned ends which is shown in Fig. Ex. 7-3.1. For this beam, or column, the stiffnesses are constant over the length of the beam. The only load is the axial force $\lambda \bar{P}$, where λ is a scalar load parameter, and the load is compressive.

The Euler Column

For this structure (7.87a) simply is

$$N' = 0 \qquad (Ex.\ 7\text{-}3.1)$$

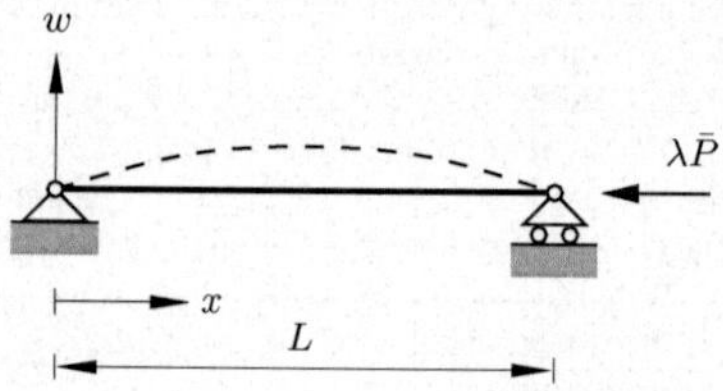

Fig. Ex. 7-3.1: *The Euler Column.*

which, together with the relevant static boundary condition (7.24b)

$$N(L) = -\lambda \bar{P} \qquad (Ex.\ 7\text{-}3.2)$$

provides

$$N(x) = -\lambda \bar{P} \qquad (Ex.\ 7\text{-}3.3)$$

Even if a transverse load, say $\bar{p}_w(x)$, had been applied, (*Ex. 7-3.3*) still holds. The static boundary conditions associated with the transverse loads follow from (7.24e) and (7.24f)

$$\begin{aligned} M(0) = 0 &\Rightarrow w''(0) = 0 \\ M(L) = 0 &\Rightarrow w''(L) = 0 \end{aligned} \qquad (Ex.\ 7\text{-}3.4)$$

where the constitutive relation connecting M and κ as well as the strain-displacement relation for κ and w'' has been exploited. If we normalize $\bar{P}$ such that

$$\bar{P} = \pi^2 \frac{EI}{L^2} \qquad (Ex.\ 7\text{-}3.5)$$

Differential equation for eigenvalue problem

(7.89b) yields

$$w^{iv} + \lambda \left(\frac{\pi}{L}\right)^2 w'' = 0 \qquad (Ex.\ 7\text{-}3.6)$$

which, together with the boundary conditions, constitutes a *linear eigenvalue problem* in λ. The solution is[7.15]

Solution

$$\left.\begin{aligned} \lambda_{(n)} &= n^2 \\ w_{(n)} &= \xi_{(n)} \sin\left(n\pi \frac{x}{L}\right) \end{aligned}\right\} n \in [1, 2, \ldots, \infty] \qquad (Ex.\ 7\text{-}3.7)$$

Buckling Load = Classical Critical Load λ_c = Euler Load λ_E Buckling mode $w_{(1)}$

where $\lambda_{(n)}$ are the eigenvalues and $\xi_{(n)}$ designates the amplitudes of the associated eigenfunctions $w_{(n)}$. As in all other eigenvalue problems the amplitudes are undetermined. The value $\lambda_{(1)} = 1$ is the *Buckling Load*, which is also called the *Classical Critical Load* λ_c or *Euler Load* λ_E of the column, and $w_{(1)}$ is usually referred to as the *Buckling Mode*. In most cases, only the lowest eigenvalue is of major interest, except possibly when a number of other eigenvalues are close to the lowest,

[7.15] If you don't believe me you might check my claim, simply insert (*Ex. 7-3.7*) in (*Ex. 7-3.6*).

which indeed is not the case for the Euler column in that the second eigenvalue is four times the first. For shells it is, however, typical that there is a cluster of buckling modes associated with almost the same buckling load. In any case, when the eigenfunctions are utilized as a basis for a series expansion all terms are needed, no matter how far the higher eigenvalues lie from the lowest.

For later purposes, see Example Ex 32-5, write the potential energy associated with the transverse displacement w for the Euler Column

$$\boldsymbol{\Pi}_{\mathrm{P}}(w) = \tfrac{1}{2}\int_0^L EI\,(w'')^2\,dx + \tfrac{1}{2}\int_0^L N\,(w')^2\,dx \qquad (Ex.\ 7\text{-}3.8)$$

Potential energy for Euler Column

where it may seem strange that we may extract only one part of the potential energy. In general, this it not admissible, but here the action in the axial and transverse directions may be separated in this respect, as may be seen from the structure of (*Ex. 7-3.1*)–(*Ex. 7-3.4*). Thus N may be regarded as a constant in (*Ex. 7-3.8*). If you still have your doubts, you may realize that you can construct (*Ex. 7-3.9*) from (*Ex. 7-3.6*) when (*Ex. 7-3.5*) is reintroduced.

When we exploit the fact that EI as well as N are constants and that N is given by (*Ex. 7-3.3*) we get

$$\boldsymbol{\Pi}_{\mathrm{P}}(w) = \tfrac{1}{2}EI\int_0^L (w'')^2\,dx - \tfrac{1}{2}\lambda\bar{P}\int_0^L (w')^2\,dx \qquad (Ex.\ 7\text{-}3.9)$$

Potential energy for Euler Column

It is easily verified that requiring that the first variation of $\boldsymbol{\Pi}_{\mathrm{P}}(w)$ vanishes provides the correct differential equation and boundary conditions, see below.

$$0 = \delta\boldsymbol{\Pi}_{\mathrm{P}}(w) = EI\int_0^L w''\delta w''dx - \lambda\bar{P}\int_0^L w'\delta w'dx \quad (Ex.\ 7\text{-}3.10)$$

or

$$\begin{aligned} 0 = {} & EI[w''\delta w']_0^l - EI\int_0^L w'''\delta w'dx \\ & - \lambda\bar{P}[w'\delta w]_0^l + \lambda\bar{P}\int_0^L w''\delta w dx \end{aligned} \qquad (Ex.\ 7\text{-}3.11)$$

and, finally

$$\begin{aligned} 0 = {} & EI[w''\delta w']_0^l - EI[w'''\delta w]_0^l + EI\int_0^L w^{\mathrm{iv}}\delta w dx \\ & - \lambda\bar{P}[w'\delta w]_0^l + \lambda\bar{P}\int_0^L w''\delta w dx \end{aligned} \qquad (Ex.\ 7\text{-}3.12)$$

which, because of the arbitrariness of δw in the field, provides the differential equation (7.89).

As regards the boundary terms their verification is left as an exercise for the reader.

7.8 Plane, Straight Elastic Timoshenko Beams

Independent of the degree of kinematic nonlinearity the linear elastic constitutive model, Hooke's "law,"[7.16] for Timoshenko beams is

Hooke's "law"

$$\begin{bmatrix} N \\ V \\ M \end{bmatrix} = \begin{bmatrix} D_{11} & 0 & 0 \\ 0 & D_{22} & 0 \\ 0 & 0 & D_{33} \end{bmatrix} \begin{bmatrix} \varepsilon \\ \varphi \\ \kappa \end{bmatrix} \tag{7.90}$$

where D_{11} denotes the axial stiffness, D_{22} the shear stiffness, and D_{33} the bending stiffness, respectively. This separation between the axial and the bending deformation is only possible when the cross-sectional properties are referred to an axis about which coupling between axial and transverse actions does not occur.

Axial stiffness EA
Effective shear stiffness GA_e
Bending stiffness EI

Introduce Young's Modulus E, the shear modulus G, the cross-sectional area A, the so-called *effective cross-sectional area* A_e, see Part III for its definition, and the *moment of inertia* I of the cross-section. For a homogeneous, isotropic cross-section we let EA, GA_e and EI denote the *axial stiffness*, the *effective shear stiffness*, and the *bending stiffness* of the beam, respectively.[7.17]

Now, (7.90) may be written

Hooke's "law"

$$\begin{bmatrix} N \\ V \\ M \end{bmatrix} = \begin{bmatrix} EA & 0 & 0 \\ 0 & GA_e & 0 \\ 0 & 0 & EI \end{bmatrix} \begin{bmatrix} \varepsilon \\ \varphi \\ \kappa \end{bmatrix} \tag{7.91}$$

As was the case for the elastic Bernoulli-Euler beams the constitutive matrix is a diagonal one, see (7.91). In order to get an idea about the importance of accounting for the shear flexibility we shall study the simple example of a clamped–free beam, see Example Ex 7-4 below.

Ex 7-4 A Cantilever Timoshenko Beam

Cantilever Timoshenko beam

The cross-sectional properties of the beam shown inFig. Ex. 7-4.1 are assumed to be independent of the axial coordinate x. The only load is the transverse force $\bar{P}$.

[7.16] Again, remember, it is a *model*—not a *law*, but this is the traditional name.

[7.17] The reason GA_e is called the *Effective Shear Stiffness* is that the real shear stiffness of a cross-section is not equal to GA because the shear strain is never constant over the cross-section. This has the inconvenient consequence that it is necessary to perform a rather complicated analysis for every kind new of cross-section. Fortunately, over the years, such analyses have been carried out for almost any conceivable cross-section, and the results may be found in a large number of standard tables. In Part III, *Beams with Cross-Sections*, we touch this subject for the rectangular cross-section.

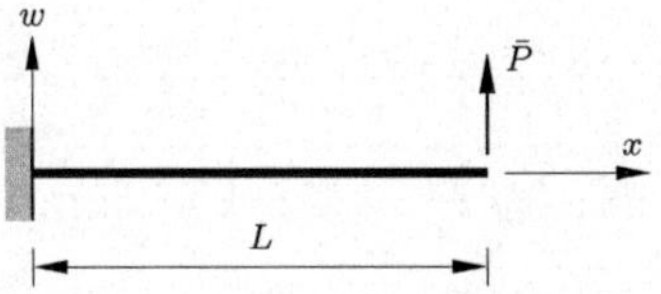

Fig. Ex. 7-4.1: *A Cantilever Beam.*

First, let us note that the structure is *statically determinate*, i.e. all generalized stresses can be found from the static conditions, including the static boundary conditions, alone. For this structure simple computations yield

$$N = 0\,,\ V = \bar{P} \text{ and } M = \bar{P}L(1-\xi)\,,\ \xi \equiv \frac{x}{L} \qquad \text{(Ex. 7-4.1)}$$

Introduce the moment-curvature relation (7.90c) and the definition (7.67) of the curvature strain κ in (*Ex. 7-4.1*c) and get

$$\omega' = (1-\xi)\frac{\bar{P}L}{EI} \qquad \text{(Ex. 7-4.2)}$$

with the solution

$$\omega = (\xi - \tfrac{1}{2}\xi^2)\frac{\bar{P}L^2}{EI} \qquad \text{(Ex. 7-4.3)}$$

Rotation ω

where we have exploited the kinematic boundary condition

$$\omega(0) = 0 \qquad \text{(Ex. 7-4.4)}$$

Now, combine ((*Ex. 7-4.1*b) with the shear strain definition (7.74) and the constitutive relation between the shear stress and the shear strain, see (7.90b)

$$\bar{P} = GA_e(w' - \omega) \qquad \text{(Ex. 7-4.5)}$$

After introduction of (*Ex. 7-4.3*) we may find

$$w' = \frac{\bar{P}}{GA_e} + \left(\xi - \tfrac{1}{2}\xi^2\right)\frac{\bar{P}L^2}{EI} \qquad \text{(Ex. 7-4.6)}$$

with the (kinematic) boundary condition[7.18]

$$w(0) = 0 \qquad \text{(Ex. 7-4.7)}$$

Thus, the solution for w is

$$w = \frac{\bar{P}L^3}{3EI}\left(\tfrac{3}{2}\xi^2 - \tfrac{1}{2}\xi^3 + \frac{3EI}{GA_eL^2}\xi\right) \qquad \text{(Ex. 7-4.8)}$$

Transverse displacement w

As an example assume that the cross-section is rectangular with depth H and width B, then $A = BH$ and $I = \frac{1}{12}BH^3$.

[7.18] I stress that this is the only kinematic boundary condition on w. Note that unlike in the case of a Bernoulli-Euler beam $w'(0) \neq 0$.

Transverse displacement w for a rectangular cross-section

When we introduce the shear modulus G given by (5.15) this becomes

$$w = \frac{\bar{P}L^3}{6EI}\left(3\xi^2 - \xi^3 + (1+\nu)\left(\frac{A}{A_e}\right)\left(\frac{H}{L}\right)^2\xi\right) \qquad (Ex.\ 7\text{-}4.9)$$

A computation, which I do not attempt to perform here, shows that for this kind of cross-section $A_e \approx \frac{5}{6}A$, but see Section 12.3 where possible values of A_e are discussed. When we further assume that $\nu = \frac{1}{4}$, which is a fairly common value, (*Ex. 7-4.8*) becomes

Displacement w for a rectangular cross-section

$$w = \frac{\bar{P}L^3}{3EI}\left(\tfrac{3}{2}\xi^2 - \tfrac{1}{2}\xi^3 + \tfrac{3}{4}\left(\frac{H}{L}\right)^2\xi\right) \qquad (Ex.\ 7\text{-}4.10)$$

The tip deflection is

Tip deflection $w(L)$

$$w(L) = \frac{\bar{P}L^3}{3EI}\left(1 + \tfrac{3}{4}\left(\frac{H}{L}\right)^2\right) \qquad (Ex.\ 7\text{-}4.11)$$

The value of $w(L)$ for the equivalent Bernoulli-Euler beam is

Tip deflection $w_{BE}(L)$ for a Bernoulli-Euler beam

$$w_{BE}(L) = \frac{\bar{P}L^3}{3EI} \qquad (Ex.\ 7\text{-}4.12)$$

and thus the second term in the parenthesis signifies the amplification due to the finite shear stiffness.

Displacements amplified in relation to the Bernoulli-Euler solution

As you can see, if the length of the beam is 5 times its depth, which is a rather stocky beam, the increase in the prediction of the tip deflection by using the Timoshenko beam theory is only 3%. You must, however, not take this as a universal truth in that e.g. for sandwich beams, which consist of a flexible core and two stiff face plates, the influence of shear flexibility must never be neglected.

Chapter 8

Plane, Curved Bernoulli-Euler Beams

Curved beams present additional problems in relation to straight ones. The reason is that the effects in the transverse direction are coupled with the effects in the axial direction—*even in the kinematically linear case.* Thus, the bending strain κ as well as the axial strain ε depends on both the axial and the transverse displacement component.

In the following we shall confine ourselves to beams that behave according to the Bernoulli-Euler assumptions, that is, they experience only axial and bending strains.

8.1 Kinematically Fully Nonlinear Curved Bernoulli-Euler Beams

8.1.1 Geometry and Kinematics

Fully nonlinear curved Bernoulli-Euler beams

Since we are dealing with Bernoulli-Euler beams we need expressions for the change of length and curvature in order to establish formulas for the two strains, namely the axial strain ε and the bending, or curvature strain κ.

Consider a part of a plane curved beam, as shown in Fig. 8.1 and let x^0_α, $\alpha = [1, 2]$[8.1] denote the coordinates of the undeformed beam and similarly let x_α designate the coordinates of the deformed beam. Then we may express the length ds^0 of the undeformed beam element by[8.2]

$$\left(ds^0\right)^2 = dx^0_\alpha dx^0_\alpha \tag{8.1}$$

and the length ds of the deformed beam element by

$$\left(ds\right)^2 = dx_\alpha dx_\alpha \tag{8.2}$$

[8.1] In some connections the upper index ^{0} is located above a symbol, e.g. $\overset{0}{\theta}$, but still it signifies a quantity that is associated with the undeformed beam. This is done in order to avoid a clutter of zeros and primes, as in $\theta^{0\prime\prime}$ versus $\overset{0}{\theta}{}''$. But, as you may see from this footnote, it may introduce extra space between text lines.

[8.2] Recall that the summation convention applied to Greek indices as well as Latin ones.

E. Byskov, *Elementary Continuum Mechanics for Everyone*,
Solid Mechanics and Its Applications 194, DOI: 10.1007/978-94-007-5766-0_8,

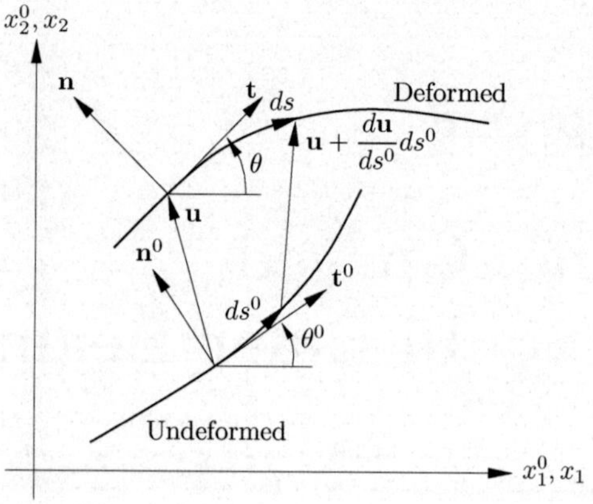

Fig. 8.1: Curved beam. The length of the undeformed and the undeformed infinitesimal elements is exaggerated.

For the sake of brevity introduce

$$( )' \equiv \frac{d( )}{ds^0} \tag{8.3}$$

and get

$$\left(x_1^0\right)' = \cos(\mathring{\theta}) \text{ and } \left(x_2^0\right)' = \sin(\mathring{\theta}) \tag{8.4}$$

where $\mathring{\theta}$ denotes the angle between the abscissa and the axis of the undeformed beam. Similarly,

$$\frac{dx_1}{ds} = \cos(\theta) \text{ and } \frac{dx_2}{ds} = \sin(\theta) \tag{8.5}$$

We intend to refer all quantities to the coordinates of the *undeformed* beam and introduce the *Stretch*[8.3] λ of the beam as the ratio between the length of the line element of the beam axis after and before deformation, respectively

Stretch λ

$$\lambda \equiv \frac{ds}{ds^0} = s' \tag{8.6}$$

Then

$$\cos(\theta) = \frac{x_1'}{\lambda} \text{ and } \sin(\theta) = \frac{x_2'}{\lambda} \tag{8.7}$$

[8.3] It is only in connection with the stretch that I allow λ to denote anything but the load factor of kinematically nonlinear problems.

In the same spirit as (7.9) we shall later define the curvature strain κ as the change in angle per *undeformed* length, see below. We shall, however, need expressions for the geometric curvature k^0 of the undeformed beam and the geometric curvature k of the deformed beam in order to express the curvature strain κ in terms of these two quantities.[8.4] The radii of curvature ρ^0 and ρ are simply

Geometric curvatures k^0 and k
Radii of curvature ρ^0 and ρ

$$\rho^0 = \frac{1}{k^0} \text{ and } \rho = \frac{1}{k} \tag{8.8}$$

For convenience, the formula (7.6) for the curvature is repeated here

$$k \equiv \frac{d\omega}{ds} = \frac{\dfrac{dx_1}{dt}\dfrac{d^2x_2}{dt^2} - \dfrac{dx_2}{dt}\dfrac{d^2x_1}{dt^2}}{\left(\left(\dfrac{dx_1}{dt}\right)^2 + \left(\dfrac{dx_2}{dt}\right)^2\right)^{\frac{3}{2}}} \tag{8.9}$$

When we choose $t = s$, then

$$k = \frac{dx_1}{ds}\frac{d^2x_2}{ds^2} - \frac{dx_2}{ds}\frac{d^2x_1}{ds^2} \tag{8.10}$$

because

$$(ds)^2 = dx_\alpha dx_\alpha \tag{8.11}$$

Then, for the undeformed beam

$$k^0 = x_1^{0\prime} x_2^{0\prime\prime} - x_2^{0\prime} x_1^{0\prime\prime} \tag{8.12}$$

When we exploit (8.6), we may get the equivalent expression for the deformed beam

$$k = \frac{x_1' x_2'' - x_2' x_1''}{\lambda^2} \tag{8.13}$$

8.1.2 Displacements and Displacement Derivatives

Let us resolve the displacement vector $\mathbf{u}$ in components u_α that are directed along the coordinate axes, see Fig. 8.2, and note that

$$x_\alpha = x_\alpha^0 + u_\alpha \tag{8.14}$$

where we, as usual, do not know u_α in advance.

Based on our experience with the straight beams, see Chapter 7, we expect that the displacement vector $\mathbf{u}$ should be resolved in the direction

[8.4] We shall see that $\kappa \neq k - k^0$.

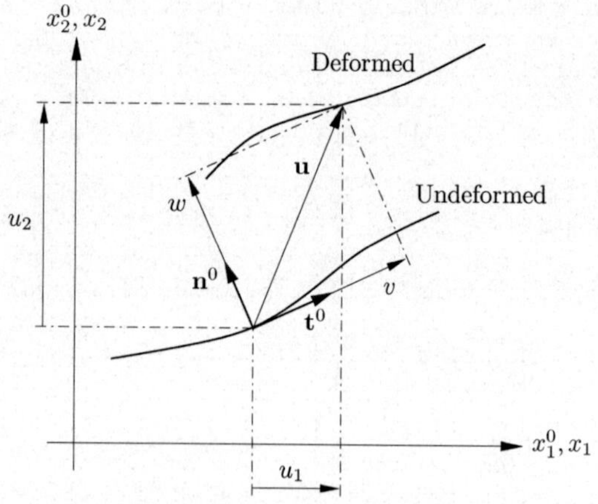

Fig. 8.2: Curved beam displacements.

of the beam and at right angles to it. The reason is that for those beams we found that the expression for the axial strain ε only entailed the first derivative of the displacement components u and w, while for the curvature strain κ the second derivative of the transverse displacement component w appeared. Thus, the displacement components u_1 and u_2, see Fig 8.2, which are directed along the coordinate axes, may be inconvenient to work with when we wish to establish expressions for the strains. Therefore, we need the relation

$$\begin{bmatrix} v \\ w \end{bmatrix} = \begin{bmatrix} +\cos(\overset{0}{\theta}) & +\sin(\overset{0}{\theta}) \\ -\sin(\overset{0}{\theta}) & +\cos(\overset{0}{\theta}) \end{bmatrix} \begin{bmatrix} u_1 \\ u_2 \end{bmatrix} \quad \text{or} \quad \mathbf{v} = \mathbf{T}\mathbf{u} \tag{8.15}$$

where v denotes the axial and w the transverse displacement component, respectively, see Fig. 8.2, and the meaning of $\mathbf{v}$, $\mathbf{T}$ and $\mathbf{u}$ ought to be clear from the equation. The inverse relation may be expressed as

$$\begin{bmatrix} u_1 \\ u_2 \end{bmatrix} = \begin{bmatrix} +\cos(\overset{0}{\theta}) & -\sin(\overset{0}{\theta}) \\ +\sin(\overset{0}{\theta}) & +\cos(\overset{0}{\theta}) \end{bmatrix} \begin{bmatrix} v \\ w \end{bmatrix} \quad \text{or} \quad \mathbf{u} = \mathbf{T}^T\mathbf{v} \tag{8.16}$$

Utilizing (8.16) we may find

$$\mathbf{u}' = \mathbf{T}^T\mathbf{v}' - \overset{0}{\theta}{}'\mathbf{T}_1\mathbf{v} \tag{8.17}$$

where

$$\mathbf{T}_1 \equiv \begin{bmatrix} +\sin(\overset{0}{\theta}) & +\cos(\overset{0}{\theta}) \\ -\cos(\overset{0}{\theta}) & +\sin(\overset{0}{\theta}) \end{bmatrix} \tag{8.18}$$

and

$$\mathbf{u}'' = \mathbf{T}^T\mathbf{v}'' - 2\overset{0}{\theta}{}'\mathbf{T}_1\mathbf{v}' - \left(\overset{0}{\theta}{}'\right)^2\mathbf{T}^T\mathbf{v} - \overset{0}{\theta}{}''\mathbf{T}_1\mathbf{v} \tag{8.19}$$

8.1.2.1 Length of the Line Element

Introduce the displacement u_α in (8.2) and get

$$(ds)^2 = (dx_\alpha + du_\alpha)(dx_\alpha + du_\alpha) \tag{8.20}$$

and utilize this expression in the definition (8.6) of the stretch λ

$$\lambda = \sqrt{1 + 2u'_\alpha {x^0_\alpha}' + u'_\alpha u'_\alpha} \tag{8.21}$$

After some tedious and rather lengthy manipulations, which involve (8.16)–(8.19), we may express the stretch in terms of the curvature $k^0 = \overset{0}{\theta}{}'$ of the undeformed beam and the two displacement components v and w

$$\lambda = \sqrt{1 + 2\left(v' - \overset{0}{\theta}{}'w\right) + \left(v' - \overset{0}{\theta}{}'w\right)^2 + \left(w' + \overset{0}{\theta}{}'v\right)^2} \tag{8.22}$$

In the interest of a shorter notation introduce [8.5]

$$\alpha \equiv v' - \overset{0}{\theta}{}'w \quad \text{and} \quad \beta \equiv w' + \overset{0}{\theta}{}'v \tag{8.23}$$

Curved beam

providing

$$\lambda = \sqrt{1 + 2\alpha + \alpha^2 + \beta^2} \tag{8.24}$$

Curved beam

Ex 8-1 *Comparison With Straight Beam*

One reason for the introduction of the quantities α and β is to show the similarity with the case of a straight beam, see e.g. (7.3), which is repeated here

$$(ds)^2 = \left((1+u')^2 + (w')^2\right)\left(ds^0\right)^2 \qquad (Ex.\ 8\text{-}1.1)$$

Straight beam

which may furnish

$$\lambda = \sqrt{1 + 2u' + (u')^2 + (w')^2} \qquad (Ex.\ 8\text{-}1.2)$$

Straight beam

Thus, for straight beams

$$\alpha = u \quad \text{and} \quad \beta = w \qquad (Ex.\ 8\text{-}1.3)$$

Straight beam

[8.5] Please do not confuse these Greek letters with the ones used to indicate the number of the coordinate axes.

8.1.2.2 Rotation of the Beam

In order to find an expression for the curvature strain κ we need formulas for the rotation ω of the beam axis. As mentioned in Section 7.2, see (7.9), we define the *Curvature Strain* κ as the change in angle per length of the undeformed beam, i.e.

Curvature strain κ

$$\kappa \equiv \theta' - \overset{0}{\theta}{}' = \lambda k - k^0 \not\equiv \frac{d\theta}{ds} - \frac{d\overset{0}{\theta}}{ds^0} = k - k^0 = \frac{1}{\rho} - \frac{1}{\rho^0} \tag{8.25}$$

We shall return to the issue of a proper definition of curvature strain, see Section 8.1.3.2.

The rotation ω, positive counterclockwise,[8.6] is

Rotation ω

$$\omega \equiv \theta - \overset{0}{\theta} \tag{8.26}$$

and thus—a little prematurely—the curvature strain κ is[8.7]

Curvature strain κ

$$\kappa = \omega' \tag{8.27}$$

When we note that[8.8]

$$\begin{bmatrix} \cos(\omega) \\ \sin(\omega) \end{bmatrix} = \begin{bmatrix} +\cos(\overset{0}{\theta}) & +\sin(\overset{0}{\theta}) \\ -\sin(\overset{0}{\theta}) & +\cos(\overset{0}{\theta}) \end{bmatrix} \begin{bmatrix} \cos(\theta) \\ \sin(\theta) \end{bmatrix} \tag{8.28}$$

where we may identify the 2×2 matrix as $\mathbf{T}$ from (8.15), and therefore may write

$$\boldsymbol{\omega} = \mathbf{T}\boldsymbol{\theta} \tag{8.29}$$

where we have introduced the (short, but maybe somewhat misleading) notation

$$\boldsymbol{\omega} \equiv \begin{bmatrix} \cos(\omega) \\ \sin(\omega) \end{bmatrix} \quad \text{and} \quad \boldsymbol{\theta} \equiv \begin{bmatrix} \cos(\theta) \\ \sin(\theta) \end{bmatrix} \tag{8.30}$$

Introduce

$$\mathbf{x} \equiv \begin{bmatrix} x_1 \\ x_2 \end{bmatrix} \quad \text{and} \quad \mathbf{x}^0 \equiv \begin{bmatrix} x_1^0 \\ x_2^0 \end{bmatrix} \tag{8.31}$$

[8.6] Maybe in a few years this concept will be meaningless, except for people wearing very expensive wristwatches.

[8.7] The word "prematurely" is present because the generalized strains are introduced later, namely in Section 8.1.3.2.

[8.8] This formula is a consequence of trigonometric identities and not particular to the problem at hand.

and recall (8.7), which may be cast in the form

$$\boldsymbol{\theta} = \frac{1}{\lambda}\mathbf{x}' \tag{8.32}$$

and find

$$\boldsymbol{\omega} = \mathbf{T}\frac{1}{\lambda}\mathbf{x}' = \frac{1}{\lambda}\mathbf{T}\left(\mathbf{x}^{0\prime} + \mathbf{u}'\right) = \frac{1}{\lambda}\mathbf{T}\left(\boldsymbol{\theta}^0 + \left(\mathbf{T}^T\mathbf{v}' - \overset{\circ}{\theta}{}'\mathbf{T}_1\mathbf{v}\right)\right) \tag{8.33}$$

Note that

$$\mathbf{T}\boldsymbol{\theta}^0 = \begin{bmatrix} 1 \\ 0 \end{bmatrix} \quad \text{and} \quad \mathbf{T}\mathbf{T}_1 = \begin{bmatrix} 0 & 1 \\ -1 & 0 \end{bmatrix} \tag{8.34}$$

to get

$$\boldsymbol{\omega} = \frac{1}{\lambda}\left(\begin{bmatrix} 1 \\ 0 \end{bmatrix} + \begin{bmatrix} v' \\ w' \end{bmatrix} - \overset{\circ}{\theta}{}'\begin{bmatrix} w \\ -v \end{bmatrix}\right) \tag{8.35}$$

and thus

$$\begin{bmatrix} \cos(\omega) \\ \sin(\omega) \end{bmatrix} = \frac{1}{\lambda}\begin{bmatrix} 1+\alpha \\ \beta \end{bmatrix} \tag{8.36}$$

In order to comfort yourself you may realize that $\cos^2(\omega)+\sin^2(\omega)$ given by (8.36) indeed equals 1.

8.1.2.3 Curvature of the Beam

When we recall the expression (8.27) for the *Curvature Strain* κ, note (8.36) and (8.23) we may realize that we can write κ in terms of α and β and therefore also in terms of the displacements v and w and of the geometry of the undeformed beam given by $\overset{\circ}{\theta}{}'$. In terms of α and β[8.9]

$$k = \frac{1}{\lambda}\overset{\circ}{\theta}{}' + \frac{1}{\lambda^3}\left((1+\alpha)\beta' - \alpha'\beta\right) \tag{8.37}$$

Curvature of the beam

This, of course, may not be the most convenient basis for expressing the curvature strain in special cases, such as beams that are initially circular, but (8.37) facilitates the discussions concerned with approximations in the kinematically moderately nonlinear case in Section 8.2 below.

[8.9] This takes some energy.

8.1.3 Generalized Strains

As is always the case for kinematically nonlinear problems there are many ways to define the generalized strains. In the present case we need an axial strain ε and a bending strain κ.

8.1.3.1 Axial Strain

An obvious measure of the axial strain is the *Engineering Strain* e

Engineering strain e

$$e \equiv \frac{ds - ds^0}{ds^0} \tag{8.38}$$

which may be expressed in terms of the stretch λ

$$\begin{aligned} e = \lambda - 1 &= \sqrt{1 + 2u'_\alpha x^{0\,\prime}_\alpha + u'_\alpha u'_\alpha} - 1 \\ &= \sqrt{1 + 2\alpha + \alpha^2 + \beta^2} - 1 \end{aligned} \tag{8.39}$$

We may also define another axial strain measure γ in the same spirit as the Lagrange strain γ_{ij}, whose definition is based on the relative change of the square of the line element, see (2.11), and may be expressed via the displacement gradients, see (2.12)

$$(ds)^2 - (ds^0)^2 = 2\gamma_{ij} dx_i dx_j \tag{8.40}$$

which may here be written in the slightly different form

Lagrange strain γ

$$\gamma \equiv \frac{(ds)^2 - (ds^0)^2}{2(ds^0)^2} = \tfrac{1}{2}\left(\lambda^2 - 1\right) = \tfrac{1}{2}(\lambda + 1)(\lambda - 1) \tag{8.41}$$

which shows that

$$\gamma \approx e \ \text{ for } \ \lambda \approx 1 \tag{8.42}$$

It is certainly the case for most engineering problems that the two axial strain measures provide almost the same value, but, of course, if you wish to analyze stretching of rubber bands you must decide whether to use γ or e because in that case $\lambda \not\approx 1$.

Note that we may express γ in terms of the quantities α and β introduced in (8.23)

$$\gamma = \alpha + \tfrac{1}{2}\alpha^2 + \tfrac{1}{2}\beta^2 \tag{8.43}$$

8.1.3.2 Curvature Strain

We have already introduced the curvature strain κ by (8.25) and note that by (8.37)

Curvature strain κ

$$\kappa = \frac{(1 + \alpha)\beta' - \alpha'\beta}{\lambda^2} \tag{8.44}$$

We shall employ (8.44) in our discussion of approximations in the kinematically moderately nonlinear theory in Section 8.2 below.

Proper Choice of Curvature Strain

You may ask why not simply define the curvature strain as the change in *geometric* curvature, which we shall denote $\hat{\kappa}$

$$\hat{\kappa} \equiv \frac{d\theta}{ds} - \frac{d\overset{0}{\theta}}{ds^0} = k - k^0 \tag{8.45}$$

Improper curvature strain $\hat{\kappa}$

since it seems somewhat arbitrary to differentiate the angle of the *deformed* beam with respect to the length of the *undeformed* beam. This may, by the way, not appear all that strange when you look at (8.27). But, let us consider a couple of consequences of using $\hat{\kappa}$ as our measure of curvature strain. The first and most important reason for not choosing $\hat{\kappa}$, is that κ satisfies the so-called rigid body criterion, which states that for a rigid body movement of the beam our strain measures must not predict straining of the beam. Therefore, in general, $\hat{\kappa}$ cannot satisfy this criterion, too.[8.10] The second reason is physical in nature. Let a circular ring be subjected to a uniform expansion and note that then $\kappa = 0$, while $\hat{\kappa} \neq 0$. Further, let us assume that the beam material is elastic. Then, according to any reasonable constitutive relation, $\hat{\kappa}$ would imply non-vanishing bending moment in the ring, and this is not to be expected. Therefore, we shall let κ be our curvature strain.

Proper choice of curvature strain important

The conclusion is that the proper choice of bending strain is

$$\kappa \equiv \frac{d\theta - d\overset{0}{\theta}}{ds^0} = \lambda k - k^0 \tag{8.46}$$

Proper curvature strain $\hat{\kappa}$

8.1.4 Equilibrium Equations

Except for Example Ex 7-2, where a fully kinematically nonlinear solution was used as basis for discussing the validity of the kinematically linear theory, and the classic example of the *Elastica*, see Example Ex 8-2 below, I do not intend to cover fully nonlinear beam theories and do not establish equilibrium equations for the general case. The reason why I stay away from this is that only for very special structures under very special circumstances is it necessary to apply a completely nonlinear theory-

8.1.5 Constitutive Relations

Since we do not cover the subject of equilibrium equations for the general case there is no reason to discuss the subject of constitutive relations in any depth. The only constitutive model we shall use for the fully kinematically nonlinear case is linear elasticity, but mention that an elastic-plastic equivalent of the Elastica, the *Plastica*, see (Yu & Johnson 1982), has been studied.

[8.10] See Example Ex 7-1 for a discussion on the subject of the degree to which this criterion must be satisfied, but bear in mind that in the present context the theory is s fully kinematically nonlinear one and therefore the criterion must be satisfied unconditionally.

Ex 8-2 The Elastica

The Elastica

Jacob (James) Bernoulli Leonhard Euler

It appears that Jacob (James) Bernoulli was the first to establish the differential equation for this problem, while the first solution was obtained by Euler (1744), see the appendix "De curvis elasticis." Although his solution is very old it covers a fully nonlinear problem with only one kinematic restriction, namely that the beam axis is inextensible. The beam is assumed to behave linearly elastic at all deformation levels which may sound like a very strong assumption, but for instance the spring in an old-fashioned wristwatch can undergo very large bending strains and still be elastic—otherwise it would not be of much value for its intended application.

I include this example for two reasons. One is that it is one of the most famous of the classics, the other is that it is often used as a benchmark for testing finite element programs. I might mention that over the years between Euler's time and today this column, often with different boundary conditions, has been the subject of numerous studies and add that there is still interest in the problem *per se*, see e.g. (Kuznetsov & Levyakov 2002). However, these studies are usually of a more scholarly kind and objective than I aim for in this book, so I shall not provide more references to such work.

Thus, although the elegant method used below is limited to a very special class of problems, I consider it worthwhile to include here.

Below, I shall follow neither Bernoulli nor Euler but—to some extent—the the derivations by Timoshenko & Gere (1961).

Ex 8-2.1 Generalized Strain

The *Elastica* is inextensible

Only one strain: the curvature strain κ

Under the assumption of inextensibility

$$\lambda = 1 \Rightarrow 2\alpha + \alpha^2 + \beta^2 = 0 \qquad (Ex.\ 8\text{-}2.1)$$

there is only one generalized strain, namely the curvature strain κ

$$\kappa = k - k^0 = k = \omega' \qquad (Ex.\ 8\text{-}2.2)$$

where the first equality sign follows from the fact that the stretch $\lambda = 1$ and therefore does not violate (8.46), the second comes about because the Elastica is straight in the undeformed configuration, and the third is another way of writing the curvature, see e.g. (8.26) and (8.27).

Ex 8-2.2 Principle of Virtual Displacements

Only one stress: the bending moment M

Let the generalized stress be denoted M and, for convenience, the only load applied between the ends of the beam be a distributed couple $\bar{c}$, which we take to be positive counter-clockwise and be measured on the undeformed beam. When we recall that the beam is assumed inextensible there is no difference between its length L_0 before and its length L after deformation, and thus we may write the principle of virtual displacements for a beam of length L as

Principle of virtual displacements

$$\int_0^L M\delta\kappa ds^0 = \int_0^L \bar{c}\,\delta\omega ds^0 \qquad (Ex.\ 8\text{-}2.3)$$

i.e.

$$0 = \int_0^L (M' + \bar{c})\, \delta\omega ds^0 \qquad (Ex.\ 8\text{-}2.4)$$

Since there is only one generalized stress we can only get one equilibrium equation, not counting static boundary conditions, from the Principle of Virtual Displacements, namely the one that expresses moment equilibrium

$$M' + \bar{c} = 0 \qquad (Ex.\ 8\text{-}2.5)$$

Only one static field equation

For the same reason as above the Principle of Virtual Displacements can only furnish possible static boundary conditions for the bending moment

$$M(0) = \bar{C}_0 \text{ or } M(L) = \bar{C}_L \qquad (Ex.\ 8\text{-}2.6)$$

Possible static boundary conditions

Thus the interpretation of M is that it is the bending moment in the beam, which is not very surprising.

Ex 8-2.3 Linear Elasticity

Linear elasticity is here given by

$$M = EI\kappa \qquad (Ex.\ 8\text{-}2.7)$$

Hooke's "law"

where, as usual, E designates Young's Modulus and I the moment of inertia.[8.11] Below we shall use this equation with the strain κ expressed in terms of the derivative of the rotation ω

$$M = EI\omega' \qquad (Ex.\ 8\text{-}2.8)$$

Ex 8-2.4 Initially Straight Elastica

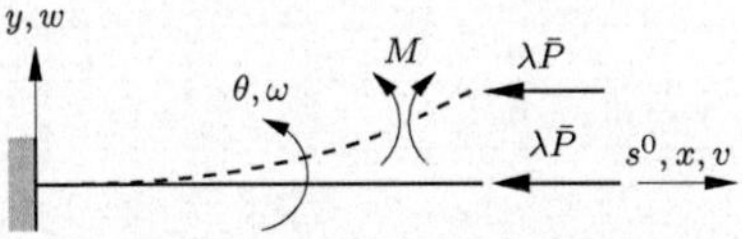

Fig. Ex. 8-2.1: *The Elastica.*

For the initially straight Elastica shown in Fig. Ex. 8-2.1 we may express moment equilibrium as

$$\lambda\bar{P}\left(w(L) - w(s^0)\right) = EI\omega' \qquad (Ex.\ 8\text{-}2.9)$$

where the assumption of linear elasticity has been utilized, λ is a scalar load parameter, and the left-hand side follows from moment equilibrium. Define the load unit $\bar{P}$ by

$$\bar{P} \equiv \left(\frac{\pi}{2L}\right)^2 EI \qquad (Ex.\ 8\text{-}2.10)$$

[8.11] Euler did not speak separately about E and I, but assumed linearity via a quantity B, which we now know to be equal to EI. By the way, his reason for calling this B was that it denoted the *Bending stiffness*—in German *Biegesteifigkeit.*

By this choice $\lambda = 1$ corresponds to initial loss of stability, as we shall see.

Because we enforce inextensibility we shall no longer differentiate between the length of the undeformed and the deformed beam and note that

Inextensibility $\Rightarrow ds = ds^0$

$$ds = ds^0 \qquad (Ex.\ 8\text{-}2.11)$$

Differentiate (*Ex. 8-2.9*) with respect to ds

$$\omega'' = -\lambda \frac{\bar{P}}{EI} w' \qquad (Ex.\ 8\text{-}2.12)$$

Note that

$$\frac{dw}{ds} = \sin(\omega) \text{ and } \frac{dx}{ds} = \cos(\omega) \qquad (Ex.\ 8\text{-}2.13)$$

and rewrite (*Ex. 8-2.11*)

Not an easy differential equation

$$\omega'' = -\lambda \frac{\bar{P}}{EI} \sin(\omega) \text{ or } \frac{d^2\omega}{ds^2} = -\lambda \frac{\bar{P}}{EI} \sin(\omega) \qquad (Ex.\ 8\text{-}2.14)$$

This differential equation does not have a simple, analytic solution and it is not an easy task to solve it by brute numerical force because the right-hand side is expressed in terms of the dependent variable. The trick here consists in multiplying both sides of (*Ex. 8-2.14*) by $d\omega$

$$\frac{d^2\omega}{ds^2}\frac{d\omega}{ds} ds = -\lambda \left(\frac{\pi}{2L}\right)^2 \sin(\omega) d\omega \qquad (Ex.\ 8\text{-}2.15)$$

which may be rewritten

$$\tfrac{1}{2} L^2 \frac{d}{ds}\left(\frac{d\omega}{ds}\right)^2 ds = -\lambda \left(\frac{\pi}{2}\right)^2 \sin(\omega) d\omega \qquad (Ex.\ 8\text{-}2.16)$$

Integration yields

$$\tfrac{1}{2} L^2 \left(\frac{d\omega}{ds}\right)^2 = +\lambda \left(\frac{\pi}{2}\right)^2 \cos(\omega) + C \qquad (Ex.\ 8\text{-}2.17)$$

Now, the only boundary condition which may be exploited here is static, namely

$$M(L) = 0 \Rightarrow \left.\frac{d\omega}{ds}\right|_L = 0 \qquad (Ex.\ 8\text{-}2.18)$$

which means that

$$C = -\lambda \left(\frac{\pi}{2}\right)^2 \cos(\omega_L) \qquad (Ex.\ 8\text{-}2.19)$$

i.e.

$$\tfrac{1}{2} L^2 \left(\frac{d\omega}{ds}\right)^2 = +\lambda \left(\frac{\pi}{2}\right)^2 \left(\cos(\omega) - \cos(\omega_L)\right) \qquad (Ex.\ 8\text{-}2.20)$$

or

$$\frac{d\omega}{ds} = \pm \frac{\pi}{L}\sqrt{\frac{\lambda}{2}\left(\cos(\omega) - \cos(\omega_L)\right)} \qquad (Ex.\ 8\text{-}2.21)$$

We choose the rotation to be positive—the case with negative rotations follows easily from this by suitable changes of sign—and may get

$$\frac{\pi}{L}\sqrt{\frac{\lambda}{2}}ds = \frac{d\omega}{\sqrt{\cos(\omega) - \cos(\omega_L)}} \qquad (Ex.\ 8\text{-}2.22)$$

Write this in terms of the half of the two angles and integrate over the entire beam

$$\frac{\pi}{L}\sqrt{\lambda}\int_0^L ds = \int_0^{\omega_L} \frac{d\omega}{\sqrt{(\sin^2(\omega_L/2) - \sin^2(\omega/2))}} \qquad (Ex.\ 8\text{-}2.23)$$

Introduce the quantities p and ϕ by

$$\sin(\omega/2) = p\sin(\phi) = \sin(\omega_L/2)\sin(\phi) \qquad (Ex.\ 8\text{-}2.24)$$

Differentiation provides

$$\frac{d\sin(\omega/2)}{d\omega} = p\cos(\phi)\frac{d\phi}{d\omega} \qquad (Ex.\ 8\text{-}2.25)$$

and thus

$$\begin{aligned} d\omega &= \frac{2p\cos(\phi)d\phi}{\cos(\omega/2)} = \frac{2p\cos(\phi)d\phi}{\sqrt{1-\sin^2(\omega/2)}} \\ &= \frac{2p\cos(\phi)d\phi}{\sqrt{1-p^2\sin^2(\phi)}} \end{aligned} \qquad (Ex.\ 8\text{-}2.26)$$

Perform the integration on the left-hand side of (*Ex. 8-2.23*) and note that by (*Ex. 8-2.24*)

$$\sin^2(\omega_L/2) - \sin^2(\omega) = p^2 - p^2\sin^2(\phi) \qquad (Ex.\ 8\text{-}2.27)$$

to get

$$\tfrac{1}{2}\pi\sqrt{\lambda} = \int_0^{\pi/2} \frac{d\phi}{\sqrt{1-p^2\sin^2(\phi)}} \qquad (Ex.\ 8\text{-}2.28)$$

where, by (*Ex. 8-2.24*) we have realized that

$$\sin(\phi_L) = 1 \Rightarrow \phi_L = \tfrac{1}{2}\pi \qquad (Ex.\ 8\text{-}2.29)$$

The right-hand side of (*Ex. 8-2.28*) is known as the *Complete Elliptic Integral of the First Kind* and is usually denoted by $K(p)$. Nowadays, values of this integral may easily be found by use of standard programs, e.g. in the *GNU Scientific Library* (Galassi, Davies, Theiler, Gough, Jungman, Alken, Booth & Rossi 2009), while in the old days, a mere 30–40 years ago, one had to resort to tables. Thus, our problem consists not so much in finding the solution of (*Ex. 8-2.28*) as interpreting the equation itself. The quantity p is the key here, and by use of (*Ex. 8-2.24*) we may get

$$\tfrac{1}{2}\pi\sqrt{\lambda} = \int_0^{\pi/2} \frac{d\phi}{\sqrt{1-\sin^2(\omega_L/2)\sin^2(\phi)}} \qquad (Ex.\ 8\text{-}2.30)$$

It may seem somewhat awkward that in order to solve this equation we must prescribe the tip rotation ω_L instead of the load level given

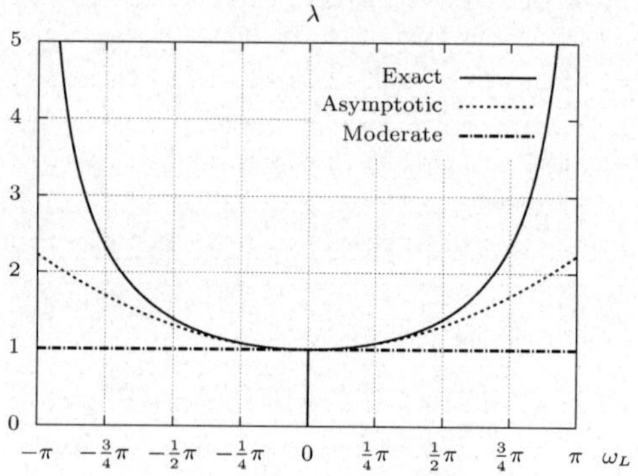

Fig. Ex. 8-2.2: *Load-displacement relation of the Elastica.*
Moderate indicates the kinematically moderately nonlinear theory of Section 7.3.

by λ. On the other hand, in many applications of the Finite Element Method this is exactly what one does. Let us investigate (*Ex. 8-2.30*) for small values of ω_L

$$\tfrac{1}{2}\pi\sqrt{\lambda} \approx \int_0^{\pi/2} \left(1 + \tfrac{1}{8}\omega_L^2 \sin^2(\phi)\right) d\phi = \tfrac{1}{2}\pi + \tfrac{1}{32}\pi\omega_L^2 \quad (Ex.\ 8\text{-}2.31)$$

i.e. an asymptotic solution for small values of ω_L is

Asymptotic solution

$$\lambda \approx 1 + \tfrac{1}{8}\omega_L^2 \text{ for } |\omega_L| \ll 1 \quad (Ex.\ 8\text{-}2.32)$$

It should be obvious from Fig. Ex. 8-2.2 that the approximation given by (*Ex. 8-2.32*) is very accurate over a large range of values of ω_L. Furthermore, the load required to provide $\omega_L = \pi$ seems to be infinitely large. This may seem surprising. However, imagine the beam stretched along its original axis, but in the opposite direction, that is, bent backwards, and realize that the load must be very great in order for this to happen.

Validity of kinematically moderately nonlinear theory

Had we applied the kinematically moderately nonlinear beam theory of Section 7.3 the postbuckling path, see Chapter 18, would have been predicted to be horizontal, see Fig. Ex. 8-2.2 and, according to that theory, the beam would thus be postbuckling neutral, not postbuckling stable. [8.12] This realization should not lead us to abandon that theory because, as you can see from Fig. Ex. 8-2.2, over an interval as large as about $(-20°, 20°)$, the exact solution is indeed almost horizontal. The Elastica theory, in particular when axial inextensibility is

[8.12] If the terms postbuckling neutral and postbuckling stable do not seem self-explanatory, you may look at Part IV, in particular Chapter 19.

not enforced, is much more difficult to apply than the kinematically moderately nonlinear one which speaks in favor of applying the latter.

Shape of the *Elastica*

In order to find the complete shape of the deformed Elastica we need to determine (v, w). A simple geometric relation provides

$$dy = \sin(\omega)ds \qquad (Ex.\ 8\text{-}2.33)$$

or, utilizing (*Ex. 8-2.22*)

$$dy = \frac{L}{\pi}\sqrt{\frac{2}{\lambda}}\frac{\sin(\omega)d\omega}{\sqrt{\cos(\omega) - \cos(\omega_L)}} \qquad (Ex.\ 8\text{-}2.34)$$

and again expressed in terms of the half angles

$$dy = \frac{L}{\pi\sqrt{\lambda}}\frac{\sin(\omega)d\omega}{\sqrt{\sin^2(\omega_L/2) - \sin^2(\omega/2)}} \qquad (Ex.\ 8\text{-}2.35)$$

Now, exploit that

$$\begin{aligned}\sin(\omega) &= 2\sin(\omega/2)\cos(\omega/2)\\ &= 2p\sin(\phi)\sqrt{1 - p^2\sin^2(\phi)}\end{aligned} \qquad (Ex.\ 8\text{-}2.36)$$

together with (*Ex. 8-2.26*) to get

$$dy = \frac{4pL}{\pi\sqrt{\lambda}}\sin(\phi)d\phi \qquad (Ex.\ 8\text{-}2.37)$$

and thus

$$y = \frac{4pL}{\pi\sqrt{\lambda}}\int_0^{\phi}\sin(t)dt \qquad (Ex.\ 8\text{-}2.38)$$

with the result

$$y(\phi; p) = \frac{4pL}{\pi\sqrt{\lambda}}(1 - \cos(\phi)) \qquad (Ex.\ 8\text{-}2.39)$$

which may seem simple, but requires that we have determined the value of p or ω_L. For $\phi = \pi/2$

$$y_L = \frac{4pL}{\pi\sqrt{\lambda}} \qquad (Ex.\ 8\text{-}2.40)$$

An expression equivalent to (*Ex. 8-2.33*) is

$$dx = \cos(\omega)ds \qquad (Ex.\ 8\text{-}2.41)$$

which by

$$\cos(\omega) = 1 - 2\sin^2(\omega/2) = 1 - 2p^2\sin^2(\phi) \qquad (Ex.\ 8\text{-}2.42)$$

may be written

$$dx = \frac{L}{\pi\sqrt{\lambda}}\frac{1 - 2p^2\sin^2(\phi)}{p\cos(\phi)}\frac{2p\cos(\phi)d\phi}{\sqrt{1 - p^2\sin^2(\phi)}} \qquad (Ex.\ 8\text{-}2.43)$$

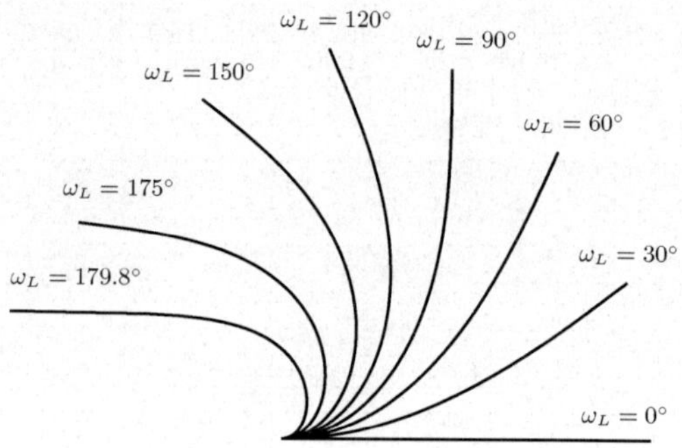

Fig. Ex. 8-2.3: *Deformation of the Elastica.*

where, once again, (*Ex. 8-2.22*) and (*Ex. 8-2.26*) have been utilized. This expression may be cast in the form

$$dx = \frac{2L}{\pi\sqrt{\lambda}}\left(2\sqrt{1-p^2\sin^2(\phi)}\,d\phi - \frac{1}{1-p^2\sin^2(\phi)}\,d\phi\right) \quad (Ex.\ 8\text{-}2.44)$$

and thus,

$$x = \frac{2L}{\pi\sqrt{\lambda}}\left(2\int_0^\phi \sqrt{1-p^2\sin^2(t)}\,dt - \int_0^\phi \frac{1}{1-p^2\sin^2(t)}\,dt\right) \quad (Ex.\ 8\text{-}2.45)$$

or

$$x(\phi;p) = \frac{2L}{\pi\sqrt{\lambda}}\left(2E(p,\phi) - F(p,\phi)\right) \quad (Ex.\ 8\text{-}2.46)$$

where $F(p,\phi)$ and $E(p,\phi)$ denote the *Incomplete Elliptic Integral* of the *First* and *Second Kind*, respectively.

With (*Ex. 8-2.39*) and (*Ex. 8-2.46*) we may plot the deflected Elastica for various values of ω_L, see Fig. Ex. 8-2.3. Note that the solution breaks down for $\omega_L = \pi = 180°$.

8.2 Kinematically Moderately Nonlinear Curved Bernoulli-Euler Beams

In this section we develop the theory that is equivalent to the one for straight Bernoulli-Euler beams, see Section 7.3. The basis is here the formulas for the fully nonlinear curved beams established in Section 8.1.

8.2.1 Kinematics

The limitations on the kinematic nonlinearity we wish to enforce here are that the length of the line element undergoes relatively small changes and that the rotation of beam is relatively limited. These are the same constraints as the ones behind the theory for the kinematically moderately nonlinear straight beam, see Section 7.3, (7.10) and (7.11).

Let us establish a Taylor expansion to second order in α and β of the stretch λ given by (8.24)

$$\lambda \approx 1 + \alpha + \tfrac{1}{2}(1 - \alpha + \alpha^2)\beta^2 \approx 1 + \alpha + \tfrac{1}{2}\beta^2 \qquad (8.47)$$

Approximate stretch

A similar expansion of $\sin(\omega)$ from (8.36) provides

$$\sin(\omega) \approx (1 - \alpha + \alpha^2)\beta \Longrightarrow \omega \approx (1 - \alpha + \alpha^2)\beta \approx \beta \qquad (8.48)$$

Approximate angle

where the same result for ω could have been obtained by expanding $\cos(\omega)$ given by (8.36). In the following we shall take (8.48) as our measure of the rotation

$$\omega = \beta = w' + \frac{v}{\rho^0} \qquad (8.49)$$

Approximate rotation

By looking at (8.47) and (8.48) you may realize that, as far as kinematic nonlinearity is concerned, α and β^2 play equivalent roles. Our constraints on the magnitude of α and β will therefore be

$$|\alpha| \ll 1 \qquad (8.50)$$

Constraint on $|\alpha|$

which is similar to the condition (7.10) on u' for straight beams, and

$$\beta^2 \ll 1 \qquad (8.51)$$

Constraint on β

which is equivalent to the condition (7.11) on w', also for straight beams. Of course, except for notational differences, for kinematically moderately nonlinear straight beams (8.50) and (8.51) express the same conditions as (7.10) and (7.11), respectively. For straight beams

$$\alpha = v' \text{ and } \beta = w' \qquad (8.52)$$

Straight beam

In plain words, the conditions on α and β say that we only allow the beams to change their length by a small amount, certainly not like a rubber band, and limit their rotation to being relatively small, say less than about $\frac{1}{10} \approx \frac{1}{30}\pi \approx 6°$, but fortunately, experience shows that the theory often holds for larger rotations.

8.2.1.1 Generalized Strains

Since we are dealing with Bernoulli-Euler beams there are only two generalized strains, namely the axial strain ε and the curvature strain κ, which we shall express in terms of the quantities α and β introduced in (8.23).

8.2.1.2 *Axial Strain*

By use of (8.50) and (8.51) we may get the expression for the axial strain, here ε, from (8.43)

Curved beam: Axial strain

$$\varepsilon = \alpha + \tfrac{1}{2}\beta^2 \tag{8.53}$$

or, in terms of the displacements v and w

Curved beam: Axial strain

$$\varepsilon = \left(v' - \overset{0}{\theta}{}'w\right) + \tfrac{1}{2}\left(w' + \overset{0}{\theta}{}'v\right)^2 = \left(v' - \frac{w}{\rho^0}\right) + \tfrac{1}{2}\left(w' + \frac{v}{\rho^0}\right)^2 \tag{8.54}$$

which for straight beams becomes

Straight beam: Axial strain

$$\varepsilon = v' + \tfrac{1}{2}(w')^2 \tag{8.55}$$

see also (7.12).

The linear part of (8.54) may deserve a comment. From the theory for straight beams we expect the term v' to appear, while the term w/ρ^0 is new. Its presence is, however, quite natural which may be seen if we imagine a beam that initially is circular, i.e. one for which $\rho^0 = \text{const.}$, and subject it to a deformation which brings it into another circular shape with $\rho = \text{const.}$ and thus also $w = \text{const.}$ Then, the beam experiences an axial strain which is linear in w. The reason for the individual terms of the quadratic term of (8.54) is less clear, except that from the straight beams we recognize the $(w')^2$, while the presence of v/ρ^0 is less obvious and requires a rather involved sketch to be substantiated. On the other hand, this is not so uncommon as you might think in that the general expressions, such as the ones of Section 8.1, are often much easier to establish than formulas for a theory with limitations on the displacements and their gradients. Having set up the general expressions it is, as in the present case, not very difficult to derive a specialized theory.

8.2.1.3 *Curvature Strain*

Here, we may make use of the expression (8.44) from Section 8.1.3.2 or cheat a little and just differentiate (8.48) to get the expression for the curvature strain κ[8.13]

Curved beam: Curvature strain

$$\kappa = \beta' \tag{8.56}$$

[8.13] As you may see, both the expressions from Section 8.1.3.2 involve the derivatives of α and β, meaning that we would have to put limitations on the magnitude of these quantities. This implies that the displacement pattern does not entail large changes in the displacement components and their derivatives up to second order. In other words, we only consider deformation patterns that are “slowly varying” in the sense that they do not entail wrinkles. Actually, if the beam is elastic, this boils down to an assumption that the energy stored is dominated by the slowly varying components of the deformation pattern and that the short-wave patterns contribute only little to the strain energy.

In terms of the displacements v and w this becomes

$$\kappa = w'' + \left(\overset{0}{\theta}'v\right)' = w'' + \left(\frac{1}{\rho^0}v\right)' \tag{8.57}$$

Curved beam: Curvature strain

which for a straight beam simplifies to

$$\kappa = w'' \tag{8.58}$$

Straight beam: Curvature strain

see also (7.14).

In (8.57) we expect the term w'', whereas the presence of $(v/\rho^0)'$ seems less obvious, see the comments above.

8.2.1.4 *Comparison Between Straight and Curved Beams*

Important differences between the strains of straight and curved beams

A comparison between the expressions for the strains of the curved and the straight beam immediately reveals a pronounced difference in that for the straight beam the curvature strain is given by only the transverse displacement component, while for the curved beam the axial displacement component also enters. For the axial strain the difference is smaller, except in the kinematically linear case, which we shall return to in Section 8.3, where the transverse displacement component does not enter for the straight beam. Thus, you may see that the strains of the curved beam are given by more complicated formulas than for the straight beam. In connection with numerical solutions this fact causes difficulties, as we shall see in Chapter 26, where we shall also see how to deal with these problems.

8.2.1.5 *Budiansky-Hutchinson Notation*

Both the axial and the curvature strain may be written in terms of the Budiansky-Hutchinson Notation, see Chapter 33,

$$\boldsymbol{\varepsilon} = \begin{bmatrix} \varepsilon \\ \kappa \end{bmatrix} = \mathbf{l}_1(\boldsymbol{u}) + \tfrac{1}{2}\mathbf{l}_2(\boldsymbol{u}) \tag{8.59}$$

where

$$\boldsymbol{u} = \begin{bmatrix} v \\ w \end{bmatrix} \tag{8.60}$$

and the linear operator $\mathbf{l}_1$ and the quadratic operator $\mathbf{l}_2$ here are given by

$$\mathbf{l}_1(\boldsymbol{u}) = \begin{bmatrix} v' - \overset{0}{\theta}'w \\ w'' + \left(\overset{0}{\theta}'v\right)' \end{bmatrix} \quad \text{and} \quad \mathbf{l}_2(\boldsymbol{u}) = \begin{bmatrix} \left(w' - \overset{0}{\theta}'v\right)^2 \\ 0 \end{bmatrix} \tag{8.61}$$

8.2.2 Equilibrium Equations

Conservative loads

As usual, we derive the equilibrium equations from the principle of virtual displacements. We shall only consider loads which are conservative in the sense that they do not change direction during deformation, see Fig. 8.4, where $\bar{p}_v$ obviously denotes the load component that is directed along the undeformed beam and $\bar{p}_w$ is the load component at right angles to this.

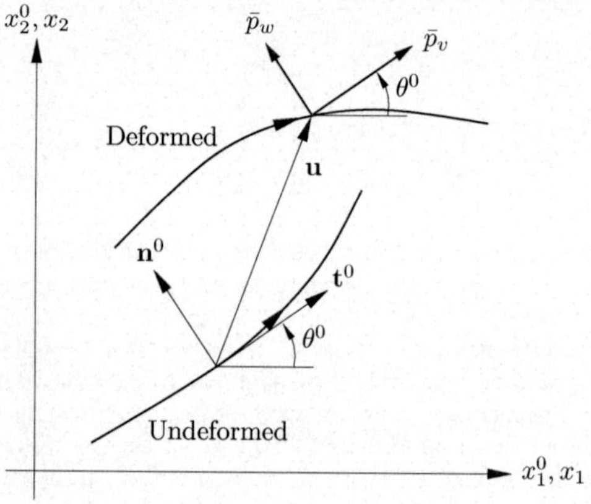

Fig. 8.4: Loads on moderately kinematically nonlinear curved beam.

Then, the principle of virtual displacements becomes

Principle of virtual displacements

$$\int_0^L (N\delta\varepsilon + M\delta\kappa)ds^0 = \int_0^L (\bar{p}_v\delta v + \bar{p}_w\delta w)ds^0 + \text{Boundary loads} \quad (8.62)$$

where the term "Boundary loads" denotes loads applied at the ends. These loads will be dealt with later.

In order to derive the equilibrium equations from (8.62) we need expressions for the variation of the strains, see (8.54) and (8.57). The variations are

Curved beam: Variation of strains

$$\begin{aligned} \delta\varepsilon &= \left(\delta v' - \frac{\delta w}{\rho^0}\right) + \left(w' + \frac{v}{\rho^0}\right)\left(\delta w' + \frac{\delta v}{\rho^0}\right) \\ \delta\kappa &= \delta w'' + \left(\frac{\delta v}{\rho^0}\right)' \end{aligned} \quad (8.63)$$

The equivalent expressions for the case of an initially straight beam, see (7.19)

Straight beam

$$\delta\varepsilon = \delta u' + w'\delta w' \text{ and } \delta\kappa = \delta w'' \quad (8.64)$$

They follow from (8.63) if we realize that for the straight beam $\rho^0 \approx \infty$ which amounts to removing all terms which include ρ^0.

After insertion of (8.63a) and (8.63b) into (8.62) and integration by parts, the left-hand side of (8.62) becomes—this takes some patience and care, but is otherwise completely straightforward

$$\begin{aligned}
&\int_0^L (N\delta\varepsilon + M\delta\kappa)ds^0 \\
&= \Big[N\delta v\Big]_0^L + \Big[N\Big(w' + \frac{v}{\rho^0}\Big)\delta w\Big]_0^L \\
&+\Big[M\delta w'\Big]_0^L - \Big[M'\delta w\Big]_0^L + \Big[M\frac{\delta v}{\rho^0}\delta w\Big]_0^L \\
&-\int_0^L \Big(N'\delta v + N\frac{\delta w}{\rho^0} + \Big(N\Big(w' + \frac{v}{\rho^0}\Big)\Big)'\delta w \\
&\qquad - N\Big(w' + \frac{v}{\rho^0}\Big)\frac{\delta v}{\rho^0}\Big)ds^0 \\
&-\int_0^L \Big(-M''\delta w + M\frac{\delta v}{\rho^0}\Big)ds^0
\end{aligned} \tag{8.65}$$

This expression does not easily lend itself to interpretation. However, when we recall (8.49)

$$\omega = \beta = w' + \frac{v}{\rho^0} \Rightarrow \delta\omega = \delta\beta = \delta w' + \frac{\delta v}{\rho^0} \tag{8.66}$$

and, analogous to (7.25), introduce the static quantity V by[8.14]

$$V \equiv -M' + N\left(w' + \frac{v}{\rho^0}\right) \tag{8.67}$$

V is not a generalized stress

we may rewrite (8.65) in the somewhat more convenient form

$$\begin{aligned}
&\int_0^L (N\delta\varepsilon + M\delta\kappa)ds^0 \\
&= \Big[N\delta v\Big]_0^L + \Big[V\delta w\Big]_0^L + \Big[M\delta\omega\Big]_0^L \\
&-\int_0^L \Big(\Big(N' - \frac{V}{\rho^0}\Big)\delta v + \Big(\Big(-V' - \frac{N}{\rho^0}\Big)\delta w\Big)ds^0
\end{aligned} \tag{8.68}$$

Now, after having equated the right-hand side of (8.68) with the right-hand side of (8.62) and exploited the arbitrariness of δv and δw, we may

[8.14] Recall that V defined by (7.25) is *not* a generalized quantity because it has no strain to do internal work with. The same is true for V defined by (8.67).

get the static field equations. They are

Static field equations including V

$$
\begin{aligned}
& N' - \frac{V}{\rho^0} + \bar{p}_v = 0 \\
& V' + \frac{N}{\rho^0} + \bar{p}_w = 0 \\
& M' + V - N\left(w' + \frac{v}{\rho^0}\right) = 0
\end{aligned}
\tag{8.69}
$$

where V has been introduced, or, apparently more complicated

Static field equations without V

$$
\begin{aligned}
& N' - N\left(w' + \frac{v}{\rho^0}\right)\frac{1}{\rho^0} + \frac{M'}{\rho^0} + \bar{p}_v = 0 \\
& M'' - \left(N\left(w' + \frac{v}{\rho^0}\right)\right)' + \frac{N}{\rho^0} - \bar{p}_w = 0
\end{aligned}
\tag{8.70}
$$

The possible static boundary conditions are

Possible static boundary conditions

$$
\begin{aligned}
& N(0) = \bar{P}_u(0), \quad N(L) = \bar{P}_u(L) \\
& V(0) = \bar{P}_w(0), \quad V(L) = \bar{P}_w(L) \\
& M(0) = \bar{C}(0), \quad M(L) = \bar{C}(L)
\end{aligned}
\tag{8.71}
$$

The similarity between these equilibrium equations and the ones derived in Section 7.3.2 ought to be clear, and we may see that for $\rho^0 \to \infty$ they become identical, except for notational differences.

8.2.3 Interpretation of Static Quantities

Unlike in Section 7.3.3 we shall not go through the entire process of interpreting the static quantities defined by the static equations above, but merely mention that N and V may be shown to be directed along the undeformed beam axis and perpendicular to it, respectively, and M is the bending moment. The reason for excluding the interpretation simply is that it takes up quite a lot of space—it is not particularly difficult to go through the derivations and, since the output is as expected, I have judged it not worthwhile to perform them.

8.3 Kinematically Linear Curved Bernoulli-Euler Beams

Kinematically linear curved B-E beams

By linearization of the formulas from Section 8.2 we may very easily get the equivalent expressions for the case of kinematic linearity. For the sake of convenience and completeness we list the most important of them below.

8.3.1 Kinematics

The rotation ω becomes, see (8.49)

$$\omega = \beta = w' + \frac{v}{\rho^0} \tag{8.72}$$

Rotation ω

and is thus the same as in the kinematically moderately nonlinear case.

8.3.1.1 Generalized Strains

Recall that we are dealing with Bernoulli-Euler beams and thus only have two generalized strains to consider, namely the axial strain ε and the curvature strain κ.

Generalized strains ε and κ

8.3.1.2 Axial Strain

The axial strain ε is, see (8.54)

$$\varepsilon = \left(v' - \frac{w}{\rho^0}\right) \tag{8.73}$$

Axial strain ε

8.3.1.3 Curvature Strain

The curvature strain κ is, see (8.57)

$$\kappa = w'' + \left(\overset{0}{\theta}{}' v\right)' = w'' + \left(\frac{v}{\rho^0}\right)' \tag{8.74}$$

Curvature strain κ

which is the same as in the kinematically moderately nonlinear case.

8.3.2 Generalized Stresses

Since the generalized strains are ε and κ the work conjugate stresses, the generalized stresses, are N and M.

Generalized stresses N and M

8.3.3 Equilibrium Equations

The formulas from Section 8.2.2 provide the static field equations

$$\begin{aligned} N' - \frac{V}{\rho^0} + \bar{p}_v &= 0 \\ V' - \frac{N}{\rho^0} + \bar{p}_w &= 0 \\ M' + V &= 0 \end{aligned} \tag{8.75}$$

Static field equations with V

where V has been introduced.

After elimination of V we may get

$$\begin{aligned} N' + \frac{M'}{\rho^0} + \bar{p}_v &= 0 \\ M'' + \frac{N}{\rho^0} - \bar{p}_w &= 0 \end{aligned} \tag{8.76}$$

Static field equations without V

These are two versions of the static field equations.

The possible static boundary conditions are

Possible static boundary conditions

$$\begin{aligned} N(0) &= \bar{P}_u(0)\,, \quad N(L) = \bar{P}_u(L) \\ V(0) &= \bar{P}_w(0)\,, \quad V(L) = \bar{P}_w(L) \\ M(0) &= \bar{C}(0)\,, \quad M(L) = \bar{C}(L) \end{aligned} \tag{8.77}$$

The example below was first examined by Brazier (1927) in a way that includes kinematic nonlinearity through a "backdoor" in the sense that from the outset his equations are all linear, but nonlinearity is introduced via coupling between axial and ring deformation.

Ex 8-3 Bending Instability of Circular Tubes

Bending instability of tubes

I have included this example for almost the same reasons as for the *Elastica*, see Example Ex 8-2. Like that example, the method used below is rather limited in scope, but the study by Brazier (1927) is in itself beautiful and provides insight into an important class of engineering problems.

Ex 8-3.1 Background

When a thin-walled beam with a circular cross-section is subjected to bending it experiences a phenomenon which, for reasons that will

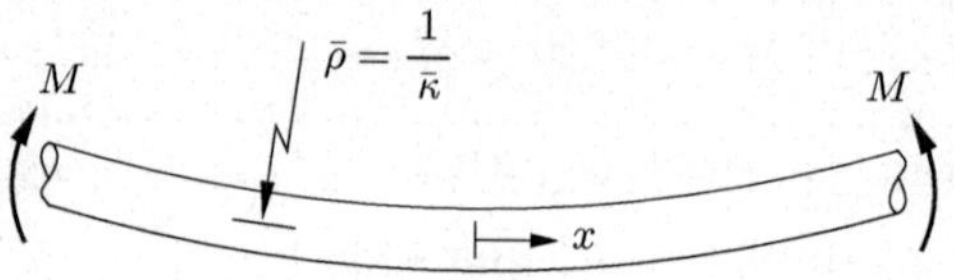

Fig. Ex. 8-3.1: *Bent tube.*

Ovalization = Brazier Effect

be clear later, is called *Ovalization*—also denoted the *Brazier Effect.* This may make the beam unstable with the result that the beam, or tube, collapses. The physics behind this is not all that difficult to understand in that the compression in the upper part of the tube, see Fig. Ex. 8-3.1, tries to push that part downwards, while the tension in the lower part makes it move upward with the result that the depth of the tubular cross-section decreases. As a result of this the tube provides a smaller bending moment than if the cross-section stayed circular. On the other hand, when bending of the tube is increased then the axial stresses are also increased which means that the tube may produce a larger bending moment. It is the competition between these two effects that may lead to a decrease in bending moment, namely when the ovalization becomes sufficiently large. To some extent, flattening of the cross-section happens in all beams that are subjected to bending, but in most cases this effect is so small that it may be neglected due to the fact that the distance between the flanges is kept almost constant by one or two webs.

Ovalization of pipelines is expensive

Ovalization has caused major financial problems in at least one important field, namely the offshore industry, in particular in connection with laying of pipelines, see e.g. (Kyriakides & Shaw 1982) and (Kyriakides & Corona 2007). The amount of money involved in losing a pipeline because of collapse is so large that it boggles the mind.[8.15]

For the sake of completeness I note that Brazier's interest in the problem of ovalization did not originate in offshore related issues, but came from problems related to aviation, see later.

Ex 8-3.2 Structural Problem—Laying of Pipelines

Laying of pipelines

We take the problem sketched in Fig. Ex. 8-3.2 as basis for our discussion of the phenomenon. During the laying processthe pipeline is bent, both at the lay vessel and at the bottom. At the lay vessel bending

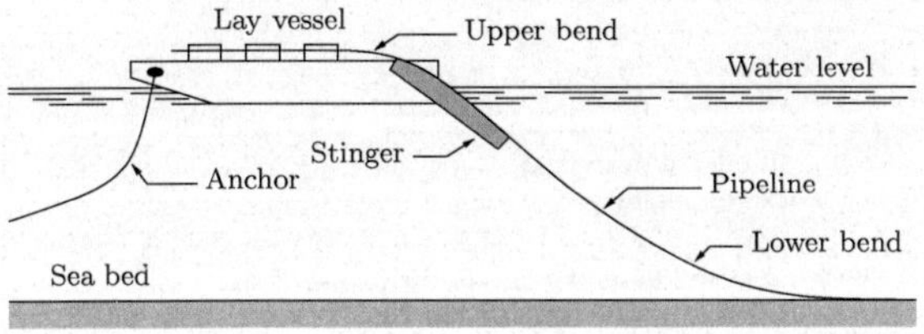

Fig. Ex. 8-3.2: *Lay vessel.*

may be controlled by the curvature of the stinger, which guides the pipeline, while at the sea bed the curvature is influenced by the tension in the anchor, the water depth, the slope of the sea bed, and other causes which are not easily controlled. In addition to bending the tube is also subjected to external pressure, which exacerbates the situation further. In the following we shall, however, only treat the pure bending situation. We shall also not consider the possibility of wrinkling and concentrated local buckling of the compressed part of the pipeline.

To some extent our analysis follows that of Brazier (1927), but we shall use a more modern formulation.

Ex 8-3.3 Geometry and Load of the Structure

Tube geometry and load on the structure

For our analysis the tube is taken to be straight before loading is applied and deformation of the cross-sections is assumed independent of the axial coordinate x of the tube, see Fig. Ex. 8-3.1. This is equivalent to applying a constant radius of curvature $\bar{\rho}$ to the tube, thus inducing a constant bending moment M. When the value of $\bar{\rho}$ is decreased, or equivalently the curvature $\bar{\kappa}$ is increased, the tube cross-section ovalizes, see Fig. Ex. 8-3.3. During this process the bending stiffness of the

8.15 At least it is almost incomprehensible compared with the salary of a University Professor.

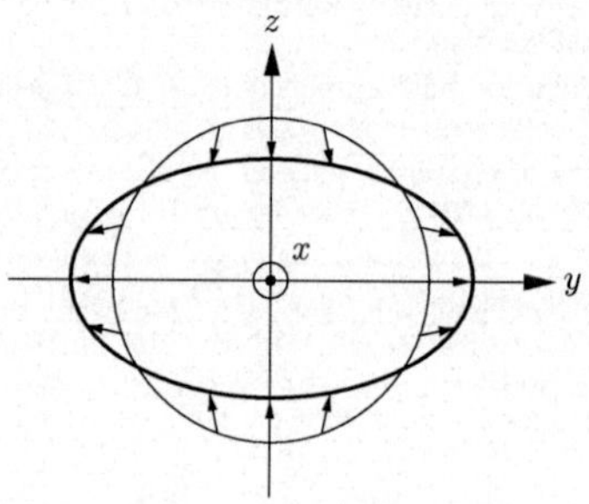

Fig. Ex. 8-3.3: Ovalized tube.

tube decreases because the depth of the cross-section decreases. While the strains and therefore the stresses of the upper and lower "fibers" of the tube increase they get a progressively smaller arm to produce bending moment through, and at some curvature the moment decreases with the result that the tube collapses. This holds true whether the material is elastic or elastic-plastic.

Ex 8-3.4 Brazier's Analysis

Brazier's Analysis

It is important to note that at Brazier's time there were no computers and no electronic calculators, implying that his analysis should be as simple as possible and involve only the absolutely necessary parameters. Solving linear—not to mention nonlinear—equations with as few as 5 unknowns was not an everyday task. Under these circumstances, in order for a simple computation to provide good results, it is extremely important to pick the right variables. Of course, performing a calculation under the assumption of plastic strain hardening was completely out of the question for the same reasons. Brazier was so intelligent—and had so much intuition—that he was able to furnish reliable results with just one unknown. That his prediction of the maximum bending moment is only about 5% higher than the correct one makes his study one of the really remarkable ones of all time. Although the problem of ovalization is inherently nonlinear, Brazier handled it as two linear problems, one on top of the other—a technique which was quite common at the time.[8.16]

Brazier's combined grasp of engineering, mechanics, and mathematics must impress everybody.

[8.16] Application of this approach requires much more insight and intuition than a more rigorous analysis. Therefore, over the years, many mistakes have been made by engineers who did not possess these qualifications.

Ex 8-3.4.1 Kinematics of the Cross-Section

Cross-section kinematics

We shall her draw on the formulas established in Section 8.3 and use the same notation as there, see Fig. Ex. 8-3.4. Later we shall see that the tube experiences loss of stability which indicates that we ought to

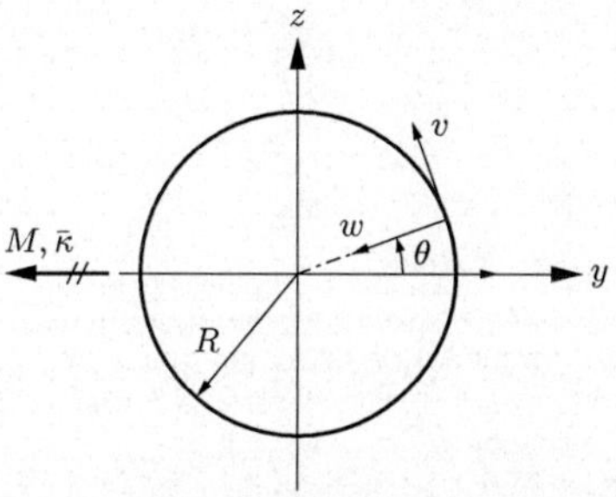

Fig. Ex. 8-3.4: *Tube cross-section.*

apply a kinematically nonlinear beam theory. The reason that this is not necessary is that nonlinearity is introduced through a backdoor, so to speak, in that the final expression for the strain ε_{xx} is nonlinear in the displacements v and w together with the applied curvature $\bar{\kappa}$, see (*Ex. 8-3.14*). Thus, Brazier neglects the nonlinearity associated with ring bending of the tube cross-section.

Ex 8-3.4.2 St. Venant Bending of the Tube

St. Venant Bending

Brazier discusses whether the experimentally observed ovalization might stem from the effect of the Poisson ratio ν. According to that theory, the so-called St. Venant bending, the axial strain ε^0_{xx} is

$$\varepsilon^0_{xx} = -\bar{\kappa} z \qquad (Ex.\ 8\text{-}3.1)$$

where ^{0} indicates that the quantity is computed according to this simple theory. Independent of the shape of the cross-section the prediction of this theory gives

$$\varepsilon^0_{yy} = -\nu\varepsilon^0_{xx}\,,\ \ \varepsilon^0_{zz} = -\nu\varepsilon^0_{xx}\,,\ \ \varepsilon^0_{yz} = 0 \qquad (Ex.\ 8\text{-}3.2)$$

since the stresses in the plane of the cross-section may be assumed to vanish. Based on (*Ex. 8-3.2*) we conclude the the displacements must be of the form

$$u^0_y = a_1 + a_2 yz\,,\ \ u^0_z = a_3 + a_4 y^2 + a_5 z^2 \qquad (Ex.\ 8\text{-}3.3)$$

where a_i are constants and the strains do not depend on a_1 and a_3. Utilizing (*Ex. 8-3.1*)–(*Ex. 8-3.1*) we may get

$$u^0_y = +\nu\bar{\kappa} yz\,,\ \ u^0_z = -\tfrac{1}{2}\left(y^2 - z^2\right) \qquad (Ex.\ 8\text{-}3.4)$$

or, more convenient for our purpose

$$v^0 = -\tfrac{1}{2}\nu\bar{\kappa}R^2\cos(\theta)\,,\quad w^0 = -\tfrac{1}{2}\nu\bar{\kappa}R^2\sin(\theta) \qquad (Ex.\ 8\text{-}3.5)$$

where the radius of the undeformed tube cross-section, which is constant, is denoted R instead of ρ^0.

Then, the strain $\varepsilon^0_{\theta\theta}$ in the circumferential direction becomes

$$\varepsilon^0_{\theta\theta} = \frac{\dot{v}^0 - w^0}{R} \qquad (Ex.\ 8\text{-}3.6)$$

where the dot $\dot{}$ denotes differentiation with respect to θ. From (*Ex. 8-3.5*) and (*Ex. 8-3.6*) we may get

$$\varepsilon^0_{\theta\theta} = \nu\bar{\kappa}R\sin(\theta) \qquad (Ex.\ 8\text{-}3.7)$$

which, by the way, might have been obtained from the expression for ε^0_{xx} in the same way as we found ε^0_{yy} and ε^0_{zz}. The reason for computing the value of $\varepsilon^0_{\theta\theta}$ is that later we shall introduce the idea of inextensibility in the plane of the cross-section.

Brazier found that for common values of material parameters and strains the absolute maximum value of v^0 (and w^0) is of the order of $R/1000$, which cannot explain the much larger ovalization found experimentally. Thus the simple, so-called technical beam theory is unable to describe ovalization.

Ex 8-3.4.3 Bending with Ovalization

Bending with ovalization

In order to establish expressions in the deformed configuration the coordinate ζ, see Fig. Ex. 8-3.5, is introduced. Brazier assumes that

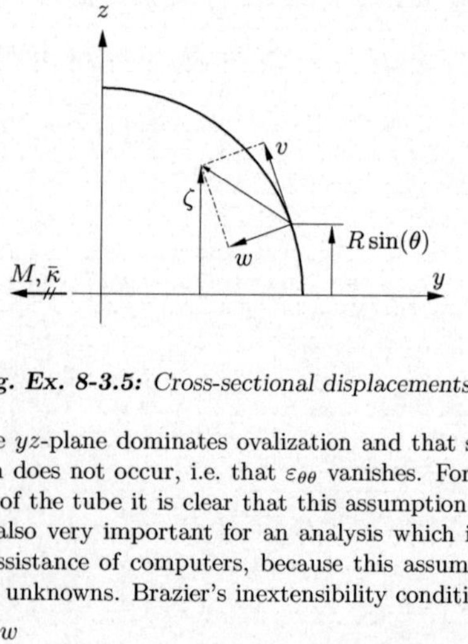

Fig. Ex. 8-3.5: *Cross-sectional displacements.*

bending in the yz-plane dominates ovalization and that stretching in the θ-direction does not occur, i.e. that $\varepsilon_{\theta\theta}$ vanishes. For a case such as ovalization of the tube it is clear that this assumption is appropriate, but it is also very important for an analysis which is performed without the assistance of computers, because this assumption halves the number of unknowns. Brazier's inextensibility condition is

Brazier's inextensibility condition is linear

$$0 = \dot{v} - w \qquad (Ex.\ 8\text{-}3.8)$$

where quantities without upper index ^{0} are associated with ovalization.

It is clear that (*Ex. 8-3.8*) is a *linear* inextensibility condition which means that for large deformations it is inadequate. Taking full nonlinearity into consideration it should be

Fully nonlinear inextensibility condition

$$0 = \frac{\dot{v} - w}{2R} + \tfrac{1}{2}\left(\frac{\dot{v} - w}{2R}\right)^2 + \tfrac{1}{2}\left(\frac{\dot{w} + v}{2R}\right)^2 \qquad (Ex.\ 8\text{-}3.9)$$

see (8.43), where the expressions (8.23) for α and β in terms of v and w have been introduced. Assuming a kinematically moderately nonlinear theory, which may be assumed valid for the present case, the inextensibility condition is

Moderately nonlinear inextensibility condition

$$0 = \frac{\dot{v} - w}{2R} + \tfrac{1}{2}\left(\frac{\dot{v} - w}{2R}\right)^2 \qquad (Ex.\ 8\text{-}3.10)$$

In the following we shall neglect all deformations from St. Venant bending. As the deformations increase, this assumption becomes more and more reasonable because bending deformation of the tube cross-section becomes more and more dominating.

Since the "load" on the tube is the prescribed curvature $\bar{\kappa}$ thepotential energy of a unit slice of the tube is

The "load" is $\bar{\kappa}$

$$\begin{aligned} &\mathbf{\Pi}_{\mathrm{P}}(v, w) \\ &= \tfrac{1}{2}R\int_0^{2\pi}\left(Et\left(\varepsilon_{xx}\right)^2 + \frac{1}{12(1-\nu^2)}Et^3\left(\kappa_{xx}\right)^2\right)d\theta \end{aligned} \qquad (Ex.\ 8\text{-}3.11)$$

where (*Ex. 8-3.8*) serves as an auxiliary condition, and where the dimension of $\mathbf{\Pi}_{\mathrm{P}}$ is correct since we are considering a slice of length 1 of the tube.[8.17] Based on Fig. Ex. 8-3.5 we may see that

$$\varepsilon_{xx} = -\bar{\kappa}\zeta \qquad (Ex.\ 8\text{-}3.12)$$

where

$$\zeta = R\sin(\theta) - u_z = (R - w)\sin(\theta) + v\cos(\theta) \qquad (Ex.\ 8\text{-}3.13)$$

i.e.

$$\varepsilon_{xx} = -\big((R - w)\sin(\theta) + v\cos(\theta)\big)\bar{\kappa} \qquad (Ex.\ 8\text{-}3.14)$$

where v and w are interconnected via the inextensibility condition (*Ex. 8-3.8*) and thus

$$\varepsilon_{xx} = -\big((R - \dot{v})\sin(\theta) + v\cos(\theta)\big)\bar{\kappa} \qquad (Ex.\ 8\text{-}3.15)$$

and

$$\delta\varepsilon_{xx} = -\big(-\delta\dot{v}\sin(\theta) + \delta v\cos(\theta)\big)\bar{\kappa} \qquad (Ex.\ 8\text{-}3.16)$$

According to (8.74) the ring bending strain κ_{xx} is

$$\kappa_{xx} = \frac{1}{R^2}(\dddot{v} + \dot{v}) \qquad (Ex.\ 8\text{-}3.17)$$

[8.17] You may be bothered by the fact that the driving "force" $\bar{\kappa}$ does not appear, but, as we shall see, it enters through ε_{xx}.

i.e.

$$\delta\kappa_{xx} = \frac{1}{R^2}(\delta\ddot{v} + \delta v) \qquad (Ex.\ 8\text{-}3.18)$$

The variation of the potential energy now is

$$\begin{aligned}\delta\mathbf{\Pi}_\mathrm{P}(v) = \bar{\kappa}^2 REt\int_0^{2\pi} &((R-\dot{v})\sin(\theta) + v\cos(\theta))\\ &(-\delta\dot{v}\sin(\theta) + \delta v\cos(\theta))d\theta\\ +\frac{E}{12(1-\nu^2)}\left(\frac{t}{R}\right)^3 &\int_0^{2\pi}(\ddot{v} + v)\\ &(\delta\ddot{v} + \delta v)\,d\theta\end{aligned} \qquad (Ex.\ 8\text{-}3.19)$$

Requiring that $\delta\mathbf{\Pi}_\mathrm{P}$ vanishes, integrating by parts and dividing by R/Et provide

$$\begin{aligned}0 = (\bar{\kappa}R)^2\int_0^{2\pi}&\Big(\tfrac{3}{2}R\sin(2\theta) - \tfrac{1}{2}(1-\cos(2\theta))\ddot{v}\\ &- \sin(\theta)\dot{v} + \tfrac{1}{2}(1+3\cos(2\theta))v\Big)\delta v d\theta\\ -\frac{1}{12(1-\nu^2)}\left(\frac{t}{R}\right)^2&\int_0^{2\pi}(v^{\mathrm{vi}} + 2v^{\mathrm{iv}} + \ddot{v}))\delta v d\theta\end{aligned} \qquad (Ex.\ 8\text{-}3.20)$$

where it has been exploited that the displacement v and its derivatives are all continuous over the the entire interval.
Since δv is arbitrary we get

$$\begin{aligned}&(v^{\mathrm{vi}} + 2v^{\mathrm{iv}} + \ddot{v})\frac{1}{12(1-\nu^2)}\left(\frac{t}{R}\right)^2\\ &= (\bar{\kappa}R)^2\Big(\tfrac{3}{2}R\sin(2\theta) - \tfrac{1}{2}(1-\cos(2\theta))\ddot{v}\\ &\qquad - \sin(\theta)\dot{v} + \tfrac{1}{2}(1+3\cos(2\theta))v\Big)\end{aligned} \qquad (Ex.\ 8\text{-}3.21)$$

The right-hand side, which may be thought of as the "load," deserves special attention because the terms between the parentheses are of different order in the displacements. While the first term is independent of the displacement v the remaining terms depend linearly on it. Assuming that the absolute value of the displacement v and its derivatives are small compared with R, then we may take the differential equation for v to be

$$v^{\mathrm{vi}} + 2v^{\mathrm{iv}} + \ddot{v} = 18(1-\nu^2)\left(\frac{R}{t}\right)^2(\bar{\kappa}R)^2R\sin(2\theta) \qquad (Ex.\ 8\text{-}3.22)$$

The only physically possible solution to (*Ex. 8-3.22*) is

Brazier's solution

$$v = -\tfrac{1}{2}(1-\nu^2)\left(\frac{R}{t}\right)^2(\bar{\kappa}R)^2R\sin(2\theta) \qquad (Ex.\ 8\text{-}3.23)$$

because all other solutions are either constant or proportional to θ, $\sin(\theta)$, $\cos(\theta)$, $\theta\sin(\theta)$, or $\theta\cos(\theta)$ and these possibilities must all be rejected since they violate the continuity condition, the (anti)symmetry condition on v, provide the St. Venant solution, or because they merely describe a rigid-body motion which is not interesting in this connection. The inextensibility condition (*Ex. 8-3.8*) and (*Ex. 8-3.22*) provide the solution for the transverse displacement component w

$$w = -(1-\nu^2)\left(\frac{R}{t}\right)^2 (\bar{\kappa}R)^2 R\cos(2\theta) \qquad (Ex.\ 8\text{-}3.24)$$

Brazier's solution

It may be worth noticing that the linear inextensibility condition (*Ex. 8-3.8*) is poorly fulfilled at the maximum on the M-$\bar{\kappa}$-curve because at that point the maximum displacements are of the order one fifth of the original radius R, but still the result for the maximum of M is only 5% too high compared with a more exact value.

We may compute the value of M by use of the Principle of Virtual Displacements with $\delta\kappa \neq 0$, or we may determine it by integration

$$M = -R\int_0^{2\pi} Et\varepsilon_{xx}\zeta d\theta = +EtR\bar{\kappa}\int_0^{2\pi}\zeta^2 d\theta \qquad (Ex.\ 8\text{-}3.25)$$

where the expression (*Ex. 8-3.12*) for ε_{xx} has been utilized. The final formula for ζ follows from (*Ex. 8-3.13*), (*Ex. 8-3.23*) and (*Ex. 8-3.24*)

$$\begin{aligned}\zeta = R\Big(&\big(1+(1+\nu^2)c^2\cos(2\theta)\big)\sin(\theta)\\ &-\tfrac{1}{2}(1-\nu^2)c^2\sin(2\theta)\cos(\theta)\Big)\end{aligned} \qquad (Ex.\ 8\text{-}3.26)$$

where the geometric parameter c is defined as

$$c \equiv (\bar{\kappa}R)\left(\frac{R}{t}\right) \qquad (Ex.\ 8\text{-}3.27)$$

Having performed the integrations in (*Ex. 8-3.25*) we find

$$M = Et^2R\pi\Big(c - \tfrac{3}{2}(1-\nu^2)c^3 + \tfrac{5}{8}(1-\nu^2)c^5\Big) \qquad (Ex.\ 8\text{-}3.28)$$

However, the last term in the parentheses is of the same order as the terms we omitted from (*Ex. 8-3.21*), and thus Brazier's result is

$$M = Et^2R\pi\Big(c - \tfrac{3}{2}(1-\nu^2)c^3\Big) \qquad (Ex.\ 8\text{-}3.29)$$

Brazier's result

The maximum value M_B, where subscript $_B$ indicates Brazier, occurs when

$$c^2 = c_B^2 = \frac{2}{9(1-\nu^2)} \qquad (Ex.\ 8\text{-}3.30)$$

which corresponds to the bending strain $\bar{\kappa}_B$

$$\bar{\kappa}_B = \frac{\sqrt{2}}{3\sqrt{(1-\nu^2)}}\left(\frac{t}{R}\right)\frac{1}{R} \qquad (Ex.\ 8\text{-}3.31)$$

and is

$$M_B = \frac{2\pi\sqrt{2}}{9\sqrt{(1-\nu^2)}}Et^2R \qquad (Ex.\ 8\text{-}3.32)$$

Brazier's maximum bending moment M_B

which is the maximum bending moment according to Brazier.

Brazier's w_B at maximum bending moment

At M_B the transverse displacement for $\theta = \pi/2$ becomes

$$w_B = \tfrac{2}{9}R \qquad (Ex.\ 8\text{-}3.33)$$

which is far from infinitesimal, not even small.

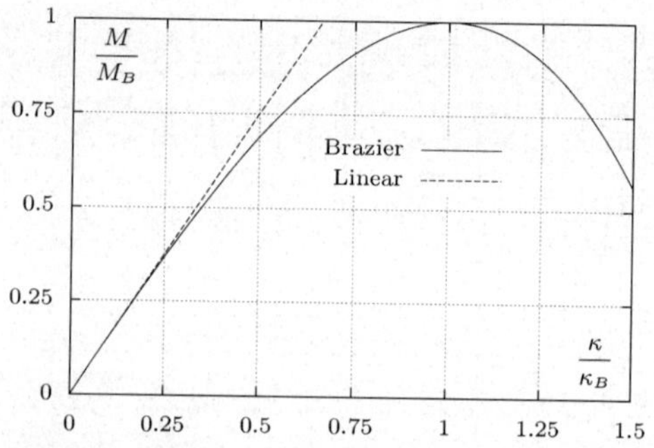

Fig. Ex. 8-3.6: *Moment-curvature relation for ovalized tube.*

Expressed as a relation between the nondimensional moment M/M_B and the nondimensional applied curvature $\bar{\kappa}/\bar{\kappa}_B$ (*Ex. 8-3.29*) becomes as simple as;

Brazier's solution

$$\frac{M}{M_B} = \tfrac{1}{2}\left(3\left(\frac{\bar{\kappa}}{\bar{\kappa}_B}\right) - \left(\frac{\bar{\kappa}}{\bar{\kappa}_B}\right)^3\right) \qquad (Ex.\ 8\text{-}3.34)$$

This relation, together with a simple linear approximation, is shown in Fig. Ex. 8-3.6 and indicates that nonlinearity does not become important until the moment reaches a value about half the maximum moment M_B or, equivalently, when the applied curvature $\bar{\kappa}$ is around half of its value at maximum bending moment

The sequence of deflections in Fig. Ex. 8-3.7 may indicate that ovalization does not become noticeable until the applied curvature indeed is rather large, maybe when $\bar{\kappa}/\bar{\kappa}_B$ is about 1.2. Around this value the cross-section begins taking the shape of a dog-bone. However, ovalization has a strong influence on the value of the bending moment much earlier in the loading process, as is clear from Fig. Ex. 8-3.7.

Like so many other great scientists, Brazier is not overly impressed by his own work.[8.18] On the other hand, I gather that Brazier must have felt great satisfaction when he had completed his analysis of the

[8.18] Personally, I have had the good fortune to know most of the best researchers of my time in structural and solid mechanics and be friends with many of them. It is my experience that the really good people do not feel the need to brag about their own achievement and belittle that of others. This, of course, does not mean that they do not take their work seriously.

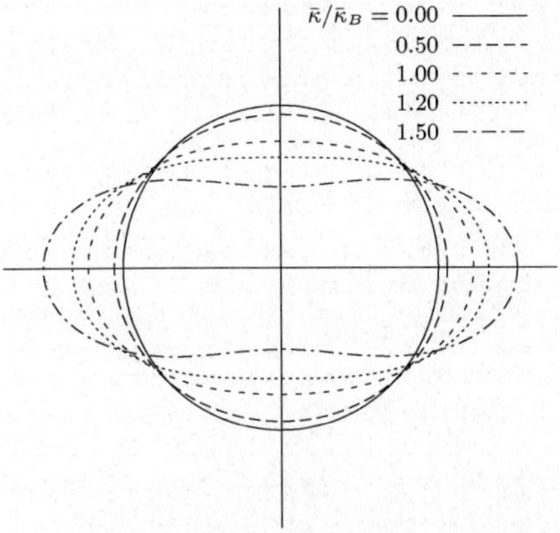

Fig. Ex. 8-3.7: *Sequence of deflections.*

ovalized tube. Yet, he is more interested in the kind of cross-sections that apparently were used in the aircraft industry of the time, see Fig. Ex. 8-3.8. He even outlines a way of handling "ovalization" of such

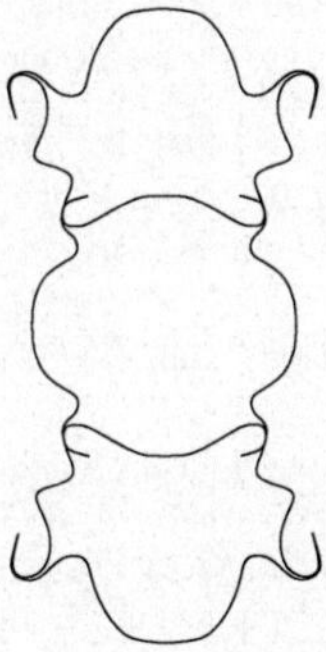

Fig. Ex. 8-3.8: *Brazier's "real" cross-section.*

cross-section. I am sure that today nobody would attempt a semi-analytic solution of this problem, but would resort to finite element methods immediately.

Ex 8-3.4.4 Another Procedure

Another procedure

Brazier might have reached a solution in a different way, once he had found the shape of the displacements v and w, either by the derivations leading to (*Ex. 8-3.23*) and (*Ex. 8-3.24*), or by considering the possible forms of v and w. He could have assumed

$$v = a\sin(2\theta) \text{ and } w = \frac{\dot{v}}{R} \tag{Ex. 8-3.35}$$

because this is the lowest order terms above the St. Venant solution. Then, he might have computed ζ from (*Ex. 8-3.13*), ε_{xx} from (*Ex. 8-3.14*), κ_{xx} from (*Ex. 8-3.17*), and finally inserted these expressions into the formula (*Ex. 8-3.11*) for the potential energy. Requiring the first variation of $\mathbf{\Pi}_{\mathrm{P}}(a)$ to vanish provides

$$a = -\frac{\frac{1}{2}(1-\nu^2)\left(\frac{R}{t}\right)^2(\bar{\kappa}R)^2 R}{1+\frac{5}{6}(1-\nu^2)\left(\frac{R}{t}\right)^2(\bar{\kappa}R)^2} \tag{Ex. 8-3.36}$$

which, when we consider the expression for (*Ex. 8-3.23*) for v, leaves the problem of whether $(R/t)^2(\bar{\kappa}R)^2$ is small compared to 1. Without any information about the geometry of the actual tube and the value of the applied curvature $\bar{\kappa}$ it is not possible to make a judgment regarding this. We may, however, cheat a little and utilize the results from Section Ex 8-3.4.3 and realize that at the maximum on the M-$\bar{\kappa}$-curve $(\bar{\kappa}R)^2(R/t)^2 \approx 0.5$, which is not small in relation to 1. However, at $\bar{\kappa}/\bar{\kappa}_B = 1$ the value of M/M_B is only about 11% higher based on (*Ex. 8-3.36*) and, considering the other approximations made above, it is difficult to decide which is the better result.

Ex 8-3.4.5 A More Accurate Procedure

Kyriakides & Shaw (1982) provide a more accurate solution, which requires the use of computers. The crux of the method is to assume

A more accurate assumption

$$v = R\sum_{j=1}^{J} b_j \sin(2j\theta) \text{ and } w = R\sum_{k=0}^{K} c_k \cos(2k\theta) \tag{Ex. 8-3.37}$$

and require that inextensibility according to the moderately nonlinear criterion (*Ex. 8-3.10*) is satisfied. Actually, the main subject of Kyriakides & Shaw (1982) is of a wider scope, namely response and stability of elastic-plastic circular pipes under combined bending and external pressure, which, of course, is a concern for real submerged pipelines.

Chapter 9

Plane Plates

Right after beams, plates are the most commonly used structural element. Plates may be loaded in their own plane only, transversely only, or in a combination of the two which, by the way, is almost always the case.[9.1] Examples of the first type of loaded plates are interior walls in buildings, while roof plates predominantly are subjected to transverse loads. Combined loads are usually more detrimental to the strength and stiffness of a plate than would be predicted by separate in-plane and transverse analyses. This is the case for outer load-carrying walls and many other structural elements. Thus, there are two basically different components of straining of plates, namely in-plane stretching and shearing, and transverse bending.

Combined loads are dangerous

In this chapter we derive the kinematic and static equations for plane plates and limit ourselves to the kinematically moderately nonlinear and to the linear case. As regards constitutive laws, we consider only linear elasticity, but emphasize that the kinematic and static equations—as is always the case—are independent of the constitutive law.

Recall: kinematics and statics always independent of constitutive law

Before we can begin establishing a theory for plane plates,such as the one shown in Fig. 9.1, we need to decide which strain measures are needed. First of all, we choose to consider the plate to be a *two-dimensional* continuum. Thus, it has no thickness, and the resistance of the plate to deformation must be computed from a three-dimensional theory, or measured in the laboratory. Next, the two most fundamental deformations that a plate can experience are in-plane stretching and shearing, and transverse bending. The theory we develop here neglects transverse shear deformations and is called a *Kirchhoff-Love* plate theory in the kinematically linear case and *von Kármán plate theory* in the kinematically moderately nonlinear case. For in-plane stretching and shearing we may rely on the strain measures from the three-dimensional theory with proper adjustments, while the bending part is new. Whether the theory is moderately nonlinear or linear we expect bending to be limited in the sense that the second derivatives of the shape

Two-dimensional continuum: "No thickness"

Kirchhoff-Love and von Kármán plates: Transverse shear strain assumed $\equiv 0$ In-plane stretching and shearing $\not\equiv 0$

[9.1] It is somewhat surprising that the English language, which has many more words than Danish, does not have a word for a plate loaded in the first two ways. In Danish the first type is called *skive* and the second *plade* with *plade* also meaning the third case. The equivalent words in German are *Scheibe* and *Platte*.

E. Byskov, *Elementary Continuum Mechanics for Everyone*,
Solid Mechanics and Its Applications 194, DOI: 10.1007/978-94-007-5766-0_9,

of the transverse displacement component of the deformed plate suffice, see (9.1b) below.[9.2] This is tantamount to keeping only the first terms of the series expansion of the curvature of the plate. The stretching and shearing measures, however, differ with the degree of nonlinearity, see (9.1a).

9.1 Kinematically Moderately Nonlinear Plates

Consider a plane plate, as shown in Fig. 9.1. By A^0 we denote the interior of the plate, while the boundary of the plate is Γ^0, which is smooth except for the point D^0, where the boundary forms a corner[9.3] with the

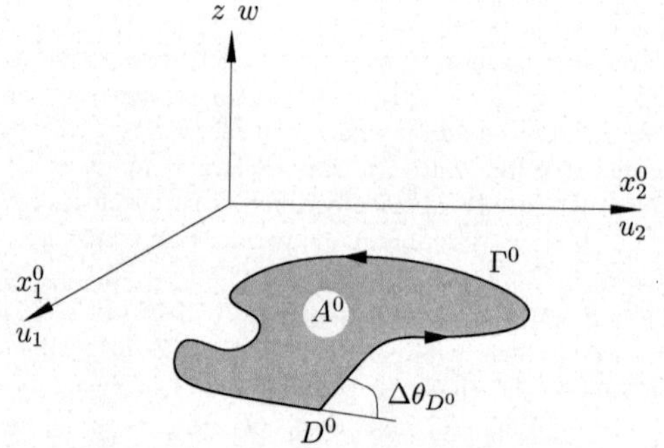

Fig. 9.1: Plane plate. Note the corner at D^0.

angle $\Delta\theta_{D^0}$.[9.4] The corner point—whether the angle $\Delta\theta_{D^0}$ is equal to 0 or not—plays an important role in connection with formulation of the static boundary conditions, see Section 9.1.4.

Later we shall integrate along the boundary Γ^0 such that the plate lies to the left of the integration path.

9.1.1 Kinematic Description

Greek indices: range [1,2]

The plate is described in a Cartesian coordinate system (x_1^0, x_2^0, z) with the axes x_β^0, $\beta = [1, 2]$ in the plane of the plate and z in the transverse direction. Here, lower-case Greek indices—as everywhere else in this book—cover the range $[1, 2]$, and the summation convention applies to repeated lower-case

[9.2] A similar assumption was introduced in the section on kinematically moderately nonlinear straight beams, see Section 7.3.

[9.3] It should be straightforward to generalize to cases with more than one corner.

[9.4] We insist that $-\pi < \Delta\theta_{D^0} < +\pi$ and emphasize that $\Delta\theta_{D^0}$ may take the value 0, in which case the notion of a corner becomes less obvious.

Greek indices similar to the way it applies to repeated lower-case Latin indices.

Assume that the kinematic nonlinearity only enters in the expression for the membrane strains $\varepsilon_{\alpha\beta}$ and that the curvature strains $\kappa_{\alpha\beta}$ are given by expressions that are linear in the displacements. Furthermore, assume that quadratic terms in the in-plane displacements u_α may be neglected, but retain quadratic terms in the transverse displacement w.[9.5] Then, the Lagrange strain measure (2.19) provides (9.1a), and a simple expansion of the geometric curvature yields (9.1b). The justification of the approximation in (9.1a) is discussed below.

$$\begin{aligned}\varepsilon_{\alpha\beta} &= \tfrac{1}{2}\left(u_{\alpha,\beta} + u_{\beta,\alpha}\right) + \tfrac{1}{2} u_{\gamma,\alpha} u_{\gamma,\beta} + \tfrac{1}{2} w_{,\alpha} w_{,\beta} \\ &\approx \tfrac{1}{2}\left(u_{\alpha,\beta} + u_{\beta,\alpha}\right) + \tfrac{1}{2} w_{,\alpha} w_{,\beta} \\ \kappa_{\alpha\beta} &= w_{,\alpha\beta}\end{aligned} \tag{9.1}$$

Membrane strains $\varepsilon_{\alpha\beta}$
Curvature strains $\kappa_{\alpha\beta}$

Generalized strains $\varepsilon_{\alpha\beta}$ and $\kappa_{\alpha\beta}$

Now, we have chosen the generalized strains, see Section 2.4.2, namely $\varepsilon_{\alpha\beta}$ and $\kappa_{\alpha\beta}$, the rest of the derivations are trivial because the *Principle of Virtual Displacements* in a sense takes over once we have set up the kinematic relations, and the equilibrium equations may be considered mere consequences. Although the procedure is systematic and therefore straightforward it may be tedious and sometimes require a fair amount of mathematical skill.

It is worthwhile emphasizing that in the expression for $\varepsilon_{\alpha\beta}$ only the nonlinear terms caused by the transverse displacement component w are kept, while the possible contributions $\frac{1}{2}u_{\gamma,\alpha}u_{\gamma,\beta}$ are left out. This is justifiable in many cases—certainly in most building structures—because the terms $\frac{1}{2}u_{1,1}u_{1,1}$ and $\frac{1}{2}u_{2,2}u_{2,2}$ loosely speaking only take significant values if the plate behaves in a rubber-like fashion. Only the terms $\frac{1}{2}u_{1,2}u_{1,2}$ and $\frac{1}{2}u_{2,1}u_{2,1}$, which describe straining from in-plane rotations, might compete with the terms $\frac{1}{2}w_{,\alpha}w_{,\beta}$. The remaining terms, $\frac{1}{2}u_{1,1}u_{1,2}$ and $\frac{1}{2}u_{2,1}u_{2,2}$ must be compared with the linear terms $u_{1,2}$ and $u_{2,1}$, respectively, and in most cases they are negligible because $|u_{1,1}| \ll 1$ and $|u_{2,2}| \ll 1$. Also, if these terms are retained while the bending measure is assumed linear the theory becomes inconsistent in the sense that terms which are assumed small in the expression for some of the strain measures are taken to be important in others. In the following, we employ the expression for $\varepsilon_{\alpha\beta}$ given by (9.1a). I do not intend to go any deeper into this, but mention that the freedom in choosing strain measures is limited, see also Chapter 7, in particular Section 7.3 where this question is discussed in connection with beam theories.

[9.5] These assumptions are analogous to the ones imposed on the beam in Section 7.3, where $\frac{1}{2}(u')^2$ is neglected, while $\frac{1}{2}(w')^2$ is kept in the expression for the axial strain ε, and a linear measure w'' for the bending strain κ. is assumed valid.

9.1.2 Budiansky-Hutchinson dot Notation

In the *Budiansky-Hutchinson dot notation*, see Chapter 33, the relevant operators $\mathbf{l}_1$, $\mathbf{l}_2$ and $\mathbf{l}_{11}$ become

Linear operator *lone* Quadratic operator $\mathbf{l}_2$

$$\mathbf{l}_1(\boldsymbol{u}) \sim \begin{bmatrix} \frac{1}{2}(u_{\alpha,\beta} + u_{\beta,\alpha}) \\ w_{,\alpha\beta} \end{bmatrix} \text{ and } \mathbf{l}_2(\boldsymbol{u}) \sim \begin{bmatrix} w_{,\alpha} w_{,\beta} \\ 0 \end{bmatrix} \tag{9.2}$$

Bilinear operator

$$\mathbf{l}_{11}(\boldsymbol{u}^a, \boldsymbol{u}^b) \sim \begin{bmatrix} \frac{1}{2}\left(w^a_{,\alpha} w^b_{,\beta} + w^b_{,\alpha} w^a_{,\beta}\right) \\ 0 \end{bmatrix} \tag{9.3}$$

where superscripts ^{a} and ^{b} indicate two different displacement fields.

9.1.3 Internal Virtual Work

The interpretation of the internal virtual work $\boldsymbol{\sigma} \cdot \delta\boldsymbol{\varepsilon}$ clearly is

Internal virtual work

$$\boldsymbol{\sigma} \cdot \delta\boldsymbol{\varepsilon} = \int_{A^0} (N_{\alpha\beta}\delta\varepsilon_{\alpha\beta} + M_{\alpha\beta}\delta\kappa_{\alpha\beta})\, dA^0 \tag{9.4}$$

Generalized stresses $N_{\alpha\beta}$ and $M_{\alpha\beta}$

where the generalized stresses $N_{\alpha\beta}$ and $M_{\alpha\beta}$ $N_{\alpha\beta}$ have not yet been given any physical meaning. But, already at this point it seems worthwhile mentioning that because (9.4) expresses work, the dimension of $N_{\alpha\beta}$ is force/length, while the dimension of $M_{\alpha\beta}$ is force×length/length. We only know that they are the generalized stresses of the plate since they work through the generalized strains defined by (9.1). For later purposes, introduce the strain-displacement relations (9.1) and get

$$\begin{aligned} \boldsymbol{\sigma} \cdot \delta\boldsymbol{\varepsilon} &= \int_{A^0} N_{\alpha\beta}\left(\tfrac{1}{2}(\delta u_{\alpha,\beta} + \delta u_{\beta,\alpha} + w_{,\alpha}\delta w_{,\beta} + w_{,\beta}\delta w_{,\alpha})\right) dA^0 \\ &\quad + \int_{A^0} M_{\alpha\beta}\delta w_{,\alpha\beta} dA^0 \\ &= \int_{A^0} \Big(N_{\alpha\beta}(\delta u_{\alpha,\beta} + w_{,\alpha}\delta w_{,\beta}) + M_{\alpha\beta}\delta w_{,\alpha\beta}\Big) dA^0 \end{aligned} \tag{9.5}$$

In order to arrive at the the last line of (9.5) we have utilized the fact that since the strain variations are symmetric in their indices $_\alpha$ and $_\beta$ then—without loss of generality—the generalized stresses $N_{\alpha\beta}$ and $M_{\alpha\beta}$ may be taken to be symmetric in $_\alpha$ and $_\beta$

$N_{\alpha\beta}$ and $M_{\alpha\beta}$ are symmetric

$$N_{\alpha\beta} = N_{\beta\alpha} \text{ and } M_{\alpha\beta} = M_{\beta\alpha} \tag{9.6}$$

9.1.4 External Virtual Work

We choose to limit the contributions to the external virtual work $\overline{\boldsymbol{T}} \cdot \delta\boldsymbol{u}$ to be composed of loads distributed over the plate—"body forces"[9.6]—and of "surface tractions,"[9.7] i.e. loads on the boundary, see Figs. 9.2 and 9.3. Thus, we have excluded point loads and line loads in the interior A^0 of the

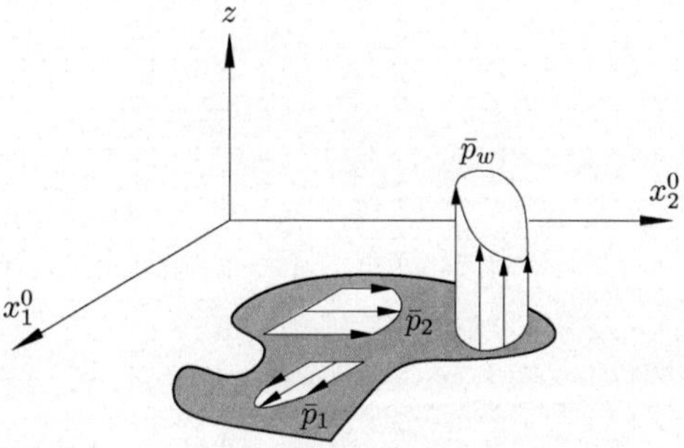

Fig. 9.2: "Body forces."

plate,[9.8] while the boundary Γ^0 may be subjected to line loads everywhere and concentrated loads at the corner.[9.9] All concentrated loads, including $\bar{P}_D$,[9.10] may be treated as limits of distributed loads or by use of the *Dirac delta Function*, if you prefer that.

"Body forces"

In the interior A^0 of the plate the distributed loads, the "body forces," consist of $\bar{p}_\alpha(x_\beta^0)$ in the plane and $\bar{p}_w(x_\beta^0)$ in the transverse direction.

"Surface loads", tractions

The boundary may be subjected to the following distributed loads—"surface loads", also called tractions—on the static boundary Γ_T^0: $\bar{p}_t(x_\beta^0)$ in the tangential direction, $\bar{p}_n(x_\beta^0)$ in the direction of the normal to the boundary, $\bar{q}(x_\beta^0)$ transverse to the plate, and the moment load $\bar{c}_n(x_\beta^0)$ directed along the *negative* tangential direction, see also Fig. 9.3. A concentrated

Concentrated boundary load

[9.6] The term "body force" is misleading unless you interpret it in a generalized sense, where the "volume" of the plate is its area. Recall that here the plate is treated as a two-dimensional body.

[9.7] Again, you must consider the dimension of the structure, here the dimension is 2, and interpret the surface as the plate boundary.

[9.8] My reason for omitting these kinds of loads, which may be important, is that the following equations already are fairly complicated as they stand. These concentrated loads may be modeled by loads whose resultant is finite while shrinking the area they attack.

[9.9] See the comments on the corner in the footnote on page 188.

[9.10] The reason for providing $\bar{P}_D$ with a special status is that—as mentioned above in Section 9.1—it plays an important role in connection with formulation of the static boundary conditions in Section 9.1.4.

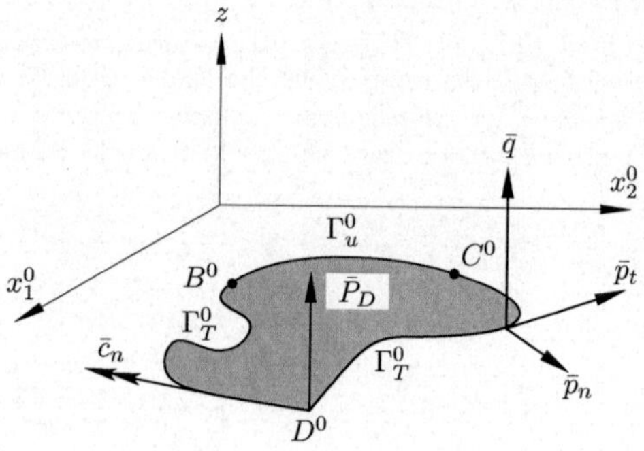

Fig. 9.3: Boundary loads—"surface tractions."
Note the corner load.
Filled circles denote ends of intervals.

load $\bar{P}_D$, which acts in the transverse direction, may be applied to the corner $\Delta\theta_{D^0}$.[9.11] The kinematic boundary Γ_u^0 will be subjected to reactions, which are denoted as the loads, except that they are not supplied with an overbar.

While the meaning of most of the load terms should be clear from Fig. 9.3, the distributed moment load $\bar{c}_n$, which is directed along the *negative* tangent t_α^0 to the undeformed plate, may be worth mentioning because it is not possible to impose a similar load in the direction of the normal n_α^0 to the undeformed plate and thus prescribe M_{nt}. This is because the variation $\delta w_{,t}$, where $_{,t}$ signifies the derivative in the direction of the tangent to the undeformed boundary, depends on δw on the boundary and therefore may not be varied independently. This fact has an important consequence as far as the static boundary conditions are concerned, see Section 9.1.6, in particular (9.25c).

In view of these loads, the external virtual work is

External virtual work

$$\begin{aligned}\overline{\boldsymbol{T}} \cdot \delta \boldsymbol{u} &= \int_{A^0} (\bar{p}_\alpha \delta u_\alpha + \bar{p}_w \delta w)\, dA^0 \\ &+ \oint_{\Gamma_T^0} (\bar{p}_t \delta u_t + \bar{p}_n \delta u_n + \bar{q}\delta w + \bar{c}_n \delta w_{,n})\, d\Gamma^0 \\ &+ \bar{P}_D \delta w_D\end{aligned} \tag{9.7}$$

9.11 If you fail to see why this load should be more prominent than other possible concentrated loads, don't feel bad. The reason for its importance and the necessity to include it follow from the static boundary conditions, see the derivations that lead to (9.21) below.

In (9.7) the interpretation of the integral over the area is straightforward, but the notation of the boundary integral requires a little explanation. In the interest of keeping the notation short we shall not explicitly indicate that the static boundary Γ_T^0 may consist of more than one part of the total boundary S^0.[9.12] However, the integral over Γ_T^0 actually designates the integral over the static boundary from its beginning, point B^0 to the point D^{0-} immediately before the corner D^0 plus the integral from the point D^{0+} immediately after the corner to its end C^0. In most terms the distinction between integrating in this sense and integrating from B^0 to C^0 is inconsequential, but for some terms, which we identify in each case below, the difference indeed is very important.

9.1.5 Principle of Virtual Displacements

From (9.5) and (9.7) the principle of virtual displacements for the plate is

$$\begin{aligned}&\int_{A^0}\Big(N_{\alpha\beta}\left(\delta u_{\alpha,\beta}+w_{,\alpha}\delta w_{,\beta}\right)+M_{\alpha\beta}\delta w_{,\alpha\beta}\Big)dA^0\\&=\int_{A^0}\left(\bar{p}_\alpha\delta u_\alpha+\bar{p}_w\delta w\right)dA^0\\&+\oint_{\Gamma_T^0}\left(\bar{p}_t\delta u_t+\bar{p}_n\delta u_n+\bar{q}\delta w+\bar{c}_n\delta w_{,n}\right)d\Gamma^0\\&+\bar{P}_D\delta w_D\end{aligned} \tag{9.8}$$

Principle of virtual displacements

In order to derive expressions—the equilibrium equations—which relate $N_{\alpha\beta}$ and $M_{\alpha\beta}$ to the applied loads it is necessary that the variational equation does not contain derivatives (with respect to the coordinates x_α^0) of the displacement variations.[9.13] Otherwise we are unable to exploit the fact that the displacement variations are arbitrary, except that they must satisfy the homogeneous kinematic boundary conditions, i.e. vanish on Γ_u^0. Therefore, we must apply the *Divergence Theorem* to the internal virtual work. First, however, terms like $N_{\alpha\beta}\delta u_{\alpha,\beta}$ must be rewritten

$$N_{\alpha\beta}\delta u_{\alpha,\beta}=\left(N_{\alpha\beta}\delta u_\alpha\right)_{,\beta}-N_{\alpha\beta,\beta}\delta u_\alpha \tag{9.9}$$

$$N_{\alpha\beta}w_{,\alpha}\delta w_{,\beta}=\left(N_{\alpha\beta}w_{,\alpha}\delta w\right)_{,\beta}-\left(N_{\alpha\beta,\beta}w_{,\alpha}\right)\delta w \tag{9.10}$$

$$\begin{aligned}M_{\alpha\beta}\delta w_{,\alpha\beta}&=\left(M_{\alpha\beta}\delta w_{,\alpha}\right)_{,\beta}-M_{\alpha\beta,\beta}\delta w_{,\alpha}\\&=\left(M_{\alpha\beta}\delta w_{,\alpha}\right)_{,\beta}-\left(M_{\alpha\beta,\beta}\delta w\right)_{,\alpha}+M_{\alpha\beta,\beta\alpha}\delta w\end{aligned} \tag{9.11}$$

[9.12] I strongly suspect—hope—that anyone should be able to generalize the formulas to cases where the static boundary is split up into several parts.

[9.13] As you will see from (9.18) there is one exception, namely $\delta w_{,n}$ on the boundary which is very much like $\delta w'$ at the ends of a beam.

An intermediate result is

Internal virtual work

$$\begin{aligned}\boldsymbol{\sigma}\cdot\delta\boldsymbol{\varepsilon} = &\oint_{\Gamma_T^0}\Big(N_{\alpha\beta}n_\beta^0\delta u_\alpha + N_{\alpha\beta}w_{,\alpha}n_\beta^0\delta w \\ &\qquad + M_{\alpha\beta}n_\beta^0\delta w_{,\alpha} - M_{\alpha\beta,\beta}n_\alpha^0\delta w\Big)d\Gamma^0 \\ &-\int_{A^0}\Big(N_{\alpha\beta,\beta}\delta u_\alpha + (N_{\alpha\beta,\beta}w_{,\alpha})\,\delta w - M_{\alpha\beta,\beta\alpha}\delta w\Big)dA^0\end{aligned} \tag{9.12}$$

We may immediately observe that the third term in the boundary integral contains $\delta w_{,\alpha}$, which cannot be varied independently of δw in the interior of the plate. However, as mentioned earlier, while the variation $\delta w_{,t}$ depends on the value of δw on the boundary, the variation $\delta w_{,n}$ does

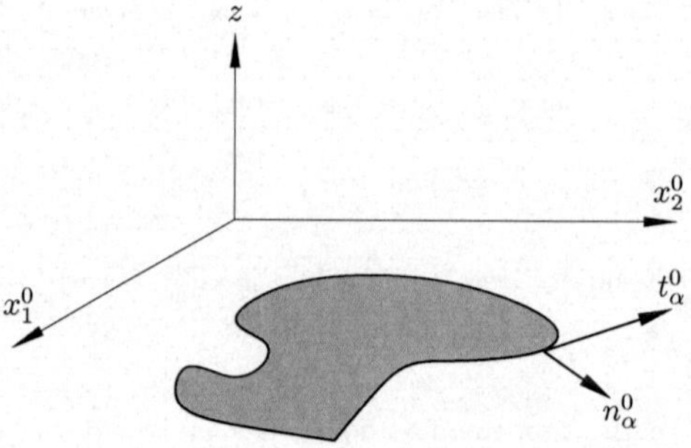

Fig. 9.4: Tangent t_α^0 and normal n_α^0 to the boundary.

not. Also, regarding formulation of the boundary conditions the coordinate system x_α^0 in general is an unnatural choice since the boundaries often lie in other directions than parallel to the axes. Therefore, it is convenient to introduce the unit normal n_α^0 and the unit tangent t_α^0 as local coordinate axes on the boundary of the *undeformed plate*. When we do that, the first integral on the right-hand side of (9.12) becomes

$$\begin{aligned}&\oint_{\Gamma_T^0}\Big(N_{\alpha\beta}n_\beta^0\delta u_\alpha + N_{\alpha\beta}w_{,\alpha}n_\beta^0\delta w \\ &\qquad + M_{\alpha\beta}n_\beta^0\delta w_{,\alpha} - M_{\alpha\beta,\beta}n_\alpha^0\delta w\Big)d\Gamma^0 \\ &= \oint_{\Gamma_T^0}\Big((N_{nn}\delta u_n + N_{tn}\delta u_t) + (N_{nn}w_{,n} + N_{tn}w_{,t})\delta w \\ &\qquad + (M_{nn}\delta w_{,n} + M_{tn}\delta w_{,t}) - (M_{nn,n} + M_{tn,t})\delta w\Big)d\Gamma^0\end{aligned} \tag{9.13}$$

where the meaning of the subscripts $_n$ and $_t$ should be obvious. Note that the summation convention does *not* apply to subscripts $_n$ and $_t$ because they indicate specific directions, namely the normal and tangential direction, respectively.

One of the above terms deserves special attention

$$\begin{aligned}
&\oint_{\Gamma_T^0} M_{tn}\delta w_{,t} d\Gamma^0 \\
&= \oint_{\Gamma_T^0} \Big((M_{tn}\delta w)_{,t} - M_{tn,t}\delta w\Big) d\Gamma^0 \\
&= [M_{tn}\delta w]_{A^0}^{D^{0-}} + [M_{tn}\delta w]_{D^{0+}}^{B^0} - \oint_{\Gamma_T^0} M_{tn,t}\delta w d\Gamma^0 \\
&= [M_{tn}\delta w]_{A^0}^{B^0} - [M_{tn}\delta w]_{D^{0-}}^{D^{0+}} - \oint_{\Gamma_T^0} M_{tn,t}\delta w d\Gamma^0
\end{aligned} \tag{9.14}$$

Here, the first term vanishes because $\delta w \equiv 0$ on the kinematic boundary Γ_u^0 and, therefore, by continuity also at the points A^0 and B^0 between the static boundary Γ_T^0 and the kinematic boundary Γ_u^0. Since we do not allow discontinuities in δw at D^0, the second term may be written

$$-[M_{tn}\delta w]_{D^{0-}}^{D^{0+}} = -\Delta M_{tn}(D^0)\delta w_D \tag{9.15}$$

where Δ indicates a difference in value. Therefore,

$$\begin{aligned}
&\oint_{\Gamma_T^0} \Big(N_{\alpha\beta} n_\beta^0 \delta u_\alpha + N_{\alpha\beta} w_{,\alpha} n_\beta^0 \delta w \\
&\qquad + M_{\alpha\beta} n_\beta^0 \delta w_{,\alpha} - M_{\alpha\beta,\beta} n_\alpha^0 \delta w \Big) d\Gamma^0 \\
&= \oint_{\Gamma_T^0} \Big(N_{nn}\delta u_n + N_{tn}\delta u_t \\
&\qquad + (N_{nn} w_{,n} + N_{tn} w_{,t} - M_{nn,n} - 2M_{tn,t})\delta w \\
&\qquad + M_{nn}\delta w_{,n} \Big) d\Gamma^0 \\
&\quad -\Delta M_{tn}(D^0)\delta w_D
\end{aligned} \tag{9.16}$$

Clearly, all terms in these integrals depend on the local coordinate axes n^0 and t^0, which change at D^0 when $\Delta\theta_{D^0} \neq 0$, and, therefore, all integrands have discontinuities at D^0. However, these jumps do not contribute terms like the last one in (9.16). The effect of the discontinuities of the integrands is merely that the integrations must be split up.

By (9.13) and (9.16) the internal virtual work (9.12) may be expressed in the following form

Internal virtual work

$$
\begin{aligned}
\boldsymbol{\sigma}\cdot\delta\boldsymbol{\varepsilon} = & \int_{A^0}\Big(- N_{\alpha\beta,\beta}\delta u_\alpha \\
& \qquad + \Big(M_{\alpha\beta,\beta\alpha} - (N_{\alpha\beta}w_{,\alpha})_{,\beta}\Big)\delta w\Big) dA^0 \\
& + \oint_{\Gamma_T^0}\Big(N_{nn}\delta u_n + N_{tn}\delta u_t \\
& \qquad + (-M_{nn,n} - 2M_{tn,n} + N_{nn}w_{,n} + N_{tn}w_{,t})\,\delta w \\
& \qquad + M_{nn}\delta w_{,n}\Big) d\Gamma^0 \\
& - \Delta M_{tn}(D^0)\delta w_D
\end{aligned}
\tag{9.17}
$$

When we utilize the expression (9.7) for the external virtual work and (9.17) for the internal virtual work, respectively, it is now possible to rewrite the principle of virtual displacements (9.8)

Principle of virtual displacements

$$
\begin{aligned}
0 = & + \int_{A^0}\Big(N_{\alpha\beta,\beta} + \bar{p}_\alpha\Big)\delta u_\alpha dA^0 \\
& - \int_{A^0}\Big(M_{\alpha\beta,\beta\alpha} - (N_{\alpha\beta}w_{,\alpha})_{,\beta} - \bar{p}_w\Big)\delta w dA^0 \\
& - \oint_{\Gamma_T^0}\Big(N_{nn} - \bar{p}_n\Big)\delta u_n d\Gamma^0 - \oint_{\Gamma_T^0}\Big(N_{tn} - \bar{p}_t\Big)\delta u_t d\Gamma^0 \\
& + \oint_{\Gamma_T^0}\Big(M_{nn,n} + 2M_{tn,n} - N_{nn}w_{,n} - N_{tn}w_{,t} + \bar{q}\Big)\delta w d\Gamma^0 \\
& - \oint_{\Gamma_T^0}\Big(M_{nn} - \bar{c}_n\Big)\delta w_{,n} d\Gamma^0 \\
& + \Big(\Delta M_{tn}(D^0) + \bar{P}_D\Big)\delta w_D
\end{aligned}
\tag{9.18}
$$

which is the expression that provides us with the static equations for kinematically moderately nonlinear plane plates.

9.1.6 Equilibrium Equations

All the equilibrium equations, which are derived below, are consequences of the arbitrariness of the variations. The static field equations follow from the fact that δu_α and δw are arbitrary in A^0

Static field equations

$$
\left.\begin{aligned}
& N_{\alpha\beta,\beta} + \bar{p}_\alpha = 0 \\
& M_{\alpha\beta,\beta\alpha} - (N_{\alpha\beta}w_{,\alpha})_{,\beta} - \bar{p}_w = 0
\end{aligned}\right\} \quad x_\alpha^0 \in A^0
\tag{9.19}
$$

and the static boundary conditions are consequences of the arbitrariness of

δu_n, δu_t, δw and $\delta w_{,n}$[9.14]

$$\left.\begin{aligned} &N_{nn} = \bar{p}_n \\ &N_{tn} = \bar{p}_t \\ &M_{nn,n} + 2M_{tn,n} - N_{nn}w_{,n} - N_{tn}w_{,t} = -\bar{q} \\ &M_{nn} = \bar{c}_n \end{aligned}\right\} x_\alpha^0 \in \Gamma_T^0 \tag{9.20}$$

Static boundary conditions

while the last condition, which may be considered a continuity or boundary condition, is given by the fact that δw_D is arbitrary[9.15]

$$\Delta M_{tn}(D^0) = -\bar{P}_D \tag{9.21}$$

Static (dis)continuity condition

As usual, we would like to be able to interpret the static quantities, i.e. be able to draw them as arrows, etc., but before we attempt to do that we introduce the variable V_α by

$$V_\alpha \equiv -\big(M_{\alpha\beta,\beta} - N_{\alpha\beta}w_{,\beta}\big)\ , \ x_\alpha^0 \in A^0 \tag{9.22}$$

Introduce shear force V_α

and similarly for the boundary

$$\left.\begin{aligned} V_n &\equiv -\big(M_{nn,n} + M_{nt,t} - N_{nn}w_{,n} - N_{nt}w_{,t}\big) \\ V_t &\equiv -\big(M_{tn,n} + M_{tt,t} - N_{tn}w_{,n} - N_{tt}w_{,t}\big) \end{aligned}\right\} x_\alpha^0 \in \Gamma_T^0 \tag{9.23}$$

Introduce V_n and V_t

These new quantities make it possible to express the equilibrium equations in shorter forms. The static field equations become

$$\left.\begin{aligned} &N_{\alpha\beta,\beta} + \bar{p}_\alpha = 0 \\ &V_{\alpha,\alpha} + \bar{p}_w = 0 \\ &M_{\alpha\beta,\beta} - N_{\alpha\beta}w_{,\beta} + V_\alpha = 0 \end{aligned}\right\} x_\alpha^0 \in A^0 \tag{9.24}$$

Static field equations

[9.14] If you do not feel comfortable handling all these variations at one time, you may proceed in the following way:

1. Let all other variations than δu_α in A^0 vanish. Still, δu_α may take sufficiently arbitrary values in A^0 with (9.19a) as the result.
2. Now that (9.19a) is satisfied the first term of (9.18) vanishes independently of the value of δu_α.
3. Repeat step 1 and step 2 for all other field terms.
4. Now, only the boundary terms remain, and we proceed in much the same way as in steps 1–3.

[9.15] It may seem strange that a jump in a bending moment can equal a force, but, as mentioned earlier, the dimension of the plate moments is force×length/length=force.

The static boundary conditions may now be expressed as

Static boundary conditions

$$\left.\begin{aligned} &N_{nn} = \bar{p}_n \\ &N_{tn} = \bar{p}_t \\ &V_n - M_{tn,t} = \bar{q} \\ &M_{nn,n} + M_{nt,t} - V_n - N_{nn}w_{,n} - N_{nt}w_{,t} = 0 \\ &M_{nn} = \bar{c}_n \end{aligned}\right\} x^0_\alpha \in \Gamma^0_T \tag{9.25}$$

while, as before, the last static condition (9.21) remains unaltered

Static continuity condition

$$\Delta M_{tn}(D^0) = -\bar{P}_D \tag{9.26}$$

The first two boundary conditions (9.25a) and (9.25b) ought not to present any problems, and the fourth condition (9.25d) expresses the transverse shear force V_n in terms of the displacements, the membrane forces and the moments.

Impossible to prescribe the boundary value of V_n independent of the derivative of M_{nt}

On the other hand, the third boundary condition (9.25c) seems somewhat suspicious at a first glance. In plain English it says that it is impossible to prescribe the boundary value of the transverse shear force V_n independent of the value of the (derivative of) the boundary value of the torsional moment M_{nt}. The reason behind this fact must be sought in our choice of curvature strain measure $\kappa_{\alpha\beta}$ as the second derivatives of the transverse displacement w, which precludes possible transverse shear strains. By expressing the curvature strains via the transverse displacement we have deprived the plate of kinematic degrees of freedom. When we limit the kinematic degrees of freedom we must expect consequences for the static quantities because they are linked together by the principle of virtual displacements.

The last condition, namely (9.26), is also a consequence of the kinematic constraints we have imposed on the plate.[9.16] We could have chosen to express the curvature strains in terms of the rotations of the plate and thus not reduce the number of kinematic degrees of freedom with the result that the transverse shear forces would have obtained the rank of generalized stresses in that the theory would have included transverse shear strains, which, of course, by their mere existence would be generalized. Such theories, the so-called *Mindlin plate theories*, are in use, in particular for thick plates and sandwich plates, but the one derived above is the most common because it is reasonably accurate for usual thin plates and because it involves fewer variables and therefore requires fewer constitutive parameters. We shall therefore not derive the Mindlin plate theories but mention that the task is somewhat easier than the one performed here.

Mindlin plate theory

[9.16] For those who are acquainted with Johansen's *Yield Line Theory*, see (Johansen 1963), this force at a corner is very familiar.

9.1.7 Interpretation of Static Quantities

Interpretation of static quantities

In order to derive the equilibrium equations in the interior of the plate we consider an element with the area (dx_1dx_2), see Fig. 9.5.[9.17]

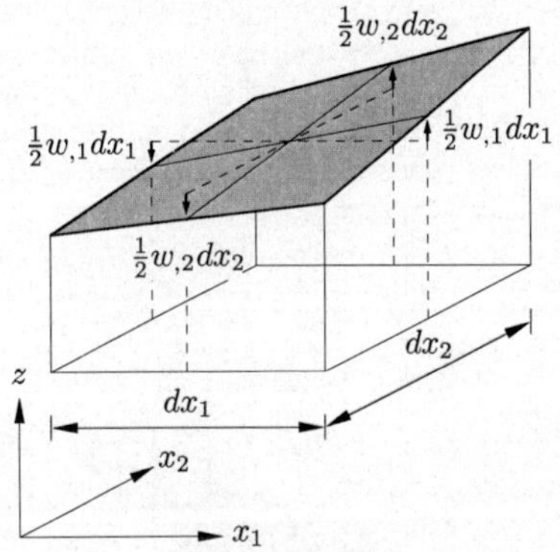

Fig. 9.5: Plane plate element.

It is vital to keep in mind the kinematic assumptions regarding the de-

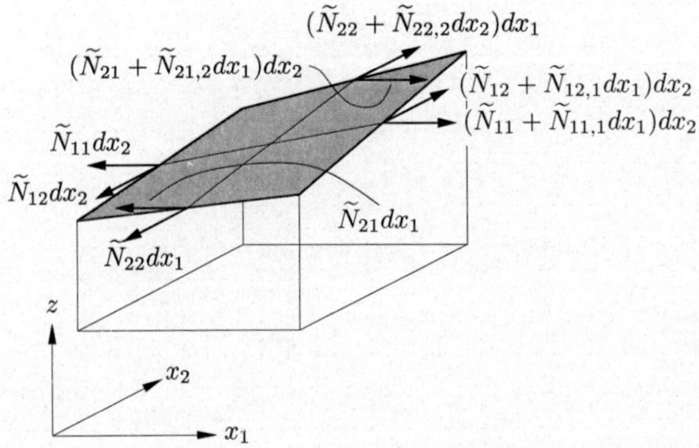

Fig. 9.6: Plane plate element with membrane forces.

gree of nonlinearity when we interpret the static quantities. Recall that the

[9.17] Alternatively, you may consider equilibrium of an arbitrary, finite part of the plate and exploit the *Divergence Theorem* to arrive at the equilibrium equations below. Whether you prefer to do it in one way or the other depends on your personal taste. In the present context I consider the derivations below more instructive, but others may disagree.

strains are small, while the rotations are assumed to be moderate. Therefore, we conclude that the projection on the x_α-plane of a deformed plate element to a good approximation has the same area as the undeformed element. This is especially important when we establish the moment equilibrium equations. All the forces and moments, which are introduced below, are assumed not to change direction with deformation of the plate, see Fig. 9.6 and Fig. 9.8. Also, they are measured on the undeformed plate, but this assumption is of less consequence because the strains are small.

Membrane forces $\widetilde{N}_{\alpha\beta}$

The membrane forces $\widetilde{N}_{\alpha\beta}$ are shown in Fig. 9.6, where it is pertinent to realize that they act on the deformed—displaced—plate element, otherwise the components of $\widetilde{N}_{\alpha\beta}$ would not contribute to the moment equilibrium

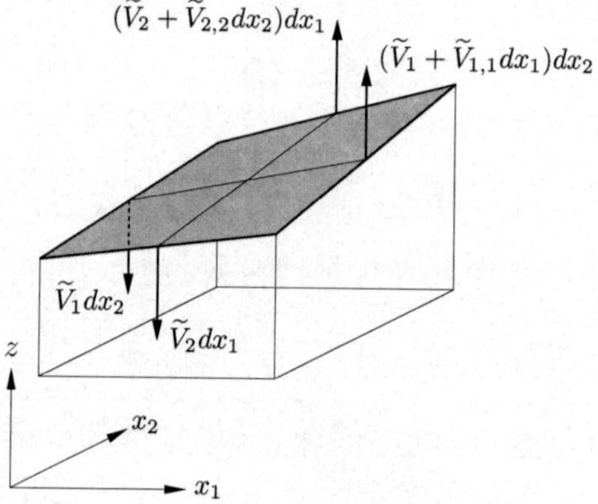

Fig. 9.7: Plane plate element with shear forces.

equations. Again, Fig. 9.6 only shows some of the forces acting on the plate element and, thus, like Fig. 9.7–Fig. 9.9, it is *not* a free-body diagram.

Transverse shear forces $\widetilde{V}_\alpha$

Further, introduce the transverse shear forces $\widetilde{V}_\alpha$ with the components shown in Fig. 9.7.

Bending and torsional moments $\widetilde{M}_{\alpha\beta}$

Finally, introduce the bending and torsional moments $\widetilde{M}_{\alpha\beta}$, whose components are shown in Fig. 9.8. The bending moments are the ones for which the indices on M are equal. The reason for the apparently awkward directions of the moments is the lack of symmetry inherent in an (x_1, x_2, z)-coordinate system. However, it may be observed that the directions of the bending and torsional moments follow a coordinate system which is rotated $\pi/2$ with respect to the x_α-system.

The contributions from the "body forces," which were introduced earlier, are shown in Fig. 9.9. When we consider the forces, including the moments

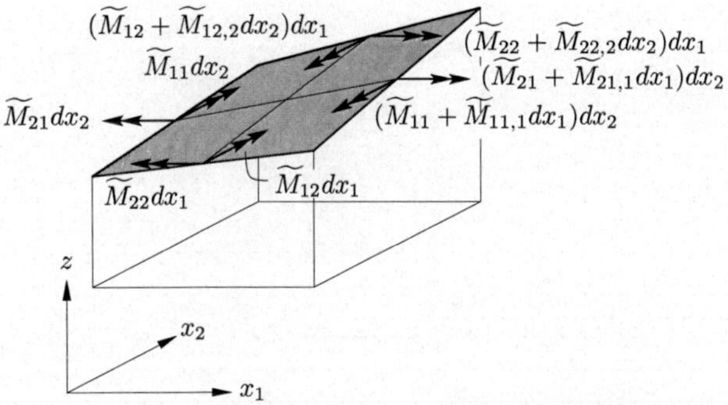

Fig. 9.8: Plane plate element with bending and torsional moments.

and loads, shown in Fig. 9.6–Fig. 9.9 moment equilibrium about the x_1-axis requires

$$\begin{aligned}0 = &-\widetilde{M}_{22}dx_1 + (\widetilde{M}_{22}+\widetilde{M}_{22,2}dx_2)dx_1\\&-\widetilde{M}_{21}dx_2 + (\widetilde{M}_{21}+\widetilde{M}_{21,1}dx_1)dx_2\\&-(\widetilde{N}_{22}+\widetilde{N}_{22,2}dx_2)dx_1 w_{,2}dx_2\\&-(\widetilde{N}_{12}+\widetilde{N}_{12,1}dx_1)dx_2 w_{,1}dx_1\\&-\widetilde{V}_1 dx_2 \tfrac{1}{2}dx_2 + (\widetilde{V}_1+\widetilde{V}_{1,1}dx_1)dx_2\tfrac{1}{2}dx_2\\&+(\widetilde{V}_2+\widetilde{V}_{2,2}dx_2)dx_1 dx_2\\&+\bar{p}_w dx_1 dx_2 \tfrac{1}{2}dx_2\\&+\bar{p}_2 dx_1 dx_2 w_{,2}dx_2\end{aligned} \tag{9.27}$$

This is rather horrendous, but the fact that dx_1 and dx_2 are small provides us with means to simplify the expression

$$\widetilde{M}_{22,2}+\widetilde{M}_{21,1}-\widetilde{N}_{22}w_{,2}-\widetilde{N}_{12}w_{,1}+\widetilde{V}_2=0 \tag{9.28}$$

which may immediately be written in the more compact form

$$\widetilde{V}_2 = -\left(\widetilde{M}_{2\beta,\beta}-\widetilde{N}_{2\beta}w_{,\beta}\right) \tag{9.29}$$

Similar manipulations on the moment equilibrium about the x_2-axis fur-

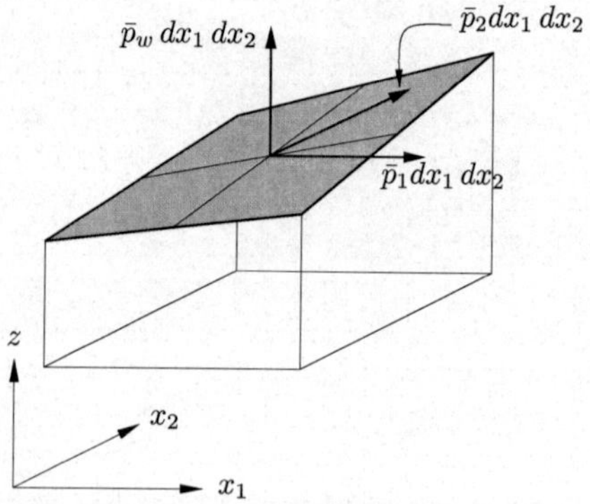

Fig. 9.9: Plane plate element with loads.

nish[9.18]

$$\widetilde{V}_1 = -\left(\widetilde{M}_{1\beta,\beta} - \widetilde{N}_{1\beta} w_{,\beta}\right) \tag{9.30}$$

The outcome of (9.29) and (9.30) is

Moment equilibrium

$$\widetilde{V}_\alpha = -\left(\widetilde{M}_{\alpha\beta,\beta} - \widetilde{N}_{\alpha\beta} w_{,\beta}\right) \tag{9.31}$$

which already strongly indicates that the quantities, which enter (9.31), may be interpreted as the same quantities without tilde, i.e. the generalized quantities $N_{\alpha\beta}$ and $M_{\alpha\beta}$ and the additional quantity V_α.

Having gone through the agony of establishing (9.31) we now turn to the much simpler task of deriving the equations that express equilibrium in the x_α-plane. Equilibrium in the x_1-direction requires

$$\begin{aligned} 0 = & -\widetilde{N}_{11} dx_2 + (\widetilde{N}_{11} + \widetilde{N}_{11,1} dx_1) dx_2 \\ & -\widetilde{N}_{21} dx_1 + (\widetilde{N}_{21} + \widetilde{N}_{21,2} dx_2) dx_1 \\ & + \bar{p}_1 dx_1 dx_2 \end{aligned} \tag{9.32}$$

i.e.

$$\begin{aligned} & \widetilde{N}_{11,1} + \widetilde{N}_{21,2} + \bar{p}_1 = 0 \\ \Rightarrow\ & \widetilde{N}_{\beta 1,\beta} + \bar{p}_1 = 0 \end{aligned} \tag{9.33}$$

9.18 The derivations are so much like the previous ones that—against my declared policy—I have decided to omit them. The only difficulties come about because of the strange directions of the components of $\widetilde{M}_{\alpha\beta}$ and $\widetilde{V}_\alpha$.

By proceeding in an equivalent way for the x_2-direction we conclude that

$$\widetilde{N}_{\beta\alpha,\beta} + \bar{p}_\alpha = 0 \text{ or } \widetilde{N}_{\alpha\beta,\beta} + \bar{p}_\alpha = 0 \tag{9.34}$$

Membrane equilibrium

where the last expression in (9.34) is a consequence of the symmetry of $\widetilde{N}_{\alpha\beta}$, which—by the way—has not been proved yet. This symmetry is a result of the fact that we have excluded moment loads acting in the z-direction. It is not very difficult to write the equation which expresses moment equilibrium about the z-axis, and the equation expressing equilibrium about the center of the element is

$$\begin{aligned} 0 = &+ \widetilde{N}_{12} dx_2 \tfrac{1}{2} dx_1 + \left(\widetilde{N}_{12} + \widetilde{N}_{12,1} dx_1\right) dx_2 \tfrac{1}{2} dx_1 \\ &- \widetilde{N}_{21} dx_1 \tfrac{1}{2} dx_2 - \left(\widetilde{N}_{21} + \widetilde{N}_{21,2} dx_2\right) dx_1 \tfrac{1}{2} dx_2 \end{aligned} \tag{9.35}$$

which yields

$$\widetilde{N}_{12} = \widetilde{N}_{21} \tag{9.36}$$

Symmetry of $N_{\alpha\beta}$

showing that $\widetilde{N}_{\alpha\beta}$ indeed is symmetric.

A comparison between the equations expressing internal equilibrium of the static quantities without and with tilde shows that we may interpret the quantities without tilde ($N_{\alpha\beta}$, V_α and $M_{\alpha\beta}$) as the equivalent quantities ($\widetilde{N}_{\alpha\beta}$, $\widetilde{V}_\alpha$ and $\widetilde{M}_{\alpha\beta}$).

Interpretation of the static boundary loads and reactions follows from derivations analogous to the ones above.

9.2 Plane Elastic Plates

The situation as regards constitutive models for plates is much the same as for beams, see Section 7.7, in that linear (hyper)elasticity and perfect plasticity are employed more than probably all other models combined. It is not my intention to derive the constitutive models for plates here, but see Chapter 15, where linearly elastic plate properties are determined. In order to establish boundary value problems for linearly elastic plates we need to recall that the generalized strains are $\varepsilon_{\alpha\beta}$ and $\kappa_{\alpha\beta}$ and that the generalized stresses are $N_{\alpha\beta}$ and $M_{\alpha\beta}$. Any linear constitutive "law," which, by proper choice of reference plane in the plate, presupposes that the in-plane, or membrane, action is decoupled from the transverse bending, may be expressed[9.19]

$$N_{\alpha\beta} = D^M_{\alpha\beta\gamma\omega} \varepsilon_{\gamma\omega} \text{ and } M_{\alpha\beta} = D^B_{\alpha\beta\gamma\omega} \kappa_{\gamma\omega} \tag{9.37}$$

Linear constitutive model

Here, superscripts ^{M} and ^{B} indicate membrane and bending, respectively. Both $D^M_{\alpha\beta\gamma\omega}$ and $D^B_{\alpha\beta\gamma\omega}$ are independent of deformation, but may vary with position in the plate.

[9.19] Membrane displacements and bending may couple through kinematical nonlinearity.

9.2.1 Generalized Quantities

Whether we limit ourselves to isotropic plates, which we actually do later, or wish to treat anisotropic plates as well (9.37) does not offer a very convenient way of writing the constitutive law for many purposes, except theoretical ones. The reason is that the strain and stress tensors are symmetric and therefore may be rearranged in a way which is similar to the one employed in Section 5.1 in order to simplify the constitutive relations. The main new feature is that, in the present connection, we let $\{\sigma\}$ comprise both the in-plane, or membrane stresses $N_{\alpha\beta}$ and the bending moments $M_{\alpha\beta}$. Therefore, $\{\varepsilon\}$ must cover the curvature strains $\kappa_{\alpha\beta}$ as well as the in-plane, or membrane strains $\varepsilon_{\alpha\beta}$.

Define the generalized stresses $\sigma_i,\ i \in [1,6]$[9.20]

Rearrangement of stresses ⇒ New generalized stresses

$$\{\sigma\} \equiv \begin{bmatrix} \sigma_1 \\ \sigma_2 \\ \sigma_3 \\ \sigma_4 \\ \sigma_5 \\ \sigma_6 \end{bmatrix} \equiv \begin{bmatrix} N_{11} \\ N_{22} \\ N_{12} \\ M_{11} \\ M_{22} \\ M_{12} \end{bmatrix} \tag{9.38}$$

Then, we are forced by the principle of virtual displacements to define the generalized strains $\varepsilon_i,\ i \in [1,6]$ as

Rearrangement of stresses ⇒ New generalized strains

$$\{\varepsilon\} \equiv \begin{bmatrix} \varepsilon_1 \\ \varepsilon_2 \\ \varepsilon_3 \\ \varepsilon_4 \\ \varepsilon_5 \\ \varepsilon_6 \end{bmatrix} \equiv \begin{bmatrix} \varepsilon_{11} \\ \varepsilon_{22} \\ 2\varepsilon_{12} \\ \kappa_{11} \\ \kappa_{22} \\ 2\kappa_{12} \end{bmatrix} \tag{9.39}$$

Stresses in Terms of Strains

The rearrangement of the strain and stress measures (9.38) and (9.39) implies that the linear constitutive relation (9.37), i.e. Hooke's "law," may be written

Hooke's "law"

$$\{\sigma\} = [D]\{\varepsilon\} \tag{9.40}$$

where $[D]$ designates the material stiffness matrix.

The constitutive relation (9.40) may be given a more explicit form which shows that the membrane strains do not produce bending moments and that

[9.20] For the same reason as in Section 5.1 we define the generalized stresses $\{\sigma\}$ before the generalized strains.

bending strains do not cause membrane stresses.

$$\begin{bmatrix} \sigma_1 \\ \sigma_2 \\ \sigma_3 \\ \sigma_4 \\ \sigma_5 \\ \sigma_6 \end{bmatrix} = \begin{bmatrix} D_{11} & D_{12} & D_{13} & 0 & 0 & 0 \\ D_{21} & D_{22} & D_{23} & 0 & 0 & 0 \\ D_{31} & D_{32} & D_{33} & 0 & 0 & 0 \\ 0 & 0 & 0 & D_{44} & D_{45} & D_{46} \\ 0 & 0 & 0 & D_{54} & D_{55} & D_{56} \\ 0 & 0 & 0 & D_{64} & D_{65} & D_{66} \end{bmatrix} \begin{bmatrix} \varepsilon_1 \\ \varepsilon_2 \\ \varepsilon_3 \\ \varepsilon_4 \\ \varepsilon_5 \\ \varepsilon_6 \end{bmatrix} \tag{9.41}$$

Hooke's "law" for plate

9.2.2 Constitutive Relations for Isotropic Plates

We shall not investigate (9.37) or (9.41) for anisotropic plates, but limit ourselves to the simpler case of isotropic plates.

Stresses in Terms of Strains

The constitutive matrix $D_{ij}, i, j \in [1, 6]$ may be written in a more convenient form

$$[D] = \begin{bmatrix} D^M_{11} & D^M_{12} & D^M_{13} & 0 & 0 & 0 \\ D^M_{21} & D^M_{22} & D^M_{23} & 0 & 0 & 0 \\ D^M_{31} & D^M_{32} & D^M_{33} & 0 & 0 & 0 \\ 0 & 0 & 0 & D^B_{11} & D^B_{12} & D^B_{13} \\ 0 & 0 & 0 & D^B_{21} & D^B_{22} & D^B_{23} \\ 0 & 0 & 0 & D^B_{31} & D^B_{32} & D^B_{33} \end{bmatrix} \tag{9.42}$$

Material stiffness matrix $[D]$

For convenience the membrane stiffness D^M_{ij} may be given by

$$\begin{bmatrix} D^M_{11} & D^M_{12} & D^M_{13} \\ D^M_{21} & D^M_{22} & D^M_{23} \\ D^M_{31} & D^M_{32} & D^M_{33} \end{bmatrix} = D^M \left[D^H \right] \tag{9.43}$$

Membrane part of material stiffness matrix D^M_{ij}

It is customary to assume a condition of *plane stress*, see Section Ex 5-3.2, as far as the membrane stress-strain relation is concerned because plates which may be assumed to bend noticeably are fairly thin. For convenience, the stress-strain relation (*Ex. 5-3.18*) for plane stress is repeated here

$$\sigma_{\alpha\beta} = \frac{E}{1+\nu}\left(\varepsilon_{\alpha\beta} + \frac{\nu}{1-\nu}\varepsilon_{\gamma\gamma}\delta_{\alpha\beta}\right) \quad \text{and} \quad \sigma_{3j} \equiv 0 \tag{9.44}$$

and, once you remember to integrate over the thickness t, it takes only a little effort to derive (9.45) and (9.46)

$$\left[D^H \right] \equiv \begin{bmatrix} 1 & \nu & 0 \\ \nu & 1 & 0 \\ 0 & 0 & \frac{1}{2}(1-\nu) \end{bmatrix} \tag{9.45}$$

and

$$D^M \equiv \frac{Et}{1-\nu^2} \tag{9.46}$$

where t is the thickness of the plate.

While the upper left-hand submatrix D^M_{ij} follows directly from the formulas for a state of plane stress determination of the expressions for the entries in the lower right-hand side submatrix D^M_{ij} requires some work, see Section 15.1 where it is shown that for the bending and torsional stiffness we may write D^B_{ij} in the form

Bending-torsion part of material stiffness matrix D^B_{ij}

$$\begin{bmatrix} D^B_{11} & D^B_{12} & D^B_{13} \\ D^B_{21} & D^B_{22} & D^B_{23} \\ D^B_{31} & D^B_{32} & D^B_{33} \end{bmatrix} = D^B \left[D^H\right] \tag{9.47}$$

where $\left[D^H\right]$ is given by (9.45), and

$$D^B \equiv \frac{Et^3}{12(1-\nu^2)} \tag{9.48}$$

Strains in Terms of Stresses

Sometimes we need the inverse relationship of (9.41). Again, we may write separate expressions connecting the membrane strains $(\varepsilon_1, \varepsilon_2, \varepsilon_3)$ with the membrane forces $(\sigma_1, \sigma_2, \sigma_3)$ and the bending strains $(\varepsilon_4, \varepsilon_5, \varepsilon_6)$ with the bending moments $(\sigma_4, \sigma_5, \sigma_6)$, respectively. By simple manipulations these equations may be derived from (9.41)–(9.48).

The results are

$$\{\varepsilon\} = [C]\{\sigma\} \tag{9.49}$$

or

$$\begin{bmatrix} \varepsilon_1 \\ \varepsilon_2 \\ \varepsilon_3 \\ \varepsilon_4 \\ \varepsilon_5 \\ \varepsilon_6 \end{bmatrix} = \begin{bmatrix} C_{11} & C_{12} & C_{13} & 0 & 0 & 0 \\ C_{21} & C_{22} & C_{23} & 0 & 0 & 0 \\ C_{31} & C_{32} & C_{33} & 0 & 0 & 0 \\ 0 & 0 & 0 & C_{44} & C_{45} & C_{46} \\ 0 & 0 & 0 & C_{54} & C_{55} & C_{56} \\ 0 & 0 & 0 & C_{64} & C_{65} & C_{66} \end{bmatrix} \begin{bmatrix} \sigma_1 \\ \sigma_2 \\ \sigma_3 \\ \sigma_4 \\ \sigma_5 \\ \sigma_6 \end{bmatrix} \tag{9.50}$$

where $[C]$ designates the material compliance matrix, or material flexibility matrix, and where the structure of $[C]$ is seen to be the same as that of $[D]$.

We may split up [C] in the same fashion that we used for [D]

$$[C] = \begin{bmatrix} C^M_{11} & C^M_{12} & C^M_{13} & 0 & 0 & 0 \\ C^M_{21} & C^M_{22} & C^M_{23} & 0 & 0 & 0 \\ C^M_{31} & C^M_{32} & C^M_{33} & 0 & 0 & 0 \\ 0 & 0 & 0 & C^B_{11} & C^B_{12} & C^B_{13} \\ 0 & 0 & 0 & C^B_{21} & C^B_{22} & C^B_{23} \\ 0 & 0 & 0 & C^B_{31} & C^B_{32} & C^B_{33} \end{bmatrix} \tag{9.51}$$

Inverse Hooke's "law" for plate

Here C^M_{ij} designates the membrane part of the material compliance matrix

$$\begin{bmatrix} C^M_{11} & C^M_{12} & C^M_{13} \\ C^M_{21} & C^M_{22} & C^M_{23} \\ C^M_{31} & C^M_{32} & C^M_{33} \end{bmatrix} = C^M \left[C^H\right] \tag{9.52}$$

Membrane part of material compliance matrix C^M_{ij}

with

$$\left[C^H\right] \equiv \begin{bmatrix} 1 & -\nu & 0 \\ -\nu & 1 & 0 \\ 0 & 0 & 2(1+\nu) \end{bmatrix} \tag{9.53}$$

and

$$C^M \equiv \frac{1}{Et} \tag{9.54}$$

Similarly, for the bending and torsional compliance

$$\begin{bmatrix} C^B_{11} & C^B_{12} & C^B_{13} \\ C^B_{21} & C^B_{22} & C^B_{23} \\ C^B_{31} & C^B_{32} & C^B_{33} \end{bmatrix} = C^B \left[C^H\right] \tag{9.55}$$

Bending-torsion part of material compliance matrix C^B_{ij}

where $\left[C^H\right]$ is given by (9.53), and

$$C^B \equiv \frac{12}{Et^3} \tag{9.56}$$

9.2.3 Differential Equations

It is our intention to write the equilibrium equation (9.19b)

$$M_{\alpha\beta,\beta\alpha} - (N_{\alpha\beta} w_{,\alpha})_{,\beta} - \bar{p}_w = 0\,,\ x^0_\alpha \in A^0 \tag{9.57}$$

Transverse equilibrium

in terms of the transverse displacement component w and the membrane forces $N_{\alpha\beta}$.

The reason why we do not attempt to express the equilibrium equation (9.19a)

In-plane equilibrium

$$N_{\alpha\beta,\beta} + \bar{p}_\alpha = 0 \, , \; x^0_\alpha \in A^0 \tag{9.58}$$

in terms of the in-plane displacement components u_α is that in many cases the membrane forces are statically determinate and therefore given by the equilibrium equations alone. Of course, if that is not the case we must invoke the displacements. The only term that needs attention is $M_{\alpha\beta,\beta\alpha}$, which we may rewrite as

$$\begin{aligned} M_{\alpha\beta,\beta\alpha} &= M_{11,11} + M_{12,12} + M_{21,21} + M_{22,22} \\ &= M_{11,11} + 2M_{12,12} + M_{22,22} \end{aligned} \tag{9.59}$$

where the symmetry of $M_{\alpha\beta}$ has been exploited.

When we utilize the definition (9.1b) of the curvature strain $\kappa_{\alpha\beta}$

$$\kappa_{\alpha\beta} = w_{,\alpha\beta} \tag{9.60}$$

and the formulas from Section 9.2.1 we may get

$$\begin{bmatrix} M_{11} \\ M_{22} \\ M_{12} \end{bmatrix} = D^B \begin{bmatrix} \kappa_{11} + \nu\kappa_{22} \\ \nu\kappa_{11} + \kappa_{22} \\ \frac{1}{2}(1-\nu)2\kappa_{12} \end{bmatrix} = D^B \begin{bmatrix} w_{,11} + \nu w_{,22} \\ \nu w_{,11} + w_{,22} \\ (1-\nu) w_{,12} \end{bmatrix} \tag{9.61}$$

Then, (9.59) may be written in terms of the displacement gradients $w_{,\alpha\beta}$

$$\begin{aligned} M_{\alpha\beta,\beta\alpha} &= D^B\big((w_{,1111} + \nu w_{,2211}) + 2(1-\nu)w_{,1212} \\ &\qquad + (\nu w_{,1122} + w_{,2222})\big) \\ &= D^B\big(w_{,1111} + 2w_{,1212} + w_{,2222}\big) \end{aligned} \tag{9.62}$$

With the definition of the biharmonic operator ∇^4

Biharmonic operator ∇^4

$$\nabla^4 f \equiv f_{,1111} + 2f_{,1212} + f_{,2222} \tag{9.63}$$

where $f = f(x_\alpha)$ designates a function defined in A^0, we may then write (9.57) as

Transverse equilibrium

$$D^B \nabla^4 w - (N_{\alpha\beta} w_{,\alpha})_{,\beta} = \bar{p}_w \, , \; x^0_\alpha \in A^0 \tag{9.64}$$

which must always be solved along with (9.58) and the boundary conditions.

9.2.4 Boundary Conditions

9.2.4.1 Kinematic Boundary Conditions

In most cases, the kinematic boundary conditions are straightforward and, therefore, we do not treat them in any detail except to say that, as far as the transverse displacement component w is concerned, the most common kinematic boundary conditions prescribe the value of w itself or its partial derivative $w_{,n}$, which is the same as the slope normal to the boundary—or both.

9.2.4.2 Static Boundary Conditions

We may formulate the static boundary conditions (9.20c) and (9.20d) in terms of the transverse displacement component w and the membrane forces N_{nn}, $N_{nt} = N_{tn}$, and thus (9.20) yields

$$\left.\begin{aligned} &N_{nn} = \bar{p}_n \\ &N_{tn} = \bar{p}_t \\ &D^B\big(w_{,nnn} + \nu w_{,ttn} + 2(1-\nu)w_{,tnn}\big) \\ &\qquad\qquad - N_{nn}w_{,n} - N_{tn}w_{,t} = -\bar{q} \\ &D^B\big(w_{,nn} + \nu w_{,tt}\big) = \bar{c}_n \end{aligned}\right\} x_\alpha^0 \in \Gamma_T^0 \tag{9.65}$$

Static boundary conditions without V_n

and, similarly, (9.25) becomes

$$\left.\begin{aligned} &N_{nn} = \bar{p}_n \\ &N_{tn} = \bar{p}_t \\ &D^B(1-\nu)w_{,ntt} + V_n = \bar{q} \\ &D^B\big(w_{,nnn} + w_{,ntt}\big) - N_{nn}w_{,n} - N_{nt}w_{,t} + V_n = 0 \\ &D^B\big(w_{,nn} + \nu w_{,tt}\big) = \bar{c}_n \end{aligned}\right\} x_\alpha^0 \in \Gamma_T^0 \tag{9.66}$$

Static boundary conditions with V_n

9.2.5 The Airy Stress Function

In some applications, when there is no in-plane loading on the plate, i.e.

$$\bar{p}_1 = 0 \text{ and } \bar{p}_2 = 0 \tag{9.67}$$

it proves feasible to introduce the *Airy Stress Function* Φ, which is defined by the following relations, see e.g. (Brush & Almroth 1975)

$$N_{11} = \Phi_{,22}\,, \quad N_{12} = -\Phi_{,12} \text{ and } N_{22} = \Phi_{,11} \tag{9.68}$$

The Airy Stress Function Φ

With Φ as given above and $\bar{p}_\alpha = 0$ (9.58) becomes

$$\Phi_{,221} + (-\Phi_{,122}) \equiv 0 \text{ and } (-\Phi_{,121}) + \Phi_{,112} \equiv 0 \tag{9.69}$$

and in-plane equilibrium is thus satisfied by Φ as defined by (9.68). This, however, does not mean that any Φ is the solution to some plate problem because the strains that follow from the stresses given by Φ cannot always be integrated to a displacement field. In other words, there is no guaranty that the stress field results in compatible displacements, see also Sections 2.2.4 and 4.2.2, where this subject is mentioned in connection with the three-dimensional continuum. The crux of the matter is that there are *three* components of strain (and stress), while there are only *two* independent displacement components, and therefore there must exist a connection between the strain components. In the present case, the compatibility condition is expressed as an equation between derivatives of the membrane strains and terms that are nonlinear in derivatives of the transverse displacements

$$\varepsilon_{11,22} + \varepsilon_{22,11} - 2\varepsilon_{12,12} = (w_{,12})^2 - w_{,11}w_{,22} \tag{9.70}$$

which, by straightforward manipulations and utilizing the strain definitions (9.1), may be proved to be correct. This is left for the reader to do. We must ensure that (9.70) is satisfied by the solutions in terms of Φ, which is not necessarily the case, since the definition of Φ only insures equilibrium of the membrane stresses. Therefore, we need to express (9.70) in terms of Φ instead of $\varepsilon_{\alpha\beta}$ and utilize strain-stress relations (9.49) or (9.50) and the defining equations (9.68) for Φ.

Utilize (9.38), (9.39), (9.49)–(9.54) and (9.68) to get

$$\begin{aligned} Et\varepsilon_{11,22} &= +\Phi_{,2222} - \nu\Phi_{,1122} \\ Et\varepsilon_{22,11} &= -\nu\Phi_{,2211} + \Phi_{,1111} \\ Et\varepsilon_{12,12} &= -(1+\nu)\Phi_{,1212} \end{aligned} \tag{9.71}$$

This, inserted in (9.70), yields the differential equation for the *Airy Stress Function* Φ

Differential equation for the Airy Φ

$$\nabla^4\Phi = Et\left((w_{,12})^2 - w_{,11}w_{,22}\right) \ , \ x^0_\alpha \in A^0 \tag{9.72}$$

This differential equation together with the differential equation (9.64) for the transverse displacement component w

Nonlinear differential equation for w

$$D^B\nabla^4 w - (N_{\alpha\beta}w_{,\alpha})_{,\beta} = \bar{p}_w \ , \ x^0_\alpha \in A^0 \tag{9.73}$$

constitutes a set of equations for description of membrane and bending of plane elastic plates under the assumption of kinematic moderate nonlinearity.

Recall that the *Permutation Symbol* $e_{\alpha\beta}$ in the two-dimensional case is defined by (31.19), which is repeated here[9.21]

Permutation symbol $e_{\alpha\beta}$

$$e_{\alpha\beta} \equiv \begin{cases} 0 \ \text{for} \ (\alpha,\beta)=(1,1) \ \text{and} \ (\alpha,\beta)=(2,2) \\ +1 \ \text{for} \ (\alpha,\beta)=(1,2) \\ -1 \ \text{for} \ (\alpha,\beta)=(2,1) \end{cases} \tag{9.74}$$

[9.21] Do not confuse the permutation symbol with a linear strain measure. As mentioned before, the many different notations for strain cause problems.

then (9.73) may be rewritten

Nonlinear differential equation for w

$$D^B \nabla^4 w - e_{\alpha\gamma} e_{\beta\delta} (\Phi_{,\delta\gamma} w_{,\alpha})_{,\beta} = \bar{p}_w \, , \; x^0_\alpha \in A^0 \tag{9.75}$$

Finally, (9.72) and (9.75) together with the relevant boundary conditions form a complete set of equations for the plate.

The in-plane boundary conditions may well warrant some comments. If the boundary conditions are static, then their formulation in terms of Φ is straightforward because the second derivatives of Φ simply are the stresses. On the other hand, since the in-plane behavior is described in terms of a *stress* function kinematic boundary conditions present difficulties, which must be handled individually, see e.g. Example Ex 18-3.

There are several advantages of the application of the *Airy Stress Function* Φ. The first is that for the in-plane behavior of the plate we are dealing with only *one* stress function, namely Φ, instead of *two* displacement functions u_1 and u_2—the disadvantage as regards boundary conditions is mentioned above. Another, less obvious benefit, is that in a number of cases, where numerical methods are applied, the use of a stress function may make the particularly annoying numerical problem of the so-called *nonlinear membrane locking*[9.22] disappear. This unwanted effect is caused by an internal mismatch between polynomial contributions from linear in-plane displacement components and nonlinear transverse components of the membrane strain. Especially in postbuckling it proves to be difficult to choose approximations to u_α and w such that the expression for $\varepsilon_{\alpha\beta}$, see (9.1), can model a constant strain which is necessary in order to avoid self-straining. In a number of buckling and postbuckling studies, e.g. (Hutchinson & Amazigo 1967), (Fitch 1968), (Stephens 1971), (Byskov & Hutchinson 1977), (Byskov & Hansen 1980) and (Byskov, Damkilde & Jensen 1988) Airy-type stress functions have been utilized with great advantage in that membrane locking has been avoided.[9.23]

Membrane locking

It may be mentioned that membrane locking may be handled by other means than stress functions. One of the more popular ones is the application of *reduced integration* which consists in applying a numerical integration scheme that has fewer points than necessary for exact integration of the terms of the membrane strain. To my taste such a procedure must make one very cautious, and I much prefer the use of *Lagrange Multipliers* which is straightforward and safe, see e.g. (Byskov 1989).

Reduced integration

Lagrange multipliers

[9.22] Personally, I don't like the term membrane locking and prefer something like *internal mismatch*, but that is not the standard nomenclature.

[9.23] None of the authors of the papers before about 1980 were aware of the problem of membrane locking, simply because it was not discovered until later. Therefore, it is a fortunate coincidence that they employed the $w - \Phi$-formulation and thus had such success with their numerical procedures.

9.2.6 Other Stress Functions

It may be relevant mentioning that the *Airy Stress Function* Φ is not the only useful stress function, see in particular (Muskhelishvili 1963). For many purposes, the formulation in terms of the complex functions by Muskhelishvili (1963) yields answers to complicated problems, for example involving cracks modeled as mathematical slits etc., in an easy way—*once you have mastered the method*, which indeed takes some time and presupposes a working knowledge of complex functional analysis. At least in the standard version the formulas of Muskhelishvili cover only kinematically linear plates with in-plane loads, which is our next subject.

9.3 Kinematically Linear Plates

Many structural analyses are performed under the assumption of kinematic linearity and, therefore, the linear version of the plate equations from Sections 9.1 and 9.2 is used widely. Since you can obtain the linear equations simply by omitting all kinematically nonlinear terms from the formulas of Section 9.1 I have judged it unnecessary to derive them separately and also to cite them all here, except for the following ones.

9.3.1 Kinematic Description

Here, (9.1) becomes

Generalized strains $\varepsilon_{\alpha\beta}$ and $\kappa_{\alpha\beta}$

$$\begin{aligned} \varepsilon_{\alpha\beta} &= \tfrac{1}{2}\left(u_{\alpha,\beta} + u_{\beta,\alpha}\right) \\ \kappa_{\alpha\beta} &= w_{,\alpha\beta} \end{aligned} \tag{9.76}$$

where the kinematic linearity is obvious.

9.3.2 Equilibrium Equations

The static field equations (9.19) simplify to

Generalized stresses $N_{\alpha\beta}$ and $M_{\alpha\beta}$

$$\left.\begin{aligned} N_{\alpha\beta,\beta} + \bar{p}_\alpha &= 0 \\ M_{\alpha\beta,\beta\alpha} - \bar{p}_w &= 0 \end{aligned}\right\} \; x^0_\alpha \in A^0 \tag{9.77}$$

and the static boundary conditions (9.20) now are

Static boundary conditions

$$\left.\begin{aligned} N_{nn} &= \bar{p}_n \\ N_{tn} &= \bar{p}_t \\ M_{nn,n} + 2M_{tn,n} &= -\bar{q} \\ M_{nn} &= \bar{c}_n \end{aligned}\right\} \; x^0_\alpha \in \Gamma^0_T \tag{9.78}$$

Static (dis)continuity condition unaltered

while the discontinuity condition (9.21) remains unaltered

$$\Delta M_{tn}(D^0) = -\bar{P}_D \tag{9.79}$$

which should not be surprising because (9.21) is linear.

The definition (9.22) of the extra static variable V_α is now

$$V_\alpha \equiv -M_{\alpha\beta,\beta}\ ,\ x^0_\alpha \in A^0 \tag{9.80}$$

Introduce shear force V_α

and on the boundary (9.23) becomes

$$\left.\begin{aligned} V_n &\equiv -\Big(M_{nn,n} + M_{nt,t} - N_{nn}w_{,n} - N_{nt}w_{,t}\Big) \\ V_t &\equiv -\Big(M_{tn,n} + M_{tt,t} - N_{tn}w_{,n} - N_{tt}w_{,t}\Big) \end{aligned}\right\}\ x^0_\alpha \in \Gamma^0_T \tag{9.81}$$

Introduce V_n and V_t

with the consequence that the static field equations (9.24) now are

$$\left.\begin{aligned} N_{\alpha\beta,\beta} + \bar{p}_\alpha &= 0 \\ V_{\alpha,\alpha} + \bar{p}_w &= 0 \\ M_{\alpha\beta,\beta} + V_\alpha &= 0 \end{aligned}\right\}\ x^0_\alpha \in A^0 \tag{9.82}$$

Static field equations

and the possible static boundary conditions (9.25) take the simpler form

$$\left.\begin{aligned} N_{nn} &= \bar{p}_n \\ N_{tn} &= \bar{p}_t \\ V_n - M_{tn,t} &= \bar{q} \\ M_{nn,n} + M_{nt,t} - V_n &= 0 \\ M_{nn} &= \bar{c}_n \end{aligned}\right\}\ x^0_\alpha \in \Gamma^0_T \tag{9.83}$$

Static boundary conditions

with the remark that it is still impossible to prescribe the boundary value of V_n independent of the derivative of M_{nt}.

9.3.3 Interpretation of Static Quantities

The interpretation of the static quantities becomes easier than for the moderately nonlinear plate because we may let the deformed plate coincide with the undeformed one in Figs. 9.5–9.9. Therefore, this task is not performed here, but is left for the reader.

Ex 9-1 Linear Plate Example

Here, we shall investigate one of the simple examples of linear bending of a plate loaded transversely. In this case we need not concern ourselves with the in-plane boundary conditions except that we must support the plate such that it does not move in its own plane. When we assume that all four sides of the plate are simply supported we get the boundary conditions shown in Fig. Ex. 9-1.1. As regards loading of the plate we shall limit ourselves to the case of a uniform load whose intensity is $\bar{p} =$ const., see Fig Ex. 9-1.2.

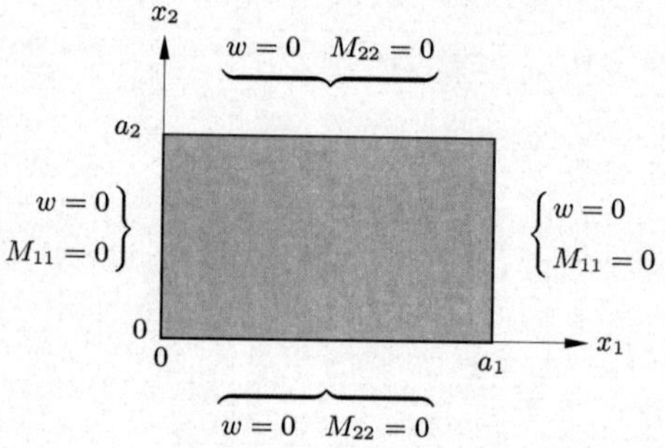

Fig. Ex. 9-1.1: *Plate geometry, coordinate system and boundary conditions.*

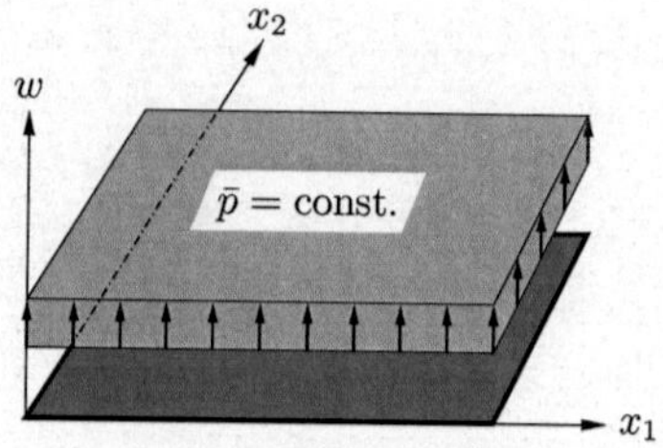

Fig. Ex. 9-1.2: *Plate and applied load.*

To make it simple, we shall assume isotropy with the bending stiffness matrix, given by the product of D^B and $[D^H]$, see (9.46) and (9.45), respectively.

Furthermore, we shall take the material properties to be independent of x_α. Then, if we use our experience from beam bending we would probably expect the solution for w to be straightforward and consist of polynomials in x_α. Even under these very simple conditions this proves to be quite far from the truth, as we shall below.

Analyses similar to the one below may be found in many other textbooks, such as the classic one by Timoshenko & Woinowsky-Krieger (1959),[9.24] although the present derivations differ somewhat.

[9.24] According to correspondence in 2011 between the bookstore at the Technical University of Denmark and McGraw-Hill publication of this book will has been discontinued. To me, this sounds improbable, but maybe nobody reads older books.

First, let us note that because of the kinematic linearity the differential equation (9.75) simplifies to

Linear differential equation for w

$$D^B \nabla^4 w = \bar{p}_w \,, \; x^0_\alpha \in A^0 \qquad (Ex.\ 9\text{-}1.1)$$

Ex 9-1.1 The Series Solution by Navier

Series solution by Navier

The sines satisfy all boundary conditions and they are mutually orthogonal ⇒ only one equation with one unknown per coefficient

The classic approach to bending of a plate with constant material properties is to expand the transverse displacement w in trigonometric functions. For the simply supported plates these functions are sines. The main reasons for doing this are that each term satisfies all boundary conditions and that when the material properties of the plate are independent of x_α, and therefore the the ensuing set of equations turn out to decouple such that the coefficients of the expansion are all given by equations with only one unknown. Especially before computers became an everyday tool this was highly appreciated.

It may not be obvious from the derivations in Ex 9-1.1.1 that the orthogonality of the sines is exploited, but it is utilized behind our backs, and the orthogonality becomes clear in Ex 9-1.1.2.

Basically, the present problem is very much like the problem of torsion of a rectangular cross-section, see Example Ex 13-6.1. Here, however, we shall discuss more possible ways of establishing the equations to determine the coefficients of the trigonometric terms.

With the definitions

$$\tilde{\jmath} \equiv \frac{j\pi}{l_1} \quad \text{and} \quad \tilde{k} \equiv \frac{k\pi}{l_2} \qquad (Ex.\ 9\text{-}1.2)$$

we may easily see that all static as well as all kinematic boundary conditions are satisfied by functions w_{jk} given by

$$w_{jk} = \sin(\tilde{\jmath} x_1) \sin(\tilde{k} x_2), \; (j,k) = 1, 2, \ldots, \infty \qquad (Ex.\ 9\text{-}1.3)$$

while they violate the differential equation (*Ex. 9-1.1*).

Navier assumes a solution in the form

$$w \approx \widetilde{w} = \sum_{j=1}^{\infty} \sum_{k=1}^{\infty} v_{jk} \sin(\tilde{\jmath} x_1) \sin(\tilde{k} x_2) \qquad (Ex.\ 9\text{-}1.4)$$

where v_{jk} denote constants which are determined by the method. If we plug the term associated with v_{jk} into the right-hand side of (*Ex. 9-1.1*) we may get

$$D^B \left(\tilde{\jmath}^2 + \tilde{k}^2\right)^2 \sin(\tilde{\jmath} x_1) \sin(\tilde{k} x_2) \qquad (Ex.\ 9\text{-}1.5)$$

Obviously, no such term matches the right-hand side of (*Ex. 9-1.1*), but there are several possibilities to cope with this problem. One is to expand the right-hand side in terms of the same sines, that is write a double *Fourier Series* of $\bar{p}$, see e.g. (Kreyszig 1993).

Ex 9-1.1.1 Solution by Use of Fourier Series

In general,

Expand the load in sines

$$\bar{p} = \sum_{j=1}^{\infty}\sum_{k=1}^{\infty} B_{jk}\sin(\tilde{\jmath}x_1)\sin(\tilde{k}x_2) \tag{Ex. 9-1.6}$$

with

$$B_{jk} = \frac{4}{l_1 l_2}\int_0^{l_2}\int_0^{l_1}\bar{p}\sin(\tilde{\jmath}x_1)\sin(\tilde{k}x_2)dx_1dx_2 \tag{Ex. 9-1.7}$$

Since $\bar{p}$ is constant in our case[9.25] this becomes[9.26]

$$B_{jk} = \frac{4\bar{p}}{\tilde{\jmath}\tilde{k}l_1l_2}\left(1-(-1)^j\right)\left(1-(-1)^k\right) \tag{Ex. 9-1.8}$$

where we may note that $B_{jk} = 0$ when j or k is (or both are) even. By the way, it is quite obvious that even terms of the expansion of $\bar{p}$ must vanish because the sines for even values are antisymmetric about the midpoint of the plate.

Insert (*Ex. 9-1.5*) and (*Ex. 9-1.6*) with (*Ex. 9-1.8*) into (*Ex. 9-1.1*) and get

$$\begin{aligned} &D^B\sum_{j=1}^{\infty}\sum_{k=1}^{\infty} v_{jk}\left(\tilde{\jmath}^2+\tilde{k}^2\right)^2\sin(\tilde{\jmath}x_1)\sin(\tilde{k}x_2) \\ &= \frac{4\bar{p}}{l_1l_2}\sum_{j=1}^{\infty}\sum_{k=1}^{\infty}\frac{1}{\tilde{\jmath}\tilde{k}}\left(1-(-1)^j\right)\left(1-(-1)^k\right)\sin(\tilde{\jmath}x_1)\sin(\tilde{k}x_2)\end{aligned} \tag{Ex. 9-1.9}$$

In order for this to hold for all values of x_α it is necessary that

$$v_{jk} = \frac{4\left(1-(-1)^j\right)\left(1-(-1)^k\right)}{\left(\tilde{\jmath}\tilde{k}\right)\left(\tilde{\jmath}^2+\tilde{k}^2\right)^2}\frac{\bar{p}}{D^Bl_1l_2} \tag{Ex. 9-1.10}$$

or, realizing that only odd terms enter

$$v_{jk} = \frac{16}{\left(\tilde{\jmath}\tilde{k}\right)\left(\tilde{\jmath}^2+\tilde{k}^2\right)^2}\frac{\bar{p}}{D^Bl_1l_2}\,,\quad (j,k) = 1,3,\ldots,\infty \tag{Ex. 9-1.11}$$

Expressed in terms of j and k only, the expression for w becomes

$$w(x_\alpha) = \frac{16\bar{p}}{\pi^6D^B}\sum_{j=1,3}^{\infty}\sum_{k=1,3}^{\infty}\frac{\sin\left(j\pi\dfrac{x_1}{l_1}\right)\sin\left(k\pi\dfrac{x_2}{l_2}\right)}{(jk)\left(\left(\dfrac{j}{l_1}\right)^2+\left(\dfrac{k}{l_2}\right)^2\right)^2} \tag{Ex. 9-1.12}$$

[9.25] Saint-Venant treats the general case of $\bar{p}$ depending of the coordinates, but our aim is more limited.

[9.26] For practical reasons we keep j and k along with $\tilde{\jmath}$ and $\tilde{k}$.

After proper differentiations of w and use of the constitutive model we may get

$$M_{11}(x_\alpha) = \frac{16\bar{p}}{\pi^4} \sum_{j=1,3}^{\infty} \sum_{k=1,3}^{\infty} \frac{\left(\frac{j}{l_1}\right)^2 + \nu\left(\frac{k}{l_2}\right)^2}{(jk)\left(\left(\frac{j}{l_1}\right)^2 + \left(\frac{k}{l_2}\right)^2\right)^2} \times \sin\left(j\pi\frac{x_1}{l_1}\right) \sin\left(k\pi\frac{x_2}{l_2}\right) \quad \text{(Ex. 9-1.13)}$$

which may provide the expression for M_{22} by moving the factor ν from the second to the first term in the numerator. For completeness the formula for M_{12} is given here

$$M_{12}(x_\alpha) = \frac{8\bar{p}(\nu-1)}{\pi^4} \sum_{j=1,3}^{\infty} \sum_{k=1,3}^{\infty} \frac{\cos\left(j\pi\frac{x_1}{l_1}\right) \cos\left(k\pi\frac{x_2}{l_2}\right)}{(jk)\left(\left(\frac{j}{l_1}\right)^2 + \left(\frac{k}{l_2}\right)^2\right)^2} \quad \text{(Ex. 9-1.14)}$$

In the present case the work done by the load $\bar{p}$ is easily computed by integration of w as given by (*Ex. 9-1.12*). It is

$$Work = \frac{64\bar{p}l_1l_2}{\pi^8 D^B} \sum_{j=1,3}^{\infty} \sum_{k=1,3}^{\infty} \frac{1}{(jk)^2\left(\left(\frac{j}{l_1}\right)^2 + \left(\frac{k}{l_2}\right)^2\right)^2} \quad \text{(Ex. 9-1.15)}$$

With (*Ex. 9-1.12*), (*Ex. 9-1.13*) and (*Ex. 9-1.15*) in hand we may compute the value of $w^M = w(l_1/2, l_2/2)$ and $M_{11}^M = M(l_1/2, l_2/2)$ as well as the work done by the for a square plate. The convergence of these three quantities is studied in Fig. Ex. 9-1.3, which shows that all three quantities as well as the value of the displacement exhibit a strong convergence and that only very few terms are needed in order to get a sufficient accuracy. It is, however, also obvious that the value of the displacement converges much faster than that of the bending moment. This is no surprise because the expression for the bending moment entails differentiating the displacement twice and thus accuracy must be lost. The work converges even stronger than the displacement because it is computed by integration of the displacement. The difference in rate of convergence is obvious from (*Ex. 9-1.12*), (*Ex. 9-1.13*) and (*Ex. 9-1.15*) where the expression for the bending moment is proportional to the square of the number of the term compared with the expression for the displacement. Conversely, the expression for the work is inversely proportional to the number of the term compared to the expression.

Very few sines are needed for engineering accuracy. Work converges faster than displacement which converges faster than moment

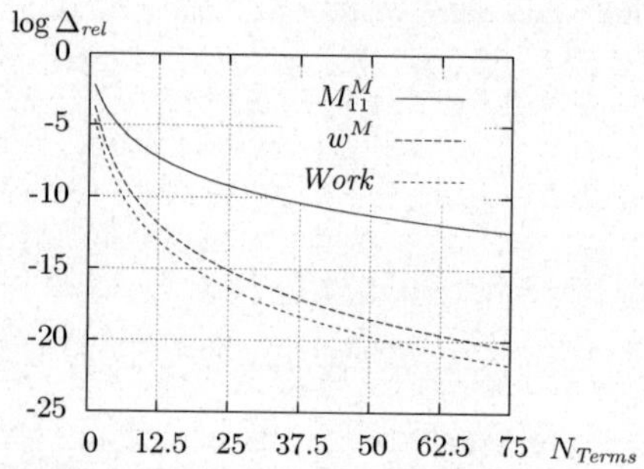

Fig. Ex. 9-1.3: *Relative error on bending moment M_{11} and of displacement w at the midpoint as well as the work done by the load of a square plate. Fourier series approximation. $\nu = 0.3$.*[a]

[a] The values obtained by use of about 10,000 terms are taken to be exact.

Ex 9-1.1.2 Solution by Use of Potential Energy

If we, instead of using Fourier Series, establish the potential energy $\Pi_P(w)$ of the plate

$$\Pi_P(w) = \tfrac{1}{2}D^B \int_0^{l_2}\int_0^{l_1}\Big(w_{,11}^2 + 2w_{,12}^2 + w_{,22}^2 + \nu(w_{,11}w_{,22} - 2w_{,12}^2)\Big)dx_2dx_1 - \int_0^{l_2}\int_0^{l_1} \bar{p}w dx_2dx_1 \qquad (Ex.\ 9\text{-}1.16)$$

we may realize that the assumption (*Ex. 9-1.4*) results in terms containing integrals of the type

$$\int_0^{l_1} \sin\left(j\pi\frac{x_1}{l_1}\right)\sin\left(m\pi\frac{x_1}{l_1}\right)dx_1 \qquad (Ex.\ 9\text{-}1.17)$$

with similar integrals in the x_2-direction. It is clear that these integrals vanish unless $j = m$, and therefore the system of equations to determine v_{jk} simplify considerably in that every v_{jk} may be determined from an equation with just one unknown. Thus, the orthogonality of the terms from (*Ex. 9-1.4*) is clear. Since the solution is the same as the one found in Example Ex 9-1.1.1 it is not rederived here, but the idea of a formulation in terms of a potential energy is utilized in Example Ex 9-1.3.

Ex 9-1.2 Single Series Approximation

Apparently Lévy (1899) was the first to apply the single series approximation

$$w(x_\alpha) = \sum_{j=1}^{\infty} Y_j(x_2) \sin\left(j\pi \frac{x_1}{a_1}\right) \qquad (Ex.\ 9\text{-}1.18)$$

to the problem of plate bending. The method is very much like the single series approximation utilized by Saint-Venant to solve the problem of torsion of a bar with rectangular section, see Ex 13-6.2.

The advantage of this method over the double series approximation is that the functions $Y_j(x_2)$ may be determined for each value of j—they turn out to entail hyperbolic sines and cosines—before the sine series is used. In the old days when the only available tools for computation of values of new expressions were tables of functions, such as logarithms and trigonometric and hyperbolic functions it was important to limit the number of numerical operation, while it mattered less if the analytical derivations became very involved. Today, this consideration is not very important because computers are very good at performing repetitive tasks. Therefore we shall not explore this method further, but you may get an idea of how it works from Ex 13-6.2 where a similar method is applied to the problem of torsion.

Few terms needed with single series approximation. Rather heavy mathematical manipulations

Ex 9-1.3 Polynomial Approximation

As mentioned above, orthogonality between terms of the displacement assumption (*Ex. 9-1.4*) is a desirable feature. On the other hand, it is a quality which comes at a price, namely that the higher the number of the term the more wavy it gets. Thus, while the exact solution is very smooth the approximate one given by sines always contains many shortwave terms. Having introduced the sine with, say 17 half-waves, part of the obligation of many of the next sines is to try to smooth the 17 half-waves, and so on *ad infinitum*. It may, therefore, be an obvious idea to look for other types of functions as basis for our approximation. The simplest functions are polynomials

Too many shortwave terms in solution in sines. Try polynomials

$$w \approx \widetilde{w} = \sum_{j=1}^{\infty}\sum_{k=1}^{\infty} v_{jk}^{s} \left(\xi_1(1-\xi_1)\right)^j \left(\xi_2(1-\xi_2)\right)^k \qquad (Ex.\ 9\text{-}1.19)$$

with

$$\xi_\alpha \equiv \frac{x_\alpha}{l_\alpha}\ , \text{ no sum over } \alpha = (1,2) \qquad (Ex.\ 9\text{-}1.20)$$

These polynomials suffer, unfortunately, from being far from orthogonal, meaning that we get a fully populated coefficient matrix of the system of equation that determines v_{jk}^{s}. If, however, we only need far fewer terms than was the case for the expansion in sines, then this does not matter.

The seemingly erratic convergence of the values of the displacement and the bending moment, see Fig. Ex. 9-1.4, does not show up in the

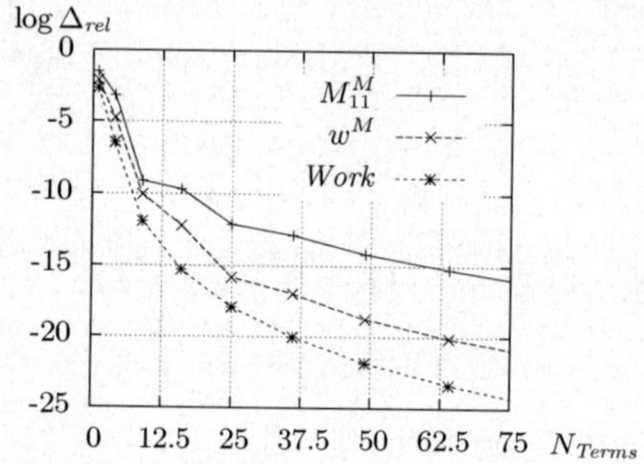

Fig. Ex. 9-1.4: *Relative error on bending moment M_{11} and displacement w at the midpoint as well as the work done by the load of a square plate. Polynomial approximation given by (Ex. 9-1.20).* $\nu = 0.3$.

curve associated with the work done by the load. The reason for this difference is that the work is a global quantity while the other two are given in a point. It is fair to say the the accuracy obtained by rather few sines is sufficient for all practical purposes and that the polynomial approximation is unable to compete in this case.

Another, potentially better, choice of polynomials satisfies the static boundary conditions of the simply supported plate is

$$w \approx \widetilde{w} = \sum_{j=1}^{\infty} \sum_{k=1}^{\infty} v_{jk}^b (\xi_1^4 - 2\xi_1^3 + \xi_1)^j (\xi_2^4 - 2\xi_2^3 + \xi_2)^k \quad \text{(Ex. 9-1.21)}$$

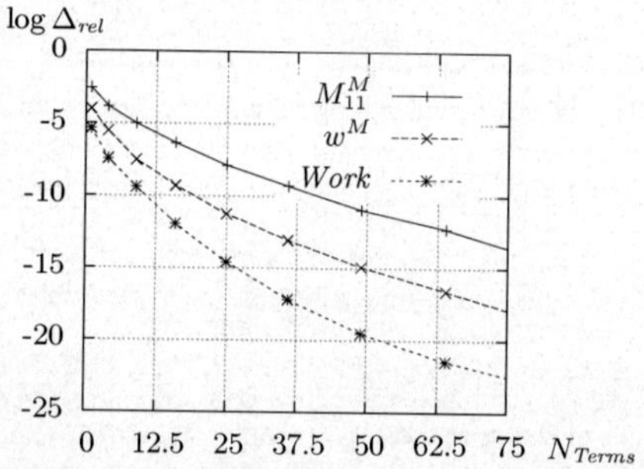

Fig. Ex. 9-1.5: *Relative error on bending moment M_{11} and displacement w at the midpoint as well as the work done by the load of a square plate. Polynomial approximation given by (Ex. 9-1.19).* $\nu = 0.3$.

There is a penalty to pay in that the terms are more complicated than in (*Ex. 9-1.20*), but much of the effort lies in solving the ensuing systems of equations, so the extra computer time may turn out to be quite small, after all.

Luckily, even for the same number of unknowns the simple polynomials provide better accuracy which may be seen by comparing Figs. Ex. 9-1.4 and Ex. 9-1.5. The reason for this is that satisfying the static boundary conditions constrains the solution too much, see also Example Ex 32-1. Given a fixed number of terms it is often the case that it does not pay to satisfy all boundary conditions exactly—here, it is, of course, mandatory that we satisfy the kinematic ones.

In general, don't satisfy all boundary conditions exactly, only the ones demanded by the variational principle

There may be a lesson to learn from this, namely that only if you have very good reasons for it should you satisfy other boundary conditions than the ones demanded by the variational principle because too much computational attention may be wasted by doing it.

9.4 Kinematically Linear vs. Nonlinear Plate Theory

Many factors determine whether it is necessary to apply a nonlinear theory

Whether you should apply a nonlinear or a linear plate theory depends on several circumstances. If the plate is subjected to a combination of transverse and in-plane loads it is usually a good idea to apply a kinematically nonlinear theory if the plate is "thin." Another factor which enters the question is the structural material of the plate. If the plate is made of steel or another metal, it is usually fairly thin, and a nonlinear theory may be needed. The same may hold in the case of wooden plates. Concrete plates are usually fairly thick and it is rarely necessary to apply a nonlinear theory for their analysis. On the other hand, today high strength concrete is used more and more frequently and plates made of such a material may be so thin that they require a nonlinear analysis.[9.28]

[9.28] Time-dependent effects such as creep and relaxation are very important for the behavior of many concrete structures and may interact heavily with kinematic nonlinearities with the outcome that a seemingly good-natured concrete structure fails some time—possibly years—after load has been applied.

Since time-dependency is not covered in this book I shall not go further into that subject but refer the reader to other literature. It may be necessary to caution that much literature on behavior of concrete is written in a way which seems alien to people used to mechanics.

Part III

Beams with Cross-Sections and Plates with Thickness

Chapter 10

Introduction to "Beams with Cross-Sections"

In Part II beams and bars were introduced as one-dimensional structural elements because in that way the theories for them could be established in a systematic way. At that point we were not concerned with whether the beams and bars had cross-sections—we just assumed that we knew the constitutive relations between the generalized stresses and the generalized strains. For a Bernoulli-Euler beam they are the generalized stresses N and M and the generalized strains ε and κ, while for a Timoshenko beam the generalized stresses are N, V and M and the generalized strains are ε, φ and κ . So, we might say that we considered beams and bars "without cross-sections." On the other hand, when we wish to analyze a "real" structure we need a way to determine the constitutive behavior of the structural elements. For linearly elastic beams this entails setting up formulas for EA, GA_e and EI for Timoshenko beams and EA and EI for Bernoulli-Euler beams. For perfectly plastic or elastic-plastic beams and bars other constitutive parameters enter the picture.

Determination of constitutive relations for beams

Since the majority of structural analyses assumes linear elasticity this will be our first topic here. It is, however, good to bear in mind that no material behaves linearly elastic and that it may be necessary to account for other effects such as plasticity or time-dependent behavior.[10.1]

Here, only linear elasticity

[10.1] In general, the more complicated and expensive the structure the more pertinent it is to use a more realistic material model. In designing a beam in a carport one can make do with linear elasticity, but in the case of a big concrete bridge that does not suffice.

E. Byskov, *Elementary Continuum Mechanics for Everyone*,
Solid Mechanics and Its Applications 194, DOI: 10.1007/978-94-007-5766-0_10,

Chapter 11

Bending and Axial Deformation of Linear Elastic Beam Cross-Sections

11.1 Linear Elastic Material

In Sections 7.7 and 7.8, which are concerned with plane, straight linearly elastic beams, the first with Bernoulli-Euler and the second with Timoshenko beams, respectively, we assumed the following constitutive relations, see (7.86) and (7.91), respectively

$$\begin{bmatrix} N \\ M \end{bmatrix} = \begin{bmatrix} EA & 0 \\ 0 & EI \end{bmatrix} \begin{bmatrix} \varepsilon \\ \kappa \end{bmatrix} \quad \text{and} \quad \begin{bmatrix} N \\ V \\ M \end{bmatrix} = \begin{bmatrix} EA & 0 & 0 \\ 0 & GA_e & 0 \\ 0 & 0 & EI \end{bmatrix} \begin{bmatrix} \varepsilon \\ \varphi \\ \kappa \end{bmatrix} \tag{11.1}$$

Hooke's "Law"

but, at that point we did not discuss the physical meaning of EA, GA_e and EI, except that they connected the generalized stresses and generalized strains. In this Chapter we shall attempt to determine A and I for a number of different beam cross-sections, and in Chapter 12 we do the same for A_e.

11.1.1 Purpose

In Part II the meaning of the generalized quantities was clear, but when we consider two- or three-dimensional bodies this is not necessarily the case, as we shall see.

11.1.2 Beam Cross-Section and Beam Fibers

Cut corners and apply simplified continuum mechanics

If we were to do things completely right, we should base our derivations on continuum mechanics as it was introduced in Part I and treat all beams as three-dimensional bodies subject to some kinematic or static constraints, possibly using the finite element method with three-dimensional elements. We shall, however, choose a much easier possibility, namely one which con-

E. Byskov, *Elementary Continuum Mechanics for Everyone*,
Solid Mechanics and Its Applications 194, DOI: 10.1007/978-94-007-5766-0_11,

siders the beam as consisting of "fibers," see Fig. 11.1, where one of these fibers is shown. When we do this, we do not account for strains perpendicular to the beam axis, making the derivations far simpler than in a a full three-dimensional analysis. Thus, we seek an engineering solution which is not the correct one, but one that, hopefully, is sufficiently accurate.

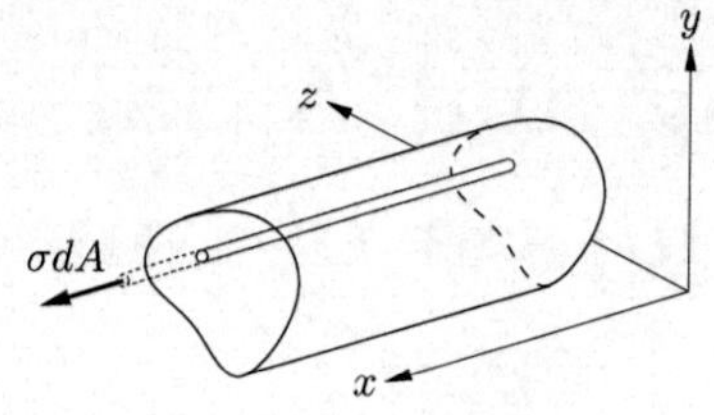

Fig. 11.1: Part of a beam with a "fiber."

Beam "fiber"

In the present connection we must not take the word "fiber" on its face value, e.g. as a wood fiber, but rather imagine that the beam consists of infinitely many, infinitely thin threads which cannot become longer or shorter without deformation of their neighbors. Before we may exploit the idea of beam fibers we must assume their constitutive behavior which we simply take as a linear relation between their axial strain ε and axial stress σ, see Fig. 11.2, where the force acting on a fiber with area dA is dP.

$$\sigma \equiv \lim_{A \to 0} \frac{P}{A} = \frac{dP}{dA} \tag{11.2}$$

Now we assume that ε

$$\sigma(x) = E(x)\varepsilon(x) \tag{11.3}$$

where it is indicated that Young's Modulus E may vary along the length of the beam. For simplicity we shall assume that E does not vary over the cross-section. Letting E vary would merely complicate the derivations and results below, but it is quite easy to account for varying Young's Moduli. In general the deformation of the fibers varies along the length

$$\sigma(x, y, z) = E(x, y, z)\varepsilon(x, y, z) \tag{11.4}$$

Assume $E =$ const. $\Rightarrow$ Reinforced concrete beams not covered

We shall, however, limit ourselves to the case of a constant Young's Modulus

$$\sigma(x, y, z) = E\varepsilon(x, y, z)\,,\ E = \text{const.} \tag{11.5}$$

thus excluding for example reinforced concrete beams since the reinforcement bars and the concrete have very different Young's Moduli.

You may have observed that the coordinates here are denoted x, y and z instead of x_j , $j = 1, 2, 3$ which is the habit in the major part of this book. The main reason for abandoning the latter notation which for most purposes is the more systematic one is that, usually the coordinate along the beam axis is called x. The advantage of using x_1 and x_2 instead of y and z, respectively, to indicate the axes perpendicular to the beam axis would be minor, as you may see from the derivations below, since we are only considering bending about the z-axis.[11.1] Then, the two coordinates in the cross-section play different roles which makes the x_α-notation less obvious. Finally, the notation used here is the traditional one—not a very strong argument in my opinion.

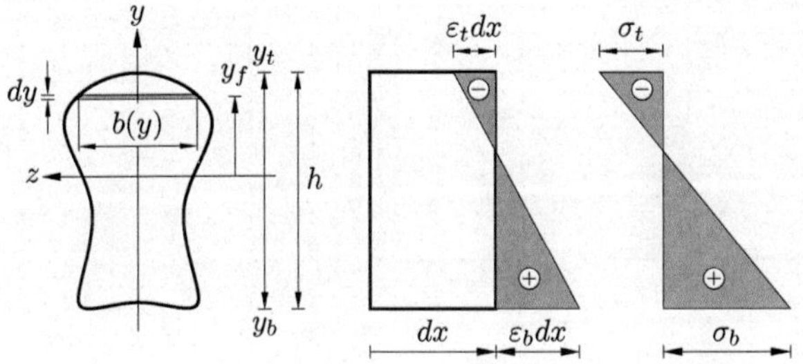

Fig. 11.2: Part of a beam with a "fiber" stretching over the width of the beam.

Behind the assumption that the beam does not deflect in the direction of the z-axis lies the requirements that the beam cross-section is symmetric with respect to the y-axis and that the loading exhibits the same symmetry.

11.1.3 Pure Axial Strain

Axial strain ε
Axial force N

Assume that all beam fibers are subjected to the same strain $\varepsilon =$ const. and that $E =$ const. everywhere, then the fiber stress σ is independent of all coordinates. Therefore, it is easy to integrate the fiber stress. We expect to find $N = EA\varepsilon$, see (11.1), and consequently we call the integral N

$$N = \int_A \sigma dA = \int_A E\varepsilon dA = \varepsilon \int_A E dA = \varepsilon E \int_A dA$$
$$N = EA\varepsilon \, , \; EA = \text{const.} \tag{11.6}$$

Thus, we have found the relationship between the axial strain and the axial force of the beam. Now, we turn to the more interesting case of curvature (bending) strain as well as axial strain.

[11.1] You might ask why I don't cover bending about two axes. The reason is that I want to address the most basic issues and Part II is only concerned with in-plane bending.

11.1.4 Both Axial and Curvature Strain in Bernoulli-Euler Beams

Axial strain ε
Curvature strain κ
Axial force N
Bending moment M

In Section 7.5 we developed the kinematically linear theory for straight Bernoulli-Euler beams. This theory entails the axial strain ε and the curvature strain κ as its generalized strains. The axial force N and the bending moment M are their work conjugate generalized stresses, see e.g. Section 7.5 and Section 7.3, which covers the more general case of moderate kinematic nonlinearity. We shall therefore seek expressions connecting κ and M in addition to the relation between ε and N. As shown in Fig. 11.2 and Fig. 11.3

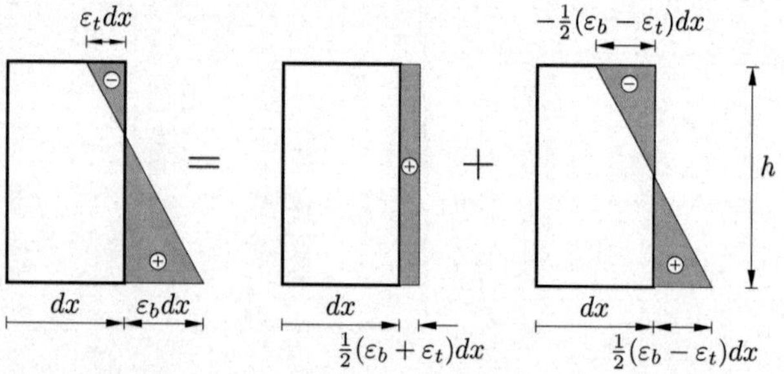

Fig. 11.3: Beam strain divided into axial and bending contributions.

we assume that a plane cross-section remains plane after deformation. It is not shown in those figures that we insist that the cross-sections are perpendicular to the beam axis, both before and after deformation, but this is a consequence of the restriction of the Bernoulli-Euler hypothesis of vanishing shear strains, see Sections 7.4 and 7.6.

Traditionally, it is presumed that the cross-sections do not shrink or expand, but this is not relevant when we base our derivations on the idea of fibers. It is, of course, important when continuum mechanics is used, see also Example Ex 12-1 where it is discussed how a Bernoulli-Euler beam might be loaded.

When we consider small deformations and observe the above limitations as regards the deformation of the cross-sections we may find that the axial displacement of the beam fibers u can be described by a bilinear expression

Linear displacement

$$u(x, y) = c_x(x)(c_0 + c_y y) \tag{11.7}$$

where we note that the axial displacement does not vary with z. Then,

Linear strain

$$\varepsilon_f(x, y) = \frac{du(x, y)}{dx} = \frac{dc_x(x)}{dx}(c_0 + c_y y) \tag{11.8}$$

where ε_f denotes the axial fiber strain.

At this point we need to relate the generalized strains ε and κ to $\varepsilon_f(x, y)$. In Fig. 11.3 the fiber strain is divided into two parts, namely one which is constant and one which covers the rest. It should come as no surprise that we intend to identify the constant part as the axial beam strain ε, while we hope that the rest is closely related to the beam curvature strain κ. Based on Fig. 11.3 we may find

$$\varepsilon = \tfrac{1}{2}(\varepsilon_b + \varepsilon_t) \tag{11.9}$$

As regards the curvature strain we must go through a somewhat more complicated path. Originally, we defined the curvature strain κ as the derivative of the change in angle of the beam axis, see Section 7.2, (7.9)

$$\kappa \equiv \frac{d\omega}{dx} \tag{11.10}$$

Nonlinear curvature measure

and in the case of a straight, linearly or moderately linear, see Section 7.3, (7.14), beam we found it to be

$$\kappa = \frac{d^2w}{dx^2} = \frac{d}{dx}\left(\frac{dw}{dx}\right) \tag{11.11}$$

Linear curvature measure

which is a linear expression for the curvature strain as the change in angle per length of the beam. The Bernoulli-Euler conditions force the angle of the cross-section to be the same as the angle of the beam axis and thus we may find the curvature strain as the change in angle of the cross-section per length of the beam. Then from Fig. 11.3

$$\kappa_{zz} = \frac{(\varepsilon_b - \varepsilon_t)dx}{dx}\frac{1}{h} = \frac{(\varepsilon_b - \varepsilon_t)}{h} \tag{11.12}$$

Curvature strain in terms of fiber strains

where we have indicated that the curvature is about the z-axis.

In (11.9)–(11.12) it is not explicitly stated that the strains may vary with the axial coordinate x. In order to determine the "fiber stress" $\sigma_f(x, y)$, and because of the later computation of the cross-sectional constants A, S and I, we wish to express the fiber strain $\varepsilon_f(x, y)$ in terms of the generalized beam strains $\varepsilon(x)$ and $\kappa(x)$

$$\varepsilon_f(x, y) = \varepsilon(x) - y\kappa_{zz}(x) \tag{11.13}$$

Fiber strain in terms of axial and bending strain

which is seen to be of the same form as (11.8). The minus sign in front of the second term is due to the fact that a positive value of κ results in a negative contribution to the fiber strain in the upper part of the cross-section.

11.1.5 Axial Force, Zeroth- and First-Order Moments

Our present assumption of linear elastic behavior implies linear relations between the fiber strain ε_f and the fiber stress σ_f

$$\sigma_f(x, y, z) = E(x, y, z)\varepsilon_f(x, y) \tag{11.14}$$

Hooke's "Law"

but symmetry about the y-axis demands that

$$E(x,y,z) = E(x,y,-z) \tag{11.15}$$

For the time being we limit ourselves to the case of a Young's Modulus E which is independent of the coordinates y and z

$$\sigma_f(x,y) = E(x)\varepsilon_f(x,y) \tag{11.16}$$

Integrate the fiber stress σ_f over the cross-section to get

$$\begin{aligned} N(x) &= \int_A \sigma_f(x,y)dA = \int_A E(x)\varepsilon_f(x,y)dA \\ &= E(x)\int_A \varepsilon_f(x,y)dA = E(x)\int_A \big(\varepsilon(x) - y\kappa_{zz}(x)\big)dA \end{aligned} \tag{11.17}$$

and thus

Axial force N

$$N(x) = E(x)\big(A(x)\varepsilon(x) - S_z(x)\kappa_{zz}(x)\big) \tag{11.18}$$

where

Cross-sectional area A

$$\text{Cross-sectional area}: \quad A(x) \equiv \int_{A(x)} dA \tag{11.19}$$

and

Static moment S_z

$$\text{Static moment about the } z\text{-axis}: \quad S_z(x) \equiv \int_{A(x)} ydA \tag{11.20}$$

Zeroth-order moment = Area
First-order moment = Static moment

For the sake of completeness it may be worth mentioning that the area A is the zeroth-order moment of the cross-section, while S_y is its first-order moment about the z-axis. By the way, some authors prefer the name *first moment* or *static moment* instead of *first-order moment.*

As (11.18) shows there are two possibilities of getting a state of pure axial force. Either $\kappa = 0$ or $S_z = 0$ (or both, of course), where the first was investigated in Section 11.1.3. The other possibility puts demands on the location of the x-axis. We shall return to that possibility in Section 11.1.8.

11.1.6 Bending Moment and Second-Order Moments

Probably it is not surprising that in the general case the stresses, given by the fiber strain in (11.13), result in a bending moment. Therefore, we compute their moment $M_z(x)$ about the z-axis

$$\begin{aligned} M_z(x) &= -\int_A y\sigma_f(x,y)dA = -\int_A yE\varepsilon_f(x,y)dA \\ &= -E\int_A y\varepsilon_f(x,y)dA = -E\int_A \big(y\varepsilon(x) - y^2\kappa_{zz}(x)\big)dA \\ &= -\varepsilon(x)E\int_A ydA + \kappa_{zz}(x)E\int_A y^2dA \end{aligned} \tag{11.21}$$

Then,

$$M_z(x) = -ES_z\varepsilon(x) + EI_{zz}\kappa_{zz}(x) \tag{11.22}$$

where

Moment of Inertia I_{zz} = Second-order moment

$$\text{Moment of inertia about the } z\text{-axis:} \quad I_{zz} \equiv \int_A y^2 dA \tag{11.23}$$

and where we observe that the minus sign on the first term on the left-hand side of (11.22) is caused by the fact that a positive value of σ_f for $y > 0$ results in a negative contribution to the bending moment M_z. The quantity I_{zz} is the second-order moment about the z-axis of the cross-sectional area.[11.2]

As seen from (11.18) and (11.22) the stress-strain relations are not of the same form as postulated in (11.1a), originally (7.86), in that the terms with S_z do not appear there. The reason is that when we introduced (11.1a) and (7.86) the beam did not have depth and therefore the only natural choice of beam axis was the position of the one-dimensional beam itself. In Section 11.1.8 we look into the issue of the position of the beam axis.

11.1.7 Summary of Linear Elastic Stress-Strain Relations

In view of the fact that we have confined ourselves to plane beams we introduce a simpler notation for the bending moment, the static moment and for the moment of inertia. The reason for using a more complicated notation was that we wished to make it clear about which axes we were taking the moments.

Therefore, introduce

Summary of cross-sectional properties and of linear elastic stress-strain relations

$$\begin{aligned}
&\text{Cross-sectional area:} && A \equiv \int_A dA \\
&\text{Static moment about the } z\text{-axis:} && S = S_z \equiv \int_A y dA \\
&\text{Moment of inertia about the } z\text{-axis:} && I = I_{zz} \equiv \int_A y^2 dA \qquad (11.24) \\
&\text{Axial force:} && N \equiv \int_A \sigma_f dA \\
&\text{Bending moment about the } z\text{-axis:} && M = M_{zz} \equiv \int_A \sigma_f y dA
\end{aligned}$$

[11.2] It ought to be obvious that the cross-section is associated with one more second-order moment similar to I_{zz}, namely the second-order moment $I_{yy} \equiv \int_A z^2 dA$ about the y-axis. In addition to this there exists a mixed second-order $I_{yz} \equiv \int_A yz dA$. Since we are only concerned with bending that is symmetric about the y-axis we do not need these other second-order moments.

With this notation we get the following constitutive conditions according to *Hooke's "Law"*

Hooke's "Law"

$$\begin{bmatrix} N \\ M \end{bmatrix} = \begin{bmatrix} EA & -ES \\ -ES & EI \end{bmatrix} \begin{bmatrix} \varepsilon \\ \kappa \end{bmatrix} \tag{11.25}$$

Invert this relationship and get

Hooke's "Law"

$$\begin{bmatrix} \varepsilon \\ \kappa \end{bmatrix} = \frac{1}{E(AI - S^2)} \begin{bmatrix} I & S \\ S & A \end{bmatrix} \begin{bmatrix} N \\ M \end{bmatrix} \tag{11.26}$$

which we intend to utilize in Section 11.1.8 when we discuss the placement of the x-axis.

11.1.8 Cross-Sectional Axes—Beam Axis and Center of Gravity

Axes of the cross-section

In the above derivations we chose the position of the x-axis completely arbitrarily. If we can choose the axis such that the first-order moment S vanishes, then Hooke's Law simplifies, see (11.25) and (11.26). More often than not, this proves to be a fairly good idea,[11.3] in part because it appears a reasonable request that it should be possible to apply an axial loading without causing bending of the beam and, in the same spirit, to apply a moment load without stretching of the beam axis. These assumptions lie behind the derivations of the strictly one-dimensional beam theory, see Sections 7.5, 8.1 and 8.2 where we, without any kind of discussion, identified the beam axis as the physical configuration of the one-dimensional beam. As we may see from (11.26) we can get both of the above wishes fulfilled if the static moment S vanishes. This condition may be used to fix the position of the beam axis with the implication that for pure bending the fiber stress σ_f of the beam axis equals zero. Then the beam axis and the so-called *neutral axis* coincide. By the same token this means that the neutral axis is the one which does not experience axial strain under pure moment loading.

Neutral axis

11.1.9 The Beam Axis at the Neutral Axis

In the following we shall discuss some examples of determining the moments of various order of some types of cross-sections. In all these cases we choose the beam axis and the neutral axis to coincide. Then,

Hooke's "Law" for beam axis = neutral axis

$$\begin{bmatrix} N \\ M \end{bmatrix} = \begin{bmatrix} EA & 0 \\ 0 & EI \end{bmatrix} \begin{bmatrix} \varepsilon \\ \kappa \end{bmatrix} \tag{11.27}$$

[11.3] Saying that it is a good idea to let the beam axis be the same as the neutral axis may be a too strong statement because in the case of interconnected beams it may indeed be better to choose a common axis which may not cause S to vanish in any of the beams.

The inverse relationship is

$$\begin{bmatrix} \varepsilon \\ \kappa \end{bmatrix} = \begin{bmatrix} 1/(EA) & 0 \\ 0 & 1/(EI) \end{bmatrix} \begin{bmatrix} N \\ M \end{bmatrix} \tag{11.28}$$

Hooke's "Law" for beam axis = neutral axis

At this point we may remark that it is quite easy to find the cross-sectional properties associated with other axes than the neutral axis once we have determined the ones related to that axis, see (11.32).

11.1.10 Independence of Results of Choice of Beam Axis

Strains and stresses must not depend on choice of beam axis

I gather that everyone would consider it unfortunate if the results as regards the stress and strain distribution over the cross-section depended on the choice of beam axis. If that were the case, then the physical results would depend on the choice of coordinate system which must be deemed unacceptable. In that case, the description would not be *objective* and the situation would be similar if the speed of a car depended on the coordinate system. Our results must be *frame indifferent*. We shall see that our description is objective and that our solutions are frame indifferent. In Fig. 11.4

Results must be frame indifferent

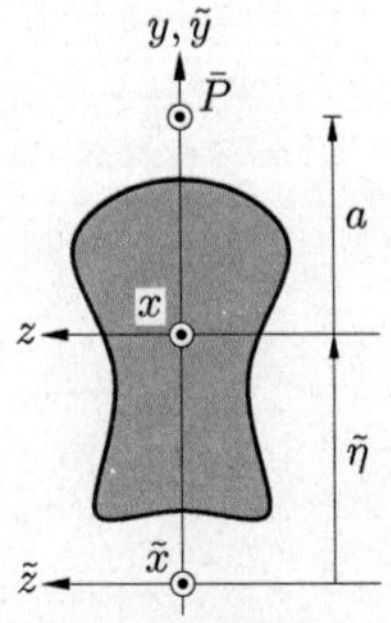

Fig. 11.4: Beam axes.

the beam axis, the $\tilde{x}$-axis lies at $\tilde{\eta} > 0$ below the neutral axis, the x-axis. As indicated, the load consists of a force $\bar{P}$ at $y = a > 0$. This means that we cover loadings that consist of a force $\bar{P}$ at the center of gravity in addition to a moment, a couple $\bar{C}$. Of course, a situation with pure bending deserves another treatment, but, as you might observe, we can describe that situation by letting $a\bar{P} = -\bar{C}$ and at the same time let $\bar{P} = 0$ in the formulas below. Applying $\bar{P}$ at $y = a$ along with $-\bar{P}$ at $y = -a$ produces a a case with no resulting axial force, but with a prescribed couple $\bar{C} = -2\bar{P}a$. Thus, we may reuse the formulas below by simple addition.

From Fig. 11.4

$$\tilde{y} = y + \tilde{\eta} \tag{11.29}$$

where tilde ($\tilde{}$) indicates that the quantity is referred to the $(\tilde{x}, \tilde{y}, \tilde{z})$-coordinate system.[11.4]

When we refer the axial force and bending moment to the two axes we may see from Fig. 11.4 that

$$\begin{aligned} N &= \bar{P}\,, \quad M = -a\bar{P} \\ \widetilde{N} &= \bar{P}\,, \quad \widetilde{M} = -(a + \tilde{\eta})\bar{P} \end{aligned} \tag{11.30}$$

Utilize (11.24) to find the cross-sectional properties that refer to the z-axis

$$\begin{aligned} A &= \int_A dA \\ S &= \int_A y dA = 0 \\ I = I_{zz} &= \int_A y^2 dA \end{aligned} \tag{11.31}$$

In the same spirit the properties associated with the $\tilde{z}$-axis

Cross-sectional properties referred to the y- and $\tilde{y}$-axes

$$\begin{aligned} \tilde{A} &= \int_A dA = A \\ \tilde{S} = S_{\tilde{z}} &= \int_A \tilde{y} dA = \int_A (y + \tilde{\eta}) dA = S + \tilde{\eta} A \\ &= +\tilde{\eta} A \\ \tilde{I} = I_{\tilde{z}\tilde{z}} &= \int_A \tilde{y}^2 dA = \int_A (y^2 + 2y\tilde{\eta} + \tilde{\eta}^2) da = I + 2\tilde{\eta} S + \tilde{\eta}^2 A \\ &= I + \tilde{\eta}^2 A \end{aligned} \tag{11.32}$$

which means that the static moment $\tilde{S}$ is equal to the distance $\tilde{\eta}$ from the neutral axis multiplied by the area A and that the moment of inertia $\tilde{I}$ is equal to the moment of inertia I about the neutral axis plus the the square of the distance $\tilde{\eta}$ from the neutral axis multiplied by the area A. Later these results prove to be very valuable. From (11.32) we observe that the moment of inertia I about the neutral axis is the smallest one possible.

[11.4] In this connection the meaning of the tilde ought to be clear and must not be confused with other cases where I have used it to indicate an approximate solution.

Once we have found A, $\tilde{S}$ and $\tilde{I}$ we may determine $\tilde{\eta}$ and I

$$\begin{aligned}\tilde{\eta} &= \frac{\tilde{S}}{a} \\ \tilde{I} &= I + \left(\frac{\tilde{S}}{A}\right)^2 A \;\Rightarrow\; I = \tilde{I} - \frac{\tilde{S}^2}{A}\end{aligned} \tag{11.33}$$

Let us insert (11.30a) in (11.28) and get

$$\begin{bmatrix}\varepsilon \\ \kappa\end{bmatrix} = \begin{bmatrix}1/(EA) & 0 \\ 0 & 1/(EI)\end{bmatrix}\begin{bmatrix}\bar{P} \\ -a\bar{P}\end{bmatrix} = \begin{bmatrix}\bar{P}/(EA) \\ -a\bar{P}/(EI)\end{bmatrix} \tag{11.34}$$

Utilize (11.13) and (11.34) and obtain the following expression for the fiber strain ε_f

$$\varepsilon_f = \varepsilon - y\kappa = \frac{\bar{P}}{EA} + y\frac{a\bar{P}}{EI} \tag{11.35}$$

In a similar fashion use (11.26) and (11.32) to get

$$\begin{bmatrix}\tilde{\varepsilon} \\ \tilde{\kappa}\end{bmatrix} = \frac{1}{EAI}\begin{bmatrix}(I + \tilde{\eta}^2 A) & +\tilde{\eta}A \\ +\tilde{\eta}A & A\end{bmatrix}\begin{bmatrix}\bar{P} \\ -(a+\tilde{\eta})\bar{P}\end{bmatrix} \tag{11.36}$$

where we have exploited the fact that

$$\tilde{A}\tilde{I} - \tilde{S}^2 = AI \tag{11.37}$$

After some trivial manipulations we may get

$$\begin{bmatrix}\tilde{\varepsilon} \\ \\ \tilde{\kappa}\end{bmatrix} = \begin{bmatrix}\dfrac{\bar{P}}{EA} + \tilde{\eta}\dfrac{a\bar{P}}{EI} \\ \\ -\dfrac{a\bar{P}}{EI}\end{bmatrix} = \begin{bmatrix}\varepsilon + \tilde{\eta}\kappa \\ \\ \kappa\end{bmatrix} \tag{11.38}$$

Relation between $(\tilde{\varepsilon}, \tilde{\kappa})$ and (ε, κ)

where (11.34) has been utilized. Finally we may use (11.38) and (11.13) with $(\varepsilon_f, \varepsilon, \kappa_{zz}, y) = (\tilde{\varepsilon}_f, \tilde{\varepsilon}, \tilde{\kappa}, \tilde{y})$ to get

$$\tilde{\varepsilon}_f = \tilde{\varepsilon} - \tilde{y}\tilde{\kappa} = (\varepsilon + \tilde{\eta}\kappa) - (y + \tilde{\eta})\kappa = \varepsilon - y\kappa = \varepsilon_f \tag{11.39}$$

Fiber strain is independent of coordinate system

and thus we have shown that the choice of beam axis does not influence the physical results. Therefore the description is objective.

11.1.11 Distribution of Axial Strain and Axial Stress

Fiber strain is linear over the cross-section

In Section 11.1.10 we concluded that the fiber strain ε_f is

$$\varepsilon_f = \varepsilon - y\kappa \tag{11.40}$$

which means that the fiber strain ε is linear over the cross-section, see Fig. 11.2. Under the assumption of linear elasticity the fiber stress σ_f will also be linearly distributed over the cross-section because the constitutive model (11.16) together with (11.40) gives:

Fiber stress is linear over the cross-section

$$\sigma_f = E(\varepsilon - y\kappa) \tag{11.41}$$

If we exploit the constitutive relations (11.27) we may express the fiber stress in terms of the axial force N and the bending moment M

Navier's formula

$$\sigma_f = \frac{N}{A} - y\frac{M}{I} \tag{11.42}$$

In the literature this formula is known as *Navier's Formula* after the French engineer *Claude Louis Marie Henri Navier.*

In order to use a simpler nomenclature we shall omit the lower index $_f$ and talk about the axial strain ε and the axial stress σ in stead of ε_f and σ_f, respectively. Later, when we wish to refer to the strain and stress of the beam axis we must indicate this explicitly. A commonly used notation uses a lower index $_0$ to signify the neutral axis. If, however, we insist on using another beam axis we need another indicator.

11.1.12 Examples of Moments of Inertia

Our first example is a rectangular cross-section which is important in itself, but the results are useful in many connections, see examples Ex 11-3 and Ex 11-4.

Ex 11-1 Rectangular Cross-Section

Rectangular cross-section

The rectangular cross-section, see Fig. Ex. 11-1.1, probably is the simplest example in this connection. By z we denote the axis about which the static moment S vanishes,[11.5] while $\tilde{z}$ is an axis that we may place as we see fit. The reason it lies at the bottom of the cross-section is that that choice makes the following derivations easy.

Utilize (11.24) to get the quantities referred to the $\tilde{z}$-axis

$$\begin{aligned}
\tilde{A} &= \int_A dA = bh \\
\tilde{S} = S_{\tilde{z}} &= \int_A \tilde{y} dA = \tfrac{1}{2}bh^2 \\
\tilde{I} = I_{\tilde{z}\tilde{z}} &= \int_A \tilde{y}^2 dA = \tfrac{1}{3}bh^3
\end{aligned} \tag{Ex. 11-1.1}$$

[11.5] You may have guessed that the figure is misleading in this respect.

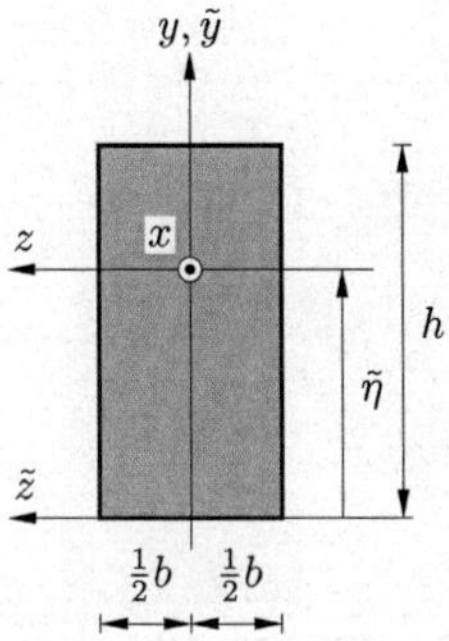

Fig. Ex. 11-1.1: *Rectangular cross-section.*

In the same fashion we may find the quantities referred to the z-axis

$$A = \int_A dA = bh$$

$$S = S_z = \int_A y dA = b \int_{-\tilde{\eta}}^{h-\tilde{\eta}} y dy = \tfrac{1}{2} b(h^2 - 2h\tilde{\eta}) \qquad \text{(Ex. 11-1.2)}$$

$$I = I_{zz} = \int_A y^2 dA = \tfrac{1}{3} b(h^3 - 3h^2\tilde{\eta} + 3h\tilde{\eta}^2 - 2\tilde{\eta}^3)$$

Area of rectangle $A = bh$

If we demand that $S = 0$, then

$$S = 0 \Rightarrow \tilde{\eta} = \tfrac{1}{2} h \qquad \text{(Ex. 11-1.3)}$$

which results in

$$I = \tfrac{1}{12} bh^3 \qquad \text{(Ex. 11-1.4)}$$

Moment if inertia of rectangle $I = \frac{1}{12} bh^3$

It is probably not surprising that in the present case the x-axis lies at the center of the cross-section, simply because the center of gravity of a rectangle is its mid-point.[11.6]

Our next example is almost as simple as Ex 11-1 and also very fundamental in that beams with circular cross-sections are used in many structures.

Ex 11-2 Circular Cross-Section

Circular cross-section

Circular cross-sections, see Fig. Ex. 11-2.1, are often used in high-voltage masts. It ought to be self-evident that the center of gravity coincides with the center of the circle. Consequently, we know immediately where we should place the x, y, z-coordinate system. We need the following simple geometric relations

$$y = R\sin\theta\,, \quad dy = R\cos\theta d\theta$$
$$dA = (2R\cos\theta)dy = 2R^2\cos^2\theta d\theta \qquad \text{(Ex. 11-2.1)}$$

[11.6] By the way, the result $I = \frac{1}{12}bh^3$ for the rectangle is assumed known to students who have passed an introductory course on statics and strength of materials.

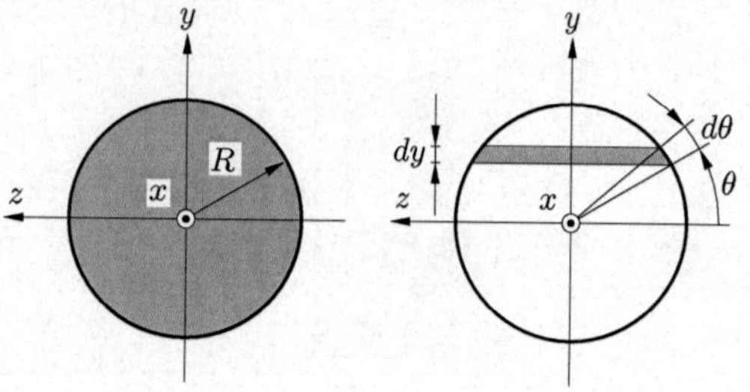

Fig. Ex. 11-2.1: *Circular cross-section.*

Once more we use (11.24) and find

Area of circle $A = \pi R^2$ Moment of inertia of circle $I = \frac{\pi}{4}R^4$

$$\begin{aligned} A &= \int_A dA = \int_{-\pi/2}^{+\pi/2} 2R^2\cos^2\theta d\theta = \pi R^2 \\ S &= \int_A y dA = \int_{-\pi/2}^{+\pi/2} 2R^3\cos^2\theta\sin\theta d\theta = 0 \\ I &= \int_A y^2 dA = \int_{-\pi/2}^{+\pi/2} 2R^4\cos^2\theta\sin^2\theta d\theta = \frac{\pi}{4}R^4 \end{aligned} \qquad (Ex.\ 11\text{-}2.2)$$

where the result (*Ex. 11-2.2a*) most likely is known in advance, and (*Ex. 11-2.2b*) is the expected.

Sometimes it is convenient to express the above results in terms of the diameter d the area A of the circle

Area of circle $A = \frac{\pi}{4}d^2$ Moment of inertia of circle $I = \frac{1}{16}Ad^2$

$$\begin{aligned} A &= \frac{\pi}{4}d^2 \\ I &= \frac{\pi}{64}d^4 = \tfrac{1}{4}AR^2 = \tfrac{1}{16}Ad^2 \end{aligned} \qquad (Ex.\ 11\text{-}2.3)$$

Our two first examples of cross-sections were characterized by their simplicity. During the analysis of the next two cross-sections we shall exploit the result from example Ex 11-1 in connection with (11.32c) to handle more complicated cases.

Ex 11-3 T-Shaped Cross-Section

T-shaped cross-section

Many structures are designed with T-shaped beam cross-sections, see Fig. Ex. 11-3.1. The reason for its popularity is that it is inexpensive and easy to inspect for corrosion, while a circular tube, for instance, may corrode from the inside without any indication of deterioration on the outside. As mentioned above, in the derivations below we shall utilize some of the result from Example Ex 11-1 and (11.32).

Obviously, there is no intuitively "correct" choice of the $(\tilde{x}, \tilde{y}, \tilde{z})$-coordinate system, except that the $\tilde{y}$-axis must respect the symmetry. However, three possibilities seem rather obvious, namely choosing the

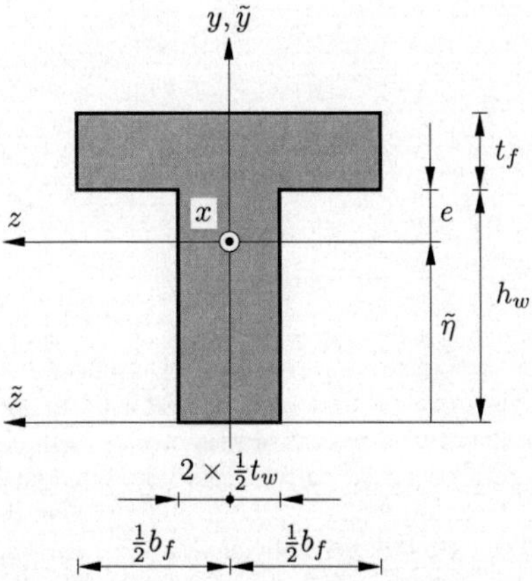

Fig. Ex. 11-3.1: *T-shaped cross-section.*

$\tilde{x}$-axis at the top of the cross-section, or, as shown in Fig. Ex. 11-3.1, at the bottom, or at the junction between web and flange.

Referring to the figure we may easily find

$$\begin{aligned} A_w &= t_w h_w \\ A_f &= t_f b_f \\ A &= A_w + A_f = t_w h_w + t_f b_f \\ S_w &= A_w\left(\tfrac{1}{2}h_w - \tilde{\eta}\right) = -h_w t_w\left(\tilde{\eta} - \tfrac{1}{2}h_w\right) \\ S_f &= A_f\left(h_w + \tfrac{1}{2}t_f - \tilde{\eta}\right) = b_f t_f\left(h_w + \tfrac{1}{2}t_f - \tilde{\eta}\right) \\ S &= S_w + S_f = A_f\left(h_w + \tfrac{1}{2}hk - \tilde{\eta}\right) + A_w\left(\tfrac{1}{2}h_w - \tilde{\eta}\right) \end{aligned} \qquad (Ex.\ 11\text{-}3.1)$$

Area of T-shaped cross-section $A = t_w h_w + t_f b_f$

where index $_w$ indicates the web and index $_f$ the flange. Demand that the static moment S about the z-axis vanishes and get

$$\begin{aligned} \tilde{\eta} &= \frac{h_w(2A_f + A_w) + t_f A_f}{2A} \\ &= \frac{t_w h_w^2 + 2b_f h_w t_f + b_f t_f^2}{2(b_f t_f + h_w t_w)} \end{aligned} \qquad (Ex.\ 11\text{-}3.2)$$

The distance e from the junction between flange and web to the centroid is then

$$e = h_w - \tilde{\eta} = \frac{h_w^2 t_w - b_f t_f^2}{2A} \qquad (Ex.\ 11\text{-}3.3)$$

and the moment of inertia I becomes

$$\begin{aligned} I = {} & A_w(\tilde{\eta} - \tfrac{1}{2}h_w)^2 + \tfrac{1}{12}t_w h_w^3 \\ & + A_f(e + \tfrac{1}{2}t_f)^2 + \tfrac{1}{12}t_f^3 b_f \end{aligned} \qquad (Ex.\ 11\text{-}3.4)$$

or

$$I = A_w(e - \tfrac{1}{2}h_w)^2 + \tfrac{1}{12}t_w h_w^3 \\ +A_f(e + \tfrac{1}{2}t_f)^2 + \tfrac{1}{12}t_f^3 b_f \qquad (Ex.\ 11\text{-}3.5)$$

or

Moment of inertia of T-shaped cross-section I

$$I = A_w(e - \tfrac{1}{2}h_w)^2 + \tfrac{1}{12}A_w h_w^2 \\ +A_f(e + \tfrac{1}{2}t_f)^2 + \tfrac{1}{12}A_f t_f^2 \qquad (Ex.\ 11\text{-}3.6)$$

Regarding the T-shaped cross-section in Example Ex 11-3 we did not assume anything about the thicknesses of the web and flange in relation to the width and depth of the cross-section. It is not difficult to derive formulas for thin-walled T-shaped cross-sections from the above formulas by realizing that in that case t_f is small compared to h_w and b_f, but I shall not do it here. It is not necessary to assume t_w small, but chances are that it is, otherwise the flanges would not contribute sufficiently.

In the next example, namely Example Ex 11-4, we shall, however, assume that the thickness of the flanges is much smaller than the depth and width of the I-shaped profile we investigate.

Ex 11-4 Thin-Walled I*-Shaped Cross-Section*

Thin-walled I-shaped cross-section

Numerous structures are designed with beams with doubly symmetric, thin-walled I-shaped cross-section because it has proved to be a rather inexpensive way of obtaining good stiffness and strength associated with bending about the z-axis. Such a cross-section has much smaller stiffness and strength associated with the y-axis, but often this is of less concern. Its torsional stiffness is also very low. Once more we uti-

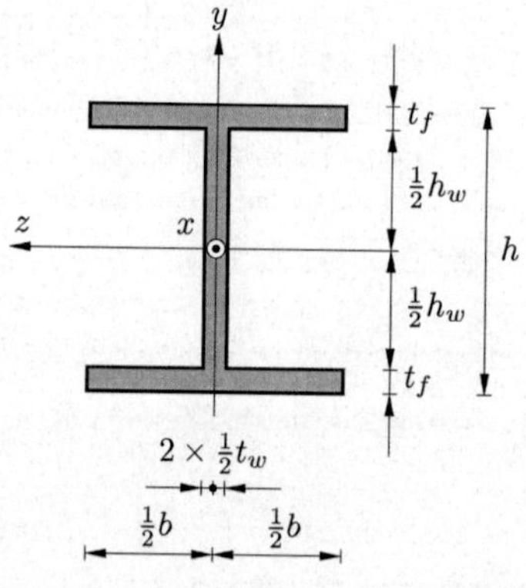

Fig. Ex. 11-4.1: I-shaped cross-section.

lize some of the results we obtained in Example Ex 11-1 and (11.32).

Without any computations we may see that the centroid must coincide with the x-axis shown in the figure. This makes the following derivations much easier. By "thin-walled" we understand that the thickness t_f of the flanges is very small in comparison with the depth h and width b of the cross-section.[11.7] In mathematical terms we require

$$t_f \ll h \Rightarrow h_w \approx h\,; \quad t_f \ll b$$
$$t_w \ll b\,; \quad t_w \ll h \qquad (Ex.\ 11\text{-}4.1)$$

Definition of "thin-walled"

Referring to Fig. Ex. 11-4.1 we may find

$$A_w = t_w h_w \approx t_w h$$
$$A_f = t_f b$$
$$A = A_w + A_f = t_w h_w + t_f b \approx t_w h + t_f b \qquad (Ex.\ 11\text{-}4.2)$$
$$I = \tfrac{1}{12} t_w h_w^3 + 2A_f\left(\tfrac{1}{2}(h_w + t_f)\right)^2 + 2\tfrac{1}{12} b t_f^3$$

Area A and moment of inertia I of thin-walled I-shaped cross-section

providing

$$I \approx \tfrac{1}{12} h_w^3 t_w + 2h_w^2 b t_f = \tfrac{1}{12} A_w h_w^2 + 2A_f h_w^2 \qquad (Ex.\ 11\text{-}4.3)$$

Moment of inertia I of thin-walled I

or

$$I \approx \left(1 + 6\frac{A_f}{A_w}\right)\tfrac{1}{12} A_w h_w^2 = \left(1 + 6\frac{A_f}{A_w}\right) I_w \qquad (Ex.\ 11\text{-}4.4)$$

Moment of inertia I of thin-walled I

where I_w denotes the moment of inertia of the web. This result may also be given in the form

$$I \approx \left(1 + \frac{A_w}{24A_f}\right) I_F \qquad (Ex.\ 11\text{-}4.5)$$

Moment of inertia I of thin-walled I

where I_F is the moment of inertia of *both* flanges. Both (*Ex. 11-4.4*) and (*Ex. 11-4.5*) show that it pays better to add material to the flanges than to the web—*provided that only the bending stiffness about the z-axis is of concern.* If the web is too thin, then it may buckle leaving the cross-section worthless, or it may not be able to carry the shear stress necessary to "transport" axial stresses between the upper and the lower flange which may be just as fatal.

Our last example deals with a circular tube. In various truss structures this profile is used very often due to the fact that it offers a feasible alternative in terms of usage of material. It is particularly advantageous with respect to torsional stiffness, while its bending stiffness is less than that of the I-profile of the same amount of material. The latter cross-section has a very low torsional stiffness seen in this light. It must be acknowledged that it makes for a somewhat uneasy feeling that circular tubes are difficult to inspect for internal corrosion.

[11.7] In order to compute the moment of inertia about the z-axis it is not necessary to put any restrictions on the thickness t_w of the web. On the other hand, it is not likely that one should choose a value of t_w that is of the same order of magnitude as h or b.

Ex 11-5 Circular Tube—Ring-Shaped Cross-Section

Circular tube. Ring-shaped cross-section

In this example we intend to determine the area and moment of inertia of a circular ring as shown in Fig. Ex. 11-5.1. Here we may exploit the

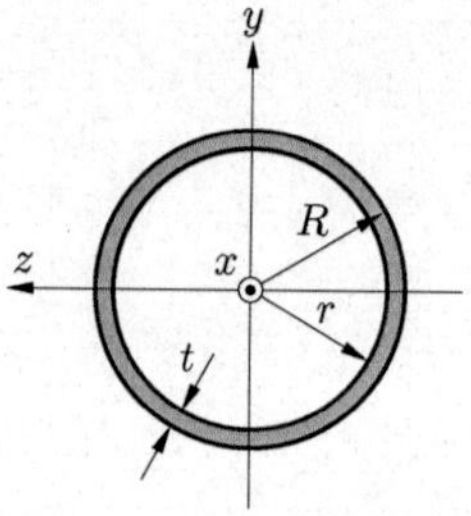

Fig. Ex. 11-5.1: *Ring-shaped, circular cross-section.*

result (*Ex. 11-2.3b*), in that we may subtract the results for a circle with radius r from results for the circle with radius R.[11.8]

Area A and moment of inertia I for a circular tube

$$\begin{aligned} &A_R = \pi R^2 \,,\ A_r = \pi r^2 \,,\ A = \pi(R^2 - r^2) \\ &I_R = \frac{\pi}{4}R^4 \,,\ I_r = \frac{\pi}{4}r^4 \,,\ I = \frac{\pi}{4}(R^4 - r^4) \end{aligned} \qquad (Ex.\ 11\text{-}5.1)$$

When we introduce

$$r = R - t \qquad (Ex.\ 11\text{-}5.2)$$

where t is the wall thickness we may find

$$A = \pi(R^2 - (R-t)^2) \text{ and } I = \frac{\pi}{4}(R^4 - (R-t)^4) \qquad (Ex.\ 11\text{-}5.3)$$

If we are dealing with thin cross-sections, i.e. when $t \ll R$, we may find the very simple result

Area A and moment of inertia I for a thin-walled circular tube

$$\begin{aligned} &A \approx 2\pi rt \\ &I \approx \pi R^3 t \text{ or } I \approx \tfrac{1}{2}AR^2 \end{aligned} \qquad (Ex.\ 11\text{-}5.4)$$

The above examples do not cover all the cross-sections that are used in structural engineering, but if you come across one which is not covered here you ought to be able to "do the sums yourself," as the British sometimes say.

[11.8] We may do this because the circles have coinciding centroids.

Chapter 12

Shear Deformation of Linear Elastic Beam Cross-Sections

In Chapter 11 on bending and axial deformation of "beams with cross-sections" we based the derivations on the concept of beams built of "fibers" which may elongate or shorten. This idea is in the spirit of Bernoulli-Euler beam theory, whose generalized strains are the axial strain ε and the curvature strain κ, see Sections 7.5, 7.3 and 7.2 and therefore these beams only have two generalized stresses, namely the axial force N and the bending moment M which are the work conjugate of ε and κ, respectively. On the other hand, Bernoulli-Euler beams do have a shear force V which does not have a work conjugate strain.[12.1] Sometimes it is said that in Bernoulli-Euler beams the shear force is the internal reaction to the assumption that the shear strain equals zero. The reason for this way of seeing things is that the reaction force at the support of a beam may be said to be a reaction to the requirement that the support does not move. In order for a shear force to exist in a Bernoulli-Euler beam it must exhibit shear stresses, although they cannot be computed from shear strains. The only way that they may be determined is then through equilibrium equations in a similar way that the shear force V was introduced in (7.25), p. 130. When we turn to the detailed study of shear stresses in the interior of "beams with cross-sections" the idea of beam fibers no longer suffices and we have to apply some kind of continuum mechanics, as we shall see below.

Concept of beam fiber not good for shear

At this point it seems reasonable to emphasize that the expressions for the shear stresses over a cross-section require much more work than what lead to the definition of the shear force V in (7.25).

In Section 7.6 we introduced Timoshenko beams which have one more generalized strain, namely the shear strain φ. This fact might make it possible to determine the shear stresses from the shear strains, but the

[12.1] Just by looking at Fig. 7.3 we can see that the shear force $\widetilde{V}$ is needed in order for equilibrium to be possible.

E. Byskov, *Elementary Continuum Mechanics for Everyone*,
Solid Mechanics and Its Applications 194, DOI: 10.1007/978-94-007-5766-0_12,

formulas we derive below are also valid for Bernoulli-Euler beams although they do not assume that the beam experiences shear strains.

12.1 Without and With a Cross-Section

To set the stage, consider a beam such as the one shown in Fig. 12.1 where, to make things simpler, we have not applied any axial load. From Chapter 11

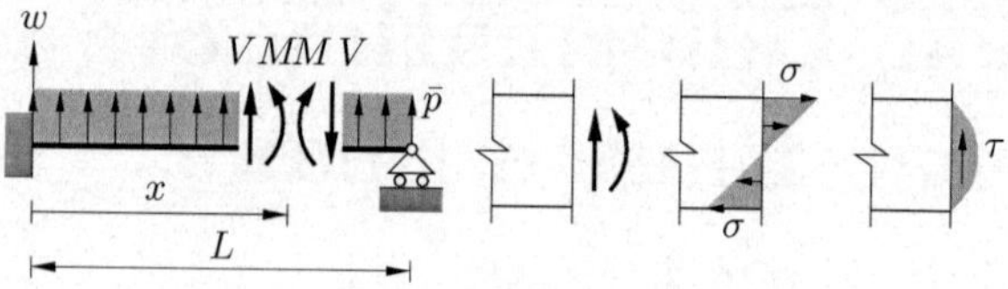

Fig. 12.1: Beam with transverse load, viewed as a one-dimensional and a two-dimensional structure.
In this figure, the beam cross-section is taken to be symmetric about the beam axis.

you ought to have a clear understanding of the meaning of the axial stress $\sigma = \sigma_{xx}$. Now our aim is to establish expressions for the other stress, the shear stress $\tau = \sigma_{xy}$ shown in the figure.[12.2]

12.2 Formulas for Shear Stresses in Beams

Shear stresses assumed constant perpendicular to the paper plane.

In Fig. 12.1 the distribution of the shear stress τ is sketched, but nothing is said about its variation perpendicular to the paper plane. Our focus shall not be on the shear strain $\tau(x,y,z) = \sigma_{xy}(x,y,z)$ but on the integral over the width of the product of τ and the width $b(x,y,z)$

$$Q(x,y) \equiv \int_{-b/2}^{b/2} \tau(x,y,z)dz \text{ implying } V(x) = \int_{h_l}^{h_u} Q(x,y)dy \quad (12.1)$$

where h_l and h_u denote the lower and upper coordinate of the cross-section. Therefore we might think of τ as independent of z and make things easier for ourselves and assume that the stresses σ_{xx}, σ_{xy} and σ_{yy} do not depend on z, see Fig. 12.2 and the derivations based on that figure. The formulas following from this assumption are valid whether the stresses depend on z or not.

12.2.1 A Little Continuum Mechanics

Derivations may look like voodoo

Some of the derivations below may seem to be the result of a cross between voodoo and divine inspiration. It is certainly true that the first time somebody has derived a set of complicated formulas it may have been possible

[12.2] The distribution of τ over the cross-section is based on conjecture at this point.

due to some kind of divine inspiration. It is often the case that, when somebody else later presents the same formulas, the other person is able to give a derivation which is much more systematic and logical—I hope that I succeed in this respect.

Later the derivations become systematic

In the present connection we only need expressions for a plate loaded in its own plane, much like the cases of Example Ex 5-3, p. 100. We shall, however, not rely heavily on the results from that but utilize the equilibrium equations (4.75), p. 76, from the kinematically linear continuum mechanics.

Consider equilibrium of an infinitesimal rectangle, see Fig. 12.2, where we note that the width b may vary with the coordinates x and y.[12.3]

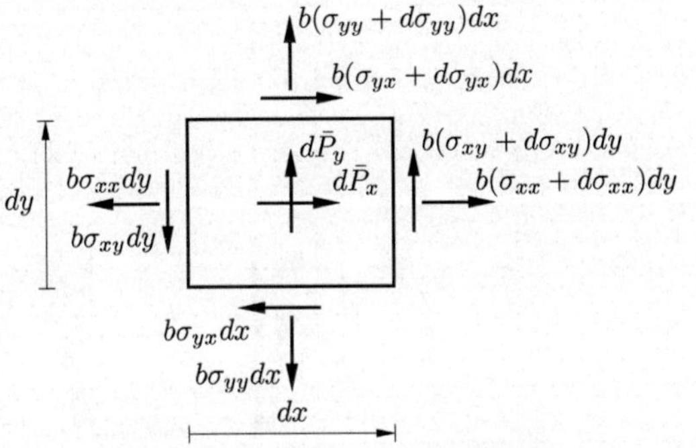

Fig. 12.2: An infinitesimal rectangle.

The load on the rectangle is $b\bar{p}_x$ per area in the x-direction and $b\bar{p}_y$ per area in the y-direction, where $\bar{p}_x$ and $\bar{p}_y$ are of dimension [force/length^3]. On the rectangle $dxdy$ this means that the loads may be given as the resultants $d\bar{P}_x$ and $d\bar{P}_y$[12.4]

$$\begin{aligned} &\text{In the } x\text{-direction}: \quad d\bar{P}_x = b\bar{p}_x dxdy \\ &\text{In the } y\text{-direction}: \quad d\bar{P}_y = b\bar{p}_y dxdy \end{aligned} \tag{12.2}$$

In order to have equilibrium the other forces shown in Fig. 12.2 are needed. Since we assume that no loads act in the z-direction we may immediately write the relevant equilibrium equations by rewriting (4.75)

$$\begin{aligned} &\text{In the } x\text{-direction}: \quad 0 = \sigma_{xx,x} + \sigma_{yx,y} + \bar{p}_x \\ &\text{In the } y\text{-direction}: \quad 0 = \sigma_{yy,y} + \sigma_{xy,x} + \bar{p}_y \end{aligned} \tag{12.3}$$

As indicated in Fig. 12.1, the symbol τ denotes the shear stress and σ

12.3 For the reason for using an (x, y, z)-coordinate, see Section 11.1.2, p. 229.

12.4 Recall the comment that the stresses may be thought of as being independent of y.

designates the axial stress. Using that notation (12.3a) may be written

$$\text{In the } x\text{-direction}: \quad 0 = \sigma_{,x} + \tau_{,y} + \bar{p}_x \tag{12.4}$$

Later we shall see that it may be easier to determine the shear stress acting on a plane perpendicular to the one we would have preferred. However, symmetry of the stress tensor, $\sigma_{xy} = \sigma_{yx}$, may then be exploited.

12.2.2 Axial and Transverse Equilibrium

It is our aim to determine the shear stresses over the cross-section, see Fig. 12.3. First, note that the shear stresses must vanish on the upper and lower surfaces because of symmetry of the stress tensor. This fact justifies the sketch of the parabolic-like distribution of τ shown in the figure.[12.5] From Navier's formula, see (11.42), we know that the axial stress is linearly distributed in the y-direction. The way to go about the task consists in

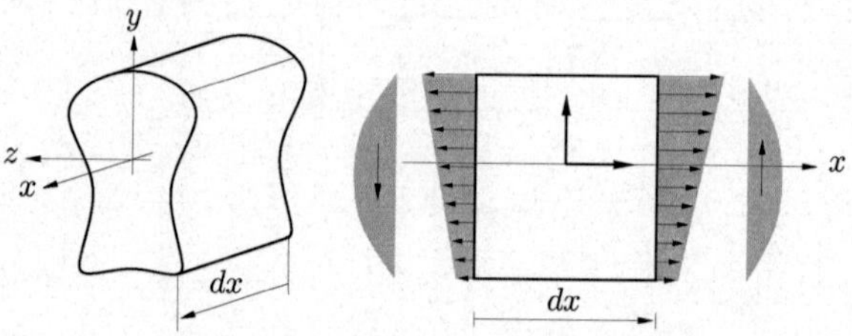

Fig. 12.3: Infinitesimal part of a beam with indication of stresses and applied loads.

establishing equilibrium in the y-direction of the upper shadowed part, see Fig. 12.4 and Fig. 12.5. First, however, we exploit Navier's formula (11.42) which, for convenience, is repeated here

Navier's formula

$$\sigma = \frac{N}{A} - y\frac{M}{I} \tag{12.5}$$

where, for simplicity, we write σ instead of σ_f.

Differentiation with respect to x provides

$$d\sigma = \frac{dN}{A} - y\frac{dM}{I} = \frac{N'}{A}dx - y\frac{M'}{I}dx = \frac{N'}{A}dx + y\frac{V}{I}dx \tag{12.6}$$

[12.5] At this moment we cannot a priori exclude the possibility that the shear stress vanishes somewhere between the upper and lower surface. On the other hand, if it looked like a third degree polynomial with the value zero in the middle, then the shear stresses would only be able to carry a small load in the y-direction which indicates that the distribution sketched in the figure is reasonable.

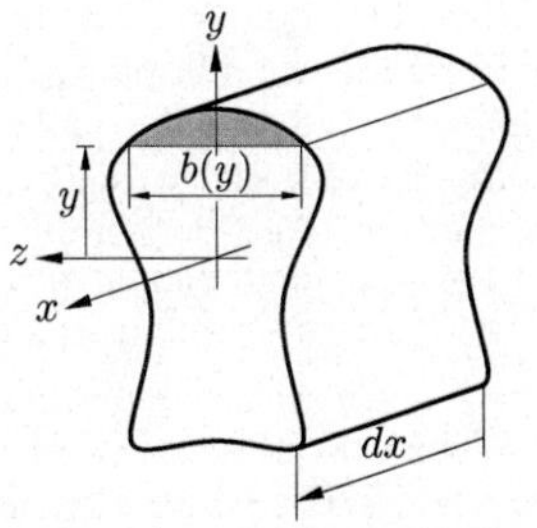

Fig. 12.4: Infinitesimal beam element with shaded upper part.

where prime ( )′ means differentiation with respect to x, and where the definition of the shear force V in terms of the derivative of the bending moment M, i.e. $V \equiv M'$, see (7.61), is introduced.

Before we establish the equilibrium equations for the upper part of the infinitesimal beam element we need to discuss how the the beam loads $\bar{p}_u$ and $\bar{p}_w$, see Fig. 7.2, p. 128, are distributed over the cross-section, i.e. we need to know how $\bar{q}_x$ and $\bar{q}_y$, see Fig. 12.5, vary with y—as you may recall,

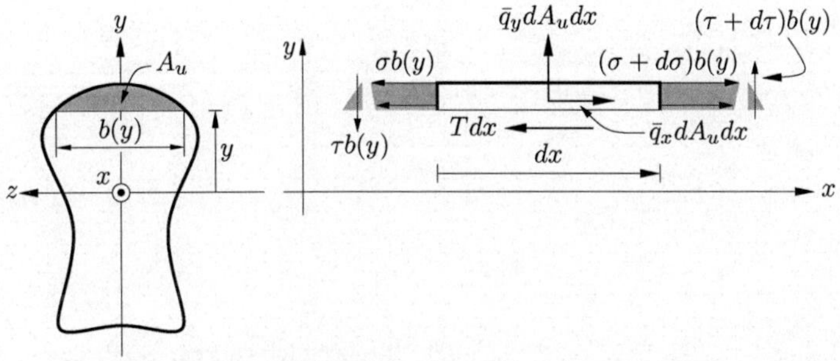

Fig. 12.5: Cross-section and free-body diagram of upper part of infinitesimal beam element.

we assume that strains and stresses are independent of z. In the same spirit as (12.1) which defined Q let us introduce T by, see Fig. 12.5

$$T(x, y) \equiv \int_{-b/2}^{b/2} \sigma_{yx}(x, y, z)dz \tag{12.7}$$

In this connection it is important to realize that we are dealing with a continuum which is subjected to severe kinematic constraints, namely the constraints inherent in the Bernoulli-Euler theory. Here, the important condition is that we assume that the cross-sections remain plane after deformation. Therefore, the only meaningful distributions of $\bar{q}_x$ are the ones that are linear in y, but the resultant of $\bar{q}_x$, namely $\bar{p}_u$, is associated with pure axial deformation which limits the possibilities to $\bar{q}_x$ being independent of the coordinates, i.e. constant over the cross-section.

When we introduced the one-dimensional beam theories of Sections 7.2–7.6 we tacitly assumed that all loads acted on the beam axis, i.e. at the centroid, simply because it was the only obvious choice and because we did not consider two- or three-dimensional cross-sections. Thus, our assumption agrees with the above statement as regards the distribution of $\bar{q}_x$, but see also Example Ex 12-1 which is intended to shed some more light on the issue of how a Bernoulli-Euler beam can be loaded.

Horizontal equilibrium of the upper part, see Fig. 12.5, requires

$$0 = -T dx + \int_{A_u} (\sigma + d\sigma) dA - \int_{A_u} \sigma dA + \left(\int_{A_u} \bar{q}_x dA \right) dx \tag{12.8}$$

or, when (12.6) is introduced

$$0 = -T dx + \left(\int_{A_u} \left(\frac{N'}{A} + y \frac{V}{I} \right) dA \right) dx + \left(\int_{A_u} \bar{q}_x dA \right) dx \tag{12.9}$$

i.e.

$$T = \frac{N'}{A} \int_{A_u} dA + \frac{V}{I} \int_{A_u} y \, dA + \bar{q}_x \int_{A_u} dA \tag{12.10}$$

Since

$$\bar{p}_u = A \bar{q}_x \tag{12.11}$$

we may get

$$T = \frac{A_u}{A} N' + \frac{S_u}{I} V + \frac{A_u}{A} \bar{p}_x \tag{12.12}$$

where the definition of S_u, see (11.19b), has been introduced. Furthermore, (7.62a)

$$N' + \bar{p}_x = 0 \tag{12.13}$$

with the the result that (12.12) may give

$$T = \frac{S_u}{I} V \tag{12.14}$$

12.2.3 Moment Equilibrium

Moment equilibrium

At this point we have exploited horizontal equilibrium and only two more relevant equilibrium equations remain. Whether we should choose vertical equilibrium or moment equilibrium is probably not obvious, but against vertical equilibrium speaks the fact that in that case we need to consider

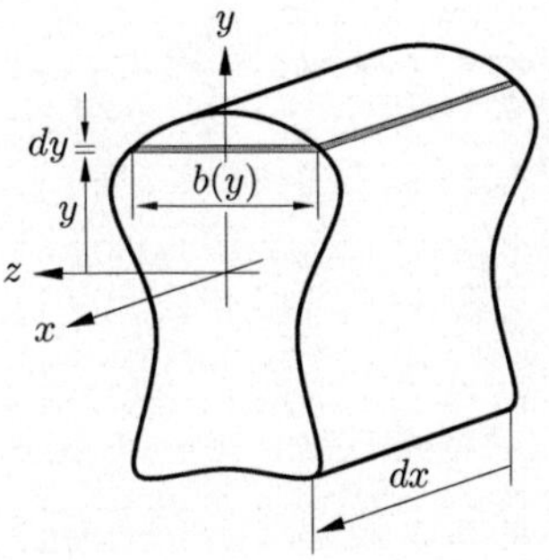

Fig. 12.6: Part of beam with a "fiber" spanning the width.

the distribution over the cross-section of the vertical load $\bar{p}_w$ and this issue is closely connected to the question of how the shear stresses vary with y. Therefore, we turn to moment equilibrium of an infinitesimal "fiber" which lies y from the centroid, see Fig. 12.6 and Fig. 12.7.

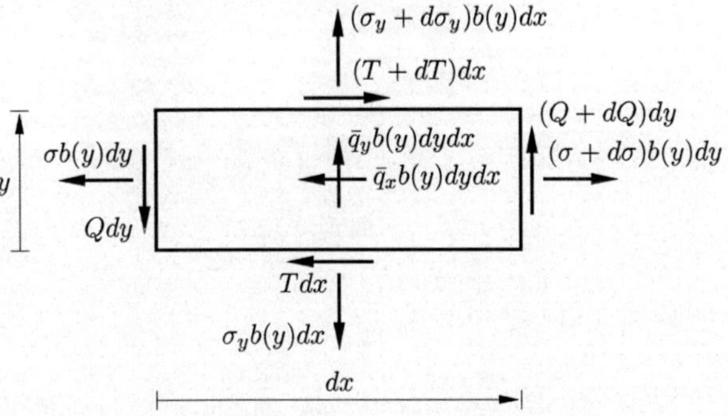

Fig. 12.7: Infinitesimal beam element.

Moment equilibrium, positive counter-clockwise, about the midpoint of the element shown in Fig. 12.7 requires

$$0 = +\tfrac{1}{2}dxQdy + \tfrac{1}{2}dx(Q + dQ)dy - \tfrac{1}{2}dyTdx - \tfrac{1}{2}dy(T + dt)dx \quad (12.15)$$

Elimination of higher order terms in the infinitesimal quantities provides

$$Q = T = \frac{S_u}{I}V \quad (12.16)$$

and thus the shear stress τ in the y-direction becomes

$$\tau(y) = \frac{V}{I} \frac{S_u}{b(y)} \tag{12.17}$$

Grashof's Formula

This formula is often called *Grashof's formula.*

Fortunately, as we predicted in Section 12.2.2, (12.17) shows that the shear stress vanishes on the upper and lower faces of the beam.

Ex 12-1 Where to Load a Beam

Consider the ring shown in Fig. Ex. 12-1.1. Clearly, if we analyze the ring as a two-dimensional structure the relative values of the two loads p_i and p_o are very important for the strains and stresses in the ring. In (Muskhelishvili 1963) you may find the formulas that can be used to solve the problems below.[12.6]

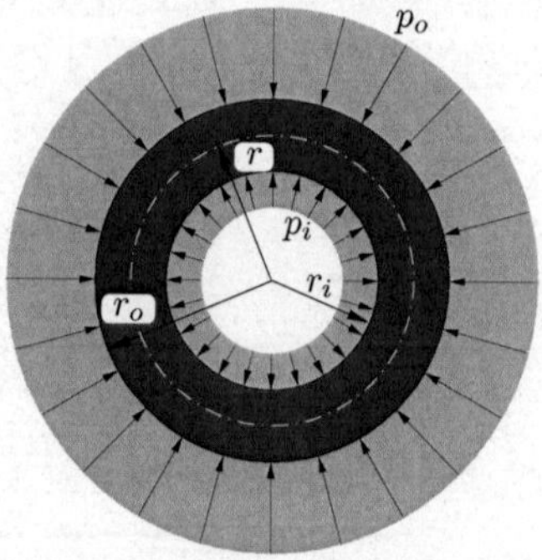

Fig. Ex. 12-1.1: *A ring beam with inner and outer loading.*

By r denote the radius of the centerline of the ring, i.e. $r = (r_i + r_o)/2$, and let h denote the depth of the ring, i.e. $h = r_o - r_i$. To make things easier, let the thickness (perpendicular to the paper plane) be 1.

Ex 12-1.1 Same Load on Inside and Outside

If we let $p = p_i = p_o$, then the situation is one of hydrostatic pressure and the stress is uniform in the ring, which simply contracts. If we regard the ring as a curved Bernoulli-Euler beam subjected to the pressure ph/r simple statics shows that the axial force N in the beam is ph and thus the axial stress is p which agrees with the two-dimensional analysis.

The reason for the contraction is that p_o acts on a larger area than p_i.

[12.6] You may have to adjust to some notations that are unusual today. For instance, X_y is used instead of σ_{yx} or σ_{21}, etc. Furthermore, in polar coordinates $\widehat{rr}$ is the same as σ_{rr}, etc.

Ex 12-1.2 Only Load on Outside

If we load the ring on the outside only and let $p_o = ph/r_o$ the total load on the ring is the same as before, but the two-dimensional strains and stresses will be different.

Ex 12-1.3 Only Load on Inside

For the case with only an inside load of $p_i = ph/r_i$ the total load is again the same as before, but the strains and stresses will be different from the ones found in the earlier cases.

So, which is the correct way to load the beam as a curved Bernoulli-Euler beam?

No simple answer

In some ways the answer is somewhat disturbing in that there is none, simply because in the one-dimensional theory no distinction is made between the two faces of the beam. One might say that the Bernoulli-Euler beam cannot feel if it is loaded on the outside, the inside or somewhere else as long as the total load is the same.

Hopefully, this example helps to clarify some aspects of the limitations of Bernoulli-Euler beam theory.[12.7]

12.2.4 Examples of Shear Stress Computations

In Section 11.1.12 we started by treating the rectangular cross-section, and we shall do the same here.

Ex 12-2 Rectangular Cross-Section

Rectangular cross-section

For the rectangular cross-section, see Fig. Ex. 12-2.1, we may determine the first-order moment S_u of the shaded area A_u

$$S_u = \int_{A_u} y\,dA = \int_y^{h/2} \eta b\,d\eta = \tfrac{1}{2} b \left(\left(\frac{h}{2}\right)^2 - y^2 \right) \qquad (Ex.\ 12\text{-}2.1)$$

and by the formula (*Ex. 11-1.4*) for the moment of inertia of the rectangle and (12.17) we may find

$$\tau(y) = \tfrac{3}{2} \left(1 - \left(\frac{y}{h/2}\right)^2 \right) \frac{V}{A} \qquad (Ex.\ 12\text{-}2.2)$$

where A is the area of the rectangle.

Maximum of the shear stress takes place at $y = 0$, i.e. at the centroid

$\tau_{\max}$

$$\tau_{\max} = \frac{3}{2} \frac{V}{A} \qquad (Ex.\ 12\text{-}2.3)$$

[12.7] Nothing prevents us from developing beam theories which entail some measures of the deformation of the beam depth. Such theories have been proposed over the years, but against the idea of including more sophisticated strain measures speaks the simple fact that theories of that kind easily get very complicated, maybe without providing much more insight into the behavior of the beam.

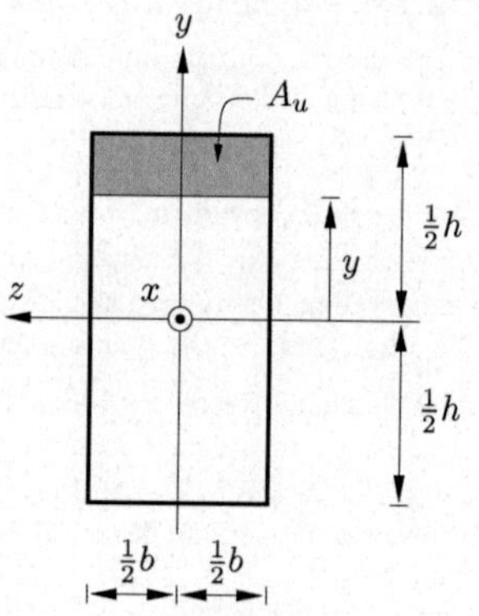

Fig. Ex. 12-2.1: *Rectangular Cross-Section.*

which shows that the maximum shear stress is equal to one and a half times its average.[12.8]

For completeness sake we ought to ensure that the resultant of the shear stresses equals the shear force V. In this case it is left for the reader to do this.[12.9]

Determination of the shear stress is only half the task in that we would also like to know the shear stiffness of the cross-section. This, however, is a topic that has no definitive answer as we shall see in Section 12.3, in particular Section 12.3.1. We shall not pursue this here but mention that the value of the *effective cross-section* A_e, see Section 7.8 for a rectangular cross-section is often taken to be $\frac{5}{6}A$. Thus, the *effective shear stiffness* is

Effective shear stiffness of rectangle

$$GA_e \approx \tfrac{5}{6}GA \qquad \text{(Ex. 12-2.4)}$$

Ex 12-3 Circular Cross-Section

Circular cross-section

Here, the computations are a little more complicated than for the rectangle, see Fig. Ex. 12-3.1. As in Example Ex 11-2 we shall use angles as variables. Here,

$$S_u = \int_{A_u} \eta \, dA \qquad \text{(Ex. 12-3.1)}$$

[12.8] Actually, the value of the factor $\frac{3}{2}$ in (*Ex. 12-2.3*) is the limit for $b/h \approx 0$. This is not obvious from our analysis which does not account for deformations in the z-directions. For finite values of b/h the value of the factor is slightly less than 1.5.

[12.9] If you know that the area under a parabola is equal to two thirds of the base multiplied by its maximum value, then the task should not be insurmountable.

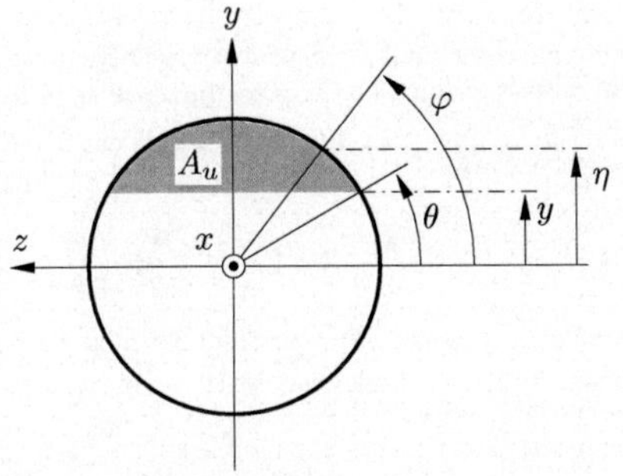

Fig. Ex. 12-3.1: Circular cross-section.

Expressed in terms of η φ in (Ex. 11-2.1)

$$\begin{aligned}\eta &= R\sin\varphi\,, \quad d\eta = R\cos\varphi d\varphi \\ dA &= (2R\cos\varphi)d\varphi = 2R^2\cos^2\varphi d\varphi\end{aligned} \qquad \text{(Ex. 12-3.2)}$$

providing

$$S_u = \int_0^{\pi/2} 2R^3\cos^2\varphi\sin\varphi\, d\varphi \qquad \text{(Ex. 12-3.3)}$$

It is a matter of taste how one performs the integration. Today, most students will do it using some kind of program like MuPAD, Maple, Mathematica or maxima, but here we do it the oldfashioned way, namely by hand.[12.10]

Introduce the variable ζ

$$\zeta \equiv \cos\varphi\,, \quad d\zeta = -\sin\varphi\, d\varphi \qquad \text{(Ex. 12-3.4)}$$

with

$$\varphi = \theta \Rightarrow \zeta = \cos\theta\,, \quad \varphi = \frac{\pi}{2} \Rightarrow \zeta = 0 \qquad \text{(Ex. 12-3.5)}$$

and thus,

$$S_u = 2R^3 \int_0^{\cos\theta} \zeta^2\, d\zeta = \tfrac{2}{3}R^3\cos^3\theta \qquad \text{(Ex. 12-3.6)}$$

Utilize (Ex. 11-2.2c) and (12.17) to get

$$\tau = \frac{V}{I}\frac{S_u}{b(\theta)} = \frac{4V}{3\pi R^2}\cos^2\theta \qquad \text{(Ex. 12-3.7)}$$

implying that the maximum of τ is

$$\tau_{\max} = \frac{4}{3\pi}\frac{V}{R^2} = \frac{4}{3}\frac{V}{A} \qquad \text{(Ex. 12-3.8)}$$

τ_{max}

12.10 To be sure I have checked the results by use of maxima.

The maximum value of the shear stress therefore is about 33% higher than its average which is a smaller increase than for the rectangle.

For a number of purposes it is more convenient to express τ by y instead of by θ

$$\tau = \frac{4V}{3\pi R^2}\cos^2(\sin^{-1}(y/R)) = \frac{4V}{3\pi R^2}\left(1 - \left(\frac{y}{R}\right)^2\right) \quad \text{(Ex. 12-3.9)}$$

Also for this cross-section we ought to have made certain that the integral of τ is V, but, again I leave it for the reader to perform the necessary computations.

The next examples deal with thin-walled cross-sections. For the solid cross-sections of the previous examples shear stresses rarely present problems regarding stability and strength. The situation is very different for the examples below in that, for example, the shear stresses can be so large that they cause buckling or rupture of the web of an I-beam.

Ex 12-4 Thin-Walled I-Shaped Cross-Section

Thin-walled I-shaped cross-section

As we shall see, the shear stresses in the web of a thin-walled I-beam are relatively high because the flanges only to a small degree participates in carrying the shear loading—their obligation is mainly to provide a bending moment.

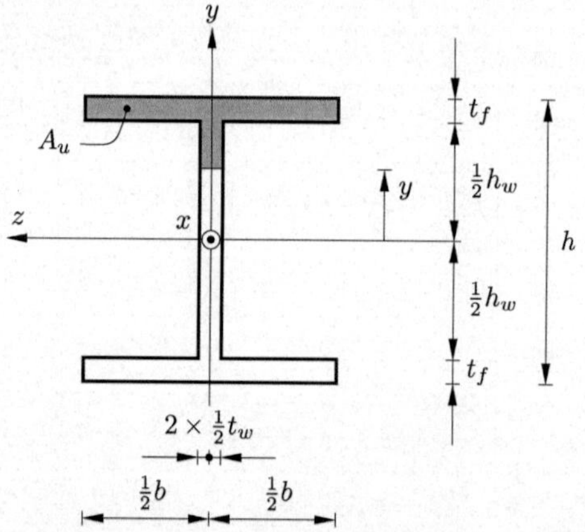

Fig. Ex. 12-4.1: I-shaped cross-section. Analysis of shear stresses in the web.

Ex 12-4.1 Shear Stresses in the Web

Here we may exploit some of the results from Example Ex 11-1 and Example Ex 11-4 in connection with (12.17). For $|y| < \frac{1}{2}h - t_f$ there are two contributions to S_u

$$\begin{aligned} S_u &= \int_y^{h/2} b(\eta)\,d\eta = \int_y^{h/2-t_f} \eta t_w\,d\eta + \int_{h/2-t_f}^{h/2} \eta b\,d\eta \\ &= \tfrac{1}{2}\left(\tfrac{1}{2}h - t_w\right)^2 t_w - \tfrac{1}{2}y^2 t_w + \tfrac{1}{8}h^2 b - \tfrac{1}{2}\left(\tfrac{1}{2}h - t_w\right)^2 b \end{aligned} \qquad (Ex.\ 12\text{-}4.1)$$

The assumption (*Ex. 11-4.1*) of thin flanges implies that we may write

$$\begin{aligned} S_u &\approx \tfrac{1}{2}t_w\left(\left(\tfrac{1}{2}h\right)^2 - y^2\right) + \tfrac{1}{2}t_f bh \\ &= \tfrac{1}{8}\left(1 - \tilde{y}^2\right)h^2 t_w + \tfrac{1}{2}bht_f \end{aligned} \qquad (Ex.\ 12\text{-}4.2)$$

where

$$\tilde{y} \equiv \frac{y}{h/2} \approx \frac{y}{h_w} \qquad (Ex.\ 12\text{-}4.3)$$

We may also find the expression for S_u without integration when we realize that the static moment of a domain is equal to its area multiplied by the distance to its centroid

$$\begin{aligned} S_u &= \left(\left(\tfrac{1}{2}h_w - y\right)t_w\right)\left(\tfrac{1}{2}\left(\tfrac{1}{2}h_w + y\right)\right) \\ &\quad + (bt_f)\left(\tfrac{1}{2}h_w + \tfrac{1}{2}t_f\right) \end{aligned} \qquad (Ex.\ 12\text{-}4.4)$$

and after utilizing that $h_w = h - 2t_f$ we may get (*Ex. 12-4.1*).
The above results combined with (*Ex. 11-4.5*) and inserted into (12.17) may provide

$$\tau \approx \frac{3}{2}\,\frac{1 - \tilde{y}^2 + 4\tilde{A}_f}{1 + 6\tilde{A}_f}\,\frac{V}{ht_w} \qquad (Ex.\ 12\text{-}4.5)$$

where $\tilde{A}_f$ is a relative measure of the material in one of the flanges

$$\tilde{A}_f \equiv \frac{bt_f}{ht_w} \approx \frac{bt_f}{h_w t_w} \qquad (Ex.\ 12\text{-}4.6)$$

If we integrate the value of τ from (*Ex. 12-4.5*) over the web, then—under the assumption of thin flanges and thin web—we may get the value V which is comforting to know.
It seems clear that there are three places where the shear stresses are particularly interesting, namely at the middle of the web and immediately under and above the joint between web and flange. The above results together with (*Ex. 11-4.5*) may be inserted in (12.17)

$\tau_{\max}$ and τ at junction between web and flange

$$\begin{aligned} y = 0: &\quad \tau = \tau_{\max} \approx \frac{3}{2}\,\frac{1 + 4\tilde{A}_f}{1 + 6\tilde{A}_f}\,\frac{V}{ht_w} \\ y = \tfrac{1}{2}h_w^-: &\quad \tau = \tau^- \approx \frac{3}{2}\,\frac{4\tilde{A}_f}{1 + 6\tilde{A}_f}\,\frac{V}{ht_w} \\ y = \tfrac{1}{2}h_w^+: &\quad \tau = \tau^+ \approx \frac{3}{2}\,\frac{4\tilde{A}_f}{1 + 6\tilde{A}_f}\,\frac{V}{hb} \end{aligned} \qquad (Ex.\ 12\text{-}4.7)$$

where τ^- and τ^+ denote the shear stress below and above the junction, respectively.

This shows that these shear stresses[12.11] are negligible in the flanges. In order to get a feeling for the magnitude of the shear stresses let us assume that $b \approx h$ and $t_f \approx t_w$

$\tau_{\max}$ and τ at junction between web and flange. Thin-walled cross-section

$$
\begin{aligned}
y = 0: &\quad \tau = \tau_{\max} \approx \frac{15}{14}\frac{V}{ht_w} = \frac{5}{7}\left(\frac{3}{2}\frac{V}{ht_w}\right) \\
y = \tfrac{1}{2}h_w^-: &\quad \tau = \tau^- \approx \frac{12}{14}\frac{V}{ht_w} \\
y = \tfrac{1}{2}h_w^+: &\quad \tau = \tau^+ \approx \frac{12}{14}\frac{V}{ht_w}\frac{t_w}{b} \approx 0
\end{aligned}
\qquad (Ex.\ 12\text{-}4.8)
$$

where the factor 5/7 in (*Ex. 12-4.8a*) indicates how much the maximum shear stress is reduced in relation to the value in a rectangular cross-section of the same size as the web, see Example Ex 12-2.

It is quite remarkable how efficiently the web carries the shear stresses in that they only vary little between centroid and flange. This may seem surprising because in a rectangle, see Example Ex 12-2, they vanish at the upper and lower edges. This difference must be caused by the existence of the flanges although their primary obligation is to carry bending moments.

Ex 12-4.2 Shear Stresses in the Flanges

Shear stresses in the flanges

The stress distribution in an I-cross-section is much more complicated

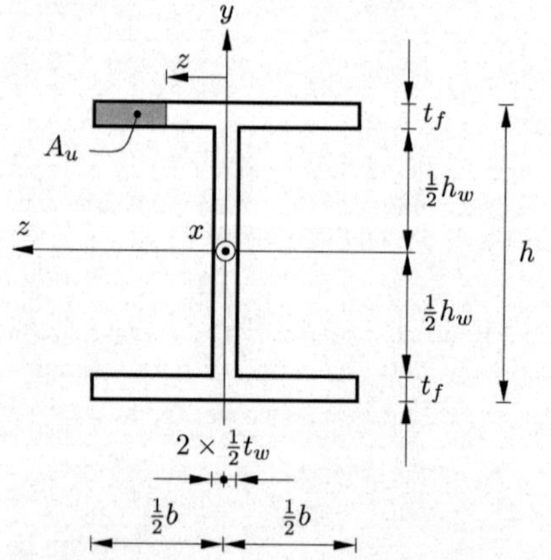

Fig. Ex. 12-4.2: *I-shaped cross-section. Analysis of shear stresses in the flanges.*

[12.11] The justification for writing "these shear stresses" is that—as we shall see later—in the flanges there are shear stresses in the z-direction.

than that in a rectangle or other simple cross-sections. The details around the concave corners alone, be they rounded or not, cannot be described by the methods we establish here. In advance it seems clear that there must be large strain gradients at the junction between web and flange, which is where the shear stress in the web must be transferred to be spread over the flange. There is another effect which has to do with the fact that there exist shear stresses in the the z-direction.

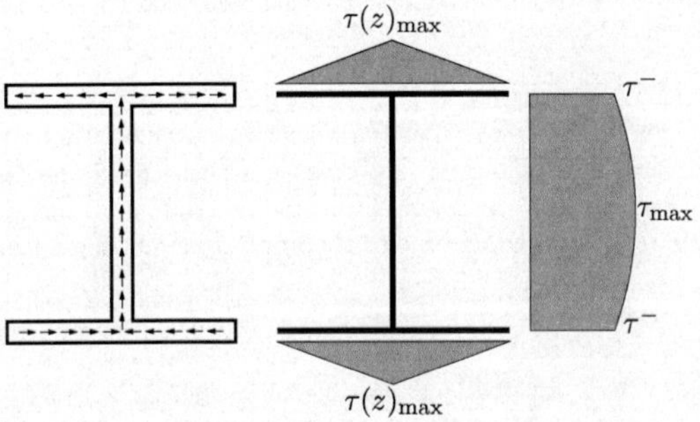

Fig. Ex. 12-4.3: *I-shaped cross-section.*
Sketch of shear stresses in web and flanges.

In Section 12.2 we considered cuts parallel with the z-axis, but here we need results that are valid for cuts parallel with the y-axis. If we go through the derivations we may see that for our present purpose we may utilize the result from Section 12.2.2 and exchange y and z. It was an important assumption that the shear stresses did not vary with z, which is equivalent to assume that in the present context they do not vary with y. Thus, we may use (12.17) with y and z interchanged to get

$$\tau(z) = \frac{V}{I}\frac{S_u}{b(z)} \qquad \text{(Ex. 12-4.9)}$$

The static moment S_u is here

$$S_u = \tfrac{1}{2}(h_w + t_f)\Big(\big(\tfrac{1}{2}b - z\big)\, t_f\Big) \qquad \text{(Ex. 12-4.10)}$$

and therefore

$$\tau(z) = 3\frac{\tilde{b}(1-\tilde{z})}{1+6\tilde{A}_f}\frac{V}{ht_w}\,,\ z \geq 0 \qquad \text{(Ex. 12-4.11)}$$

where

$$\tilde{z} \equiv \frac{z}{b/2} \qquad \text{(Ex. 12-4.12)}$$

and

$$\tilde{b} \equiv \frac{b}{h} \qquad (Ex.\ 12\text{-}4.13)$$

are introduced.

The maximum value $\tau(z)_{\max}$ of $\tau(z)$ occurs at the web, i.e. for $z = 0$

$\tau(z)_{\text{max}}$ in flange at web

$$\tau(z)_{\max} = 3\frac{\tilde{b}}{1+6\tilde{A}_f}\frac{V}{ht_w} \qquad (Ex.\ 12\text{-}4.14)$$

Relative value of τ at junction between web and flange

That this value may not be neglected may be seen from the fact that

$$\frac{\tau(z)_{\max}}{\tau^-} = \frac{t_w}{2t_f} \qquad (Ex.\ 12\text{-}4.15)$$

The geometry of our last example is much simpler than that of the thin-walled I-beam of Example Ex 12-4. On the other hand, we shall investigate shear stresses whose direction follows the circle instead of the (y, z)-coordinate system which makes the analysis somewhat more complicated.

Ex 12-5 Circular Tube—Ring-Shaped Cross-Section

Circular tube. Ring-shaped cross-section

Here, we shall compute the shear stresses that act on an end of the circular ring and are perpendicular to the radii, see Fig. Ex. 12-5.1.

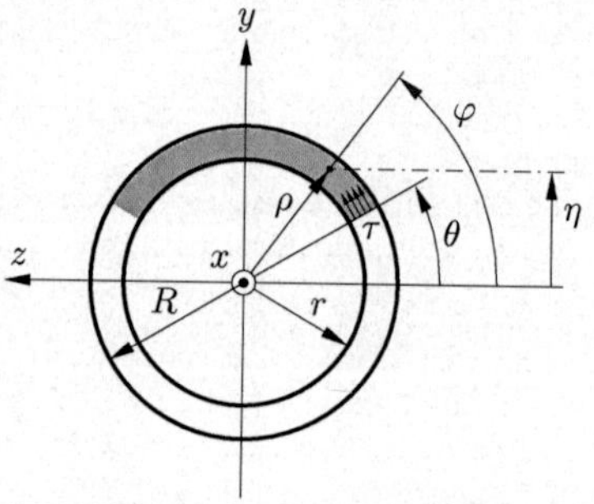

Fig. Ex. 12-5.1: *Circular ring.*

We may reason in the same way as in Example Ex 12-4.2 when we derived the formula (*Ex. 12-4.11*) for the shear stress $\tau(z)$ in the flange. In the same spirit as before we assume that the shear stresses do not vary along the radius and that they are perpendicular to the radius. Once more (12.17)

$$\tau(y) = \frac{V}{I}\frac{S_u}{b(y)} \qquad (Ex.\ 12\text{-}5.1)$$

applies. Therefore, we shall determine the static moment S_u of the shaded area and interpret the meaning of the width which earlier was denoted $b(y)$. In order to determine the static moment S_u we need the area element dA. With the variables shown in Fig. Ex. 12-5.1 we may get

$$dA = \rho\, d\varphi\, d\rho \tag{Ex. 12-5.2}$$

and the ordinate η

$$\eta = \rho \sin\varphi \tag{Ex. 12-5.3}$$

i.e.

$$\begin{aligned} S_u &= \int_{A_u} \eta\, dA = \int_\theta^{\pi-\theta}\int_r^R (\rho\sin\varphi)(\rho\, d\varphi\, d\rho) \\ &= \int_\theta^{\pi-\theta} \sin\varphi\, d\varphi \int_r^R \rho^2\, d\rho \end{aligned} \tag{Ex. 12-5.4}$$

and thus,

$$S_u = \tfrac{1}{3}\left(R^3 - r^3\right)\cos\theta \tag{Ex. 12-5.5}$$

From Example Ex 11-5 we have the formula (*Ex. 11-5.1*)

$$I = \frac{\pi}{4}\left(R^4 - r^4\right) \tag{Ex. 12-5.6}$$

which we need below.

Instead of the thickness, which we called $b(y)$, we need to introduce the thickness t of the wall of the tube

$$b(y) \sim t = R - r \tag{Ex. 12-5.7}$$

i.e.

$$\tau = \frac{4}{3\pi}\frac{\left(R^3 - r^3\right)}{\left(R^4 - r^4\right)}\frac{1}{(R-r)}\cos(\theta)V \tag{Ex. 12-5.8}$$

which, after a simple rewriting, may provide

$$\tau = \frac{4}{3\pi}\frac{\left(R^2 + Rr + r^2\right)}{\left(R^4 - r^4\right)}\cos(\theta)V \tag{Ex. 12-5.9}$$

The area A of the tube wall

$$A = \pi\left(R^2 - r^2\right) \tag{Ex. 12-5.10}$$

may be utilized to rewrite (*Ex. 12-5.9*)

$$\tau = \frac{4}{3}\frac{\left(R^2 + Rr + r^2\right)}{\left(R^2 + r^2\right)}\cos(\theta)\frac{V}{A} \tag{Ex. 12-5.11}$$

From (*Ex. 12-5.9*) and (*Ex. 12-5.11*) it is quite obvious that the maximum value $\tau_{\max}$ of the shear stress occurs for $\theta = 0$, i.e. at the z-axis, which seems intuitively clear

$$\tau_{\max} = \frac{4}{3}\frac{\left(R^2 + Rr + r^2\right)}{\left(R^2 + r^2\right)}\frac{V}{A} \tag{Ex. 12-5.12}$$

$\tau_{\max}$

If we imagine that the cross-section carried the shear force V by a constant shear stress τ_{uniform}

$$\tau_{\text{uniform}} = \frac{V}{A} \qquad (Ex.\ 12\text{-}5.13)$$

then we may see that the maximum shear stress $\tau_{\max}$ is a factor k larger than if it were uniformly distributed, where

$$k = \frac{4}{3}\frac{(R^2 + Rr + r^2)}{(R^2 + r^2)} \to 2 \text{ for } r \to R \qquad (Ex.\ 12\text{-}5.14)$$

The largest relative increase in maximum shear stress occurs for $r \to R$. Therefore, introduce the wall thickness t in (*Ex. 12-5.12*) and expand in t to get

$\tau_{\max}$

$$\tau_{\max} \approx 2\left(1 - \tfrac{1}{6}\left(\frac{t}{R}\right)^2\right)\frac{V}{A} + O\left((t/R)^3\right) \qquad (Ex.\ 12\text{-}5.15)$$

which furnishes a another insight into the value of k.

12.3 Shear Stiffness

Shear stiffness: a difficult question

Effective shear stiffness GA_e

Strangely enough, the question of beam shear stiffness is not an easy one to answer. At least, one would imagine that determination of the effective shear stiffness of a rectangular cross-section, given by the value of the "Effective Area" A_e, see Section 7.8, would be a topic in introductory courses on structural mechanics, but that is far from the truth as we shall see. In fact, papers on the subject are still published in international journals. As an example, see Renton (1991) who discusses various ways of obtaining A_e and proposes new ones.

12.3.1 Rectangular Cross-Section

Let us here consider a case of bending with a fair amount of shearing, namely the "beam" of thickness 1 shown in Fig. 12.2. In both cases of

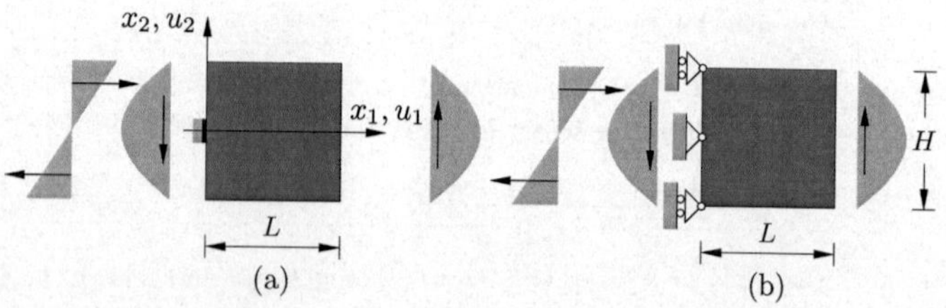

Fig. 12.2: "Beam" clamped in two different ways. The loading is defined in Fig. Ex. 27-1.1.

support the exact solution is known, see (Timoshenko & Goodier 1970), where the solution for the case (a) is derived under the assumption of plane stress.[12.12] In case (a) clamping is enforced by requiring that the rotation at the left midpoint vanishes, while in case (b) it is given by preventing horizontal translation of the three points as shown in the figure.[12.13] The solution for case (b) follows from case (a) by rotating the beam the proper angle. I shall not derive the solutions but merely provide them here.[12.14]

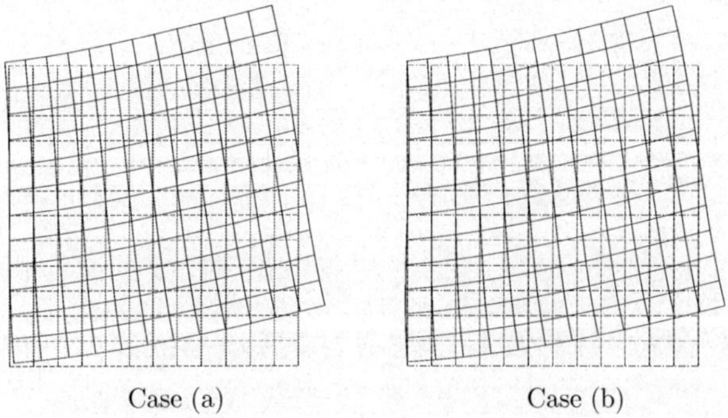

Fig. 12.3: Deformation patterns for $H = L$.

By comparing the solutions of the two cases with that found by application of Timoshenko beam theory we may find estimates of the value of A_e.

The deformation patterns for both cases are shown in Fig. 12.3 where you may see that—to the naked eye—they look very similar. In particular, note that the rotation at $(x_1, x_2) = (0, 0)$ in case (b) is very small indicating that we may expect the results from the two cases to be close. In spite of this, the value of the *Effective Area* A_e computed on the basis of the two ways of clamping differ more than just a little, as we shall see. The values of the tip displacement $w_{tip} \equiv w(1)$ do not agree, either. For convenience introduce the nondimensional coordinates ξ_α and ξ by

$$\begin{aligned} &\xi_1 \equiv x_1/L\,,\ \xi_1 \in [0,1] \ \text{ and } \ \xi_2 \equiv x_2/H\,,\ \xi_2 \in [-\tfrac{1}{2}, \tfrac{1}{2}] \\ &\xi \equiv \xi_1 \end{aligned} \tag{12.18}$$

[12.12] Timoshenko & Goodier (1970) treat a case which may be found as the mirror image of the one shown here as (a) in Fig. 12.2.
[12.13] A finite element computation shows that preventing the entire left-hand side of the "beam" from translating horizontally results in very small differences from case (b).
[12.14] The reason why case (b) and not case (a) is utilized in Example Ex 27-1 is that prescribing an angle is not possible with the finite elements applied there.

As you may realize, the theory for in-plane deformation of plates, which is developed and used in this book, does not entail rotations as primary variables meaning that prescribing a rotation is usually not possible.

and the "beam" deflection $w(\xi)$ by

$$w(\xi) \equiv u_2(\xi, 0) \tag{12.19}$$

12.3.2 Case (a). Solution by Timoshenko & Goodier

Timoshenko & Goodier (1970)

$$u_1^{(a)} = \frac{\bar{P}L^3}{6EI}\left(3\xi_1^2\xi_2 - 6\xi_1\xi_2 - (2+\nu)\xi_2^3\left(\frac{H}{L}\right)^2\right)\left(\frac{H}{L}\right)$$

$$u_2^{(a)} = \frac{\bar{P}L^3}{6EI}\left(3\xi_1^2 - \xi_1^3 + \left(\tfrac{3}{2}\left(1+\nu-2\nu\xi_2^2\right)\xi_1 + 3\nu\xi_2^2\right)\left(\frac{H}{L}\right)^2\right) \tag{12.20}$$

$$w^{(a)} = \frac{\bar{P}L^3}{6EI}\left(3\xi^2 - \xi^3 + \tfrac{3}{2}\left(1+\nu\right)\xi\left(\frac{H}{L}\right)^2\right)$$

12.3.3 Case (b). Solution of Rotated "Beam"

Rotated "beam"

$$u_1^{(b)} = \frac{\bar{P}L^3}{6EI}\left(3\xi_1^2\xi_2 - 6\xi_1\xi_2 - (2+\nu)\left(\xi_2^3 - \tfrac{1}{4}\xi_2\right)\left(\frac{H}{L}\right)^2\right)\left(\frac{H}{L}\right)$$

$$u_2^{(b)} = \frac{\bar{P}L^3}{6EI}\left(3\xi_1^2 - \xi_1^3 + \left(\left(1+\tfrac{5}{4}\nu - 3\nu\xi_2^2\right)\xi_1 + 3\nu\xi_2^2\right)\left(\frac{H}{L}\right)^2\right) \tag{12.21}$$

$$w^{(b)} = \frac{\bar{P}L^3}{6EI}\left(3\xi^2 - \xi^3 + \left(1+\tfrac{5}{4}\nu\right)\xi\left(\frac{H}{L}\right)^2\right)$$

12.3.4 Timoshenko Beam Theory

In Example Ex 7-4 the present example is analyzed by use of Timoshenko beam theory with the result given in (*Ex. 7-4.9*)

Timoshenko beam theory

$$w^{Timo\ beam} = \frac{\bar{P}L^3}{6EI}\left(3\xi^2 - \xi^3 + (1+\nu)\left(\frac{A}{A_e}\right)\left(\frac{H}{L}\right)^2\xi\right) \tag{12.22}$$

12.3.5 Values of the Effective Area

By comparing (12.22) with (12.20c) we may get the first estimate of the value of the effective area A_e

A_e from case (a)

$$A_e^{(a)} = \tfrac{2}{3}A \approx 0.667A \tag{12.23}$$

Given that the whole exercise centers on determining the influence of the value of Poisson's Ratio ν, it is somewhat surprising that the value of $A_e^{(a)}$ does not depend on that quantity. Using (12.22) and (12.21c) we may get the second estimate

$$A_e^{(b)} = \frac{4+4\nu}{4+5\nu}A \tag{12.24}$$

A_e from case (b)

This implies that the value of A_e/A should lie in the interval $[12/13, 1] \approx [0.92, 1]$ where the first value is associated with incompressibility, i.e. with $\nu = 1/2$ and the second with $\nu = 0$. It should come as no surprise that the beam of case (a) is more flexible than that of case (b) because it is less clamped. As mentioned in Example Ex 7-4 the value of A_e/A which appears in most tables is $5/6 \approx 0.83$, see e.g. (Sundström 2010), which lies between the values found above.

12.3.6 A Simple Lower Bound

Lower bound of Shear stiffness

If we exploit our knowledge about the distribution of the shear stresses we may derive a lower bound of the shear stiffness by use of the *Principle of Minimum Complementary Energy.*

Consider a short part of a solid beam subjected to shear at the right-hand end, see Fig. 12.4. From the previous investigations we know that the plate shear stress $N_{12}(L, x_2) = \sigma_{12}(L, x_2)t(x_2)$ must vary as a parabola as indicated in the figure. As in the previous derivations, we shall assume that the plate is in plane stress. In order to exclude bending and obtain a state of pure shear we shall let $L \to 0$ later. Application of the above mentioned

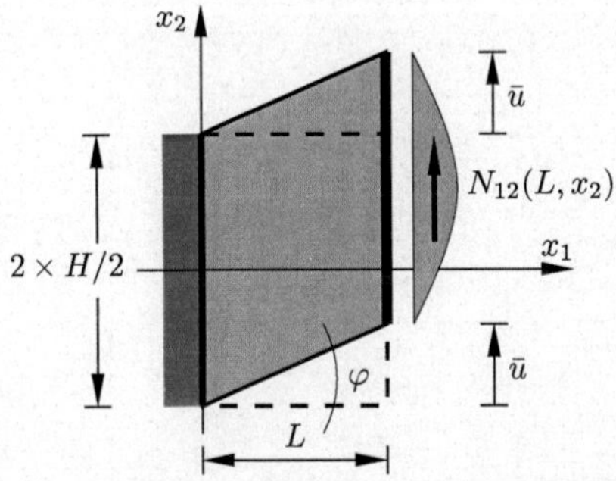

Fig. 12.4: A small part of a beam subjected to shearing.

variational principle requires that we can establish a stress field $\tilde{\boldsymbol{\sigma}}$ which satisfies all equilibrium equations, see Section 33.5. The requirement that the variation $\delta\boldsymbol{\Pi}_C(\tilde{\boldsymbol{\sigma}})$ of the complementary energy $\boldsymbol{\Pi}_C(\tilde{\boldsymbol{\sigma}})$ vanishes implies (33.46), which is repeated here

$$\mathbf{C}(\tilde{\boldsymbol{\sigma}}) \cdot \delta\tilde{\boldsymbol{\sigma}} = \delta\widetilde{\boldsymbol{T}} \cdot \bar{\boldsymbol{u}} \tag{12.25}$$

$\delta\boldsymbol{\Pi}_C$

Here, the field $(\tilde{\boldsymbol{\sigma}}, \widetilde{\boldsymbol{T}})$ must satisfy all equilibrium equations, the stress variation $\delta\tilde{\boldsymbol{\sigma}}$ must satisfy the homogeneous equilibrium equations, $\mathbf{C}$ denotes the material flexibility, and $\bar{\boldsymbol{u}}$ designates the prescribed displacements.

In the examples below, we limit ourselves to a stress field with only one parameter, but employ the Finite Element formalism used in this book and denote it $\{v_\beta\}$. Then,

Assumed equilibrium stress field

$$\tilde{\boldsymbol{\sigma}} \sim \{\sigma\} = \begin{bmatrix} N_{11} \\ N_{22} \\ N_{12} \end{bmatrix} = [N_\beta]\{v_\beta\} = \begin{bmatrix} 8\dfrac{x_1 x_2}{H^2} \\ 0 \\ \left(1 - 4\left(\dfrac{x_2}{H}\right)^2\right) \end{bmatrix} \{v_\beta\} \qquad (12.26)$$

where it is easy to see that the field equilibrium equations $N_{\alpha\beta,\alpha} = 0$ and the stress boundary conditions at $x_2 = \pm H/2$ are satisfied.

The tractions $\widetilde{\boldsymbol{T}}$ at $x_1 = L$ are

Tractions at $x_1 = L$

$$\widetilde{\boldsymbol{T}} \sim \{T\} = \begin{bmatrix} T_1 \\ T_2 \end{bmatrix} = \begin{bmatrix} N_{11}(L, x_2) \\ N_{12}(L, x_2) \end{bmatrix} \{v_\beta\} = [N_T]\{v_\beta\} \qquad (12.27)$$

In the present case we enforce the following kinematic boundary conditions

Kinematic boundary conditions at $x_1 = (0, L)$

$$u_\alpha(0, x_2) = 0\,,\ u_2(L, x_2) = \bar{u} \qquad (12.28)$$

or

$$\{\bar{u}\} = \begin{bmatrix} 0 \\ \bar{u} \end{bmatrix} = \begin{bmatrix} 0 \\ 1 \end{bmatrix} \bar{u} = [N_u]\{\bar{v}_u\} \qquad (12.29)$$

With the above formulas in hand we may write (12.25) as

$$\begin{aligned} &\{\delta v_\beta\} \int_A [N_\beta]^T [C][N_\beta] dA \{v_\beta\} \\ &= \{\delta v_\beta\} \int_{-H/2}^{H/2} [N_T]^T [N_u] dx_2 \{\bar{v}_u\}\ \forall \delta\beta \end{aligned} \qquad (12.30)$$

where $[C]$ is the material flexibility matrix of the plate

Material flexibility matrix $[C]$

$$[C] = \frac{1}{Et} \begin{bmatrix} 1 & -\nu & 0 \\ -\nu & 1 & 0 \\ 0 & 0 & 2\,(\nu + 1) \end{bmatrix} \qquad (12.31)$$

which, as indicated, depends on the plate thickness and thus is not necessarily constant. In a more compact notation (12.30) becomes

$$\{\delta v_\beta\}[K_{\beta\beta}]\{v_\beta\} = \delta\beta\{R_\beta\}\, \forall \{\delta v_\beta\} \Rightarrow [K_{\beta\beta}]\{v_\beta\} = \{R_\beta\} \qquad (12.32)$$

which may be solved analytically.

Ex 12-6 Rectangular Cross-Section

Rectangular cross-section

After some straightforward computations we may get the expression for $\{v_\beta\}$

$$\{v_\beta\} = \left[\frac{15EtH^2}{8\,(5L^2 + 3\nu H^2 + 3H^2)}\right]\varphi \qquad (Ex.\ 12\text{-}6.1)$$

which for $L = 0$ becomes

$$\{v_\beta\} = \left[\frac{5Et}{8(1+\nu)}\right]\varphi \qquad (Ex.\ 12\text{-}6.2)$$

Integration at $(x_1 = L)$ provides the value of the shear force V

$$V = \int_{-H/2}^{H/2} [N_T]\,\{v_\beta\}dx_2 = \frac{5EtH}{12\,(1+\nu)}\varphi \qquad (Ex.\ 12\text{-}6.3)$$

Therefore,

Effective Area A_e of Rectangular Cross-Section

$$A_e = \frac{5EtH}{12\,(1+\nu)\,G} = \tfrac{5}{6}Ht = \tfrac{5}{6}A \qquad (Ex.\ 12\text{-}6.4)$$

This value is the same as the one given in most tables, as mentioned above. It is, however, important to notice that it a lower bound and that the "true" value may be higher, see e.g. (12.24). Then, we have also shown that the coefficient $\frac{2}{3}$ found in (12.23) is too small.

Ex 12-7 Circular Cross-Section

Circular cross-section

Since the derivations for the case of a circular cross-section are very similar to the previous ones they are not shown here, while the final result for the effective area is noted

Effective Area A_e of Circular Cross-Section

$$A_e = \tfrac{9}{10}A \qquad (Ex.\ 12\text{-}7.1)$$

which agrees with Sundström (2010).

12.3.7 Concluding Remarks

Of course, the above study is not the final word on how to compute the shear stiffness, in part because three-dimensional effects have not been taken into account. For the circular cross-section this is more important than for the rectangular one, and in the case of a T-shaped or an I-shaped cross-section the assumption that it may be treated as a plate it is even more problematic. Still the lowers bounds obtained here seem justified.

Analysis ought to have been three-dimensional

Furthermore, for many structures the value of A_e/A may not be all that important. For instance, in Example Ex 7-4, the difference between the results for the tip displacement for $A_e/A = 9/10$ and $A_e/A = 5/6$ differ as little as about 1% for a beam whose length is only twice its depth, which is a *very* short beam. Even taking $A_e = A$ implies a difference as little as about 2.5%.

Results fairly insensitive to value of A_e/A

Maybe this insensitivity to values of A_e/A is the reason that most introductory courses on beam theory barely mention the issue of shear stiffness.

Chapter 13

Unconstrained Torsion

13.1 Introduction

In Sections 7.1–8.3 we have only treated beams that deform in a plane, but did not consider the phenomenon of torsion of beams and bars although plane beams and bars under torsion often remain in the plane. In principle, torsion of straight, one-dimensional members is relatively easy and therefore we did not devote space for that subject but introduce it here through the example below.

Ex 13-1 One-Dimensional Torsion

For the case shown in Fig. Ex. 13-1.1 the analysis is rather easy because the torsional moment is constant throughout the entire member. Let

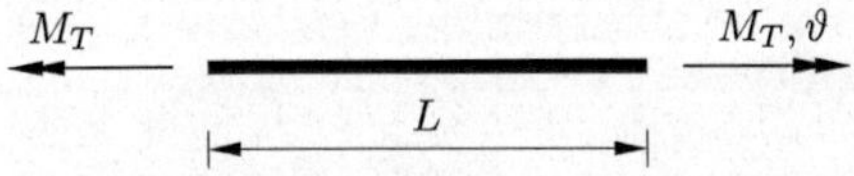

Fig. Ex. 13-1.1: *One-dimensional body subjected to a torsional moment M_T.*

us demand that the left end of the bar is prevented from rotating. Then, the obvious question is how much does the right end rotates. If the constitutive properties are constant over the entire member the solution to the problem of the rotation ϑ of the right end is not difficult if we assume linear elasticity. Then, it seems reasonable that ϑ is proportional to the applied torsional moment M_T and to the length L, but inversely proportional to the shear modulus G and some cross-sectional property I_T associated with torsion. This constant is often called the *Torsional Constant* and is sometimes denoted by J. For now we don't need to view these two entities separately but may talk about their product as the torsional stiffness of the cross-section.

Torsional constant I_T

$$\vartheta = \frac{M_T L}{G I_T} \qquad \textit{(Ex. 13-1.1)}$$

A major part of this chapter is concerned with determining the cross-sectional properties relevant for linear elastic torsion. Among the results are formulas for I_T for a number of commonly used cross-sections.

E. Byskov, *Elementary Continuum Mechanics for Everyone*,
Solid Mechanics and Its Applications 194, DOI: 10.1007/978-94-007-5766-0_13,

Unconstrained torsion
Warping allowed

In the following we shall deal with the problem of free, i.e. unconstrained torsion of cylindrical bars. The meaning of the term unconstrained is that the cross-sections of the bar are free to warp. A quite natural constraint consists, however, in clamping the cross-section at one end leading to at least local disturbances in relation to a solution to the unconstrained problem. Thus, if we glue the end of a bar to some stiff foundation the theory presented below is not valid—at least close to the supported end. Fortunately, it is such that constraining one end often only results in a local effect, a boundary layer, which has little effect at a finite distance from the support.

Since the present chapter is aimed at presenting unconstrained torsion of three-dimensional bodies in as simple a fashion as possible I have chosen to utilize the *Index Notation* with the *Summation Convention*. I have done this in part because this makes it much easier to recognize terms that constitute a divergence than by use of the old-fashioned xyz-notation, but the main reason is that all the derivations become much easier this way.

In many textbooks soap film analogies are utilized to handle the problem of linear elastic torsion. This may have been a good idea before the advent of computers, but nowadays seems like an unnecessary detour. In the same spirit, I compute the torsional moment directly by integrating the effect of the shear strains instead of appealing to contour lines of the stress function for that purpose. Thus, as I usually recommend, I try to let physical insight into the problems rather than mathematical tricks take the leading role.[13.1]

Whenever possible we shall utilize some variational principle as the basis for constructing finite elements. And, if possible, I prefer the variational principle to express some kind of energy, such as the potential Energy $\mathbf{\Pi}_\mathrm{P}$, the complementary energy $\mathbf{\Pi}_\mathrm{C}$, or some modified version of one of these "real" energies. Alternatively, we may base our finite elements on some principle of virtual work, either the principle of virtual displacements, the principle of virtual forces or some modified version of either one. But, in a number of cases, no such energy or (virtual) work based principle exists, or they may not furnish appealing equations and we must try alternative formulations of the finite element method. There are many types of alternative finite element methods, for example ones that are based on a Galerkin principle or on a weighted residual, see e.g. (Cook, Malkus & Plesha 2002).

[13.1] Originally, I had intended to write a little section on how to construct a basis for finite element analysis of unconstrained torsion showing that one could use the differential equation for the stress function directly instead of an energy functional or a principle of virtual work. This would have taken up something like the six pages it does in the present note. Unfortunately, I discovered that the texts available to my students constituted a very poor foundation for my purpose and, therefore, I felt that most of the other twenty-five or so pages were needed. Once I got started, I felt a need for redoing and, in some cases, expanding the classic examples of the circular cross-section, the ring, etc., with the result that the section got a lot bigger than was my original intention.

Although the problem may be written in terms of the axial displacement, which may be associated with a potential energy, I prefer to formulate it using a stress function.

13.2 Structural Problem

The problem of unconstrained[13.2] torsion of a cylindrical bar is sketched in Figs. 13.2 and 13.3, where the latter shows that the bar may contain holes. As regards the material of the bar we shall assume that the constitutive

Cross-sections free to warp

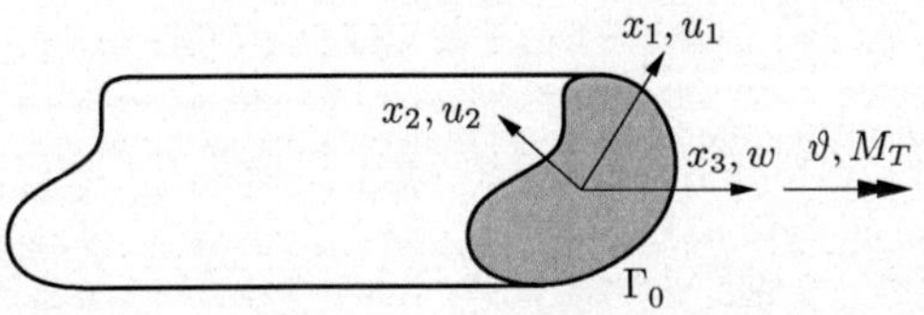

Fig. 13.2: Cylindrical body subjected to a torsional moment M_T.

properties are independent of the axial coordinate x_3. Also, the torsional moment M_T is assumed to be independent of the axial coordinate x_3.[13.3] As you will see, here the coordinate axis in the cross-section are denoted $x_\alpha, \alpha = 1, 2$, which is in contrast to our choice in Chapter 12 because here the index notation offers great advantages. The rotation of the cross-sections is denoted ϑ.

Torsional moment M_T
Rotation of cross-section: $\vartheta(x_3)$

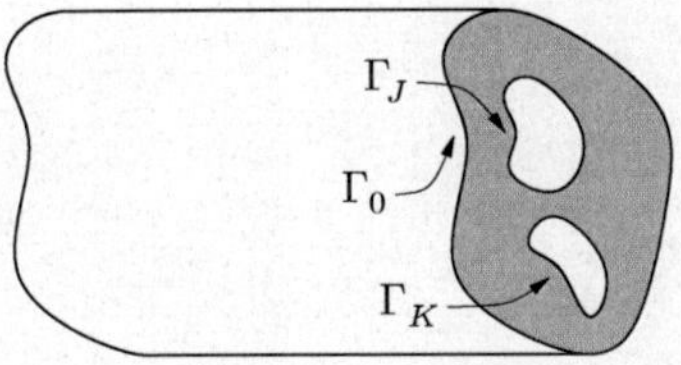

Fig. 13.3: Cylindrical body with holes.

Below, we derive the necessary formulas below in a way that serves the purpose of introducing a notation that is particularly convenient for our purpose of setting up finite element equations.

[13.2] The meaning of the term unconstrained is explained further subsequently, but here it suffices to know that the cross-sections are free to warp.

[13.3] The quantity M_T which I denote torsional moment is often called the twisting moment or the torque.

13.3 Geometry

Referring to Figs. 13.2 and 13.3, the outer boundary of the cross-section of the bar is denoted Γ_0, while boundaries of possible cylindrical holes are designated $\Gamma_J, J = 1, 2, \ldots, M$, where M is the number of holes. The cross-section is independent of the axial coordinate x_3. The area covered by material is A_0, while A_J designates the area of hole J, and the entire area, including $A_J, J = 0, 1, \ldots, M$, inside Γ_0 is A.

13.4 Kinematics

We shall assume that the displacement field entails rotation $\vartheta(x_3)$ of the cross-sections as rigid bodies, except that the cross-sections are free to warp which is the reason for the term unconstrained. The warping of the cross-sections is assumed to be independent of x_3. These assumptions have been verified by experiments as early as the Nineteenth Century. Any displacement field u_i which obeys these constraints may be written:[13.4]

First assumed displacement field

$$u_\alpha(x_i) = e_{\beta\alpha}(x_\beta - x_\beta^0)\vartheta(x_3)\,, \quad u_3(x_i) = \psi(x_\alpha)\vartheta_{,3}(x_3) \tag{13.1}$$

Warping function ψ

where, as usual, Greek indices cover the range $[1, 2]$, while Latin indices take the values $[1, 2, 3]$. By $\psi(x_\alpha)$ denote the warping function, and let x_α^0 signify the coordinates of the axis of twist. Summation over lower-case indices is implied in the following. As usual, the permutation symbol $e_{\beta\alpha}$ is defined by

Permutation symbol $e_{\beta\alpha}$

$$e_{\beta\alpha} = \begin{cases} +1 \text{ for } \beta = 1, \alpha = 2 \\ -1 \text{ for } \beta = 2, \alpha = 1 \\ 0 \text{ for } \beta = \alpha \end{cases} \tag{13.2}$$

We shall further restrict the angle of twist ϑ such that is varies linearly with the axial coordinate x_3

Further restriction on displacements

$$\vartheta(x_3) = \theta x_3 \tag{13.3}$$

A consequence of this assumption is that the displacements are

Final assumed displacement field

$$u_\alpha(x_i) = e_{\beta\alpha}(x_\beta - x_\beta^0)\theta x_3 \text{ and } u_3(x_i) = \psi(x_\alpha)\theta \tag{13.4}$$

where we may choose $x_3 = 0$ at a place that suits our purpose, probably at one end of the bar.

[13.4] Note that, as usual, the description of kinematics of the body entails neither information about statics, nor of constitutive models. Thus, the results below for kinematics are also valid for plasticity.

13.4.1 Strains

As a result of the displacement assumptions of (13.1) and (13.2) the only non-vanishing strains are $\varepsilon_{\alpha 3}$

$$\begin{aligned}\varepsilon_{\alpha\beta} &= \tfrac{1}{2}\left(e_{\gamma\alpha}\delta_{\gamma\beta} + e_{\gamma\beta}\delta_{\gamma\alpha}\right) = \tfrac{1}{2}(e_{\beta\alpha} + e_{\alpha\beta}) = 0 \\ \varepsilon_{\alpha 3} &= \varepsilon_{3\alpha} = \tfrac{1}{2}\left(e_{\beta\alpha}(x_\beta - x^0_\beta) + \psi_{,\alpha}\right)\vartheta_{,3} \\ &= \tfrac{1}{2}\left(e_{\beta\alpha}(x_\beta - x^0_\beta) + \psi_{,\alpha}\right)\theta \\ \varepsilon_{33} &= \psi\vartheta_{,33} = 0\end{aligned} \tag{13.5}$$

Strains

where the Kronecker delta, as usual, is defined by

$$\delta_{\alpha\beta} = \begin{cases} +1 \text{ for } \beta = \alpha \\ \;\;0 \text{ for } \beta \neq \alpha \end{cases} \tag{13.6}$$

Kronecker delta $\delta_{\alpha\beta}$

13.5 Statics

In the same spirit as the one behind the kinematic assumptions we shall presuppose that the stress state is independent of the axial coordinate x_3. For almost any conceivable constitutive relation $\sigma_{\alpha\beta}$ and σ_{33} vanish because the strains $\varepsilon_{\alpha\beta}$ and ε_{33} are all equal to zero. Then, the only non-vanishing stresses are $\sigma_{\alpha 3}(x_\gamma)$, where we have indicated that the stress state does not vary with the axial coordinate. Thus, the only equilibrium equations that are not satisfied a priori are

Stress state assumed independent of axial coordinate x_3

$$\sigma_{3\alpha,\alpha} = 0 \tag{13.7}$$

Only surviving equilibrium equation

Once we have decided on the constitutive relation we have established a boundary value problem consisting of the above kinematic and static relations in connection with the constitutive relation. As we shall see, for linear elasticity and, by the way also for perfect plasticity, there is a convenient way of formulating the boundary value problem in terms of a stress function T, which will be defined later.

13.6 Stress Function

We intend to formulate equilibrium in terms of a stress function T, see Section 13.6.1 below. First, however, for convenience, let

$$\tau_\alpha \equiv \sigma_{3\alpha} = \sigma_{\alpha 3} \tag{13.8}$$

New notation $\tau_\alpha \equiv \sigma_{3\alpha} = \sigma_{\alpha 3}$

Then, we may write (13.7) as

$$\tau_{\alpha,\alpha} = 0 \tag{13.9}$$

Equilibrium equation

We could, of course, have continued using $\sigma_{3\alpha}$ and $\sigma_{\alpha 3}$, but the following expressions would have looked more complicated.

13.6.1 Equilibrium

Introduce the stress function $T(x_\alpha)$ by

Stress function T

$$\tau_\alpha = e_{\alpha\beta} T_{,\beta} \tag{13.10}$$

In terms of this function the equilibrium equation (13.9) becomes

Equilibrium equation

$$e_{\alpha\beta} T_{,\beta\alpha} = 0 \tag{13.11}$$

which is always satisfied because $e_{\alpha\beta}$ is antisymmetric, while $T_{,\beta\alpha}$ is symmetric in its indices.[13.5]

Are the strains compatible?

Already at this point we mention that, as always when a formulation is in terms of a stress function, there is a problem regarding compatibility. Satisfaction of the static equations does not guaranty that the strains are compatible, i.e. that they may be integrated to yield a displacement field, see also Section 4.2.2. We shall deal with this issue later.

On all boundaries, the ones at the holes as well as the outer one, the static boundary condition is

Static boundary conditions

$$\sigma_{3\alpha} n_\alpha = 0 \text{ or } \tau_\alpha n_\alpha = 0 \tag{13.12}$$

where n_α is the (outward) normal of the surface. Written in terms of the stress function T this becomes

Static boundary conditions

$$e_{\alpha\beta} T_{,\beta} n_\alpha = 0 \tag{13.13}$$

The normal n_α to a boundary $\Gamma_J, J = 0, 1, \ldots, M$ may easily be expressed in terms of the tangent t_β

n_α in terms of t_α

$$n_\alpha = e_{\alpha\beta} t_\beta \tag{13.14}$$

or similarly

t_α in terms of n_α

$$t_\beta = e_{\alpha\beta} n_\alpha \tag{13.15}$$

With this formula in hand we may rewrite the boundary condition (13.13)

Static boundary conditions

$$T_{,\beta} t_\beta = 0 \tag{13.16}$$

which is the same as

Static boundary conditions

$$T_{,s} = 0 \tag{13.17}$$

where $_{,s}$ means differentiation along the boundary. Therefore, independent

[13.5] Prandtl (1904) gave the function $T(x_\alpha)$ a "physical" interpretation in that he regarded it as a stress surface which may be thought of as a soap film which suffers small displacements caused by atmospheric pressure. Although this interpretation for some purposes proves to be convenient I shall not go into any details in this matter.

of the constitutive relation, the stress function is constant along all boundaries, both the outer and the inner ones

Stress function $T = $ const. on all boundaries

$$T = T_J,\ x_\alpha \in \Gamma_J\,,\ J = 0, 1, \ldots, M \tag{13.18}$$

where T_J denote constants whose values are determined by the solution to the boundary value problem, which is derived below. One of these values must be fixed in advance, and we shall always take the value of T at the outer boundary to be zero

Choose $T = T_0 = 0$ on outer boundary

$$T_0 = 0 \tag{13.19}$$

13.6.2 Compatibility

Ensure that $\psi_{,\alpha}$ can be integrated

In general, compatibility equations may be derived in a number of ways. Here, we ensure that the derivatives $\psi_{,\alpha}$ can be integrated to provide the warping function $\psi(x_\alpha)$ in an unambiguous way.

Rewrite (13.5b)

$$\psi_{,\alpha}\theta = 2\varepsilon_{\alpha 3} - e_{\beta\alpha}(x_\beta - x_\beta^0)\theta \tag{13.20}$$

Prove that ψ is single-valued

In order that u_3 and thereby ψ is single-valued we integrate along a closed curve Γ and require that the integral of $\psi_{,\alpha}t_\alpha$ vanishes[13.6]

$$0 = \oint_\Gamma \psi_{,\alpha}t_\alpha d\Gamma \tag{13.21}$$

i.e. that

$$0 = \oint_\Gamma (2\varepsilon_{\alpha 3} - e_{\beta\alpha}(x_\beta - x_\beta^0)\theta))t_\alpha d\Gamma \tag{13.22}$$

When we introduce (13.14) we may get

$$\oint_\Gamma 2\varepsilon_{\alpha 3}t_\alpha d\Gamma = \oint_\Gamma (x_\beta - x_\beta^0)n_\beta\theta d\Gamma \tag{13.23}$$

We may apply the divergence theorem to the right-hand side of (13.23)

$$\oint_\Gamma 2\varepsilon_{\alpha 3}t_\alpha d\Gamma = \int_{A_\Gamma} (x_{\beta,\beta} - x_{\beta,\beta}^0)\theta dA \tag{13.24}$$

where A_Γ denotes the area inside Γ. Since x_β^0 denotes the coordinates to a fixed point the second term in the parenthesis is zero. Further,

$$x_{\beta,\beta} = \delta_{\beta\beta} = 2 \tag{13.25}$$

[13.6] The curve Γ may be one of the previously introduced Γ_J, but is not confined to the boundary of any hole or the outer boundary Γ_0.

We may now conclude that the strains are compatible when

Compatibility equation

$$\oint_{\Gamma} \varepsilon_{\alpha 3} t_\alpha d\Gamma = A_\Gamma \theta \tag{13.26}$$

Compatibility equation independent of constitutive relation

It is worthwhile noting that this result is independent of the constitutive relation.[13.7] On the other hand, since we wish to formulate it in terms of the stress function T we need to introduce the constitutive relation which, in the cases below, is taken to be linear elastic.

If the curve Γ is associated with an inner boundary Γ_J we may see that (13.21) and therefore (13.26) introduces a kinematic condition on all inner boundaries

Kinematic condition on inner boundaries

$$\oint_{\Gamma_J} \varepsilon_{\alpha 3} t_\alpha d\Gamma = A_J \theta \tag{13.27}$$

13.6.3 Torsional Moment

Torsional moment M_T important

The value of the torsional moment M_T is, of course, one of the most important quantities. It is not necessary to divide the following derivations into two parts, namely one which is concerned with simply connected regions, i.e. bodies without holes, and one dealing with multiply connected regions. However, in order to facilitate understanding, we begin with simply connected regions.

13.6.3.1 Torsional Moment—Simply Connected Region

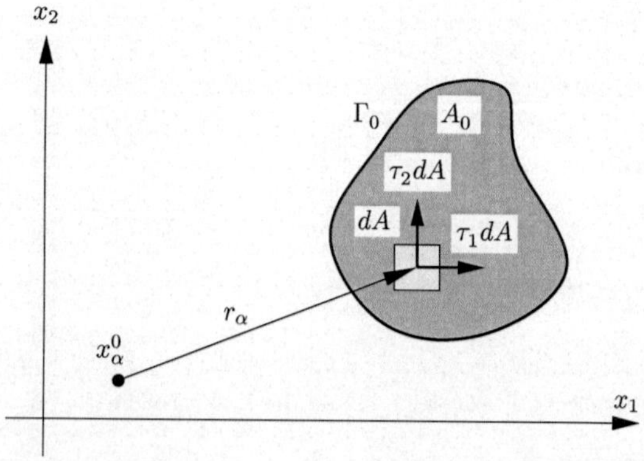

Fig. 13.4: Simply connected region with shear stresses.

[13.7] This fact should not come as a surprise since the condition is about compatibility of a *kinematic* quantity, namely the strain.

The torsional moment M_T may be computed as

$$M_T = \int_{A_0} e_{\alpha\beta} r_\alpha \tau_\beta dA = e_{\alpha\beta} e_{\beta\gamma} \int_{A_0} r_\alpha T_{,\gamma} dA$$
$$= e_{\alpha\beta} e_{\beta\gamma} \int_{A_0} ((r_\alpha T)_{,\gamma} - r_{\alpha,\gamma} T) dA \qquad (13.28)$$

where, see Fig. 13.4

$$r_\alpha = x_\alpha - x_\alpha^0 \qquad (13.29)$$

Apply the divergence theorem to the first term in the integral and realize that

$$r_{\alpha,\gamma} = \delta_{\alpha\gamma} \qquad (13.30)$$

to get

$$M_T = e_{\alpha\beta} e_{\beta\gamma} \oint_{\Gamma_0} r_\alpha n_\gamma T d\Gamma - e_{\alpha\beta} e_{\beta\gamma} \delta_{\alpha\gamma} \int_{A_0} T dA \qquad (13.31)$$

When we note that

$$e_{\alpha\beta} e_{\beta\alpha} = -2 \qquad (13.32)$$

and exploit (13.14) we may get

$$M_T = e_{\alpha\beta} \oint_{\Gamma_0} r_\alpha (-t_\beta) T d\Gamma - e_{\alpha\beta} e_{\beta\alpha} \int_{A_0} T dA \qquad (13.33)$$

or

$$M_T = -T_0 \oint_{\Gamma_0} r_\alpha n_\alpha d\Gamma + 2 \int_{A_0} T dA \qquad (13.34)$$

where we have utilized the fact that T is constant on Γ_0, see (13.18).

Apply the divergence theorem to the first term and get

$$M_T = -T_0 \int_{A_0} r_{\alpha,\alpha} dA + 2 \int_{A_0} T dA \qquad (13.35)$$

Rearrange terms and utilize that

$$r_{\alpha,\alpha} = \delta_{\alpha\alpha} = 2 \qquad (13.36)$$

to arrive at the following result for the torsional moment expressed in terms of the stress function T

$$M_T = 2 \int_{A_0} T dA - 2 T_0 A_0 \qquad (13.37)$$

Torsional moment M_T

The last term has no influence on the computed value of the torsional moment M_T and in almost all cases we choose $T_0 = 0$, see (13.19).

Thus, we may get the following simple expression

Torsional moment M_T for $T_0 = 0$

$$M_T = 2\int_{A_0} T dA \tag{13.38}$$

As we may see, the torsional moment M_T is equal to twice the volume of the stress function T—*independent of the constitutive relation.*

It is worth noticing that the above derivation also shows that the resultant of τ_α vanishes

Resultant of τ_α vanishes

$$\int_{A_0} \tau_\alpha dA = 0 \tag{13.39}$$

otherwise a resulting shear force would have made the value of M_T dependent of the choice of x_α^0 which cannot be not the case.

13.6.3.2 Torsional Moment—Multiply Connected Region.

When the cylinder contains holes and therefore the cross-section is a multiply connected region the computation of the torsional moment M_T becomes more complicated. This is mainly due to the integrations involved, see below.

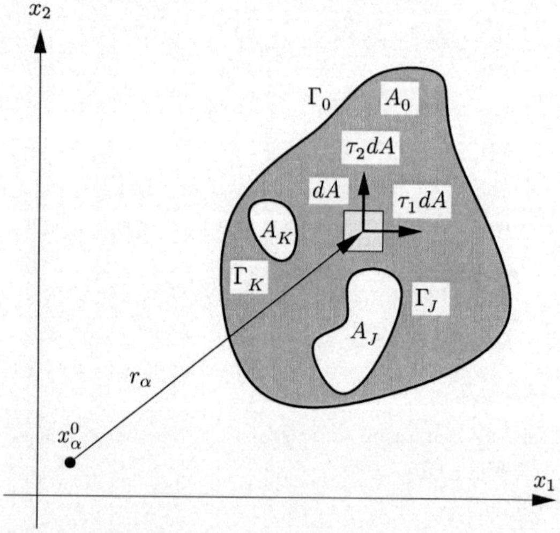

Fig. 13.5: Multiply connected region with shear stresses.

As before, let A denote the entire area inside the outer boundary Γ_0, A_0 designate the area covered by the material, and let A_J be the area of hole number J, $J = 1, 2, \ldots, M$ where M is the number of holes. Then, we note that

$$A = A_0 + \sum_{J=1}^{M} A_J \tag{13.40}$$

The torsional moment M_T may still be computed as, see Fig. 13.5

$$\begin{aligned} M_T &= \int_{A_0} e_{\alpha\beta} r_\alpha \tau_\beta dA = e_{\alpha\beta} e_{\beta\gamma} \int_{A_0} r_\alpha T_{,\gamma} dA \\ &= e_{\alpha\beta} e_{\beta\gamma} \int_{A_0} ((r_\alpha T)_{,\gamma} - r_{\alpha,\gamma} T) dA \end{aligned} \tag{13.41}$$

where the only contribution clearly comes from the area A_0 since it is the only area covered by material.

Problems with divergence theorem for multiply connected region

In the case of the simply connected region, we applied the divergence theorem. We shall also do that here, but we must be very careful because of the fact that the region is multiply connected. We shall exploit a trick when we convert a surface integral to a boundary integral in that we connect the outer boundary Γ_0 with the boundaries Γ_J of the holes through paths which go back and forth between Γ_0 and Γ_J, $J = 1, 2, \ldots, M$, see Fig. 13.6. The path going from Γ_0 to Γ_J coincides with the path going back, but in the drawing the two paths are shown as if they were separate. In this way we have converted the multiply connected region into a simply connected one. The reason why this is permissible is that the contributions to the integrals when we travel from Γ_0 to Γ_J is exactly the same as the one we get on the return trip but with the opposite sign. Therefore, the two contributions cancel out and the only contributions to the boundary integral come from the outer and inner boundaries. Note that the inner boundaries Γ_J are traveled in the opposite direction of the outer boundary Γ_0 meaning that the contributions from Γ_J have the opposite sign of the contribution from Γ_0. With these considerations in mind (13.41) provides

Trick: go from Γ_0 to Γ_J and back the same way

$$M_T = e_{\alpha\beta} e_{\beta\gamma} \left(\oint_{\Gamma_0} r_\alpha T n_\gamma d\Gamma - \sum_{J=1}^{M} \oint_{\Gamma_J} r_\alpha T n_\gamma d\Gamma - \delta_{\alpha\gamma} \int_{A_0} T dA \right) \tag{13.42}$$

Then,

$$M_T = e_{\alpha\beta} e_{\beta\gamma} \left(T_0 \oint_{\Gamma_0} r_\alpha n_\gamma d\Gamma - \sum_{J=1}^{M} T_J \oint_{\Gamma_J} r_\alpha n_\gamma d\Gamma \right) + 2 \int_{A_0} T dA \tag{13.43}$$

where we have exploited the fact that the stress function T is constant on all boundaries.

In applying the divergence theorem to the first term in the parenthesis it is important to note that the area is the *total area* A, *not* A_0. We apply

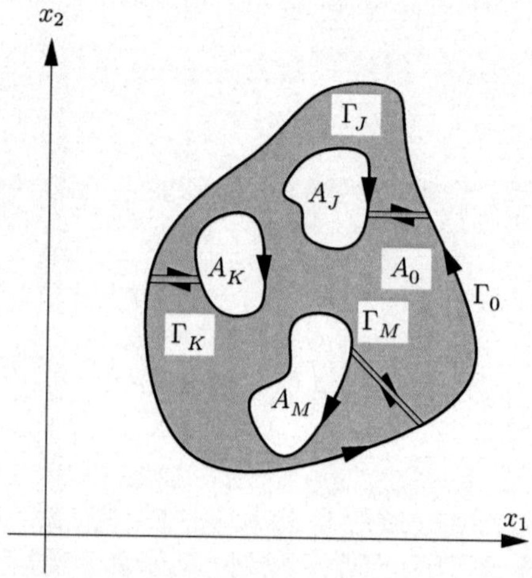

Fig. 13.6: Multiply connected region with modified integration path.

the divergence theorem to all boundary integrals and get

$$M_T = -2T_0 A + 2\sum_{J=1}^{M} T_J A_J + 2\int_{A_0} T dA \tag{13.44}$$

or, considering (13.40):

$$M_T = 2\int_{A_0} T dA + 2\sum_{J=1}^{M} (T_J - T_0) A_J - 2T_0 A_0 \tag{13.45}$$

If, as usual, we choose $T_0 = 0$ the result becomes simpler

Torsional moment M_T for $T_0 = 0$

$$M_T = 2\int_{A_0} T dA + 2\sum_{J=1}^{M} T_J A_J \tag{13.46}$$

As was the case of the simply connected region the torsional moment M_T is equal to twice the volume of the stress function—*independent of the constitutive relation*, but note that the stress function itself clearly depends on that relation.

We may extend the definition of T such that it covers the holes and takes the values T_J, $J = 1, 2, \ldots, M$ in the holes. Then, (13.46) takes a simpler form, namely

$$M_T = 2\int_A T dA \tag{13.47}$$

Torsional moment M_T for $T_0 = 0$

13.7 Linear Elasticity

Under the assumption of linear elasticity the torsional moment M_T must be proportional to the shear modulus G since the only non-vanishing strains are the shear strains $e_{\alpha\beta}$. Therefore, we may postulate the following relationship between M_T and the angle of twist θ

$$M_T = G I_T \theta \tag{13.48}$$

where I_T is the *Torsional Moment of Inertia* which, except for circular cross-sections, is not the same as the polar moment of inertia I_p, see the following examples.[13.8]

Usually: $I_T \neq I_p$

In order to describe torsion for a variety of cross-sections we need a differential equation for the warping function ψ, but first we derive a formula which results from the compatibility condition in the case of linear elasticity.

13.7.1 Compatibility

In this case we may easily express the strains in terms of the stresses and therefore also in terms of the stress function T

$$2\varepsilon_{\alpha 3} = \frac{\tau_\alpha}{G} = \frac{1}{G} e_{\alpha\beta} T_{,\beta} \tag{13.49}$$

Strains and stresses in terms of T

which is inserted in (13.24) with (13.25)

$$\oint_\Gamma \frac{1}{G} e_{\alpha\beta} T_{,\beta} t_\alpha d\Gamma = \int_{A_\Gamma} 2\theta dA \tag{13.50}$$

where A_Γ denotes the area inside the boundary Γ. When we exploit (13.14) we may get

$$-\oint_\Gamma \frac{1}{G} T_{,\beta} n_\beta d\Gamma = \int_{A_\Gamma} 2\theta dA \tag{13.51}$$

By use of the divergence theorem and rearrangement of terms we may get

$$\int_{A_\Gamma} \left(\frac{1}{G} T_{,\beta\beta} + 2\theta \right) dA = 0 \tag{13.52}$$

which must hold for all closed paths Γ.

[13.8] In a principle of virtual work the torsional moment M_T and the angle of twist θ may serve as generalized stress and strain, respectively.

Therefore, we may get the simple result that

(13.53) together with

$$T_{,\beta\beta} = -2G\theta \tag{13.53}$$

ensures compatibility for the case of linear elasticity when the uniqueness condition (13.51), which for $\Gamma = \Gamma_J$ may be written

(13.54) ensures compatibility

$$-\oint_{\Gamma_J} \frac{1}{G} T_{,\beta} n_\beta d\Gamma = 2\theta A_J \tag{13.54}$$

also is satisfied on all Γ_J.

13.7.2 Warping Function

The equilibrium equation (13.9), which, for convenience, is repeated here

Equilibrium equation

$$\tau_{\alpha,\alpha} = 0 \tag{13.55}$$

may be expressed in terms of the strains, see (13.5b) and the constitutive relation (13.49)

Equilibrium equation

$$0 = G\left(e_{\beta\alpha}\delta_{\beta\alpha} + \psi_{,\alpha\alpha}\right)\theta \tag{13.56}$$

with the result that the warping function ψ satisfies

DE for warping function ψ

$$\psi_{,\alpha\alpha} = 0 \tag{13.57}$$

which shows that the warping function ψ is a harmonic function. This expression is, however, not sufficient to determine ψ—usually we would need boundary conditions in order to have a well-posed boundary value problem for ψ. Here, we go about the task in a different way, as shown below, in that we utilize the requirement that the strains are compatible. The compatibility condition (13.22) was based on (13.20)

$$\psi_{,\alpha}\theta = 2\varepsilon_{\alpha 3} - e_{\beta\alpha}(x_\beta - x_\beta^0)\theta \tag{13.58}$$

Introduce the constitutive relation (13.49) which connects the shear strains $\varepsilon_{\alpha\beta}$ and the derivatives $T_{,\beta}$ of the stress function T and get

DE for warping function ψ

$$\psi_{,\alpha} = e_{\alpha\beta}\left(\frac{1}{G\theta}T_{,\beta} + \left(x_\beta - x_\beta^0\right)\right) \tag{13.59}$$

where we have exploited the fact that $e_{\alpha\beta} = -e_{\beta\alpha}$. Note that the expressions (13.59) and (13.57) may be utilized to determine the shape of warping function ψ to within a constant. This, however, does not present any severe problem because we may add any constant to ψ and still get the same strains and stresses—a rigid body translation of the bar does, of course, not influence the strains and stresses.

(13.59) and (13.57) determine ψ to within a constant

13.7.3 Examples of Elastic Torsion

Below, we explore examples of torsion of bars with cross-sections of various complexity. We begin with the simplest case, namely the circle.

Ex 13-2 Circular Cross-Section

The simplest example in this context is a bar with a circular cross-section, see Fig. Ex. 13-2.1. In this case we can find the shape of the

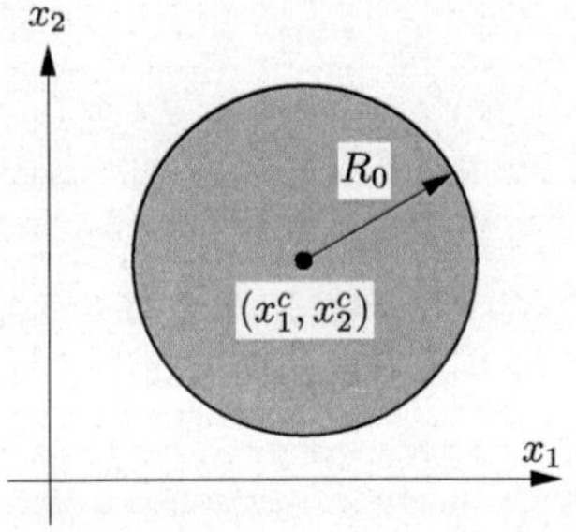

Fig. Ex. 13-2.1: *Circular cross-section.*

stress function T in a very straightforward manner because we know that it must satisfy (13.53) and take a constant value T_0 on the boundary. When we, as always, choose $T_0 = 0$ the following expression, where C denotes a constant which is determined later, is a good candidate for T

In this case we may guess the shape of T

$$T = C\Big(\big(R_0\big)^2 - \big(x_\alpha - x_\alpha^c\big)\big(x_\alpha - x_\alpha^c\big)\Big) \qquad \text{(Ex. 13-2.1)}$$

because it takes the value 0 on the boundary, and its second derivatives are constants

$$T_{,\beta} = -2C\delta_{\alpha\beta}\big(x_\alpha - x_\alpha^c\big) = -2C\big(x_\beta - x_\beta^c\big) \qquad \text{(Ex. 13-2.2)}$$

i.e.

$$T_{,\beta\beta} = -4C \qquad \text{(Ex. 13-2.3)}$$

where, according to (13.53)

$$T_{,\beta\beta} = -2G\theta \qquad \text{(Ex. 13-2.4)}$$

and thus

$$T = \tfrac{1}{2}G\theta\Big(\big(R_0\big)^2 - \big(x_\alpha - x_\alpha^c\big)\big(x_\alpha - x_\alpha^c\big)\Big) \qquad \text{(Ex. 13-2.5)}$$

satisfies the field equation and boundary condition on T and therefore constitutes the solution.

The relation between the torsional moment M_T and the twist θ is given by (13.38), which here becomes

$$\begin{aligned} M_T &= 2\int_{A_0} T dA = G\theta \int_{A_0} \Big(\big(R_0\big)^2 - r^2 \Big) dA \\ &= +G\theta\pi\big(R_0\big)^4 - G\theta \int_0^{R_0}\int_0^{2\pi} r^2 d\varphi r dr \\ &= +G\theta\pi\big(R_0\big)^4 - G\theta 2\pi \tfrac{1}{4}\big(R_0\big)^4 \end{aligned} \qquad (Ex.\ 13\text{-}2.6)$$

The final result for M_T is

Torsional moment M_T

$$M_T = \tfrac{1}{2}\pi G\theta \big(R_0\big)^4 \qquad (Ex.\ 13\text{-}2.7)$$

and thus

Torsional moment of inertia I_T

$$I_T = \tfrac{1}{2}\pi\big(R_0\big)^4 \qquad (Ex.\ 13\text{-}2.8)$$

Only for circular cross-sections: $I_T = I_p$

We may see that $\pi/2(R_0)^4$ is the *Polar Moment of Inertia* I_P of the circular region. It is, however, important to notice that the polar moment of inertia only for circular cross-sections, including circular rings, plays this crucial role and that in all other cases the resistance to twist may not be expressed in terms of I_P. Furthermore, as we shall see later, the circular cross-section is a very special one in that it is the only one which does not warp.

The maximum shear stress is, of course, one of the major features of a problem involving torsion. In the present case, T is symmetric about the center of the circle which means that we may focus on the stresses for $x_2 = x_2^c$. Here, the expression for T is

$$T = \tfrac{1}{2}G\theta\Big(\big(R_0\big)^2 - \big(x_1 - x_1^c\big)^2\Big) \qquad (Ex.\ 13\text{-}2.9)$$

According to (13.10) we get

$$\begin{aligned} \tau_1 &= e_{12}T_{,2} = 0 \\ \tau_2 &= e_{21}T_{,1} = G\theta x_1 \end{aligned} \qquad (Ex.\ 13\text{-}2.10)$$

This means that the maximum shear stress $\max(\tau)$ is

$\max(\tau)$

$$\max(\tau) = G\theta R_0 = \frac{2M_T}{\pi\big(R_0\big)^3} \qquad (Ex.\ 13\text{-}2.11)$$

Based on (*Ex. 13-2.9*) it is tempting to conclude that the maximum shear stress always must occur as far as possible from the center of the cross-section, but this is not a generally valid conclusion—as a matter of fact, the opposite is usually the case, see e.g. Example Ex 13-3 below.

It is of major interest to know the warping—if any such phenomenon occurs—of the circular cross-section. We shall utilize (13.59) as basis for this purpose. Determine

$$T_{,\beta} = -G\theta\delta_{\beta\alpha}\big(x_\alpha - x_\alpha^c\big) = -G\theta\big(x_\beta - x_\beta^c\big) \qquad (Ex.\ 13\text{-}2.12)$$

which is inserted into (13.59)

$$\psi_{,\alpha} = \Big(-(x_\beta - x^c_\beta) + (x_\beta - x^0_\beta)\Big)e_{\alpha\beta} \qquad (Ex.\ 13\text{-}2.13)$$

i.e.

$$\psi_{,\alpha} = (x^c_\beta - x^0_\beta)e_{\alpha\beta} \qquad (Ex.\ 13\text{-}2.14)$$

or

$$\psi_{,1} = +(x^c_2 - x^0_2) \quad \text{and} \quad \psi_{,2} = -(x^c_1 - x^0_1) \qquad (Ex.\ 13\text{-}2.15)$$

which provides

No warping of circular cross-section

$$\psi = +(x^c_2 - x^0_2)x_1 - (x^c_1 - x^0_1)x_2 \qquad (Ex.\ 13\text{-}2.16)$$

which shows that after deformation the cross-section is still a plane and, if $x^0_\alpha = x^c_\alpha$, then ψ vanishes. Whether or not the axis of torsion coincides with the centroid of the cross-section no warping occurs. This is a feature that is particular to circular cross-sections.

The next example, which is concerned with torsion of bars with elliptic cross-sections, is more complicated and sheds some light on the issues of maximum shear stress and of warping.

Ex 13-3 Elliptic Cross-Section

Although the elliptic cross-section, see Fig. Ex. 13-3.1, does not look much more complicated than the circular one there are important differences, as we shall see.

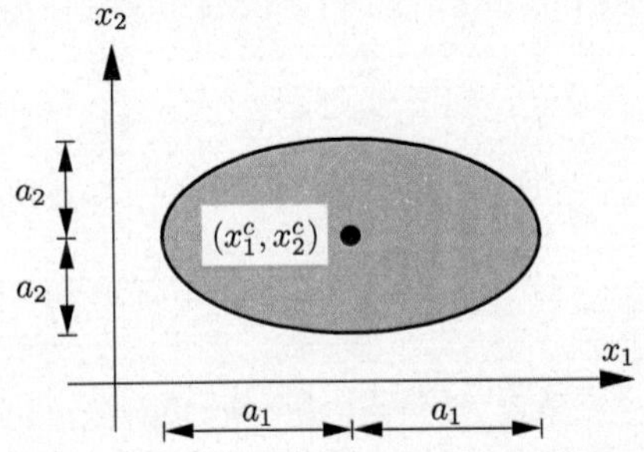

Fig. Ex. 13-3.1: *Elliptic cross-section.*

From the outset, we proceed as we did in Example Ex 13-2, but, as we shall see soon, the analysis and the conclusions differ substantially from the ones for the circular cross-section. In the spirit of (*Ex. 13-2.1*) we assume the following expression for T

Assumed stress function

$$T = C\left(1 - \left(\frac{x_1 - x^c_1}{a_1}\right)^2 - \left(\frac{x_2 - x^c_2}{a_2}\right)^2\right) \qquad (Ex.\ 13\text{-}3.1)$$

which provides

$$T_{,\beta} = -2C\left(\frac{\delta_{1\beta}(x_1 - x_1^c)}{(a_1)^2} + \frac{\delta_{2\beta}(x_2 - x_2^c)}{(a_2)^2}\right) \qquad (Ex.\ 13\text{-}3.2)$$

and thus

$$T_{,\beta\beta} = -2C\left(\frac{\delta_{1\beta}\delta_{1\beta}}{(a_1)^2} + \frac{\delta_{2\beta}\delta_{2\beta}}{(a_2)^2}\right) = -2C\left(\frac{1}{(a_1)^2} + \frac{1}{(a_2)^2}\right) (Ex.\ 13\text{-}3.3)$$

i.e.

$$T_{,\beta\beta} = -2C\frac{(a_1)^2(a_2)^2}{(a_1)^2 + (a_2)^2} \qquad (Ex.\ 13\text{-}3.4)$$

and therefore

$$T = G\theta\frac{(a_1)^2(a_2)^2}{(a_1)^2 + (a_2)^2}\left(1 - \left(\frac{x_1 - x_1^c}{a_1}\right)^2 - \left(\frac{x_2 - x_2^c}{a_2}\right)^2\right) (Ex.\ 13\text{-}3.5)$$

which agrees with (*Ex. 13-2.9*) for the circle.

In order to facilitate the computation of the torsional moment M_T introduce another set of coordinates y_α, where

$$y_\alpha = x_\alpha - x_\alpha^c \qquad (Ex.\ 13\text{-}3.6)$$

Then, (*Ex. 13-3.5*) becomes

Stress function T

$$T = G\theta\frac{(a_1)^2(a_2)^2}{(a_1)^2 + (a_2)^2}\left(1 - \left(\frac{y_1}{a_1}\right)^2 - \left(\frac{y_2}{a_2}\right)^2\right) \qquad (Ex.\ 13\text{-}3.7)$$

and the torsional moment M_T is

$$\begin{aligned} M_T &= 2G\theta\frac{(a_1)^2(a_2)^2}{(a_1)^2 + (a_2)^2}\int_{A_0}\left(1 - \left(\frac{y_1}{a_1}\right)^2 - \left(\frac{y_2}{a_2}\right)^2\right) dA \\ &= 2G\theta\frac{(a_1)^2(a_2)^2}{(a_1)^2 + (a_2)^2}\left(A_0 - \frac{I_{22}}{(a_1)^2} - \frac{I_{11}}{(a_2)^2}\right) \end{aligned} \qquad (Ex.\ 13\text{-}3.8)$$

where A_0 is the area of the ellipse, while I_{11} and I_{22} are its second-order moments

$$A_0 = \pi a_1 a_2\,,\ I_{11} = \frac{\pi}{4}a_1(a_2)^3\,,\ I_{22} = \frac{\pi}{4}(a_1)^3 a_2 \qquad (Ex.\ 13\text{-}3.9)$$

Therefore,

$$M_T = 2G\theta\frac{(a_1)^2(a_2)^2}{(a_1)^2 + (a_2)^2}\left(\pi a_1 a_2 - \frac{\pi}{4}a_1 a_2 - \frac{\pi}{4}a_1 a_2\right) (Ex.\ 13\text{-}3.10)$$

The final result for M_T is

Torsional moment M_T for ellipse

$$M_T = \pi G\theta\frac{(a_1)^3(a_2)^3}{(a_1)^2 + (a_2)^2} \qquad (Ex.\ 13\text{-}3.11)$$

which agrees with (*Ex. 13-2.7*) for the circular cross-section when $a_1 = a_2 = R_0$. For I_T we therefore find

Torsional moment of inertia I_T for ellipse

$$I_T = \pi\frac{(a_1)^3(a_2)^3}{(a_1)^2 + (a_2)^2} \qquad (Ex.\ 13\text{-}3.12)$$

which also may be seen to agree with (*Ex. 13-2.8*) for the circular cross-section when we let $a_1 = a_2 = R_0$.

In terms of the torsional moment M_T the stress function may be written

$$T = \frac{1}{\pi}\frac{M_T}{a_1 a_2}\left(1 - \left(\frac{y_1}{a_1}\right)^2 - \left(\frac{y_2}{a_2}\right)^2\right) \qquad (Ex.\ 13\text{-}3.13)$$

Stress function T for ellipse

We may write the expression for M_T in another form

$$M_T = \frac{1}{4\pi^2}\frac{{A_0}^4}{I_P} G\theta\Big((a_1)^2 + (a_2)^2\Big) \qquad (Ex.\ 13\text{-}3.14)$$

which shows that for the elliptic bar the torsional moment indeed is *not* proportional to the polar moment of inertia.

For elliptic cross-section: $I_T \neq I_p$

Also in this case we compute the maximum shear stress, and for that purpose we need the partial derivatives of T

$$\begin{aligned} T_{,1} &= -2G\theta\frac{(a_2)^2}{(a_1)^2 + (a_2)^2}\,y_1 = -\frac{2}{\pi}\frac{M_T}{(a_1)^3\,a_2}\,y_1 \\ T_{,2} &= -2G\theta\frac{(a_1)^2}{(a_1)^2 + (a_2)^2}\,y_2 = -\frac{2}{\pi}\frac{M_T}{a_1\,(a_2)^3}\,y_1 \end{aligned} \qquad (Ex.\ 13\text{-}3.15)$$

and thus

$$\begin{aligned} \tau_1 &= -2G\theta\frac{(a_1)^2}{(a_1)^2 + (a_2)^2}\,y_2 \\ \tau_2 &= +2G\theta\frac{(a_2)^2}{(a_1)^2 + (a_2)^2}\,y_1 \end{aligned} \qquad (Ex.\ 13\text{-}3.16)$$

From (*Ex. 13-3.16*) it is clear that the maximum shear stress $\max(\tau)$ must occur at the end of one of the axes. If, as indicated in Fig. Ex. 13-3.1, $a_1 > a_2$ the maximum absolute value of the shear stress is found at $(y_1, y_2) = (0, \pm a_2)$

$$\max(|\tau|) = |\tau_1(0, \pm a_2)| \text{ for } a_1 > a_2 \qquad (Ex.\ 13\text{-}3.17)$$

$\max(\tau)$ occurs at end of *minor* axis

i.e.

$$\max(|\tau|) = 2G\theta\frac{(a_1)^2\,a_2}{(a_1)^2 + (a_2)^2} = \frac{2M_T}{\pi a_1\,(a_2)^2} \qquad (Ex.\ 13\text{-}3.18)$$

$\max(\tau)$ for $a_1 > a_2$

which for $a_1 = a_2 = R_0$ provides (*Ex. 13-2.11*).

It sounds counter-intuitive that the maximum shear stress does not occur as far away from the center of the cross-section as possible. And, this is even more surprising since the maximum shear stress indeed is found at the boundary. Researchers as great as Coulomb (1787) and Navier (1827) seem to have reasoned wrongly along such lines, and even Cauchy (1829) did not find the correct solution. It appears that Saint-Venant (1855) was the first to establish a valid theory for unconstrained torsion.

Why does $\max(|\tau|)$ not occur as far as possible from the centroid?

The issue of possible warping of the cross-section is important, and we proceed in much the same way as in Example Ex 13-2. Utilizing (13.59) and (*Ex. 13-3.15*), we get

$$\begin{aligned}\psi_{,1} &= +\left(-\frac{2M_T}{\pi G\theta a_1 (a_2)^3}+1\right) y_2 = \frac{(a_2)^2-(a_1)^2}{(a_1)^2+(a_2)^2}\, y_2 \\ \psi_{,2} &= -\left(-\frac{2M_T}{\pi G\theta (a_1)^3 a_2}+1\right) y_1 = \frac{(a_2)^2-(a_1)^2}{(a_1)^2+(a_2)^2}\, y_1\end{aligned} \qquad (Ex.\ 13\text{-}3.19)$$

Fig. Ex. 13-3.2: *Ellipsoidal cross-section, the warping function ψ.*

which shows that the formula for the warping function ψ must have the following form

$$\psi = \frac{(a_2)^2-(a_1)^2}{(a_1)^2+(a_2)^2}\left(x_1 - x_1^c\right)\left(x_2 - x_2^c\right) + C_w \qquad (Ex.\ 13\text{-}3.20)$$

where C_w denotes a constant which depends on the choice of coordinate system and on the axis of torsion. Independent of the values of x_α^c and C_w, except for terms that are linear in the coordinates, the shape of the warping is that of a hyperbolic paraboloid. When the axis of twist coincides with the centroid of the ellipse we get

The elliptic cross-section warps

$$\psi = \frac{(a_2)^2-(a_1)^2}{(a_1)^2+(a_2)^2}\, x_1 x_2 \qquad (Ex.\ 13\text{-}3.21)$$

Clearly, when the two axes are equal we recover the result for a circular cross-section.

Elastic torsion of a bar with ring-shaped cross-section, which is our next example, constitutes the simplest example of a multiply connected region.

Ex 13-4 Circular Tube—Ring-Shaped Cross-Section

The simplest example of a multiply connected cross-section is the circular ring, see Fig. Ex. 13-4.1. Here, we may exploit some of the results

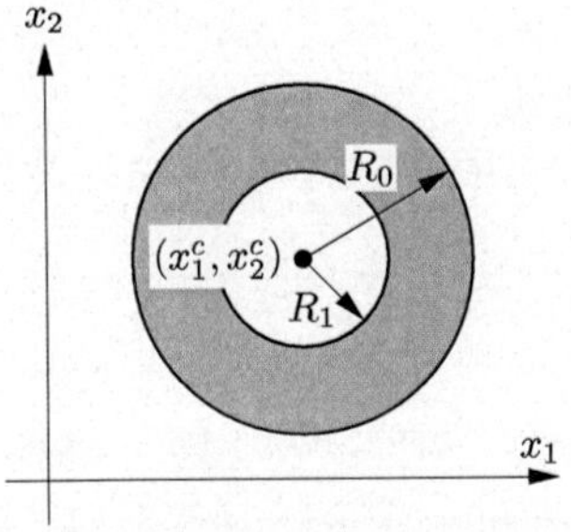

Fig. Ex. 13-4.1: *Ring cross-section.*

from Section Ex 13-2. We may see that the expression (*Ex. 13-2.5*) also is valid in the present case

$$T = \tfrac{1}{2}G\theta\Big(\big(R_0\big)^2 - \big(x_\alpha - x_\alpha^c\big)\big(x_\alpha - x_\alpha^c\big)\Big) \qquad (Ex.\ 13\text{-}4.1)$$

T from circle results

because it is 0 on the outer boundary, is constant on the inner one, has constant second derivatives, and that the uniqueness condition (13.54) is satisfied

$$\begin{aligned} -\oint_{\Gamma_1} \frac{1}{G} T_{,\beta} n_\beta d\Gamma &= -\oint_{\Gamma_1} \frac{1}{G}\big(-G\theta(x_\alpha - x_\alpha^0)\delta_{\alpha\beta}\big) n_\beta d\Gamma \\ &= \theta \oint_{\Gamma_1} (x_\beta - x_\beta^0) n_\beta d\Gamma = \theta \int_{A_1} \delta_{\beta\beta} d\Gamma \qquad (Ex.\ 13\text{-}4.2) \\ &= 2\theta A_J \end{aligned}$$

Cast in a different form (*Ex. 13-4.1*) is

$$T = \tfrac{1}{2}G\theta\Big(\big(R_0\big)^2 - r^2\Big) \qquad (Ex.\ 13\text{-}4.3)$$

Stress function T

where r denotes the distance from the center.
According to (13.46), the torsional moment M_T is

$$M_T = 2\int_{A_0} T dA + 2T_1 A_1 \qquad (Ex.\ 13\text{-}4.4)$$

The first term on the right-hand side is

$$\begin{aligned} 2\int_{A_0} T dA &= G\theta \int_0^{2\pi}\int_{R_1}^{R_0} \Big(\big(R_0\big)^2 - r^2\Big) r dr d\varphi \\ &= \frac{\pi}{2} G\theta \Big(\big(R_0\big)^2 - \big(R_1\big)^2\Big)^2 \end{aligned} \qquad (Ex.\ 13\text{-}4.5)$$

and the second term becomes

$$\begin{aligned} 2T_1 A_1 &= G\theta\Big(\big(R_0\big)^2 - \big(R_1\big)^2\Big)\pi\big(R_1\big)^2 \\ &= \pi G\theta\Big(\big(R_0\big)^2\big(R_1\big)^2 - \big(R_1\big)^4\Big) \end{aligned} \qquad (Ex.\ 13\text{-}4.6)$$

Torsional moment M_T for circular ring

Torsional moment of inertia I_T for circular ring

with the result that

$$M_T = \frac{\pi}{2} G\theta\Big((R_0)^4 - (R_1)^4\Big) \qquad \text{(Ex. 13-4.7)}$$

and

$$I_T = \frac{\pi}{2}\Big((R_0)^4 - (R_1)^4\Big) \qquad \text{(Ex. 13-4.8)}$$

Oftentimes the circular cylinder is thin-walled

$$\Big(t \ll (R_0 - R_1) \ll R_0\Big) \Rightarrow \Big(R_0 + R_1 \approx 2R_0\Big) \qquad \text{(Ex. 13-4.9)}$$

which may be utilized in rewriting (*Ex. 13-4.7*)

Torsional moment M_T for thin-walled ring

$$M_T = \frac{\pi}{2} G\theta\Big((R_0)^2 + (R_1)^2\Big)\Big(R_0 + R_1\Big)\Big(R_0 - R_1\Big)$$
$$\Rightarrow M_T \approx 2\pi G\theta (R_0)^3 t\,,\ t \ll R_0 \qquad \text{(Ex. 13-4.10)}$$

In Section Ex 13-2 we saw that the *polar moment of inertia* I_P of the circular region played an important role. The same is true for the ring-shaped cross-section, but we emphasize that this is a very special feature of these cross-sections.

Also in the present case the maximum shear stress is found at the outer boundary. It is

max(τ)

$$\max(\tau) = G\theta R_0 \qquad \text{(Ex. 13-4.11)}$$

i.e. the same result as for the solid circular cross-section when it is subjected to the same twist. This is not so surprising as it may seem at a first glance since both cross-sections in this case experience the same strains. If, on the other hand, we compute the maximum shear strain for an applied torsional moment M_T we may get

max(τ) for thick-walled ring

$$\max(\tau) = \frac{2M_T R_0}{\pi\Big((R_0)^4 - (R_1)^4\Big)} \qquad \text{(Ex. 13-4.12)}$$

and for thin-walled rings

max(τ) for thin-walled ring

$$\max(\tau) \approx \frac{M_T}{2\pi (R_0)^3}\,, \quad t \ll R_0 \qquad \text{(Ex. 13-4.13)}$$

For a given value of the applied moment M_T both values are greater than the value for the solid circular cross-section.

It seems somewhat surprising—at least to me—that a cross-section with corners could be susceptible to an analytic treatment. The rectangular cross-section, see Example Ex 13-6, indeed requires a quite heavy analysis involving either a double series expansion or a single series expansion of a sophisticated kind. However, the problem of torsion of a bar with a cross-section shaped as an equilateral triangle, see Fig. Ex. 13-5.1, has an exact solution, as we shall see below.

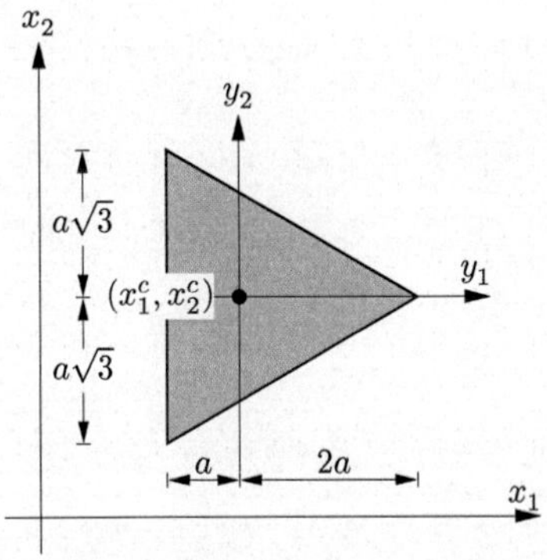

Fig. Ex. 13-5.1: *Equilateral triangle cross-section.*

Ex 13-5 Equilateral Triangle Cross-Section

I this example we do not so much derive the solution as make an educated guess, which we then prove to be correct. We begin by noting the equations of the sides of the equilateral triangle

$$\begin{aligned} 0 &= y_1 + a \\ 0 &= \sqrt{3}y_1 + 3y_2 - 2\sqrt{3}a \\ 0 &= \sqrt{3}y_1 - 3y_2 - 2\sqrt{3}a \end{aligned} \qquad \text{(Ex. 13-5.1)}$$

Since we may set the value of the stress function T equal to zero at the entire boundary we may try the following expression

Also here we guess the shape of T

$$T = C\Big(y_1+a\Big)\Big(\sqrt{3}y_1+3y_2-2\sqrt{3}a\Big)\Big(\sqrt{3}y_1-3y_2-2\sqrt{3}a\Big) \qquad \text{(Ex. 13-5.2)}$$

where C is a constant to be determined by the differential equation (13.53)

$$T_{,\beta\beta} = -2G\theta \qquad \text{(Ex. 13-5.3)}$$

We compute $T_{,11}$ and $T_{,22}$

$$T_{,11} = +18C(y_1 - a)\,,\ T_{,22} = -18C(y_1 + a) \qquad \text{(Ex. 13-5.4)}$$

and thus

Stress function T for equilateral triangle

$$T = \frac{G\theta}{6a}\Big(y_1 + a\Big)\Big(y_1^2 - 4y_1a + 4a^2 - 3y_2^2\Big) \qquad \text{(Ex. 13-5.5)}$$

This expression for the stress function seems to lack the three-fold symmetry, which it obviously must have since we are dealing with an equilateral triangle. However, if we express T in terms of the radius r and the angle φ from the y_1-axis

Stress function T for equilateral triangle

$$T = \tfrac{1}{6}G\theta a^2\left(4 - 3\left(\frac{r}{a}\right)^2 + \left(\frac{r}{a}\right)^3\cos(3\varphi)\right) \qquad \text{(Ex. 13-5.6)}$$

then, symmetry is revealed.

Torsional moment M_T for equilateral triangle

Integration over the area and multiplication by 2 provides the torsional moment M_T

$$M_T = \frac{9\sqrt{3}}{5} G\theta a^4 \qquad \text{(Ex. 13-5.7)}$$

which also may be expressed in terms of the side length s of the equilateral triangle

Torsional moment M_T for equilateral triangle

$$M_T = \frac{\sqrt{3}}{80} G\theta s^4 \qquad \text{(Ex. 13-5.8)}$$

and

Torsional moment of inertia I_T for equilateral triangle

$$I_T = \frac{9\sqrt{3}}{5} a^4 = \frac{\sqrt{3}}{80} s^4 \qquad \text{(Ex. 13-5.9)}$$

In order to find the warping function ψ we need $T_{,\beta}$

$$\begin{aligned} T_{,1} &= -\tfrac{1}{2} G\theta a \left(2\left(\frac{y_1}{a}\right) - \left(\frac{y_1}{a}\right)^2 + \left(\frac{y_2}{a}\right)^2 \right) \\ T_{,2} &= -G\theta a \left(1 + \left(\frac{y_1}{a}\right)\right) \left(\frac{y_2}{a}\right) \end{aligned} \qquad \text{(Ex. 13-5.10)}$$

With these two expressions and (13.59) in hand it is a minor task to find the formula for the warping function ψ

Warping function ψ for equilateral triangle

$$\psi = -\tfrac{1}{6} a^2 \left(3\left(\frac{y_1}{a}\right)^2 \left(\frac{y_2}{a}\right) - \left(\frac{y_2}{a}\right)^3 \right) \qquad \text{(Ex. 13-5.11)}$$

In terms of r and φ this is

Warping function ψ for equilateral triangle

$$\psi = -\tfrac{1}{6} a^2 \left(\frac{r}{a}\right)^3 \sin(3\varphi) \qquad \text{(Ex. 13-5.12)}$$

which exhibits the necessary symmetries of ψ.

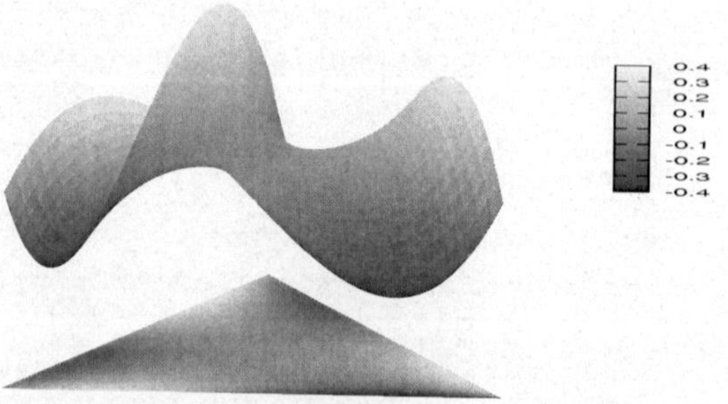

Fig. Ex. 13-5.2: *Equilateral triangle, the warping function ψ.*

The maximum shear stress $\max(|\tau_\alpha|)$ occurs at the midpoint of the sides.

It is easy to compute this value at the side $y_1 = -a$. It is

$$\max(|\tau_\alpha|) = \tfrac{1}{2} G\theta a \qquad (Ex.\ 13\text{-}5.13)$$

or, in terms of the torsional moment M_T

$$\max(|\tau_\alpha|) = \frac{5}{18\sqrt{3}} \frac{M_T}{a^3} \qquad (Ex.\ 13\text{-}5.14)$$

The warping function is shown in Fig. Ex. 13-5.2.

max($|\tau_\alpha|$) for equilateral triangle

max($|\tau_\alpha|$) for equilateral triangle

Relatively few cross-sections have an exact solution, and earlier one would have to resort to semi-analytic solutions as the ones given in the next example, while today the finite element method presents a viable alternative. On the other hand, the finite element method can only provide solutions in terms of *numbers*, not *formulas*, and is in that respect inferior to some approximative methods. For instance, the next example, which is concerned with the important issue of torsion of a bar with a solid rectangular cross-section, lends itself to semi-analytic methods which provide insight into the analytic nature of the asymptotic results for very slender rectangles. This can never be obtained by application of finite element programs. Sure, you may run your finite element program a vast number of times and plot the results, but, unless you know the asymptotic behavior in advance, you will never be able to extract this behavior analytically from your numerical results.

The finite element method cannot provide formulas—only number

It may be mentioned that, in addition to the examples above, among the cross-sections which have exact solutions are a circle with a circular indent and all cross-sections whose boundary may be mapped onto the unit circle by use of complex functional analysis, e.g. hypocycloids and epicycloids.

Ex 13-6 Rectangular Cross-Section

The problem of torsion of a bar with rectangular cross-section,see Fig. Ex. 13-6.1, cannot be solved exactly. In this example we shall

Rectangular cross-section

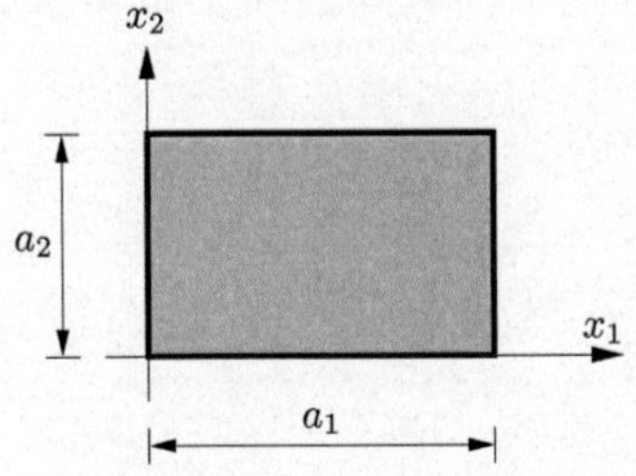

Fig. Ex. 13-6.1: *Rectangular cross-section.*

outline two such methods, one which is very straightforward, but tedious in execution, the other less obvious but providing results in a simpler form. The first is originally due to Navier and later reworked by Saint-Venant in 1883, the latter established by Saint-Venant.

Ex 13-6.1 The Series Solution by Navier

Any trial function T_{jk} according to

$$T_{jk} = \sin\left(j\pi\frac{x_1}{a_1}\right)\sin\left(k\pi\frac{x_2}{a_2}\right),\ (j,k) = 1,2,\ldots,\infty \qquad (Ex.\ 13\text{-}6.1)$$

satisfies the boundary conditions but is seen to violate the differential equation (13.53), which, for convenience. is given here

$$T_{,\beta\beta} = -2G\theta \qquad (Ex.\ 13\text{-}6.2)$$

Navier assumes a solution in the form of

$$T \approx \widetilde{T} = \sum_{j=1}^{\infty}\sum_{k=1}^{\infty} v_{jk}\sin\left(j\pi\frac{x_1}{a_1}\right)\sin\left(k\pi\frac{x_2}{a_2}\right) \qquad (Ex.\ 13\text{-}6.3)$$

where v_{jk} denote constants which are determined by the method. Then he applies a Ritz method in that he, except for the factor $\frac{1}{2}$ on the right-hand side and notational differences, constructs the functional $\mathbf{\Pi}$

$$\mathbf{\Pi}(v_{mn}) = \tfrac{1}{2}\int_{A_0}\left(\widetilde{T}_{,\beta\beta}(v_{jk}) + 2G\theta\right)^2 dA \qquad (Ex.\ 13\text{-}6.4)$$

The idea is now to make $\mathbf{\Pi}$ stationary with respect to all v_{mn} and in that fashion apply the method of least squares. After many, rather tedious, manipulations the result for v_{mn} is

$$v_{jk} = G\theta\frac{32}{\pi^4 jk}\frac{1}{\left(\dfrac{j}{a_1}\right)^2 + \left(\dfrac{k}{a_2}\right)^2} \qquad (Ex.\ 13\text{-}6.5)$$

$$(j,k) = 1,3,\ldots,\infty$$

In this way the differential equation is satisfied in an approximate way. The result for the torsional moment M_T is

Torsional moment M_T for rectangle

$$M_T = G\theta\frac{256}{\pi^6}\sum_{j=1,3}^{\infty}\sum_{k=1,3}^{\infty}\frac{1}{(jk)^2}\frac{a_1 a_2}{\left(\dfrac{j}{a_1}\right)^2 + \left(\dfrac{k}{a_2}\right)^2} \qquad (Ex.\ 13\text{-}6.6)$$

Sines with high numbers have wrinkles

The reason why the trial functions of (*Ex. 13-6.1*) have been so popular in the past is that the sine terms are mutually orthogonal which means that it is only necessary to solve a (vast) number of equations with just one unknown each to get the solution (*Ex. 13-6.5*). There is, however, a problem with using trigonometric terms in that the higher the number of the term the higher its frequency and, since the solution does not have wrinkles, many of the following terms must be used to smooth out the solution. Thus, the trade-off is the number of terms that is necessary to obtain a decent accuracy. Today, when solving systems of equations is an everyday task it is more appealing to use trial functions which are all smooth

$$\widetilde{T} = \sum_{j=1}^{\infty}\sum_{k=1}^{\infty}\left(\frac{x_1}{a_1}\left(1-\frac{x_1}{a_1}\right)\right)^j\left(\frac{x_2}{a_2}\left(1-\frac{x_2}{a_2}\right)\right)^k v_{jk} \quad \text{(Ex. 13-6.7)}$$

Polynomial terms are smooth

While it is true that this assumption results in a set of equations for v_{jk} whose coefficient matrix is fully populated the outcome is that we need far fewer terms in order to achieve a satisfactory accuracy.

Polynomial terms result in fully populated coefficient matrix

As explained in Section 28 the functional (*Ex. 13-6.4*) is not the best one in this connection. It is preferable to base the computations on the functional $\mathbf{\Pi_T}$, see (28.1)

$$\mathbf{\Pi_T}(T) = \tfrac{1}{2}\int_{A_0} T_{,\alpha}T_{,\alpha}dA - 2\int_{A_0} G\theta T dA \quad \text{(Ex. 13-6.8)}$$

because it entails derivatives of a lower order than (*Ex. 13-6.4*). The trial functions of (*Ex. 13-6.7*) may still be used.

Ex 13-6.2 The Single Series Solution by Saint-Venant

The method of Saint-Venant is very much different from Navier's. He expands not only T but also the right-hand side $-2G\theta$, which, does not seem like an obvious idea. The reason for doing this is that in this way he is able to simplify the ensuing problem such that it entails a single series instead of Navier's double series.[13.9] The Fourier series for the right-hand side is

Expand both sides of equation

$$-2G\theta = \sum_{j=1}^{\infty} v_j \sin\left(j\pi\frac{x_1}{a_1}\right) \quad \text{(Ex. 13-6.9)}$$

while the expansion for the stress function T is assumed to be

$$T = \sum_{j=1}^{\infty} X_j(x_2)\sin\left(j\pi\frac{x_1}{a_1}\right) \quad \text{(Ex. 13-6.10)}$$

where $X_j(x_2)$ are functions which depends only on x_2, as indicated. When these two expansions are inserted into the partial differential equation for T a set of ordinary differential equations for $X_j(x_2)$ results. Furthermore, demanding that all $X_j(x_2)$ satisfy the boundary conditions at $x_2 = (0, a_2)$

Ordinary differential equations for $X_j(x_2)$

$$X_j(0) = X_j(a_2) = 0 \quad \text{(Ex. 13-6.11)}$$

provides well-posed boundary value problems for $X_j(x_2)$. As was the case for Navier's method a fairly long sequence of manipulations are

[13.9] These old-timers had great command of mathematical methods, we have to admit. Today, very few would be able to compete with them in that respect.

needed in order to arrive at the final results. Among these are the expression for the coefficients $X_j(x_2)$

$$X_j(x_2) = G\theta \frac{8(a_1)^2}{(j\pi)^3} \left(1 - \cosh\left(j\pi \frac{x_2}{a_1}\right) - \frac{1 - \cosh\left(j\pi \frac{a_2}{a_1} i\right)}{\sinh\left(j\pi \frac{a_2}{a_1} i\right)} \sinh\left(j\pi \frac{x_2}{a_1}\right) \right) \quad \text{(Ex. 13-6.12)}$$

the formula for the torsional moment M_T

$$M_T = G\theta \tfrac{1}{3} (a_1)^3 a_2 \times \left(1 - 3\left(\frac{2}{\pi}\right)^5 \left(\frac{a_1}{a_2}\right) \sum_{j=1,3}^{\infty} \frac{1}{j^5} \tanh\left(j\pi \frac{a_2}{2a_1}\right)\right) \quad \text{(Ex. 13-6.13)}$$

the stress τ_1

$$\tau_1 = G\theta \frac{8a_1}{\pi^2} \times \sum_{j=1,3}^{\infty} \frac{1}{j^2} \frac{\sinh\left(j\pi \frac{a_2 - 2x_2}{2a_1}\right)}{\cosh\left(j\pi \frac{a_2}{2a_1}\right)} \sin\left(j\pi \frac{x_1}{a_1}\right) \quad \text{(Ex. 13-6.14)}$$

the stress τ_2

$$\tau_2 = -G\theta \left(a_1 - 2x_1 - \frac{8a_1}{\pi^2} \sum_{j=1,3}^{\infty} \frac{1}{j^2} \frac{\cosh\left(j\pi \frac{a_2 - 2x_2}{2a_1}\right)}{\cosh\left(j\pi \frac{a_2}{2a_1}\right)} \cos\left(j\pi \frac{x_1}{a_1}\right) \right) \quad \text{(Ex. 13-6.15)}$$

The general expression for τ_α in terms of the torsional moment M_T is judged to be so complicated that it is not worth to write it here. In any case, if you want to compute it you will probably use a computer, and, in that case, it is not difficult to achieve the result from the above formulas..

Analogous with the elliptic cross-section the maximum shear stress occurs at the midpoint of the long side of the rectangle. An approximation obtained from (Ex. 13-6.13)–(Ex. 13-6.15) is

$$\max(|\tau_\alpha|) \approx 4.8 \frac{M_T}{b^3}\ , \quad b = \min(a_1, a_2) \quad \text{(Ex. 13-6.16)}$$

The two expansions give identical results

A final note concerning Saint-Venant's solution may be in order in that the above results appear to suffer from a lack of symmetry with respect to the axes. That this was not the case was proved by Lévy (1899), who showed that the double series of Navier, which indeed possesses this symmetry, may be converted into the single series by Saint-Venant.

13.8 Concluding Remarks

As regards the old French and German literature I have a confession to make: I have not read the original sources and therefore have relied on the book by Frandsen (1946) and, to a lesser extent the book by Timoshenko & Goodier (1970).

Chapter 14

Introduction to "Plates with Thickness"

In Part II we treated plates as two-dimensional structural elements and could therefore establish theories in a consistent, straightforward way. Clearly, at that point the linear elastic constutive model connecting the generalized stresses $N_{\alpha\beta}$ and $M_{\alpha\beta}$ with the generalized strains $\varepsilon_{\alpha\beta}$ and $\kappa_{\alpha\beta}$[14.1] was merely postulated. Here, we shall follow the ideas from Chapter 11. In some respects the task i easier because we are only concerned with isotropic homogeneous linear elastic plates with vanishing transverse shear strains,[14.2] as we shall see.

[14.1] Note, that in many cases — for instance in establishing plate finite elements — the generalized membrane and bending strains usually are taken to be $(\varepsilon_{11}, \varepsilon_{22}, 2\varepsilon_{12})$ and $(\kappa_{11}, \kappa_{22}, 2\kappa_{12})$, respectively. That is, don't forget the factor 2 where it applies.
[14.2] We do account for the shear strain ε_{12} in the plane of the plate

E. Byskov, *Elementary Continuum Mechanics for Everyone*,
Solid Mechanics and Its Applications 194, DOI: 10.1007/978-94-007-5766-0_14,

Chapter 15

Bending and In-Plane Deformation of Linear Elastic Plates

15.1 Linear Elastic Plates

In Section 9.2 we assumed linear relations between the two-dimensional generalized plate strains $(\varepsilon^M_{\alpha\beta}, \kappa^M_{\alpha\beta})$ and their work conjugate (generalized) plate stresses $(N^M_{\alpha\beta}, M^M_{\alpha\beta})$, see (9.37), where superscript ^{M} is introduced here to indicate the mid-plane. If we, as we did in Section 9.2, wish to work with the generalized, rearranged plate quantities $(\varepsilon^M_j, \kappa^M_j, \sigma^M_j, M^M_j)$, then footnote 14.1, p. 299, concerning the factor 2 in the strain measures is important to note. In the present connection, our most basic kinematic

Two-dimensional plate strain and stress measures

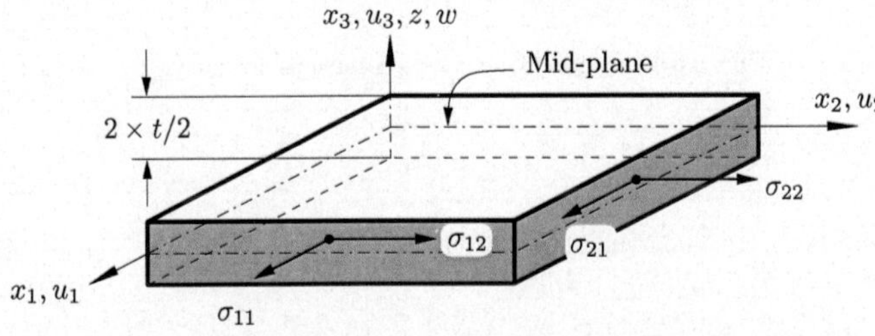

Fig. 15.1: Three-dimensional plane plate with stresses $\sigma_{\alpha\beta}$.

assumptions for the plate as a two-dimensional body are given by (9.1), which, in a slightly different notation, are

$$\varepsilon^M_{\alpha\beta} = \tfrac{1}{2}\left(u^M_{\alpha,\beta} + u^M_{\beta,\alpha}\right) + \tfrac{1}{2} w^M_{,\alpha} w^M_{,\beta} \quad \text{and} \quad \kappa^M_{\alpha\beta} = w^M_{,\alpha\beta} \tag{15.1}$$

Membrane strains $\varepsilon^M_{\alpha\beta}$
Curvature strains $\kappa^M_{\alpha\beta}$

but these formulas do not describe the strains in a plate as a three-dimensional body, see Fig. 15.1, and in order to derive relations between

E. Byskov, *Elementary Continuum Mechanics for Everyone*,
Solid Mechanics and Its Applications 194, DOI: 10.1007/978-94-007-5766-0_15,

$(N^M_{\alpha\beta}, M^M_{\alpha\beta})$ and $(\varepsilon^M_{\alpha\beta}, \kappa^M_{\alpha\beta})$ we need assumptions regarding the displacements throughout the thickness of the plate. It is worth noticing that the following derivations concern homogeneous plates, only.[15.1]

The idea of "fibers" does not apply here

The situation is different from the one we discussed in Chapter 11 in that we need to consider the shear stresses $\sigma_{12} = \sigma_{21}$ in addition to the normal stresses σ_{11} and σ_{22} and thus the idea of "fibers" does not work well here. As regards the stresses $\sigma_{\alpha 3} = \sigma_{3\alpha}$ they are internal reactions to the assumption that the shear strains $\varepsilon_{\alpha 3} = \varepsilon_{3\alpha}$ vanish. Therefore they do not enter through any constitutive relation in the same way as the shear force V cannot be determined through constitutive relations in Bernoulli-Euler beams. The remaining stress component σ_{33} is taken to be zero because the plate is considered to be in plane stress due to the condition that it is thin, see Example Ex 5-3.2.

The plate is in plane stress: $\sigma_{33} \equiv 0$

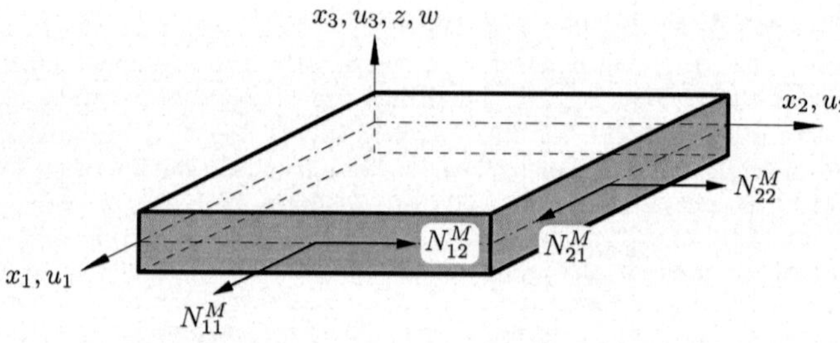

Fig. 15.2: Plane plate with membrane forces $N^M_{\alpha\beta}$.

15.1.1 Outline of Procedure

We wish constitutive relations between two-dimensional stress and strain measures

In order to arrive at the constitutive relation connecting the generalized plate stresses $(N^M_{\alpha\beta}, M^M_{\alpha\beta})$ with their work conjugate strains $(\varepsilon^M_{\alpha\beta}, \kappa^M_{\alpha\beta})$ we need to express the strains $\varepsilon_{\alpha\beta}$ at any point over the thickness in terms of the displacements $(u^M_{\alpha,\beta}, w^M_{,\alpha\beta})$ of the mid-plane. Then, we can use the three-dimensional stress-strain relations to compute the stresses $\sigma_{\alpha\beta}$ at the corresponding points and, finally, we integrate over the thickness to get expressions for $N^M_{\alpha\beta}$ and $M^M_{\alpha\beta}$ in terms of $(\varepsilon^M_{\alpha\beta}, \kappa^M_{\alpha\beta})$.

[15.1] In many cases the kinematic assumptions may be taken to be independent of the material, but for plates—and beams—the deformation of the cross-section depends strongly on the material and, for instance sandwich plates are not covered here because the below assumption that the cross-sections remain plane under deformation is seldom valid for such structures.

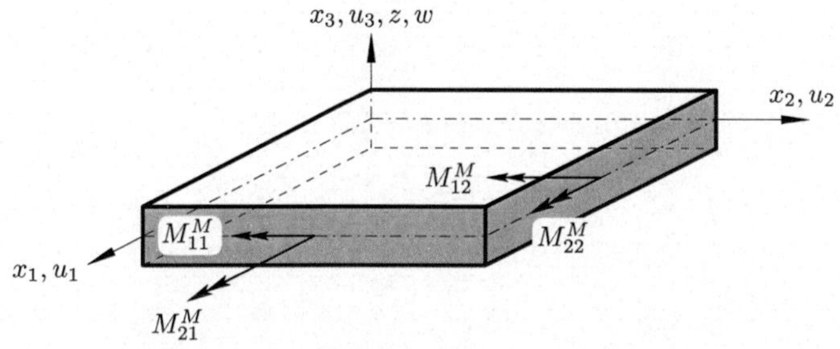

Fig. 15.3: Plane plate with bending and torsional moments $M_{\alpha\beta}^M$.

15.1.2 Kinematic Relations

We shall assume that the cross-sections remain plane during deformation and may therefor write the displacements (u_α, w) of an arbitrary point as

Displacements (u_α, w) of arbitrary point

$$u_\alpha = u_\alpha^M + x_3\,\omega_\alpha^M \quad \text{and} \quad w = w^M \tag{15.2}$$

where the rotations ω^M are given by

Rotations ω_α^M of cross-section

$$\omega_\alpha^M = w_{,\alpha}^M \tag{15.3}$$

The strains $\varepsilon_{\alpha\beta}$ of an arbitrary point are then

Strains $\varepsilon_{\alpha\beta}$ of arbitrary point

$$\varepsilon_{\alpha\beta} = \tfrac{1}{2}\left(u_{\alpha,\beta} + u_{\beta,\alpha}\right) + x_3\,\omega_{\alpha,\beta} + \tfrac{1}{2}w_{,\alpha}w_{,\beta} \tag{15.4}$$

which, utilizing (15.1)–(15.4), may be expressed in terms of the strains of the mid-plane

Strains $\varepsilon_{\alpha\beta}$ of arbitrary point

$$\varepsilon_{\alpha\beta} = \varepsilon_{\alpha\beta}^M + x_3\,\kappa_{\alpha\beta}^M \tag{15.5}$$

15.1.3 Three-Dimensional Constitutive Relations

Since the plate is assumed thin we may utilize the expressions in (*Ex. 5-3.18*) which are valid for the plane stress situation, namely

Three-dimensional constitutive relations

$$\begin{aligned} \sigma_{\alpha\beta} &= \frac{E}{1+\nu}\left(\varepsilon_{\alpha\beta} + \frac{\nu}{1-\nu}\varepsilon_{\gamma\gamma}\delta_{\alpha\beta}\right) \\ \sigma_{3j} &\equiv 0 \end{aligned} \tag{15.6}$$

15.1.4 Constitutive Relations for Two-Dimensional Plate

Finally, with the above formulas in hand we may integrate over the plate thickness and get the relations between the mid-plane generalized stresses

$(N^M_{\alpha\beta}, M^M_{\alpha\beta})$ and their work conjugate (generalized) strains $(\varepsilon^M_{\alpha\beta}, \kappa^M_{\alpha\beta})$

Two-dimensional constitutive relations

$$
\begin{aligned}
N^M_{\alpha\beta} &= \int_{-t/2}^{t/2} \sigma_{\alpha\beta} dx_3 = \frac{Et}{1+\nu}\left(\varepsilon^M_{\alpha\beta} + \frac{\nu}{1-\nu}\varepsilon^M_{\gamma\gamma}\delta_{\alpha\beta}\right) \\
M^M_{\alpha\beta} &= \int_{-t/2}^{t/2} \sigma_{\alpha\beta} x_3 dx_3 = \frac{Et^3}{12(1+\nu)}\left(\kappa^M_{\alpha\beta} + \frac{\nu}{1-\nu}\kappa^M_{\gamma\gamma}\delta_{\alpha\beta}\right)
\end{aligned}
\tag{15.7}
$$

It takes only little effort to rewrite (15.7) to derive the stress-strain relations (9.42)–(9.48) from Section 9.2.2 get

$$
\begin{aligned}
N^M_{11} &= \frac{Et}{1+\nu}\left(\varepsilon^M_{11} + \frac{\nu}{1-\nu}\left(\varepsilon^M_{11} + \varepsilon^M_{22}\right)\right) \\
N^M_{12} &= \frac{Et}{1+\nu}\varepsilon^M_{12} \\
N^M_{22} &= \frac{Et}{1+\nu}\left(\varepsilon^M_{22} + \frac{\nu}{1-\nu}\left(\varepsilon^M_{22} + \varepsilon^M_{11}\right)\right) \\
M^M_{11} &= \frac{Et^3}{12(1+\nu)}\left(\kappa^M_{11} + \frac{\nu}{1-\nu}\left(\kappa^M_{11} + \nu\kappa^M_{22}\right)\right) \\
M^M_{12} &= \frac{Et^3}{12(1+\nu)}\kappa^M_{12} \\
M^M_{22} &= \frac{Et^3}{12(1+\nu)}\left(\kappa^M_{22} + \frac{\nu}{1-\nu}\left(\kappa^M_{22} + \nu\kappa^M_{11}\right)\right)
\end{aligned}
\tag{15.8}
$$

and in a more convenient form

Two-dimensional constitutive relations between generalized strains $\varepsilon_{\alpha\beta}$ and $\kappa_{\alpha\beta}$ and generalized stresses $N_{\alpha\beta}$ and $M_{\alpha\beta}$

$$
\begin{aligned}
N^M_{11} &= \frac{Et}{1-\nu^2}\left(\varepsilon^M_{11} + \nu\varepsilon^M_{22}\right) \\
N^M_{12} &= \frac{Et}{1-\nu^2}\left(1-\nu\right)\varepsilon^M_{22} \\
N^M_{22} &= \frac{Et}{1-\nu^2}\left(\varepsilon^M_{22} + \nu\varepsilon^M_{11}\right) \\
M^M_{11} &= \frac{Et^3}{12(1-\nu^2)}\left(\kappa^M_{11} + \nu\kappa^M_{22}\right) \\
M^M_{12} &= \frac{Et^3}{12(1-\nu^2)}(1-\nu)\kappa^M_{12} \\
M^M_{22} &= \frac{Et^3}{12(1-\nu^2)}\left(\kappa^M_{22} + \nu\kappa^M_{11}\right)
\end{aligned}
\tag{15.9}
$$

Again, I emphasize that when you reorder the above quantities which have double lower indices such that they only have one lower index, then you must be certain to introduce the factor 2, see footnote 14.1, p. 299.

Part IV
Buckling

Chapter 16

Stability—Buckling

Many of the most devastating structural failures have been caused by instability, that is, buckling in one form or another, see e.g. Levy & Salvadori (1992), who claim that most of the important structural catastrophes throughout the history were due to loss of stability. Therefore, and because the subject brings out the advantages of the *Budiansky-Hutchinson notation*, see Chapter 33, very clearly, I feel that an introduction to the subject is called for. As we shall see below there are various ways that a structure may experience instability depending on its design and purpose. Often the most dramatic instabilities happen when the engineer has designed a structure, or a structural element, to carry the load in pure compression. Given this fact you might wonder why an engineer would want to design a structure that may fail due to this type of instability, which is called bifurcation buckling. The reason is that—*provided no instability occurs*—the most efficient and therefore cheapest way to carry loads is in tension or compression. For example, many early steel bridges were designed as trusses with their individual structural members mainly in tension or compression.

Worst structural failures caused by instability

Tension and compression are the cheapest ways to carry loads—*provided no instability occurs*

Fig. 16.1: Simple examples of stable and unstable states.

Below, we concentrate on the case of elastic buckling and cover geometrically perfect structures, but discuss the very important subject of buckling of geometrically imperfect structures and present results for one particular structure, namely a truss column. Instability associated with plasticity will only be covered through the famous *Shanley Model Column*.

Just from looking at Fig. 16.1 most of us have a clear feeling that the ball on the left is in a *stable state* because we may disturb it and it will seek to return to its position in the middle of the valley. Given friction,

E. Byskov, *Elementary Continuum Mechanics for Everyone*,
Solid Mechanics and Its Applications 194, DOI: 10.1007/978-94-007-5766-0_16,

eventually it will end up there after a number of travels up and down the sides of the valley. If we push the ball in the middle it will begin rolling in the direction of the shove, and were there no friction it would continue to roll with a constant speed. This ball is in a *neutral state.* Friction or no friction, it takes only a little disturbance to make the ball on the right fall off the ridge, and this ball is therefore in an *unstable state.*

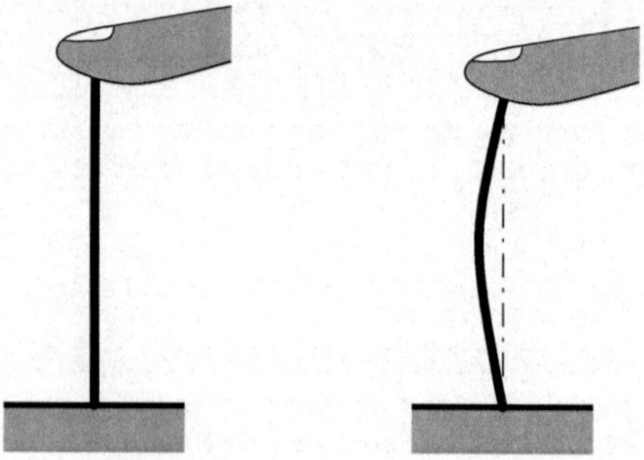

Fig. 16.2: Simple example of column instability—axial compression of a plastic ruler.
Comment: Your finger and the friction between the table and the ruler keep the ruler form moving sideways.

Instructive as the example of Fig. 16.1 may be it can only serve as an illustration of the concept of stability—not of the kind of stability, namely static structural stability, that we cover below. You may experience structural instability of the kind that is called bifurcation instability, see Chapter 17, if you position a plastic ruler vertically with one end on a table surface and you push downward on the other, see Fig. 16.2. If you are able to align the force that you exert with the ruler, then, for small forces it will stay straight, but when the force reaches a certain value, the buckling load, then the ruler bends without you having pushed it transverse to its direction.

Chapter 17

Stability Concepts

Although it is no easy task to formalize the concept of stability, several different stability criteria have been established. In our context we are concerned with structural stability, mainly stability of structures that are subjected to static loads, and this makes the task somewhat easier. However, at this point we shall not try to establish a formal foundation but introduce the concept of structural stability through simple examples. Static instability can be divided into two classes, namely *limit load buckling* and *bifurcation buckling*.

Stability criteria

Static loads

17.1 Static Stability and Instability Phenomena

Consider a structure subjected to a load whose intensity can be increased. If, at some certain load intensity, the displacements tend to infinity (possibly in an asymptotic sense), or the displacement pattern changes abruptly, then we say that the structure has lost its stability.The first phenomenon is termed *snap through* or *limit load instability*, while the second is known as *bifurcation buckling*. Below, we introduce these two phenomena through two simple examples.

Snap buckling = Limit load instability

Bifurcation buckling

17.1.1 Limit Load Buckling—Snap-Through

As an example of *limit load buckling* consider the structure shown in Fig. 17.1,

Limit load buckling

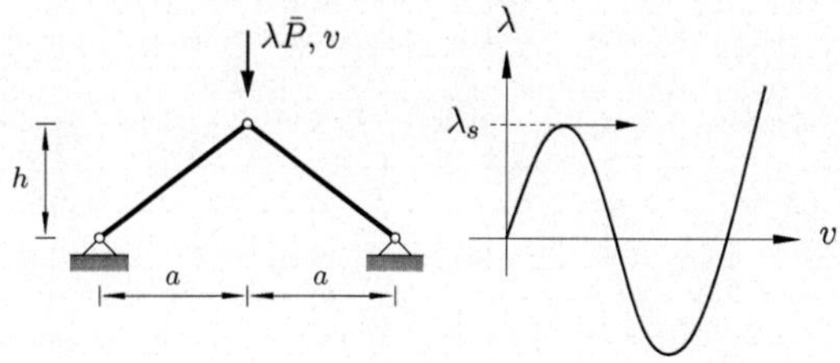

Fig. 17.1: Simple example of limit load instability.

E. Byskov, *Elementary Continuum Mechanics for Everyone*,
Solid Mechanics and Its Applications 194, DOI: 10.1007/978-94-007-5766-0_17,

which comprises two symmetric bars that are inflexible, but can undergo large axial deformations. The telescopes of a motorcycle front may serve as an example of such structural elements. If the structure is sufficiently shallow it can only undergo symmetric deflection patterns when it is loaded symmetrically.

Snap-through at snap buckling load λ_s

The structure is loaded vertically by $\lambda\bar{P}$, where $\bar{P}$ is a load unit, and the scalar load parameter λ is a measure of the load intensity.[17.1] When λ is increased the vertical deflection v increases until λ reaches a maximum, the *Snap Buckling Load* λ_s, and the structure experiences a dynamic *Snap-Through* from a position where the loaded joint is above the supports to one where it lies below, see the schematic plot in Fig. 17.1. Although the displacements increase abruptly their shape is basically the same throughout the entire loading history.

17.1.2 Bifurcation Buckling—Classical Critical Load

The bar of the model column in Fig. 17.2 is inflexible, but able to compress or extend axially. Before any load is applied the bar is vertical. The load

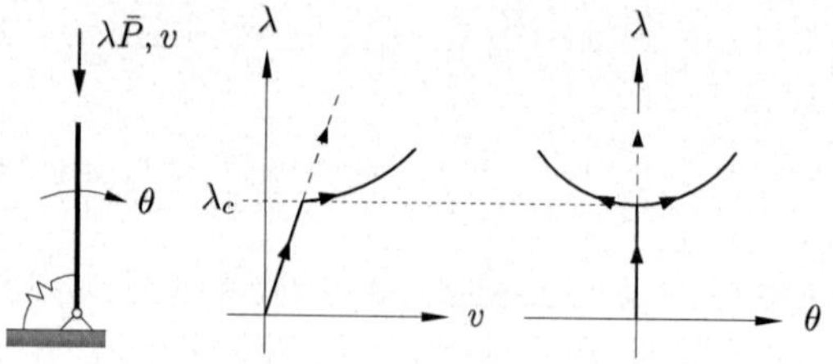

Fig. 17.2: Simple example of bifurcation instability.

$\lambda\bar{P}$ is kept vertical at all times.

Classical critical load λ_c

The bar stays vertical until the load reaches a certain value, given by $\lambda = \lambda_c$, where λ_c is the *Classical Critical Load.* At λ_c the bar suddenly deflects either to the right or the left, and the deformation pattern changes from a purely vertical one to one which also entails a rotation about the support. Thus, the deformation pattern has changed its basic character and the structure is said to have suffered *Bifurcation Buckling* in that the equilibrium paths, i.e. the paths connecting λ with the displacements v and θ, bifurcate at $\lambda = \lambda_c$. If the spring is linearly elastic this column can support loads in excess of λ_c, but because of the abrupt change in deformation pattern λ_c is considered a critical value—the classical critical load.

Bifurcation buckling

[17.1] Recall that throughout this book an overbar signifies that the quantity is prescribed.

Examples akin to the ones introduced here will be discussed in some detail later.

17.1.3 Further Comments

Even if the s-shape of the equilibrium path as shown in Fig. 17.1 is very small and therefore of a local nature, see Fig. 17.3a, we shall still consider λ_s the stability limit. In the same spirit, λ_c in Fig. 17.3b is taken as the

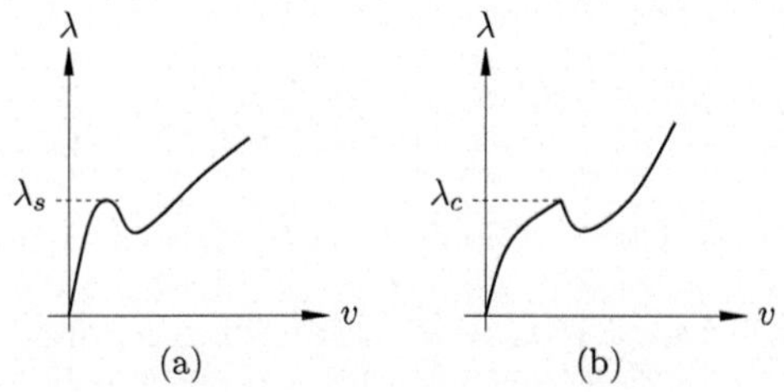

Fig. 17.3: Local limit load instability and local bifurcation instability.

classical critical load and, therefore, as an expression of the load-carrying capacity of the structure. The reason for this is that, although the structure can support loads in excess of λ_s or λ_c, respectively, its behavior after either of these loads is reached has changed significantly, and the stiffness of the structure has dropped. Furthermore, both at λ_s and at λ_c the structure experiences dynamic effects which may cause it to rupture.

17.2 Criteria and Methods for Determination of Stability and Instability

Three criteria for instability

Here, we consider three different criteria for instability, namely a static neighbor equilibrium criterion, an energy based criterion, and a dynamical criterion which will be formulated loosely here. Before doing so, we mention the intuitive criterion which states that a structure is stable if the load-displacement relation, i.e. the equilibrium path, is such that an increase in load is followed by an increase in displacement of the load, see Figs. 17.1 and 17.2.

17.2.1 Static Neighbor Equilibrium Stability Criterion

Neighbor equilibrium

This criterion states that a structure is unstable if it is possible to find two or more equilibrium configurations that are associated with the same value of the load, and the method consists in finding such neighboring equilibrium states and the load(s) at which this is possible. If a neighbor equilibrium state is possible, then the structure is unstable and will experience bifurcation buckling at the associated load, the bifurcation load λ_c.

17.2.2 Energy-Based Static Stability Criterion

The sign of the second variation of Π_P determines stability

The potential energy Π_P of a structure which is in an equilibrium state is stationary with respect to the displacements. If the potential energy has a minimum for the equilibrium displacements, then the structure is known to be stable, otherwise it is unstable.

17.2.3 Dynamic Stability Criterion

Dynamical method: oscillatory motion means stability, divergent motion instability

If we subject a structure to some small disturbance and the structure responds with an oscillatory motion, then we consider the structure to be stable in the dynamic sense. Conversely, if the structure exhibits a divergent motion, that is, the displacements increase, it is said to be unstable.

Already at this stage we point out that these three criteria do not always give the same answer, simply because they are valid for different situations. For example, the static criteria cannot—at least usually not with success—be applied to problems that entail non-conservative loads,[17.2] and the potential energy is only applicable to elastic problems.

Conservative loads Non-conservative loads

Since we shall not concern ourselves with non-conservative loads later it might have sufficed to mention that the work of a conservative load is

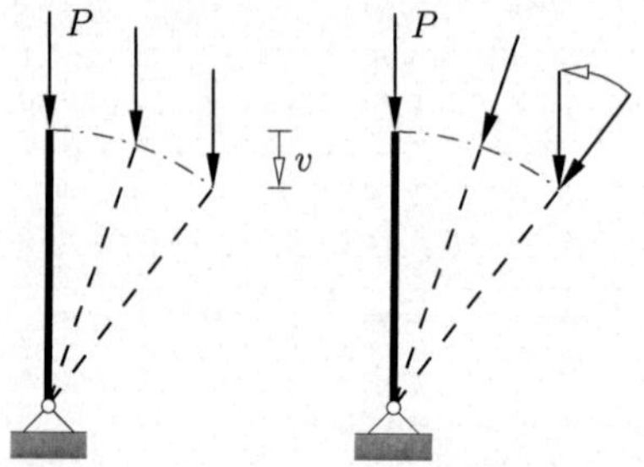

Fig. 17.4: Conservative load (a).
Non-conservative load (b).

independent of the path it travels. On the other hand, it may be instructive to view the two situations sketched in Fig. 17.4. In case (a) the load P remains vertical throughout the entire history with the result that the work

[17.2] Mount a jet engine on the top of a flagpole such that its thrust is compressive and follows the tangent to the top of the column. Then the column experiences a non-conservative load, and the buckling load can only be determined by the dynamic stability criterion.

it produces is always equal to Pv. In case (b) the load rotates with the rod until the final geometrical configuration, whereupon it is rotated back such that it becomes vertical. Through that process the load has not done any work. Although the final picture is the same in both cases, only the conservative load of case (a) has done any work, and it is obvious that the work is Pv independent of the path the load may have traveled between its original and final position. A conservative load may be derived from a load potential, here Pv, but see also the term $\overline{\boldsymbol{T}} \cdot \boldsymbol{u}$ in (33.26).

17.3 Introductory Example

Instead of trying to establish a general foundation we begin by considering a simple example which displays many of the basic features of stability and instability. In this example we shall take several postulates at face value and postpone all proofs till later chapters.

Ex 17-1 Model Column

Model column

Consider the model structure shown in Fig. Ex. 17-1.1. If the column is exactly vertical and the load is exactly vertical and applied at the

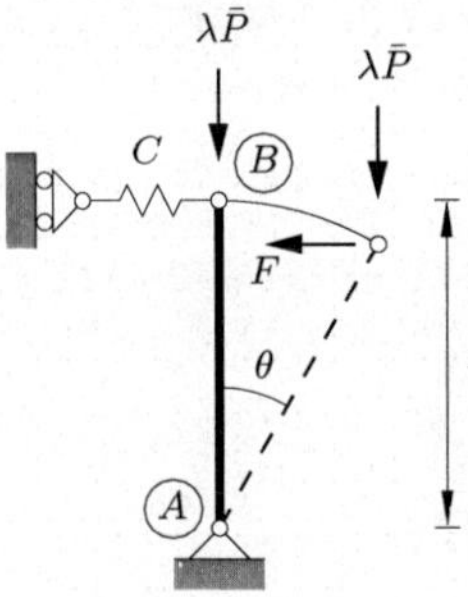

Fig. Ex. 17-1.1: *Perfect model column.*

top of the column we are dealing with a perfect structure. If, on the other hand, the column suffers from a (slight) inclination it is said to be geometrically imperfect. If the load is tilted or applied with some eccentricity the structure is subjected to load imperfections. Below, we first investigate the perfect realization of the model column and conclude with an analysis of its geometrically imperfect counterpart.

Ex 17-1.1 Perfect Model Column

Geometrically perfect model column

In order to elucidate some of the aspects of the methods mentioned in Section 17.2.2, we analyze the behavior of the model column shown in Fig. Ex. 17-1.1. The structure consists of a rigid vertical bar of length L which is supported by a spring with stiffness C that can slide vertically and therefore always produces a horizontal force F. The bar is loaded by a force $\lambda\bar{P}$ which always points downwards. This means that the load is conservative and that we may employ either of the methods mentioned in Section 17.2.2.

For convenience,[17.3] the load unit $\bar{P}$ is given by

$$\bar{P} = CL \qquad (Ex.\ 17\text{-}1.1)$$

Ex 17-1.2 *The Method of Neighboring Equilibrium*

Neighbor equilibrium

Moment equilibrium requires, see Fig. Ex. 17-1.1

$$\lambda\bar{P}L\sin\theta = FL\cos\theta \qquad (Ex.\ 17\text{-}1.2)$$

Constitutive relation for spring

where the constitutive law for the spring gives

$$F = CL\sin\theta \qquad (Ex.\ 17\text{-}1.3)$$

and thus

$$\lambda\bar{P}L\sin\theta = CL^2\sin\theta\cos\theta \qquad (Ex.\ 17\text{-}1.4)$$

or in view of (*Ex. 17-1.1*)

$$\lambda\sin\theta = \sin\theta\cos\theta \qquad (Ex.\ 17\text{-}1.5)$$

For $\theta \neq p\pi$ $(p = 0, \pm1, \pm2, \ldots)$ this provides the equilibrium path after bifurcation, the postbuckling path

Postbifurcation

$$\lambda = \cos\theta \qquad (Ex.\ 17\text{-}1.6)$$

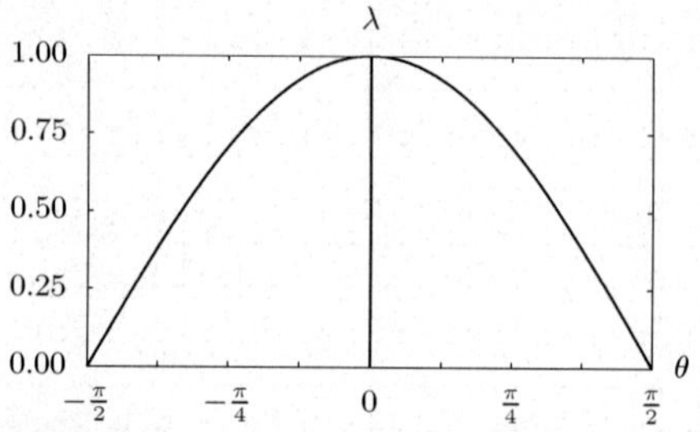

Fig. Ex. 17-1.2: *Perfect model column. Equilibrium paths.*

The relation (*Ex. 17-1.6*), along with the prebuckling path, i.e. for $\theta = 0$, is shown in Fig. Ex. 17-1.2.

Ex 17-1.2.1 *Bifurcation*

Bifurcation

At bifurcation:

$$|\theta| \to 0 \;\Rightarrow\; \cos\theta \to 1 \qquad (Ex.\ 17\text{-}1.7)$$

and therefore

$$\lambda_c = 1 \qquad (Ex.\ 17\text{-}1.8)$$

[17.3] That this choice is convenient is revealed below, see (*Ex. 17-1.8*).

Ex 17-1.3 Potential Energy

In this book we shall always denote a potential $\mathbf{\Pi}$, and in particular designate the potential energy $\mathbf{\Pi}_\mathrm{P}$. For this structure $\mathbf{\Pi}_\mathrm{P}$ is given by

Potential energy

$$\mathbf{\Pi}_\mathrm{P}(\theta) = \tfrac{1}{2}C(L\sin\theta)^2 - \lambda\bar{P}\big(L(1-\cos\theta)\big) \qquad (Ex.\ 17\text{-}1.9)$$

Potential energy $\mathbf{\Pi}_\mathrm{P}(\theta)$

where the first term on the right-hand side is the elastic energy and the second the potential of the load.

The equilibrium equation is found by requiring that the first variation $\delta\mathbf{\Pi}_\mathrm{P}$ of $\mathbf{\Pi}_\mathrm{P}$ vanishes

$$\delta\mathbf{\Pi}_\mathrm{P} = 0 \Rightarrow \frac{\partial\mathbf{\Pi}_\mathrm{P}}{\partial\theta}\,\delta\theta = 0 \,\forall\, \delta\theta \qquad (Ex.\ 17\text{-}1.10)$$

Equilibrium

which gives the same equilibrium equation (*Ex. 17-1.4*) as the static method.

Ex 17-1.3.1 Stability of the Solution

Stability is given by the sign of the second variation $\delta^2\mathbf{\Pi}_\mathrm{P}$ of $\mathbf{\Pi}_\mathrm{P}$. If $\delta^2\mathbf{\Pi}_\mathrm{P} > 0$ the structure is stable, otherwise it is unstable. The second variation is

Stability of solution

$$\delta^2\mathbf{\Pi}_\mathrm{P} = \frac{\partial^2\mathbf{\Pi}_\mathrm{P}}{\partial\theta^2}\,(\delta\theta)^2 \qquad (Ex.\ 17\text{-}1.11)$$

Second variation $\delta^2\mathbf{\Pi}_\mathrm{P}$

or

$$\begin{aligned}\delta^2\mathbf{\Pi}_\mathrm{P} &= \big(CL^2(\cos^2\theta - \sin^2\theta) - \lambda\bar{P}L\cos\theta\big)(\delta\theta)^2 \\ &= -CL^2\sin^2\theta(\delta\theta)^2\end{aligned} \qquad (Ex.\ 17\text{-}1.12)$$

where (*Ex. 17-1.1*) and (*Ex. 17-1.6*) have been exploited. Clearly, for $\theta \neq 0$ the second variation $\delta^2\mathbf{\Pi}_\mathrm{P}$ is never positive. The bifurcated path is therefore always unstable in agreement with the earlier findings.

Ex 17-1.3.2 Bifurcation

If our only interest is the determination of the bifurcation load, we may take $\mathbf{\Pi}_\mathrm{P}$ in a linearized form, i.e. with $\sin\theta \approx \theta$ and $\cos\theta \approx (1 - \frac{1}{2}\theta^2)$

$$\mathbf{\Pi}_\mathrm{P} \approx \tfrac{1}{2}CL^2\theta^2 - \tfrac{1}{2}\lambda\bar{P}L\theta^2 = \tfrac{1}{2}CL^2(1-\lambda)\theta^2 \qquad (Ex.\ 17\text{-}1.13)$$

Linearized $\mathbf{\Pi}_\mathrm{P}$

which gives $\delta\mathbf{\Pi}_\mathrm{P}$

$$\delta\mathbf{\Pi}_\mathrm{P} \approx CL^2\theta(1-\lambda)\delta\theta = 0 \,\forall\, \delta\theta \qquad (Ex.\ 17\text{-}1.14)$$

Bifurcation by linearized $\mathbf{\Pi}_\mathrm{P}$

with the result $\lambda_c = 1$ in accordance with the previous results, see Figs. Ex. 17-1.2 and (*Ex. 17-1.8*).

If we would have liked to determine whether the column is postbuckling stable or not we could not have used (*Ex. 17-1.13*), but must expand $\mathbf{\Pi}_\mathrm{P}$ to fourth order in θ.

Ex 17-1.4 Dynamical Method

For the present purpose we define the quantity I as the moment of inertia of the rigid bar about its support. Note that I is *not* associated with bending stiffness.

Moment of inertia I about the support

Ex 17-1.4.1 Bifurcation

Over-dot indicates differentiation with respect to time

According to a standard convention, an over-dot denotes the derivative with respect to time t.

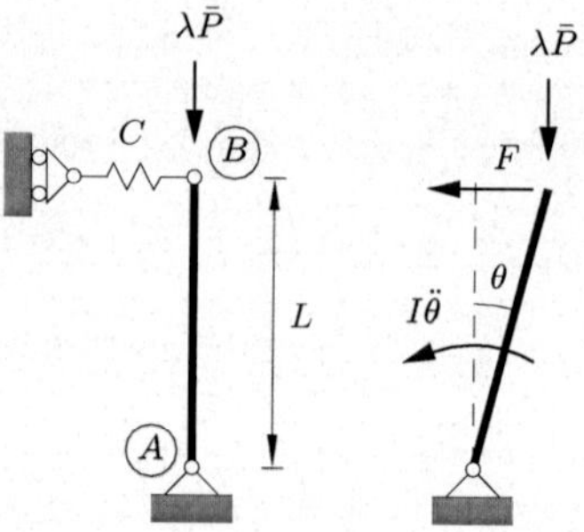

Fig. Ex. 17-1.3: *Perfect model column. Dynamic bifurcation.*

We shall only consider small disturbances and may therefore utilize (*Ex. 17-1.7*) and write the equation of motion

Small disturbances

$$\lambda \bar{P} L \theta - C L^2 \theta - I \ddot{\theta} = 0 \qquad (Ex.\ 17\text{-}1.15)$$

or, in view of (*Ex. 17-1.1*)

$$I\ddot{\theta} + CL^2(1-\lambda)\theta = 0 \qquad (Ex.\ 17\text{-}1.16)$$

For convenience, introduce $\tilde{I}$ by

$$\tilde{I} \equiv \frac{I}{CL^2} \qquad (Ex.\ 17\text{-}1.17)$$

to get

Dimensionless equation of motion

$$\tilde{I}\ddot{\theta} + (1-\lambda)\theta = 0 \qquad (Ex.\ 17\text{-}1.18)$$

with the characteristic equation

Characteristic equation

$$\tilde{I}R^2 + (1-\lambda) = 0 \qquad (Ex.\ 17\text{-}1.19)$$

Obviously, there are 3 possibilities which determine the character of the solution.

They are

1. $\lambda < 1$
   Here, $R^2 < 0$ which implies that θ is expressed in terms of exponential functions of imaginary arguments in time. The motion is therefore oscillatory and the configuration is stable.
2. $\lambda = 1$
   In this case, $R^2 = 0$, and θ is a linear function of time, and the motion is divergent with the result that the configuration is unstable.

3. $\lambda > 1$
   In this range, $R^2 > 0$ and θ is given in terms of exponential functions of real arguments in time, and consequently the motion is divergent and the configuration is unstable.

Thus, for this example we may see that the dynamic criterion gives the same instability load as the static criterion.

Ex 17-1.4.2 Dynamic Check of Stability of (Static) Postbuckling Path

Dynamic check of stability of solution

Assume that the structure is deflected an angle $\theta = \theta_0$ and is in equilibrium. We add a small disturbance, given by the additional angle

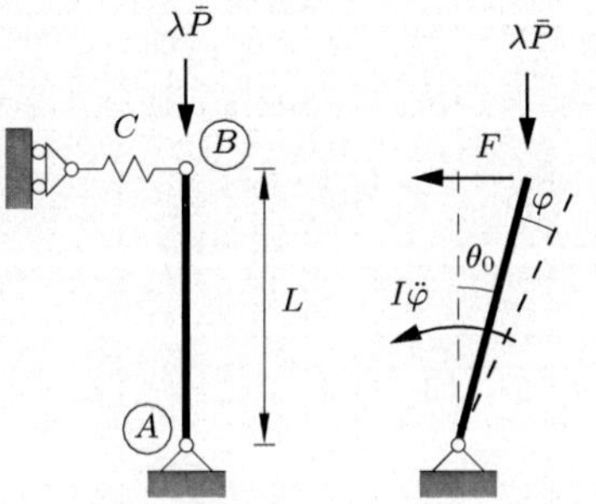

Fig. Ex. 17-1.4: *Perfect model column. Dynamic check of stability of postbuckling path.*

φ, where $|\varphi| \ll 1$. Noting that $\ddot{\theta} = \ddot{\varphi}$, the equation of motion is now altered to, see Fig. Ex. 17-1.4

$$\lambda\bar{P}L\sin\theta - CL^2\sin\theta\cos\theta - I\ddot{\theta} = 0 \qquad \text{(Ex. 17-1.20)}$$

where the total angle θ is

$$\theta = \theta_0 + \varphi, \; |\varphi| \ll 1 \qquad \text{(Ex. 17-1.21)}$$

Because of the assumption that φ is small we may write

$$\sin\theta \approx \sin\theta_0 + \varphi\cos\theta_0\,, \quad \cos\theta \approx \cos\theta_0 - \varphi\sin\theta_0 \qquad \text{(Ex. 17-1.22)}$$

with the result that the equation of motion is

$$\begin{aligned} 0 &= \lambda(\sin\theta_0 + \varphi\cos\theta_0) \\ &\quad -\big(\sin\theta_0\cos\theta_0 + \varphi\cos(2\theta_0)\big) - \tilde{I}\ddot{\varphi} \end{aligned} \qquad \text{(Ex. 17-1.23)}$$

Utilize the equilibrium equation (*Ex. 17-1.5*) with θ_0 substituted for θ and get

$$0 = \tilde{I}\ddot{\varphi} + (\cos(2\theta_0) - \lambda\cos\theta_0)\varphi \qquad \text{(Ex. 17-1.24)}$$

The analysis is analogous to the one performed above when bifurcation was the issue. Since the case $\theta_0 = 0$ was covered above, it is only interesting to treat the case $\theta_0 \neq 0$, where we can exploit (*Ex. 17-1.6*) to get the following condition for oscillatory motion

$$\cos(2\theta_0) - \cos^2\theta_0 > 0 \Rightarrow \cos^2\theta_0 > 1 \qquad (Ex.\ 17\text{-}1.25)$$

which is never satisfied. The postbuckling path is therefore always unstable, which agrees well with the results derived in the part dealing with the equilibrium method, see Fig. Ex. 17-1.2, and with the results obtained by application of the potential energy, see Example Ex 17-1.3. Without any proof it is important to mention that this finding is valid because we are dealing with a conservative load. This excludes loads that change direction during the loading process.

Ex 17-1.5 Imperfect Model Column

Geometrically imperfect model column

It is never possible to make a completely straight column which is exactly vertical and subjected to a load that is strictly vertical. As a result of this it is important to be able to analyze an imperfect structure

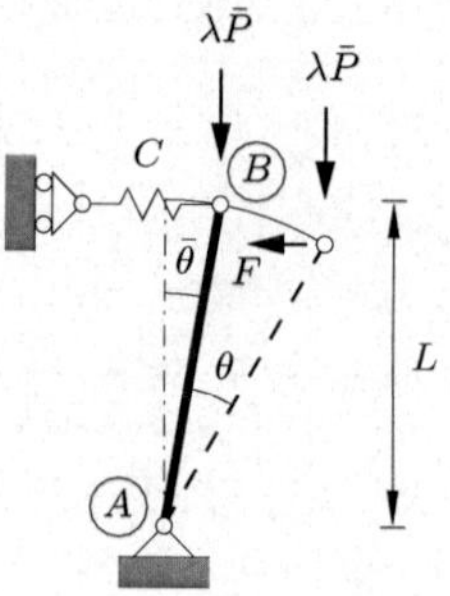

Fig. Ex. 17-1.5: *Imperfect model column.*

and to determine the influence of the imperfections. In the present case it suffices to consider the model column with an imperfection that consists in the unloaded bar being tilted an angle $\bar{\theta}$ (positive in the same direction as θ), while the load is vertical. Whether or not the bar is exactly straight does not matter since it is rigid.

It is important to note that θ signifies the *additional* rotation in contrast with the notation of some other authors who let θ denote the *total* rotation. Both notations have their virtues, but I prefer the former.

Ex 17-1.5.1 Equilibrium Analysis

Equilibrium analysis

Moment equilibrium about the support, see Fig. Ex. 17-1.5, requires—note that it is the additional rotation θ which causes the spring to produce a force

$$\lambda\bar{P}L\sin(\bar{\theta}+\theta) = CL\big(\sin(\bar{\theta}+\theta) - \sin\bar{\theta}\big)\cos(\bar{\theta}+\theta) \qquad (Ex.\ 17\text{-}1.26)$$

which for $\theta \neq -\bar{\theta} + p\pi \; (p = 0, \pm 1, \pm 2, ...)$ provides

$$\begin{aligned} \lambda &= \left(1 - \frac{\sin\bar{\theta}}{\sin(\bar{\theta}+\theta)}\right)\cos(\bar{\theta}+\theta) \\ \Rightarrow \lambda &= \cos(\bar{\theta}+\theta) - \sin\bar{\theta}\cot(\bar{\theta}+\theta) \end{aligned} \qquad (Ex.\ 17\text{-}1.27)$$

The plots in Fig. Ex. 17-1.6 show equilibrium paths for a number of positive imperfections. Negative imperfections result in paths that

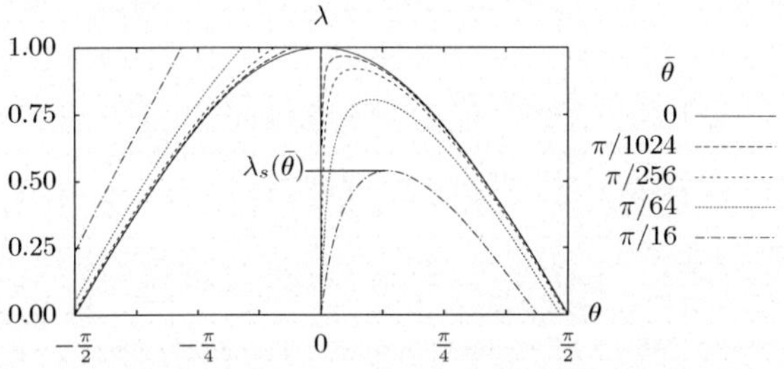

Fig. Ex. 17-1.6: *Imperfect model column. Maxima on the equilibrium paths:* $\lambda_s(\bar{\theta})$.

are symmetrical about the λ-axis with the ones shown. We shall in particular focus on the equilibrium paths that emanate from the origin, $(\theta, \lambda) = (0, 0)$, since these paths are the only ones that can be reached in a load-controlled device. It is clear from the plots that imperfections cause a decrease in the load-carrying capacity as given by the maximum on the equilibrium paths.

The equilibrium paths to the left of the λ-axis are all unstable and may only be realized in a displacement-controlled device. They are, in this example, less important.

Ex 17-1.5.2 Potential Energy

We may apply the principle of minimum potential energy to the imperfect model column. The potential energy is

$$\begin{aligned} \Pi_P = &+ \tfrac{1}{2}CL^2\big(\sin(\bar{\theta}+\theta) - \sin\bar{\theta}\big)^2 \\ &- \lambda\bar{P}L\big((1-\cos(\bar{\theta}+\theta)) - (1-\cos\bar{\theta})\big) \end{aligned} \qquad (Ex.\ 17\text{-}1.28)$$

Potential energy Π_P

where the constant term $(1 - \cos\bar{\theta})$ is inconsequential and is only present in order to make the analogy with (*Ex. 17-1.9*) clear. The first variation of (*Ex. 17-1.28*) is

$$\begin{aligned} \delta\Pi_P = &\Big(CL^2\big(\sin(\bar{\theta}+\theta) - \sin\bar{\theta}\big)\cos(\bar{\theta}+\theta) \\ &- \lambda\bar{P}L\sin(\bar{\theta}+\theta)\Big)\delta\theta = 0 \;\forall\; \delta\theta \end{aligned} \qquad (Ex.\ 17\text{-}1.29)$$

First variation of Π_P

which may give (*Ex. 17-1.26*) and (*Ex. 17-1.27*).

In order to explore the stability of the equilibrium paths of the imperfect column we take the second variation of $\mathbf{\Pi}_{\rm P}$

Second variation of $\mathbf{\Pi}_{\rm P}$

$$\delta^2\mathbf{\Pi}_{\rm P} = CL^2\big(\cos(2(\bar{\theta}+\theta)) + \sin(\bar{\theta}+\theta)\sin\bar{\theta} - \lambda\cos(\bar{\theta}+\theta)\big)(\delta\theta)^2 \qquad (Ex.\ 17\text{-}1.30)$$

Stability requires that $\delta^2\mathbf{\Pi}_{\rm P} > 0 \;\forall\; \delta\theta$, i.e. that

$$\lambda\cos(\bar{\theta}+\theta) < \cos(2(\bar{\theta}+\theta)) + \sin(\bar{\theta}+\theta)\sin\bar{\theta} \qquad (Ex.\ 17\text{-}1.31)$$

or, in view of (*Ex. 17-1.27*)

$$\left(1 - \frac{\sin\bar{\theta}}{\sin(\bar{\theta}+\theta)}\right)\cos^2(\bar{\theta}+\theta) < \cos(2(\bar{\theta}+\theta)) + \sin\bar{\theta}\sin(\bar{\theta}+\theta) \qquad (Ex.\ 17\text{-}1.32)$$

or

Criterion for stability of imperfect column

$$\frac{\sin\bar{\theta}}{\sin^3(\bar{\theta}+\theta)} > 1 \qquad (Ex.\ 17\text{-}1.33)$$

which is always fulfilled for small $|\theta|$ on the paths going through the origin. Thus, a geometrically imperfect realization of the model column is stable at small load levels, which agrees with the plots in Fig. Ex. 17-1.6.

The critical states, i.e. the ones where snap buckling occurs, are reached when $\theta = \theta_S$

Criterion for snap-through

$$\sin^3(\bar{\theta}+\theta_S) = \sin\bar{\theta} \qquad (Ex.\ 17\text{-}1.34)$$

Introduce this into (*Ex. 17-1.27*) and get an expression for the snap buckling load λ_s as a function of the imperfection $\bar{\theta}$

Snap buckling load λ_s

$$\lambda_s = \big(1 - (\sin\bar{\theta})^{2/3}\big)^{3/2} \qquad (Ex.\ 17\text{-}1.35)$$

For small $|\bar{\theta}|$ this may be expanded to give

Snap buckling load λ_s for small values of $|\bar{\theta}|$

$$\lambda_s \approx 1 - \tfrac{3}{2}\bar{\theta}^{2/3} \qquad (Ex.\ 17\text{-}1.36)$$

Imperfection sensitivity

The cusp on both curves in Fig. Ex. 17-1.7, which is typical for all elastic imperfection sensitive structures, shows that small imperfections may cause a severe loss of load-carrying capacity. Actually, much of modern stability theory is devoted to asymptotic methods that may be used to compute λ_s as a function of $\bar{\theta}$. But, already at this point, we mention that the exponent in the asymptotic expression for λ_s is 2/3 or 1/2 for *all* linearly elastic structures.

Asymptotic value of exponent: always 2/3 or 1/2 for linearly elastic structures

In the same spirit that lead to (*Ex. 17-1.36*) expand both sides of (*Ex. 17-1.34*)

$$(\bar{\theta}+\theta_S)^3 \approx \bar{\theta} - \tfrac{1}{6}\bar{\theta}^3 \qquad (Ex.\ 17\text{-}1.37)$$

where we have truncated after terms of power 3. From (*Ex. 17-1.37*) we get the explicit expression for θ_S

$$\theta_S \approx \sqrt[3]{\bar{\theta} - \tfrac{1}{6}\bar{\theta}^3} - \bar{\theta} \qquad (Ex.\ 17\text{-}1.38)$$

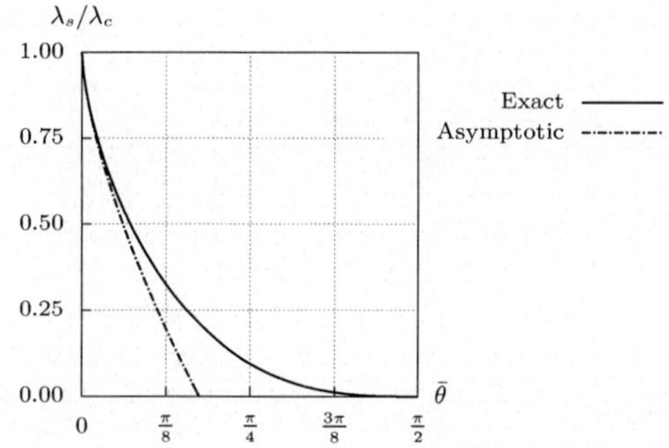

Fig. Ex. 17-1.7: *Imperfection sensitivity of model column.*

which may be compared to the exact formula that is derived from (*Ex. 17-1.34*)

$$\theta_S = \sin^{-1}\left(\sqrt[3]{\sin\bar{\theta}}\right) - \bar{\theta} \qquad (Ex.\ 17\text{-}1.39)$$

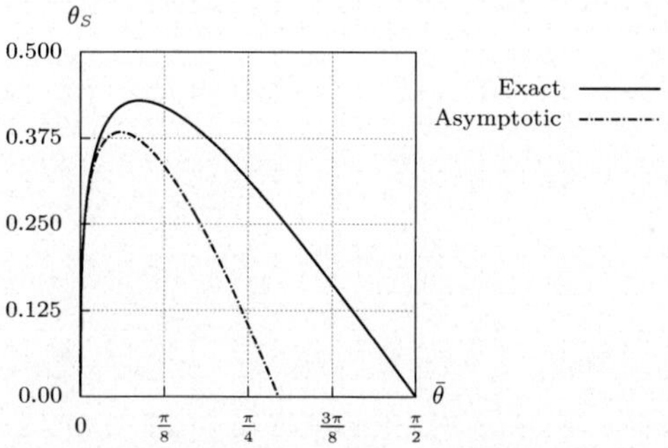

Fig. Ex. 17-1.8: *Imperfect model column. Rotation θ_S at maximum load as a function of imperfection $\bar{\theta}$.*

The plots in Fig. Ex. 17-1.7 show that for all realistic imperfections (*Ex. 17-1.38*) approximates (*Ex. 17-1.39*) sufficiently well.

We may identify the points at which $\delta^2\mathbf{\Pi}_{\mathrm{P}}$ vanishes as the points with zero slope on the equilibrium paths. Differentiate λ, given by (*Ex. 17-1.27*), with respect to θ

$$\frac{\partial\lambda}{\partial\theta} = -\sin(\bar{\theta}+\theta) + \frac{\sin\bar{\theta}}{\sin^2(\bar{\theta}+\theta)} \qquad (Ex.\ 17\text{-}1.40)$$

A comparison with (*Ex. 17-1.34*) shows that the slope is zero at (θ_S, λ_s) determined by the requirement that the second variation $\delta^2\mathbf{\Pi}_{\mathrm{P}}$ of the potential energy Π_P vanishes.

Ex 17-1.5.3 Dynamical Method

Dynamical method

A dynamic analysis of the imperfect model column follows the pattern of the dynamic analysis of the perfect model column. Since such an analysis does not provide new results in relation to the ones obtained by the two static methods we do not perform it here.

Chapter 18

Elastic Buckling Problems with Linear Prebuckling

Buckling comprises so many important different subtopics[18.1] that it is impossible to cover all of them in any detail here. Therefore, we concentrate on some of the problems which are characterized by at least two qualities, namely that they reveal important features of buckling and that they can be treated by analytical means. Among the topics are elastic buckling problems with linear prebuckling, see Section 18.3, including postbuckling behavior, see Chapter 19, the influence of geometric imperfections, see Chapter 20, in particular Example Ex 20-2, and elastic-plastic stability, which is dealt with in Chapter 21.

Buckling comprises many important different subtopics

Although bifurcation buckling never occurs in a real structure it is customary to consider this phenomenon rather than the more appropriate limit load buckling. There are reasons of varying relevance for doing this. First of all, by nature limit load buckling is nonlinear and therefore results in mathematical problems that are more difficult to solve, and before the advent of computers only very simple cases of limit load buckling could be handled; so, by tradition, these problems have received less attention. Another, better, reason is that many structures experience limit load buckling only because of imperfections, i.e. because of deviations from the perfect geometry and perfect load configuration, and that the bifurcation buckling load in some cases is still a good approximation to the real load-carrying capacity. Thirdly, for structures that are very imperfection sensitive asymptotic methods, which only entail quantities computed at the bifurcation load, may provide good estimates of the limit load. These asymptotic methods have been studied since 1945 and are the subject of many books and papers, see e.g. (Koiter 1945), (Budiansky 1966), (Thompson & Hunt 1973), (Budiansky 1974), (Byskov & Hutchinson 1977),

Bifurcation buckling never occurs in real structures

Asymptotic methods may give good estimates of the real load-carrying capacity

[18.1] Among the very important ones are: postbuckling behavior and imperfection sensitivity, see e.g. (Koiter 1945), (Thompson & Hunt 1973) and (Budiansky 1974); influence of plasticity, see e.g. (Hutchinson 1974b); mode interaction, see e.g. (Koiter & Kuiken 1971), (Tvergaard 1973a), and (Byskov & Hutchinson 1977), etc. Time-dependent problems are not covered.

E. Byskov, *Elementary Continuum Mechanics for Everyone*,
Solid Mechanics and Its Applications 194, DOI: 10.1007/978-94-007-5766-0_18,

and (Budiansky & Hutchinson 1979). Even in the elastic-plastic case asymptotic expansions have been developed and applied with success, see in particular (Hutchinson 1974b). Because of their complexity none of these methods will be introduced in this book, but some indication of their usefulness may be seen from Example Ex 17-1.5.

In the following we rely on the general equations for the three-dimensional continuum derived in Chapter 2, the beam equations from Chapter 7, and the plate equations Chapter 9, which may all be rewritten using the *Budiansky-Hutchinson Notation*, see Chapter 33. However, in order to make the exposition self-contained some of the basic equations are also given below.

18.1 Nonlinear Prebuckling

Nonlinear prebuckling

Here, we shall be concerned with *elastic* bifurcation buckling only and limit ourselves to the important case of *linear prebuckling.* First, however, consider bifurcation buckling with a nonlinear prebuckling path, see Fig. 18.1. As indicated in Fig. 18.1, the displacements $\boldsymbol{u}_0$ on the prebuckling path as

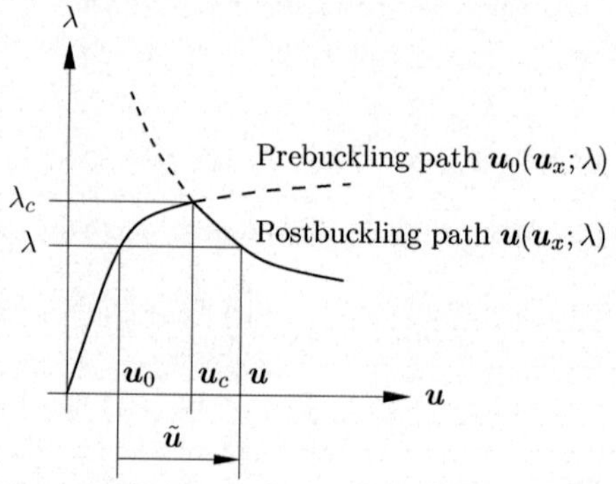

Fig. 18.1: Bifurcation buckling with nonlinear prebuckling.

well as the displacements $\boldsymbol{u}$ on the postbuckling path depend nonlinearly on the scalar load parameter λ

$$\boldsymbol{u}_0 = \boldsymbol{u}_0(\boldsymbol{x}; \lambda) \quad \text{and} \quad \boldsymbol{u} = \boldsymbol{u}(\boldsymbol{x}; \lambda) \tag{18.1}$$

where also the dependence on the spatial coordinates $\boldsymbol{x}$ is indicated.

Little bending: Linear prebuckling often valid

However, in many cases a structure responds in a way suggesting that its prebuckling path can be considered linear in the load parameter λ. This is the case when the prebuckling state entails little or no bending. If bending

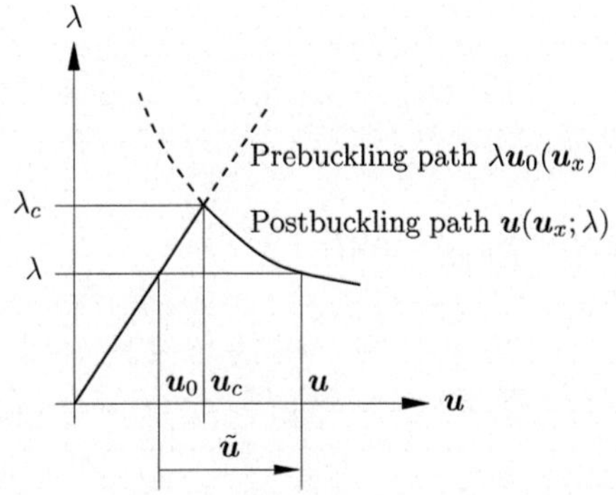

Fig. 18.2: Bifurcation buckling with linear prebuckling.

strains are important in the prebuckling state, linearity of the prebuckling path is certainly not a good assumption. But, in most cases the engineer tries to design a structure, which may experience buckling, in such a way that the prebuckling state is axial or in-plane and is therefore approximately linear. One reason for doing this is that pure compression—or tension—is a very efficient way to carry loads.

18.2 Some Prerequisites

As before, see Section 33, the generalized strains $\boldsymbol{\varepsilon}$ are derived from the generalized displacements $\boldsymbol{u}$ according to

$$\boldsymbol{\varepsilon} = \mathbf{l}_1(\boldsymbol{u}) + \tfrac{1}{2}\mathbf{l}_2(\boldsymbol{u}) \tag{18.2}$$

Quadratic strain

where $\mathbf{l}_1$ and $\mathbf{l}_2$ denote a linear and a quadratic operator, respectively. A bilinear operator $\mathbf{l}_{11}$ is defined by the following equation

$$\mathbf{l}_2(\boldsymbol{u}_a + \boldsymbol{u}_b) = \mathbf{l}_2(\boldsymbol{u}_a) + 2\,\mathbf{l}_{11}(\boldsymbol{u}_a, \boldsymbol{u}_b) + \mathbf{l}_2(\boldsymbol{u}_b) \tag{18.3}$$

Quadratic operator $\mathbf{l}_2$

where $\boldsymbol{u}_a$ and $\boldsymbol{u}_b$ are two independent displacement fields. Obviously

$$\mathbf{l}_{11}(\boldsymbol{u}_a, \boldsymbol{u}_b) = \mathbf{l}_{11}(\boldsymbol{u}_b, \boldsymbol{u}_a) \quad \text{and} \quad \mathbf{l}_{11}(\boldsymbol{u}, \boldsymbol{u}) = \mathbf{l}_2(\boldsymbol{u}) \tag{18.4}$$

Bilinear operator $\mathbf{l}_{11}$

The constitutive law is taken to be linearly hyperelastic

$$\boldsymbol{\sigma} = \mathbf{H}(\boldsymbol{\varepsilon}) \tag{18.5}$$

Hooke's "law"

where $\boldsymbol{\sigma}$ designates the generalized stresses and $\mathbf{H}$ is a linear operator. We assume that the reciprocal relation

$$\mathbf{H}(\boldsymbol{\varepsilon}_a)\boldsymbol{\varepsilon}_b = \mathbf{H}(\boldsymbol{\varepsilon}_b)\boldsymbol{\varepsilon}_a \tag{18.6}$$

Reciprocal relation

holds for all strain fields. Note that this is a point-wise condition because the (18.6) does not entail a dot and therefore an inner product is not implied. This relation is not as restrictive as you may think, which you may realize if you consider a number of typical elastic structures.

18.3 Linear Prebuckling

Linear prebuckling

We restrict ourselves to cases with linear prebuckling in the sense that the displacement, strain and stress fields on the fundamental path are given by

Linear prebuckling

$$\boldsymbol{u}(\boldsymbol{x};\lambda) = \lambda \boldsymbol{u}_0(\boldsymbol{x})\,,\ \boldsymbol{\varepsilon}(\boldsymbol{x};\lambda) = \lambda \boldsymbol{\varepsilon}_0(\boldsymbol{x}) \text{ and } \boldsymbol{\sigma}(\boldsymbol{x};\lambda) = \lambda \boldsymbol{\sigma}_0(\boldsymbol{x}) \qquad (18.7)$$

where subscript $_0$ indicates that a quantity is computed on the prebuckling path, $\boldsymbol{x}$ designates the coordinates, and λ is a scalar load parameter. There are other, more general, ways to define linear prebuckling, e.g. assuming that the bending energy associated with prebuckling is much smaller than the axial or membrane energy, but this is the simplest way, which is in most cases satisfactory. In order to keep the notation less complicated, we do not indicate the dependence on $\boldsymbol{x}$ below.

We impose the additional restriction that

No prebuckling bending

$$\mathrm{l}_{11}(\boldsymbol{u}_0, \boldsymbol{v}) = 0\ \forall\, \boldsymbol{v} \qquad (18.8)$$

In plain language this condition implies that all prebuckling bending effects can be ignored, as hinted at above. Thus, this restriction is not as limiting as it may seem at a first glance. As consequences of (18.8) we get;

$$\mathrm{l}_2(\boldsymbol{u}_0) = 0 \quad \text{and} \quad \boldsymbol{\varepsilon}_0 = \mathrm{l}_1(\boldsymbol{u}_0) \qquad (18.9)$$

18.3.1 Principle of Virtual Displacements

Load distribution $\overline{\boldsymbol{T}}$
Load intensity λ

In the *Principle of Virtual Displacements*, see (33.14), the load is given as $\overline{\boldsymbol{T}}$, but here it proves convenient to denote it by $\lambda\overline{\boldsymbol{T}}$, where $\overline{\boldsymbol{T}}$ signifies the load *distribution* and the scalar load parameter λ expresses the load *intensity*. Thus, the principle now is

Principle of virtual displacements

$$\boldsymbol{\sigma} \cdot \delta\boldsymbol{\varepsilon} = \lambda \overline{\boldsymbol{T}} \cdot \delta \boldsymbol{u} \qquad (18.10)$$

which is valid on the postbuckling as well as on the prebuckling path since both paths are equilibrium states. In (18.10) $\delta\boldsymbol{\varepsilon}$ is the variation of strains, $\delta\mathbf{u}$ the variation of displacements, and the dot implies an inner product, i.e. the proper integral over the whole structure or a scalar vector product.

The strain variation $\delta\boldsymbol{\varepsilon}$ is defined by

Strain variation $\delta\boldsymbol{\varepsilon}$

$$\boldsymbol{\varepsilon}(\boldsymbol{u} + \epsilon\delta\boldsymbol{u}) = \boldsymbol{\varepsilon}(\boldsymbol{u}) + \epsilon\delta\boldsymbol{\varepsilon}(\boldsymbol{u}) + O(\epsilon^2)\,,\ |\epsilon| \ll 1 \qquad (18.11)$$

"American epsilon" ϵ

where ϵ often is called the "American epsilon."

Now, (18.2)–(18.4) provide us with

$$\varepsilon(\boldsymbol{u}+\epsilon\delta\boldsymbol{u}) = \varepsilon(\boldsymbol{u}) + \epsilon\{\mathbf{l}_1(\delta\boldsymbol{u}) + \mathbf{l}_{11}(\boldsymbol{u},\delta\boldsymbol{u})\} + O(\epsilon^2) \tag{18.12}$$

and comparison with (18.11) furnishes

$$\delta\varepsilon(\boldsymbol{u}) = \mathbf{l}_1(\delta\boldsymbol{u}) + \mathbf{l}_{11}(\boldsymbol{u},\delta\boldsymbol{u}) \tag{18.13}$$

Strain variation $\delta\varepsilon$

which on the prebuckling path simply is

$$\delta\varepsilon_0 = \mathbf{l}_1(\delta\boldsymbol{u}) \tag{18.14}$$

Strain variation in prebuckling $\delta\varepsilon_0$

because of (18.8).

Above we have exploited the fact that the displacement variation $\delta\boldsymbol{u}$ may be taken to be the same for all states. Then, on the prebuckling path the *Principle of Virtual Displacements* is

$$\boldsymbol{\sigma}_0 \cdot \mathbf{l}_1(\delta\boldsymbol{u}) = \overline{\boldsymbol{T}} \cdot \delta\boldsymbol{u} \tag{18.15}$$

Principle of virtual displacements in prebuckling

and the constitutive relation (18.5), i.e. linear hyperelasticity, provides

$$\boldsymbol{\sigma}_0 = \mathbf{H}(\varepsilon_0) \tag{18.16}$$

Hooke's "law" in prebuckling

18.3.2 Bifurcation Buckling—Classical Critical Load

On the postbuckling path we write the total displacements $\boldsymbol{u}$ as the sum of a contribution $\lambda\boldsymbol{u}_0$ from prebuckling and an increment $\tilde{\boldsymbol{u}}$, see Fig. 18.2, which are both associated with the same value of the load parameter, as seen from Fig. 18.2

$$\boldsymbol{u} = \lambda\boldsymbol{u}_0 + \tilde{\boldsymbol{u}} \tag{18.17}$$

Postbuckling displacement

and similarly

$$\varepsilon = \lambda\varepsilon_0 + \tilde{\varepsilon} \quad \text{and} \quad \boldsymbol{\sigma} = \lambda\boldsymbol{\sigma}_0 + \tilde{\boldsymbol{\sigma}} \tag{18.18}$$

Postbuckling strain and stress

As mentioned above, we focus attention on bifurcation buckling which is assumed to take place at the load level λ_c or λ_1. The bifurcation buckling load is also known as the classical critical load, which indicates that it is associated with bifurcation buckling and not limit load buckling of a structure, be it geometrically perfect or not. Below, we assume that there is only one buckling mode $\boldsymbol{u}_1$ associated with λ_c.

Bifurcation buckling Classical critical load = Bifurcation buckling load

From (18.13), (18.14) and (18.17) we get

$$\delta\varepsilon = \delta\varepsilon_0 + \mathbf{l}_{11}(\tilde{\boldsymbol{u}},\delta\boldsymbol{u}) \tag{18.19}$$

Postbuckling strain variation

In postbuckling the principle of virtual displacements (18.10) provides the following expression

Principle of virtual displacements in postbuckling

$$\lambda\boldsymbol{\sigma}_0 \cdot \mathbf{l}_1(\delta\boldsymbol{u}) + \tilde{\boldsymbol{\sigma}} \cdot \mathbf{l}_1(\delta\boldsymbol{u}) + \lambda\boldsymbol{\sigma}_0 \cdot \mathbf{l}_{11}(\tilde{\boldsymbol{u}}, \delta\boldsymbol{u}) + \tilde{\boldsymbol{\sigma}} \cdot \mathbf{l}_{11}(\tilde{\boldsymbol{u}}, \delta\boldsymbol{u}) = \lambda\boldsymbol{T} \cdot \delta\boldsymbol{u} \tag{18.20}$$

where (18.18) and (18.19) have been utilized.

From (18.2)–(18.4), (18.8)–(18.14) and (18.18) we may get

$$\tilde{\boldsymbol{\varepsilon}} = \mathbf{l}_1(\tilde{\boldsymbol{u}}) + \tfrac{1}{2}\mathbf{l}_2(\tilde{\boldsymbol{u}}) \tag{18.21}$$

With (18.5), (18.16) and (18.18b) we note that

$$\tilde{\boldsymbol{\sigma}} = \mathbf{H}(\tilde{\boldsymbol{\varepsilon}}) \tag{18.22}$$

When we subtract (18.15) multiplied by λ from (18.20), we may get

$$\tilde{\boldsymbol{\sigma}} \cdot \mathbf{l}_1(\delta\boldsymbol{u}) + \lambda\boldsymbol{\sigma}_0 \cdot \mathbf{l}_{11}(\tilde{\boldsymbol{u}}, \delta\boldsymbol{u}) + \tilde{\boldsymbol{\sigma}} \cdot \mathbf{l}_{11}(\tilde{\boldsymbol{u}}, \delta\boldsymbol{u}) = 0 \tag{18.23}$$

It may immediately be observed that the third term is quadratic in $\tilde{\boldsymbol{u}}$ while the first two terms are linear. In order to discover bifurcation let $\lambda \to \lambda_c$ which at the same time means that any suitable norm $||\tilde{\boldsymbol{u}}||$ of $\tilde{\boldsymbol{u}}$ goes to zero

$$\lim_{\lambda \to \lambda_c} ||\tilde{\boldsymbol{u}}|| = 0 \tag{18.24}$$

Let $\boldsymbol{u}_1$ designate $\tilde{\boldsymbol{u}}$ for $\lambda = \lambda_c$, then (18.21) gives us

$$\boldsymbol{\varepsilon}_1 = \mathbf{l}_1(\boldsymbol{u}_1) \tag{18.25}$$

where $\boldsymbol{\varepsilon}_1$ signifies the strain associated with λ_c and $\boldsymbol{u}_1$.

Further, note that

$$\boldsymbol{\sigma}_1 = \mathbf{H}(\boldsymbol{\varepsilon}_1) \tag{18.26}$$

Linear eigenvalue problem
Classical critical load λ_c
Buckling mode $\boldsymbol{u}_1$

is linear in $\boldsymbol{u}_1$. Then (18.23) provides

$$\boldsymbol{\sigma}_1 \cdot \mathbf{l}_1(\delta\boldsymbol{u}) + \lambda_c\boldsymbol{\sigma}_0 \cdot \mathbf{l}_{11}(\boldsymbol{u}_1, \delta\boldsymbol{u}) = 0 \tag{18.27}$$

This is a *Linear Eigenvalue Problem* which determines the *Classical Critical Load* λ_c, (often denoted λ_1), and its associated *Buckling Mode* $\boldsymbol{u}_1$. The shape of the buckling mode is only determined to within a factor, as can be easily seen from (18.27).

Note that if the buckling mode $\boldsymbol{u}_1$ has been determined, then an expression for the value of the classical critical load λ_c follows from (18.27) when $\boldsymbol{u}_1$ is used instead of $\delta\boldsymbol{u}$

$$\lambda_c = -\frac{\boldsymbol{\sigma}_1 \cdot \mathbf{l}_1(\boldsymbol{u}_1)}{\boldsymbol{\sigma}_0 \cdot \mathbf{l}_2(\boldsymbol{u}_1)} = -\frac{\mathbf{H}(\mathbf{l}_1(\boldsymbol{u}_1)) \cdot \mathbf{l}_1(\boldsymbol{u}_1)}{\boldsymbol{\sigma}_0 \cdot \mathbf{l}_2(\boldsymbol{u}_1)} \tag{18.28}$$

In many of the following derivations we shall use this formula, which is the basis for the very useful *Rayleigh Quotient*, see Section 18.3.5 below.

18.3.3 Higher Bifurcation Loads

Higher bifurcation loads

Repeating the above analysis, we can get a linear eigenvalue problem for each higher buckling mode $\boldsymbol{u}_J$ associated with the bifurcation load λ_J, where $J = 2, 3, ...$

Higher bifurcation loads

$$\boldsymbol{\sigma}_J \cdot \mathbf{l}_1(\delta\boldsymbol{u}) + \lambda_J \boldsymbol{\sigma}_0 \cdot \mathbf{l}_{11}(\boldsymbol{u}_J, \delta\boldsymbol{u}) = 0 \tag{18.29}$$

Since all buckling modes are kinematically admissible in the strictest sense, i.e. they satisfy the homogeneous kinematic boundary conditions, we may take $\delta\boldsymbol{u} = \boldsymbol{u}_K$ in (18.29)

$$\boldsymbol{\sigma}_J \cdot \mathbf{l}_1(\boldsymbol{u}_K) + \lambda_J \boldsymbol{\sigma}_0 \cdot \mathbf{l}_{11}(\boldsymbol{u}_J, \boldsymbol{u}_K) = 0 \tag{18.30}$$

Similarly, we may write a relation analogous to (18.29), but with λ_K and $\boldsymbol{u}_K$ instead of λ_J and $\boldsymbol{u}_J$ and introduce $\delta\boldsymbol{u} = \boldsymbol{u}_J$ to get

$$\boldsymbol{\sigma}_K \cdot \mathbf{l}_1(\boldsymbol{u}_J) + \lambda_K \boldsymbol{\sigma}_0 \cdot \mathbf{l}_{11}(\boldsymbol{u}_K, \boldsymbol{u}_J) = 0 \tag{18.31}$$

Relations similar to (18.25) hold for $\boldsymbol{\varepsilon}_J$ and $\boldsymbol{\varepsilon}_K$. Further, the reciprocal relation (18.6) gives us

$$\boldsymbol{\sigma}_J \cdot \boldsymbol{\varepsilon}_K = \boldsymbol{\sigma}_K \cdot \boldsymbol{\varepsilon}_J \tag{18.32}$$

When we subtract (18.31) from (18.30) we get

$$(\lambda_J - \lambda_K)\boldsymbol{\sigma}_0 \cdot \mathbf{l}_{11}(\boldsymbol{u}_J, \boldsymbol{u}_K) = 0 \tag{18.33}$$

where the symmetry of $\mathbf{l}_{11}$ has been exploited.

Equation (18.33) shows that each pair of buckling modes $\boldsymbol{u}_J$ and $\boldsymbol{u}_K$ with $\lambda_J \neq \lambda_K$ are *orthogonal* in the so-called *energy sense*

Orthogonality of buckling modes

$$\boldsymbol{\sigma}_0 \cdot \mathbf{l}_{11}(\boldsymbol{u}_J, \boldsymbol{u}_K) = 0\,, \quad \lambda_J \neq \lambda_K \tag{18.34}$$

For a pair of buckling modes $\boldsymbol{u}_M$ and $\boldsymbol{u}_N$ that are associated with $\lambda_M = \lambda_N$ we have the freedom to choose them to be orthogonal in the sense of (18.34). For convenience, consistency, and continuity with the case of separate buckling loads we shall take all buckling modes to be orthogonal in the following fashion:[18.2]

Orthogonality of buckling modes

$$\boldsymbol{\sigma}_0 \cdot \mathbf{l}_{11}(\boldsymbol{u}_J, \boldsymbol{u}_K) = 0\,, \quad J \neq K \tag{18.35}$$

In most cases we wish to normalize the buckling modes in a particular way, sometimes in a geometrical way such that the maximum displacement is equal to a plate thickness or a beam depth, but under other circumstances

[18.2] It would seem unreasonable if two buckling modes with coincident buckling were to be taken to be orthogonal in some other sense than (18.34), while two buckling modes associated with buckling loads that are only an *iota* apart indeed are orthogonal according to (18.34).

in the following energy sense such that

$$\sigma_0 \cdot \mathrm{l}_2(\boldsymbol{u}_J) \tag{18.36}$$

takes a certain value for all $\boldsymbol{u}_J$.

The first example below is concerned with buckling of the *Euler Column*, which was treated in Example Ex 7-3, but here is given as an example of interpretation of the above formulas, which are in the *Budiansky-Hutchinson Notation*.

Ex 18-1 The Euler Column

The Euler Column

Let us reconsider the Euler Column analyzed in Example Ex 7-3 in light of the present theory and notation.

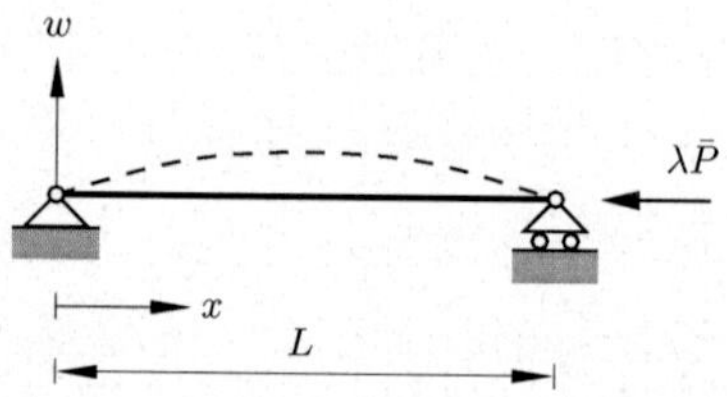

Fig. Ex. 18-1.1: *The Euler Column.*

Before we proceed it is necessary to investigate whether an assumption of linear prebuckling is valid. If we consider a column made of a common structural material such as steel, wood, concrete, or even some kind of plastic, it seems obvious that when an axial load is applied as shown in Fig. Ex. 18-1.1 the displacements are purely axial and very small until buckling occurs.[18.3] Provided that this is true, we may assume that prebuckling is linear. Also, the fact that prebuckling clearly does not involve bending in this case supports the postulate of linear prebuckling.

Ex 18-1.1 Generalized Quantities and Interpretations of the Budiansky-Hutchinson Notation

First we need the generalized quantities, i.e. the displacements, strains and stresses according to the kinematically moderately nonlinear Bernoulli-Euler beam theory.[18.4]

The generalized displacements are the axial displacement component u and the transverse displacement component w, and thus

[18.3] You may try the experiment with a plastic ruler shown in Fig. 16.2. Probably you will realize that the axial deformation is extremely small before bifurcation, i.e. buckling, takes place.

[18.4] In Example Ex 18-2 we choose another beam theory, namely the Timoshenko beam theory, which also accounts for shear strains.

$$\boldsymbol{u} = \begin{bmatrix} u_1 \\ u_2 \end{bmatrix} \equiv \begin{bmatrix} u \\ w \end{bmatrix} \qquad (Ex.\ 18\text{-}1.1)$$

Generalized displacements

The generalized strains are the axial strain ε and the bending strain κ

$$\boldsymbol{\varepsilon} = \begin{bmatrix} \varepsilon_1 \\ \varepsilon_2 \end{bmatrix} \equiv \begin{bmatrix} \varepsilon \\ \kappa \end{bmatrix} = \begin{bmatrix} u' + \frac{1}{2}(w')^2 \\ w'' \end{bmatrix} \qquad (Ex.\ 18\text{-}1.2)$$

Generalized strains

where prime indicates differentiation with respect to the axial coordinate x. Furthermore, see (7.12) and (7.14), or Section 7.7 where the following interpretations are also given.

Now, (*Ex. 18-1.2*) in connection with (18.3) shows that we may interpret $\mathbf{l}_1$ and $\mathbf{l}_2$ as

$$\mathbf{l}_1(\boldsymbol{u}) = \begin{bmatrix} u' \\ w'' \end{bmatrix} \text{ and } \mathbf{l}_2(\boldsymbol{u}) = \begin{bmatrix} (w')^2 \\ 0 \end{bmatrix} \qquad (Ex.\ 18\text{-}1.3)$$

Linear and quadratic operators

Similarly, the interpretation of $\mathbf{l}_{11}$ is

$$\mathbf{l}_{11}(\boldsymbol{u}_a, \boldsymbol{u}_b) = \begin{bmatrix} w'_a w'_b \\ 0 \end{bmatrix} \qquad (Ex.\ 18\text{-}1.4)$$

Bilinear operator

where subscripts $_a$ and $_b$ indicate two different displacement fields.

According to this theory, the generalized stresses are the axial force N and the bending moment M

$$\boldsymbol{\sigma} = \begin{bmatrix} \sigma_1 \\ \sigma_2 \end{bmatrix} \equiv \begin{bmatrix} N \\ M \end{bmatrix} \qquad (Ex.\ 18\text{-}1.5)$$

Generalized stresses

Finally, the constitutive operator $\mathbf{H}$ is

$$\mathbf{H} = \begin{bmatrix} EA & 0 \\ 0 & EI \end{bmatrix} \qquad (Ex.\ 18\text{-}1.6)$$

Hooke's "law"

where EA designates the axial and EI the bending stiffness, respectively. Later we restrict ourselves to the case where the beam properties are constant along the length.

When we have established the boundary conditions and with the above formulas in hand, we may easily write the variational equations for the prebuckling state and for buckling.

Ex 18-1.2 Boundary Conditions

As usual, we need the boundary conditions for the structural problem. Here they are

$$\begin{aligned} &u(0) = 0\,, \quad w(0) = 0\,,\ M(0) = 0 \ \Rightarrow\ w''(0) = 0 \\ &N(L) = -\bar{P}\,,\ w(L) = 0\,,\ M(L) = 0 \ \Rightarrow\ w''(L) = 0 \end{aligned} \qquad (Ex.\ 18\text{-}1.7)$$

Boundary conditions

where we have written the static boundary conditions on M in terms of the second derivative of the displacement.

Below, the indication of prebuckling and buckling by subscripts $_0$ and $_1$, respectively, is changed such that the numbers are placed on top of the quantities, e.g. $\overset{0}{N}$ instead of N_0.[18.5] The reason for doing this may not be obvious in this example, but in Example Ex 18-3 the original notation would result in a cluttering of subscripts designating prebuckling and buckling in combination with subscripts indicating coordinate axes.

Ex 18-1.3 Prebuckling

The variational statement (18.15) becomes

Prebuckling

$$\int_0^L \left(\overset{0}{N} \delta u' + \overset{0}{M} \delta w'' \right) dx = (-\bar{P}) \delta u(L) \qquad \text{(Ex. 18-1.8)}$$

which provides the solution[18.6]

Prebuckling stresses

$$\overset{0}{N} = -\bar{P} \text{ and } \overset{0}{M} = 0 \qquad \text{(Ex. 18-1.9)}$$

and

Prebuckling displacements

$$\overset{0}{u} = -\frac{\bar{P}}{EA} x \text{ and } \overset{0}{w} = 0 \qquad \text{(Ex. 18-1.10)}$$

Ex 18-1.4 Buckling

Here, (18.27) becomes

Buckling

$$\begin{aligned} 0 = & \int_0^L \left(\overset{1}{N} \delta u' + \overset{1}{M} \delta w'' \right) dx \\ & + \lambda_c \int_0^L \left(\overset{0}{N} \overset{1}{w}' \delta w' \right) dx \end{aligned} \qquad \text{(Ex. 18-1.11)}$$

or, because of (*Ex. 18-1.9*)

$$\begin{aligned} 0 = & \int_0^L \left(\overset{1}{N} \delta u' + \overset{1}{M} \delta w'' \right) dx \\ & - \lambda_c \bar{P} \int_0^L \left(\overset{1}{w}' \delta w' \right) dx \end{aligned} \qquad \text{(Ex. 18-1.12)}$$

Integration by parts provides

$$\begin{aligned} 0 = & \left[\overset{1}{N} \delta u \right]_0^L + \left[\overset{1}{M} \delta w' \right]_0^L - \left[(\overset{1}{M}' + \lambda_c \bar{P} w') \delta w \right]_0^L \\ & - \int_0^L \overset{1}{N}' \delta u dx \\ & + \int_0^L \left(\overset{1}{M}'' + \lambda_c \bar{P} \overset{1}{w}'' \right) \delta w dx \end{aligned} \qquad \text{(Ex. 18-1.13)}$$

Buckling axial displacement $\overset{1}{u} = 0$

The axial component $\overset{1}{u}$ of the buckling mode $\boldsymbol{u}_1$ is seen to vanish. Then its transverse component $\overset{1}{w}$ may be determined by the relatively simple

[18.5] You might ask why don't I always use numbers on top to indicate prebuckling and buckling (and postbuckling, if relevant). The reason is that I find the use of subscripts more aesthetically pleasing.

[18.6] Since the prebuckling state is (assumed to be) linear only one solution is possible.

differential equation in $\overset{1}{w}$ below, which is a linear eigenvalue problem in $\overset{1}{w}$, where it is assumed that the bending stiffness is independent of the axial coordinate

$$\overset{1}{M}'' + \lambda_c \bar{P}\overset{1}{w}'' = 0 \Rightarrow EI\overset{1}{w}^{iv} + \lambda_c \bar{P}\overset{1}{w}'' = 0 \qquad (Ex.\ 18\text{-}1.14)$$

Buckling eigenvalue problem

with boundary conditions

$$\overset{1}{w}(0) = 0 \,, \; \overset{1}{w}(L) = 0 \,, \; \overset{1}{w}''(0) = 0 \,, \; \overset{1}{w}''(L) = 0 \qquad (Ex.\ 18\text{-}1.15)$$

Boundary conditions

Together, the equations (*Ex. 18-1.14*) and (*Ex. 18-1.15*) constitute a linear eigenvalue problem with the solution[18.7]

$$\left.\begin{aligned} u_J(x) &= 0 \\ w_J(x) &= \xi_J \sin\left(J\frac{\pi x}{L}\right) \end{aligned}\right\} \; J = 1, 2, 3, \ldots \qquad (Ex.\ 18\text{-}1.16)$$

Displacements of the solution

where the amplitudes ξ_J are undetermined, as usual, and

$$\lambda_J = J^2 \frac{\pi^2 EI}{\bar{P}L^2} \,, \; J = 1, 2, 3, \ldots \qquad (Ex.\ 18\text{-}1.17)$$

Buckling loads

The smallest of these values, λ_1, is the *Classical Critical Load*, also called the *Euler Load*, λ_c

$$\lambda_c = \frac{\pi^2 EI}{\bar{P}L^2} \qquad (Ex.\ 18\text{-}1.18)$$

Classical critical load = Euler buckling load

with the associated buckling mode $\overset{1}{w}$

$$\overset{1}{w} = \xi_1 \sin(\pi x / L) \qquad (Ex.\ 18\text{-}1.19)$$

Buckling mode

The expressions (*Ex. 18-1.16*)–(*Ex. 18-1.19*) probably constitute the most famous of all solutions to elastic buckling problems.

It is a characteristic feature of *elastic* structures with infinitely many degrees of freedom that bifurcation is possible at infinitely many *distinct* values of the load parameter. The reason for emphasizing the words elastic and distinct is that structures which experience bifurcation in the *plastic* regime may bifurcate over *intervals* of the load, see e.g. Chapter 21.

In other structures, such as shells and long plates the bifurcation loads are usually not as well separated as is the case for the Euler Column. Without going into any details, it may be worth mentioning that this can lead to *Mode Interaction*, see Chapter 20, with the consequence that the structure becomes strongly imperfection sensitive. Example Ex 20-2 provides some insight into this phenomenon.

[18.7] You know this from your calculus course, but if you have forgotten the answer you can just insert the postulated solution into the differential equation and the boundary conditions to verify it.

Ex 18-2 A Pinned-Pinned Column Analyzed by Timoshenko Theory

A Timoshenko Column

As an example of the influence of the shear flexibility on the buckling load of a column we consider the same column that was analyzed in Examples Ex 7-3 and Ex 18-1, but perform the analysis using Timoshenko beam theory and refer to Fig. Ex. 18-1.1. We shall employ the same procedure as in Example Ex 18-1, but shorten it somewhat.

Ex 18-2.1 Generalized Quantities and Interpretations of the Budiansky-Hutchinson Notation

Here, the generalized displacements are the axial displacement component u, the transverse displacement component w, and the rotation ω

Generalized displacements

$$\boldsymbol{u} = \begin{bmatrix} u_1 \\ u_2 \\ u_3 \end{bmatrix} \equiv \begin{bmatrix} u \\ w \\ \omega \end{bmatrix} \qquad \text{(Ex. 18-2.1)}$$

and the generalized strains are the axial strain ε, the shear strain φ and the bending strain κ, i.e.

Generalized strains

$$\boldsymbol{\varepsilon} = \begin{bmatrix} \varepsilon_1 \\ \varepsilon_2 \\ \varepsilon_3 \end{bmatrix} \equiv \begin{bmatrix} \varepsilon \\ \varphi \\ \kappa \end{bmatrix} = \begin{bmatrix} u' + \frac{1}{2}(w')^2 \\ w' - \omega \\ \omega' \end{bmatrix} \qquad \text{(Ex. 18-2.2)}$$

The interpretation of $\mathbf{l}_1$ and $\mathbf{l}_2$ is

Linear and quadratic operators

$$\mathbf{l}_1(\boldsymbol{u}) = \begin{bmatrix} u' \\ w' - \omega \\ \omega' \end{bmatrix} \quad \text{and} \quad \mathbf{l}_2(\boldsymbol{u}) = \begin{bmatrix} (w')^2 \\ 0 \\ 0 \end{bmatrix} \qquad \text{(Ex. 18-2.3)}$$

where prime, as before, indicates differentiation with respect to the axial coordinate x. Similarly, the interpretation of $\mathbf{l}_{11}$ is

Bilinear operator

$$\mathbf{l}_{11}(\boldsymbol{u}_a, \boldsymbol{u}_b) = \begin{bmatrix} w'_a w'_b \\ 0 \\ 0 \end{bmatrix} \qquad \text{(Ex. 18-2.4)}$$

where subscripts $_a$ and $_b$ indicate two different displacement fields.

The generalized stresses are the axial force N, the shear force V and the bending moment M

Generalized stresses

$$\boldsymbol{\sigma} = \begin{bmatrix} \sigma_1 \\ \sigma_2 \\ \sigma_3 \end{bmatrix} \equiv \begin{bmatrix} N \\ V \\ M \end{bmatrix} \qquad \text{(Ex. 18-2.5)}$$

Finally, the constitutive operator $\mathbf{H}$ is

Hooke's "law"

$$\mathbf{H} = \begin{bmatrix} EA & 0 & 0 \\ 0 & GA_e & 0 \\ 0 & 0 & EI \end{bmatrix} \qquad \text{(Ex. 18-2.6)}$$

where we in this example assume that the beam properties are constant along the length.

Ex 18-2.2 Boundary Conditions

The boundary conditions are

$$\begin{aligned} &u(0)=0\,, \qquad w(0)=0\,,\ M(0)=0\ \Rightarrow\ \omega'(0)=0 \\ &N(L)=-\bar{P}\,,\ w(L)=0\,,\ M(L)=0\ \Rightarrow\ \omega'(L)=0 \end{aligned} \qquad \text{(Ex. 18-2.7)}$$

Boundary conditions

As in Example Ex 18-2 prebuckling and buckling are denoted by $_0$ and $_1$, respectively, e.g. $\overset{0}{N}$.

Ex 18-2.3 Prebuckling

In view of our experience from Example Ex 18-2 we postulate

$$\begin{aligned} &\overset{0}{N}=-\bar{P}\,,\ \overset{0}{V}=0\ \text{ and }\ \overset{0}{M}=0 \\ &\overset{0}{u}=-\frac{\bar{P}}{EA}\,,\ \overset{0}{\omega}=0\ \text{ and }\ \overset{0}{w}=0 \end{aligned} \qquad \text{(Ex. 18-2.8)}$$

Prebuckling stresses. Prebuckling displacements

which are seen to fulfill (7.48) in prebuckling.

Ex 18-2.4 Buckling

Here, (18.27) becomes

$$\begin{aligned} 0=&\int_0^L \left(\overset{1}{N}\delta u'+\overset{1}{V}(\delta w'-\delta\omega)+\overset{1}{M}\delta\omega'\right)dx \\ &+\lambda_c\int_0^L \overset{0}{N}\overset{1}{w}{}'\delta w'\,dx \end{aligned} \qquad \text{(Ex. 18-2.9)}$$

or, because of (Ex. 18-2.8a)

$$\begin{aligned} 0=&\int_0^L \left(\overset{1}{N}\delta u'+\overset{1}{V}\left(\delta\overset{1}{w}{}'-\delta\overset{1}{\omega}\right)+\overset{1}{M}\delta\omega'\right)dx \\ &-\lambda_c\bar{P}\int_0^L \overset{1}{w}{}'\delta w'\,dx \end{aligned} \qquad \text{(Ex. 18-2.10)}$$

Buckling problem

Here, only one term entails δu implying that

$$\overset{1}{N}\equiv 0 \qquad \text{(Ex. 18-2.11)}$$

which we shall exploit in the following. Now, integration by parts provides

$$\begin{aligned} 0=&+\left[\left(\overset{1}{V}-\lambda_c\bar{P}\overset{1}{w}{}''\right)\delta w\right]_0^L+\left[\overset{1}{M}\delta\omega\right]_0^L \\ &-\int_0^L \left(\overset{1}{M}{}'+\overset{1}{V}\right)\delta\omega\,dx \\ &-\int_0^L \left(\overset{1}{V}{}'-\lambda_c\bar{P}\overset{1}{w}{}''\right)\delta w\,dx \end{aligned} \qquad \text{(Ex. 18-2.12)}$$

The components $\overset{1}{w}$ and $\overset{1}{\omega}$ of the buckling mode may be determined by the relatively simple differential equations below

Buckling problem

$$\begin{aligned} \overset{1}{V}' - \lambda_c \bar{P} \overset{1}{w}'' &= 0 \\ \overset{1}{M}' + \overset{1}{V} &= 0 \end{aligned} \qquad (Ex.\ 18\text{-}2.13)$$

With the strain-displacement relations and the constitutive relations inserted this becomes

Buckling problem

$$\begin{aligned} GA_e(\overset{1}{w}'' - \overset{1}{\omega}') - \lambda_c \bar{P} \overset{1}{w}'' &= 0 \\ EI\overset{1}{\omega}'' + GA_e(\overset{1}{w}' - \overset{1}{\omega}) &= 0 \end{aligned} \qquad (Ex.\ 18\text{-}2.14)$$

The boundary conditions are

Buckling boundary conditions

$$\begin{aligned} &\overset{1}{w}(0) = 0 \\ &\overset{1}{w}(L) = 0 \\ &\overset{1}{M}(0) = 0 \ \Rightarrow\ \overset{1}{\omega}'(0) = 0 \\ &\overset{1}{M}(L) = 0 \ \Rightarrow\ \overset{1}{\omega}'(L) = 0 \end{aligned} \qquad (Ex.\ 18\text{-}2.15)$$

Together, the equations (*Ex. 18-2.14*) and (*Ex. 18-2.15*) constitute a linear eigenvalue problem. In the present context we shall only be concerned with the one associated with the lowest eigenvalue, namely λ_c. When we utilize that $\overset{0}{N}$ is constant and equal to $-\bar{P}$ we may find the solution

Solution to buckling problem

$$\begin{aligned} \overset{1}{w}(x) &= \xi_1 \sin\left(\pi \frac{x}{L}\right) \\ \overset{1}{\omega}(x) &= \xi_1 \frac{1}{1 + \bar{P}/(GA_e)} \left(\frac{\pi}{L}\right) \cos\left(\pi \frac{x}{L}\right) \\ &= \xi_1 \lambda_c \left(\frac{\pi}{L}\right) \cos\left(\pi \frac{x}{L}\right) \\ \lambda_c &= \frac{1}{1 + \bar{P}/(GA_e)} = \left(1 + \pi^2 \frac{EI}{GA_e L^2}\right)^{-1} \end{aligned} \qquad (Ex.\ 18\text{-}2.16)$$

When we recognize that $\lambda = 1$ corresponds to the buckling load according to the Bernoulli-Euler theory we may see that shear flexibility lowers the value. If, for example, the length of a column with a solid cross-section is about 5 times the depth of a square cross-section, the load-carrying capacity according to the Timoshenko theory is about 7.5% lower than the prediction from using Bernoulli-Euler theory.[18.8] For a sandwich column the values computed by these theories lie even further.

The second type of examples of interpretation and application of the above general formulas is concerned with plate buckling, which is somewhat more

[18.8] For such a stubby column plasticity may very well enter making the above elastic analysis problematic. On the other hand, sandwich columns with a soft core, which is the common design, may very well buckle without any trace of plasticity.

complicated than column buckling if for no other reason then because there are two spatial coordinates instead of one. It may be worth mentioning that, while the *Euler Column* is postbuckling neutral, i.e. after buckling the load stays the same independently of the total compression,[18.9] plates exhibit stable postbuckling behavior in that the plate is able to support greater loads after buckling has occurred. One consequence is that after buckling the plate still possesses stiffness towards in-plane compression, while the column is infinitely flexible with respect to additional axial compression.

Ex 18-3 Buckling of an Elastic Plate

Plate buckling

As mentioned in Chapter 9 plates may carry transverse as well as in-plane loads, and here we concentrate on the case of in-plane load

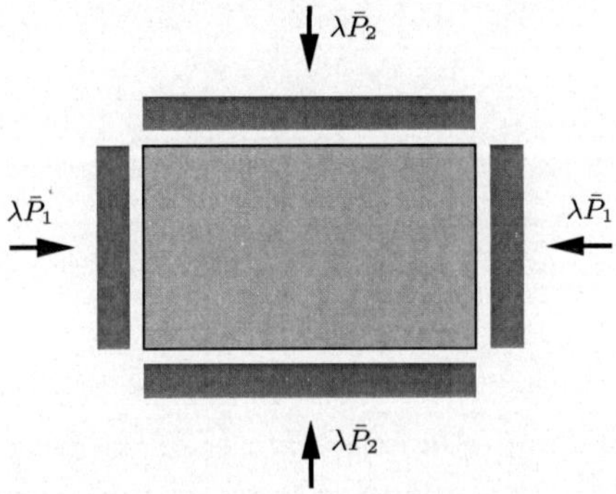

Fig. Ex. 18-3.1: *Simply supported plate with lubricated grips.*

on elastic plates. If the load is predominantly transverse the plate may fail because the stresses become too large, but if the load is in-plane the plate may buckle at fairly low stresses in much the same way as the *Euler Column*. To be specific, consider the simply supported plate shown in Fig. Ex. 18-3.1. In the transverse direction the plate is simply supported, i.e. the transverse displacement component w and the bending moment M_{nn} vanish on the boundary. In its own plane the sides of the plate remain straight, but are subjected to loads through infinitely rigid lubricated grips, see Fig. Ex. 18-3.1, and thus the in-plane shear stress N_{nt} and the displacement derivative $u_{n,n}$ vanish on the boundary. The boundary conditions are given in Fig. Ex. 18-3.2, while the load is shown in Fig. Ex. 18-3.1.

18.9 This is not completely true, but a consequence of the beam theory used here. On the other hand, the rotations of the column have to be large before a more exact theory gives answers that differ noticeably from the ones provided by the present theory, see Examples Ex 8-2 and Ex 7-2.

It may be worth mentioning that the *static* boundary conditions, which are quite obvious, also follow from the variational statement, i.e. the principle of virtual work for the plate. If we so choose, we may derive the *static* boundary conditions separately for the prebuckling and buckling problems from their associated variational statements.

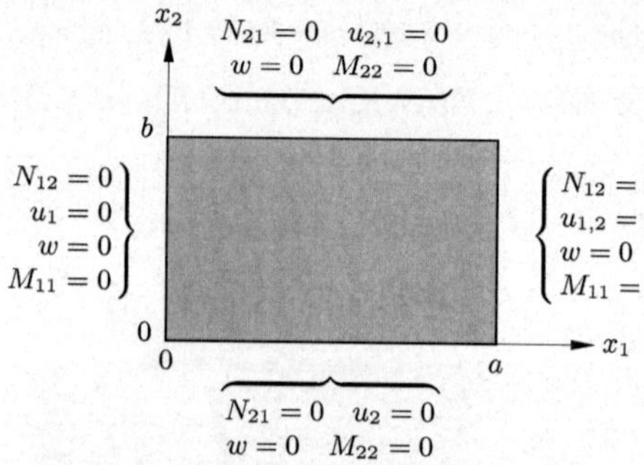

Fig. Ex. 18-3.2: *Plate geometry, coordinate system and boundary conditions.*

Linear prebuckling reasonable

As far as possible, we follow the same procedure below as in Example Ex 18-1 and begin by realizing that, for similar reasons as for the *Euler Column*, an assumption of linear prebuckling is valid here, too.

We assume that the plate is isotropic and homogeneous, i.e. that its properties are independent of direction and coordinate and may therefore exploit the expressions derived in Section 9.2. The plate thickness is t, ν designates *Poisson's Ratio*, and E denotes *Young's Modulus*. The plate is rectangular with sides a and b, as shown in Fig. Ex. 18-3.2.

Ex 18-3.1 Generalized Quantities and Interpretations of the Budiansky-Hutchinson Notation

As in Example Ex 18-1 we need the generalized displacements, strains and stresses, here according to the *von Kármán* plate theory developed in Chapter 9, particularly in Sections 9.2 and 9.2.5.

Here, the generalized displacements are the in-plane displacement components u_α and the transverse displacement component w, and thus

Generalized displacements

$$\boldsymbol{u} = \begin{bmatrix} u_1 \\ u_2 \\ u_3 \end{bmatrix} \equiv \begin{bmatrix} u_1 \\ u_2 \\ w \end{bmatrix} \qquad (Ex.\ 18\text{-}3.1)$$

It proves convenient to choose the generalized strains as ε_j , $j = 1, \ldots, 6$, see (*Ex. 18-3.2*) below, instead of the the membrane strains $\varepsilon_{\alpha\beta}$ together with the bending strains $\kappa_{\alpha\beta}$. The main reason for not making the latter choice is that it would have been at the expense of having a fourth order constitutive tensor $D_{\alpha\beta\gamma\delta}$, resulting in more complicated equations than the ones below.

$$\boldsymbol{\varepsilon} = \begin{bmatrix} \varepsilon_1 \\ \varepsilon_2 \\ \varepsilon_3 \\ \varepsilon_4 \\ \varepsilon_5 \\ \varepsilon_6 \end{bmatrix} \equiv \begin{bmatrix} \varepsilon_{11} \\ \varepsilon_{22} \\ 2\varepsilon_{12} \\ \kappa_{11} \\ \kappa_{22} \\ 2\kappa_{12} \end{bmatrix} = \begin{bmatrix} u_{1,1} \\ u_{2,2} \\ u_{1,2}+u_{2,1} \\ w_{,11} \\ w_{,22} \\ 2w_{,12} \end{bmatrix} + \frac{1}{2} \begin{bmatrix} (w_{,1})^2 \\ (w_{,2})^2 \\ 2w_{,1}w_{,2} \\ 0 \\ 0 \\ 0 \end{bmatrix} \quad (Ex.\ 18\text{-}3.2)$$

Generalized strains

Then, the interpretation of the operators $\mathbf{l}_1$, $\mathbf{l}_2$ and $\mathbf{l}_{11}$ is

$$\mathbf{l}_1(\boldsymbol{u}) = \begin{bmatrix} u_{1,1} \\ u_{2,2} \\ u_{1,2}+u_{2,1} \\ w_{,11} \\ w_{,22} \\ 2w_{,12} \end{bmatrix} \quad \text{and} \quad \mathbf{l}_2(\boldsymbol{u}) = \begin{bmatrix} (w_{,1})^2 \\ (w_{,2})^2 \\ 2w_{,1}w_{,2} \\ 0 \\ 0 \\ 0 \end{bmatrix} \quad (Ex.\ 18\text{-}3.3)$$

Linear and quadratic operators

and

$$\mathbf{l}_{11}(\boldsymbol{u}_a, \boldsymbol{u}_b) = \begin{bmatrix} w^a_{,1}w^b_{,1} \\ w^a_{,2}w^b_{,2} \\ w^a_{,1}w^b_{,2} + w^a_{,2}w^b_{,1} \\ 0 \\ 0 \\ 0 \end{bmatrix} \quad (Ex.\ 18\text{-}3.4)$$

Bilinear operator

where subscripts $_a$ and $_b$ and superscripts ^{a} and ^{b} indicate two different displacement fields.

The generalized stresses are the membrane stresses $N_{\alpha\beta}$ and the bending and torsional moments $M_{\alpha\beta}$ arranged according to

$$\boldsymbol{\sigma} = \begin{bmatrix} \sigma_1 \\ \sigma_2 \\ \sigma_3 \\ \sigma_4 \\ \sigma_5 \\ \sigma_6 \end{bmatrix} \equiv \begin{bmatrix} N_{11} \\ N_{22} \\ N_{12} \\ M_{11} \\ M_{22} \\ M_{12} \end{bmatrix} \quad (Ex.\ 18\text{-}3.5)$$

Generalized stresses

which means that the constitutive operator $\mathbf{H}$ is

$$\mathbf{H} = \left[\begin{array}{c|c} D^M\left[D^H\right] & [0] \\ \hline [0] & D^B\left[D^H\right] \end{array}\right] \quad (Ex.\ 18\text{-}3.6)$$

Hooke's "law"

where $[D^H]$, D^M and D^B are given by (9.45), (9.46) and (9.48), respectively

$$\left[D^H\right] \equiv \begin{bmatrix} 1 & \nu & 0 \\ \nu & 1 & 0 \\ 0 & 0 & \frac{1}{2}(1-\nu) \end{bmatrix} \tag{Ex. 18-3.7}$$

$$D^M \equiv \frac{Et}{1-\nu^2} \tag{Ex. 18-3.8}$$

$$D^B \equiv \frac{Et^3}{12(1-\nu^2)} \tag{Ex. 18-3.9}$$

and the matrix [0] is

$$[0] \equiv \begin{bmatrix} 0 & 0 & 0 \\ 0 & 0 & 0 \\ 0 & 0 & 0 \end{bmatrix} \tag{Ex. 18-3.10}$$

In order to formulate variational equations for the prebuckling and buckling states we need the boundary conditions, which are given below.

Ex 18-3.2 Boundary Conditions

As usual, we need the boundary conditions for the structural problem. They are already noted in Fig. Ex. 18-3.2, but are recapitulated below for completeness

Boundary conditions

$$\begin{aligned} &w=0\,,\ M_{11}=0\,,\ u_1=0\,, \quad N_{12}=0 \text{ for } x_1=0 \\ &w=0\,,\ M_{11}=0\,,\ u_{1,2}=0\,,\ N_{12}=0 \text{ for } x_1=a \\ &w=0\,,\ M_{22}=0\,,\ u_2=0\,, \quad N_{12}=0 \text{ for } x_2=0 \\ &w=0\,,\ M_{22}=0\,,\ u_{2,1}=0\,,\ N_{12}=0 \text{ for } x_2=b \end{aligned} \tag{Ex. 18-3.11}$$

In addition to these boundary conditions we must fulfill the integral conditions

Loads = integral of stresses on boundary

$$\begin{aligned} \lambda\bar{P}_1 &= \int_0^b N_{11} dx_2 \quad \text{for } x_1=0 \text{ and } x_1=a \\ \lambda\bar{P}_2 &= \int_0^a N_{22} dx_1 \quad \text{for } x_2=0 \text{ and } x_2=b \end{aligned} \tag{Ex. 18-3.12}$$

From (*Ex. 18-3.11*) we conclude

$$\begin{aligned} &w=0\,,\ w_{,2}=0\,,\ w_{,22}=0 \text{ at } (x_1=0)\vee(x_1=a) \\ &w=0\,,\ w_{,1}=0\,,\ w_{,11}=0 \text{ at } (x_2=0)\vee(x_2=b) \end{aligned} \tag{Ex. 18-3.13}$$

and, in view of the definition (9.1b) of the bending strain and the constitutive relation given by e.g. (9.61), we may therefore rewrite (*Ex. 18-3.11*)

Boundary conditions

$$\begin{aligned} &w=0\,,\ w_{,11}=0\,,\ u_1=0\,, \quad N_{12}=0 \text{ for } x_1=0 \\ &w=0\,,\ w_{,11}=0\,,\ u_{1,2}=0\,,\ N_{12}=0 \text{ for } x_1=a \\ &w=0\,,\ w_{,22}=0\,,\ u_2=0\,, \quad N_{12}=0 \text{ for } x_2=0 \\ &w=0\,,\ w_{,22}=0\,,\ u_{2,1}=0\,,\ N_{12}=0 \text{ for } x_2=b \end{aligned} \tag{Ex. 18-3.14}$$

Ex 18-3.2.1 Boundary Conditions and the Airy Stress Function

The Airy Stress Function Φ

If we wish to employ the *Airy Stress Function* Φ, see Section 9.2.5, the boundary conditions on the zeroth and first derivatives of the in-plane displacements deserve special care because the stress function is associated with the *second* derivatives. In the following we derive boundary conditions at $x_1 = a$ and mention that the equivalent conditions at the other edges follow by symmetry. The first boundary condition, however, is straightforward and follows directly from the fact that the in-plane shear stress vanishes on the boundary

In-plane shear stress on boundary

$$N_{12} = 0 \Rightarrow \Phi_{,12} = 0 \text{ on } x_1 = a \qquad (Ex.\ 18\text{-}3.15)$$

In order to arrive at the next boundary condition we compute[18.10]

$$N_{21,2} = -\Phi_{,122} = -\Phi_{,221} \text{ on } x_1 = a \qquad (Ex.\ 18\text{-}3.16)$$

and

$$\begin{aligned} 0 = \varepsilon_{12,2} &= \tfrac{1}{2}\left(u_{1,22} + u_{2,12} + w_{,12}w_{,2} + w_{,1}w_{,22}\right) \\ &= \tfrac{1}{2}\left(0 + u_{2,12} + 0 + 0\right) \text{ on } x_1 = a \end{aligned} \qquad (Ex.\ 18\text{-}3.17)$$

Further, compute

$$\varepsilon_{22,1} = u_{2,21} + w_{,21}w_{,2} = 0 + 0 \text{ on } x_1 = a \qquad (Ex.\ 18\text{-}3.18)$$

Also,

$$\begin{aligned} 0 = \varepsilon_{22,1} &= \frac{1}{E(1-\nu^2)}\left(N_{22,1} + \nu N_{11,1}\right) \\ \Rightarrow\ 0 &= \Phi_{,111} + \nu\Phi_{,221} = \Phi_{,111} + \nu\Phi_{,122} \\ &= \Phi_{,111} + 0 \\ \Rightarrow\ \Phi_{,111} &= 0 \text{ on } x_1 = a \end{aligned} \qquad (Ex.\ 18\text{-}3.19)$$

The remaining boundary condition on Φ at $x_1 = a$ comes from the condition (*Ex. 18-3.12*a)

$$\begin{aligned} \lambda\bar{P}_1 &= \int_0^b N_{11} dx_2 = \int_0^b \Phi_{,22} dx_2 \\ \Rightarrow \lambda\bar{P}_1 &= \Phi_{,2}(a,b) - \Phi_{,2}(a,0) \end{aligned} \qquad (Ex.\ 18\text{-}3.20)$$

To summarize, the boundary conditions on Φ are

Boundary conditions on Φ

$$\begin{aligned} &\Phi_{,12} = 0,\ \Phi_{,111} = 0 \text{ at } x_1 = 0 \text{ and at } x_1 = a \\ &\Phi_{,12} = 0,\ \Phi_{,222} = 0 \text{ at } x_2 = 0 \text{ and at } x_2 = b \\ &\Phi_{,2}(a,b) - \Phi_{,2}(a,0) = \lambda\bar{P}_1 \\ &\Phi_{,1}(a,b) - \Phi_{,1}(0,b) = \lambda\bar{P}_2 \end{aligned} \qquad (Ex.\ 18\text{-}3.21)$$

[18.10] If the following derivations seem less than self-evident to you, you are not the first one. On the other hand, due to the nature of Φ, it must be clear that displacement boundary conditions must be converted into conditions on stress and after that into strains and, finally, into derivatives of Φ.

Ex 18-3.3 Prebuckling

Recall (*Ex. 18-3.11*) and let Δ_α indicate the homogeneous in-plane displacements at the edges $x_1 = a$ and $x_2 = b$, respectively. Then, $\delta u_1(a, x_2) = \delta\Delta_1$ and $\delta u_2(x_1, b) = \delta\Delta_2$, and the variational statement (18.15) becomes

Principle of virtual displacements in prebuckling

$$\int_0^a\int_0^b\left(\overset{0}{N}_{\alpha\beta}\tfrac{1}{2}\left(\delta u_{\alpha,\beta}+\delta u_{\beta,\alpha}\right)+\overset{0}{M}_{\alpha\beta}\delta w_{,\alpha\beta}\right)dx_1dx_2 = -\bar{P}_1\delta\Delta_1-\bar{P}_2\delta\Delta_2 \qquad (Ex.\ 18\text{-}3.22)$$

or

$$\int_0^a\int_0^b\left(\overset{0}{N}_{\alpha\beta}\delta u_{\alpha,\beta}+\overset{0}{M}_{\alpha\beta}\delta w_{,\alpha\beta}\right)dx_1dx_2 = -\bar{P}_1\delta\Delta_1-\bar{P}_2\delta\Delta_2 \qquad (Ex.\ 18\text{-}3.23)$$

because of the symmetry of $N_{\alpha\beta}$ and therefore also of $\overset{0}{N}_{\alpha\beta}$.

Rewriting of the left-hand side of (*Ex. 18-3.23*) followed by application of the *Divergence Theorem* yields

$$\begin{aligned}
&\int_0^a\int_0^b\left(\overset{0}{N}_{\alpha\beta}\delta u_{\alpha,\beta}+\overset{0}{M}_{\alpha\beta}\delta w_{,\alpha\beta}\right)dx_1dx_2\\
&=\int_0^a\int_0^b\left(\left(\overset{0}{N}_{\alpha\beta}\delta u_\alpha\right)_{,\beta}-\overset{0}{N}_{\alpha\beta,\beta}\delta u_\alpha\right)dx_1dx_2\\
&+\int_0^a\int_0^b\left(\left(\overset{0}{M}_{\alpha\beta}\delta w_{,\alpha}\right)_{,\beta}-\left(\overset{0}{M}_{\alpha\beta,\beta}\delta w\right)_{,\alpha}\right.\\
&\qquad\left.+\overset{0}{M}_{\alpha\beta,\beta\alpha}\delta w\right)dx_1dx_2 \qquad (Ex.\ 18\text{-}3.24)\\
&=\oint_\Gamma\overset{0}{N}_{\alpha\beta}n_\alpha\delta u_\alpha d\Gamma-\int_0^a\int_0^b\overset{0}{N}_{\alpha\beta,\beta}\delta u_\alpha dx_1dx_2\\
&+\oint_\Gamma\overset{0}{M}_{\alpha\beta}n_\beta\delta w_{,\alpha}d\Gamma-\oint_\Gamma\overset{0}{M}_{\alpha\beta,\beta}n_\alpha\delta w d\Gamma\\
&+\int_0^a\int_0^b\overset{0}{M}_{\alpha\beta,\beta\alpha}\delta w dx_1dx_2
\end{aligned}$$

The first integral vanishes on $x_1 = 0$ and on $x_2 = 0$, and the remaining contributions are

$$\begin{aligned}
&\int_0^b\overset{0}{N}_{11}\delta u_1dx_2\Big|_{x_1=a}+\int_a^0\overset{0}{N}_{22}\delta u_2(-dx_1)\Big|_{x_2=b}\\
&=\delta\Delta_1\int_0^b\overset{0}{N}_{11}dx_2\Big|_{x_1=a}+\delta\Delta_2\int_0^a\overset{0}{N}_{22}dx_1\Big|_{x_2=b}
\end{aligned} \qquad (Ex.\ 18\text{-}3.25)$$

which must balance the contributions from the applied loads, i.e. the right-hand side of (*Ex. 18-3.23*). This, however, does not provide any useful statements regarding the boundary conditions on the membrane stresses $N_{\alpha\beta}$, only on the resultants.

Since the variations δu_α of the in-plane displacements u_α are arbitrary in the field the second term provides the field equations for in-plane equilibrium

Prebuckling in-plane static field equation

$$\overset{0}{N}_{\alpha\beta,\beta} = 0 \qquad (Ex.\ 18\text{-}3.26)$$

At the faces $\delta w_{,n}$ is arbitrary, while $\delta w_{,t} = 0$, and thus the third term yields the boundary conditions on the edge moments $\overset{0}{M}_{nn}$

Prebuckling boundary condition on moment

$$\overset{0}{M}_{nn} = 0\,,\ x_1 = (0, a) \text{ and } x_2 = (0, b) \qquad (Ex.\ 18\text{-}3.27)$$

The fourth term vanishes because $\delta w = 0$ on all sides and thus it furnishes no information.

There is no transverse load on the plate and, therefore, the fifth term results in a statement of the transverse equilibrium

Prebuckling transverse static field equation

$$\overset{0}{M}_{\alpha\beta,\beta\alpha} = 0 \qquad (Ex.\ 18\text{-}3.28)$$

Of course, we could have obtained these equilibrium equations from the general expressions of Section 9.1.6 by linearization and after a slight change of notation, but I judge it instructive to furnish a specific example.

A possible solution to (*Ex. 18-3.25*)–(*Ex. 18-3.28*) is[18.11]

Prebuckling solution

$$\begin{aligned} &\overset{0}{N}_{11} = -\frac{\bar{P}_1}{b}\,,\ \overset{0}{N}_{22} = -\frac{\bar{P}_2}{a}\,,\ \overset{0}{N}_{12} = 0 \\ &\overset{0}{M}_{\alpha\beta} = 0\,, \quad \overset{0}{w} = 0 \end{aligned} \qquad (Ex.\ 18\text{-}3.29)$$

which may be verified by insertion into the field equations and the boundary conditions when the constitutive relations are observed.

From (*Ex. 18-3.29*) it is obvious that the prebuckling state is in-plane, as expected.

Ex 18-3.3.1 The Airy Stress Function

A formulation in terms of the *Airy Stress Function* Φ has its advantages, and below we give the relevant equations for the prebuckling state. After linearization, the equations derived in Sections Ex 18-3.2.1 and 9.2.5 may provide the expressions below. The field equations become, see (9.72) and (9.73)

Prebuckling static field equations

$$\nabla^4 \overset{0}{\Phi} = 0 \text{ and } D^B \nabla^4 \overset{0}{w} = 0 \qquad (Ex.\ 18\text{-}3.30)$$

and the in-plane boundary conditions follow from (*Ex. 18-3.21*)

Prebuckling in-plane static boundary conditions

$$\begin{aligned} &\overset{0}{\Phi}_{,12} = 0\,,\ \overset{0}{\Phi}_{,111} = 0 \text{ at } (x_1 = 0) \vee (x_1 = a) \\ &\overset{0}{\Phi}_{,12} = 0\,,\ \overset{0}{\Phi}_{,222} = 0 \text{ at } (x_2 = 0) \vee (x_2 = b) \\ &\overset{0}{\Phi}_{,2}(a, b) - \overset{0}{\Phi}_{,2}(a, 0) = \bar{P}_1 \\ &\overset{0}{\Phi}_{,1}(a, b) - \overset{0}{\Phi}_{,1}(0, b) = \bar{P}_2 \end{aligned} \qquad (Ex.\ 18\text{-}3.31)$$

[18.11] Usually, I like to *derive* solutions rather than rely on guesswork, which is then proved correct, but in the present case it seems justified to me.

Prebuckling boundary conditions on transverse displacement

The boundary conditions on the transverse displacement component $\overset{0}{w}$ are, see (*Ex. 18-3.11*)

$$\begin{aligned} &\overset{0}{w}=0\,,\ \overset{0}{w}_{,11}=0 \ \text{ for } \ x_1=0 \\ &\overset{0}{w}=0\,,\ \overset{0}{w}_{,11}=0 \ \text{ for } \ x_1=a \\ &\overset{0}{w}=0\,,\ \overset{0}{w}_{,22}=0 \ \text{ for } \ x_2=0 \\ &\overset{0}{w}=0\,,\ \overset{0}{w}_{,22}=0 \ \text{ for } \ x_2=b \end{aligned} \qquad (Ex.\ 18\text{-}3.32)$$

Apart from inconsequential linear contributions to $\overset{0}{\Phi}$ the solution in terms of $\overset{0}{\Phi}$ and $\overset{0}{w}$ is

Prebuckling solution

$$\overset{0}{\Phi}=-\frac{\bar{P}_1}{2b}\left(x_2\right)^2-\frac{\bar{P}_2}{2a}\left(x_1\right)^2 \ \text{ and } \ \overset{0}{w}=0 \qquad (Ex.\ 18\text{-}3.33)$$

which, of course, agrees with the solution (*Ex. 18-3.29*).

Ex 18-3.4 Buckling

As was the case for the prebuckling state we formulate and solve the governing equations both as variational statements and in terms of w and Φ. Here, the eigenvalue problem(18.27) becomes

Principle of virtual displacements in buckling Eigenvalue problem

$$\begin{aligned} 0=&\int_0^a\int_0^b\left(\overset{1}{N}_{\alpha\beta}\delta u_{\alpha,\beta}+\overset{1}{M}_{\alpha\beta}\delta w_{,\alpha\beta}\right)dx_1dx_2 \\ &+\lambda_c\int_0^a\int_0^b\overset{0}{N}_{\alpha\beta}\overset{1}{w}_{,\alpha}\delta w_{,\beta}dx_1dx_2 \end{aligned} \qquad (Ex.\ 18\text{-}3.34)$$

where the symmetry of $\overset{0}{N}_{\alpha\beta}$ has been exploited.
The first integral may be handled in the same way as the integral on the right-hand side of (*Ex. 18-3.23*). Therefore, we concentrate on the second integral. By use of the *Divergence Theorem* we get

$$\begin{aligned} &\lambda_c\int_0^a\int_0^b\overset{0}{N}_{\alpha\beta}\overset{1}{w}_{,\alpha}\delta w_{,\beta}dx_1dx_2 \\ &=\lambda_c\int_0^a\int_0^b\left(\left(\overset{0}{N}_{\alpha\beta}\overset{1}{w}_{,\alpha}\delta w\right)_{,\beta}-\left(\overset{0}{N}_{\alpha\beta}\overset{1}{w}_{,\alpha}\right)_{,\beta}\delta w\right)dx_1dx_2 \qquad (Ex.\ 18\text{-}3.35) \\ &=\lambda_c\oint_\Gamma\overset{0}{N}_{\alpha\beta}\overset{1}{w}_{,\alpha}n_\beta\delta w d\Gamma-\lambda_c\int_0^a\int_0^b\left(\overset{0}{N}_{\alpha\beta}\overset{1}{w}_{,\alpha}\right)_{,\beta}\delta w dx_1dx_2 \end{aligned}$$

Here, the first integral vanishes because $\delta w=0$ on the boundary. The second integral and the second term of the first integral in (*Ex. 18-3.34*) must balance each other since these are the only field terms that entail δw

$$\begin{aligned} 0=&\int_0^a\int_0^b\overset{1}{M}_{\alpha\beta,\beta\alpha}\delta w dx_1dx_2 \\ &-\lambda_c\int_0^a\int_0^b\left(\overset{0}{N}_{\alpha\beta}\overset{1}{w}_{,\alpha}\right)_{,\beta}\delta w dx_1dx_2=0 \end{aligned} \qquad (Ex.\ 18\text{-}3.36)$$

Since δw is arbitrary over the interior of the plate this may provide the differential equation, the eigenvalue problem

Buckling static field equation. Eigenvalue problem

$$\overset{1}{M}_{\alpha\beta,\beta\alpha} - \lambda_c \left(\overset{0}{N}_{\alpha\beta} \overset{1}{w}_{,\alpha} \right)_{,\beta} = 0 \qquad (Ex.\ 18\text{-}3.37)$$

When we utilize the ∇-notation we may rewrite the first term, see also (*Ex. 18-3.30*) and (9.73), and by use of the prebuckling solution (*Ex. 18-3.29*) the second term is simplified with the result

Buckling problem

$$D^B \nabla^4 \overset{1}{w} + \lambda_c \left(\frac{\bar{P}_1}{b} \overset{1}{w}_{,11} + \frac{\bar{P}_2}{a} \overset{1}{w}_{,22} \right) = 0 \qquad (Ex.\ 18\text{-}3.38)$$

It may be verified that the in-plane displacements and forces associated with buckling all vanish

In-plane buckling displacements or forces vanish

$$\overset{1}{u}_{\alpha} = 0 \text{ and } \overset{1}{N}_{\alpha\beta} = 0 \qquad (Ex.\ 18\text{-}3.39)$$

Thus, we are left with the task of solving (*Ex. 18-3.38*) with its associated boundary conditions

Buckling boundary conditions on transverse displacement

$$\begin{aligned} \overset{1}{w} &= 0\,,\ \overset{1}{w}_{,11} = 0 \text{ for } x_1 = 0 \\ \overset{1}{w} &= 0\,,\ \overset{1}{w}_{,11} = 0 \text{ for } x_1 = a \\ \overset{1}{w} &= 0\,,\ \overset{1}{w}_{,22} = 0 \text{ for } x_2 = 0 \\ \overset{1}{w} &= 0\,,\ \overset{1}{w}_{,22} = 0 \text{ for } x_2 = b \end{aligned} \qquad (Ex.\ 18\text{-}3.40)$$

which follow directly from (*Ex. 18-3.14*) or may be derived from the variational statement (*Ex. 18-3.34*) of the buckling problem. Since the ∇-term as well as the $\overset{1}{w}_{,\alpha\beta}$-terms involve differentiation of only even order, a solution in terms of trigonometric functions presents an obvious possibility. The boundary conditions preclude cosines, and therefore we investigate whether

Buckling solution?

$$\overset{1}{w} = \xi \sin\left(m\pi \frac{x_1}{a}\right) \sin\left(n\pi \frac{x_2}{b}\right) \qquad (Ex.\ 18\text{-}3.41)$$

which satisfies all boundary conditions also satisfies the differential equation (*Ex. 18-3.38*). In (*Ex. 18-3.40*) ξ is an amplitude, m and n are positive integers. After rearrangement of terms (*Ex. 18-3.40*) and (*Ex. 18-3.38*) yield

$$\begin{aligned} 0 = \Bigg(& D^B \left(\left(m\frac{\pi}{a}\right)^4 + 2\left(m\frac{\pi}{a}\right)^2 \left(n\frac{\pi}{b}\right)^2 + \left(n\frac{\pi}{b}\right)^4 \right) \\ & - \lambda_c \left(\frac{\bar{P}_1}{b} \left(m\frac{\pi}{a}\right)^2 + \frac{\bar{P}_2}{a} \left(n\frac{\pi}{b}\right)^2 \right) \Bigg) \\ & \times \xi \sin\left(m\pi\frac{x_1}{a}\right) \sin\left(n\pi\frac{x_2}{b}\right) \end{aligned} \qquad (Ex.\ 18\text{-}3.42)$$

Nontrivial solutions, i.e. solutions for $\xi \neq 0$, require that the coefficient of the sines vanish with the result that

Bifurcation loads

$$\lambda_c(m,n) = D^B \frac{\left(\left(m\frac{\pi}{a}\right)^2 + \left(n\frac{\pi}{b}\right)^2 \right)^2}{\frac{\bar{P}_1}{b}\left(m\frac{\pi}{a}\right)^2 + \frac{\bar{P}_2}{a}\left(n\frac{\pi}{b}\right)^2} \qquad (Ex.\ 18\text{-}3.43)$$

All these values of λ_c denote bifurcation loads, but in general we are only interested in the lowest one, which—maybe surprisingly—is not always the value for $(m, n) = (1, 1)$, as we shall see below.

Ex 18-3.4.1 *Specific Plate*

As an example, let $a = 2b$, $\bar{P}_1 = \bar{P}$ and $\bar{P}_2 = 0$ and introduce

$$\widetilde{m} \equiv \left(\frac{m}{n}\right) \qquad (Ex.\ 18\text{-}3.44)$$

and get

$$\lambda_c(\widetilde{m}, n) = \frac{\pi^2 D^B}{4\bar{P}b} \frac{\left(\widetilde{m}^2 + 4\right)^2 n^2}{\widetilde{m}^2} \qquad (Ex.\ 18\text{-}3.45)$$

Clearly, the larger n the larger the value of $\lambda_c(\widetilde{m}, n)$ for a fixed value of $\widetilde{m}$. Therefore, we minimize $\lambda_c(\widetilde{m}, n)$ with respect to $\widetilde{m}$, which requires

$$\begin{aligned} & 0 = 4\widetilde{m}^3\left(\widetilde{m}^2 + 4\right) - 2\widetilde{m}\left(\widetilde{m}^2 + 4\right)^2 \\ \Rightarrow\ & 0 = (\widetilde{m} - 2)(\widetilde{m} + 2)(\widetilde{m}^2 + 4)\widetilde{m} \ \Rightarrow\ \widetilde{m} = 2 \\ \Rightarrow\ & m = 2n \end{aligned} \qquad (Ex.\ 18\text{-}3.46)$$

Since the lowest possible value of n is 1 we get the minimum of $\lambda_c(m, n)$ for[18.12]

$$m = 2 \quad \text{and} \quad n = 1 \qquad (Ex.\ 18\text{-}3.47)$$

with the associated value

Buckling load

$$\lambda_c = \lambda_c(2, 1) = \frac{4\pi^2 D^B}{\bar{P}b} \qquad (Ex.\ 18\text{-}3.48)$$

and buckling mode

Buckling mode

$$\overset{1}{w} = \xi \sin\left(\pi\frac{x_1}{b}\right) \sin\left(\pi\frac{x_2}{b}\right) \qquad (Ex.\ 18\text{-}3.49)$$

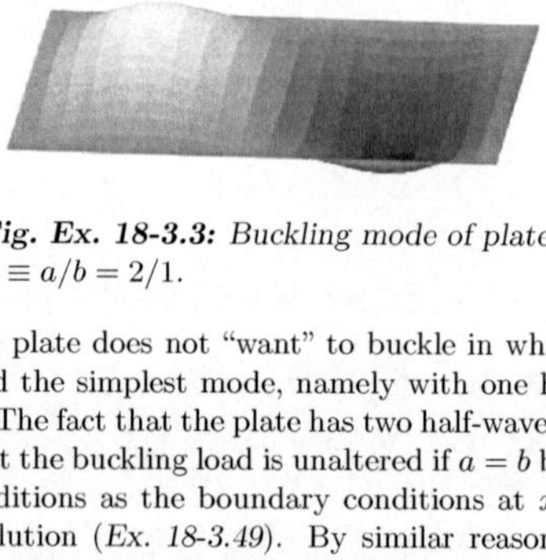

Fig. Ex. 18-3.3: *Buckling mode of plate.* $\alpha \equiv a/b = 2/1$.

The buckling mode is not the one that appears to be the simplest possible

Thus, the plate does not "want" to buckle in what may otherwise be considered the simplest mode, namely with one half-wave in both directions. The fact that the plate has two half-waves in the x_1-directions shows that the buckling load is unaltered if $a = b$ because at $x_1 = b$ the same conditions as the boundary conditions at $x_1 = 2b$ are satisfied by the solution (*Ex. 18-3.49*). By similar reasoning we may further conclude that as long as the length of the plate is an integer multiple of its width the buckling load is given by (*Ex. 18-3.48*).

[18.12] Note that both n and m are positive integers, as required.

Ex 18-3.4.2 Long Plates

In view of the results for the plate that is twice as long as it is wide it may be interesting to investigate the buckling load of longer and shorter plates. In order to do so, let us normalize the buckling loads on the buckling load of the square plate with side length b, namely λ_c of (*Ex. 18-3.48*). We shall only consider load in the x_1-direction and may therefore utilize (*Ex. 18-3.43*) with $n = 1$ because any larger value of n would increase λ_c. When we normalize in the way indicated above, the only parameters left are the ratio $\alpha \equiv a/b$ of the lengths of the two faces and the number of axial half-waves m with the result that the normalized buckling load $\tilde{\lambda}_c$ becomes

Normalized buckling load of long plate

$$\tilde{\lambda}_c = \left(\frac{\alpha^2 + m^2}{2\alpha m} \right)^2 , \ \alpha \equiv \frac{a}{b} \qquad (Ex.\ 18\text{-}3.50)$$

From Fig. *Ex. 18-3.50* it is clear that that the buckling load increases dramatically when a/b is lower than, say 2/3. On the other hand, when

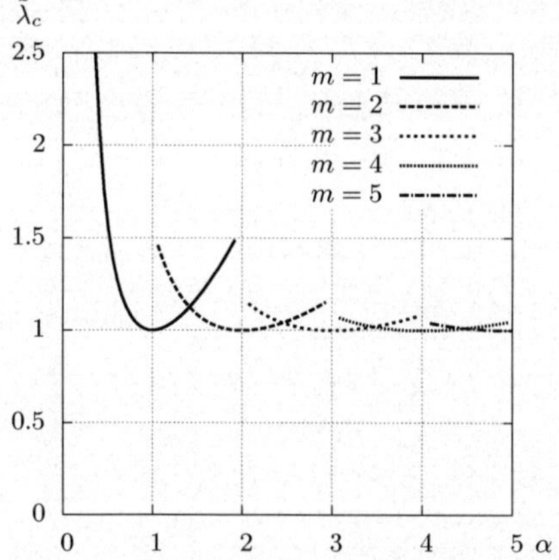

Fig. Ex. 18-3.4: *Normalized buckling load of long plates.*
The parameter $\alpha \equiv a/b$.

a/b is increased from the value 1, then the buckling load becomes more and more independent of a/b.

Ex 18-3.4.3 The Airy Stress Function

The w-Φ-notation is also applicable to the buckling problem. In line with the procedure from Section 18.3.2 we may write

w-Φ-notation

$$\Phi = \lambda \overset{0}{\Phi} + \tilde{\Phi} \text{ and } w = \lambda \overset{0}{w} + \tilde{w} \qquad (Ex.\ 18\text{-}3.51)$$

From (9.75) we may then get

$$0 = D^B \nabla^4(\lambda \overset{0}{w} + \widetilde{w}) - e_{\alpha\gamma} e_{\beta\delta} \left(\left(\lambda \overset{0}{\Phi} + \widetilde{\Phi} \right)_{,\delta\gamma} \left(\lambda \overset{0}{w} + \widetilde{w} \right) \right)_{,\beta} \tag{Ex. 18-3.52}$$

When we exploit our knowledge of $\overset{0}{\Phi}$ and $\overset{0}{w}$ (*Ex. 18-3.52*) becomes

$$D^B \nabla^4 \widetilde{w} - e_{\alpha\gamma} e_{\beta\delta} \left(\left(\lambda \overset{0}{\Phi} + \widetilde{\Phi} \right)_{,\delta\gamma} \widetilde{w} \right)_{,\beta} = 0 \tag{Ex. 18-3.53}$$

and because $\widetilde{\Phi}$ as well as $\widetilde{w}$ is small

$$D^B \nabla^4 \widetilde{w} - \lambda e_{\alpha\gamma} e_{\beta\delta} \left(\overset{0}{\Phi}_{,\delta\gamma} \widetilde{w} \right)_{,\beta} = 0 \tag{Ex. 18-3.54}$$

When we insert (*Ex. 18-3.50*) in (9.72) we may get

$$\nabla^4 \left(\lambda \overset{0}{\Phi} + \widetilde{\Phi} \right) = Et \left(\left(\lambda \overset{0}{w} + \widetilde{w} \right)_{,12}^2 - \left(\lambda \overset{0}{w} + \widetilde{w} \right)_{,11} \left(\lambda \overset{0}{w} + \widetilde{w} \right)_{,22} \right) \tag{Ex. 18-3.55}$$

Because of the smallness of $\widetilde{\Phi}$ and $\widetilde{w}$ and our knowledge that $\overset{0}{w}$ vanishes

$$\nabla^4 \widetilde{\Phi} = 0 \tag{Ex. 18-3.56}$$

For buckling, the in-plane boundary conditions are

Buckling in-plane boundary conditions

$$\begin{aligned} &\overset{1}{\Phi}_{,12} = 0, \ \overset{1}{\Phi}_{,111} = 0 \ \text{ at } \ x_1 = 0 \ \text{ and at } \ x_1 = a \\ &\overset{1}{\Phi}_{,12} = 0, \ \overset{1}{\Phi}_{,222} = 0 \ \text{ at } \ x_2 = 0 \ \text{ and at } \ x_2 = b \\ &\overset{1}{\Phi}_{,2}(a,b) - \overset{1}{\Phi}_{,2}(a,0) = 0 \\ &\overset{1}{\Phi}_{,1}(a,b) - \overset{1}{\Phi}_{,1}(0,b) = 0 \end{aligned} \tag{Ex. 18-3.57}$$

We may readily verify that the solution found above satisfies the equations given in the w-Φ-notation.

18.3.4 Expansion Theorem

Expansion theorem

From calculus it is known that any function $\boldsymbol{w}$ can be expanded in terms of eigenfunctions, here the buckling modes $\boldsymbol{u}_j$

Number of modes M

$$\boldsymbol{w} = \xi_j \boldsymbol{u}_j \,, \ \text{sum over } \ j = 1, \ldots, M \tag{18.37}$$

where the number of nodes M may be infinite.

The amplitudes ξ_J may be determined from (18.37) in conjunction with the orthogonality condition (18.35). Simply write

$$\boldsymbol{\sigma}_0 \cdot \mathbf{l}_{11}(\boldsymbol{w}, \boldsymbol{u}_J) = \boldsymbol{\sigma}_0 \cdot \mathbf{l}_{11}(\xi_j \boldsymbol{u}_j, \boldsymbol{u}_J) \tag{18.38}$$

and exploit the fact that the buckling modes are—or may be taken to be—

mutually orthogonal in the sense of (18.35) to get

$$\xi_J = \frac{\sigma_0 \cdot \mathrm{l}_{11}(\boldsymbol{w}, \boldsymbol{u}_J)}{\sigma_0 \cdot \mathrm{l}_2(\boldsymbol{u}_J)} \tag{18.39}$$

Participation ξ_J of $\boldsymbol{u}_J$ in $\boldsymbol{w}$

where ξ_J is the participation of $\boldsymbol{u}_J$ in $\boldsymbol{w}$.

18.3.5 Numerical and Approximate Solutions, the Rayleigh Quotient

Approximate values of buckling load Rayleigh Quotient and Rayleigh-Ritz Procedure

In many cases it is either impossible or not worth the effort to try to determine the exact value of λ_c and the exact shape of $\boldsymbol{u}_1$. More often than not, we shall therefore resort to numerical or approximate solutions. The *Rayleigh Quotient* is a very strong tool in this connection, in particular when used in the *Rayleigh-Ritz Procedure* which we establish below.

The *Rayleigh Quotient* $\boldsymbol{\Lambda}[\boldsymbol{\phi}]$, which is a functional of the kinematically admissible displacement field $\boldsymbol{\phi}(\boldsymbol{x})$, is here defined by

$$\boldsymbol{\Lambda}[\boldsymbol{\phi}] \equiv -\frac{\mathbf{H}(\mathrm{l}_1(\boldsymbol{\phi})) \cdot \mathrm{l}_1(\boldsymbol{\phi})}{\sigma_0 \cdot \mathrm{l}_2(\boldsymbol{\phi})} \tag{18.40}$$

Rayleigh Quotient $\boldsymbol{\Lambda}[\boldsymbol{\phi}]$

A comparison between (18.40) and (18.27) with $\delta\boldsymbol{u} = \boldsymbol{u}_1$ shows that

$$\boldsymbol{\Lambda}[\boldsymbol{u}_1] = \lambda_c \tag{18.41}$$

$\boldsymbol{\Lambda}[\boldsymbol{u}_1] = \lambda_c$

If we are able to guess a function $\boldsymbol{\phi}$ that is close to $\boldsymbol{u}_1$ in shape it may therefore seem reasonable that the *Rayleigh Quotient* provides us with a good estimate of λ_c. It is our intention to show that this is true.

18.3.6 Stationarity of the Rayleigh Quotient

Let us investigate $\boldsymbol{\Lambda}[\boldsymbol{\phi}]$ in the neighborhood of $\boldsymbol{u}_1$. In order to do this write

$$\boldsymbol{\phi} = \boldsymbol{u}_1 + \epsilon \boldsymbol{w}, \quad |\epsilon| \ll 1 \tag{18.42}$$

and insert it into the Rayleigh Quotient to get

$$\boldsymbol{\Lambda}[\boldsymbol{\phi}] = -\frac{\mathbf{H}(\mathrm{l}_1(\boldsymbol{u}_1 + \epsilon\boldsymbol{w})) \cdot \mathrm{l}_1(\boldsymbol{u}_1 + \epsilon\boldsymbol{w})}{\sigma_0 \cdot \mathrm{l}_2(\boldsymbol{u}_1 + \epsilon\boldsymbol{w})} \tag{18.43}$$

which may be differentiated with respect to ϵ to furnish

$$\begin{aligned}\frac{\partial\boldsymbol{\Lambda}}{\partial\epsilon} = \; & \big(2\sigma_0 \cdot \mathrm{l}_{11}(\boldsymbol{u}_1 + \epsilon\boldsymbol{w}, \boldsymbol{w})\,\mathbf{H}(\mathrm{l}_1(\boldsymbol{u}_1 + \epsilon\boldsymbol{w})) \cdot \mathrm{l}_1(\boldsymbol{u}_1 + \epsilon\boldsymbol{w}) \\ & - 2\sigma_0 \cdot \mathrm{l}_2(\boldsymbol{u}_1 + \epsilon\boldsymbol{w})\,\mathbf{H}(\mathrm{l}_1(\boldsymbol{u}_1 + \epsilon\boldsymbol{w})) \cdot \mathrm{l}_1(\boldsymbol{w})\big) \\ & \times \big(\sigma_0 \cdot \mathrm{l}_2(\boldsymbol{u}_1 + \epsilon\boldsymbol{w})\big)^{-2}\end{aligned} \tag{18.44}$$

For $\epsilon = 0$ the value of this is

$$\left.\frac{\partial\boldsymbol{\Lambda}}{\partial\epsilon}\right|_{\epsilon=0} = \frac{2\sigma_0 \cdot \mathrm{l}_{11}(\boldsymbol{u}_1, \boldsymbol{w})\,\sigma_1 \cdot \varepsilon_1 - 2\sigma_0 \cdot \mathrm{l}_2(\boldsymbol{u}_1)\,\sigma_1 \cdot \mathrm{l}_1(\boldsymbol{w})}{\big(\sigma_0 \cdot \mathrm{l}_2(\boldsymbol{u}_1)\big)^2} \tag{18.45}$$

From (18.27) with $\delta\boldsymbol{u} = \boldsymbol{u}_1$ we get

$$\boldsymbol{\sigma}_1 \cdot \boldsymbol{\varepsilon}_1 = -\lambda_c \boldsymbol{\sigma}_0 \cdot \mathbf{l}_2(\boldsymbol{u}_1) \tag{18.46}$$

where (18.25) has been introduced.

Now, (18.45) yields

The Rayleigh Quotient $\boldsymbol{\Lambda}[\boldsymbol{\phi}]$ is stationary for $\boldsymbol{\phi} = \boldsymbol{u}_1$

$$\left.\frac{\partial \boldsymbol{\Lambda}}{\partial \epsilon}\right|_{\epsilon=0} = -2\,\frac{\boldsymbol{\sigma}_1 \cdot \mathbf{l}_1(\boldsymbol{w}) + \lambda_c \boldsymbol{\sigma}_0 \cdot \mathbf{l}_{11}(\boldsymbol{u}_1, \boldsymbol{w})}{\boldsymbol{\sigma}_0 \cdot \mathbf{l}_2(\boldsymbol{u}_1)} = 0 \tag{18.47}$$

The numerator is the eigenvalue problem (18.27) for $\boldsymbol{u}_1$ and λ_c where $\boldsymbol{w}$ has been substituted for $\delta\boldsymbol{u}$. Therefore, the right-hand side of (18.47) vanishes, and the *Rayleigh Quotient* $\boldsymbol{\Lambda}[\boldsymbol{\phi}]$ is therefore stationary for $\boldsymbol{\phi} = \boldsymbol{u}_1$.

18.3.7 Minimum Property of the Rayleigh Quotient?

Is $\Lambda(\boldsymbol{\phi})$ minimum for $\boldsymbol{\phi} = \boldsymbol{u}_1$?

The stationarity of the *Rayleigh Quotient* is, in itself, an important property, but if we were able to show that the *Rayleigh Quotient* attains a minimum for $\boldsymbol{\phi} = \boldsymbol{u}_1$ this tool becomes much stronger because, given two different approximations $\boldsymbol{\phi}_a$ and $\boldsymbol{\phi}_b$, then we could immediately tell which is the better. In order to prove this we could compute $\partial^2\boldsymbol{\Lambda}/\partial\epsilon^2$ for $\epsilon = 0$ and see if this is always positive. A rather lengthy computation would provide us with

$$\left.\frac{\partial^2 \boldsymbol{\Lambda}}{\partial \epsilon^2}\right|_{\epsilon=0} = 2\frac{\lambda_c}{\boldsymbol{\Lambda}[\boldsymbol{w}]}\,\frac{\mathbf{H}(\mathbf{l}_1(\boldsymbol{w})) \cdot \mathbf{l}_1(\boldsymbol{w})}{\boldsymbol{\sigma}_1 \cdot \boldsymbol{\varepsilon}_1}(\boldsymbol{\Lambda}[\boldsymbol{w}] - \lambda_1) \tag{18.48}$$

which is not very helpful in itself. We shall therefore employ a more direct approach. Omitting an unessential factor (recall that the buckling mode is determined to within an arbitrary factor), any kinematically admissible field $\boldsymbol{\phi}$ can be written as

Number of modes M

$$\boldsymbol{\phi} = \boldsymbol{u}_1 + a_j \boldsymbol{u}_j\,, \text{ sum over } j = 2, 3, \ldots, M \tag{18.49}$$

where M, as mentioned above, may be infinite, a_j are coefficients, which are independent of the spatial coordinates, and $\boldsymbol{u}_1, \boldsymbol{u}_2, \ldots, \boldsymbol{u}_M$ are the buckling modes associated with $\lambda_1 (= \lambda_c), \lambda_2, \ldots, \lambda_M$ ordered such that

$$\lambda_{j+1} \geq \lambda_j \geq \lambda_{j-1} \geq \ldots \geq \lambda_1 > 0 \tag{18.50}$$

Problems when some eigenvalues are negative and others are positive

If all eigenvalues are negative the ordering must be changed accordingly, and $\boldsymbol{\Lambda}[\boldsymbol{\phi}]$ will prove to take its *maximum* value for $\boldsymbol{\phi} = \boldsymbol{u}_1$. Cases with a number of negative eigenvalues, while the rest are positive, are not covered by the following, and we shall assume that all eigenvalues are of the same sign, in particular positive, as indicated in (18.50). Actually, cases with both positive and negative eigenvalues need not be contrived since the load on a structure may very well induce compression in parts of the structure for $\lambda > 0$ and tension in other parts.

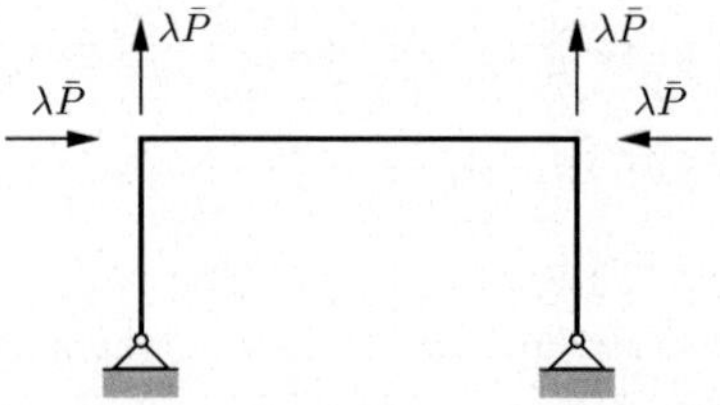

Fig. 18.5: Frame with positive and negative buckling loads.

As an example of this, consider the frame in Fig. 18.5 where it must be clear that it may buckle for some negative value of λ as well as for a positive one because for $\lambda < 0$ the columns which were in tension for $\lambda > 0$ now experience compression while the horizontal member then is in tension.[18.13]

The numerator on the right-hand side of (18.40) becomes

$$\mathbf{H}(\mathbf{l}_1(\boldsymbol{\phi})) \cdot \mathbf{l}_1(\boldsymbol{\phi}) = \mathbf{H}(\mathbf{l}_1(\boldsymbol{u}_1 + a_j \boldsymbol{u}_j)) \cdot \mathbf{l}_1(\boldsymbol{u}_1 + a_k \boldsymbol{u}_k) \tag{18.51}$$

Sum from index = 2, not 1

where a repeated lower-case index indicates summation from 2 to M. This notation is employed throughout the rest of this proof.

The orthogonality condition (18.34) which can also be written as

$$\boldsymbol{\sigma}_J \cdot \boldsymbol{\varepsilon}_K = 0\,, \quad J \neq K \tag{18.52}$$

Buckling modes are orthogonal

and simplifies (18.51) significantly

$$\mathbf{H}(\mathbf{l}_1(\boldsymbol{\phi})) \cdot \mathbf{l}_1(\boldsymbol{\phi}) = \boldsymbol{\sigma}_1 \cdot \boldsymbol{\varepsilon}_1 + \sum_{J=2}^{M} a_J^2 \boldsymbol{\sigma}_J \cdot \boldsymbol{\varepsilon}_J \tag{18.53}$$

Utilize (18.29) with $\delta\boldsymbol{u} = \boldsymbol{u}_1, \boldsymbol{u}_2, \ldots, M$ to get

$$\mathbf{H}(\mathbf{l}_1(\boldsymbol{\phi})) \cdot \mathbf{l}_1(\boldsymbol{\phi}) = -\lambda_1 \boldsymbol{\sigma}_0 \cdot \mathbf{l}_2(\boldsymbol{u}_1) - \sum_{J=2}^{M} a_J^2 \lambda_J \boldsymbol{\sigma}_0 \cdot \mathbf{l}_2(\boldsymbol{u}_J) \tag{18.54}$$

Normalize all the eigenfunctions such that

$$\boldsymbol{\sigma}_0 \cdot \mathbf{l}_2(\boldsymbol{u}_J) = q\,, \quad J = 1, 2, \ldots, M \tag{18.55}$$

[18.13] For what it is worth, to my experience, these cases seldom present severe problems as regards the usefulness of the Rayleigh Quotient.

Then, if q is some positive, arbitrary, but fixed number, (18.54) becomes

$$\mathbf{H}(\mathbf{l}_1(\boldsymbol{\phi})) \cdot \mathbf{l}_1(\boldsymbol{\phi}) = -\left(\lambda_1 + \sum_{J=2}^{M} a_J^2 \lambda_J\right) q \tag{18.56}$$

The denominator of (18.40) is

$$\boldsymbol{\sigma}_0 \cdot \mathbf{l}_2(\boldsymbol{\phi}) = \boldsymbol{\sigma}_0 \cdot \mathbf{l}_{11}(\boldsymbol{u}_1 + a_j \boldsymbol{u}_j, \boldsymbol{u}_1 + a_k \boldsymbol{u}_k) \tag{18.57}$$

The orthogonality condition and the normalization make it possible to write (18.57) as

$$\boldsymbol{\sigma}_0 \cdot \mathbf{l}_2(\boldsymbol{\phi}) = q\left(1 + \sum_{J=2}^{M} a_J^2\right) \tag{18.58}$$

When we rewrite (18.56)

$$\mathbf{H}(\mathbf{l}_1(\boldsymbol{\phi})) \cdot \mathbf{l}_1(\boldsymbol{\phi}) = -\lambda_1 \left(1 + \sum_{J=2}^{M} a_J^2 \frac{\lambda_J}{\lambda_1}\right) q \tag{18.59}$$

we may get

$$\boldsymbol{\Lambda}[a_p] = -\frac{-\lambda_1 \left(1 + \displaystyle\sum_{J=2}^{M} a_J^2 \frac{\lambda_J}{\lambda_1}\right) q}{\left(1 + \displaystyle\sum_{J=2}^{M} a_J^2\right) q} \tag{18.60}$$

Recall that

$$\lambda_J \geq \lambda_1 \tag{18.61}$$

When all $\lambda_j > 0$: $\boldsymbol{\Lambda}[\boldsymbol{\phi}]$ is minimum for $\boldsymbol{\phi} = \boldsymbol{u}_1$

This expression, in conjunction with (18.60), shows that

$$\boldsymbol{\Lambda}[\boldsymbol{\phi}] \geq \lambda_1 \tag{18.62}$$

Thus, we have proved that the *Rayleigh Quotient*— provided that all buckling loads are positive—is minimum for the buckling mode $\boldsymbol{u}_1$ associated with the lowest buckling load λ_1, or λ_c.

18.3.8 The Rayleigh-Ritz Procedure

The Rayleigh-Ritz Procedure works with more than one trial function

Instead of working with *one* assumed displacement field $\boldsymbol{\phi}(\boldsymbol{x})$ we may write it as a sum of a number, say N, assumed functions, or trial functions, $\boldsymbol{\phi}_j(\boldsymbol{x})$, $j = 1, 2, \ldots, N$, that are all kinematically admissible

$$\boldsymbol{\phi}(\boldsymbol{x}) = v_j \boldsymbol{\phi}_j(\boldsymbol{x}), \text{ sum over } j = 1, 2, \ldots, N \tag{18.63}$$

where we return to the usual habit of letting a repeated lower-case index indicate summation from 1 to N, v_j are coefficients, which do not depend on the spatial coordinates, and where all the *Trial Functions* $\boldsymbol{\phi}_j(\boldsymbol{x})$ are chosen once and for all. Below, we omit the indication that the fields $\boldsymbol{\phi}_j$ and $\boldsymbol{\sigma}_0$ depend on the spatial coordinate $\boldsymbol{x}$. The expression (18.40) for the *Rayleigh Quotient* now is

Trial functions $\boldsymbol{\phi}_j(\boldsymbol{x})$

$$\boldsymbol{\Lambda}[v_p] = -\frac{\mathbf{H}(\mathbf{l}_1(v_j\boldsymbol{\phi}_j)) \cdot \mathbf{l}_1(v_k\boldsymbol{\phi}_k)}{\boldsymbol{\sigma}_0 \cdot \mathbf{l}_2(v_m\boldsymbol{\phi}_m)} \tag{18.64}$$

or, because v_j are constants

$$\boldsymbol{\Lambda}[v_p] = -\frac{v_j v_k \mathbf{H}(\mathbf{l}_1(\boldsymbol{\phi}_j)) \cdot \mathbf{l}_1(\boldsymbol{\phi}_k)}{v_m v_n \boldsymbol{\sigma}_0 \cdot \mathbf{l}_{11}(\boldsymbol{\phi}_m, \boldsymbol{\phi}_n)} \tag{18.65}$$

Note that, in general, $\boldsymbol{\phi}_m$ and $\boldsymbol{\phi}_n$ are not orthogonal.

For simplicity introduce the "stiffness matrix" K_{jk}

"Stiffness matrix" K_{jk}

$$K_{jk} = \mathbf{H}(\mathbf{l}_1(\boldsymbol{\phi}_j)) \cdot \mathbf{l}_1(\boldsymbol{\phi}_k) \tag{18.66}$$

That K_{jk} is closely connected with the stiffness of the structure may be seen from reading Part V.

Further, define the so-called *Geometric (Stiffness) Matrix*[18.14] K^G_{mn}

"Geometric (stiffness) matrix" K^G_{mn}

$$K^G_{mn} = \boldsymbol{\sigma}_0 \cdot \mathbf{l}_{11}(\boldsymbol{\phi}_m, \boldsymbol{\phi}_n) \tag{18.67}$$

which expresses the (de)stabilizing effect of the in-plane or membrane loads—stabilizing when the load is predominantly dominated by tension.

Then,

$$\boldsymbol{\Lambda}[v_p] = -\frac{K_{jk} v_j v_k}{K^G_{mn} v_m v_n} \tag{18.68}$$

We know that the best approximation to λ_1 we can obtain from (18.68) is the lowest and therefore we make $\boldsymbol{\Lambda}$ stationary with respect to all v_p, i.e. we require that

Demand that $\Lambda(v_p)$ is stationary

$$\frac{\partial \boldsymbol{\Lambda}}{\partial v_p} = 0 \tag{18.69}$$

This gives us

$$K_{pk} v_k K^G_{mn} v_m v_n - K_{jk} v_j v_k K^G_{pn} v_n = 0 \tag{18.70}$$

[18.14] Here, we have utilized a common *Finite Element Method* nomenclature, see Part V, according to which K_{jk} is termed the *Stiffness Matrix*, and K^G_{mn} is the *Geometric (Stiffness) Matrix*. It is immediately observed that both matrices K_{jk} and K^G_{mn} are symmetric in their indices.

With a change of dummy indices

$$(K^G_{mn} v_m v_n K_{pj} - K_{mn} v_m v_n K^G_{pj}) v_j = 0 \tag{18.71}$$

Buckling ~ matrix eigenvalue problem

Utilize (18.68) to get a matrix eigenvalueproblem

$$\left(K_{pj} + \mathbf{\Lambda}[v_m] K^G_{pj}\right) v_j = 0 \tag{18.72}$$

which provides us with N eigenvalues λ_J and their associated eigenvectors $v_j^{(J)}$. The lowest of these eigenvalues is Λ_1 which is our best approximation to λ_1 because

Lowest eigenvalue $\Lambda_1 \gtrapprox \lambda_1$ with eigenvector $v_j^{(1)}$

$$\Lambda_1 \geq \lambda_1 \tag{18.73}$$

and

$$\Lambda_J \geq \Lambda_1\,, \quad J = 2, 3, ..., N \tag{18.74}$$

In numerical applications it is usually such that Λ_1 determined from (18.72) is not the best estimate we can get. The value is improved if we compute $\mathbf{\Lambda}[v_p]$ from (18.68) on the basis of the values of v_p found from (18.72).

The Rayleigh-Ritz procedure can be extended to cover the higher order buckling modes and their associated buckling loads. We do not, however, intend to include this here.

18.3.9 Another Finite Element Notation

Finite element notation

Occasionally, we shall employ a notation that is more in accord with another of the most common ones in *Finite Element* literature. According to that notation a column matrix with elements v_j is denoted $\{v\}$, a row matrix with elements B_j by $[B]$, and a two-dimensional matrix with elements K_{jk} is given by $[K]$. Since this notation is less explicit than the one used above it gives less information but, on the other hand, the reader is not bothered by information that sometimes is so detailed that it obscures the main message of an equation. Further, in finite element literature, there is a standard set of matrix names, which we introduce below through rewriting some of the above formulas.

Less information, but no unnecessary details

Equation (18.63), which gives the displacement interpolation, becomes

Displacement interpolation matrix $[N]$

$$\{u\} = [N]\{v\} \tag{18.75}$$

while the *linear* strain contribution $\{e\}$ is given by

Strain distribution matrix $[B]$

$$\{e\} = [B]\{v\} \tag{18.76}$$

where the strain distribution matrix $[B]$ is derived from $[N]$ by proper (and obvious) differentiations according to the linear part of the strain-displacement relation, i.e. from $\mathbf{l}_1(\boldsymbol{u})$. Note that $[N]$ and $\{v\}$ correspond to $\boldsymbol{\phi}_j$ and v_j, respectively.

The linear constitutive operator $\mathbf{H}$ is replaced by a constitutive matrix $[D]$ such that the *linear* part $\{s\}$ of the stress vector $\{\sigma\}$ is

$$\{s\} = [D]\{e\} = [D][B]\{v\} \tag{18.77}$$

Constitutive matrix $[D]$

The geometric matrix $[G]$ is given by the nonlinear term of the strain-displacement relation, i.e. by $\mathbf{l}_2(\boldsymbol{u})$, and is therefore derived from $[N]$ by proper differentiations. Thus, we can write the total strain matrix $\{\varepsilon\}$

$$\{\varepsilon\} = [B]\{v\} + \tfrac{1}{2}\sum_k \{j_k\}\Big(\{v\}^T[G]^T[G]\{v\}\Big)_k \tag{18.78}$$

Geometric matrix $[G]$

where the meaning of $\{j_k\}$ follows from Examples Ex 18-4.1 and Ex 18-4.2. In the second term of the right-hand side of (18.78) we sum over the nonlinear strain components, whose number usually is smaller than the total number of strain components. The above mentioned examples ought to clarify the way this term may be interpreted for theoretical purposes.

Second term of (18.78) explained later

Ex 18-4 Interpretation of $\sum_k\Big(\{j_k\}\{v\}^T[G]^T[G]\{v\}\Big)_k$

Ex 18-4.1 Bernoulli-Euler Beam

Consider a Bernoulli-Euler beam and recall (7.12) and (7.14) which may be collected to

$$\{\varepsilon\} = \begin{bmatrix}\varepsilon\\ \kappa\end{bmatrix} = \begin{bmatrix}u'\\ w''\end{bmatrix} + \tfrac{1}{2}\begin{bmatrix}(w')^2\\ 0\end{bmatrix} \tag{Ex. 18-4.1}$$

Generalized strains $\{\varepsilon\}$

Utilizing the finite element notation this may be written

$$\{\varepsilon\} = [N]\{v\} + \tfrac{1}{2}\begin{bmatrix}1\\ 0\end{bmatrix}\{v\}^T[G]^T[G]\{v\} \tag{Ex. 18-4.2}$$

Generalized strains $\{\varepsilon\}$

Ex 18-4.2 von Kármán Plate

For a *von Kármán plate* there are 6 generalized strains, namely 3 membrane strains and 3 bending strains, but only the membrane strains contain nonlinear terms. Thus, the situation becomes more complicated. Recall (9.1) or, more convenient for our present purpose, (*Ex. 18-3.2*) which, with a slight change of notation is

$$\{\varepsilon\} = \begin{bmatrix}\varepsilon_1\\ \varepsilon_2\\ \varepsilon_3\\ \varepsilon_4\\ \varepsilon_5\\ \varepsilon_6\end{bmatrix} \equiv \begin{bmatrix}\varepsilon_{11}\\ \varepsilon_{22}\\ 2\varepsilon_{12}\\ \kappa_{11}\\ \kappa_{22}\\ 2\kappa_{12}\end{bmatrix} = \begin{bmatrix}u_{1,1}\\ u_{2,2}\\ u_{1,2}+u_{2,1}\\ w_{,11}\\ w_{,22}\\ 2w_{,12}\end{bmatrix} + \tfrac{1}{2}\begin{bmatrix}(w_{,1})^2\\ (w_{,2})^2\\ 2w_{,1}w_{,2}\\ 0\\ 0\\ 0\end{bmatrix} \tag{Ex. 18-4.3}$$

Generalized strains $\{\varepsilon\}$

Define $[G_\alpha]$, $\alpha = [1, 2]$ by

$$w_{,1} = [G_1]\{v\} \quad \text{and} \quad w_{,2} = [G_2]\{v\} \tag{Ex. 18-4.4}$$

then we may write (*Ex. 18-4.3*) in the following form

$$\begin{aligned}\{\varepsilon\} = [N]\{v\} &+ \tfrac{1}{2}\{j_1\}\{v\}^T[G_1]^T[G_1]\{v\} \\ &+ \tfrac{1}{2}\{j_2\}\{v\}^T[G_2]^T[G_2]\{v\} \\ &+ \tfrac{1}{2}\{j_3\}\{v\}^T[G_1]^T[G_2]\{v\} \\ &+ \tfrac{1}{2}\{j_3\}\{v\}^T[G_2]^T[G_1]\{v\}\end{aligned} \qquad (Ex.\ 18\text{-}4.5)$$

where

$$\begin{aligned}\{j_1\}^T &\equiv [\,1,\ 0,\ 0,\ 0,\ 0,\ 0\,] \\ \{j_2\}^T &\equiv [\,0,\ 1,\ 0,\ 0,\ 0,\ 0\,] \\ \{j_3\}^T &\equiv [\,0,\ 0,\ 1,\ 0,\ 0,\ 0\,]\end{aligned} \qquad (Ex.\ 18\text{-}4.6)$$

Based on these examples interpretation of $\sum_k \{j_k\}\{v\}^T[G]^T[G]\{v\}$ for other types of structures ought to be fairly easy.

Now, the stress vector $\{\sigma\}$ is

Stress vector $\{\sigma\}$

$$\{\sigma\} = [D]\{\varepsilon\} = [D]\Big([B]\{v\} + \tfrac{1}{2}\sum_k \left(\{j_k\}\{v\}^T[G]^T[G]\{v\}\right)_k\Big) \qquad (18.79)$$

Again, the nonlinear term must be handled the way described above.

With the above formulas in hand, we can compute the (linear) stiffness matrix $[K]$

Linear stiffness matrix $[K]$

$$[K] = [B]^T \cdot ([D][B]) \qquad (18.80)$$

Geometric (stiffness) matrix $[K_G]$

The geometric (stiffness) matrix $[K_G]$ is

$$[K_G] = \boldsymbol{\sigma}_0 \cdot ([G]^T[G]) \qquad (18.81)$$

where the dot implies integration over the structure, and in (18.80) and (18.81) we have taken the liberty of mixing the *Budiansky-Hutchinson Notation* with the *Finite Element Notation.*

Then, the *Rayleigh Quotient* (18.68) becomes

The Rayleigh Quotient $\boldsymbol{\Lambda}[\{v\}]$

$$\boldsymbol{\Lambda}[\{v\}] = -\frac{\{v\}^T[K]\{v\}}{\{v\}^T[K_G]\{v\}} \qquad (18.82)$$

and the eigenvalue problem (18.72) is

Matrix eigenvalue problem

$$([K] + \boldsymbol{\Lambda}_m[K_G])\{v\} = \{0\} \qquad (18.83)$$

Practical Implementation of $\sum_k \Big(\{j_k\}\{v\}^T[G]^T[G]\{v\}\Big)_k$

Don't use $\{j_k\}$ in implementations

As often is the case interpretations the above kind are useful in theoretical context, but rather inappropriate in connection with practical implementation in a computer program, see also page 446 and page 449, because we don't want to multiply by zero if it can be avoided.

18.3.10 A Word of Caution

In applications you should avoid establishing $\mathbf{\Lambda}$ as a function of v_j and then differentiate with respect to v_p. It is always easier and more efficient to solve the eigenvalue problem (18.72) or (18.83).

18.3.11 Examples of Application of the Rayleigh Quotient and the Rayleigh-Ritz Procedure

In most cases it is not possible to find an exact value of the classical critical load λ_c, and in general we must therefore resort to some numerical or approximate method.[18.15] By far the most commonly used method is to exploit the minimum property of the *Rayleigh Quotient*,[18.16] in particular in connection with the *Rayleigh-Ritz Procedure*. However, in order to determine the validity of the results obtained in this way, we study examples for which exact solutions are known.

It may be worth mentioning that today the *Finite Element Method* is usually applied when the geometry of the structure in question is complicated. For structures of the kind investigated below, the *Rayleigh-Ritz Procedure*, with assumed displacement fields that are continuous everywhere, suffices. On the other hand, the study of *Roorda's Frame* in Example Ex 18-5 entails a simple application of the *Finite Element Method*, and in all examples the notation of that method is employed because of its clarity and ease.

For complicated geometries, use the Finite Element Method

Ex 18-5 Roorda's Frame—Application of the Rayleigh-Ritz Procedure

Rayleigh-Ritz Procedure applied to Roorda's frame

As an example of application of the *Rayleigh Quotient* in connection with the *Rayleigh-Ritz Procedure* let us consider the famous *Roorda's Frame*, see Fig. Ex. 18-5.1, whose elastic imperfection sensitivity was determined experimentally by Roorda (1965). Later its postbuckling behavior and imperfection sensitivity were analyzed by Koiter (1966), and its elastic-plastic postbuckling and imperfection sensitivity were investigated by Byskov (1989). In the present analysis—as in Koiter's—we shall assume that the members of the frame are inextensible with the result that the prebuckling path entails no displacements but only axial forces

Inextensibility $\Rightarrow$ linear prebuckling

$$N^0_{AB} = -\bar{P} = -\frac{EI}{L^2} \text{ and } N^0_{BC} = 0 \qquad (Ex.\ 18\text{-}5.1)$$

Linear prebuckling

Thus, the prebuckling state is linear in the strictest sense. As a reference, we shall occasionally cite Koiter's results (Koiter 1966), but we do not include his analysis here—beautiful as it is.

18.15 By an approximate method I mean one that furnishes formulas, while a numerical method results in numbers.

18.16 Recall that minimum is guarantied when all buckling loads are positive, otherwise Rayleigh's Quotient is only stationary.

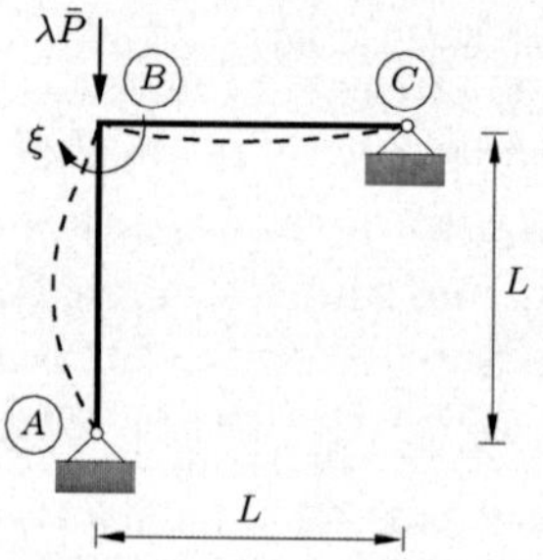

Fig. Ex. 18-5.1: *Roorda's Frame.*

The *Rayleigh Quotient* (18.40) is

Rayleigh quotient $\Lambda[\phi]$

$$\Lambda[\phi] = -\frac{\mathbf{H}(\mathbf{l}_1(\phi)) \cdot \mathbf{l}_1(\phi)}{\boldsymbol{\sigma}_0 \cdot \mathbf{l}_2(\phi)} \qquad \text{(Ex. 18-5.2)}$$

In the present case

$$\mathbf{H}(\mathbf{l}_1(\phi)) \cdot \mathbf{l}_1(\phi) = \int_A^B EI(w_1'')^2 dx + \int_B^C EI(w_1'')^2 dx \qquad \text{(Ex. 18-5.3)}$$

and

$$\begin{aligned} \boldsymbol{\sigma}_0 \cdot \mathbf{l}_2(\phi) &= \int_A^B N_{AB}^0 (w_1')^2 dx + \int_B^C N_{BC}^0 (w_1')^2 dx \\ &= \int_A^B N_{AB}^0 (w_1')^2 dx \end{aligned} \qquad \text{(Ex. 18-5.4)}$$

The denominator in the Rayleigh Quotient does not contain contributions from the axial displacement component u_1, only from the transverse component w_1. If we allowed the frame to be extensible in buckling, then the numerator would entail a term that is quadratic in u_1 and, since the Rayleigh Quotient attains a minimum for the exact solution, it is immediately seen that u_1 must vanish and that the only displacement component of the buckling mode is w_1. Therefore, we may write the *Rayleigh Quotient* as

Only transverse displacements in buckling

Rayleigh quotient $\Lambda[\widetilde{w}]$

$$\Lambda[\widetilde{w}] = -\frac{\displaystyle\int_A^B EI(\widetilde{w}'')^2 dx + \int_B^C EI(\widetilde{w}'')^2 dx}{\displaystyle\int_A^B N_{AB}^0 (\widetilde{w}')^2 dx} \qquad \text{(Ex. 18-5.5)}$$

where $\widetilde{w}$ signifies the approximation to the transverse component of the buckling mode $\boldsymbol{u}_1$.[18.17]

[18.17] Don't let the minus sign bother you, recall that $N_{AB}^0 < 0$.

Ex 18-5.1 One-Parameter Solution

One-Parameter Solution

In AB and in BC the simplest admissible displacement fields, which also satisfy all static boundary conditions,[18.18] namely that the bending moments vanish at the supports, are given by

Trial function which satisfy the static boundary conditions

$$w_{AB} = L\left(-\tfrac{1}{2}\zeta^3 + \tfrac{1}{2}\zeta\right)\xi$$
$$w_{BC} = L\left(-\tfrac{1}{2}\zeta^3 + \tfrac{3}{2}\zeta^2 - \zeta\right)\xi \qquad (Ex.\ 18\text{-}5.6)$$

where

$$\zeta \equiv \frac{x}{L} \qquad (Ex.\ 18\text{-}5.7)$$

With (*Ex. 18-5.6*) in hand we could utilize (*Ex. 18-5.5*) directly, but because we later increase the number of fields we use the finite element notation given in Section 18.3.9.

By comparing (*Ex. 18-5.6*) with (18.75) we may make the following identifications

$$\{v_{(1)}\} = \{\xi\} \qquad (Ex.\ 18\text{-}5.8)$$

where subscript $_{(1)}$ indicates that this is our first approximation, and

Displacement interpolation matrices $[N]$

$$[N_{AB}] = \left[L\left(-\tfrac{1}{2}\zeta^3 + \tfrac{1}{2}\zeta\right)\right]$$
$$[N_{BC}] = \left[L\left(-\tfrac{1}{2}\zeta^3 + \tfrac{3}{2}\zeta^2 - \zeta\right)\right] \qquad (Ex.\ 18\text{-}5.9)$$

Compare (18.78) with the expression for the axial strain ε, see (7.12)

$$\varepsilon \equiv u' + \tfrac{1}{2}\left(w'\right)^2 \qquad (Ex.\ 18\text{-}5.10)$$

and get

Geometric matrix $[G]$

$$[G_{AB}] = \frac{1}{L}\frac{d[N_{AB}]}{d\zeta} = [-\tfrac{3}{2}\zeta^2 + \tfrac{1}{2}]$$
$$[G_{BC}] = \frac{1}{L}\frac{d[N_{BC}]}{d\zeta} = [-\tfrac{3}{2}\zeta^2 + 3\zeta - 1] \qquad (Ex.\ 18\text{-}5.11)$$

Since the only non-vanishing buckling strain is the bending strain κ which is defined by, see (7.14)

Only bending strain in buckling

$$\kappa \equiv w'' \qquad (Ex.\ 18\text{-}5.12)$$

we find

Strain distribution matrix $[B]$

$$[B_{AB}] = \frac{1}{L^2}\frac{d^2[N_{AB}]}{d\zeta^2} = \frac{1}{L}[-3\zeta]$$
$$[B_{BC}] = \frac{1}{L^2}\frac{d^2[N_{BC}]}{d\zeta^2} = \frac{1}{L}[-3\zeta + 3] \qquad (Ex.\ 18\text{-}5.13)$$

[18.18] It is not necessary that the trial functions satisfy *any* static conditions, but, sometimes, with a given number of trial functions we get a better solution if the static boundary conditions are fulfilled, but see Example Ex 32-4, in particular the comments page 571 and Example Ex 32-1, page 551 where a similar situation is discussed. In the present case, it is easy to find such functions, but for other structures it may not be.

Constitutive matrix $[D]$

The constitutive matrix $[D]$ is

$$[D] = [EI] \qquad (Ex.\ 18\text{-}5.14)$$

Now, it is easily found that

Contributions to stiffness matrix

$$[K_{AB}] = \left[+3\frac{EI}{L}\right] = \frac{EI}{L}[+3]$$
$$[K_{BC}] = \left[+3\frac{EI}{L}\right] = \frac{EI}{L}[+3] \qquad (Ex.\ 18\text{-}5.15)$$

and that

Contributions to geometric (stiffness) matrix

$$[K^G_{AB}] = \left[-\bar{P}\frac{L}{5}\right] = \frac{EI}{L}\left[-\tfrac{1}{5}\right]$$
$$[K^G_{BC}] = [0] \qquad (Ex.\ 18\text{-}5.16)$$

For the whole structure

Eigenvalue problem

$$([K] + \lambda[K_G])\{v_{(1)}\} = \{0\} \qquad (Ex.\ 18\text{-}5.17)$$

with

Stiffness matrix $[K]$ and geometric (stiffness) matrix $[K_G]$

$$[K] = \frac{EI}{L}[+6]$$
$$[K_G] = \left[-\bar{P}\frac{L}{5}\right] = \frac{EI}{L}[-\tfrac{1}{5}] \qquad (Ex.\ 18\text{-}5.18)$$

The solution is

$\lambda_{(1)} \gg \lambda_{ex}$

$$\lambda_{(1)} = 30 \qquad (Ex.\ 18\text{-}5.19)$$

while the exact value—to 14 significant digits—is given by

λ_{ex}

$$\lambda_{ex} = 13.885\,942\,905\,965 \qquad (Ex.\ 18\text{-}5.20)$$

The value of $\lambda_{(1)}$, given by (*Ex. 18-5.19*), is clearly not satisfactory. The reason is that the assumption for $[N_{AB}]$ given by (*Ex. 18-5.9*a) does not account for the change in displacement field caused by the influence of the compressive axial force in AB. Therefore, the expression for $[K_{AB}]$, see (*Ex. 18-5.15*a), is inaccurate, while the expression for $[K_{BC}]$ given by (*Ex. 18-5.15*b) is exact.

The presence of the beam BC has increased the buckling load by around 40% over the buckling load of the Euler Column which is given by the value $\lambda_c^{Euler} = \pi^2$.

Ex 18-5.2 Two-Parameter Solution

Two-Parameter Solution

Since the displacement field utilized above is exact in BC, we focus attention on AB. There are, of course, many possible ways to improve the displacement field, e.g. dividing AB into a number of finite elements. We shall, however, pursue a different course in that we still let AB be an undivided part of the frame, but increase the number of displacement functions in that column. This may be done in many different ways. However, because of the inextensibility of the beams the additional term(s) in $[N_{AB}]$ must fulfill the condition that the buckling mode entails vanishing displacements at A and at B. Furthermore, we choose the additional term(s) in $[N_{AB}]$ such that they provide no additional rotation at B. This may seem like an arbitrary choice, but the reason for doing this is that the buckling mode amplitude is then easily identified as the displacement field parameter associated with the first term in $[N_{AB}]$. Especially in the computation of the postbuckling coefficient a, see Example Ex 19-1 this proves very convenient.

Finally, if we can satisfy the static boundary condition that the bending moment M_A at A vanishes, we may expect a good approximation to the column displacements. It is easily verified that the following assumption conforms to our requirements

Displacement interpolation matrix $[N]$

$$[N_{AB}] = L[-\tfrac{1}{2}\zeta^3 + \tfrac{1}{2}\zeta\,;\; +\zeta^4 - \tfrac{3}{2}\zeta^3 + \tfrac{1}{2}\zeta] \qquad \text{(Ex. 18-5.21)}$$

with $[N_{BC}]$ unchanged. Differentiations yield

Geometric matrix $[G]$

$$[G_{AB}] = [N'_{AB}] = [-\tfrac{3}{2}\zeta^2 + \tfrac{1}{2}\,;\; +4\zeta^3 - \tfrac{9}{2}\zeta^2 + \tfrac{1}{2}] \qquad \text{(Ex. 18-5.22)}$$

and further

Strain distribution matrix $[B]$

$$[B_{AB}] = \frac{1}{L}[-3\zeta\,;\quad +12\zeta^2 - 9\zeta] \qquad \text{(Ex. 18-5.23)}$$

Compute the stiffness matrix $[K_{AB}]$

$$\begin{aligned}[K_{AB}] &= \int_A^B EI[B_{AB}]^T[B_{AB}]dx \\ &= \frac{EI}{L^2}L\int_0^1 [B_{AB}]^T[B_{AB}]d\zeta \\ &= \frac{EI}{L}\int_0^1 \begin{bmatrix} -3\zeta \\ +12\zeta^2 - 9\zeta \end{bmatrix} \big[-3\zeta\,;\, +12\zeta^2 - 9\zeta\big]d\zeta \end{aligned} \qquad \text{(Ex. 18-5.24)}$$

with the result

Stiffness matrix $[K]$

$$[K_{AB}] = \frac{EI}{L}\begin{bmatrix} +3 & 0 \\ 0 & +\frac{9}{5} \end{bmatrix} \qquad \text{(Ex. 18-5.25)}$$

It may be worthwhile mentioning that the expression for $[K_{AB}]$ does not entail off-diagonal terms, which means that the two elements in $[N_{AB}]$ are mutually orthogonal. This is desirable in any approximation because it results in numerically stable systems of equations and less work.

In a similar way, we compute the geometric (stiffness) matrix $[K^G_{AB}]$

$$\begin{aligned}[K^G_{AB}] &= \int_A^B N^0_{AB}[G_{AB}]^T[G_{AB}]dx \\ &= -\bar{P}L\int_0^1 [G_{AB}]^T[G_{AB}]d\zeta \\ &= -\bar{P}L\int_0^1 \begin{bmatrix} -\frac{3}{2}\zeta^2 + \frac{1}{2} \\ +4\zeta^3 - \frac{9}{2}\zeta^2 + \frac{1}{2} \end{bmatrix} \\ &\qquad \big[-\tfrac{3}{2}\zeta^2 + \tfrac{1}{2}\,;\; +4\zeta^3 - \tfrac{9}{2}\zeta^2 + \tfrac{1}{2}\big]\,d\zeta \end{aligned} \qquad \text{(Ex. 18-5.26)}$$

Contribution to geometric (stiffness) matrix $[K^G_{AB}]$

The result is

$$[K^G_{AB}] = -\frac{EI}{L}\begin{bmatrix} +\frac{1}{5} & +\frac{1}{10} \\ +\frac{1}{10} & +\frac{3}{35} \end{bmatrix} . \qquad (Ex.\ 18\text{-}5.27)$$

As before, the first element in $[K]$ is twice that of $[K_{AB}]$ because the beam BC also contributes stiffness against rotation of B. Then, the eigenvalue problem is still given in the form of (*Ex. 18-5.17*), which results in the equation

$$+\frac{1}{140}\lambda^2 - \frac{153}{175}\lambda + \frac{54}{5} = 0 \qquad (Ex.\ 18\text{-}5.28)$$

with the solution

$\lambda_{(2)} \approx \lambda_{ex}$

$$\begin{aligned} \lambda_{(2)} &= \min\left(+\tfrac{306}{5} \pm \tfrac{6}{5}\sqrt{1551}\right) = 13.94071520 \\ \text{cf. } \lambda_{ex} &= 13.8859735 \end{aligned} \qquad (Ex.\ 18\text{-}5.29)$$

where, as indicated, only the smallest solution is of any interest in the present connection. The relative error on the approximate solution is only 0.4%, which is sufficiently accurate for almost any purpose. The eigenvector, i.e. the approximation to the buckling mode, associated with $\lambda_{(2)}$ is $\{v\}_{(2)}$, with

buckling mode

$$\{v_{AB}\}_{(2)} = \begin{bmatrix} 1 \\ (9+\sqrt{1551})/21 \end{bmatrix} = \begin{bmatrix} 1 \\ 2.303\,939\,873 \end{bmatrix} \qquad (Ex.\ 18\text{-}5.30)$$

where we have normalized the buckling mode such that $\xi = 1$ corresponds to a rotation of one radian of point B. For BC

$$\{v_{BC}\}_{(2)} = [1] \qquad (Ex.\ 18\text{-}5.31)$$

It is, of course, possible to continue the above process, and for the sake of completeness I give the expression for $[N_{AB}]$

More terms in $[N_{AB}]$

$$\begin{aligned} [N_{AB}] = [\,&-\tfrac{1}{2}\zeta^3 + \tfrac{1}{2}\zeta;\ \zeta^4 - \tfrac{3}{2}\zeta^3 + \tfrac{1}{2}\zeta;\ \ldots; \\ &\ldots;\ \zeta^{j+2} - \tfrac{j+1}{j}\zeta^{j+1} + \tfrac{1}{j}\zeta;\ \ldots] \end{aligned} \qquad (Ex.\ 18\text{-}5.32)$$

and the values of the approximations to λ_c are

$$\begin{aligned} \lambda_{(1)} &= 30\,, & \tilde{\lambda}_{(1)} &= 2.160\,458 \\ \lambda_{(2)} &= 13.940\,715\,20\,, & \tilde{\lambda}_{(2)} &= 1.003\,944 \\ \lambda_{(3)} &= 13.921\,293\,88\,, & \tilde{\lambda}_{(3)} &= 1.002\,545 \\ \lambda_{(4)} &= 13.886\,049\,19\,, & \tilde{\lambda}_{(4)} &= 1.000\,007 \\ \lambda_{(5)} &= 13.885\,973\,47\,, & \tilde{\lambda}_{(5)} &= 1.000\,002 \\ \lambda_{(6)} &= 13.885\,973\,47\,, & \tilde{\lambda}_{(6)} &= 1.000\,002 \end{aligned} \qquad (Ex.\ 18\text{-}5.33)$$

where $\tilde{\lambda}_{(j)} \equiv \lambda_{(j)}/\lambda_{ex}$.

The effort involved in establishing and solving the eigenvalue problem grows geometrically with the number of terms and, even with as few as three terms in $[N_{AB}]$ it is wise to resort to some analytic manipulation program, such as maxima, Mathematica, Maple, MuPAD, or the older MuMath.

In view of our success with Roorda's Frame we may try the Rayleigh-Ritz procedure on another kind of structure, namely a plate. The geometry is simpler than that of Roorda's Frame, but, as we have seen in Example Ex 18-3, the buckling mode of the plate depends strongly on the ratio between its two sides. Therefore, if we utilized the buckling mode for a square plate to analyze a long plate we would get very erroneous results because it cannot model a buckling mode with many half-waves. From Fig. Ex. 18-3.4 it must be clear that such an assumption would predict much too high values of the buckling load. For instance, if the plate is twice as long as it is wide the overshoot would be about 50%, see the curve for $m = 1$. Even if we extended the investigation to many more terms than the ones used below and saw convergence of the estimates of the buckling load, we would still not get close to the correct result if we did not include terms that allowed for more buckles. As always, engineering insight into the nature of the problem at hand is pertinent.

Engineering insight is important

Ex 18-6 Plate Buckling—Rayleigh-Ritz Procedure

As our second example of the *Rayleigh Quotient* and the *Rayleigh-Ritz Procedure*we return to the problem of buckling of a plane plate, see Example Ex 18-3, and exploit the information about the prebuckling state given there, see (*Ex. 18-3.29*)

Rayleigh-Ritz Procedure

$$\begin{aligned} &\overset{0}{N}_{11} = -\frac{\bar{P}_1}{b}\,,\ \overset{0}{N}_{22} = -\frac{\bar{P}_2}{a}\,,\ \overset{0}{N}_{12} = 0 \\ &\overset{0}{M}_{\alpha\beta} = 0\,,\ \overset{0}{w} = 0 \end{aligned} \qquad (Ex.\ 18\text{-}6.1)$$

We shall also rely on the interpretation of the *Budiansky-Hutchinson* notation given in Example Ex 18-3.1. In order to keep the analysis simple we limit ourselves to the case of a square plate with equal compression in both in-plane directions, i.e.

$$b = a \text{ and } \bar{P}_1 = \bar{P}_2 = \bar{P} \qquad (Ex.\ 18\text{-}6.2)$$

Square plate Equal compression

As in the example that was concerned with *Roorda's Frame*, see Example Ex 18-5, the *Rayleigh Quotient* (18.40) is

$$\Lambda[\phi] = -\frac{\mathbf{H}(\mathbf{l}_1(\phi))\cdot\mathbf{l}_1(\phi)}{\sigma_0\cdot\mathbf{l}_2(\phi)} \qquad (Ex.\ 18\text{-}6.3)$$

Rayleigh Quotient

where,

$$\begin{aligned} \mathbf{H}(\mathbf{l}_1(\phi))\cdot\mathbf{l}_1(\phi) &= \int_A D^B_{\alpha\beta\gamma\delta}\overset{1}{\tilde{\kappa}}_{\alpha\beta}\overset{1}{\tilde{\kappa}}_{\delta\gamma}dA \\ \sigma_0\cdot\mathbf{l}_2(\phi) &= \int_A \overset{0}{N}_{\alpha\beta}\overset{1}{\tilde{w}}_{,\alpha}\overset{1}{\tilde{w}}_{,\beta}dA \end{aligned} \qquad (Ex.\ 18\text{-}6.3)$$

and tilde (˜) indicates that the field is an approximate one.

When we introduce the strain-displacement relation (9.1b) the *Rayleigh*

Quotient may be written in terms of displacement derivatives $\overset{1}{\widetilde{w}}_{,\alpha}$ and $\overset{1}{\widetilde{w}}_{,\alpha\beta}$ only. The expression then is

Rayleigh Quotient

$$\Lambda[\widetilde{w}] = -\frac{\displaystyle\int_A D^B_{\alpha\beta\gamma\delta}\overset{1}{\widetilde{w}}_{,\alpha\beta}\overset{1}{\widetilde{w}}_{,\delta\gamma}dA}{\displaystyle\int_A \overset{0}{N}_{\alpha\beta}\overset{1}{\widetilde{w}}_{,\alpha}\overset{1}{\widetilde{w}}_{,\beta}dA} \qquad (Ex.\ 18\text{-}6.4)$$

Ex 18-6.1 ***One-Parameter Solution***

One-Parameter solution

Recall that the approximating displacement field does not need to satisfy the static boundary conditions and choose the simplest nondimensional field possible

Assumed w_1 need not satisfy any static conditions

$$\overset{1}{\widetilde{w}} = 16\zeta_1(1-\zeta_1)\zeta_2(1-\zeta_2) \qquad (Ex.\ 18\text{-}6.5)$$

where

$$\zeta_\alpha \equiv \frac{x_\alpha}{a},\ \alpha = (1,2) \qquad (Ex.\ 18\text{-}6.6)$$

We may cast this in the *Finite Element* notation and write the displacement as

$$\widetilde{w} = [N]\{v\} \qquad (Ex.\ 18\text{-}6.7)$$

Displacement interpolation matrix $[N]$

with the displacement interpolation matrix $[N]$[18.19]

$$[N] = a[16\zeta_1(1-\zeta_1)\zeta_2(1-\zeta_2)] \qquad (Ex.\ 18\text{-}6.8)$$

which provides the geometric matrices $[G_1]$ and $[G_2]$

Geometric matrices $[G]_\alpha$

$$\begin{aligned} [G_1] &= [N]_{,1} = [16\zeta_2\,(-1+\zeta_2)\,(-1+2\zeta_1)] \\ [G_2] &= [N]_{,2} = [16\zeta_1\,(-1+2\zeta_2)\,(-1+\zeta_1)] \end{aligned} \qquad (Ex.\ 18\text{-}6.9)$$

and the strain distribution matrix $[B]$

Strain distribution matrix $[B]$

$$[B] = \begin{bmatrix} [N]_{,11} \\ [N]_{,22} \\ 2[N]_{,12} \end{bmatrix} = \frac{1}{a}\begin{bmatrix} 32\zeta_2\,(-1+\zeta_2) \\ 32\zeta_1\,(-1+\zeta_1) \\ (-32+64\zeta_2)\,(-1+2\zeta_1) \end{bmatrix} \qquad (Ex.\ 18\text{-}6.10)$$

where the strains and stresses are generalized in the same fashion as in Example Ex 18-3.1.

Then, the stiffness matrix $[K]$ is

Stiffness matrix $[K]$

$$[K] = \int_A D^B[B]^T\left[D^H\right][B]dA = \left[\tfrac{5632}{45}D^B\right] \qquad (Ex.\ 18\text{-}6.11)$$

where the expressions for D^B and $\left[D^H\right]$ may be found in Example Ex 18-3.1, and where the geometric (stiffness) matrix becomes

Geometric (stiffness) matrix $[K_G]$

$$[K_G] = \int_A \overset{0}{N}_{\alpha\beta}[G_\alpha]^T[G_\beta]dA = \left[-\tfrac{256}{45}\,a\bar{P}\right] \qquad (Ex.\ 18\text{-}6.12)$$

18.19 The term “Displacement interpolation matrix” may be a little misleading here in that the displacement is not interpolated between nodes, but, still, the meaning ought to be clear.

Since the eigenvalue problem that determines the buckling load is given as

$$\Big([K] + \lambda[K^G]\Big)\,\{v_1\} = \{0\} \tag{Ex. 18-6.13}$$

Eigenvalue problem

it is easily found that the first approximation $\lambda_{(1)}$ to λ_c is

$$\lambda_{(1)} = \frac{5632}{256}\frac{D^B}{a\bar{P}} = 22\,\frac{D^B}{a\bar{P}}$$
$$\text{cf.}\quad \lambda_{ex} = 2\pi^2\,\frac{D^B}{a\bar{P}} = 19.739\,209\,\frac{D^B}{a\bar{P}} \tag{Ex. 18-6.14}$$

$\lambda_{(1)}$ off by 11%

Thus, the error on $\lambda_{(1)}$ is about 11%. This is half the error of the one-parameter solution to the *Euler Column* in Example Ex 32-5.1, which at first may be surprising because the assumed displacement fields are of the same order in both cases. However, compare (*Ex. 18-6.10*) with (*Ex. 32-5.5*b) and realize that while the latter predicts a constant bending moment the former shows that the displacement assumption (*Ex. 18-6.5*) results in bending and torsional moments that vary over the plate, although κ_{11} is constant in the x_1-direction (the ζ_1-direction) and κ_{22} is constant in the x_2-direction. This means that neither of the bending moments M_{11} and M_{22} nor the torsional moment M_{12} vanishes independently of the value of ν. It is thus reasonable that the displacement assumption for the plate provides a somewhat better approximation to the classical critical load.

Ex 18-6.2 Two-Parameter Solution

Two-Parameter Solution

An error of 11% is probably more than an engineer is willing to accept, so we try to improve the solution by adding one more term to the displacement approximation

$$[N] = a\big[\,16\zeta_1(1-\zeta_1)\zeta_2(1-\zeta_2)\;;\; 256\,(\zeta_1)^2\,(1-\zeta_1)^2\,(\zeta_2)^2\,(1-\zeta_2)^2\big] \tag{Ex. 18-6.15}$$

Displacement interpolation matrix $[N]$

Now, the elements of $[G_1]$ are

$$\begin{aligned} G_1(1) &= N(1)_{,1} = 16\zeta_2\,(-1+\zeta_2)\,(-1+2\zeta_1)\\ G_1(2) &= N(2)_{,1} = 512\,(\zeta_2)^2\,\zeta_1\,(-1+\zeta_2)^2\\ &\qquad\times(-1+2\zeta_1)\,(-1+\zeta_1)\end{aligned} \tag{Ex. 18-6.16}$$

Geometric matrix $[G]$

and the elements of $[G_2]$ become

$$\begin{aligned} G_2(1) &= N(2)_{,1} = 16\zeta_1\,(-1+2\zeta_2)\,(-1+\zeta_1)\\ G_2(2) &= N(2)_{,2} = 512\,(\zeta_1)^2\,\zeta_2\,(-1+2\zeta_2)\\ &\qquad\times(-1+\zeta_2)\,(-1+\zeta_1)^2\end{aligned} \tag{Ex. 18-6.17}$$

Geometric matrix $[G]$

Similarly, the elements of the first row of $[B]$ are

$$\begin{aligned} B(1,1) &= N(1)_{,11} = \frac{1}{a}32\zeta_2\,(-1+\zeta_2)\\ B(1,2) &= N(2)_{,11} = \frac{1}{a}512\,(\zeta_2)^2\,(-1+\zeta_2)^2\\ &\qquad\times\big(1+6\,(\zeta_1)^2-6\zeta_1\big)\end{aligned} \tag{Ex. 18-6.18}$$

First row of strain distribution matrix $[B]$

The elements of the second row of $[B]$ are

Second row of strain distribution matrix $[B]$

$$B(2,1) = N(1)_{,22} = \frac{1}{a}32\zeta_1(-1+\zeta_1)$$
$$B(2,2) = N(2)_{,22} = \frac{1}{a}512(\zeta_1)^2(-1+\zeta_1)^2 \times \left(1-6\zeta_2+6(\zeta_2)^2\right) \qquad \text{(Ex. 18-6.19)}$$

The elements of the third row of $[B]$ are

Third row of strain distribution matrix $[B]$

$$B(3,1) = 2N(1)_{,12} = \frac{1}{a}(-32+64\zeta_2)(-1+2\zeta_1)$$
$$B(3,2) = 2N(2)_{,12} = \frac{1}{a}2048\zeta_1\zeta_2 \times(-1+2\zeta_2)(-1+\zeta_2) \times(-1+2\zeta_1)(-1+\zeta_1) \qquad \text{(Ex. 18-6.20)}$$

With the above expressions for $[G_1]$, $[G_2]$ and $[B]$ in hand it is a fairly straightforward, but tedious,[18.20] process to compute $[K]$ and $[K_G]$, and only the final results for these matrices are given below

Stiffness matrix $[K]$

$$[K] = D^B \begin{bmatrix} \frac{5632}{45} & \frac{8192}{225} \\ \frac{8192}{225} & \frac{262144}{1225} \end{bmatrix} \qquad \text{(Ex. 18-6.21)}$$

and

Geometric (stiffness) matrix $[K_G]$

$$[K_G] = -\bar{P}a \begin{bmatrix} \frac{256}{45} & \frac{2048}{525} \\ \frac{2048}{525} & \frac{131072}{33075} \end{bmatrix} \qquad \text{(Ex. 18-6.22)}$$

The ensuing eigenvalue problem is

Eigenvalue problem

$$\left([K]+\lambda[K_G]\right)\{v\} = \{0\} \qquad \text{(Ex. 18-6.23)}$$

see also (18.83). In this case the easiest way to get the solution is to determine the roots of the determinant of the coefficient matrix to $\{v\}$

$$\left|[K]+\lambda[K_G]\right| = 0 \qquad \text{(Ex. 18-6.24)}$$

which becomes

$$\frac{631494441024}{2480625} - \frac{10636754944}{7441875}\left(\frac{\lambda a\bar{P}}{D^B}\right) + \frac{54525952}{7441875}\left(\frac{\lambda a\bar{P}}{D^B}\right)^2 = 0 \qquad \text{(Ex. 18-6.25)}$$

with the solution

$$\lambda = \frac{D^B}{a\bar{P}}\frac{4}{13}\begin{bmatrix} 317-\sqrt{63790} \\ 317+\sqrt{63790} \end{bmatrix} = \frac{D^B}{a\bar{P}}\begin{bmatrix} 19.825\,593 \\ 175.251\,330 \end{bmatrix} \qquad \text{(Ex. 18-6.26)}$$

Here, we are only interested in the smaller of these eigenvalues, which is our second approximation to λ_c

[18.20] I left this task for maxima to handle.

The error on our final approximation, namely

$$\lambda_{(2)} = 19.825\,593\,\frac{D^B}{aP} \qquad (Ex.\ 18\text{-}6.27)$$

$\lambda_{(2)} \approx \lambda_c$

is as little as 0.4%, which must be deemed acceptable for all practical purposes.

18.3.12 Concluding Comments on the Examples Above

It seems obvious that the results obtained in the preceding examples are very accurate although the effort involved was moderate. One might object that the analytical derivations involved in the plate examples, in particular in Example Ex 18-6.2, are lengthy. However, left to some analytic manipulation program, such as maxima, Mathematica, Maple, MuPAD, or even the old MuMath, the two-parameter solution in Example Ex 18-6.2 only requires a program of about 50–100 lines including print-out commands. Furthermore, with only minor modifications the program can be used for more terms with the comment that the run-times may become prohibitively long.

Chapter 19

Initial Postbuckling with a Unique Buckling Mode

Here, we consider problems that belong to the category of Chapter 18, namely problems with *Linear Prebuckling*, but take a broader view in that

Linear prebuckling

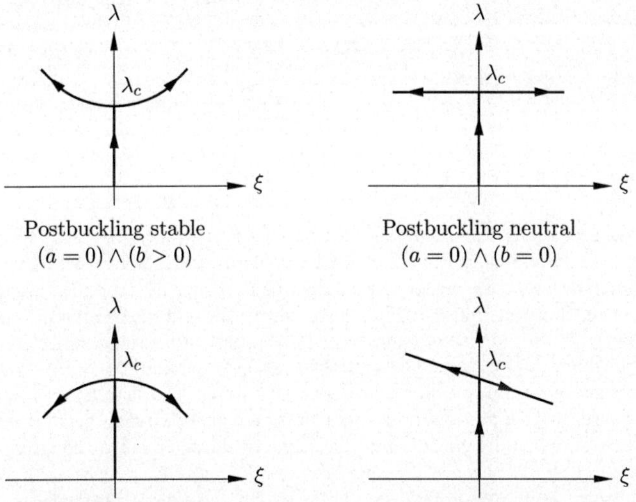

Fig. 19.1: Types of initial postbuckling behavior: $\lambda/\lambda_c \approx 1 + a\xi + b\xi^2$. The first and second postbuckling constants are a and b, respectively. The buckling mode amplitude is ξ.

we establish a procedure for investigating the initial behavior after buckling—the *Postbuckling Behavior*. Already at this point we mention that the formu-

Postbuckling behavior

E. Byskov, *Elementary Continuum Mechanics for Everyone*,
Solid Mechanics and Its Applications 194, DOI: 10.1007/978-94-007-5766-0_19,

The size of a "small" neighborhood is difficult to predict

las are only valid in a "small" neighborhood around the classical critical load λ_c and its associated displacements $\boldsymbol{u}_c$. What is meant by "small" depends strongly on the problem. In some cases the procedure yields results that are valid in a much larger neighborhood of $(\boldsymbol{u}_c, \lambda_c)$ than might be expected, in other cases the range of validity is smaller. This, however, is less important than you might think, in part because our interest lies in characterizing the type of postbuckling behavior of the structure in that a structure may be postbuckling stable, neutral, unstable, or postbuckling stable and unstable depending on the direction of the displacement in postbuckling, see Fig. 19.1. Thus, it is a question whether the structure is capable of supporting an additional load after buckling or if it becomes unstable immediately at bifurcation. In Fig. 19.1, as in the remaining part of this chapter, ξ denotes the amplitude of the buckling mode. In Example Ex 17-1.5 the rotation θ served as the buckling mode amplitude.

Postbuckling stability

Imperfection sensitivity

But, even more important than the determination of the postbuckling behavior itself is the question of imperfection sensitivity which is closely connected to the kind of postbuckling behavior, as you may infer from Fig. Ex. 17-1.5 of Example Ex 17-1.5. As it turns out, the real, asymptotic load-carrying capacity λ_s of a geometrically imperfect structure may be determined once the classical critical load λ_c and a constant characterizing the postbuckling behavior of the geometrically perfect structure have been determined, see Chapter 20.

W.T. Koiter

W.T. Koiter (1945) was the first to establish a theory for initial postbuckling of geometrically perfect elastic structures and imperfection sensitivity of geometrically imperfect elastic structures. Koiter's work must be considered one of the major landmarks in the theory of structural analysis and is certainly the most highly regarded contribution to the area of elastic stability.[19.1] In the following, an asymptotic theory for postbuckling behavior of geometrically perfect structures in the spirit of Koiter is derived using a different notation and point of departure. While Koiter uses the potential energy the Harvard School takes the principle of virtual displacements as their basis, and I shall follow their line of development although some of the details below are different. To some extent the derivations in this chapter are based on unpublished notes by Hutchinson (1974a),[19.2] but the presentation given below differs in many respects from Hutchinson's.[19.3]

Load-carrying capacity of geometrically imperfect structures

In Chapter 20 a set of formulas to determine the load-carrying capacity of geometrically imperfect structures are cited, but not derived because I consider the theory too difficult to follow in the present context. The ensuing formulas are, however, so important that they deserve to be given

[19.1] Personally, I have always found Koiter's derivations and notation hard to follow and therefore I have chosen to follow the Harvard School of Budiansky and Hutchinson.

[19.2] Reading Hutchinson's notes during a sabbatical leave at Harvard University was my entry into the wonderful world of postbuckling and imperfection sensitivity analysis.

[19.3] I think that some of the crucial derivations are performed in a new, more direct way.

here. For derivations of the formulas you may consider the original paper by Budiansky (1974) and the more recent one by Christensen & Byskov (2010), where the theory is extended.

Nonlinear prebuckling

In addition to problems involving linear prebuckling Fitch (1968) and Budiansky (1974) treat the more difficult case of postbuckling of structures with nonlinear prebuckling, and Byskov, Christensen & Jørgensen (1996) have extended the analysis to cover very general auxiliary conditions.

Mode interaction

Since about 1970, beginning with the work by A. van der Neut (1969), the subject of interaction between simultaneous and nearly simultaneous buckling modes, i.e. buckling modes that are associated with the same or nearly the same buckling loads, has been studied by, among others, Koiter & Kuiken (1971), Tvergaard (1973a), Tvergaard (1973b),Thompson & Hunt (1973), Crawford & Hedgepeth (1975), Koiter (1976), Byskov & Hutchinson (1977), Byskov (1979), Koiter & van der Neut (1979), Byskov & Hansen (1980), Byskov (1983), Byskov (1987–88), Byskov et al. (1988), Møllmann & Goltermann (1989a) and Møllmann & Goltermann (1989b).[19.4]

The paper by Peek & Kheyrkhahan (1993) is also worth mentioning because it extends the scope to cover nonlinear prebuckling.[19.5]

Survey papers on buckling

The survey papers by Hutchinson & Koiter (1970), by Tvergaard (1976), by Budiansky & Hutchinson (1979), as well as the paper by Budiansky (1974) mentioned above, are valuable as introductions to the general topic of elastic stability, including postbuckling and imperfection sensitivity, while the survey paper by Byskov (2004) is solely concerned with mode interaction.

19.1 Selected Formulas from Chapter 18

Initial postbuckling behavior

We wish to be able to predict the behavior on the postbuckling path immediately after buckling, that is, in a close neighborhood of $(\lambda_c, \boldsymbol{u}_c)$ in load-displacement space, see Fig. 18.2 or, which is the same, Fig. 19.2 below.

For convenience, here I repeat some of the most fundamental formulas from Part 33 and Chapter 18 and note that subscript $_0$ designates prebuckling, while subscripts $_1$ and $_c$ indicate buckling.[19.6]

[19.4] The works mentioned here are concerned with mode interaction caused by designs which result in close buckling loads. This is basically different than mode interaction in shells where mode interaction is an inherent property of the structure. That kind of mode interaction had been studied earlier

[19.5] I am sure that most engineers will find this article very difficult because of its mathematical nature.

[19.6] Note that subscripts $_1$ and $_{11}$ on $\mathbf{l}_1$ and $\mathbf{l}_{11}$, respectively, do *not* indicate buckling, but linearity of the operator.

19.1.1 General Formulas

The expressions (19.1)—(19.4) are not associated with prebuckling, buckling or postbuckling, but with the kind of continuum mechanics we utilize.

$$\boldsymbol{\varepsilon} = \mathbf{l}_1(\boldsymbol{u}) + \tfrac{1}{2}\mathbf{l}_2(\boldsymbol{u}) \;, \quad \delta\boldsymbol{\varepsilon} = \mathbf{l}_1(\delta\boldsymbol{u}) + \mathbf{l}_{11}(\boldsymbol{u}, \delta\boldsymbol{u}) \tag{19.1}$$

$$\mathbf{l}_2(\boldsymbol{u}+\boldsymbol{v}) = \mathbf{l}_2(\boldsymbol{u}) + 2\mathbf{l}_{11}(\boldsymbol{u}, \boldsymbol{v}) + \mathbf{l}_2(\boldsymbol{v}) \tag{19.2}$$

$$\boldsymbol{\sigma} \cdot \delta\boldsymbol{\varepsilon} = \overline{\boldsymbol{T}} \cdot \delta\boldsymbol{u} \tag{19.3}$$

$$\boldsymbol{\sigma} = \mathbf{H}(\boldsymbol{\varepsilon}) \,, \; \mathbf{H}(\boldsymbol{\varepsilon}_a)\boldsymbol{\varepsilon}_b = \mathbf{H}(\boldsymbol{\varepsilon}_b)\boldsymbol{\varepsilon}_a \tag{19.4}$$

19.1.2 Fundamental Path—Prebuckling

Linearity of the prebuckling path is given by, and results in, (18.7)–(18.9)

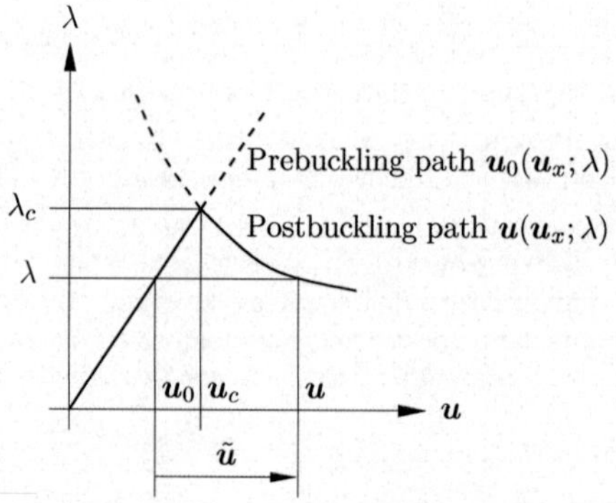

Fig. 19.2: Postbuckling with linear prebuckling.

which are reproduced here

$$\boldsymbol{u}(\boldsymbol{x};\lambda) = \lambda\boldsymbol{u}_0(\boldsymbol{x}) \,, \; \boldsymbol{\varepsilon}(\boldsymbol{x};\lambda) = \lambda\boldsymbol{\varepsilon}_0(\boldsymbol{x}) \text{ and } \boldsymbol{\sigma}(\boldsymbol{x};\lambda) = \lambda\boldsymbol{\sigma}_0(\boldsymbol{x}) \tag{19.5}$$

$$\mathbf{l}_{11}(\boldsymbol{u}_0, \boldsymbol{v}) = 0 \;\forall \boldsymbol{v} \tag{19.6}$$

$$\mathbf{l}_2(\boldsymbol{u}_0) = 0 \,, \; \boldsymbol{\varepsilon}_0 = \mathbf{l}_1(\boldsymbol{u}_0) \text{ and } \delta\boldsymbol{\varepsilon}_0 = \mathbf{l}_1(\delta\boldsymbol{u}) \tag{19.7}$$

$$\boldsymbol{\sigma}_0 \cdot \mathbf{l}_1(\delta\boldsymbol{u}) = \overline{\boldsymbol{T}} \cdot \delta\boldsymbol{u} \text{ with } \boldsymbol{\sigma}_0 = \mathbf{H}(\boldsymbol{\varepsilon}_0) \tag{19.8}$$

19.1.3 Buckling—Bifurcation

Other consequences of the linear prebuckling are

$$\boldsymbol{\varepsilon}_1 = \mathbf{l}_1(\boldsymbol{u}_1) \text{ and } \delta\boldsymbol{\varepsilon}_1 = \mathbf{l}_1(\delta\boldsymbol{u}) \tag{19.9}$$

$$\boldsymbol{\sigma}_1 \cdot \mathbf{l}_1(\delta\boldsymbol{u}) + \lambda_c\boldsymbol{\sigma}_0 \cdot \mathbf{l}_{11}(\boldsymbol{u}_1, \delta\boldsymbol{u}) = 0 \text{ with } \boldsymbol{\sigma}_1 = \mathbf{H}(\boldsymbol{\varepsilon}_1) \tag{19.10}$$

In applications of the theory which is derived below we shall assume that we can solve both the prebuckling problem given by (19.8) and the buckling problem (19.10), either in closed form or in some approximate way, but mention that this may constitute a problem in itself. In most situations in engineering practice we may need to apply some finite element method which usually is straightforward. However, when we are concerned with solving the *second-order problem* (19.34) we must be particularly careful, see e.g. (Byskov 1989).

19.1.4 Initial Postbuckling

On the bifurcated path we may always write the displacement field as

$$\begin{aligned} \boldsymbol{u}(\boldsymbol{x};\lambda) &= \lambda \boldsymbol{u}_0(\boldsymbol{x}) + \xi(\lambda)\boldsymbol{u}_1(\boldsymbol{x}) + \tilde{\tilde{\boldsymbol{u}}}(\boldsymbol{x};\lambda) \\ &= \lambda \boldsymbol{u}_0(\boldsymbol{x}) + \xi_1(\lambda)\boldsymbol{u}_1(\boldsymbol{x}) + \sum_{j=2}^{\infty} \xi_j(\lambda)\boldsymbol{u}_j(\boldsymbol{x}) \end{aligned} \tag{19.11}$$

Bifurcated path

where $\xi_1(\lambda) = \xi(\lambda)$ denotes the denotes the amplitude of the buckling mode associated with the lowest bifurcation load $\lambda_c = \lambda_1$, and $\xi_j(\lambda)$ is the amplitude of the jth buckling mode $\boldsymbol{u}_j(\boldsymbol{x})$ associated with the buckling load λ_j, where $\lambda_1 \leq \lambda_2 \leq \cdots$. The second line of (19.11) follows from the fact that any function over the domain—the structure—may be expanded in terms of the eigenfunctions, i.e. the buckling modes.

Amplitude of buckling mode $\xi_1(\lambda) = \xi(\lambda)$

In Chapter 18 we have proved that the buckling modes satisfy the orthogonality condition (18.35)[19.7]

Buckling modes are orthogonal

$$\boldsymbol{\sigma}_0 \cdot \mathbf{l}_{11}(\boldsymbol{u}_J, \boldsymbol{u}_K) = 0\,, \;\; J \neq K (J,K) \in [1;\infty[\tag{19.12}$$

in particular

$$\boldsymbol{\sigma}_0 \cdot \mathbf{l}_{11}(\boldsymbol{u}_1, \boldsymbol{u}_J) = 0\,, \;\; J = [2;\infty[\tag{19.13}$$

Koiter (1945) suggested that, at least in initial postbuckling, i.e. for small values of ξ, $\tilde{\tilde{\boldsymbol{u}}}$ can be approximated by an asymptotic expansion in ξ such that

Initial postbuckling behavior

$$\begin{aligned} \boldsymbol{u}(\boldsymbol{x};\lambda) = \lambda \boldsymbol{u}_0(\boldsymbol{x}) &+ \xi(\lambda)\boldsymbol{u}_1(\boldsymbol{x}) \\ &+ \xi(\lambda)^2\boldsymbol{u}_{11}(\boldsymbol{x}) + \xi(\lambda)^3\boldsymbol{u}_{111}(\boldsymbol{x}) + O(\xi^4) \end{aligned} \tag{19.14}$$

with similar expansions for $\boldsymbol{\varepsilon}$ and $\boldsymbol{\sigma}$

$$\begin{aligned} \boldsymbol{\varepsilon}(\boldsymbol{x};\lambda) = \lambda \boldsymbol{\varepsilon}_0(\boldsymbol{x}) &+ \xi(\lambda)\boldsymbol{\varepsilon}_1(\boldsymbol{x}) \\ &+ \xi(\lambda)^2\boldsymbol{\varepsilon}_{11}(\boldsymbol{x}) + \xi(\lambda)^3\boldsymbol{\varepsilon}_{111}(\boldsymbol{x}) + O(\xi^4) \end{aligned} \tag{19.15}$$

and

$$\begin{aligned} \boldsymbol{\sigma}(\boldsymbol{x};\lambda) = \lambda \boldsymbol{\sigma}_0(\boldsymbol{x}) &+ \xi(\lambda)\boldsymbol{\sigma}_1(\boldsymbol{x}) \\ &+ \xi(\lambda)^2\boldsymbol{\sigma}_{11}(\boldsymbol{x}) + \xi(\lambda)^3\boldsymbol{\sigma}_{111}(\boldsymbol{x}) + O(\xi^4) \end{aligned} \tag{19.16}$$

[19.7] If the structure only has a finite number of degrees of freedom, say M, the substitute M for ∞ below.

Postbuckling fields $\boldsymbol{u}_{11}$, $\boldsymbol{u}_{111}$, ε_{11}, ε_{111}, $\boldsymbol{\sigma}_{11}$, $\boldsymbol{\sigma}_{111}$

The fields $\boldsymbol{u}_{11}$ and $\boldsymbol{u}_{111}$ are the (first and second) postbuckling displacement fields, while ε_{11} and ε_{111} denote the (first and second) postbuckling strain fields, and $\boldsymbol{\sigma}_{11}$ and $\boldsymbol{\sigma}_{111}$ are the (first and second) postbuckling stress fields, respectively. The notation for the postbuckling fields used here differs from Koiter's and from the one employed by the Harvard School of Budiansky and Hutchinson, see e.g. (Hutchinson 1974a), (Fitch 1968) and (Budiansky 1974), while it is in accord with Byskov & Hutchinson (1977). The reason for using the present notation stems from a wish to avoid confusion between the higher buckling modes $\boldsymbol{u}_J$ and the postbuckling fields, which are denoted $\boldsymbol{u}_2$ and $\boldsymbol{u}_3$ by some of the authors mentioned above.

Since $\boldsymbol{u}_{11}$ and $\boldsymbol{u}_{111}$ are part of the sum in (19.11), whose elements are all orthogonal to $\boldsymbol{u}_1$ in the sense of (19.13), we may conclude that similar conditions apply to the fields $\boldsymbol{u}_{11}$ and $\boldsymbol{u}_{111}$

$\boldsymbol{u}_1$ orthogonal on higher order modes

$$\begin{aligned} \boldsymbol{\sigma}_0 \cdot \mathbf{l}_{11}(\boldsymbol{u}_1, \boldsymbol{u}_{11}) &= 0 \\ \boldsymbol{\sigma}_0 \cdot \mathbf{l}_{11}(\boldsymbol{u}_1, \boldsymbol{u}_{111}) &= 0 \end{aligned} \tag{19.17}$$

Higher order modes: not mutually orthogonal

It is very important to note that an orthogonality condition analogous to (19.12) does not apply between $\boldsymbol{u}_{11}$ and $\boldsymbol{u}_{111}$ because, loosely speaking, both of these fields contain components from *all* buckling modes higher than the first one.

On the bifurcated path we have utilized ξ as a perturbation parameter, and, for consistency, we expand the load parameter λ itself in ξ

Expansion of λ

$$\frac{\lambda}{\lambda_c} = 1 + a\xi + b\xi^2 + c\xi^3 + O(\xi^4) \tag{19.18}$$

which can be viewed as a shift from load control to displacement control, or, as a mere substitution of variables.

Postbuckling behavior
Postbuckling constants a and b

If we are able to determine the value of the postbuckling behavior in the immediate neighborhood of the bifurcation point, given by the value of the postbuckling constants a and b, then we know whether the structure is postbuckling stable, neutral or unstable, see Fig. 19.1.

As usual, we express the strain quantities, here $\varepsilon_1, \varepsilon_{11}$ and ε_{111}, in terms of the displacements—both on the prebuckling path and on the postbuckling path—and introduce (19.14) in (19.1). The two terms in (19.1) are

$$\begin{aligned} \mathbf{l}_1(\boldsymbol{u}) &= \lambda \mathbf{l}_1(\boldsymbol{u}_0) + \xi \mathbf{l}_1(\boldsymbol{u}_1) + \xi^2 \mathbf{l}_1(\boldsymbol{u}_{11}) + \xi^3 \mathbf{l}_1(\boldsymbol{u}_{111}) + O(\xi^4) \\ &= \lambda_c \mathbf{l}_1(\boldsymbol{u}_0) \ + \xi\big(\mathbf{l}_1(\boldsymbol{u}_1) + a\lambda_c \mathbf{l}_1(\boldsymbol{u}_0)\big) \\ &\qquad + \xi^2\big(\mathbf{l}_1(\boldsymbol{u}_{11}) + b\lambda_c \mathbf{l}_1(\boldsymbol{u}_0)\big) \\ &\qquad + \xi^3\big(\mathbf{l}_1(\boldsymbol{u}_{111}) + c\lambda_c \mathbf{l}_1(\boldsymbol{u}_0)\big) \\ &\qquad + O(\xi^4) \end{aligned} \tag{19.19}$$

and, when we exploit the linearity, see (18.8) or (19.6), of the prebuckling state

$$\begin{aligned}\tfrac{1}{2}\mathbf{l}_2(\boldsymbol{u}) &= \tfrac{1}{2}\mathbf{l}_2(\lambda\boldsymbol{u}_0+\xi\boldsymbol{u}_1+\xi^2\boldsymbol{u}_{11}+\xi^3\boldsymbol{u}_{111})+O(\xi^4)\\ &= \tfrac{1}{2}\xi^2\mathbf{l}_2(\boldsymbol{u}_1)+\xi^3\mathbf{l}_{11}(\boldsymbol{u}_1,\boldsymbol{u}_{11})+O(\xi^4)\end{aligned} \tag{19.20}$$

Thus, the strain $\boldsymbol{\varepsilon}$ may be expressed as

$$\begin{aligned}\boldsymbol{\varepsilon} = \lambda_c\mathbf{l}_1(\boldsymbol{u}_0) &+ \xi\big(\mathbf{l}_1(\boldsymbol{u}_1)+a\lambda_c\mathbf{l}_1(\boldsymbol{u}_0)\big)\\ &+\xi^2\big(\mathbf{l}_1(\boldsymbol{u}_{11})+\tfrac{1}{2}\mathbf{l}_2(\boldsymbol{u}_1)+b\lambda_c\mathbf{l}_1(\boldsymbol{u}_0)\big)\\ &+\xi^3\big(\mathbf{l}_1(\boldsymbol{u}_{111})+\mathbf{l}_{11}(\boldsymbol{u}_1,\boldsymbol{u}_{11})+c\lambda_c\mathbf{l}_1(\boldsymbol{u}_0)\big)\\ &+O(\xi^4)\end{aligned} \tag{19.21}$$

Expansion of strain $\boldsymbol{\varepsilon}$

or, using (19.15) and (19.18)

$$\begin{aligned}\boldsymbol{\varepsilon} = \lambda_c\boldsymbol{\varepsilon}_0 &+ \xi\big(\boldsymbol{\varepsilon}_1+a\lambda_c\boldsymbol{\varepsilon}_0\big)+\xi^2\big(\boldsymbol{\varepsilon}_{11}+b\lambda_c\boldsymbol{\varepsilon}_0\big)\\ &+\xi^3\big(\boldsymbol{\varepsilon}_{111}+c\lambda_c\boldsymbol{\varepsilon}_0\big)+O(\xi^4)\end{aligned} \tag{19.22}$$

Expansion of strain $\boldsymbol{\varepsilon}$

When we compare (19.21) with (19.22) we may get the following expressions for $\boldsymbol{\varepsilon}_0$–$\boldsymbol{\varepsilon}_{111}$

$$\begin{aligned}\boldsymbol{\varepsilon}_0 &= \mathbf{l}_1(\boldsymbol{u}_0)\\ \boldsymbol{\varepsilon}_1 &= \mathbf{l}_1(\boldsymbol{u}_1)\\ \boldsymbol{\varepsilon}_{11} &= \mathbf{l}_1(\boldsymbol{u}_{11})+\tfrac{1}{2}\mathbf{l}_2(\boldsymbol{u}_1)\\ \boldsymbol{\varepsilon}_{111} &= \mathbf{l}_1(\boldsymbol{u}_{111})+\mathbf{l}_{11}(\boldsymbol{u}_1,\boldsymbol{u}_{11})\end{aligned} \tag{19.23}$$

Prebuckling, buckling and postbuckling displacement-strain relations

As expected, the expressions for $\boldsymbol{\varepsilon}_0$ and $\boldsymbol{\varepsilon}_1$ agree with (19.7b) and (19.9a), respectively, as well with (18.9) and (18.25), respectively.

The constitutive model, see 19.4 or (18.5), provides

$$\boldsymbol{\sigma}_{11} = \mathbf{H}(\boldsymbol{\varepsilon}_{11}) \ \text{ and } \ \boldsymbol{\sigma}_{111} = \mathbf{H}(\boldsymbol{\varepsilon}_{111}) \tag{19.24}$$

Postbuckling stress-strain relations

In order to apply the principle of virtual displacements (19.3) we need the variation $\delta\boldsymbol{\varepsilon}$ of the strain field $\boldsymbol{\varepsilon}$. By use of the expansion of λ (19.18) and the linearity condition (19.6) the expression (19.1b) for the strain variation becomes

$$\begin{aligned}\delta\boldsymbol{\varepsilon} &= \mathbf{l}_1(\delta\boldsymbol{u})+\mathbf{l}_{11}\big(\lambda\boldsymbol{u}_0+\xi\boldsymbol{u}_1+\xi^2\boldsymbol{u}_{11}+\xi^3\boldsymbol{u}_{111},\delta\boldsymbol{u}\big)\\ &= \mathbf{l}_1(\delta\boldsymbol{u})+\xi\mathbf{l}_{11}(\boldsymbol{u}_1,\delta\boldsymbol{u})+\xi^2\mathbf{l}_{11}(\boldsymbol{u}_{11},\delta\boldsymbol{u})+\xi^3\mathbf{l}_{11}(\boldsymbol{u}_{111},\delta\boldsymbol{u})\end{aligned} \tag{19.25}$$

Strain variation

Above, as in the following derivations, we have omitted all high-order terms $O(\xi^4)$.

The principle of virtual displacements (19.3) now becomes

Principle of virtual displacements

$$\begin{aligned} &\lambda\overline{\boldsymbol{T}}\cdot\delta\boldsymbol{u} \\ &= \big(\lambda\boldsymbol{\sigma}_0+\xi\boldsymbol{\sigma}_1+\xi^2\boldsymbol{\sigma}_{11}+\xi^3\boldsymbol{\sigma}_{111}\big) \\ &\quad\cdot\big(\mathbf{l}_1(\delta\boldsymbol{u})+\xi\mathbf{l}_{11}(\boldsymbol{u}_1,\delta\boldsymbol{u})+\xi^2\mathbf{l}_{11}(\boldsymbol{u}_{11},\delta\boldsymbol{u}) \\ &\qquad+\xi^3\mathbf{l}_{11}(\boldsymbol{u}_{111},\delta\boldsymbol{u})\big) \end{aligned} \tag{19.26}$$

Before we collect terms of like powers in ξ we may note that the principle of virtual displacements on the fundamental path, (19.8), may be utilized to rewrite (19.26)

$$\begin{aligned} &\lambda\boldsymbol{\sigma}_0\cdot\mathbf{l}_1(\delta\boldsymbol{u}) \\ &= \big(\lambda\boldsymbol{\sigma}_0+\xi\boldsymbol{\sigma}_1+\xi^2\boldsymbol{\sigma}_{11}+\xi^3\boldsymbol{\sigma}_{111}\big) \\ &\quad\cdot\big(\mathbf{l}_1(\delta\boldsymbol{u})+\xi\mathbf{l}_{11}(\boldsymbol{u}_1,\delta\boldsymbol{u})+\xi^2\mathbf{l}_{11}(\boldsymbol{u}_{11},\delta\boldsymbol{u}) \\ &\qquad+\xi^3\mathbf{l}_{11}(\boldsymbol{u}_{111},\delta\boldsymbol{u})\big) \end{aligned} \tag{19.27}$$

with the result

$$\begin{aligned} 0 = \;&\xi\big(\boldsymbol{\sigma}_1\cdot\mathbf{l}_1(\delta\boldsymbol{u})+\lambda_c\boldsymbol{\sigma}_0\cdot\mathbf{l}_{11}(\boldsymbol{u}_1,\delta\boldsymbol{u})\big) \\ &+\xi^2\big(\boldsymbol{\sigma}_{11}\cdot\mathbf{l}_1(\delta\boldsymbol{u})+\lambda_c\boldsymbol{\sigma}_0\cdot\mathbf{l}_{11}(\boldsymbol{u}_{11},\delta\boldsymbol{u}) \\ &\qquad+\boldsymbol{\sigma}_1\cdot\mathbf{l}_{11}(\boldsymbol{u}_1,\delta\boldsymbol{u})+a\lambda_c\boldsymbol{\sigma}_0\cdot\mathbf{l}_{11}(\boldsymbol{u}_1,\delta\boldsymbol{u})\big) \\ &+\xi^3\big(\boldsymbol{\sigma}_{111}\cdot\mathbf{l}_1(\delta\boldsymbol{u})+\lambda_c\boldsymbol{\sigma}_0\cdot\mathbf{l}_{11}(\boldsymbol{u}_{111},\delta\boldsymbol{u}) \\ &\qquad+\boldsymbol{\sigma}_1\cdot\mathbf{l}_{11}(\boldsymbol{u}_{11},\delta\boldsymbol{u})+\boldsymbol{\sigma}_{11}\cdot\mathbf{l}_{11}(\boldsymbol{u}_1,\delta\boldsymbol{u}) \\ &\qquad+b\lambda_c\boldsymbol{\sigma}_0\cdot\mathbf{l}_{11}(\boldsymbol{u}_1,\delta\boldsymbol{u})+a\lambda_c\boldsymbol{\sigma}_0\cdot\mathbf{l}_{11}(\boldsymbol{u}_{11},\delta\boldsymbol{u})\big) \end{aligned} \tag{19.28}$$

We require that (19.28) holds for all (small) values of ξ, but, before we proceed with this endeavor, we need to pay attention to the variations. As usual, the variations must fulfill the kinematic conditions, which in the present case includes fulfillment of the homogeneous boundary conditions, i.e. the same boundary conditions which apply to the buckling modes $\boldsymbol{u}_j$ and therefore also to $\boldsymbol{u}_1$, $\boldsymbol{u}_{11}$ and $\boldsymbol{u}_{111}$. In the following Sections 19.1.5, 19.1.6 and 19.1.7 we let the variation $\delta\boldsymbol{u}$ be

$$\delta\boldsymbol{u} = \delta\hat{\boldsymbol{u}}+\boldsymbol{u}_1\delta\alpha \tag{19.29}$$

where $\delta\hat{\boldsymbol{u}}$ is orthogonal to $\boldsymbol{u}_1$ in the usual sense

$$\boldsymbol{\sigma}_0\cdot\mathbf{l}_{11}(\boldsymbol{u}_1,\delta\hat{\boldsymbol{u}}) = 0 \tag{19.30}$$

and $\delta\alpha$ is a scalar.

19.1.5 First-Order Problem—Buckling Problem

The coefficient to ξ in (19.28) yields

First-order problem = Buckling problem

$$\begin{aligned} 0 = &\big(\boldsymbol{\sigma}_1\cdot\mathbf{l}_1(\delta\hat{\boldsymbol{u}})+\lambda_c\boldsymbol{\sigma}_0\cdot\mathbf{l}_{11}(\boldsymbol{u}_1,\delta\hat{\boldsymbol{u}})\big) \\ &+\delta\alpha\big(\boldsymbol{\sigma}_1\cdot\mathbf{l}_1(\boldsymbol{u}_1)+\lambda_c\boldsymbol{\sigma}_0\cdot\mathbf{l}_2(\boldsymbol{u}_1)\big) \\ &\forall(\delta\hat{\boldsymbol{u}},\delta\alpha) \end{aligned} \tag{19.31}$$

and thus we have recovered (19.10) and (18.28)—it would have been very unfortunate if we had got other results.

19.1.6 Second-Order Problem

We may derive the second-order problem, which is also called the first postbuckling problem, from the coefficient to ξ^2 in (19.28) and, as we shall see, at the same time get a formula that determines the first-order postbuckling constant a.

$$\begin{aligned} 0 = &\big(\boldsymbol{\sigma}_{11} \cdot \mathrm{l}_1(\delta\hat{\boldsymbol{u}}) + \lambda_c \boldsymbol{\sigma}_0 \cdot \mathrm{l}_{11}(\boldsymbol{u}_{11}, \delta\hat{\boldsymbol{u}}) \\ &+ \boldsymbol{\sigma}_1 \cdot \mathrm{l}_{11}(\boldsymbol{u}_1, \delta\hat{\boldsymbol{u}}) + a\lambda_c \boldsymbol{\sigma}_0 \cdot \mathrm{l}_{11}(\boldsymbol{u}_1, \delta\hat{\boldsymbol{u}})\big) \\ &+ \delta\alpha\big(\boldsymbol{\sigma}_{11} \cdot \mathrm{l}_1(\boldsymbol{u}_1) + \lambda_c \boldsymbol{\sigma}_0 \cdot \mathrm{l}_{11}(\boldsymbol{u}_{11}, \boldsymbol{u}_1) \\ &\quad + \boldsymbol{\sigma}_1 \cdot \mathrm{l}_{11}(\boldsymbol{u}_1, \boldsymbol{u}_1) + a\lambda_c \boldsymbol{\sigma}_0 \cdot \mathrm{l}_{11}(\boldsymbol{u}_1, \boldsymbol{u}_1)\big) \\ &\forall(\delta\hat{\boldsymbol{u}}, \delta\alpha) \end{aligned} \tag{19.32}$$

or, in view of the orthogonality conditions

$$\begin{aligned} 0 = &\big(\boldsymbol{\sigma}_{11} \cdot \mathrm{l}_1(\delta\hat{\boldsymbol{u}}) + \lambda_c \boldsymbol{\sigma}_0 \cdot \mathrm{l}_{11}(\boldsymbol{u}_{11}, \delta\hat{\boldsymbol{u}}) \\ &+ \boldsymbol{\sigma}_1 \cdot \mathrm{l}_{11}(\boldsymbol{u}_1, \delta\hat{\boldsymbol{u}})\big) \\ &+ \delta\alpha\big(\boldsymbol{\sigma}_{11} \cdot \mathrm{l}_1(\boldsymbol{u}_1) \\ &\quad + \boldsymbol{\sigma}_1 \cdot \mathrm{l}_2(\boldsymbol{u}_1) + a\lambda_c \boldsymbol{\sigma}_0 \cdot \mathrm{l}_2(\boldsymbol{u}_1)\big) \\ &\forall(\delta\hat{\boldsymbol{u}}, \delta\alpha) \end{aligned} \tag{19.33}$$

Second-order problem = First postbuckling problem

The first term yields a variational statement of the second-order boundary value problem

Second-order problem boundary value problem

$$\boldsymbol{\sigma}_{11} \cdot \mathrm{l}_1(\delta\hat{\boldsymbol{u}}) + \lambda_c \boldsymbol{\sigma}_0 \cdot \mathrm{l}_{11}(\boldsymbol{u}_{11}, \delta\hat{\boldsymbol{u}}) = -\boldsymbol{\sigma}_1 \cdot \mathrm{l}_{11}(\boldsymbol{u}_1, \delta\hat{\boldsymbol{u}}) \tag{19.34}$$

while the second term provides us with a formula to determine the first-order postbuckling constant a. But, note that the second term involves $\boldsymbol{\sigma}_{11} \cdot \mathrm{l}_1(\boldsymbol{u}_1)$, where $\boldsymbol{\sigma}_{11}$ belongs to the second-order problem. Fortunately, we are able to express $\boldsymbol{\sigma}_{11} \cdot \mathrm{l}_1(\boldsymbol{u}_1)$ in terms of quantities of the first order problem alone

$$\begin{aligned} \boldsymbol{\sigma}_{11} \cdot \mathrm{l}_1(\boldsymbol{u}_1) &= \boldsymbol{\sigma}_{11} \cdot \boldsymbol{\varepsilon}_1 = \boldsymbol{\sigma}_1 \cdot \boldsymbol{\varepsilon}_{11} = \boldsymbol{\sigma}_1 \cdot \big(\mathrm{l}_1(\boldsymbol{u}_{11}) + \tfrac{1}{2}\mathrm{l}_2(\boldsymbol{u}_1)\big) \\ &= \tfrac{1}{2}\boldsymbol{\sigma}_1 \cdot \mathrm{l}_2(\boldsymbol{u}_1) \end{aligned} \tag{19.35}$$

where we have exploited the reciprocal relation (18.6) and where $\boldsymbol{\sigma}_1 \cdot \mathrm{l}_1(\boldsymbol{u}_{11})$ vanishes because of the orthogonality between $\boldsymbol{u}_1$ and $\boldsymbol{u}_{11}$ in that the first-order problem (19.31) with $\delta\hat{\boldsymbol{u}} = \boldsymbol{u}_{11}$ becomes

$$\boldsymbol{\sigma}_1 \cdot \mathrm{l}_1(\boldsymbol{u}_{11}) + \lambda_c \boldsymbol{\sigma}_0 \cdot \mathrm{l}_{11}(\boldsymbol{u}_1, \boldsymbol{u}_{11}) = 0 \tag{19.36}$$

where the last term vanishes because of the orthogonality condition (19.17a). Therefore, we may express the first-order postbuckling constant a solely in terms of quantities from the first-order problem

First postbuckling constant a

$$a = -\frac{3}{2}\frac{\boldsymbol{\sigma}_1 \cdot \mathrm{l}_2(\boldsymbol{u}_1)}{\lambda_c \boldsymbol{\sigma}_0 \cdot \mathrm{l}_2(\boldsymbol{u}_1)} = +\frac{3}{2}\frac{\boldsymbol{\sigma}_1 \cdot \mathrm{l}_2(\boldsymbol{u}_1)}{\boldsymbol{\sigma}_1 \cdot \boldsymbol{\varepsilon}_1} \tag{19.37}$$

and thus if $a \neq 0$ we are not forced to solve the second-order problem in order to determine the postbuckling behavior to lowest order.

19.1.7 Third-Order Problem

We do not intend to solve the third-order problem, which is the same as the second postbuckling problem, because we shall never attempt to go beyond the term $b\xi^2$ in the expansion for λ (19.18). This problem is the coefficient to ξ^3 in (19.27)

Third-order problem = Second postbuckling problem

$$\begin{aligned}
0 = \big(&\boldsymbol{\sigma}_{111} \cdot \mathrm{l}_1(\delta\hat{\boldsymbol{u}}) + \lambda_c \boldsymbol{\sigma}_0 \cdot \mathrm{l}_{11}(\boldsymbol{u}_{111}, \delta\hat{\boldsymbol{u}}) \\
&+ \boldsymbol{\sigma}_1 \cdot \mathrm{l}_{11}(\boldsymbol{u}_{11}, \delta\hat{\boldsymbol{u}}) + \boldsymbol{\sigma}_{11} \cdot \mathrm{l}_{11}(\boldsymbol{u}_1, \delta\hat{\boldsymbol{u}}) \\
&+ b\lambda_c \boldsymbol{\sigma}_0 \cdot \mathrm{l}_{11}(\boldsymbol{u}_1, \delta\hat{\boldsymbol{u}}) + a\lambda_c \boldsymbol{\sigma}_0 \cdot \mathrm{l}_{11}(\boldsymbol{u}_{11}, \delta\hat{\boldsymbol{u}})\big) \\
&+ \delta\alpha\big(\boldsymbol{\sigma}_{111} \cdot \mathrm{l}_1(\boldsymbol{u}_1) + \lambda_c \boldsymbol{\sigma}_0 \cdot \mathrm{l}_{11}(\boldsymbol{u}_{111}, \boldsymbol{u}_1) \\
&\quad + \boldsymbol{\sigma}_1 \cdot \mathrm{l}_{11}(\boldsymbol{u}_{11}, \boldsymbol{u}_1) + \boldsymbol{\sigma}_{11} \cdot \mathrm{l}_2(\boldsymbol{u}_1) \\
&\quad + b\lambda_c \boldsymbol{\sigma}_0 \cdot \mathrm{l}_2(\boldsymbol{u}_1) + a\lambda_c \boldsymbol{\sigma}_0 \cdot \mathrm{l}_{11}(\boldsymbol{u}_{11}, \boldsymbol{u}_1)\big) \\
&\forall(\delta\hat{\boldsymbol{u}}, \delta\alpha)
\end{aligned} \tag{19.38}$$

Proceeding along much the same lines as in Section 19.1.6 we get the third-order boundary value problem

Third-order problem boundary value problem

$$\begin{aligned}
&\boldsymbol{\sigma}_{111} \cdot \mathrm{l}_1(\delta\hat{\boldsymbol{u}}) + \lambda_c \boldsymbol{\sigma}_0 \cdot \mathrm{l}_{11}(\boldsymbol{u}_{111}, \delta\hat{\boldsymbol{u}}) \\
&= -\big(\boldsymbol{\sigma}_1 \cdot \mathrm{l}_{11}(\boldsymbol{u}_{11}, \delta\hat{\boldsymbol{u}}) + \boldsymbol{\sigma}_{11} \cdot \mathrm{l}_{11}(\boldsymbol{u}_1, \delta\hat{\boldsymbol{u}}) \\
&\quad + a\lambda_c \boldsymbol{\sigma}_0 \cdot \mathrm{l}_{11}(\boldsymbol{u}_{11}, \delta\hat{\boldsymbol{u}})\big)
\end{aligned} \tag{19.39}$$

If we were to compute the third-order fields we would usually not do it unless both a and b were equal to zero and in that case (19.39) simplifies somewhat, as you can see.

The expression for b follows from the second term of (19.38)

$$\begin{aligned}
0 = &\boldsymbol{\sigma}_{111} \cdot \mathrm{l}_1(\boldsymbol{u}_1) + \boldsymbol{\sigma}_1 \cdot \mathrm{l}_{11}(\boldsymbol{u}_{11}, \boldsymbol{u}_1) + \boldsymbol{\sigma}_{11} \cdot \mathrm{l}_2(\boldsymbol{u}_1) \\
&+ b\lambda_c \boldsymbol{\sigma}_0 \cdot \mathrm{l}_2(\boldsymbol{u}_1)
\end{aligned} \tag{19.40}$$

Derivations analogous to those leading to the result of (19.35) furnish

$$\begin{aligned}
\boldsymbol{\sigma}_{111} \cdot \mathrm{l}_1(\boldsymbol{u}_1) &= \boldsymbol{\sigma}_{111} \cdot \boldsymbol{\varepsilon}_1 = \boldsymbol{\sigma}_1 \cdot \boldsymbol{\varepsilon}_{111} \\
&= \boldsymbol{\sigma}_1 \cdot \big(\mathrm{l}_1(\boldsymbol{u}_{111}) + \mathrm{l}_{11}(\boldsymbol{u}_1, \boldsymbol{u}_{11})\big) \\
&= \boldsymbol{\sigma}_1 \cdot \mathrm{l}_{11}(\boldsymbol{u}_1, \boldsymbol{u}_{11})
\end{aligned} \tag{19.41}$$

Now, we may get rid of the third-order term $\boldsymbol{\sigma}_{111} \cdot \mathrm{l}_1(\boldsymbol{u}_1)$ and find the final result for the second-order postbuckling constant b

$$b = -\frac{\boldsymbol{\sigma}_{11} \cdot \mathrm{l}_2(\boldsymbol{u}_1) + 2\boldsymbol{\sigma}_1 \cdot \mathrm{l}_{11}(\boldsymbol{u}_{11}, \boldsymbol{u}_1)}{\lambda_c \boldsymbol{\sigma}_0 \cdot \mathrm{l}_2(\boldsymbol{u}_1)} = +\frac{\boldsymbol{\sigma}_{11} \cdot \mathrm{l}_2(\boldsymbol{u}_1) + 2\boldsymbol{\sigma}_1 \cdot \mathrm{l}_{11}(\boldsymbol{u}_{11}, \boldsymbol{u}_1)}{\boldsymbol{\sigma}_1 \cdot \boldsymbol{\varepsilon}_1} \tag{19.42}$$

Second postbuckling constant b

19.1.8 Solubility Conditions on the Second- and Third-Order Problems

Second- and third-order problems are singular: Invoke orthogonality

By comparing the first-order problem with the left-hand side of the second- and third-order problems you may immediately observe that the latter are singular. Therefore, the second- and third-order problems cannot be solved without imposing an auxiliary condition, namely the orthogonality conditions given by (19.17a) and (19.17b)

$$\begin{aligned} \boldsymbol{\sigma}_0 \cdot \mathrm{l}_{11}(\boldsymbol{u}_1, \boldsymbol{u}_{11}) &= 0 \\ \boldsymbol{\sigma}_0 \cdot \mathrm{l}_{11}(\boldsymbol{u}_1, \boldsymbol{u}_{111}) &= 0 \end{aligned} \tag{19.43}$$

which in variational form both become

$$\boldsymbol{\sigma}_0 \cdot \mathrm{l}_{11}(\boldsymbol{u}_1, \delta \boldsymbol{u}) = 0 \quad \text{or} \quad \boldsymbol{\sigma}_0 \cdot \mathrm{l}_{11}(\boldsymbol{u}_1, \delta \hat{\boldsymbol{u}}) = 0 \tag{19.44}$$

in accord with (19.30).

Ex 19-1 Postbuckling of Roorda's Frame and the First-Order Postbuckling Coefficient

Our success in Example Ex 18-5 with the computation of λ_c for Roorda's Frame by the Rayleigh-Ritz procedure makes it an obvious idea to compute the first order postbuckling coefficient a based on the approximate solution from that example. In the present case, where the prebuckling state is linear, the formula for the first order postbuckling constant a is, see (19.37)

$$a = -\frac{3}{2}\frac{\boldsymbol{\sigma}_1 \cdot \mathrm{l}_2(\boldsymbol{u}_1)}{\lambda_c \boldsymbol{\sigma}_0 \cdot \mathrm{l}_2(\boldsymbol{u}_1)} = +\frac{3}{2}\frac{\boldsymbol{\sigma}_1 \cdot \mathrm{l}_2(\boldsymbol{u}_1)}{\boldsymbol{\sigma}_1 \cdot \boldsymbol{\varepsilon}_1} \tag{Ex. 19-1.1}$$

where,

$$\boldsymbol{\sigma}_1 \cdot \mathrm{l}_2(\boldsymbol{u}_1) = \int_A^B N_{AB}^1 \left(w'_{AB}\right)^2 dx + \int_B^C N_{BC}^1 \left(w'_{BC}\right)^2 dx \tag{Ex. 19-1.2}$$

There is a small problem in that we do not know the value of N_{AB}^1 and N_{BC}^1 and, since we have imposed inextensibility, we cannot employ the constitutive law for this purpose—recall that $\varepsilon \equiv 0$, while $N \not\equiv 0$. As usual in these cases, we can think of the axial force as an internal reaction to the inextensibility condition. Therefore, we must invoke the principle of virtual displacements, or a set of equilibrium equations. We

$\varepsilon \equiv 0$, but $N \not\equiv 0$. N is a reaction to the inextensibility condition choose the latter possibility and compute the bending moment $M^1_{B,BC}$ at point B in BC

$$M^1_{B,BC} = \frac{EI}{L}[B_{BC}(0)]\{v^1_{BC}\} = +3\frac{EI}{L} \tag{Ex. 19-1.3}$$

We may also compute the value of the bending moment $M^1_{B,AB}$ at point B in AB

$$\begin{aligned} M^1_{B,AB} &= -\left([3\,;0] - \lambda_{(2)}\left[\tfrac{1}{5}\,;\tfrac{1}{10}\right]\right)\{v^1_{AB}\}\frac{EI}{L} \\ &= +3\frac{EI}{L} \end{aligned} \tag{Ex. 19-1.4}$$

where we have exploited the stiffness matrix and the geometric (stiffness) matrix for AB in the computation, and where the first minus sign comes from the fact that $M^1_{B,AB}$ is positive in the opposite direction of the rotation ξ of B. Note that it is much better to compute $M^1_{B,AB}$ as the nodal force from element AB, i.e. by use of the stiffness matrix and the geometric (stiffness) matrix, rather than via $[B_{AB}(1)]$ and $\{v^1_{AB}\}$. The latter procedure involves differentiation of an approximate solution, while the former is based on integration and is an application of the principle of virtual displacements to determine a force quantity, and this is always more robust. The second possibility would have given us the value $M^1_{B,AB} = \frac{1}{7}(-12+\sqrt{1551})EI/L = 3.92EI/L$, which clearly is unsatisfactory.

Now, moment equilibrium of BC gives the vertical reaction R^1_C at C

$$R^1_C = +3\frac{EI}{L^2} \tag{Ex. 19-1.5}$$

and vertical equilibrium then provides N^1_{AB}

$$N^1_{AB} = +3\frac{EI}{L^2} \tag{Ex. 19-1.6}$$

In the same way, we can get the value of N^1_{BC} as

$$N^1_{BC} = +3\frac{EI}{L^2} \tag{Ex. 19-1.7}$$

For the computation of the numerator in a we need

$$\begin{aligned} &\boldsymbol{\sigma}_1 \cdot \mathrm{l}_2(\boldsymbol{u}_1) \\ &= +\int_A^B N^1_{AB}\{v^1_{AB}\}^T[G_{AB}]^T[G_{AB}]\{v^1_{AB}\}dx \\ &\quad +\int_B^C N^1_{BC}\{v^1_{BC}\}^T[G_{BC}]^T[G_{BC}]\{v^1_{BC}\}dx \\ &= +3\frac{EI}{L^2}L\{v^1_{AB}\}^T\int_0^1[G_{AB}]^T[G_{AB}]d\zeta\{v^1_{AB}\} \\ &\quad +3\frac{EI}{L^2}L\{v^1_{BC}\}^T\int_0^1[G_{BC}]^T[G_{BC}]d\zeta\{v^1_{BC}\} \\ &= \frac{4131+67\sqrt{1551}}{1715}\frac{EI}{L} \end{aligned} \tag{Ex. 19-1.8}$$

Similarly, for the denominator in a

$$\begin{aligned}
&\lambda_{(2)}\boldsymbol{\sigma}_0 \cdot \mathbf{l}_2(\boldsymbol{u}_1) \\
&= \lambda_{(2)} \int_A^B N_{AB}^0 \{v_{AB}^1\}^T [G_{AB}]^T [G_{AB}] \{v_{AB}^1\} dx \\
&= -\lambda_{(2)} \bar{P} L \{v_{AB}^1\}^T \int_0^1 [G_{AB}]^T [G_{AB}] d\zeta \{v_{AB}^1\} \qquad \text{(Ex. 19-1.9)} \\
&= -\lambda_{(2)} \bar{P} L \frac{6}{245} \left(517 + 3\sqrt{1551}\right) \frac{EI}{L} \\
&= -\frac{6\left(517 + 3\sqrt{1551}\right)}{245} \frac{EI}{L}
\end{aligned}$$

Then, the value of a is found to be

Approximate value of a for Roorda's frame

$$a_{(2)} = \frac{1}{72380}\left(18612 + 227\sqrt{1551}\right) = 0.380\,656 \qquad \text{(Ex. 19-1.10)}$$

cf. the exact value found by Koiter (1966)[19.8]

Exact value of a for Roorda's frame

$$a_{exact} = 0.380\,520 \qquad \text{(Ex. 19-1.11)}$$

The relative error on $a_{(2)}$ is thus about 0.035% and, therefore, even smaller than the error on $\lambda_{(2)}$.

As can be inferred from the above derivations, asymmetry of load or geometry will usually result in asymmetric postbuckling behavior as we saw in Example Ex 19-1. If both geometry and load are symmetric we expect symmetric postbuckling behavior. So, if we we change the direction of the load on Roorda's Frame such that it is symmetric the postbuckling behavior will probably change from asymmetric to symmetric, as we shall see in the next example, Example Ex 19-2, which is concerned with postbuckling of a symmetric two-bar frame and determination of its second-order postbuckling coefficient.

In this connection it is less important that we can perform an exact analysis instead of an approximate one.

Ex 19-2 Postbuckling of Symmetric Two-Bar Frame. Second-Order Postbuckling Coefficient

This example may be viewed as Roorda's frame with the load rotated $\pi/4$ to get the symmetric two-bar frame shown in Fig. Ex. 19-2.1. Again, we assume that the members of frame are inextensible and, as was the case in Example Ex 18-5 and Example Ex 19-1, the consequence is that the only non-vanishing prebuckling state field quantities are the axial forces. Define the load unit $\bar{P}$ according to and the quantity P_0

Load unit $\bar{P}$

$$\bar{P} = \pi^2 \frac{EI}{L^2} \quad \text{and} \quad P_0 = \frac{\bar{P}}{\sqrt{2}} = \frac{\pi^2}{\sqrt{2}} \frac{EI}{L^2} \qquad \text{(Ex. 19-2.1)}$$

respectively.

[19.8] Koiter did not give the result with this many digits.

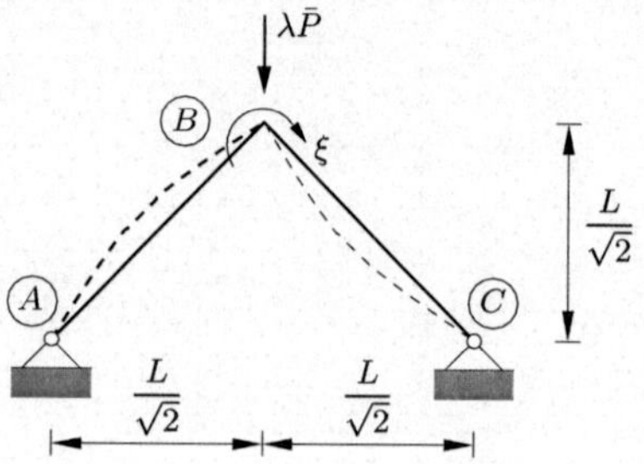

Fig. Ex. 19-2.1: *A Symmetric Two-Bar Frame*

Then,

Prebuckling stresses

$$N^0_{AB} = N^0_{BC} = -\frac{\bar{P}}{\sqrt{2}} = -\frac{\pi^2}{\sqrt{2}}\frac{EI}{L^2} = -P_0 \qquad \text{(Ex. 19-2.2)}$$

Contrary to our approach for Roorda's Frame we shall solve the buckling and postbuckling problems exactly—in part because here it proves easier than doing it in an approximate way. Since the two structural members are equal and carry the same prebuckling axial force one cannot "help" the other carry the load at buckling. The outcome of this is that in *buckling* the two members might as well have been hinged at point B. Therefore, we know the buckling load immediately

Buckling load λ_c

$$\lambda_c = \sqrt{2} \text{ and } \lambda_c P_0 = \pi^2 \frac{EI}{L^2} \qquad \text{(Ex. 19-2.3)}$$

with the buckling mode

Buckling mode w_1

$$w_1 = w^1_{AB} = -w^1_{BC} = \frac{L}{\pi}\sin\left(\pi\frac{x}{L}\right) \qquad \text{(Ex. 19-2.4)}$$

and

Buckling axial force N^1

$$N_1 = N^1_{AB} = N^1_{BC} = 0 \qquad \text{(Ex. 19-2.5)}$$

where we have normalized the buckling mode such that a rotation of 1 radian corresponds to $\xi = 1$, and where the values of the buckling axial forces follow from equilibrium equations for the two members noting that $M^1_B = 0$. The first order-postbuckling constant a vanishes because of symmetry of geometry and load on the two-bar frame, and the fact that the buckling axial force N_1 vanishes proves this. Therefore, we must compute the second-order postbuckling constant b, see (19.42). Then, we need to determine the postbuckling fields and, in order to so we must exploit the inextensibility condition several times and, at the same time, observe that since the buckling mode is *antisymmetric*, the postbuckling state is *symmetric*.

Because of inextensibility

$$\begin{aligned} &\varepsilon_{11} \equiv 0 \Rightarrow u'_{11} = -\tfrac{1}{2}\left(w'_1\right)^2 \\ &\Rightarrow u_{11} = \tfrac{1}{4}\left(-\left(\frac{L}{2\pi}\right)\sin\left(2\pi\frac{x}{L}\right) - x\right) \end{aligned} \tag{Ex. 19-2.6}$$

Postbuckling axial displacement u_{11}

where we have exploited the fact that u_{11} vanishes at the supports. Symmetry implies that

$$u_B^{11} = w_B^{11}\,,\ \text{where}\ u_B^{11} = -\tfrac{1}{4}L \tag{Ex. 19-2.7}$$

and

$$\delta u_B^{11} = \delta w_B^{11} \tag{Ex. 19-2.8}$$

Here, the second order problem is, see (19.34)

$$\boldsymbol{\sigma}_{11} \cdot \mathbf{l}_1(\delta \boldsymbol{u}_{11}) + \lambda_c \boldsymbol{\sigma}_0 \cdot \mathbf{l}_{11}(\boldsymbol{u}_{11}, \delta \boldsymbol{u}_{11}) = 0 \tag{Ex. 19-2.9}$$

because

$$N_1 \equiv 0 \tag{Ex. 19-2.10}$$

Thus, in AB and BC

$$\begin{aligned} 0 = &\int_0^L M_{11}\delta w''_{11} dx + \int_0^L N_{11}\delta u'_{11} dx \\ &+\lambda_c \int_0^L N_0 w'_{11}\delta w'_{11} dx \end{aligned} \tag{Ex. 19-2.11}$$

Integration by parts provides

$$\begin{aligned} 0 = &\left[M_{11}\delta w'_{11}\right]_0^L - \left[M'_{11}\delta w_{11} - \lambda_c N_0 w'_{11}\delta w_{11}\right]_0^L \\ &+ \left[N_{11}\delta u_{11}\right]_0^L \\ &+ \int_0^L \left\{M''_{11} - \lambda_c N_0 w''_{11}\right\}\delta w_{11} dx \\ &- \int_0^L N'_{11}\delta u_{11} dx \end{aligned} \tag{Ex. 19-2.12}$$

From (*Ex. 19-2.12*) and the symmetry condition we get the differential equations

$$\begin{aligned} &N'_{11} = 0 \\ &w_{11}^{\mathrm{iv}} + \left(\frac{\pi}{L}\right)^2 w''_{11} = 0 \end{aligned} \tag{Ex. 19-2.13}$$

Postbuckling differential equations

and the boundary conditions

$$\begin{aligned} &M_{11}(0) = 0 \Rightarrow w''_{11}(0) = 0 \\ &N_{11}(L) - EIw'''_{11}(L) + \lambda_c N_0 w'_{11}(L) = 0 \\ &w'_{11}(L) = 0 \Rightarrow N_{11}(L) - EIw'''_{11}(L) = 0 \end{aligned} \tag{Ex. 19-2.14}$$

Postbuckling differential boundary conditions

which together determine the postbuckling fields.

Second-order postbuckling constant b

Since $N_1 = 0$ the expression for b is

$$b = -\frac{\boldsymbol{\sigma}_{11} \cdot \mathbf{l}_2(\boldsymbol{u}_1)}{\lambda_c \boldsymbol{\sigma}_0 \cdot \mathbf{l}_2(\boldsymbol{u}_1)} \qquad (Ex.\ 19\text{-}2.15)$$

and, apparently, we do not need to determine w_{11}. However, we must determine N_{11} and, since the constitutive equation is of no use to us here because of inextensibility, we are forced to compute N_{11} through equilibrium equations which involve M_{11} and, therefore, also w_{11}. From (*Ex. 19-2.13b*) and (*Ex. 19-2.14*) we get the solution

Postbuckling transverse displacement w_{11}

$$w_{11} = -\frac{x}{4} - \frac{L}{4\pi} \sin\left(\pi \frac{x}{L}\right) \qquad (Ex.\ 19\text{-}2.16)$$

Utilize (*Ex. 19-2.16*) together with (*Ex. 19-2.13a*) and (*Ex. 19-2.14c*) to get

Postbuckling axial force N_{11}

$$N_{11} = -\frac{\pi^2}{4} \frac{EI}{L^2} \qquad (Ex.\ 19\text{-}2.17)$$

Then,

$$\boldsymbol{\sigma}_{11} \cdot \mathbf{l}_2(\boldsymbol{u}_1) = 2 \int_0^L N_{11} \left(w_1'\right)^2 dx = -\frac{\pi^2}{4} \frac{EI}{L} \qquad (Ex.\ 19\text{-}2.18)$$

and

$$\lambda_c \boldsymbol{\sigma}_0 \cdot \mathbf{l}_2(\boldsymbol{u}_1) = -2\lambda_c \int_0^L N_0 \left(w_1'\right)^2 dx = -\pi^2 \frac{EI}{L} \qquad (Ex.\ 19\text{-}2.19)$$

where the factor 2 stems from the fact that both AB and BC contribute to the value of b. The outcome is that

Second-order postbuckling constant b

$$b = -\tfrac{1}{4} \qquad (Ex.\ 19\text{-}2.20)$$

which shows that the two-bar frame is postbuckling unstable and, therefore, also imperfection sensitive.

A final comment is that an analysis based on fully nonlinear beam equations gives $b = -\frac{1}{8}$. The reason for the difference is that the fully nonlinear "Elastica" equations, see Example Ex 8-2, predict that the Euler Column is postbuckling stable with $b = +\frac{1}{8}$ while the moderately nonlinear beam theory implies that the Euler Column is postbuckling neutral, i.e. that $b = 0$. In any case, both values of b are so small that we might as well have assumed a neutral postbuckling behavior because even with a rotation as large as $5°$, which is as big as the theory may cover, the decrease in load-carrying capacity is as low about 2% according to our above analysis and only half of that by the Elastica theory.

Chapter 20

Imperfection Sensitivity

Imperfection sensitivity

While derivation of the asymptotic theory of elastic postbuckling of structures with a distinct lowest buckling load is in principle fairly straightforward, albeit somewhat lengthy, see Chapter 19, a similar theory valid for the behavior of geometrically imperfect structures is much more complicated, and only the most important formulas, see e.g. (Budiansky 1974), (Fitch 1968) or (Christensen & Byskov 2010), are given below. In the following it is assumed that the geometric imperfection has the same shape as the buckling mode $\boldsymbol{u}_1$ because this is usually the most detrimental kind. Thus, the imperfection is taken to be $\bar{\xi}\boldsymbol{u}_1$, where $\bar{\xi}$ denotes its amplitude.

Imperfection assumed proportional to buckling mode $\boldsymbol{u}_1$

Before we cite the formulas it may be justified to mention some of the problems associated with derivation of a theory for slightly imperfect structures. As may be noted from Fig. Ex. 17-1.5 of Example Ex 17-1.5 and Fig. 20.2, the equilibrium path of the geometrically imperfect structure lies very close to the equilibrium path of the perfect structure for very small imperfections. Therefore, it seems like an obvious idea to develop the necessary asymptotic expansions using the equilibrium path of the perfect structure as a basis. This is, as we may realize intuitively, not very easy. The reason is that, while the equilibrium path of the imperfect structures may be assumed to be described by a single formula over the entire range of ξ, the equilibrium paths of the geometrically perfect structures really consists of *two* parts, namely of a straight line, the λ-axis, and a smooth curve, the postbuckling path, intersecting each other. Thus, the transition from the formulas for the perfect case to the imperfect one must be singular and, therefore, probably difficult.

Equilibrium path of geometrically imperfect structure described by *one* formula. For the geometrically perfect structure it consists of *two* intersecting parts $\Rightarrow$ singular behavior for small imperfections

Budiansky (1974) imagines that in the space spanned by $(\bar{\xi}, \xi, \lambda)$ the values of λ may be represented by a surface and that $\bar{\xi} = \alpha\xi^\gamma$ for small values of ξ, $\bar{\xi}$ and $|\lambda - \lambda_c|$, where α and γ are scalars and the postbuckling path of the geometrically perfect structure is given by $\alpha = 0$, see Fig. 20.1. The exponent γ may be chosen in a way that suits our purpose best. By choosing the values of α and γ appropriately we may reach any point in the $(\bar{\xi}, \xi)$-plane, especially the point associated with $\lambda_{\max}$. The process for doing this is, however, very far from straightforward and the reader is referred to (Budiansky 1974) or (Christensen & Byskov 2010).

E. Byskov, *Elementary Continuum Mechanics for Everyone*,
Solid Mechanics and Its Applications 194, DOI: 10.1007/978-94-007-5766-0_20,

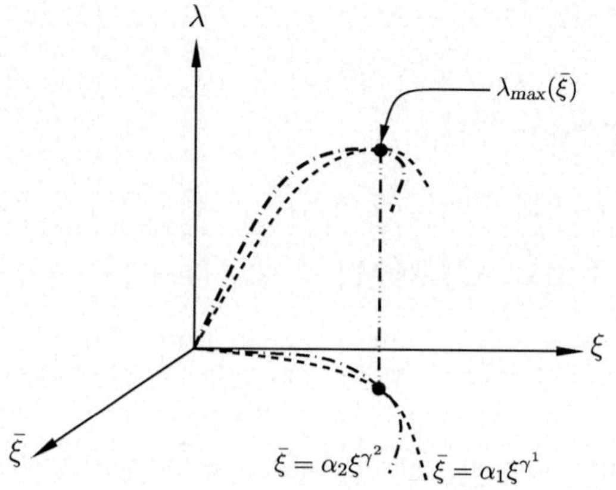

Fig. 20.1: The construct of Budiansky (1974).

20.1 Imperfection Sensitivity and a Single Buckling Mode

In our treatment in Chapter 19 of postbuckling behavior of geometrically perfect structures we needed to distinguish between symmetric and non-symmetric structures. The same holds true when we are dealing with imperfect structures.

20.1.1 Non-Vanishing First Order Postbuckling Constant

Asymmetric structure

For the first postbuckling constant $a \neq 0$ equilibrium of the geometrically imperfect structure is governed by

Equilibrium path

$$\left(1 - \frac{\lambda}{\lambda_c}\right)\xi + a\xi^2 \approx \frac{\lambda}{\lambda_c}\bar{\xi} \tag{20.1}$$

The factor λ/λ_c on the right-hand side does not follow directly from the above mentioned analyses, but is usually introduced in order to insure that the equilibrium path emanates from $(\xi, \lambda) = (0, 0)$.

Maximum load λ_s occurs at ξ_s. To lowest order

Asymmetric structures: Displacement ξ_s at limit load λ_s

$$\xi_s \approx \left(\pm\frac{\bar{\xi}}{a}\right)^{1/2}, \; a \gtrless 0 \;\text{ and }\; \frac{\lambda_s}{\lambda_c} \approx 1 - 2\left(\pm\bar{\xi}a\right)^{1/2}, \; a \gtrless 0 \tag{20.2}$$

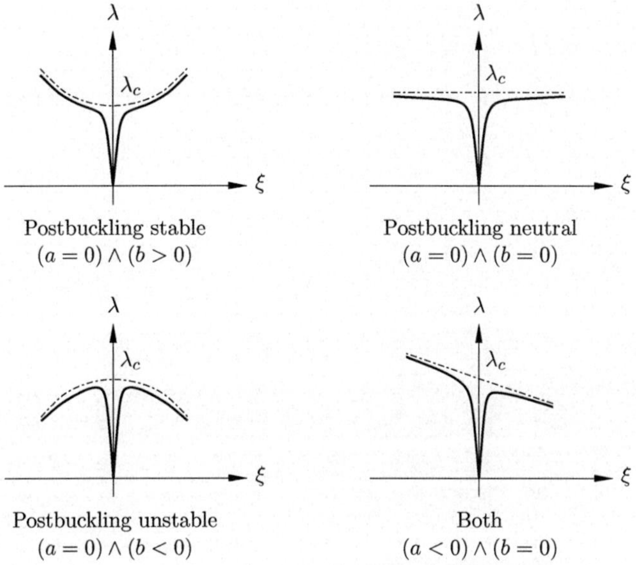

Fig. 20.2: Imperfection sensitivity depending on type of initial postbuckling behavior: $(1 - \lambda/\lambda_c)\xi + a\xi^2 + b\xi^3 \approx \lambda/\lambda_c\bar{\xi}$. The imperfection amplitude is $\bar{\xi}$.

20.1.2 Vanishing First Order Postbuckling Constant, Non-Vanishing Second Order Postbuckling Constant

When the first postbuckling constant $a = 0$ and the second order postbuckling constant $b \neq 0$ equilibrium of the geometrically imperfect structure is governed by

Symmetric structure

$$\left(1 - \frac{\lambda}{\lambda_c}\right)\xi + b\xi^3 \approx \frac{\lambda}{\lambda_c}\bar{\xi} \tag{20.3}$$

Equilibrium path

Also in this case, the factor λ/λ_c on the right-hand side is introduced in order to insure that the equilibrium path emanates from $(\xi, \lambda) = (0, 0)$.

Only when $b < 0$ does the equilibrium path display a maximum, where the expressions to lowest order are

$$\xi_s \approx \left(-\frac{\bar{\xi}}{2b}\right)^{1/3}, \; b < 0 \text{ and } \frac{\lambda_s}{\lambda_c} \approx 1 - 3(-b)^{1/3}\left(\tfrac{1}{2}\bar{\xi}\right)^{2/3} \tag{20.4}$$

Symmetric structures, $b < 0$: Displacement ξ_s at limit load λ_s

20.1.3 Is the Imperfection Detrimental?

Both (20.2) and (20.4) show that the loss of load-carrying capacity is very important for small values of the imperfection amplitude $\bar{\xi}$,

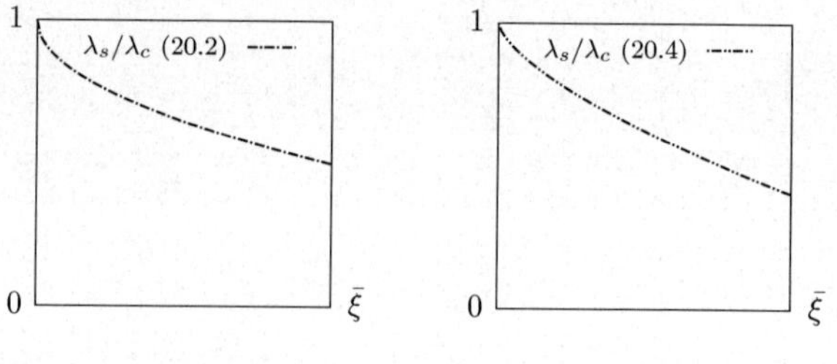

Fig. 20.3: Sketch of imperfection sensitivity according to the approximate expressions (20.2) and (20.4), respectively.

which becomes especially clear when the λ_s-$\bar{\xi}$ relation is plotted, see e.g. Fig. 20.3 and Fig. Ex. 17-1.7. Of course, the loss is greater the larger the imperfection amplitude, but the fact is that the loss for a shell structure may be as large as 75–80% for an imperfection amplitude less than the thickness of the shell skin.

Ex 20-1 Geometrically Imperfect Euler Column

Euler Column: Not imperfection sensitive by itself

One of the simplest examples of a geometrically imperfect structure is the imperfect Euler Column shown in Fig. Ex. 20-1.1. It is not imperfection sensitive by itself, but if it is used as a structural component

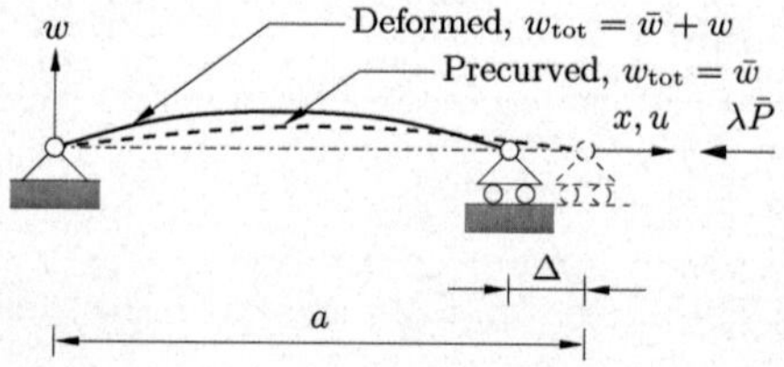

Fig. Ex. 20-1.1: *Geometrically imperfect Euler Column.*

in more complex structures the characteristics of its geometrically imperfect realization may play an important role for the behavior of the total structure, as we shall see in Example Ex 20-2 below. Of course, its behavior is by itself also interesting and deserves attention. In particular, the influence of a geometric imperfection on its axial stiffness is important.

In order to analyze the behavior of the geometrically imperfect Euler Column shown in Fig. Ex. 20-1.1 it would be correct to measure all quantities on the basis of the geometrically *imperfect* realization, and thus we ought to use strain measures for a (slightly) curved beam, whose configuration is given by a preexisting deflection $\bar{u}, \bar{w}$.[20.1] However, since we are dealing with a slightly imperfect column we are only interested in cases where the geometric imperfection is small and may refer all quantities to the perfect realization of the beam of Fig. Ex. 20-1.1.[20.2] Let subscript $_{\text{tot}}$ denote total quantities, i.e. components measured from the x-axis, and let the real displacement components measured from the geometrically imperfect shape be u and w.

Then,

Total displacements, imperfections and displacements

$$u_{\text{tot}} = \bar{u} + u\,, \quad w_{\text{tot}} = \bar{w} + w \qquad (Ex.\ 20\text{-}1.1)$$

According to the kinematically moderately nonlinear beam theory developed in Section 7.3 the generalized strains of a perfectly straight beam are, see (7.12) and (7.14)

$$\varepsilon = u' + \tfrac{1}{2}(w')^2\,, \quad \kappa = w'' \qquad (Ex.\ 20\text{-}1.2)$$

If the beam originally were straight the strains would be

"Total" strains

$$\varepsilon_{\text{tot}} = u'_{\text{tot}} + \tfrac{1}{2}\left(w'_{\text{tot}}\right)^2\,, \quad \kappa_{\text{tot}} = w''_{\text{tot}} \qquad (Ex.\ 20\text{-}1.3)$$

or

"Total" strains

$$\varepsilon_{\text{tot}} = (\bar{u} + u)' + \tfrac{1}{2}\left((\bar{w} + w)'\right)^2\,, \quad \kappa_{\text{tot}} = (\bar{w} + w)'' \qquad (Ex.\ 20\text{-}1.4)$$

Similarly, if the beam were straight in its virgin state and experienced a displacement field $(u, w) = (\bar{u}, \bar{w})$ the strains would be

"Strains" from imperfection

$$\bar{\varepsilon} = \bar{u} + \tfrac{1}{2}(\bar{w}')^2\,, \quad \bar{\kappa} = \bar{w}'' \qquad (Ex.\ 20\text{-}1.5)$$

The idea is now to assume that the stress-inducing strain can be computed as the difference between ε_{tot} and $\bar{\varepsilon}$ and between κ_{tot} and $\bar{\kappa}$, respectively

Stress-inducing strains

$$\varepsilon \approx \varepsilon_{\text{tot}} - \bar{\varepsilon}\,, \quad \kappa \approx w''_{\text{tot}} - \bar{w}'' \qquad (Ex.\ 20\text{-}1.6)$$

and thus we take the strain measures for the geometrically imperfect beam to be

Stress-inducing strains

$$\varepsilon = u' + \tfrac{1}{2}(w')^2 + \bar{w}'w'\,, \quad \kappa = w'' \qquad (Ex.\ 20\text{-}1.7)$$

With these strain measures and the principle of virtual displacements in hand we may derive the equilibrium equations for the slightly crooked beam. We may, however, also utilize the equilibrium equations (7.87) from Section 7.7, which were

Equilibrium equations. No imperfections

$$\left.\begin{array}{l} N' + \bar{p}_u = 0 \\ (EI\kappa)'' - (Nw')' - \bar{p}_w = 0 \end{array}\right\} \; x \in]0, a[\qquad (Ex.\ 20\text{-}1.8)$$

[20.1] The axial imperfection $\bar{u}$ seems somewhat difficult to define and, fortunately, it vanishes from the final expressions.

[20.2] I think that the procedure in the following is intuitively reasonable, but mention that Koiter has proved that it entails only very small errors which we may neglect.

and realize that the distributed loads vanish and that the second term of the second equation is associated with the contribution to the bending moment from the displaced axial force and therefore the w' should be changed to $(\bar{w}+w)'$ in that term. Then,[20.3]

Equilibrium equations for geometrically imperfect column

$$\left.\begin{aligned} N' &= 0 \\ (EIw'')'' - N\,(\bar{w}' + w')' &= 0 \end{aligned}\right\}\; x \in]0,a[\qquad (Ex.\ 20\text{-}1.9)$$

We consider only an axial compressive load $\lambda\bar{P}$

Only an axial load $\lambda\bar{P}$

$$N = -\lambda\bar{P} \qquad (Ex.\ 20\text{-}1.10)$$

When we assume constant elastic properties and define

Definition of load unit $\lambda\bar{P}$

$$\bar{P} \equiv \pi^2\frac{EI}{a^2} \qquad (Ex.\ 20\text{-}1.11)$$

we may get

Differential equation for imperfect column

$$w^{iv} + \lambda\left(\frac{\pi}{a}\right)^2 w'' = -\lambda\left(\frac{\pi}{a}\right)^2 \bar{w}'' \qquad (Ex.\ 20\text{-}1.12)$$

Later, we shall limit ourselves to geometric imperfections in the shape of the buckling mode w_1 of the equivalent perfect Euler Column, but for the time being assume a general shape of the imperfection which, according to the expansion theorem mentioned in Section 18.3.4, may be given by

Arbitrary imperfection $\bar{w}$

$$\bar{w} = \sum_{j=1}^{\infty}\bar{\xi}_j w_j = \sum_{j=1}^{\infty}\bar{\xi}_j \sin(j\pi x/a) \qquad (Ex.\ 20\text{-}1.13)$$

where w_j is the jth buckling mode and $\bar{\xi}_j$ the component of the imperfection in the "direction" of w_j. Because of the normalization (*Ex. 20-1.11*) the classical critical value associated with the jth buckling mode is j^2. Insert (*Ex. 20-1.13*) into (*Ex. 20-1.12*) and get

$$\begin{aligned} &+\sum_{j=1}^{\infty} j^4\left(\frac{\pi}{a}\right)^4 \xi_j \sin(j\pi x/a) - \lambda\sum_{j=1}^{\infty} j^2\left(\frac{\pi}{a}\right)^4 \xi_j \sin(j\pi x/a) \\ &= +\lambda\sum_{j=1}^{\infty} j^2\left(\frac{\pi}{a}\right)^4 \bar{\xi}_j \sin(j\pi x/a) \end{aligned} \qquad (Ex.\ 20\text{-}1.14)$$

Since all terms satisfy all boundary conditions we may get the exact result

Coefficients ξ_j of sines

$$\xi_j = \frac{\lambda}{j^2-\lambda}\bar{\xi}_j \qquad (Ex.\ 20\text{-}1.15)$$

First component w_1 of w dominates

It may easily be seen from (*Ex. 20-1.15*) that when λ is increased from 0 and approaches 1 the first component of w, namely w_1, becomes dominating. This, incidentally, is the main reason why the imperfection is

[20.3] As usual, you should be very careful and check that the strain and stress measures are generalized, i.e. work conjugate. The proof of this is in this case left to the reader to carry out. You must derive the equilibrium equations from the principle of virtual displacements and note that the ensuing equations are the same ones as below.

usually taken to be in the shape of the buckling mode associated with the lowest buckling load $\lambda_1 = \lambda_c$. In the following we shall adopt this idea and limit ourselves to the imperfection

Imperfection in shape of the buckling mode

$$\bar{w} = \bar{\xi} w_1 = \bar{\xi} \sin(\pi x/a) \qquad (Ex.\ 20\text{-}1.16)$$

and thus

Coefficient $\xi_1 = \xi$ of buckling mode

$$\xi = \frac{\lambda}{1-\lambda} \bar{\xi} \qquad (Ex.\ 20\text{-}1.17)$$

For later purposes, see Example Ex 20-2, we normalize the transverse displacements on the *Radius of Gyration* r of the cross-section. The reason for this normalization is that r is a natural measure of the cross-section and that the amplitude of an imperfection often depends on the size of the cross-section. The column length a may be another candidate for measuring the imperfection amplitude, but the values of these then become so small that it is somewhat difficult to relate to them. We indicate our normalized quantities by a tilde

Radius of gyration

Normalized coefficients

$$\tilde{\xi} \equiv \frac{\xi}{r} \text{ and } \tilde{\bar{\xi}} \equiv \frac{\bar{\xi}}{r} \qquad (Ex.\ 20\text{-}1.18)$$

Using these normalizations, we may rewrite (*Ex. 20-1.17*)

Relation between normalized coefficients

$$\tilde{\xi} = \frac{\lambda}{1-\lambda} \tilde{\bar{\xi}} \qquad (Ex.\ 20\text{-}1.19)$$

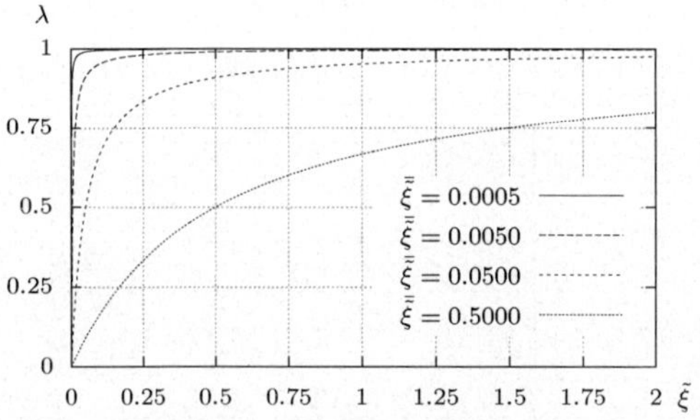

Fig. Ex. 20-1.2: *Normalized load-displacement relation for geometrically imperfect Euler Columns.*

This relation is plotted in Fig. Ex. 20-1.2 which indicates that the transverse displacements increase very sharply for a slightly imperfect column when the axial load approaches the lowest buckling load.

For the example of mode interaction in a truss column, see Example Ex 20-2 below, we need the axial stiffness of the imperfect Euler Column. In order to find this, we compute the axial shortening Δ of the column. This is done by use of a trick that may seem awkward in that we determine Δ by integration of the axial displacement component u

Axial shortening Δ

$$\Delta = -\int_0^a u' dx = -\int_0^a \left(\varepsilon - \tfrac{1}{2}(w')^2 - \bar{w}' w'\right) dx \qquad (Ex.\ 20\text{-}1.20)$$

and introduce the axial strain through the constitutive relation

$$N = EA\varepsilon \qquad (Ex.\ 20\text{-}1.21)$$

to get

Axial shortening Δ

$$\Delta = \int_0^a \left(-\frac{N}{EA} + \tfrac{1}{2}(w')^2 + \bar{w}'w' \right) dx \qquad (Ex.\ 20\text{-}1.22)$$

By use of (*Ex. 20-1.10*), (*Ex. 20-1.11*), (*Ex. 20-1.16*), (*Ex. 20-1.17*), and the equivalent of (*Ex. 20-1.16*) for w we may rewrite (*Ex. 20-1.22*) and get

$$\Delta = \int_0^a \left(+\lambda \left(\frac{\pi}{a}\right)^2 \left(\frac{I}{A}\right) + \tfrac{1}{2}\left(\frac{\pi}{a}\right)^2 \xi^2 \cos^2(\pi x/a) + \left(\frac{\pi}{a}\right)^2 \bar{\xi}\xi \cos^2(\pi x/a) \right) dx \qquad (Ex.\ 20\text{-}1.23)$$

When we introduce the relations (*Ex. 20-1.17*) and (*Ex. 20-1.18*) and carry out the integrations we may get

Shortening Δ

$$\Delta = a\left(\frac{\pi}{a}\right)^2 r^2 \left(\lambda + \left(\tfrac{1}{4}\left(\frac{\lambda}{1-\lambda}\right)^2 + \tfrac{1}{2}\left(\frac{\lambda}{1-\lambda}\right)\right) \bar{\bar{\xi}}^2 \right) \qquad (Ex.\ 20\text{-}1.24)$$

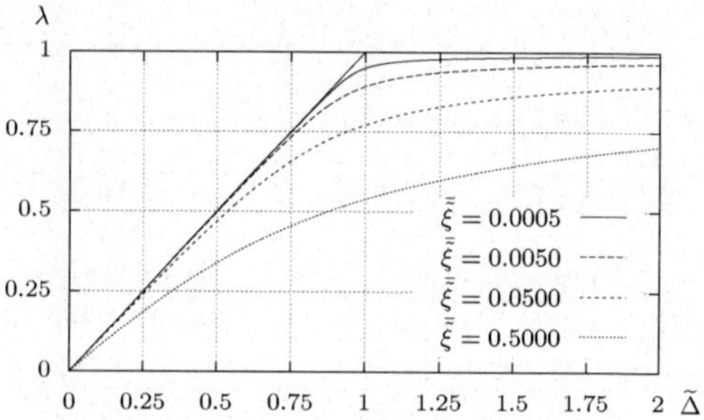

Fig. Ex. 20-1.3: *Normalized load-shortening relation for geometrically imperfect Euler Columns.*

For convenience, introduce the normalized shortening $\widetilde{\Delta}$ by

Definition of normalized shortening $\widetilde{\Delta}$

$$\widetilde{\Delta} \equiv \frac{1}{a}\left(\frac{a}{r}\right)^2 \frac{1}{r^2}\Delta \qquad (Ex.\ 20\text{-}1.25)$$

which, together with (*Ex. 20-1.18*), may be used to cast (*Ex. 20-1.24*) in the nondimensional form

Nondimensional shortening $\widetilde{\Delta}$

$$\widetilde{\Delta} = \lambda + \left(\tfrac{1}{4}\left(\frac{\lambda}{1-\lambda}\right)^2 + \tfrac{1}{2}\left(\frac{\lambda}{1-\lambda}\right)\right) \bar{\bar{\xi}}^2 \qquad (Ex.\ 20\text{-}1.26)$$

which is better suited for plotting than (*Ex. 20-1.24*), see Fig. Ex. 20-1.3, and clearly shows that the shortening increases dramatically when

$\lambda \to 1$ and that the axial stiffness of the imperfect column decreases at the same time. One reason for the interest in the change in apparent axial stiffness is that it plays an important role in determining the load-carrying capacity of a geometrically imperfect truss column, see Example Ex 20-2. At this point we merely note that the truss column only "feels" that the axial stiffness of an imperfect flange is lower than that of a perfect one—the lower stiffness might as well have been caused by a nonlinear constitutive relation—with the result that the load-carrying capacity of the truss column is smaller than if it were geometrically perfect. Therefore, we need the "apparent tangent modulus of elasticity" $E_{\tan}$ of the (slightly) crooked column, but first we compute the change $F_{\tan}$ in the axial force F due to an incremental change in the shortening Δ

"Apparent tangent modulus of elasticity " $E_{\tan}$

$$F_{\tan} \equiv \frac{dF}{d\Delta} \qquad \text{(Ex. 20-1.27)}$$

or

$$F_{\tan} = \frac{E_{\tan} A}{a} \qquad \text{(Ex. 20-1.28)}$$

and thus

$$E_{\tan} = \frac{a}{A}\frac{dF}{d\Delta} = \frac{1}{\pi^2}\left(\frac{a}{r}\right)^2 \frac{\lambda_c \bar{P}}{A}\frac{d\lambda}{d\widetilde{\Delta}} = E\frac{d\lambda}{d\widetilde{\Delta}} \qquad \text{(Ex. 20-1.29)}$$

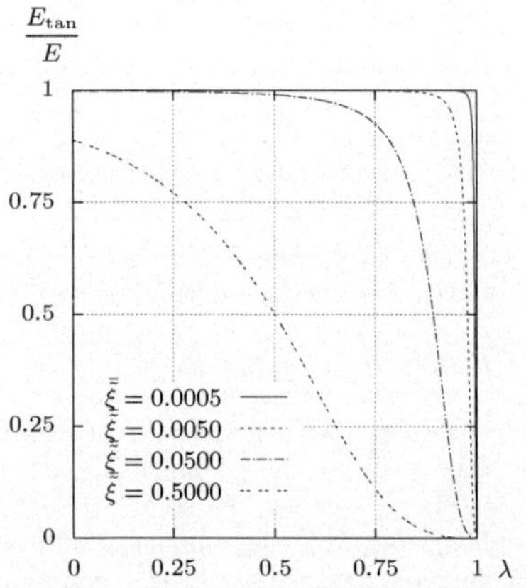

Fig. Ex. 20-1.4: Normalized apparent tangent modulus $E_{\tan}/E$ for geometrically imperfect Euler Columns.

Differentiation of (Ex. 20-1.26) provides

$$\frac{d\lambda}{d\widetilde{\Delta}} = 1 + \frac{1}{2(1-\lambda)^3}\bar{\bar{\xi}}^2 \qquad \text{(Ex. 20-1.30)}$$

Combine (Ex. 20-1.29) and (Ex. 20-1.30) to provide the desired expres-

"Apparent tangent modulus" of elasticity $E_{\tan}$

sion for $E_{\tan}$

$$E_{\tan} = \frac{1}{1 + \dfrac{1}{2(1-\lambda)^3}\bar{\bar{\xi}}^2} E \qquad (Ex.\ 20\text{-}1.31)$$

This relation is shown in Fig. Ex. 20-1.4 which proves that the value of the imperfect column as a structural member decreases strongly when $\lambda \to 1$. It may also be observed that this phenomenon gets relatively worse when the imperfection is small because then the change in stiffness is more abrupt.

20.2 Mode Interaction and Geometric Imperfections

Mode interaction

As mentioned in Chapter 19, in particular Section 20.2, interaction between buckling modes associated with the same or nearly the same classical critical load becomes particularly important for geometrically imperfect structures. This phenomenon, which is known as *Mode Interaction* has been studied intensely since about 1970.

The following formulas, derived by Byskov & Hutchinson (1977), are valid for cases with simultaneous or nearly simultaneous buckling modes provided that their wavelengths are of the same order. If, however, the wavelength of the local buckling modes is very small compared with that of the overall mode Koiter's *Slowly Varying Local Mode Amplitude Method*, see (Koiter & Kuiken 1971) and (Koiter 1976), provides more reliable results than the method by Byskov & Hutchinson (1977), see e.g. (Byskov 1979).

The basic assumption of Byskov & Hutchinson (1977) is that the displacement field $\boldsymbol{u}$ on the postbuckling path may be written

Displacements in postbuckling

$$\boldsymbol{u} = \lambda \boldsymbol{u}_0 + \xi_i \boldsymbol{u}_i + \xi_i \xi_j \boldsymbol{u}_{ij} + \cdots, \quad (i,j) = (1 \ldots M) \qquad (20.5)$$

where, as in all expressions below, sum over $(1, \ldots, M)$ is implied for lower-case indices, λ denotes a scalar load parameter, ξ_i is the amplitude of the ith buckling mode $\boldsymbol{u}_i$, $\boldsymbol{u}_{ij}$ designates the second-order field associated with buckling modes i and j, and M is the number of interacting modes. When similar expansions are assumed valid for the strains and stresses and are inserted into the principle of virtual displacements the following set of problems are generated.[20.4]

Linear boundary-value problem for the prebuckling path

Prebuckling

$$\boldsymbol{\sigma}_0 \cdot \mathrm{l}_1(\delta \boldsymbol{u}) = \overline{\boldsymbol{T}} \cdot \delta \boldsymbol{u} \qquad (20.6)$$

see also (18.15).

[20.4] As is clear from (Byskov & Hutchinson 1977), the procedure is not trivial which is the reason why their derivations are not repeated here.

Linear eigenvalue problems for the buckling modes

$$\boldsymbol{\sigma}_J \cdot \mathbf{l}_1(\delta \boldsymbol{u}) + \lambda_J \boldsymbol{\sigma}_0 \cdot \mathbf{l}_{11}(\boldsymbol{u}_J, \delta \boldsymbol{u}) = 0\,, \qquad J = (1, \ldots, M) \tag{20.7}$$

Buckling

cf. (18.29).

Linear boundary-value problems for the second-order fields

$$\begin{aligned} &\boldsymbol{\sigma}_{JK} \cdot \mathbf{l}_1(\delta \boldsymbol{u}) + \lambda_c \boldsymbol{\sigma}_0 \cdot \mathbf{l}_{11}(\boldsymbol{u}_{JK}, \delta \boldsymbol{u}) \\ &= -\tfrac{1}{2}\big(\boldsymbol{\sigma}_J \cdot \mathbf{l}_{11}(\boldsymbol{u}_K, \delta \boldsymbol{u}) + \boldsymbol{\sigma}_K \cdot \mathbf{l}_{11}(\boldsymbol{u}_J, \delta \boldsymbol{u})\big)\,, \\ &(J, K) = (1, \ldots, M) \end{aligned} \tag{20.8}$$

Postbuckling

is the equivalent of (19.34).

The postbuckling strain-displacement relation is

$$\varepsilon_{ij} = \varepsilon_{ji} = \mathbf{l}_1(\boldsymbol{u}_{ij}) + \tfrac{1}{2}\mathbf{l}_{11}(\boldsymbol{u}_i, \boldsymbol{u}_j) \tag{20.9}$$

Postbuckling strains ε_{ij}

and the stress-strain relation for quantities associated with the second-order fields is

$$\boldsymbol{\sigma}_{ij} = \mathbf{H}(\varepsilon_{ij}) \tag{20.10}$$

Postbuckling constitutive relation

It is important to note the orthogonality condition (18.35)

$$\boldsymbol{\sigma}_0 \cdot \mathbf{l}_{11}(\boldsymbol{u}_J, \boldsymbol{u}_K) = 0\,, \quad J \neq K \tag{20.11}$$

Buckling modes are orthogonal

The equilibrium path of a geometrically imperfect structure is then determined by

$$\left(1 - \frac{\lambda}{\lambda_L}\right)\xi_L + a_{ijL}\xi_i\xi_j + b_{ijkL}\xi_i\xi_j\xi_k = \left(\frac{\lambda}{\lambda_L}\right)\bar{\xi}_L\,, \qquad L = (1, \ldots, M) \tag{20.12}$$

Equilibrium path of imperfect structure

where, as in the following formulas, sum over upper-case indices is not implied unless explicitly indicated, ξ_L is the amplitude of the Lth buckling mode, $\bar{\xi}_L$ is the amplitude of the imperfection in the direction of the Lth buckling mode, a_{ij} denote first-order postbuckling constants, and b_{ijk} are second-order postbuckling constants, which are given by

$$a_{ijL} = \frac{\boldsymbol{\sigma}_L \cdot \mathbf{l}_{11}(\boldsymbol{u}_i, \boldsymbol{u}_j) + 2\boldsymbol{\sigma}_i \cdot \mathbf{l}_{11}(\boldsymbol{u}_j, \boldsymbol{u}_L)}{2\boldsymbol{\sigma}_L \cdot \varepsilon_L} \tag{20.13}$$

First-order postbuckling constants a_{ijL}

and

$$\begin{aligned} b_{ijkL} = \big(&+\boldsymbol{\sigma}_{Li} \cdot \mathbf{l}_{11}(\boldsymbol{u}_j, \boldsymbol{u}_k) + \boldsymbol{\sigma}_{ij} \cdot \mathbf{l}_{11}(\boldsymbol{u}_k, \boldsymbol{u}_L) \\ &+\boldsymbol{\sigma}_L \cdot \mathbf{l}_{11}(\boldsymbol{u}_i, \boldsymbol{u}_{jk}) + \boldsymbol{\sigma}_i \cdot \mathbf{l}_{11}(\boldsymbol{u}_L, \boldsymbol{u}_{jk}) \\ &+2\boldsymbol{\sigma}_i \cdot \mathbf{l}_{11}(\boldsymbol{u}_j, \boldsymbol{u}_{kL})\big) \\ &\big/(2\boldsymbol{\sigma}_L \cdot \varepsilon_L) \end{aligned} \tag{20.14}$$

second-order postbuckling constants b_{ijkL}

respectively.

Note that for all indices equal, the above expressions provide the formulas for the single-mode case of Chapter 18 and Section 20.1.

The following Example Ex 20-2 serves to illustrate the phenomenon of mode interaction in a simple structure. We shall, however, not exploit the formulas derived above, but perform an *ad hoc* analysis. If you wish to see asymptotic expansions applied to this problem I refer you to (Byskov 1979), where the *ad hoc* analysis, the asymptotic expansion of Byskov & Hutchinson (1977) and the slowly varying local mode amplitude method developed by Koiter (1976) are utilized. Only the results obtained in (Byskov 1979) by application of the three methods are compared in Example Ex 20-2.

Ex 20-2 Mode Interaction in a Truss Column

Mode interaction studies begin in around 1970

The first analysis of a structure exhibiting mode interaction in the spirit of the present chapter seems to be the study by van der Neut (1969) on a model of a thin-walled column. We shall, however, focus on the simpler problem of a truss column which was first analyzed by

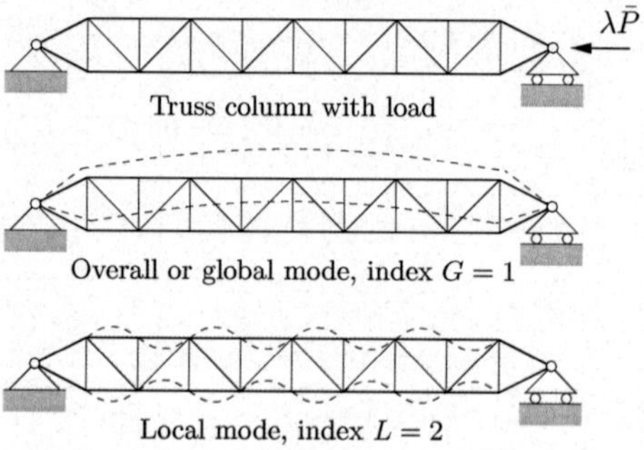

Fig. Ex. 20-2.1: *Truss column with global and local mode.*

Thompson & Hunt (1973) and, apparently independently, by Crawford & Hedgepeth (1975), both applying the *ad hoc* method described below. Later, Byskov (1979) utilized the asymptotic expansion derived above as well as Koiter's *Slowly Varying local Mode Amplitude*, see (Koiter & Kuiken 1971) and (Koiter 1976), and results from these analyses are given here.

The Quebec Bridge

It may be worth noticing that mode interaction in truss columns probably was one of the contributing causes of the famous collapse of the Quebec Bridge in 1907, which was one of the worst building disasters of the twentieth century in that 70 people were killed by the collapse which occurred during construction of the bridge.

Ex 20-2.1 A Geometrically Perfect Truss Column

Geometrically perfect truss column: Global and local instability

A geometrically perfect realization of the truss column shown in the above figure, Fig. Ex. 20-2.1, may experience loss of stability by bifurcating into an overall, or global, buckling mode, or into a local buckling mode, as shown. Already at this point we mention that both these modes are *Euler Column Buckling Modes* which are postbuckling neutral and therefore not imperfection sensitive by themselves. Our analysis below will, however, show that when the classical critical loads associated with the two modes are close the structure is imperfection sensitive due to nonlinear interaction between the overall and the local mode.

Let L denote the length of the truss, n the number of bays, a the bay length, and EI and EA the flange bending and axial stiffness, respectively. Then, we get the bifurcation load $2\lambda_L \bar{P}$ associated with local buckling of the flanges

$$2\lambda_L \bar{P} = 2\pi^2 \frac{EI}{(L/n)^2} = 2\pi^2 \frac{EI}{a^2} \qquad (Ex.\ 20\text{-}2.1)$$

For convenience we normalize the load such that

$$\lambda_L \equiv 1 \Rightarrow \bar{P} = \pi^2 \frac{EI}{a^2} \qquad (Ex.\ 20\text{-}2.2)$$

The global bifurcation load $2\lambda_G \bar{P}$ is

$$2\lambda_G \bar{P} = \pi^2 \frac{EI_G}{L^2} \qquad (Ex.\ 20\text{-}2.3)$$

where I_G designates the global moment of inertia.

We may here note that λ_G plays the role of a design parameter because if we keep the flanges untouched, but move them further apart the local buckling load λ_L stays the same, but the global buckling load λ_G is increased, meaning that the flanges have become relatively thinner.

Ex 20-2.2 A Geometrically Imperfect Truss Column

Geometrically imperfect truss column

Here, we shall employ the simple approach of Thompson & Hunt (1973) and Crawford & Hedgepeth (1975) because it seems intuitively reasonable. On the other hand, the results found by Byskov (1979) using the strictly asymptotic method by Byskov & Hutchinson (1977), which is outlined in Section 20.2, and the slowly varying local mode amplitude method of Koiter (1976) will be discussed at the end of this example.

In order to perform the following ad hoc analysis of instability in a geometrically imperfect realization of the the truss column shown in Fig. Ex. 20-2.1 we need the formulas for geometrically imperfect beams from Example Ex 20-1. In the following we shall only consider the case where the flanges of the column are geometrically imperfect in the same shape as the local buckling mode. Then, they act with the apparent tangent modulus $E_{\tan}$ found in Example Ex 20-1. Using the so-called *Shanley Criterion*, see Chapter 21, the reduced load-carrying capacity $2\lambda_R \bar{P}$ of the locally imperfect truss column can be determined. The idea behind this criterion and the analysis below very simply is that when a column of a nonlinear material bifurcates from its original straight shape the column material provides a restoring stress that is given by its tangent stiffness. Thus,

Reduced load-carrying capacity λ_R

$$2\lambda_R\bar{P} = \pi^2\frac{E_{\tan}I_G}{L^2} = \frac{E_{\tan}}{E}2\lambda_G\bar{P} \qquad (Ex.\ 20\text{-}2.3)$$

with

$$\lambda_R = \frac{E_{\tan}}{E}\lambda_G \qquad (Ex.\ 20\text{-}2.3)$$

Using the expression (*Ex. 20-1.31*) we may then get

Result from *ad hoc* analysis

$$\lambda_R = \frac{1}{1+\dfrac{1}{2(1-\lambda_R)^3}\bar{\bar{\xi}}^2}\lambda_G \qquad (Ex.\ 20\text{-}2.4)$$

which determines the load-carrying capacity of the locally imperfect truss column according to the *ad hoc* analysis. In order to compare with the formulas found by the more sophisticated methods mentioned above, see (Byskov 1979), we shall recast (*Ex. 20-2.4*) in a somewhat different form and get

Comparison between results

$$\begin{aligned} &\text{Ad hoc:} && (\lambda_G-\lambda_R)(1-\lambda_R)^3 = \tfrac{1}{2}\lambda_R\bar{\bar{\xi}}^2 \\ &\text{Asymptotic:} && (\lambda_G-\lambda_R)(1-\lambda_R)^3 = \frac{4}{\pi^2}\lambda_G\lambda_R^2\bar{\bar{\xi}}^2 \\ &\text{Slowly varying:} && (\lambda_G-\lambda_R)(1-\lambda_R)^3 = \tfrac{1}{2}\lambda_R^3\bar{\bar{\xi}}^2 \end{aligned} \qquad (Ex.\ 20\text{-}2.5)$$

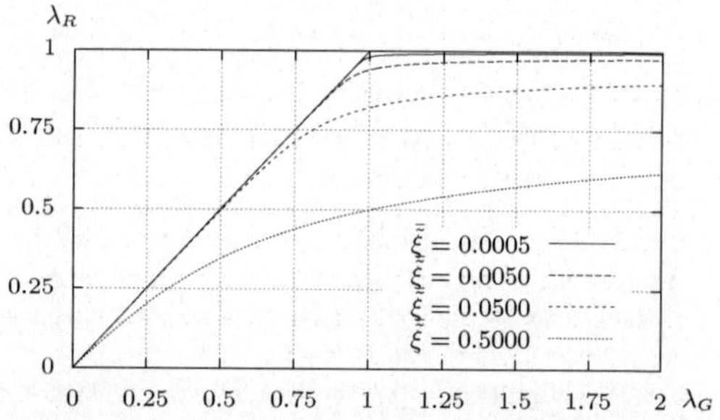

Fig. Ex. 20-2.2: *Normalized load-carrying capacity of geometrically imperfect truss columns according to* ad hoc *analysis.*

As you can see, the main structure of the expressions in (*Ex. 20-2.5*) looks almost the same, although the theories behind differ. For reasonable designs the predictions by the three formulas agree so much

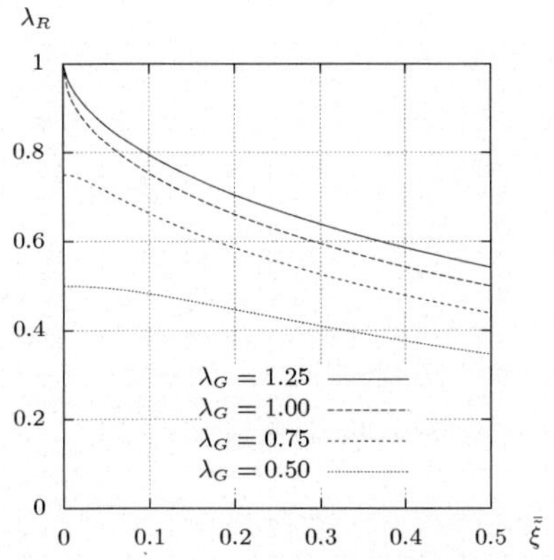

Fig. Ex. 20-2.3: *Imperfection sensitivity of truss columns according to ad hoc analysis.*

that plots of the results are almost indistinguishable. Fig. Ex. 20-2.2 shows that the sensitivity to geometric imperfection is moderate when $\lambda_G \lesssim 0.75$, i.e. for rather thick flanges. It is also clear that the most imperfection sensitive case is the one for which $\lambda_G = 1$, which corresponds to the so-called "naive optimum" where the buckling loads λ_G and λ_L are equal. From a design point of view this is indeed the optimum provided that the influence of imperfections on the load-carrying capacity is taken into account. This, by the way, seems to be a universally true statement. In the case of the old Quebec Bridge the design has probably been one for which $\lambda_G \gtrsim 1$ with the catastrophic consequences mentioned above.

"Naive optimum" is dangerous

The picture becomes even clearer when we plot the normalized load-carrying capacity against the normalized imperfection amplitude, see Fig. Ex. 20-2.3, which shows that the decrease in capacity is much greater for cases with $\lambda_G \gtrsim 1$ than otherwise. This means that the imperfection sensitivity is worse for columns with relatively thin flanges than for designs with more stocky flanges.

Thin flanges ⇒ worse imperfection sensitivity

Traditionally, engineers have preferred designs with rather thick flanges and have therefore avoided imperfection sensitive truss columns. But, because they did not know the above or a similar analysis, they could not have had imperfection sensitivity in mind.

Ex 20-2.2.1 Elastic-Plastic Truss Columns

A stocky structure is in general more likely to experience plasticity than a slender one. If plasticity and kinematic nonlinearity interact you would expect severe imperfection sensitivity. Thus, you might think that a truss column with thick flanges would exhibit strong imperfection sensitivity. This is, as shown by Byskov (1982) not the case, simply because a truss column with thick flanges bifurcates into the global mode at such low flange stresses that plasticity does not occur. For cases with imperfect, thin flanges, however, plasticity plays a more significant role.

Combinations of plasticity and geometric imperfections provide a dangerous cocktail

Other structures may buckle in the plastic regime with an increase in imperfection sensitivity as a result. Over the years many studies have shown this to be true and among those I mention (Hutchinson 1973a), (Hutchinson 1974b), (Tvergaard & Needleman 1975), (Needleman & Tvergaard 1976), (Needleman & Tvergaard 1982), (Byskov 1982–83), and (Wunderlich, Obrecht & Schrödter 1986).

Chapter 21

Elastic-Plastic Buckling The Shanley Column

21.1 Introduction

For more than 200 years it has been known that the so-called *Classical Critical Load*, also know as the *Euler Load*, constitutes a good measure of the load-carrying capacity of perfectly straight linearly elastic columns, see e.g. (Brush & Almroth 1975), (Timoshenko & Gere 1961), Example Ex 7-3, Example Ex 18-1, and Example Ex 32-5. Until the pioneering work by Shanley (1947) about 50 years ago clarified this issue there was no generally accepted equivalent definition of the load-carrying capacity of columns made of elastic-plastic materials. On the contrary, almost all professors of structural mechanics had their own elastic-plastic column formula. In retrospect it is clear that these formulas all represented some weighted mean of the *Tangent Modulus Load* λ_T, the *Reduced Modulus Load* λ_R, and the *Euler Load* $\lambda_E = \lambda_c$, see Sections 21.2.2 and 21.2.3 and Example Ex 7-3, respectively.

Important quantities:
Tangent Modulus Load λ_T
Reduced Modulus Load λ_R
Euler Load λ_c

In this chapter we discuss why buckling of elastic-plastic columns is so much more complicated an issue than that of the elastic ones and therefore remained an unsolved problem for so many years. In the process, in addition to the already well-known *Euler Buckling Load* $\lambda_E = \lambda_c$, we derive the formulas for the *Tangent Modulus Load* λ_T and for the *Reduced Modulus Load* λ_R, which are central to the understanding of the ideas of Shanley (1947). Although the concept of *Geometrical Imperfections* is crucial to Shanley's reasoning we shall not attempt to cover the work by Duberg & Wilder (1952), which is an extension of Shanley's analysis into the region of finite displacements. Nor shall we try to introduce the ingenious and very important work of Hutchinson (1974b), who presented a general theory of geometrically perfect and imperfect structures of elastic-plastic materials.

In the following we do not employ the original formulations of *Engesser*, see Sections 21.2.2, 21.2.3, 21.3.5 and 21.4.3, and of *Shanley*, see Section 21.4, but one that agrees better with the modern way of treating stability.

E. Byskov, *Elementary Continuum Mechanics for Everyone*,
Solid Mechanics and Its Applications 194, DOI: 10.1007/978-94-007-5766-0_21,

21.2 Columns of Elastic-Plastic Materials

Elastic-plastic columns

When we consider deformation of linearly elastic beams, columns, and frames we only need the value of some cross-sectional properties, such as the area and the moment of inertia, which remain constant during any loading process. In the case of elastic-plastic structures these quantities either lose their meaning or must at least be reinterpreted and recomputed at each load step. This is because in general the "fibers" of the beam experience different strains and, since the stress-strain curve is nonlinear, the stress distribution over the cross-sections not only varies in magnitude, but also in shape. Therefore, the cross-sectional properties are no longer independent of the loading on the structure.

Il'yushin theory

At a first glance, the so-called *Il'yushin* theories, see e.g. (Kachanov 1974), which try to avoid computation of the fiber strains and stresses in favor of expressing the axial force and the bending moment directly in terms of the axial strain and the bending strain seem very appealing. Unfortunately, these theories do not to work satisfactorily and have therefore in general been abandoned in favor of the above "fiber" approach.

21.2.1 Constitutive Model

Before we can perform an analysis of an elastic-plastic column we must decide which stress-strain relation to employ. As usual, this implies a competition between simplicity and reality. The simplest elastic-plastic consti-

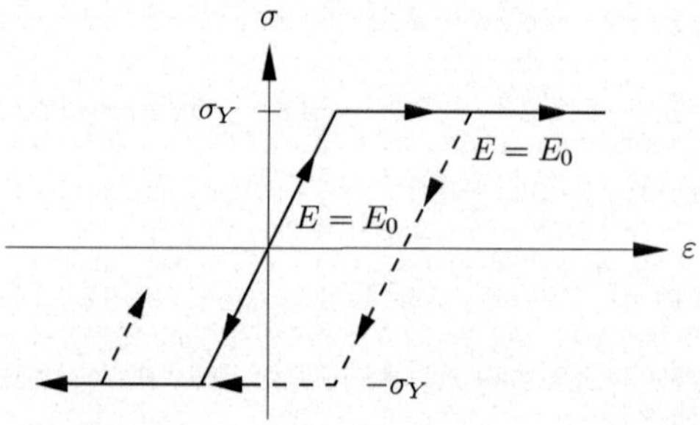

Fig. 21.1: Linear elastic-perfectly plastic material model.

tutive model is probably the linearly elastic-perfectly plastic one shown in Fig. 21.1. For reasons that should be clear later we must discard this model in the present connection because of its perfectly plastic branch, see Sections 21.2.2 and 21.2.3 below. We must therefore utilize a model which entails *Strain Hardening*, see Figs. 21.2(a) and 21.2(b). The following quan-

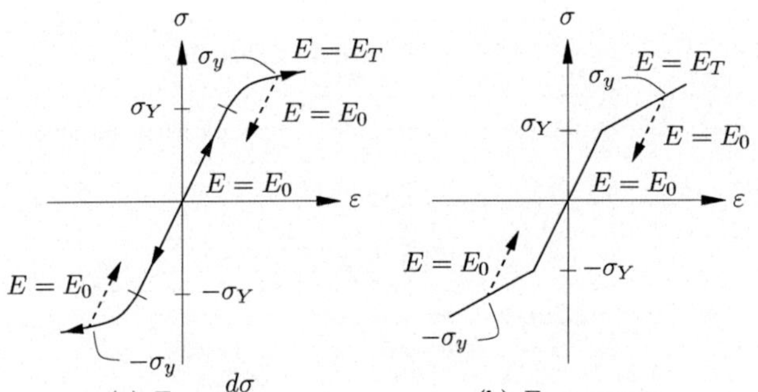

(a) $E_T = \dfrac{d\sigma}{d\varepsilon}$ (b) $E_T = \text{const.}$

Fig. 21.2: Linear elastic-hardening plastic material models.

tities defined by

$$\begin{aligned}
&\text{Fiber strain}: && \varepsilon \\
&\text{Fiber stress}: && \sigma \\
&\text{Initial modulus of elasticity}: && E_0 \\
&\text{Tangent modulus (of elasticity)}: && E_T \equiv \frac{d\sigma}{d\varepsilon} \\
&\text{Initial yield stress}: && \sigma_Y \\
&\text{Current yield stress}: && \sigma_y
\end{aligned} \tag{21.1}$$

will be used repeatedly below. In the present context we are concerned with stability of columns and may therefore without loss of generality take the stress-strain relation to be symmetric about the origin—as a matter of fact, we only need to consider the third quadrant, that is, the one with $(\varepsilon, \sigma) < (0, 0)$. As is common in theories of plasticity we assume that unloading occurs at the initial modulus of elasticity E_0 which is greater than the tangent modulus E_T.[21.1]

Nonlinear hardening

While the model of Fig. 21.2(a) displaying nonlinear hardening is more realistic than the one with linear hardening, see Fig. 21.2(b), the latter is often used in computations because it is better suited for calculations by hand. In numerical analyses by computer the sharp bends at $\pm\sigma_Y$ may cause difficulties and a constitutive model such as the one sketched in Fig. 21.2(a) is usually preferred. In Section 21.4 we shall utilize the constitutive model of Fig. 21.2(b) in order to get explicit result for the possible bifurcation loads of the *Shanley Model Column*.

[21.1] If $E_0 = E_T$ we are dealing with a linearly elastic material.

21.2.2 Engesser's First Proposal (1889)

The idea behind Engesser's first formula is quite obvious: In the prebuckling state, where the column stays straight, all fibers work with the same mod-

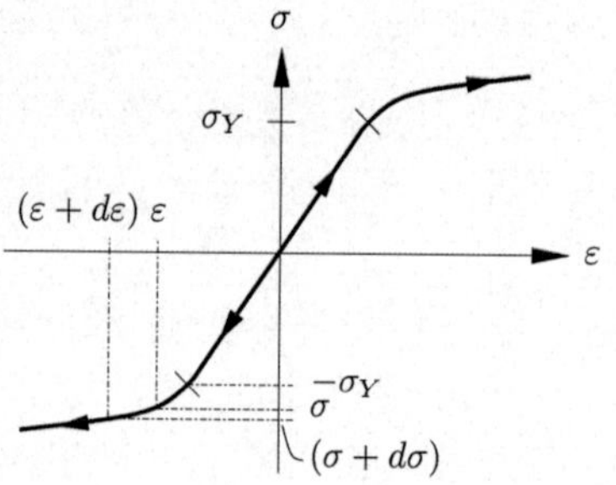

Fig. 21.3: Material with nonlinear strain hardening. Strain and stress increments are indicated.

ulus of elasticity E, namely the *Tangent Modulus* E_T for additional strains $d\sigma$, see Figs. 21.2(a), 21.2(b) and 21.3

Tangent Modulus E_T

$$d\sigma = E_T d\varepsilon \tag{21.2}$$

Now, since all fibers work with the same modulus of elasticity we can compute the buckling load as if we had a column of an elastic material with *Young's Modulus* E equal to the tangent modulus E_T. Thus, the load-carrying capacity $\lambda_T \bar{P}$ according to Engesser's first proposal, which is also known as the *Tangent Modulus Load*,[21.2] of a simply supported elastic-plastic *Euler Column* is

Tangent Modulus Load $\lambda_T \bar{P}$

$$\lambda_T \bar{P} = \pi^2 \frac{E_T(\sigma_T) I}{L^2} \tag{21.3}$$

where $\bar{P}$ is a load unit, which we may choose to suit our purpose, λ is the associated scalar load parameter, I designates the moment of inertia, and L denotes the column length, and it is indicated that the value of the tangent modulus E_T depends on the strain level and thereby on the stress σ_T at instability. When A is the area of the column cross-section and $r = \sqrt{I/A}$ is its radius of gyration (21.3) provides the *Critical Stress* σ_T

Critical stress σ_T

$$\sigma_T = -\pi^2 \frac{E_T(\sigma_T) I}{A L^2} = -\pi^2 \frac{E_T(\sigma_T)}{(L/r)^2} \tag{21.4}$$

where, L/r is the slenderness ratio.

[21.2] We shall also let λ_T and similar values of the scalar load parameter λ designate the load and thus refer to λ_T itself as the *Tangent Modulus Load*.

Note that E_T is a function of σ and therefore the determination of σ_T and of λ_T must, in general, be done iteratively. However, we may always express λ_T as follows

$$\lambda_T = \frac{E_T}{E_0}\lambda_E \tag{21.5}$$

Tangent modulus load λ_T

where λ_E is the *(Elastic) Euler Load* of the column, given by

$$\lambda_E = \pi^2 \frac{E_0 I}{\bar{P} L^2} \tag{21.6}$$

Euler load λ_E

with the associated *Euler Stress*

$$\sigma_E = -\pi^2 \frac{E_0}{(L/r)^2} \tag{21.7}$$

Euler stress σ_E

21.2.3 Engesser's Second Proposal

Engesser's first proposal was met with criticism in that it was objected that some of the fibers must unload during buckling, see Fig. 21.4. When that is the case, it is no longer justified to assume that all fibers work with the same modulus. As a result of this, Engesser put forth his *Second Proposal.*

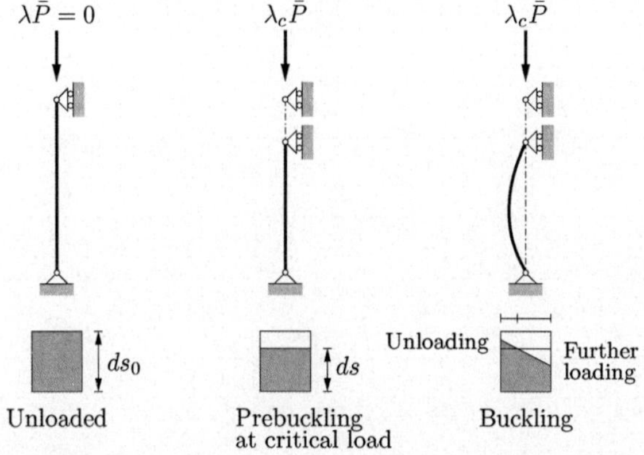

Fig. 21.4: A column and its cross-section in the unloaded state, in prebuckling, and in buckling according to Engesser's second proposal. All displacements are exaggerated.

When κ denotes the bending strain, the additional strain in relation to the prebuckling state at λ_c is

$$\Delta\varepsilon = \kappa y \tag{21.8}$$

The column cross-section is shown in Fig. 21.5, where we note that the fibers on the convex side work with the modulus E_0, while the fibers on the concave side work with E_T. We emphasize that the modulus on the concave

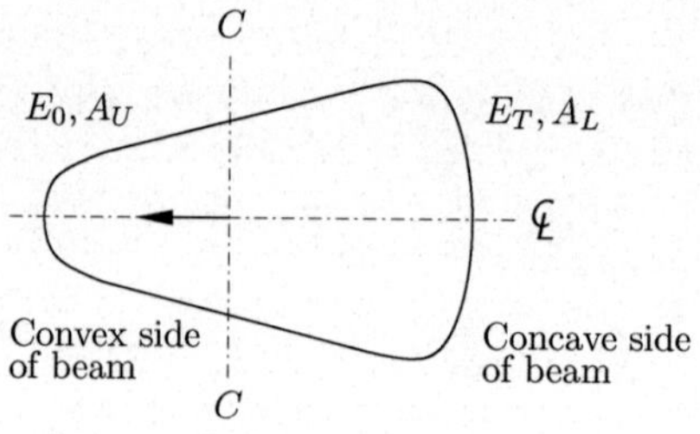

Fig. 21.5: Column cross-section

side is the same for *all* fibers, but depends on the (unknown) critical stress σ_c. The change in load between prebuckling at λ_c and at buckling therefore is

$$\begin{aligned} \Delta\lambda\bar{P} &= \int_{A_U} E_0 \Delta\varepsilon \, dA + \int_{A_L} E_T \Delta\varepsilon \, dA \\ &= \int_{A_U} E_0 \kappa y \, dA + \int_{A_L} E_T \kappa y \, dA \\ &= \kappa \left(E_0 \int_{A_U} y \, dA + E_T \int_{A_L} y \, dA \right) \end{aligned} \tag{21.9}$$

Introduce the static moments of A_U and A_L about the line C-C

$$S_U = \int_{A_U} y \, dA \quad \text{and} \quad S_L = \int_{A_L} y \, dA \tag{21.10}$$

Now, since the load does not change during buckling[21.3]

$$0 = E_0 S_U + E_T S_L \tag{21.11}$$

which determines the position of C-C. The computation must be performed iteratively in most cases.

Introduce the reduced effective Young's Modulus E_R defined by

Reduced Young's modulus E_R

$$M = \kappa E_R I \tag{21.12}$$

where I is the usual moment of inertia.

[21.3] That this is a very crucial—and erroneous–statement was not recognized for about the 50 years that passed until Shanley made his important contribution.

Express the moment M via the stress changes over the cross-section

$$M = \int_{A_U} E_0 \kappa y^2 dA + \int_{A_L} E_T \kappa y^2 dA \tag{21.13}$$

Let I_U and I_L be defined by

$$I_U \equiv \int_{A_U} y^2 dA \quad \text{and} \quad I_L \equiv \int_{A_L} y^2 dA \tag{21.14}$$

to get

$$M = \kappa \left(E_0 I_U + E_T I_L \right) \tag{21.15}$$

Then,

$$E_R = \frac{I_U}{I} E_0 + \frac{I_L}{I} E_T \tag{21.16}$$

Reduced Young's modulus E_R

We can now determine the load-carrying capacity, denoted the *Reduced Modulus Load* $\lambda_R \bar{P}$, according to Engesser's second proposal

$$\lambda_R \bar{P} = -\pi^2 \frac{E_R(\sigma_R) I}{(L/r)^2} \tag{21.17}$$

Reduced Modulus Load $\lambda_R \bar{P}$

and its associated critical stress σ_R

$$\sigma_R = -\pi^2 \frac{E_R(\sigma_R)}{(L/r)^2} \tag{21.18}$$

It is clear from (21.16) that the E_R of Engesser's second proposal is a weighted mean of the values of E_0 and E_T, and since $E_0 \geq E_T$

$$E_0 \geq E_R \geq E_T \;\Rightarrow\; -\sigma_E \geq -\sigma_R \geq -\sigma_T \;\Rightarrow\; \lambda_E \geq \lambda_R \geq \lambda_T \tag{21.19}$$

As is the case for λ_T and σ_T, in most cases the values of λ_R or σ_R can only be found by iteration.

21.2.4 Which Load Is the Critical One?

The question whether σ_T or σ_R, if any of these two, is the correct critical stress of the elastic-plastic Euler Column cannot be answered solely on the basis of the above analyses. We shall, however, compute these two "critical" loads for a specific example because we need these values for a later discussion.

21.2.5 Shanley's Experiments and Observations (1947)

Shanley's experiments

Shanley performed experiments on simply supported aluminum columns with tubular cross-sections. The buckling mode in the plastic range looked very much like the one shown in Fig. 21.6 (a) in that most bending strains occurred over a very short length at mid-span. On the basis of his exper-

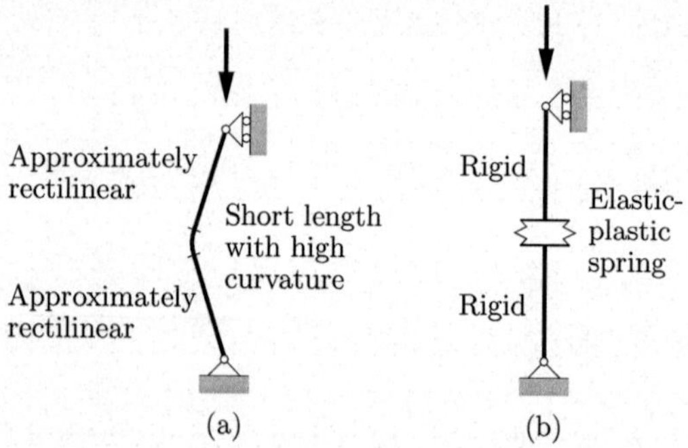

Fig. 21.6: Shanley's experiment and model.

iments Shanley suggested a column model shown in Fig. 21.6 (b), i.e. a structure consisting of two rigid parts, each forming a T, connected by an elastic-plastic element, which models the highly strained middle of the col-

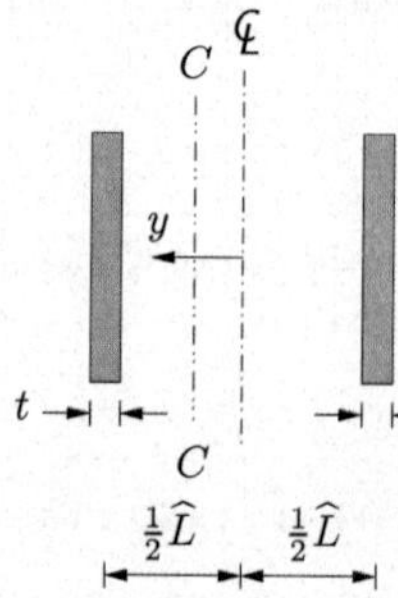

Fig. 21.7: Simple beam cross-section.

umn and is simplified such that the material of the cross-section is assumed to be concentrated in two flanges, see Fig. 21.7. In order that the column does not buckles perpendicular to the y-direction and that the cross-section is thin-walled we demand that

$$B \gg \tfrac{1}{2}\widehat{L} \gg t \tag{21.20}$$

Before we turn to Shanley's analysis we need the *Elastic Euler Load* λ_E, the *Tangent Modulus Load* λ_T, and the *Reduced Modulus Load* λ_R of Shanley's model column.

21.3 The Shanley Model Column

Shanley Model Column

In the spirit of Shanley's own observations, see Section 21.2.5 and in particular Figs. 21.6 and 21.7, he modeled the two flanges as two elastic-plastic springs and neglected the actual distribution of material. Since the columns did not bend much between the supports and the short, plastically deforming section at the middle, Shanley suggested the model shown in Fig. 21.8, which consists of a rigid T loaded by $\lambda\bar{P}$ at the top and supported on the two springs which furnish the forces F_1 and F_2.

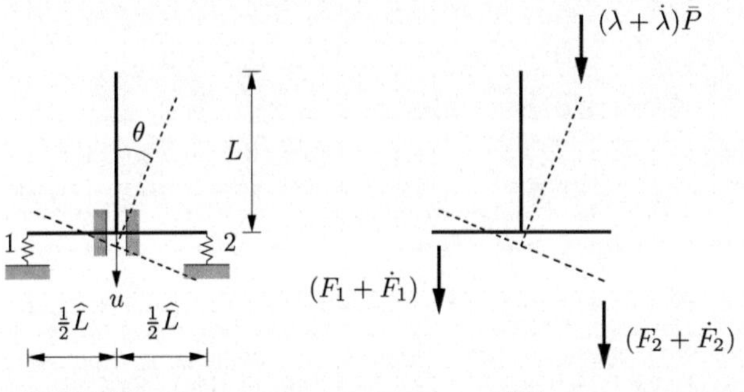

Fig. 21.8: The Shanley model column. Kinematics and statics.

Below, we establish the kinematic, the static, and the constitutive relations for the model column, both in the elastic and in the elastic-plastic regime.

21.3.1 Kinematic Relations

Kinematic relations

The number of degrees of freedom of the model column is clearly 2, and we choose them to be the downward vertical displacement u and the clockwise rotation θ. Then, the elongations of the springs, which we take to be the generalized strains $\varepsilon_\alpha, \alpha = 1, 2$, are

Spring elongations

$$\varepsilon_1 = +\tfrac{1}{2}\,\widehat{L}\sin(\theta) - u \text{ and } \varepsilon_2 = -\tfrac{1}{2}\,\widehat{L}\sin(\theta) - u \tag{21.21}$$

Note that these generalized strains have the dimension length.[21.4]

[21.4] That generalized strains have a dimension is not as uncommon as you might think, for instance curvature strains have the dimension length^{-1}.

For small values of θ, i.e. $|\theta| \ll 1$, this provides

$$\varepsilon_1 = +\tfrac{1}{2}\,\widehat{L}\theta - u \quad \text{and} \quad \varepsilon_2 = -\tfrac{1}{2}\,\widehat{L}\theta - u \tag{21.22}$$

21.3.2 Static Relations

Vertical equilibrium requires

Vertical equilibrium

$$F_1 + F_2 = -\bar{P}\lambda \tag{21.23}$$

and moment equilibrium demands that

Moment equilibrium

$$\tfrac{1}{2}\,(F_1 - F_2)\,\widehat{L} = \lambda \bar{P} L \sin(\theta) \tag{21.24}$$

and for small values of θ

Moment equilibrium for small rotations

$$\tfrac{1}{2}\,(F_1 - F_2)\,\widehat{L} = \lambda \bar{P} L \theta \tag{21.25}$$

21.3.3 Constitutive Relations

Constitutive relations

Since we are dealing with an elastic-plastic material we may only formulate the constitutive relations in terms of increments, not total quantities. Let an over-dot $(\dot{\ })$ denote an infinitesimal increment. Then the stress-strain relation for the springs may be written

$$\dot{F}_\alpha = E_\alpha A \dot{\varepsilon}_\alpha\,, \text{ no sum over } \alpha = 1, 2 \tag{21.26}$$

where it is important to note that the "area" A has dimension length—not length squared—because of the definition of the generalized strains, see (21.21).

For the linearly elastic material model E_α is always equal to E_0, while for an elastic-plastic model it is given by

Tangent moduli E_α

$$E_\alpha = \begin{cases} E_T & \text{for plastic loading} \\ E_0 & \text{otherwise} \end{cases} \tag{21.27}$$

For the results and discussions in the following we shall occasionally employ the simplest strain hardening elastic-plastic material model, namely one that entails linear hardening, see Fig. 21.2(b). Sometimes, the relation between the tangent modulus E_T and the initial modulus E_0 is introduced by the *Hardening Parameter* n

Hardening parameter n

$$E_T = \frac{1}{n}\,E_0 \tag{21.28}$$

where $n = 1 \Rightarrow$ linear elastic material, while $n = \infty \Rightarrow$ linear elastic-perfectly plastic behavior. In our analysis of the *Shanley Model Column*, see Section 21.4, we shall postpone the application of this very special constitutive assumption for as long as possible in order to render the exposition as general as possible.

21.3.4 Elastic Model Column

For the elastic realization of the model column we may compute the *Euler Buckling Load* $\lambda_E \bar{P}$ by use of the idea of neighboring equilibrium states. As usual in current buckling analyses we divide all fields in two parts, namely one which characterizes the *Prebuckling Path*, superscript ^{0}, and another, which is associated with the *Buckling Mode*, superscript ^{1}. As the buckling mode amplitude, the perturbation parameter, we shall use θ. Then, to first order in θ,[21.5] we may assume that the total kinematic and static fields may be written

Euler Buckling Load $\lambda_E \bar{P}$

$$\begin{aligned} u &= \lambda u^0 + u^1\theta \\ \varepsilon_\alpha &= \lambda\varepsilon_\alpha^0 + \varepsilon_\alpha^1\theta\,, \quad \alpha = 1,2 \\ F_\alpha &= \lambda F_\alpha^0 + F_\alpha^1\theta\,, \alpha = 1,2 \end{aligned} \tag{21.29}$$

Total fields

where

$$\lambda = \lambda_E + \lambda^1\theta \tag{21.30}$$

Total load parameter

and λ_E is the *Elastic Euler Buckling Load.*

Euler load λ_E

Prebuckling State

We may easily determine the prebuckling state, where $\theta = 0$

$$F_\alpha^0 = -\tfrac{1}{2}\bar{P} \text{ and } u^0 = \frac{\bar{P}}{2E_0A} \tag{21.31}$$

Prebuckling spring forces
Prebuckling "axial" displacement

Buckling

A small perturbation θ implies the following (small) buckling strain increments

$$\varepsilon_1^1\theta = +\tfrac{1}{2}\widehat{L}\theta \text{ and } \varepsilon_2^1\theta = -\tfrac{1}{2}\widehat{L}\theta \tag{21.32}$$

Buckling strain increments

and the constitutive relation then provides the (small) force increments associated with buckling

$$F_1^1\theta = +\tfrac{1}{2}E_0A\widehat{L}\theta \text{ and } F_2^1\theta = -\tfrac{1}{2}E_0A\widehat{L}\theta \tag{21.33}$$

Buckling force increments

For the *total* state Moment equilibrium (21.25) requires

$$\begin{aligned} &\tfrac{1}{2}\Big(\big((\lambda_E + \lambda^1\theta)F_1^0 + F_1^1\theta\big) - \big((\lambda_E + \lambda^1\theta)F_2^0 + F_2^1\theta\big)\Big)\widehat{L} \\ &= \big(\lambda_E + \lambda^1\theta\big)\bar{P}L\theta \end{aligned} \tag{21.34}$$

[21.5] As usual, in a buckling analysis, which only covers *Buckling* and does not include the *Postbuckling Path*, we may take $O(\theta^2) \ll O(\theta)$.

When we note that $F_1^0 = F_2^0$, see (21.31a), and take advantage of the smallness of θ (21.34) furnishes

$$\tfrac{1}{2}(F_1^1 - F_2^1)\theta\widehat{L} = \lambda_E \bar{P} L\theta \tag{21.35}$$

Introduce (21.33) and find

$$\tfrac{1}{2}E_0 A\widehat{L}^2\theta = \lambda_E \bar{P} L\theta \tag{21.36}$$

and, therefore

$$\lambda_E = \frac{E_0 A\widehat{L}^2}{2\bar{P}L} \tag{21.37}$$

For convenience we define $\bar{P}$ as follows

Load unit $\bar{P}$

$$\bar{P} \equiv \frac{E_0 A\widehat{L}^2}{2L} \tag{21.38}$$

and thus the *Euler Buckling Load* λ_E is

Euler Buckling Load λ_E

$$\lambda_E = 1 \tag{21.39}$$

21.3.5 Tangent Modulus Load

Tangent Modulus Load λ_T

As was the case for the the more general case of Section 21.2.2, we can conclude that we may get the value of the *Tangent Modulus Load* λ_T from the elastic counterpart by substituting E_T for E_0, where, once again, we emphasize that E_T, in general, depends on the stress $|\sigma_T|$. Since we are only concerned with cases for which $|\sigma_T|$ is greater than σ_Y, we get

Tangent Modulus Load λ_T

$$\lambda_T = \frac{E_T}{E_0}\lambda_E \tag{21.40}$$

21.3.6 Reduced Modulus Load

The computation of the *Reduced Modulus Load* λ_R is postponed till Section 21.4.3.

21.4 Shanley's Analysis and Proposal

Shanley's analysis and proposal

Clearly, the kinematic and static relations do not depend on the constitutive model. Therefore, (21.21–21.22) and (21.23–21.25) still hold, while the constitutive relations are now given by (21.26).

21.4.1 Prebuckling

Of course, the equilibrium equations (21.31a) are still correct, while (21.31b), which we do not need here, is no longer valid because we assume that the springs are loaded into the plastic regime.

Prebuckling

21.4.2 Bifurcation

One of the important contributions by Shanley was to realize that the tacit assumption behind Engesser's second proposal, see footnote 21.3 on page 406, is not necessarily valid when the column bifurcates in the plastic regime.

Bifurcation

In Section 21.3.4 superscript [1] denoted increments, i.e. quantities associated with the buckling mode. In the present, elastic-plastic case we indicate increments by an over-dot $(\dot{\ })$ in order to distinguish from the purely elastic case.[21.6]

21.4.2.1 Kinematic Relations

Since it is the strain *increments*, here the difference between the unbuckled and the buckled states, that determine whether a spring unloads or loads further, we need the incremental version of (21.22)

Kinematic relations

$$\dot{\varepsilon}_1 = +\tfrac{1}{2}\,\widehat{L}\dot{\theta} - \dot{u} \quad \text{and} \quad \dot{\varepsilon}_2 = -\tfrac{1}{2}\,\widehat{L}\dot{\theta} - \dot{u} \tag{21.41}$$

Strain increments

21.4.2.2 Static Relations

Similarly, we need the incremental static relations which—by change of notation—follow from (21.23) and (21.25)

Static relations

$$\dot{F}_1 + \dot{F}_2 = -\bar{P}\dot{\lambda}$$

and

$$\tfrac{1}{2}\Big(\big((\lambda_B + \dot{\lambda})F_1^0 + \dot{F}_1\big) - \big((\lambda_B + \dot{\lambda})F_2^0 + \dot{F}_2\big)\Big)\widehat{L} = (\lambda_B + \dot{\lambda})\bar{P}L\dot{\theta} \tag{21.42}$$

Incremental vertical equilibrium and moment equilibrium

where λ_B denotes the *Elastic-Plastic Bifurcation Load*, and where we have not anticipated that all increments are proportional to $\dot{\theta}$. Also in this case we may note that the *Prebifurcation Forces* F_1^0 and F_2^0 are equal with the result that

$$\tfrac{1}{2}(\dot{F}_1 - \dot{F}_2)\widehat{L} = (\lambda_B + \dot{\lambda})\bar{P}L\dot{\theta} \tag{21.43}$$

Moment equilibrium

21.4.2.3 Constitutive Relations

Here, we assume (21.26–21.27) in general and (21.28) for our analysis of the particular example to be valid.

Constitutive relations

[21.6] Since we are only concerned with bifurcation of the column we do not need a way to indicate postbifurcation. If we were to explore postbifurcation the over-dot probably would be an inconvenient indicator.

21.4.3 Reduced Modulus Load

Reduced Modulus Load λ_R

In view of the important assumption behind the computation of the *Reduced Modulus Load* λ_R, namely that the load does not vary immediately at bifurcation, implying that $\dot{\lambda} = 0$, see footnote 21.3 on page 406, the expressions of Section 21.4.2 simplify somewhat. Also in this case, the prebifurcation forces F_1^0 and F_2^0 are equal such that (21.43) holds, and when we assume that spring 1 unloads and spring 2 loads further, we get from (21.42)

$$E_0 A \left(+\tfrac{1}{2}\widehat{L}\dot{\theta} - \dot{u}\right) + E_T^R A \left(-\tfrac{1}{2}\widehat{L}\dot{\theta} - \dot{u}\right) = 0 \tag{21.44}$$

where we have inserted the constitutive relations and by the notation E_T^R signify that the tangent modulus of the compressed spring at the reduced modulus load is different from the tangent modulus E_T at the tangent modulus load. Solving for $\dot{u}$ we get

$$\dot{u} = \frac{E_0 - E_T^R}{2(E_0 + E_T^R)} \widehat{L}\dot{\theta} \tag{21.45}$$

Now, the other equilibrium equation (21.43) yields

$$\tfrac{1}{2}A\widehat{L}\left(\left(E_0 + E_T^R\right)\tfrac{1}{2}\widehat{L}\dot{\theta} - \left(E_0 - E_T^R\right)\dot{u}\right) = \lambda_R \bar{P} L \dot{\theta} \tag{21.46}$$

When we insert $\dot{u}$ from (21.45) in (21.46) the following result may be obtained after some trivial manipulations

Reduced modulus load λ_R

$$\lambda_R = \frac{2E_T^R}{E_0 + E_T^R} \frac{E_0 A \widehat{L}^2}{2\bar{P}L} \tag{21.47}$$

If we assume the bilinear constitutive relation of Fig. 21.2(b), then $E_T^R = E_T$, and (21.45) may be written

$$\dot{u} = \frac{\lambda_E - \lambda_T}{2(\lambda_E + \lambda_T)} \widehat{L}\dot{\theta} \tag{21.48}$$

and (21.47) becomes

$$\lambda_R = \frac{2E_T}{E_0 + E_T} \frac{E_0 A \widehat{L}^2}{2\bar{P}L} \tag{21.49}$$

Reduced modulus load λ_R for bilinear stress-strain relation

with the result that the reduced modulus load is

$$\lambda_R = \frac{2\lambda_T \lambda_E}{\lambda_E + \lambda_T} \tag{21.50}$$

where the definition (21.37) of λ_E has been utilized, but the fact that $\lambda_E = 1$ has not been exploited.

21.4.4 Possible Bifurcations

Because of the nonlinearity of the constitutive relation we must consider 4 possible ways that the column may bifurcate from the vertical unbuckled state, namely

Possible bifurcations

1. Both springs load further, i.e. $\dot{\varepsilon}_1 < 0$ and $\dot{\varepsilon}_2 < 0$,
2. Spring 1 unloads, i.e. $\dot{\varepsilon}_1 > 0$, Spring 2 loads further, i.e. $\dot{\varepsilon}_2 < 0$,
3. Spring 1 loads further, i.e. $\dot{\varepsilon}_1 < 0$, Spring 2 unloads i.e. $\dot{\varepsilon}_2 > 0$,
4. Both springs unload, i.e. $\dot{\varepsilon}_1 > 0$ and $\dot{\varepsilon}_2 > 0$.

where it is obvious that 2 and 3 are equivalent because of symmetry.

21.4.4.1 Both Springs Load Further

Both springs work with E_T. Then, from the constitutive relations (21.26) with the kinematic relations (21.41) inserted

Both springs load further

$$\begin{aligned} \dot{F}_1 &= \left(+\tfrac{1}{2}\widehat{L}\dot{\theta} - \dot{u}\right) AE_T^R \\ \dot{F}_2 &= \left(-\tfrac{1}{2}\widehat{L}\dot{\theta} - \dot{u}\right) AE_T^R \end{aligned} \tag{21.51}$$

Plug these expressions into the equilibrium equation (21.42) with the outcome that

$$-2E_T^R A\dot{u} = -\bar{P}\dot{\lambda} \tag{21.52}$$

and into the equilibrium equation (21.43) to get

$$\tfrac{1}{2}\widehat{L}^2 E_T^R A\dot{\theta} = (\lambda_B + \dot{\lambda})\bar{P}L\dot{\theta} \tag{21.53}$$

or

$$\dot{\theta}\left((\lambda_B + \dot{\lambda})\bar{P}L - \tfrac{1}{2}\widehat{L}^2 E_T^R A\right) = 0 \tag{21.54}$$

We wish to find solutions to (21.52) together with (21.54) for small increments. A trivial solution is

$$(\dot{\theta}, \dot{u}) = \left(0, \frac{\bar{P}}{2E_T^R A}\dot{\lambda}\right) \Longrightarrow \text{No bifurcation} \tag{21.55}$$

Another solution that follows from (21.54) is

$$\dot{\theta} = \text{anything and } \lambda_B + \dot{\lambda} = \frac{E_T^R A\widehat{L}^2}{2\bar{P}L} \tag{21.56}$$

meaning that

$$\lambda_B = \frac{E_T^R A\widehat{L}^2}{2\bar{P}L} \text{ and } \dot{\lambda} = 0 \tag{21.57}$$

but according to (21.52) $(\dot{\lambda} = 0) \Rightarrow (\dot{u} = 0)$. However, under these circumstances, the only displacement component, which may not vanish, is $\dot{\theta}$, but if $\dot{\theta} \neq 0$ one spring unloads in contradiction with our initial assumption. Therefore, only the trivial solution is possible, and both springs do not load further at bifurcation.

21.4.4.2 Both Springs Unload

Both springs unload

This possibility leads to the same equations as the former, except that E_0 is substituted for E_T. Therefore, also in this case, bifurcation is not possible.

21.4.4.3 Spring 1 Unloads, Spring 2 Loads Further

Spring 1 Unloads, Spring 2 Loads Further

Again, as in Section 21.4.4.1 we utilize the constitutive relations (21.26) with the kinematic relations (21.41) inserted and get

$$\dot{F}_1 = \left(+\tfrac{1}{2}\widehat{L}\dot{\theta} - \dot{u}\right) AE_0 \quad\text{and}\quad \dot{F}_2 = \left(-\tfrac{1}{2}\widehat{L}\dot{\theta} - \dot{u}\right) AE_T^R \tag{21.58}$$

where the occurrence of E_0 is due to the fact that spring 1 unloads elastically. Now, the static relations (21.42) and (21.43) provide

$$(E_0 - E_T^R)\tfrac{1}{2}\widehat{L}A\dot{\theta} - (E_0 + E_T^R)A\dot{u} = -\bar{P}\dot{\lambda} \tag{21.59}$$

and

$$(E_0 + E_T^R)\left(\tfrac{1}{2}\widehat{L}\right)^2 A\dot{\theta} - (E_0 - E_T^R)\tfrac{1}{2}\widehat{L}A\dot{u} = (\lambda_B + \dot{\lambda})\bar{P}L\dot{\theta} \tag{21.60}$$

respectively. For convenience introduce

Normalized bifurcation mode

$$\dot{v} \equiv 2\frac{L}{\widehat{L}^2}\,\dot{u} \quad\text{and}\quad \dot{\psi} \equiv \frac{L}{\widehat{L}}\,\dot{\theta} \tag{21.61}$$

Assume bilinear stress-strain relation

and utilize the definition (21.37) of λ_E, but do not exploit the fact that $\lambda_E = 1$. In order to proceed and get explicit results we assume the bilinear constitutive model of Fig. 21.2 (b) valid and recall that then $E_T^R = E_T$. The relation (21.5) between λ_T and λ_E may now be used to rewrite (21.59)

$$(\lambda_E - \lambda_T)\dot{\psi} - (\lambda_E + \lambda_T)\dot{v} = -\dot{\lambda} \tag{21.62}$$

In the same fashion (21.60) now becomes

$$(\lambda_E + \lambda_T)\dot{\psi} - (\lambda_E - \lambda_T)\dot{v} = 2(\lambda_B + \dot{\lambda})\dot{\psi} \tag{21.63}$$

or, noting that $|\dot{\lambda}| \ll \lambda_B$

$$\big((\lambda_E + \lambda_T) - 2\lambda_B\big)\dot{\psi} = (\lambda_E - \lambda_T)\dot{v} \tag{21.64}$$

The solution to the system of equations which governs the present case, i.e. (21.62) together with (21.64), is

$$\begin{aligned} \dot{\psi} &= \frac{(\lambda_E - \lambda_T)}{2(2\lambda_E\lambda_T - \lambda_B\lambda_E - \lambda_B\lambda_T)}\,\dot{\lambda} \\ \dot{v} &= \frac{(\lambda_E - 2\lambda_B + \lambda_T)}{2(2\lambda_E\lambda_T - \lambda_B\lambda_E - \lambda_B\lambda_T)}\,\dot{\lambda} \end{aligned} \tag{21.65}$$

For the sake of the discussion below we introduce the value of the Reduced Modulus Load λ_R given by (21.50).

The result is

$$\dot{\psi} = \frac{(\lambda_E - \lambda_T)}{2(\lambda_E + \lambda_T)(\lambda_R - \lambda_B)} \dot{\lambda} \text{ and } \dot{v} = \frac{(\lambda_E + \lambda_T) - 2\lambda_B}{2(\lambda_E + \lambda_T)(\lambda_R - \lambda_B)} \dot{\lambda} \qquad (21.66)$$

Normalized bifurcation mode

Before we discuss this solution in detail we emphasize that the assumption behind (21.66) is

$$(\dot{\varepsilon}_1 \geq 0) \wedge (\dot{\varepsilon}_2 \leq 0) \qquad (21.67)$$

Assumption behind this case

Recall the kinematic relations (21.41), but for convenience introduce

$$\dot{\hat{\varepsilon}}_\alpha \equiv 2\frac{L}{\hat{L}^2}\dot{\varepsilon}_\alpha \qquad (21.68)$$

and thus

$$\dot{\hat{\varepsilon}}_1 = +\dot{\psi} - \dot{v} \text{ and } \dot{\hat{\varepsilon}}_2 = -\dot{\psi} - \dot{v} \qquad (21.69)$$

and therefore

$$\dot{\hat{\varepsilon}}_1 = \frac{(\lambda_B - \lambda_T)}{(\lambda_B - \lambda_R)} \frac{\dot{\lambda}}{(\lambda_E + \lambda_T)} \text{ and } \dot{\hat{\varepsilon}}_2 = \frac{(\lambda_B - \lambda_E)}{(\lambda_B - \lambda_R)} \frac{\dot{\lambda}}{(\lambda_E + \lambda_T)} \qquad (21.70)$$

Strain increments

The inequalities (21.67) for $\dot{\varepsilon}_\alpha$ result in the requirements

$$\frac{(\lambda_B - \lambda_T)}{(\lambda_B - \lambda_R)} \frac{\dot{\lambda}}{(\lambda_E + \lambda_T)} \leq 0 \text{ and } \frac{(\lambda_B - \lambda_E)}{(\lambda_B - \lambda_R)} \frac{\dot{\lambda}}{(\lambda_E + \lambda_T)} \geq 0 \qquad (21.71)$$

Recall that

$$0 < \lambda_T < \lambda_R < \lambda_E < \infty \qquad (21.72)$$

Four intervals to investigate

and there are now 4 intervals to investigate.

First case: $0 < \lambda_B < \lambda_T$

First case

Here (21.71a) gives $\dot{\lambda} \leq 0$, but (21.71b) gives $\dot{\lambda} \geq 0$ and, therefore

$$\dot{\lambda} = 0 \qquad (21.73)$$

which in itself is all right, but (21.66) then results in

$$\dot{\psi} = 0 \text{ and } \dot{v} = 0 \qquad (21.74)$$

which means that the situation is unaltered, i.e. no transverse displacements occur, and the *structure does not experience bifurcation buckling.*

Bifurcation not possible

Second case $\lambda_T < \lambda_B < \lambda_R$

Second case

Now (21.71a) requires $\dot{\lambda} \geq 0$ and so does (21.71b). From (21.66) we see that both $\dot{\psi}$ and $\dot{v}$ are nonzero. Thus, *bifurcation is possible*, and the buckling

Bifurcation possible

mode not only has a transverse component, but also an axial component. This is a feature that is unique to *plastic* columns and similar structures.

Third case

Third case $\lambda_R < \lambda_B < \lambda_E$

Now (21.71a) enforces $\dot{\lambda} \leq 0$ and (21.71b) does the same. Again, $\dot{\lambda} \neq 0$ gives nonzero values of $\dot{\psi}$ and $\dot{v}$, and *bifurcation is possible.*

Bifurcation possible

Fourth case

Fourth case $\lambda_E < \lambda_B < \infty$

Still (21.71a) gives $\dot{\lambda} \leq 0$, but (21.71b) requires $\dot{\lambda} \geq 0$ and we are back to a situation that is similar to the first case with the result that *bifurcation is not possible.*

Bifurcation not possible

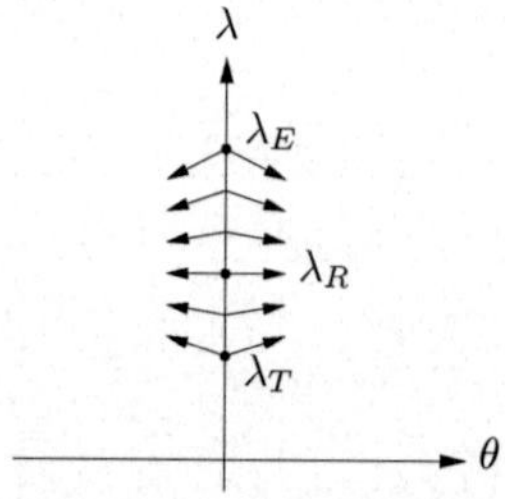

Fig. 21.9: Bifurcations of the Shanley Model Column.

Thus, to sum up:	$0 \leq \lambda \leq \lambda_T$:	No bifurcation.
	$\lambda_T \leq \lambda \leq \lambda_R$:	Bifurcation is possible for *increasing load.*
	$\lambda_R \leq \lambda \leq \lambda_E$:	Bifurcation is possible for *decreasing load.*
	$\lambda_E \leq \lambda \leq \infty$:	No bifurcation.

Although we have only been concerned with bifurcation—not with post-bifurcation, we are justified when we make the plot in Fig. 21.9.

Plastic bifurcation is possible over an interval—not only at discrete values of λ

We can see that, contrary to the elastic case, bifurcation is possible over an interval of λ, not just at discrete values. This is very remarkable since the present structure which only has two degrees of freedom meaning that its elastic counterpart only has one possible bifurcation mode.

Note that at λ_R the bifurcation takes place under *constant* load. This is consistent with the assumption behind Engesser's Second Proposal. Therefore, Engesser's Second Proposal was based on an assumption that is too restrictive for elastic-plastic buckling.

Which is the correct value of λ_B?

The remaining question now is: which is the correct value of λ_B at buckling in the sense that it is associated with the real load-carrying capacity? It seems fairly obvious that the interval $\lambda_R < \lambda_B < \lambda_E$ must be excluded because here bifurcation takes place under *decreasing load* with the result that λ_B in this interval always is too large. On the other hand, it may seem too pessimistic to pick λ_T as the buckling load, especially because *all* post-buckling curves—at least for the linear strain hardening model—may be shown to have $\lambda = \lambda_R$ as an asymptote. However, θ increases very rapidly even for a small increase in λ for the equilibrium paths that emanate from $\lambda_B \gtrsim \lambda_T$. Furthermore, an analysis including geometric imperfections provides a plot that looks like the one in Fig. 21.10. This plot indicates that the

Is λ_T too pessimistic?

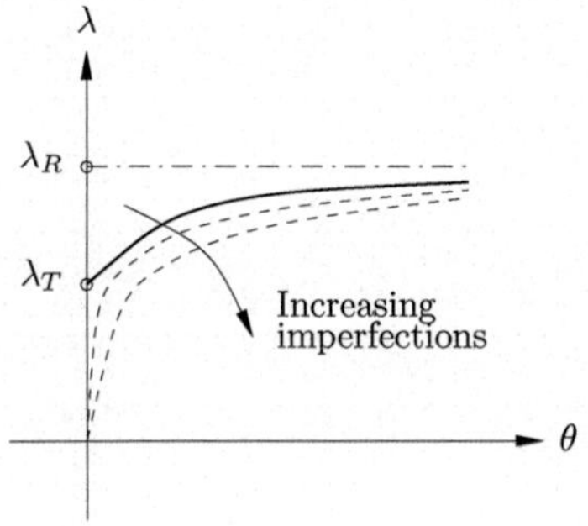

Fig. 21.10: Equilibrium paths of geometrically imperfect Shanley Model Columns.

displacements of the geometrically imperfect column increase very rapidly for load levels around λ_T.

Take λ_T as the real value of λ_B

Finally, for some cases, see (Duberg & Wilder 1952), the equilibrium paths for the imperfect structure do not have λ_R as an asymptote, but experience a maximum before that.

Shanley's conclusion (he did not know the work by Duberg & Wilder (1952) simply because it was done 3 years later than his own) was that λ_T is the best estimate we can get for the buckling load in the plastic range. His computation was, admittedly, only valid for a model column, but later studies support his conclusion. The situation may even be worse than predicted by Shanley's model in that elastic-plastic structures are often extremely sensitive to geometric imperfections, see e.g. (Byskov 1982–83) for a simple example that supports this statement.

Shanley's conclusion: The tangent modulus load λ_T is the best measure of a plastic critical load

As groundbreaking as it was, for many years Shanley's work did not receive the attention it deserved—at least it appears that few engineers outside academia knew about it.

It seems fair to state that the next breakthrough in theories for elastic-plastic buckling was due to Hutchinson, see (Hutchinson 1972), (Hutchinson 1973b), (Hutchinson 1973a), and the paper which sums up most of his work in this area, namely (Hutchinson 1974b). In our present context it is worth mentioning the model column which now often is referred to as the "Shanley-Hutchinson Model Column," see (Hutchinson 1974b),[21.7] which is similar to the model column of Shanley, except that the model entails a continuous set of springs instead of the two discrete ones shown in Fig. 21.8. Based on his model column Hutchinson developed an asymptotic theory of postbifurcation and imperfection sensitivity in the elastic-plastic regime. Christensen & Byskov (2008) reconsidered Hutchinson's asymptotic expansion and, by use of other terms in the expansion increased its validity significantly.

[21.7] Hutchinson did not give that name to the column.

Part V

Introduction to the Finite Element Method

Chapter 22

About the Finite Element Method

Finite Element Method, FEM

Among the vast number of numerical tools that can be used to solve boundary-value or eigenvalue problems[22.1] the *Finite Element Method*, usually abbreviated *FEM*, has been the most popular over the past four to five decades. This fact alone justifies that the method is introduced here, but more important to me is that the derivation of the method involves many of the basic concepts and topics of continuum mechanics, which may make the usefulness of learning continuum mechanics even more obvious—at least this is my hope.

The *Finite Element Method* comes in many different flavors which, to some extent, depend on the application, but also on the taste of the author. Like the rest of this book, the present Part V is concerned with continuum mechanics for solids and structures, covering specialized continua, namely beams, frames and plates. Therefore, we shall refrain from discussing other fields, such as fluid mechanics and heat flow.[22.2] Furthermore, dynamic problems are also left out because this introduction is meant to be short.

In connection with teaching the Finite Element Method I prefer a variational formulation, either the *Principle of Virtual Work* or *Stationarity of the Potential Energy* over the many other possibilities, such as the so-called *Weak Form*, the *"Direct Formulation,"* and the *Galerkin Formulation*, see e.g. (Ottosen & Petersson 1992). The reason that I like both variational formulations as a basis for establishing the Finite Element Method for problems in structural or solid mechanics is that they entail terms that may be interpreted in a physical way which appeals very much to me. The "Direct Formulation," is, as indicated by its name, also intuitively clear, but of much more narrow applicability.

As you will see, my approach to the finite element method is that of an engineer rather than a mathematician. There are at least two good reasons

[22.1] After some nudging, the *Finite Element Method* may also be used to solve *initial-value problems*.
[22.2] I am far from being an expert in these fields, although I have taught the finite element method for heat flow.

E. Byskov, *Elementary Continuum Mechanics for Everyone*,
Solid Mechanics and Its Applications 194, DOI: 10.1007/978-94-007-5766-0_22,

for this, one is that I am an engineer, the other is that the mathematicians' way of introducing the method seems too abstract to me. I am sure that beginning by talking about whether a solution lies in a *Hilbert*, *Banach*, or *Sobolev* space, see e.g. (Strang & Fix 1973), makes little sense to an engineering student.[22.3] On the other hand, for purposes of proving convergence, etc., the mathematical procedure has proved to be extremely fruitful.

In the interest of brevity, this Part is only concerned with the most fundamental problems of finite element analysis,[22.4] first of all, formulation of simple finite elements for bars, beams, and plates subjected to in-plane loading. But more advanced problems are also treated, e.g. the so-called locking phenomenon, while many other interesting and important problems, such as the ones associated with deriving plate bending elements, elements with curved edges and so forth will not be treated. There are many good books which cover these topics, e.g. (Bathe 1996), (Cook et al. 2002), (Zienkiewicz & Taylor 2000a), or (Zienkiewicz & Taylor 2000b).

[22.3] Some teachers may not share this view. For instance, one of my former colleagues once stated that the right way to introduce the principle of virtual displacements to sophomores was to explain that it was nothing more than a projection of the equilibrium equation on a subspace.

Needless to say, he has never been nominated for "teacher of the year."

[22.4] This means: the problems that *I* consider the most fundamental. Others may disagree.

Chapter 23

An Introductory Example in Several Parts

The history of the development of the *Finite Element Method* may be enlightening, but as an introduction to the subject a simple example may be more instructive. Below, we shall treat a beam problem in great detail and use it as our basis for "inventing" the method.

Since the example is rather long, I have broken it up by comments, set in a *slanted* type, which summarize the previous manipulations. Especially in Examples 23.3—23.6 you may not need to go through all derivations, but be sure to understand why particular procedures work or do not work.

The example below is very long—do not read all the details, unless you wish to check something.

The structure shown in Fig. 23.1 consists of a kinematically linear Bernoulli-Euler beam ABC and two springs, which support the beam at points B and C. At point C the structure is loaded by an upward force $\bar{P}$. For our

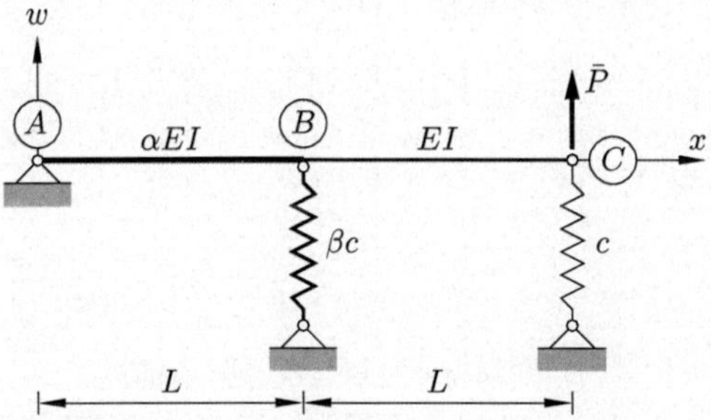

Fig. 23.1: Beam supported on two springs.

present purpose, let us acknowledge the fact that for the load $\bar{P}$ the beam only experiences transverse displacements $w(x)$, where x is the coordinate along the beam axis, w is positive upwards, and x is positive from left to right. The stiffnesses of the beams and springs are given in the figure. In the following the values of α and β are used to change the properties of the

E. Byskov, *Elementary Continuum Mechanics for Everyone*,
Solid Mechanics and Its Applications 194, DOI: 10.1007/978-94-007-5766-0_23,

structure in ways that are used to elucidate various important conditions for the solution, in particular which conditions must and must not be fulfilled by an approximate or numerical solution.

23.1 Generalized Quantities and Potential Energy

Generalized strains and stresses

In Section 2.4.2 the terms generalized strains and stresses are defined, see also Section 33.2, but let us mention that these strain and stress terms are the ones that produce (virtual) work in the *Principle of Virtual Displacements* or other versions of the *Principle of Virtual Work.*

For the structure shown in Fig. 23.1 note

1. The generalized strains of the beams and of the springs are the curvature strain κ and the elongation $\Delta\ell_S$, respectively.

Strain-displacement relations

2. The strain-displacement relations are

$$\kappa = w'' , \;\; \Delta\ell_B = w_B , \;\; \Delta\ell_C = w_C \tag{23.1}$$

where $\Delta\ell_B$ and $\Delta\ell_C$ are the elongation of the springs, and prime ($'$) denotes differentiation with respect to the physical coordinate x.

3. The beam is assumed to be linear elastic with the bending stiffness αEI in AB and EI in BC. Both springs are taken to perform linearly elastic with the stiffness βc at B and c at C, see Fig. 23.1. The value of c is given as

Spring constant c

$$c = \frac{EI}{L^3} \tag{23.2}$$

4. Although the generalized stresses do not enter in the *Potential Energy* we may note that they are the bending moment M in the beam and the forces N_B and N_C in the springs.

5. The *Potential Energy* $\mathbf{\Pi}_\mathrm{P}$ of the structure is

Potential energy $\mathbf{\Pi}_\mathrm{P}(w)$

$$\begin{aligned}\mathbf{\Pi}_\mathrm{P}(w) = &+\tfrac{1}{2}\int_A^B \alpha EI(\kappa(w))^2 dx + \tfrac{1}{2}\int_B^C EI(\kappa(w))^2 dx \\ &+\tfrac{1}{2}\beta c(\Delta\ell_B)^2 + \tfrac{1}{2}c(\Delta\ell_C)^2 - \bar{P}\Delta\ell_C\end{aligned} \tag{23.3}$$

6. Before we apply the principle of stationarity of the *Potential Energy* $\mathbf{\Pi}_\mathrm{P}$ we must identify the conditions, including boundary and continuity conditions that must be satisfied by an assumed displacement field and its variation. Both the transverse displacement w and the variation δw must be once differentiable (the slope, as well as the displacement itself, must be continuous everywhere) and satisfy the *kinematic* boundary conditions

Kinematic boundary conditions

$$w(A) = 0 \;\text{ and }\; \delta w(A) = 0 \tag{23.4}$$

They must also satisfy the *kinematic* continuity conditions

$$\begin{aligned} w'_{B-} &= w'_{B+}\,, \quad \delta w'_{B-} = \delta w'_{B+} \\ w_B &= \Delta\ell_B\,, \quad w_C = \Delta\ell_C \end{aligned} \tag{23.5}$$

Kinematic continuity conditions

Note that a similar continuity condition is not valid for w'', $\delta w''$ and higher derivatives. This is a general statement for Bernoulli-Euler beams, but in the present case it is obvious that, although the bending *moment* M is continuous at B, the bending *strain* κ must be discontinuous there if $\alpha \neq 1$ meaning that there is a jump in the bending stiffness at point B. Even in the case of $\alpha = 1$ the third derivative w''' of w is discontinuous at B because of the force in the spring at point B, which we may see from the definition (7.61)

No kinematic continuity conditions on higher derivatives of w than w'

$$V \equiv -M' \tag{23.6}$$

Definition of shear force V

of the shear force V combined with the constitutive relation (7.86b)

$$M = EI\kappa \tag{23.7}$$

Jump in EI $\Rightarrow$ jump in κ

in connection with the strain-displacement relation (7.14) or (7.88)

$$\kappa = w'' \tag{23.8}$$

Jump in κ $=$ jump in w''

and

$$V = -(EIw'')' \tag{23.9}$$

For $EI = $ const. this finally provides

$$w''' = -\frac{V}{EI} \tag{23.10}$$

Jump in V $\Rightarrow$ jump in w'''

which shows that a jump in the shear force V implies a jump in the third derivative of w, unless the bending stiffness EI has an equivalent jump.

The above comments about the continuity and discontinuity of various strain and stress quantities are extremely important in connection with establishing a finite element method. As we shall see below, if we choose displacement fields which are too continuous we must expect solutions which converge very slowly. In the worst cases, a solution may very well converge in an overall sense, but still be completely wrong at a particular point, see e.g. Figs. 23.3 and 23.6. Most of the finite elements that we derive below are based on the principle of minimum of the potential energy $\mathbf{\Pi}_\mathrm{P}$. Only generalized strains appear in $\mathbf{\Pi}_\mathrm{P}$ and, in general, we must therefore not demand continuity of the generalized stresses. Otherwise we have constructed an "over-compatible" method and are almost certain to experience the problems mentioned above.

Too continuous displacement assumptions cause slow convergence at best.
Avoid such "over-compatible" displacement assumptions

23.2 The Exact Solution

In order to assess the accuracy of the approximate results we find below we note some of the quantities of the exact solution to the present beam problem, when $\beta = 1$

α	1	3
w_B	$\frac{2}{9}\frac{\bar{P}L^3}{EI}$	$\frac{6}{25}\frac{\bar{P}L^3}{EI}$
w'_B	$\frac{1}{3}\frac{\bar{P}L^2}{EI}$	$\frac{7}{25}\frac{\bar{P}L^2}{EI}$
w_C	$\frac{2}{3}\frac{\bar{P}L^3}{EI}$	$\frac{16}{25}\frac{\bar{P}L^3}{EI}$
w'_C	$\frac{1}{2}\frac{\bar{P}L^2}{EI}$	$\frac{23}{50}\frac{\bar{P}L^2}{EI}$
N_B	$\frac{2}{3}\bar{P}$	$\frac{18}{25}\bar{P}$
N_C	$\frac{2}{3}\bar{P}$	$\frac{16}{25}\bar{P}$
M_B	$\frac{1}{3}\bar{P}L$	$\frac{9}{25}\bar{P}L$

(23.11)

23.3 The Simplest Approximation

Let us choose a displacement field $(\widetilde{w}, \widetilde{\Delta\ell}_S)$, which is as simple as possible and use it in the potential energy. The simplest field that satisfies all kinematic conditions, which are given above, is

Simplest displacement assumption

$$\widetilde{w}(\xi) = \tfrac{1}{2}a\xi \text{ with } \xi \equiv \frac{x}{L} \text{ and } a = \widetilde{w}_C = \widetilde{\Delta\ell}_C \tag{23.12}$$

where a is to be determined by requiring that the variation of the potential energy $\mathbf{\Pi}_\mathrm{P}$ vanishes, and where the factor $\frac{1}{2}$ is introduced for convenience.

With $\widetilde{w}$ as given by (23.12) we compute the associated curvature strain $\tilde{\kappa}$ and extension of the spring $\widetilde{\Delta\ell}_B$

Generalized strains

$$\tilde{\kappa} = \widetilde{w}'' = 0 \text{ and } \widetilde{\Delta\ell}_B = \tfrac{1}{2}a \tag{23.13}$$

which is not a good approximation to the distribution of the bending strain κ because the bending moment M varies linearly in AB and in BC and vanishes at A and C, which means that the bending strain cannot vanish over the entire beam. So, we must expect a poor quality of the results based on the choice (23.12).

The approximate field $\widetilde{w}$ given by (23.12) furnishes the approximate potential energy

$$\begin{aligned}\mathbf{\Pi}_{\mathrm{P}}(\widetilde{w}) = \mathbf{\Pi}_{\mathrm{P}}(a) &= \tfrac{1}{2}\beta c\left(\tfrac{1}{2}a\right)^2 + \tfrac{1}{2}ca^2 - \bar{P}a \\ &= \tfrac{1}{2}\left(1+\tfrac{1}{4}\beta\right)\frac{EI}{L^3}a^2 - \bar{P}a\end{aligned} \tag{23.14}$$

Approximate potential energy $\mathbf{\Pi}_{\mathrm{P}}(\widetilde{w})$

with the variation

$$\delta\mathbf{\Pi}_{\mathrm{P}}(a) = \delta a\left(\left(1+\frac{1}{4}\beta\right)\frac{EI}{L^3}a - \bar{P}\right) = 0\ \forall\,\delta a \tag{23.15}$$

Variation of approximate potential energy, $\delta\mathbf{\Pi}_{\mathrm{P}}(a)$

The approximate solution is then

$$\begin{aligned}&\widetilde{w}_C = a = \frac{1}{1+\frac{1}{4}\beta}\frac{\bar{P}L^3}{EI} = \frac{4}{7}\frac{\bar{P}L^3}{EI} \quad \text{for } \beta = 3 \\ &\widetilde{M} = 0 \quad \text{independent of the value of } \beta\end{aligned} \tag{23.16}$$

Approximate solution

The relative error on $\widetilde{w}_C$, which is 14% for $\alpha = 1$ and 11% for $\alpha = 3$, is too large for any purpose, and the error on the bending moment $\widetilde{M}$ is completely unsatisfactory. Note that the absolute value of the "characteristic" displacement $\widetilde{w}_C$, i.e. the displacement in the direction of the load $\bar{P}$, is smaller than the exact value, see Section 33.4.3.

"Characteristic" displacement $\widetilde{w}_C < w_{\text{exact}}$

23.4 More Terms?

Already when we saw the result (23.13) we might have looked for a better field than the one given in (23.12). The question now is how to find an improved field. One possibility is to proceed along the same lines as in Example Ex 32-1 and Example Ex 32-4.2 and add more terms that are kinematically admissible and proceed along the same lines as above. A possible set of functions is

$$\widetilde{w}(\xi) = \left(\frac{\xi}{2}\right)a_1 + \left(\frac{\xi}{2}\right)^2 a_2 + \cdots + \left(\frac{\xi}{2}\right)^j a_j + \cdots \tag{23.17}$$

More terms

where the factor $\frac{1}{2}$ is introduced for convenience. We shall not go into details of this computation, but merely cite some results below.

23.4.1 The case $\alpha = 3$

The case $\alpha = 3$ presents some additional difficulties in relation to the case $\alpha = 1$, and I wish to address these first. The variation of the tip deflection

w_C, see Fig. 23.2, as a function of the number of terms shows that, although the value of w_C indeed converges, the rate is so slow that this way of improving the results must be considered far from optimal.[23.1] Usually, we would

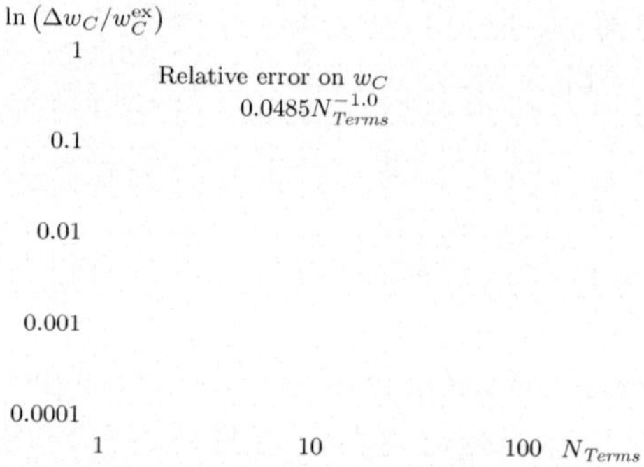

Fig. 23.2: Relative error on w_c for $\alpha = 3$. Number of terms: N_{Terms}.

The displacement assumption is too continuous causing low convergence

expect a much better rate of convergence than 1, but the present problem is bothered by the fact that the moment of inertia has a jump at point B, and thus the second derivative of the exact displacement is discontinuous at B, although the bending moment M is continuous. If the moment of inertia varies in a smooth fashion or is constant the rate of convergence is much higher. All assumed fields given by (23.17) have derivatives that are continuous everywhere, and one consequence of this is that the approximate bending moment has a jump at B, see Fig. 23.3. So, in a way, you might say that the fields of the approximate solution behave in the opposite way of those of the exact solution in that the jump in w''_{exact} is exchanged with a jump in $\widetilde{M}$.

In some cases, exploit that beams are internally statically determinate, but not here

One way of improving the values of the bending moment is to exploit the fact that a beam is *internally* statically determinate. Once we know the value of w_C we can compute the force in the spring at point C, and then a better value of M_B can be determined by equilibrium of BC. It is, however, a peculiarity of beams that they are internally statically determinate. A similar trick does not work for solids, plates and shells and may therefore be considered illegal—at least not part of a proper finite element procedure. It is clear from Fig. 23.3 that, even with 60 terms, which is a much greater number of terms than reasonable for this simple problem, the approximate

[23.1] As regards the rate of convergence, given by the exponent on the number of terms N, which is here found by experimentation with the plotting program to be -1.0, see Strang & Fix (1973) who discuss the topic of rate of convergence of finite elements in detail.

bending moment $\widetilde{M}$ is unacceptable, and it is especially unfortunate that the error on the maximum moment is the largest. Note, however, that $\widetilde{M}$ tries to approximate the exact triangular M-curve. Thus, for a very large number of terms we may expect that the $\widetilde{M}$-curve lies close to the exact one,

$\widetilde{M}$ converges very slowly

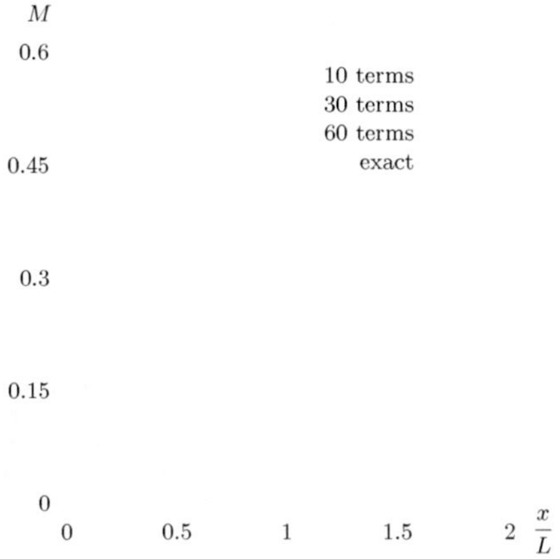

Fig. 23.3: Approximate and exact bending moment M for $\alpha = 3$.

except in a small neighborhood of point B. The accuracy of displacements is better, but not impressive. We must therefore try other possibilities if we wish to base our solution on the stationarity of the potential energy—or the *Principle of Virtual Displacements*.

23.4.2 The case $\alpha = 1$

For the case of $\alpha = 1$ the discontinuity in EI is absent and the rate of convergence of $\widetilde{w}_C$ is much better than for the case of $\alpha = 3$, see Fig. 23.4, in that the rate of convergence is now given by the exponent -3.0 instead of -1.0.

The behavior of the approximate bending moment $\widetilde{M}$ is also much better, see Fig. 23.5, but the shear force $\widetilde{V}$ is continuous at point B, while V has a jump at B equal to the spring force N_B, see Fig. 23.6, which shows that the number of terms needed to acquire a decent accuracy on $\widetilde{V}$ is excessive.

So, it seems that we ought to search for other types of fields.

Comment: Other Possibilities

One possibility is to combine the terms of (23.17) such that the new terms satisfy the static boundary conditions, also. On the other hand, whether we choose to do this or employ the functions given by (23.17) the approximating

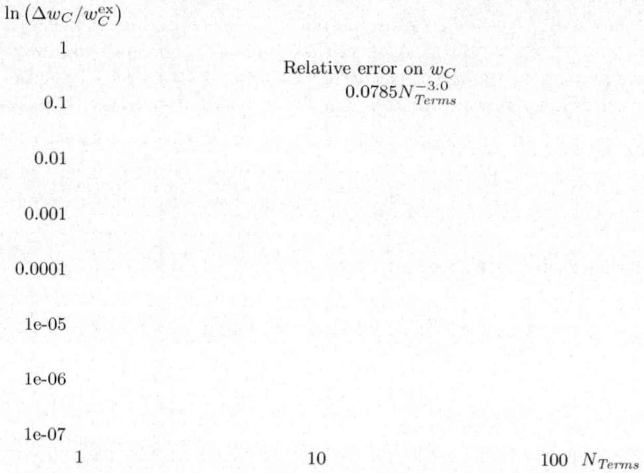

Fig. 23.4: Relative error on w_c for $\alpha = 1$. Number of terms: N_{Terms}.

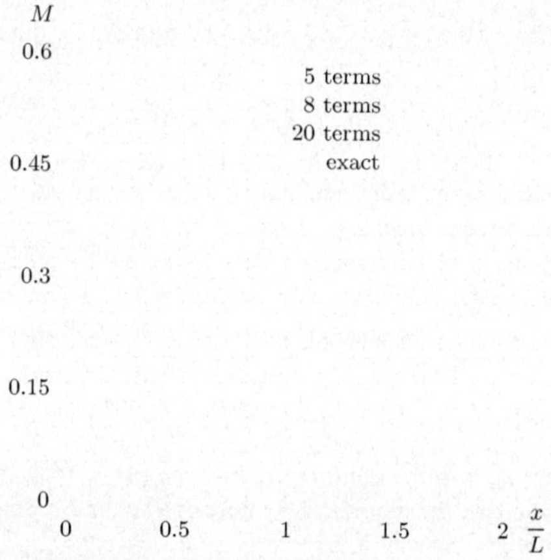

Fig. 23.5: Approximate and exact bending moment M for $\alpha = 1$.

fields all suffer from the same severe deficiency, namely that all terms are too differentiable, or over-compatible, in that their second and higher derivatives are continuous at point B. Indeed, the procedure will work, but many terms will be wasted on the problems at B. So, the obvious idea is to utilize a

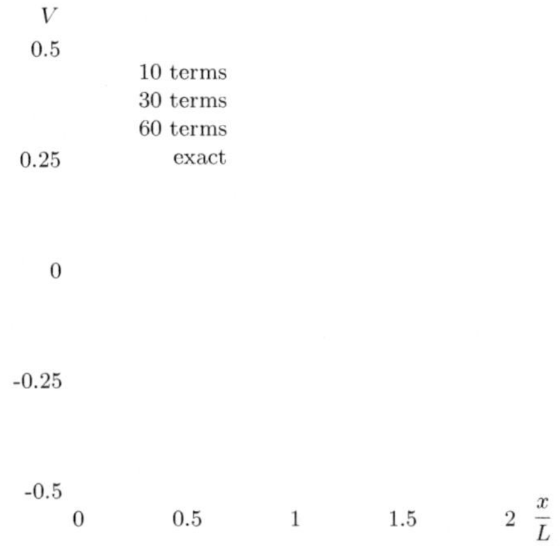

Fig. 23.6: Approximate and exact shear force V for $\alpha = 1$.

set of functions that do not exhibit over-compatibility, which is one of the central features of the Finite Element Method.[23.2]

Conclusion: Establish fields which are sufficiently continuous over proper intervals, the finite elements, and are sufficiently discontinuous between them.

In view of the limited success of the fields utilized in Section 23.4 above we shall choose fields that are differentiable over certain intervals, which are called the finite elements[23.3] *of the Finite Element Method and continuous up to and including a certain derivative, whose order is given by the requirements of the conditions that must be fulfilled by the fields used in the principle of stationary potential energy. Therefore, in the present case, we must insure that the fields are continuous up to and including the first derivative w' of w at the junction between the elements. Otherwise, we may choose the fields quite freely. The most common, and convenient, set of functions to be used is polynomials, and we shall stick with this choice in the following.*

23.5 Focus on the System, Simple Elements

In order to describe the displacement field we need at least the following parameters

[23.2] It may, however, be mentioned that if the solution is sufficiently differentiable the use of over-compatible elements may prove to be a good idea. One disadvantage is though that if we apply the program to a slightly different problem whose solution has a (small) discontinuity of the higher gradients, say because plasticity develops, then the over-compatibility is certain to cause trouble. Such a situation is likely to arise if you have not written the finite element program yourself.

[23.3] The reason for calling these intervals elements should be clear later. At present, we may just think of intervals.

1. The (transverse) displacement w_B in order to insure continuity of the displacement itself at point B.
2. The first derivative w'_B to fulfill the continuity of the rotation at point B
3. The displacement w_C because otherwise the displacement field would not be completely determined.

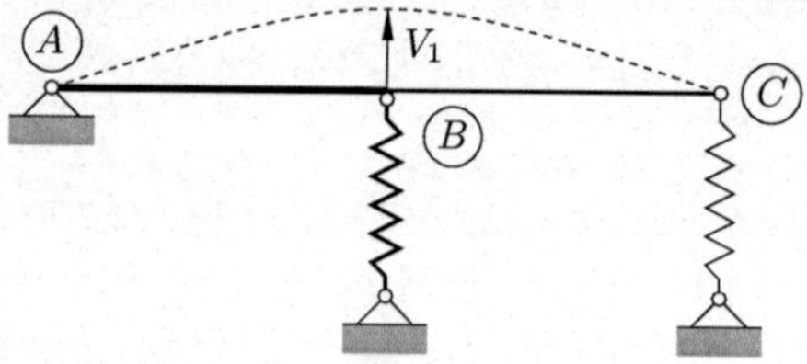

Fig. 23.7: Displacement field for $V_1 = 1$.

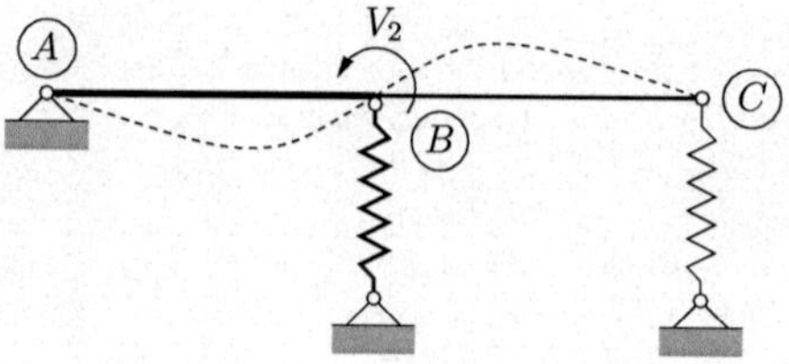

Fig. 23.8: Displacement field for $V_2 = 1$.

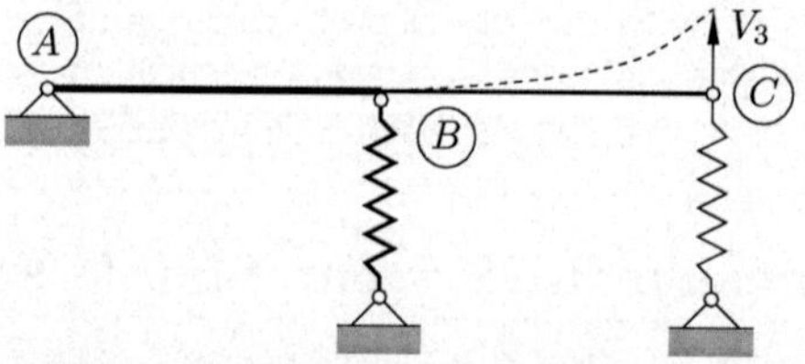

Fig. 23.9: Displacement field for $V_3 = 1$.

System displacement vector $\{V\}$

Before we go any further, we introduce a *System Displacement Vector* $\{V\}$[23.4]

$$\{V\}^T = [V_1 \,;\, V_2 \,;\, V_3] = [w_B \,;\, w'_B \,;\, w_C] \tag{23.18}$$

[23.4] We could have chosen more intervals than AB and BC at the cost of more V_j than 3. At each junction, say D, between intervals we will then have to demand that w_D and w'_D are continuous.
We shall discuss this possibility later.

see Figs. 23.7–23.9. As usual, superscript T denotes the matrix transpose. We wish to write the problem in terms of matrices since later we encounter problems so large that they can only be handled in that way.[23.5] The lowest-degree displacement fields then are

$$\begin{aligned} \widetilde{w}_1(\xi) &= (-\xi^2 + 2\xi) \qquad \text{for } \xi \in [0,2] \\ \widetilde{w}_2(\xi) &= \begin{cases} L(\xi^2 - \xi) & \text{for } \xi \in [0,1] \\ L(-\xi^2 + 3\xi - 2) & \text{for } \xi \in [1,2] \end{cases} \\ \widetilde{w}_3(\xi) &= \begin{cases} 0 & \text{for } \xi \in [0,1] \\ (\xi^2 - 2\xi + 1) & \text{for } \xi \in [1,2] \end{cases} \end{aligned} \tag{23.19}$$

Lowest-degree assumption

where, as before, see (23.12a), ξ is defined as

$$\xi \equiv \frac{x}{L} \tag{23.20}$$

with the comment that later we define ξ separately for each interval or element. When we arrange the fields $\widetilde{w}_1$–$\widetilde{w}_3$ in a row vector $[N]$, which is called the *Displacement Interpolation Vector* or the *Displacement Interpolation Matrix*

Displacement interpolation matrix $[N]$

$$[N(\xi)] = [N_1(\xi) \; ; \; N_2(\xi) \; ; \; N_3(\xi)] = [\widetilde{w}_1(\xi) \; ; \; \widetilde{w}_2(\xi) \; ; \; \widetilde{w}_3(\xi)] \tag{23.21}$$

we can write the total displacement field $\widetilde{w}(\xi)$ as

Total displacement $\widetilde{w}$

$$\widetilde{w}(\xi) = [N(\xi)]\{V\} \tag{23.22}$$

where it is implied that $[N(\xi)]$ covers $N_1(\xi)$–$N_3(\xi)$.

Comment: Weaknesses of the Chosen Fields

Second derivatives in (23.17) are constant, in the exact solution they are not

You may already have observed that the fields given by (23.19) suffer from one of the same weaknesses as the fields in (23.17), namely that the second derivative of all elements of $[N]$ are constant, implying that the bending moment is constant within AB and BC. Similar troubles occur very often when we establish the finite element method for a particular type of structure in that we are forced to make assumptions regarding the displacements which imply constant or linearly varying stresses over each element. The crux of the finite element is that, given a "sufficiently" large number of elements the solution usually proves to be a satisfactory approximation.

Let us close one eye and proceed on the basis of (23.21). In order to write the potential energy we need the curvature strain $\widetilde{\kappa}$, i.e. the second derivative $\widetilde{w}''$ of $\widetilde{w}$. Using matrix symbolism we may write

$$\widetilde{\kappa}(\xi) = \widetilde{w}''(\xi) = [B(\xi)]\{V\} \tag{23.23}$$

where the *Strain Distribution Matrix* $[B(\xi)]$ here follows from the displacement

[23.5] Actually, the same is necessary when the number of terms in (23.17) is larger than two or maybe three.

Strain distribution matrix $[B]$

interpolation matrix $[N(\xi]$ in the following way[23.6]

$$[B(\xi)] = [N''(\xi)] \tag{23.24}$$

and thus

$$B_1(\xi) = -2\frac{1}{L^2} \qquad \text{for } \xi \in [0,2]$$

$$B_2(\xi) = \begin{cases} +2\dfrac{1}{L} & \text{for } \xi \in [0,1] \\ -2\dfrac{1}{L} & \text{for } +\xi \in [1,2] \end{cases} \tag{23.25}$$

$$B_3(\xi) = \begin{cases} 0 & \text{for } \xi \in [0,1] \\ +2\dfrac{1}{L^2} & \text{for } \xi \in [1,2] \end{cases}$$

Already at this point we may get the feeling that the expression for $B_3(\xi)$ does not have to cover the beam AB because no matter the value of V_3 it has no influence on the situation in that beam. As discussed later, if the beam were supported on a large number of springs it would not be a good idea to let all $B_j(\xi)$ cover the entire structure because most of the entries in $B_j(\xi)$ would be equal to zero, at least implying a vast number of unnecessary multiplications and other manipulations.

For computation of the strain energy in the springs we need their elongations

Springs elongations $\widetilde{\Delta\ell}_B$ and $\widetilde{\Delta\ell}_C$

$$\widetilde{\Delta\ell}_B = [N(1)]\{V\} \;\text{ and }\; \widetilde{\Delta\ell}_C = [N(2)]\{V\} \tag{23.26}$$

Finally, we cast the load in vector form

Load vector $\{\bar{R}\}$

$$\{\bar{R}\}^T = \begin{bmatrix} R_1 \,;\, R_2 \,;\, R_3 \end{bmatrix} = \begin{bmatrix} 0 \,;\, 0 \,;\, \bar{P} \end{bmatrix} \tag{23.27}$$

where superscript ^{T} (again) indicates the matrix transpose.

With (23.18)–(23.27) in hand we may write the potential energy in the following form

Potential energy $\mathbf{\Pi}_\mathrm{P}$

$$\begin{aligned} \mathbf{\Pi}_\mathrm{P}(\{V\}) = &+\tfrac{1}{2}L\int_0^1 \{V\}^T[B(\xi)]^T\alpha EI[B(\xi)]\{V\}d\xi \\ &+\tfrac{1}{2}L\int_1^2 \{V\}^T[B(\xi)]^T EI[B(\xi)]\{V\}d\xi \\ &+\tfrac{1}{2}\{V\}^T[N(1)]^T\beta c[N(1)]\{V\} \\ &+\tfrac{1}{2}\{V\}^T[N(2)]^T c[N(2)]\{V\} \\ &-\{V\}^T\{\bar{R}\} \end{aligned} \tag{23.28}$$

Then, since $\{V\}$ is independent of the coordinate ξ we may write this

[23.6] As must be obvious, for other kinds of structures than frames $[B]$ must be derived from $[N]$ in other ways, see e.g. Section 24.2.

the more compact form

$$\begin{aligned}\mathbf{\Pi}_{\mathrm{P}}(\{V\}) = &+\tfrac{1}{2}\{V\}^T[K]_{AB}\{V\}d\xi \\ &+\tfrac{1}{2}\{V\}^T[K]_{BC}\{V\}d\xi \\ &+\tfrac{1}{2}\{V\}^T[K]_{B}\{V\}d\xi \\ &+\tfrac{1}{2}\{V\}^T[K]_{C}\{V\}d\xi \\ &-\{V\}^T\{\bar{R}\}\end{aligned} \tag{23.29}$$

Potential energy $\mathbf{\Pi}_{\mathrm{P}}$

where the so-called *Element Stiffness Matrices*[23.7] $[K]_{AB}$, $[K]_{BC}$, $[K]_B$, and $[K]_C$ at the *System Level*, see also Section 23.8, are defined by

$$\begin{aligned}[K]_{AB} &\equiv L\int_0^1 [B(\xi)]^T \alpha EI[B(\xi)]d\xi \\ [K]_{BC} &\equiv L\int_1^2 [B(\xi)]^T EI[B(\xi)]d\xi \\ [K]_B &\equiv [N(1)]^T \beta c[N(1)] \\ [K]_C &\equiv [N(2)]^T c[N(2)]\end{aligned} \tag{23.30}$$

Element stiffness matrices $[K]_{AB}$, $[K]_{BC}$, $[K]_B$ and $[K]_C$ at system level

with the result that

$$[K]_{AB} = 4\alpha\frac{EI}{L^3}\begin{bmatrix} +1 & -L & 0 \\ -L & +L^2 & 0 \\ 0 & 0 & 0 \end{bmatrix} \tag{23.31}$$

$$[K]_{BC} = 4\frac{EI}{L^3}\begin{bmatrix} +1 & +L & -1 \\ +L & +L^2 & -L \\ -1 & -L & +1 \end{bmatrix} \tag{23.32}$$

$$[K]_{B} = 3\frac{EI}{L^3}\begin{bmatrix} 1 & 0 & 0 \\ 0 & 0 & 0 \\ 0 & 0 & 0 \end{bmatrix} \tag{23.33}$$

and

$$[K]_{C} = \frac{EI}{L^3}\begin{bmatrix} 0 & 0 & 0 \\ 0 & 0 & 0 \\ 0 & 0 & 1 \end{bmatrix} \tag{23.34}$$

[23.7] For an explanation of the term stiffness matrix, see p. 438.

Finally the *System Stiffness Matrix* $[K]$ is

System stiffness matrix $[K]$

$$[K] = \frac{EI}{L^3}\begin{bmatrix} +(4\alpha+7) & -4(\alpha-1)L & -4 \\ -4(\alpha-1)L & +4(\alpha-1)L^2 & -4L \\ -4 & -4L & +5 \end{bmatrix} \tag{23.35}$$

Comment: All Diagonal Elements of the Stiffness Matrices are Positive

For stiffness matrices based on potential energy all diagonal elements are > 0

Note that all diagonal elements of the stiffness matrices are positive as a consequence of the fact that the strain energy function $W(\kappa) = \frac{1}{2}EI\kappa^2$ *of the beam and* $W(\Delta\ell) = \frac{1}{2}c(\Delta\ell)^2$ *of the springs are positive semi-definite, see Section 2.6.1, in particular (2.101).*

The potential energy $\mathbf{\Pi}_{\mathrm{P}}$ may now be written

Potential energy $\mathbf{\Pi}_{\mathrm{P}}$

$$\mathbf{\Pi}_{\mathrm{P}}(\{V\}) = \tfrac{1}{2}\{V\}^T[K]\{V\} - \{V\}^T\{\bar{R}\} \tag{23.36}$$

The justification of the term *Stiffness Matrix* may be seen from (23.36) in that the term containing $[K]$ is the strain energy of the system, which, for a given set of system displacements $\{V\}$, is greater the stiffer the structure.

Since we may vary V_1 independently of V_2 and V_3, etc. the solution can be found by requiring that the variation of the potential energy vanishes

$$\delta\mathbf{\Pi}_{\mathrm{P}}(\{V\}) = \{\delta V\}^T\Big([K]\{V\} - \{\bar{R}\}\Big) = 0 \;\forall\, \{\delta V\} \tag{23.37}$$

and thus

Matrix equation to determine $\{V\}$

$$[K]\{V\} = \{\bar{R}\} \tag{23.38}$$

This is the finite element system of equations for our structure—the contents of $[K]$ and $\{\bar{R}\}$ depend on the choice of displacement interpolation.

23.5.1 Interpretation of the Elements of the System Stiffness Matrix

Interpretation of the elements of the system stiffness matrix $[K]$

It seems worthwhile spending a little time on the contents of (23.37) and (23.38). In Sections 2.7 and 33.4.1 it is shown that the requirement that the (first) variation of the potential energy vanishes is equivalent to the principle of virtual displacements with the stresses expressed in terms of the strains and the constitutive relation. Furthermore, in Section 2.4 the principle of virtual displacements is derived from the equilibrium equations and is shown to be just another, sometimes convenient, way of expressing equilibrium. Therefore, (23.38) represents the equilibrium of the finite element realization of our structure which only possesses the three degrees of freedom given by $\{V\}$. Having noted this, we may interpret the elements of the system stiffness matrix $[K]$. For instance, if we let $V_1 = \overline{V}_1 = 1$, $V_2 = \overline{V}_2 = 0$ and $V_3 = \overline{V}_3 = 0$ the matrix product $[K]\{V\}$ provides

$$\{Q\} = [K]\{\overline{V}\} = \begin{bmatrix} K_{11} \\ K_{21} \\ K_{31} \end{bmatrix} \quad \text{for } \{\overline{V}\} = \begin{bmatrix} 1 \\ 0 \\ 0 \end{bmatrix} \tag{23.39}$$

Elements of columns of $[K]$ are reactions to prescribed displacements

where the element Q_j of the result $\{Q\}$—not to be confused with $\{\bar{R}\}$—is the reaction in the direction of V_j caused by the prescribed displacements given by $\{\overline{V}\}$. Thus, the elements of $[K]$ may be interpreted as the reactions associated with the prescribed displacement vector $\overline{V}$ of (23.39) or, alternatively, as the forces needed to produce the displacements $V_j = 1$ and all other $V_k = 0$.[23.8]

In this simple case with only three unknowns we can find an analytic expression for the solution

$$\{V\} = \frac{\bar{P}L^3}{EI} \begin{bmatrix} \dfrac{8\alpha}{(31\alpha+3)} \\[2ex] \dfrac{(8\alpha+3)}{(31\alpha+3)} \dfrac{1}{L} \\[2ex] \dfrac{(19\alpha+3)}{(31\alpha+3)} \end{bmatrix} \tag{23.40}$$

which for $\alpha = 1$ yields

$$\{V\} = \frac{\bar{P}L^3}{EI} \begin{bmatrix} \dfrac{4}{17} \\[2ex] \dfrac{11}{34} \dfrac{1}{L} \\[2ex] \dfrac{11}{17} \end{bmatrix} = \frac{\bar{P}L^3}{EI} \begin{bmatrix} 0.2352 \\[2ex] 0.3235 \dfrac{1}{L} \\[2ex] 0.6471 \end{bmatrix} \quad \text{cf.} \quad \frac{\bar{P}L^3}{EI} \begin{bmatrix} 0.2222 \\[2ex] 0.3333 \dfrac{1}{L} \\[2ex] 0.6667 \end{bmatrix} \tag{23.41}$$

and for $\alpha = 3$

$$\{V\} = \frac{\bar{P}L^3}{EI} \begin{bmatrix} \dfrac{1}{4} \\[2ex] \dfrac{9}{32} \dfrac{1}{L} \\[2ex] \dfrac{5}{8} \end{bmatrix} = \frac{\bar{P}L^3}{EI} \begin{bmatrix} 0.2500 \\[2ex] 0.2812 \dfrac{1}{L} \\[2ex] 0.6250 \end{bmatrix} \quad \text{cf.} \quad \frac{\bar{P}L^3}{EI} \begin{bmatrix} 0.2400 \\[2ex] 0.2800 \dfrac{1}{L} \\[2ex] 0.6400 \end{bmatrix} \tag{23.42}$$

where the exact solution is given by (23.11).

[23.8] For completeness, note that the equivalence between equilibrium and the principle of virtual displacements is utilized several times in Part II to find the equilibrium equations of specialized continua.

The displacements approximation is remarkably good in view of the fact that $\widetilde{w}'' \equiv 0 \Rightarrow \widetilde{M} \equiv 0$

The error on the characteristic displacement $V_3 = w_C$ is about 2–3% in both cases, and the error on the other displacement components is of the same order of magnitude, which is very good considering that by the assumption (23.19) the bending moment is constant in both beam AB and beam BC. Since our approximate solution is based on the stationarity of the potential energy the value of the characteristic displacement is smaller than the exact one, see Section 33.4.3. In contrast to the case of the series solutions, see (23.17), the bending moment is now continuous at point B independent of the value of α. On the other hand, the values of the bending moment are

The bending moment approximation is far from good

$$\begin{aligned} \alpha = 1: \quad & \widetilde{M} = \tfrac{3}{17}\bar{P}L = 0.1764\bar{P}L \ \text{ cf. } \ M_B = 0.3333\bar{P}L \\ \alpha = 3: \quad & \widetilde{M} = \tfrac{3}{16}\bar{P}L = 0.1875\bar{P}L \ \text{ cf. } \ M_B = 0.3600\bar{P}L \end{aligned} \tag{23.43}$$

Comment: The Error on the Approximation of the Mean Stress Is Small

Clearly, the value of $\widetilde{M}$ is a poor approximation to M_B, but note that in both cases the value of $\widetilde{M}$ (if referred to the middle of the two beams, that is, at mid-span) is about one half of M_B and thus it represents a fair approximation to the mean value of the true bending moment, which is linear in both AB and BC with zero values at the ends A and C. This is not a coincidence but a characteristic of the finite element method, namely that it makes the mean error of the stress components as small as possible, given the assumed displacement field.

The finite element method minimizes the mean error on stresses

"Real" finite element analyses focus on the elements before the system

Above, we have focused directly on the behavior of the system and based our analysis on fields that span all beams. The final aim of the finite element method is, of course, also to investigate the behavior of the structure, but, as we shall see in Section 23.6 and 23.7 below, it focuses on the behavior of the individual "elements," here the beams AB and BC and the springs at B and C.

Comment: Letting the Trial Functions Span the Entire Structure May Be a Bad Idea

Let us reconsider what we did in Section 23.5.

In (23.19) we began by introducing assumed displacement fields, also called trial functions, whose shapes are given by $\widetilde{w}_1$–$\widetilde{w}_3$, which we collected in $[N]$, and whose amplitudes are given by the Nodal Displacements *V_1–V_3, i.e. by the* system displacement vector *$\{V\}$.*

Nodal displacements given by $\{V\}$

Then, we found an expression for the potential energy by evaluating and summing it member by member of the structure, i.e. the two beams AB and BC and the two springs at points B and C. In this connection it is important to note that all these matrices covered the entire structure, and, as a consequence of this, the stiffness matrices $[K]_{AB}$, $[K]_{BC}$, $[K]_B$ and $[K]_C$, see (23.31)–(23.34), all have the dimension 3×3, although some of *them contain many more zeros than non-vanishing elements.*

For our small problem, which only has 3 unknowns, it does not matter much that we found the total stiffness matrix $[K]$ by a direct sum of the individual stiffness matrices mentioned above. If, for example, the beam were supported on 30 springs, then it would be rather foolish to store the contribution from one spring, which is only a single number, in a matrix of the order 60×60. Therefore, it seems reasonable to look for other ways to establish the system stiffness matrix $[K]$.

All stiffness matrices $[K]_{AB}$, $[K]_{BC}$, $[K]_B$ and $[K]_C$ had the same dimension as the system stiffness matrix $[K]$. Probably not a good idea

Realizing that the total potential energy of the structure is the sum of the potential energy of the members we may attack the problem differently and proceed member by member.

23.6 Focus on the Simple Elements

If we divide our structure into elements that are disconnected we get the assembly shown in Fig. 23.10. Note that the simple elements shown in the

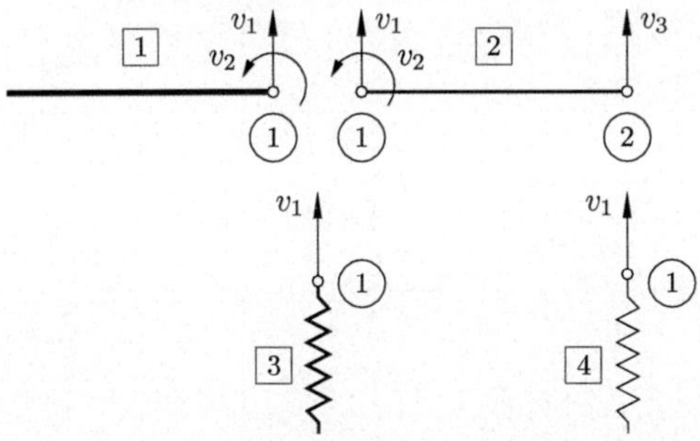

Fig. 23.10: Beam divided into disconnected simple elements. Numbers in squares denote elements, numbers in circles indicate nodes. Node and displacement numbers are local to the element.

figure are consistent with the previous analysis. We shall, however, see that there are better alternatives.

Comment: Discussion of the simple elements

Before we proceed with the process of subdividing the structure, we may discuss the contents of Fig. 23.10. Although the two beam elements have different bending stiffness, they model parts of the continuous beam that are very similar, except for the boundary conditions. If we supply element 1 with two more degrees of freedom at its left-hand end and element 2 with one more degree of freedom at its right-hand end, then the displacement

The two beams were modeled in different ways, but they are very much alike *interpolation matrix may be chosen identically in both elements. Later, we must suppress the vertical degree of freedom at the left-hand end of element* 1*, but this is a simple task.*

As mentioned above, the simple elements cannot satisfy the static boundary conditions at the supports, which is a further reason to abandon them.

23.7 Focus on the "Real" Elements

When we subdivide the continuous beam according to the ideas discussed above, we get the system shown in Fig. 23.11.

23.7.1 Beam Elements 1 and 2

In order to compute the potential energy of one of the beam finite elements

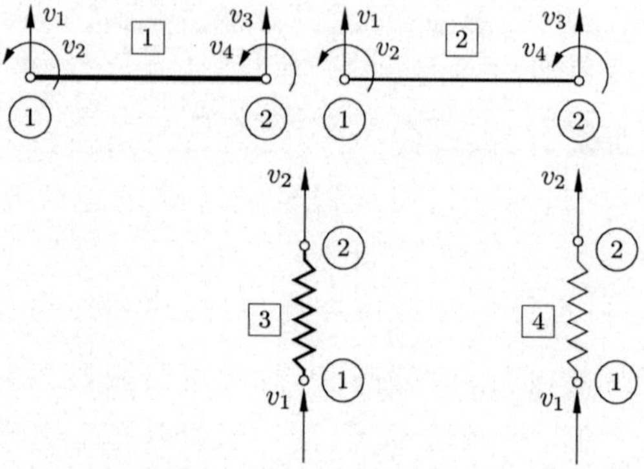

Fig. 23.11: Beam divided into disconnected "real" elements. Numbers in squares denote elements, numbers in circles indicate nodes. Node and displacement numbers are local to the element.

we must make an assumption regarding the displacement fields associated with each of the displacement components v_1–v_4, see Fig. 23.12. We may write

Displacement interpolation

$$w(\xi) = [N(\xi)]\{v\} \tag{23.44}$$

where we have used

$$\xi \equiv \frac{x}{L}, \ \xi \in [0;1] \tag{23.45}$$

instead of the physical coordinate $x \in [0;L]$ as the independent variable. It is usually a good idea to introduce nondimensional variables when you are developing finite elements—and in many other cases too, for that matter. We assume polynomial interpolations, and in order that the displacement field associated with degree j fulfills the conditions $v_j = 1$ and all other $v_k = 0$ we get the following expression for the *Displacement Interpolation Matrix* $[N]$

Displacement interpolation matrix $[N]$

$$[N(\xi)] = \big[\,(2\xi^3 - 3\xi^2 + 1)\ ;\ (\xi^3 - 2\xi^2 + \xi)L\ ;\ (-2\xi^3 + 3\xi^2)\ \ ;\ (\xi^3 - \xi^2)L\,\big] \tag{23.46}$$

This type of interpolation is called *Hermitian*, see e.g (Cook et al. 2002), and is characterized by continuity of not only the value of the function itself,

Fig. 23.12: Beam element.

but also of its first derivative. If only the value of the function is continuous, then the interpolation is of the *Lagrange* type, see e.g (Cook et al. 2002). The first kind of continuity is known as C^1- and the second as C^0-continuity.

The displacement fields associated with the four degrees of freedom, i.e. the elements of $[N(\xi)]$ are plotted in Fig. 23.13. If we inspect (23.46)

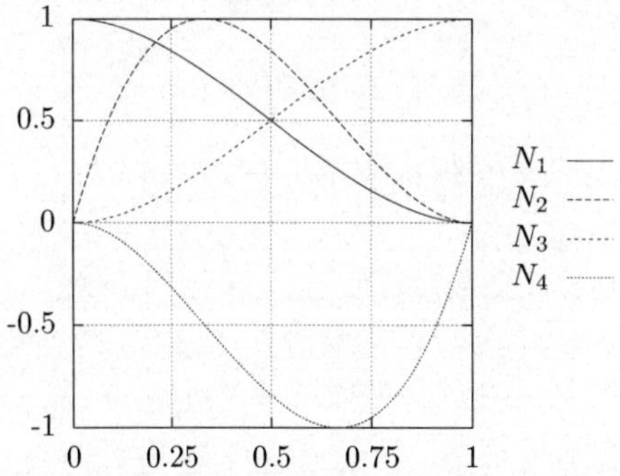

Fig. 23.13: Unit displacement fields, the components of $[N]$. The fields N_2 and N_4 are scaled such that the absolute value of their amplitudes is 1.

we may see that the elements of $[N(\xi)]$ are exact expressions for various displacement fields of a beam with constant bending stiffness. For example, the second element, $N_2(\xi)$ of $[N(\xi)]$ is the displacement field when the beam is subjected to the end displacement vector $\{v\} = [0, 1, 0, 0]^T$.

We need an expression for the (generalized) strain $\kappa(x) = w''(x)$

$$\kappa(\xi) = w''(\xi) = [B(\xi)]\{v\} \tag{23.47}$$

where we must remember that prime indicates differentiation with respect to the physical coordinate x. and where the *Strain Distribution Matrix* $[B(\xi)]_1$ is given by

Strain distribution Matrix $[B]$

$$[B(\xi)] \equiv \frac{d^2[N(\xi)]}{dx^2} = \frac{1}{L^2}\frac{d^2[N(\xi)]}{d\xi^2} \tag{23.48}$$

Thus,

Strain distribution Matrix $[B]$

$$\begin{aligned}[B(\xi)] = \Big[\frac{1}{L^2}(12\xi - 6) \;\; ; \frac{1}{L}(6\xi - 4) \; ; \\ \frac{1}{L^2}(-12\xi + 6) \; ; \frac{1}{L}(6\xi - 2) \Big]\end{aligned} \tag{23.49}$$

In the present example the *Material Stiffness Matrix* $[D(\xi)]_j$ does not vary along the length of the beams and is given by

Material stiffness matrix $[D(\xi)]_j$

$$[D(\xi)]_j \equiv [\gamma EI] \begin{cases} \gamma = \alpha & \text{for } j = 1 \\ \gamma = 1 & \text{for } j = 2 \end{cases} \tag{23.50}$$

With (23.46)—(23.50) in hand we may compute the strain energy W_j of element j

Strain energy of element j

$$\begin{aligned} W_j &= \tfrac{1}{2}L\int_0^1 \{v\}_j^T[B(\xi)]_j^T[D(\xi)]_j[B(\xi)]_j\{v\}_j d\xi \\ &= \tfrac{1}{2}\{v\}_j^T\left(L\int_0^1 [B(\xi)]_j^T[D(\xi)]_j[B(\xi)]_j d\xi\right)\{v\}_j \\ &= \tfrac{1}{2}\{v\}_j^T[k]_j\{v\}_j \end{aligned} \tag{23.51}$$

where the *Element Stiffness Matrix* $[k]_j$ is:

Beam element stiffness matrix $[k]_j$

$$\begin{aligned} [k]_j &= L\int_0^1 [B(\xi)]_j^T[D(\xi)]_j[B(\xi)]_j d\xi \\ &= \gamma\frac{EI}{L^3}\begin{bmatrix} 12 \; ; & 6L \; ; & -12 \; ; & 6L \\ 6L \; ; & 4L^2 \; ; & -6L \; ; & 2L^2 \\ -12 \; ; & -6L \; ; & 12 \; ; & -6L \\ 6L \; ; & 2L^2 \; ; & -6L \; ; & 4L^2 \end{bmatrix} \begin{cases} \gamma = \alpha & \text{for } j = 1 \\ \gamma = 1 & \text{for } j = 2 \end{cases} \end{aligned} \tag{23.52}$$

Comment: The Finite Element Stiffness Matrix Is Symmetric

You may readily observe that the finite element stiffness matrices $[k]_j$ are symmetric, and after a little work you may realize that $[k]_j$ indeed expresses the exact values of the stiffnesses of the element with constant γEI. The symmetry is not a coincidence, but a consequence of using the potential energy as the basis for $[k]$—had we used the Principle of Virtual Displacements or a Galerkin Formulation as our point of departure we might easily have got an element stiffness matrix that was non-symmetric.

Finite element stiffness matrices are symmetric

23.7.2 Interpretation of the Elements of the Element Stiffness Matrix

The elements of the stiffness matrix $[k]$ may be interpreted in the same way as was done in Example 23.5.1 for the elements of the system stiffness matrix $[K]$. If, for example, we prescribe the elements of the nodal displacement vector $\{v\}$ such that $v_4 = 1$ and all other $v_j = \bar{v}_j = 0$, then the matrix product

Interpretation of the elements of the element stiffness matrix $[k]$

$$\{q\} = [k]\{\bar{v}\} = \begin{bmatrix} k_{14} \\ k_{24} \\ k_{34} \\ k_{44} \end{bmatrix} \quad \text{for} \ \{\bar{v}\} = \begin{bmatrix} 0 \\ 0 \\ 0 \\ 1 \end{bmatrix} \tag{23.53}$$

where element q_j of the result $\{q\}$ is the reaction in the direction of v_j caused by the prescribed displacements given by $\{\bar{v}\}$. Note that $\{q\}$ is not the same as the element load vector $\{\bar{r}\}$, see (23.68). Thus, the elements of column j of the *Nodal Force Vector* $\{q\}$ are the reactions to the prescribed displacements $\{\bar{v}\}$ and, as you may see, these values are all exact for beam elements with constant bending stiffness EI. If the bending stiffness varies along the length of the beam the elements of $\{q\}$ are not the exact, but approximate values of these reactions. Whether or not the stiffness matrix is exact, its elements all have physical interpretations, either as forces or as bending moments. Already at this point we may mention that a similar interpretation is not possible for finite elements that are applied to problems of plane strain and plane stress, see Section 24.1, as well as for other types of two- or three-dimensional structures.

Nodal force vector $\{q\} \sim$ reactions to prescribed displacements $\{\bar{v}\}$

23.7.3 Spring Elements 3 and 4

For the spring elements we may immediately find the strain energy $W_j(\{v\}_j)$

$$W_j(\{v\}_j) = \tfrac{1}{2}\{v\}_j^T[k]_j\{v\}_j \tag{23.54}$$

where the element stiffness matrix $[k]_j$ is

$$[k]_j = \gamma\frac{EI}{L^3}\begin{bmatrix} 1\ ; & -1 \\ -1\ ; & 1 \end{bmatrix} \begin{cases} \gamma = \beta \ \text{for} \ j = 3 \\ \gamma = 1 \ \text{for} \ j = 4 \end{cases} \tag{23.55}$$

Spring element stiffness matrix $[k]_j$

Comment: We Need to Assemble the Finite Elements

System stiffness matrix $[K]$

So far, the finite elements of Sections 23.7.1–23.7.3 are disconnected, see also Fig. 23.11, and we need some way to assemble them in order to compute the System Stiffness Matrix $[K]$.

Don't vary the element nodal displacements independently—the beam structure would break

Looking at (23.52) and (23.54) it appears that there are $4\times2+2\times2=12$ *independent nodal displacements, which certainly is not the case because the finite elements are interconnected at the nodes. Thus, the number of independent nodal displacements is smaller than 12, and in the case of our continuous beam it is 8, see Fig. 23.14. If we varied the nodal displacements for all elements independently we would violate the continuity of the beam structure and thus have broken it up into 4 disconnected elements, which is not what we want.*

In real life, don't assemble the element stiffness matrices as is done in the following—use bookkeeping

In other words, we need to enforce the continuity conditions at the nodes A, B *and* C. *There are, at least, two different ways to do this. The one, which is described below, is particularly well suited for theoretical derivations but is not recommended as a basis for computer programs for reasons that should be obvious soon. The other method, which simply consists in bookkeeping, may be programmed quite easily but does not present a good basis for theoretical studies.*

By the way, we need to insure that the proper displacements at the supports vanish, but let us forget this problem for a while.

23.8 Assembling of the Element Matrices

We choose the *System Displacement Vector* $\{V\}$ shown in Fig. 23.14 and

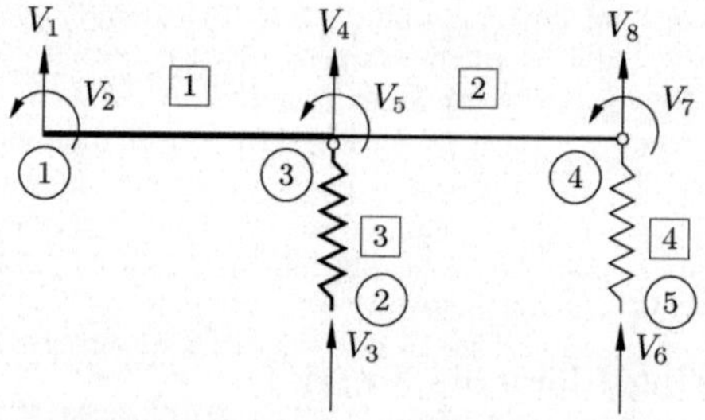

Fig. 23.14: System displacement vector $\{V\}$.

must establish the connection between the element displacement vector $\{v\}_j, j \in [1,4]$ and the system displacement vector $\{V\}$. The way we do it *here*[23.9] is to utilize an *Element Transformation Matrix* $[T]_j$ to obtain a

[23.9] I emphasize the word *here* because using element transformation matrices in real problems with many degrees of freedom is not a good idea, see the above comment.

matrix relation for each element

$$\{v\}_j = [T]_j\{V\} \tag{23.56}$$

Element transformation matrix $[T]_j$

where, for this continuous beam

$$[T]_1 = \begin{bmatrix} 1 & 0 & 0 & 0 & 0 & 0 & 0 & 0 \\ 0 & 1 & 0 & 0 & 0 & 0 & 0 & 0 \\ 0 & 0 & 0 & 1 & 0 & 0 & 0 & 0 \\ 0 & 0 & 0 & 0 & 1 & 0 & 0 & 0 \end{bmatrix} \tag{23.57}$$

$$[T]_2 = \begin{bmatrix} 0 & 0 & 0 & 1 & 0 & 0 & 0 & 0 \\ 0 & 0 & 0 & 0 & 1 & 0 & 0 & 0 \\ 0 & 0 & 0 & 0 & 0 & 0 & 0 & 1 \\ 0 & 0 & 0 & 0 & 0 & 0 & 1 & 0 \end{bmatrix} \tag{23.58}$$

$$[T]_3 = \begin{bmatrix} 0 & 0 & 1 & 0 & 0 & 0 & 0 & 0 \\ 0 & 0 & 0 & 1 & 0 & 0 & 0 & 0 \end{bmatrix} \tag{23.59}$$

and

$$[T]_4 = \begin{bmatrix} 0 & 0 & 0 & 0 & 0 & 1 & 0 & 0 \\ 0 & 0 & 0 & 0 & 0 & 0 & 0 & 1 \end{bmatrix} \tag{23.60}$$

and thus we may write the strain energy of the system as

$$\begin{aligned} W(\{V\}) &= \sum_{j=1}^{N_{El}} W_j(\{v\}_j) \\ &= \tfrac{1}{2}\sum_{j=1}^{N_{El}} \Big(\{V\}^T[T]_j^T[k]_j[T]_j\{V\}\Big) \\ &= \tfrac{1}{2}\{V\}^T\sum_{j=1}^{N_{El}} \Big([T]_j^T[k]_j[T]_j\Big)\{V\} \\ &= \tfrac{1}{2}\{V\}^T\sum_{j=1}^{N_{El}} [K]_j\{V\} \\ &= \tfrac{1}{2}\{V\}^T[K]\{V\} \end{aligned} \tag{23.61}$$

Strain energy of system

where $N_{El} = 4$ is the number of elements, and where the *Element Stiffness Matrix at System Level* is $[K]_j$

$$[K]_j \equiv [T]_j^T[k]_j[T]_j \tag{23.62}$$

Element stiffness matrix at system level $[K]_j$

of element j, see also (23.30).

The *System Stiffness Matrix* $[K]$ is given by

System stiffness matrix $[K]$

$$[K] = \sum_{j=1}^{N_{El}} [K]_j \tag{23.63}$$

see also (23.35). Note the comments on pages 446 and 449 concerning assembling in practice.

23.9 Right-Hand Side Vector

In the present case, determination of the right-hand side vector $\{\bar{R}\}$ of the system of equations does not present difficulties because the load $\bar{P}$ acts at a node

System right-hand side vector $\{\bar{R}\}$

$$\{\bar{R}\}^T = [\,0\,;0\,;0\,;0\,;0\,;0\,;\bar{P}\,;0] \tag{23.64}$$

because the load term of the potential energy $\mathbf{\Pi}_\text{P}(\{V\})$ then is

Load term of $\mathbf{\Pi}_\text{P}$

$$\{V\}^T\{\bar{R}\} = V_7 R_7 \tag{23.65}$$

as expected. Again, note the comment on page 449 concerning assembling in practice.

23.9.1 Right-Hand Side Vector for Distributed Loads

When the load is distributed over the elements $\{\bar{R}\}$ must be determined by considering the load term of the potential energy in more detail. As an example, let the beams be subjected to a vertical load, whose intensity is $\bar{p}_j(\xi), j \in [1,2]$, then the load term of element j is

Load term of element j

$$\begin{aligned} L_j \int_0^1 \bar{p}_j(\xi) w(\xi) d\xi &= L_j \int_0^1 \{v\}_j^T [N(\xi)]_j^T \bar{p}(\xi) d\xi \\ &= \{v\}_j^T L_j \int_0^1 [N(\xi)]_j^T \bar{p}(\xi) d\xi \\ &= \{v\}_j^T \{\bar{r}\}_j = \{V\}^T [T]_j^T \{\bar{r}\}_j \\ &= \{V\}^T \{\bar{R}\}_j \end{aligned} \tag{23.66}$$

and thus the *System Load Term* is

System load term $\{\bar{R}\}$

$$\{V\}^T\{\bar{R}\} = \{V\}^T \sum_1^{N_{El}} \{\bar{R}\}_j \tag{23.67}$$

where $\{\bar{r}\}_j$ is the *Element Load Vector* of element j

Element load vector $\{\bar{r}\}_j$

$$\{\bar{r}\}_j = L_j \int_0^1 [N(\xi)]_j^T \bar{p}(\xi) d\xi \tag{23.68}$$

where $\{\bar{R}\}_j$ denotes the *Element Load Vector at System Level* of element j

and is given by

$$\{\bar{R}\}_j = \sum_{1}^{N_{El}} [T]_j^T \{\bar{r}\}_j \tag{23.69}$$

Element load vector at system level $\{\bar{R}\}_j$

and where the *System Right-Hand Side Vector* $\{\bar{R}\}$ is

$$\{\bar{R}\} = \sum_{j=1}^{N_{El}} \{\bar{R}\}_j \tag{23.70}$$

System right-hand side vector $\{\bar{R}\}$

Comment: Do Not Construct the System Matrices by Matrix Products Involving Element Transformation Matrices

In the present case the above procedure for constructing the system stiffness matrix from the element stiffness matrices may not seem impractical, but if the number of system displacements were large, say 10,000, then all the $[T]_j$-matrices would have 10,000 columns, but only 2 or 4 nonzero elements. In those cases, the computational effort, both regarding time and storage, involved in the establishing of $[K]$ would be much larger than needed and, as mentioned above, simple bookkeeping[23.10] presents a more feasible alternative.

Never construct $[K]$ from $[K]_j$ and $\{\bar{R}\}$ from $\{\bar{R}\}_j$ by matrix products

23.10 Potential Energy, System of Finite Element Equations

We may now write the potential energy $\Pi_P(\{V\})$ in the same form as before, see (23.36)

$$\Pi_P(\{V\}) = \tfrac{1}{2}\{V\}^T[K]\{V\} - \{V\}^T\{\bar{R}\} \tag{23.71}$$

Potential energy of system Π_P

which, by variation with respect to $\{V\}$, yields the *Finite Element Equations*, cf. (23.38)

$$\begin{aligned} &\{\delta V\}^T\big([K]\{V\} - \{\bar{R}\}\big) = 0 \,\forall\, \{\delta V\} \\ \Rightarrow\ &[K]\{V\} = \{\bar{R}\} \end{aligned} \tag{23.72}$$

Finite element equations

where $\forall\delta\{V\}$ must be interpreted as "$\forall$ kinematically admissible $\{\delta V\}$," i.e. variations that satisfy the kinematic continuity conditions and the homogeneous kinematic boundary conditions. The solution follows from (23.72)

$$\{V\} = [K]^{-1}\{\bar{R}\} \tag{23.73}$$

Solution

Regarding finding the solution to (23.72) you must not take the factor $[K]^{-1}$ in (23.73) literally and find the inverse of $[K]$, the meaning of (23.73) is

Do not invert $[K]$

[23.10] In common computer languages, such as PL/I, Fortran. Pascal and C++, the number of statements to do this is on the order of, say 20 at the most.

that the solution $\{V\}$ satisfies this relation, but in reality you should find the solution by *Gaussian Elimination* or by use of *Cholesky Decomposition*. The description of these methods belong to the area of numerical analysis and not in this book, but see e.g. (Cook, Malkus & Plesha 1989) or (Bathe 1996).

In the above derivations it is tacitly assumed that the system stiffness matrix $[K]$ is stored as a square matrix. In the real world this is never done. The simplest of the more efficient ways is to store only the diagonal and upper half of $[K]$, but still the ensuing matrix may contain a vast number of zero elements. Therefore, the concept of *Sparse Matrix Methods*, where only the non-zero elements are stored, have been popular for quite some time.

23.11 Element Displacement Vectors

Use bookkeeping to find $\{v\}_j$ from $\{V\}$

In principle, the element displacement vectors $\{v\}_j$ are found from the system displacement vector $\{V\}$ by use of (23.56), but also here a bookkeeping scheme proves to be much more efficient.

23.12 Generalized Strains and Stresses

In our example the generalized strains are the bending strain κ in the beams and the elongation $\Delta\ell$ of the springs, and the work conjugate generalized stresses are the bending moment M and the force N in the springs, respectively. The bending strain is given by (23.47), and the bending moment in element j follows from

Bending moment in beams

$$M_j(\xi) = [D(\xi))]_j [B(\xi)]_j \{v\}_j \tag{23.74}$$

while in spring j the force is determined by

Axial force in springs

$$N_j = [k]_j \{v\}_j \tag{23.75}$$

Comment: Concluding Comments

The finite element method is a *numerical* method—not an analytic approximation method

In any case of practical interest the number of unknowns is so large that all quantities in the finite element expressions should be numerical. In our case it means that we must fix the values of α, β, EI, and L. Even with the 8 unknowns of the structure of our example it is not feasible to try to establish and solve the finite element equations in analytic form. The finite element method may be viewed as a member of a subclass of methods of approximation, namely the numerical methods.

In our example, the four finite elements furnish the exact solution because there is no distributed loads on the beams AB and BC and because the bending stiffness does not vary along the length. Therefore, there is no reason to employ more elements, but if the bending stiffness varies we may need to subdivide each beam into 2 or more elements depending on *the variation. The integral in the formula (23.52) for the element stiffness matrix $[k]_j$ is the same, but the actual numbers will be different.*

23.13 Summary of the Procedure of the Introductory Example 23

Let us summarize the procedure that provided the element stiffness matrices and the system of finite element equations, see Example 23.7–23.12.

Summary of Example 23.7

1. **Choice of Element Nodal Displacement Vectors $\{v\}_j$ and the Displacement Interpolation Matrices $[N(\xi)_i]_j$**
   The crux of the whole procedure is the proper choice of the finite element nodal displacement vectors $\{v\}_j$ and the associated displacement interpolation matrices $[N(\xi)]_j$, or in two- or three-dimensional cases $[N(\xi_i)]_j$, where ξ_i denotes the spatial coordinates. We must insure that the internal compatibility of the structure is not violated, i.e. that no holes or kinks develop between the elements, no matter what system displacements the structure is subjected to. This is an easy task regarding beam elements of the kind discussed above, but for solids, plates, and shells this condition often presents difficulties, as we shall see later.
   But, once we have chosen $[N]_j$ the rest is pretty much a routine task.

   Our solution must not create kinks and holes

2. **Computation of the Strain Distribution Matrices $[B(\xi)]_j$**
   The next step is to compute the strain distribution matrices $[B(\xi)]_j$, or $[B(\xi)_i]_j$, which follows directly from the strain-displacement relation.

   $[B(\xi)_i]_j$ follows strain definition

3. **Choice of the Material Stiffness Matrices $[D(\xi)]_j$**
   The strain distribution matrix $[D(\xi)]_j$ is, in a way, given once we have decided which structures to analyze. However, in the real structure the bending stiffness or the plate thickness may vary in a way that we are not willing to accommodate, e.g. as sines, and then we must make a choice as to the approximation, usually a polynomial one, we consider satisfactory.

   $[D(\xi)]_j$ is usually easy to establish

4. **Computation of the Element Stiffness Matrices $[k]_j$**
   With $[B(\xi)]_j$ and $[D(\xi)]_j$ in hand we can compute $[k]_j$ by integration of the product $[B(\xi)]_j^T[D(\xi)]_j[B(\xi)]_j$. In many cases we must resort to *Numerical Integration*, often by use of *Gaussian Quadratures*. This is, on the other hand, a subject that lies outside the scope of this book, but see e.g. (Cook et al. 1989) or (Bathe 1996).

   $[k]_j$ follows from $[B(\xi)]_j$ and $[D(\xi)]_j$ Numerical integration is often necessary

5. **Computation of the Element Transformation Matrices $[T]_j$**
   If we were to assemble the system stiffness matrix by use of matrix multiplications we would have to compute the element transformation matrices $[T]_j$, but, as I pointed out above, this is not recommended. So,

   Don't assemble the element stiffness matrices by matrix multiplications in all practical circumstances we shall utilize a bookkeeping scheme. This is another subject that is not treated here, but it is fairly easy to write the necessary computer code.

For the completeness of the procedure this step is included, albeit under an alias.

6. **Assembling of the Element Stiffness Matrices $[k]_j$**
   The system stiffness matrix $[K]$ is in principle given by the sum over all elements of the matrix product $[T]_j^T[k]_j[T]_j$.
7. **Assembling of the Element Right-Hand Side Vectors $\{\bar{r}\}_j$**
   The system right-hand side vector $\{\bar{R}\}$ is, in principle, given by the sum over all elements of the matrix product $[T]_j^T\{\bar{r}\}_j$.
8. **Solution of the System Finite Element Equations $[K]\{V\} = \{\bar{R}\}$**
   Don't invert $[K]$
   The system finite element equations are given by (23.72b), and the solution $\{V\}$ is found by use of elimination techniques, not by inversion of $[K]$.
9. **Determination of the Element Displacement Vectors $\{v\}_j$**
   Don't find $\{v\}_j$ by matrix multiplications
   The element displacements vectors $\{v\}_j$ are, in principle, determined by the expression (23.56), but a bookkeeping scheme proves to be much more efficient.
10. **Determination of the Generalized Strains κ and $\Delta\ell$**
    The element strains follow from $[B]_j\{v\}_j$
    The generalized strains are usually given by relations of like (23.47), but in the case of springs the generalized strain, i.e. the elongation, is simply the difference between the displacements of its two ends.
11. **Determination of the Generalized Stresses M and N**
    The element stresses follow from $[D]_j[B]_j\{v\}_j$
    In most cases, the generalized stress follows from a relation of the same type as (23.74), but in the simple case of springs it is given by (23.75).

Chapter 24

Plate Finite Elements for In-Plane States

24.1 Introduction

While the *Finite Element Method* indeed is a valuable tool for analysis of continuous beams and frames its real strength lies in the fact that it can be applied to structures made of two- and three-dimensional continua. Analysis of such structures as plates, either subjected to in-plane or transverse loads, shells, and solids is difficult, in part because the number of independent variables is large. While beam problems may be formulated in terms of *ordinary* differential equations the other types of structures are governed by *partial* differential equations which are much more demanding. Although the finite element method itself does not attack differential equations directly this fact may indicate the various levels of difficulty.

The real strength of the Finite Element Method is its capability to analyze plates, shells and 3-D solids, etc.

For beam problems computation of the stiffness matrix is comparatively straightforward, and some exact beam finite element stiffness matrices do exist, as we have seen in Section 23.7. On the other hand, sometimes we prefer an approximate one, for example for circular beam elements, see Chapter 26, because the exact stiffness matrix may exhibit unwanted features such as becoming ill-conditioned for small element lengths. Another important fact is that interpretation of the *Nodal Force Vector* $\{q\}$, which is a result of subjecting the finite element to a prescribed nodal displacement vector $\bar{v}$, see (23.53), for beams may be given a physical interpretation, see Example 23.7.2, but this is not possible for plates and similar structures.

For beams: Sometimes exact stiffness matrices

There is a conceptual problem in making the plate analogy of the beam finite elements in that when we go from a one-dimensional body to a two-dimensional one we would expect a line segment of the plate to correspond to a point of the beam, and thus the equivalent of a nodal point of a beam would be a line segment of a plate. This angle of attack has not been useful, so we use the concept of a nodal point even in two- and three-dimensional bodies. But, then another problem associated with interpretation of the nodal forces appears since, according to the theory of linear elasticity the displacement of an in-plane point force applied to a plate is infinite, see

For two- and three-dimensional bodies: No meaningful exact stiffness matrix

E. Byskov, *Elementary Continuum Mechanics for Everyone*,
Solid Mechanics and Its Applications 194, DOI: 10.1007/978-94-007-5766-0_24,

e.g. (Muskhelishvili 1963). Thus, the exact stiffness associated with any point of the plate is zero with the consequence that we may conclude that the search for an exact plate finite element stiffness matrix will be futile. Yet, a procedure similar to the one described in Chapter 23, see especially the summary Section 23.13, also proves to work for all other structures than beams.

As a simple example of a plate divided into finite element, see the wall in Fig. 24.1. In order to accommodate the slopes at the roof it is necessary

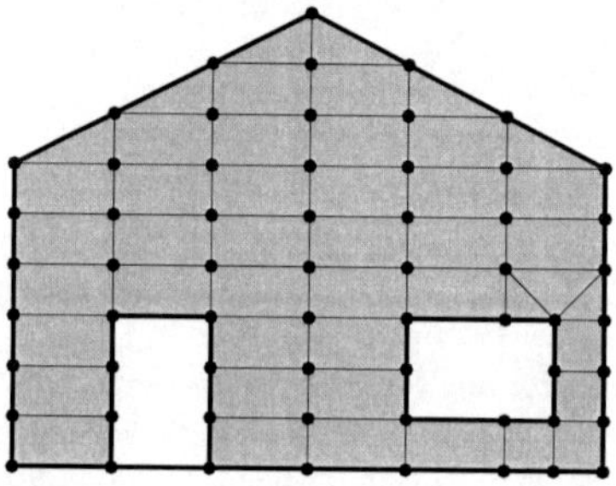

Fig. 24.1: A wall divided into rectangular and triangular shaded finite elements. Dots indicate nodal points.

to employ triangular finite elements in addition to the rectangular ones used over most of the plate.[24.1] Note also the triangular elements at the upper right-hand corner of the window. They are used to connect the more narrow rectangular elements to the right of the window opening and the rectangular elements used in most of the plate. In the following we develop the rectangular element but, mention at this point that the triangular one is also easy to construct.

24.2 A Rectangular Plate Finite Element for In-Plane States

Melosh element

In this Section we develop the very simple and yet quite useful rectangular plate finite element first derived by Melosh (1963) and shown in Fig. 24.2. The first step in the process is to realize which kinematic continuity conditions apply to a plate loaded in-plane. According to Chapter 9, see (9.1), the strains associated with in-plane deformations only are

Strain definition

$$\varepsilon_{\alpha\beta} = \frac{1}{2}\left(u_{\alpha,\beta} + u_{\beta,\alpha}\right) \tag{24.1}$$

and thus, the displacements u_α must be continuous, while the strains must

[24.1] We could, of course, have used other elements with four nodes, but of a more general shape than rectangular, and in that may have taken account of the sloping roof.

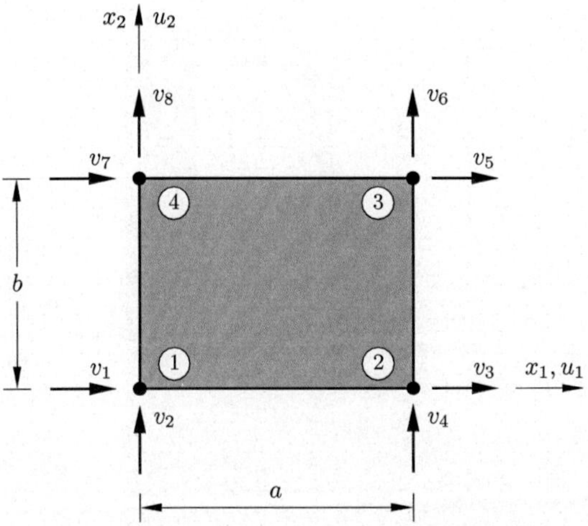

Fig. 24.2: Rectangular plate finite element.

not be assumed continuous over the element boundaries, since the strain may have jumps because of change in plate thickness or due to application of a line load $\bar{p}_\alpha$, see Fig. 9.2.

24.2.1 Displacement Field

In order to construct a finite element model of a plane plate using rectangular building blocks like the one shown in Fig. 24.2, which only has nodal displacements at the corners, we can conclude that the sides of the elements must remain straight after deformation to avoid overlapping of the elements or creation of gaps between neighboring elements, see e.g. Figs. 24.3 and 24.4. Therefore, the displacements must vary linearly over the sides of the elements, and the simplest assumption entails that they do that inside the elements, too.[24.2] We shall employ a method akin to the one utilized later in Chapter 25, in particular Section 25.4, to construct the elements of the displacement interpolation matrix $[N(x_\alpha)]$. For the displacement field associated with $v_5 = 1$ and all other $v_j = 0$, see Fig. 24.3, compatibility between finite elements requires that the sides $x_1 = 0$ and $x_2 = 0$ do not experience any displacements, and therefore we may write

Gaps or overlaps between elements are illegal

$$N_{15}(x_1, x_2) = C x_1 x_2 \tag{24.2}$$

which is linear in both x_1 and x_2. The coefficient C is determined the

[24.2] The procedure described below does not work if the elements are general quadrangles because then the sides do not remain straight if the displacement fields are constructed in that way.

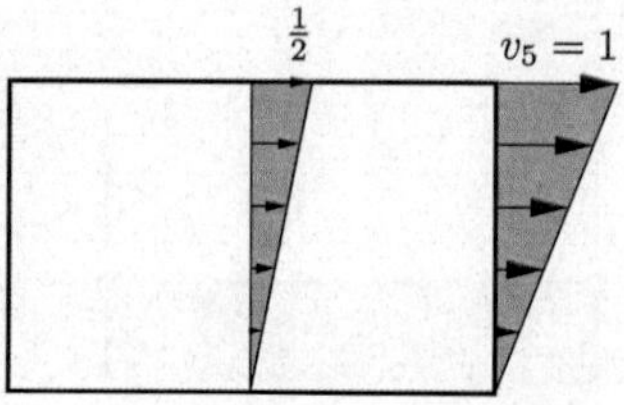

Fig. 24.3: Displacements associated with $v_5 = 1$.

demand that $N_{15}(a, b) = 1$, where a denotes the horizontal dimension of the finite element and b the vertical one. Thus,

$$N_{15}(a, b) = Cab = 1 \Rightarrow C = \frac{1}{ab} \tag{24.3}$$

The element N_{25} must vanish because of the compatibility requirement.

For the case $v_8 = 1$ and all other $v_j = 0$ the requirement is that the displacements vanish at $x_1 = a$ and $x_2 = 0$, which suggests

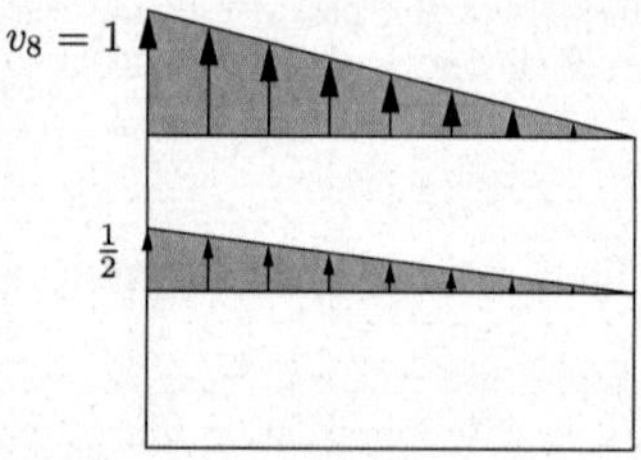

Fig. 24.4: Displacements associated with $v_8 = 1$.

$$N_{28}(x_1, x_2) = C(a - x_1)x_2 \tag{24.4}$$

The requirement that $N_{28}(0, b) = 0$ means that

$$N_{28}(0, b) = Cab = 1 \Rightarrow C = \frac{1}{ab} \tag{24.5}$$

By proceeding in this fashion we can build the displacement interpolation matrix $[N(x_\alpha)]$, where the non-zero elements are

Non-zero elements of displacement interpolation matrix $[N(x_\alpha)]$

$$\begin{aligned} N_{11} = N_{22} &= \frac{(a - x_1)(b - x_2)}{ab}, & N_{13} = N_{24} &= \frac{x_1(b - x_2)}{ab} \\ N_{15} = N_{26} &= \frac{x_1 x_2}{ab}, & N_{17} = N_{28} &= \frac{(a - x_1)x_2}{ab} \end{aligned} \tag{24.6}$$

24.2.2 Strain Distribution

In order to compute the element stiffness matrix we need to be able to write the matrix product $[B]^T[D][B]$, and for that we must rearrange the strains in a similar way to that of Section 5.1

$$\{\varepsilon(x_\alpha)\} \equiv \begin{bmatrix} \varepsilon_1 \\ \varepsilon_2 \\ \varepsilon_3 \end{bmatrix} \equiv \begin{bmatrix} \varepsilon_{11} \\ \varepsilon_{22} \\ 2\varepsilon_{12} \end{bmatrix} = \begin{bmatrix} u_{1,1} \\ u_{2,2} \\ u_{1,2} + u_{2,1} \end{bmatrix} \tag{24.7}$$

Generalized plate strains $\{\varepsilon(x_\alpha)\}$

The strain distribution matrix $[B(x_\alpha)]$ follows from the displacement assumption and is

$$[B(x_\alpha)] = \begin{bmatrix} B_{1j} \\ B_{2j} \\ B_{3j} \end{bmatrix} = \begin{bmatrix} N_{1j,1} \\ N_{2j,2} \\ N_{1j,2} + N_{2j,1} \end{bmatrix} \tag{24.8}$$

Strain distribution matrix $[B(x_\alpha)]$

where the non-zero elements of $[B]$ are

$$\begin{aligned} B_{11} &= -B_{13} = B_{32} = -B_{34} = -\frac{b - x_2}{ab} \\ B_{15} &= -B_{17} = B_{36} = -B_{38} = \frac{x_2}{ab} \\ B_{22} &= -B_{28} = B_{31} = -B_{37} = -\frac{a - x_1}{ab} \\ B_{24} &= -B_{26} = B_{33} = -B_{35} = -\frac{x_1}{ab} \end{aligned} \tag{24.9}$$

Non-zero elements of strain distribution matrix $[B(x_\alpha)]$

24.2.3 Constitutive Assumption

At this time we must decide on the material parameters given by the material stiffness matrix $[D]$, i.e. whether the material is isotropic, orthotropic or anisotropic, and whether it is in plane stress or plane strain, see Sections Ex 5-3.2 and Ex 5-3.1, respectively. Another point to discuss is whether we insist on getting an analytic solution for the stiffness matrix or if we prefer to compute it numerically. In many cases, it is not worth the effort to derive the analytic solution, but in the present case it may be a matter of taste if we choose one or the other since this finite element is very regular in shape with the result that many of the elements of the stiffness matrix are the same.

Is our material isotropic, orthotropic or anisotropic?
Is the plate in plane stress or plane strain?
Do we insist on analytic derivations?

24.2.4 Stiffness Matrix for Isotropy

For the sake of completeness, the elements of $[k]$ for an isotropic material are listed below, but recognizing the symmetry $k_{ji} = k_{ij}$ only the elements in the main diagonal and the elements above it are shown.

Diagonal and non-zero elements above the main diagonal of the element stiffness matrix $[k]$

Diagonal elements and non-zero elements above the main diagonal of plate element stiffness matrix $[k]$

$$
\begin{aligned}
k_{11} &= k_{33} = k_{55} = k_{77} = \frac{b^2 D_{11} + a^2 D_{33}}{3ab} \\
k_{12} &= -k_{34} = k_{56} = -k_{78} = -k_{16} = -k_{25} = k_{38} \\
&= k_{47} = \frac{D_{12}}{4} + \frac{D_{33}}{4} \\
k_{13} &= k_{57} = \frac{-2b^2 D_{11} + a^2 D_{33}}{6ab} \\
k_{14} &= k_{27} = k_{36} = k_{58} = -k_{18} = -k_{23} = -k_{45} \\
&= -k_{67} = \frac{D_{12}}{4} - \frac{D_{33}}{4} \\
k_{15} &= k_{37} = \frac{-b^2 D_{11} - a^2 D_{33}}{6ab} \\
k_{17} &= -k_{35} = \frac{b^2 D_{11} - 2a^2 D_{33}}{6ab} \\
k_{22} &= k_{44} = k_{66} = k_{88} = \frac{a^2 D_{11} + b^2 D_{33}}{3ab} \\
k_{24} &= k_{68} = \frac{a^2 D_{11} - 2b^2 D_{33}}{6ab} \\
k_{26} &= k_{48} = \frac{-a^2 D_{11} - b^2 D_{33}}{6ab} \\
k_{28} &= k_{46} = \frac{-2a^2 D_{11} + b^2 D_{33}}{6ab}
\end{aligned}
\tag{24.10}
$$

where we note that the elements D_{ij} of $[D]$ are proportional to the plate thickness and for the case of plane stress and isotropy are given by (9.42)–(9.46).

Some of the symmetries—determination of the signs may require special attention—are quite obvious, e.g. $k_{13} = k_{57}$, while others depend on the isotropy of the material in a more involved way.

24.2.5 A Deficiency of the Melosh Element

Deficiency of the Melosh element

Melosh element deficient in in-plane bending

Although the above finite element complies with all the basic rules for developing in-plane plate elements it is not necessarily perfect in all respects which we show below. If we wish to model a deep beam under bending using an in-plane plate element it would be desirable if the element could deform as shown to the left in Fig. 24.5, but the Melosh element is only able to deform in ways which keep its sides straight as shown in the right part of the figure. On the other hand, if we employ a large number of elements

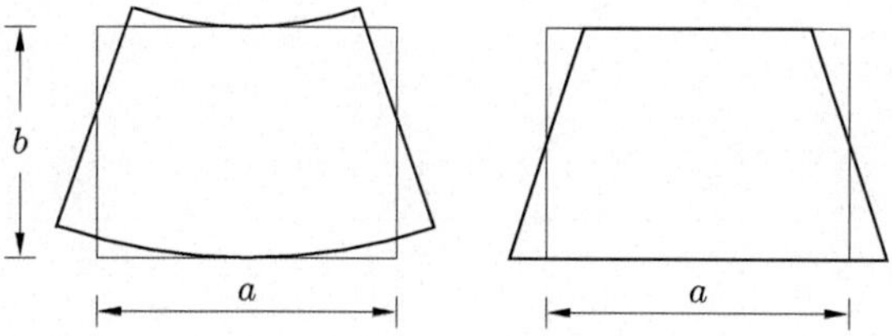

Fig. 24.5: In-plane bending of continuum and of Melosh element.

we may expect that this deficiency becomes less and less noticeable which proves to be true, see Example Ex 27-1, where its performance is compared with that of a stress hybrid finite element which looks like Melosh's but works much better in bending. The basic problem with the Melosh element in connection with bending is that it predicts shear strains associated with the displacement field shown to the right in the figure, while pure bending entails only normal strains in the x-direction. The spurious shear strain causes the Melosh element to be too stiff in cases which contain an amount of in-plane bending.[24.3] Cook et al. (2002) computes the difference in strain energy between the two types of displacement shown in the figure. In part because of the investigation in Example Ex 27-1 I consider it less pertinent for our purposes to go through Cook's derivations.

[24.3] This effect is called *shear locking*—a term which I don't like, but must accept as the standard.

Chapter 25

Internal Nodes and Their Elimination

This chapter has a two-fold purpose. The first consists in showing how internal nodes may be handled, the other is to introduce programs for manipulations of analytic expressions. For the latter I have chosen the program maxima—not because it is the best one available, but because it is free and may be downloaded for Linux, Mac and Windows. Actually, I used MuPAD[25.1] for a long time and liked it very much, but it became very expensive. To be quite honest, it took me some time to learn the little I know about maxima because its syntax is very different from MuPAD's and the syntax of most of the programming languages I have used over the past many years, namely Algol, Fortran, Pascal, C++ and the one I liked the best, PL/I.

Elimination of internal nodes
Programs to do analytic manipulations

The derivations take a bar with internal nodes as their point of departure, but the ensuing matrix algebra is generally valid. I am certain that you are able to differentiate the general expressions from the special ones.

The derivations below are of a general nature

25.1 Introduction

The derivation of finite element stiffness matrices often involves a great deal of analytic manipulations, which *can* be done by hand. However, as most people know, manipulations by hand are tedious and usually provide results that must be checked numerous times before you may trust them.

Below you will find an example of how maxima may be used for the purpose mentioned above. You may judge the example to be very pedestrian, but note that much of the procedure may be taken over almost directly when you do another example.

25.2 Structural Problem

Let us consider the problem of establishing the stiffness matrix of a linearly elastic bar element with cross-sectional properties that vary along the

Bar finite element

[25.1] A MuPAD-program looks very much like one written in Maple.

E. Byskov, *Elementary Continuum Mechanics for Everyone*,
Solid Mechanics and Its Applications 194, DOI: 10.1007/978-94-007-5766-0_25,

Many, simple or fewer, more accurate elements?

length.[25.2] We may, of course, subdivide the structure into many short elements at the cost of a larger system of equations, or we may derive stiffness matrices that are more accurate and use fewer elements, which is our intention here.

25.3 Nondimensional Quantities

The physical length of the bar element is L, and the physical axial coordinate is x. Neither of these quantities is convenient to work with, so we introduce the nondimensional axial coordinate ξ, where

Nondimensional coordinate

$$\xi \equiv \frac{x}{L}, \quad \xi \in [0,1] \tag{25.1}$$

We assume that the axial stiffness EA varies along the bar

Varying axial stiffness

$$EA(\xi) = (EA)_0 \widetilde{EA}(\xi) \tag{25.2}$$

where $\widetilde{EA}(\xi)$ is nondimensional and is given as input to the maxima program.

The reason why I use a formulation in terms of nondimensional quantities is that otherwise the expressions below become cluttered by many occurrences of L and EA. In the present example it is quite straightforward to use the resulting expressions and introduce the physical quantities.

25.4 Displacement and Displacement Interpolation

Let $u(\xi)$ be the nondimensional axial displacement and let N_{No} denote the number of nodes, which in this case is also the number of degrees of freedom of the element, then

Displacement interpolation

$$u(\xi) = [N(\xi)]\{v\} \tag{25.3}$$

where $[N(\xi)]$ is the displacement interpolation matrix and $\{v\}$ denotes the nodal displacement vector, and both contain N_{No} elements. Note that in the present case $[N]$ only has one row and therefore could be considered a row vector. This observation is not crucial to the derivation below, and we shall refer to $[N]$ as a matrix.

In some cases, such as the present one,[25.3] we can construct $[N]$ in a simple manner. Among the properties of $[N]$ are the following: element number i of $[N]$, i.e. N_i takes the value 1 at node number i and vanishes at

[25.2] The element derived below is not necessarily important in itself, but I use it as an example because it is simple and yet displays all the central features of the procedure of eliminating internal nodes.

[25.3] The plate finite element discussed in Section 24.2 was established in a manner that is as simple and based on similar observations.

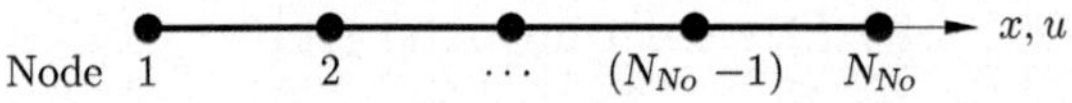

Fig. 25.1: Bar finite element. Nodes $2, \ldots, (N_{No}-)1$ are internal.

all other nodes. If $(\xi_j - \xi)$ is a factor of N_i, where ξ_j is the nondimensional coordinate of node number j, then $N_j = 0$, $j \neq i$. Therefore, we can write

$N_i(\xi)$ vanishes at all other nodes than number i

$$N_i(\xi) = C_i \prod_{j \neq i} (\xi_j - \xi) \tag{25.4}$$

where C_i is a factor determined by the requirement that $N_i(x_i) = 1$. We shall take care of this in a very simple manner. First we compute

$N^{(h)}(\xi)$ vanishes at *all* nodes

$$N^{(h)}(\xi) = \prod_{j=1}^{N_{No}} (\xi_j - \xi) \tag{25.5}$$

which vanishes at *all* nodes. Then we can construct N_i as follows

$N_i^{(h)}(\xi)$ vanishes at all other nodes than number i

$$N_i^{(h)}(\xi) = \frac{N^{(h)}}{(\xi_i - \xi)} \tag{25.6}$$

where we note that division by $(\xi_i - \xi)$ is permitted as long as we do it analytically and insure that the numerator in (25.6) is divided out. Except for the factor C_i the right-hand side of (25.6) is equal to $N_i(\xi)$. By dividing $N_i^{(h)}(\xi)$ by $N_i^{(h)}(\xi_i)$ we get the desired result

The final formula for $N_i(\xi)$

$$N_i(\xi) = \frac{N_i^{(h)}(\xi)}{N_i^{(h)}(\xi_i)} \tag{25.7}$$

A comment about the above manipulations seems relevant. If we were doing the analysis by hand it would be right out stupid to find the elements of $[N]$ in the way we followed. On the other hand, the procedure is very systematic and therefore easy to program, as we shall see.

If we take $N_{No} = 3$, maxima provides

$$[N] = \left[2\,(\xi - 1)\,(\xi - 1/2)\,;\; -4\xi\,(\xi - 1)\,;\; 2\xi\,(\xi - 1/2) \right] \tag{25.8}$$

which indeed has the properties we expect—you may easily check the expressions for the elements of $[N]$ by hand. For clarity, I have inserted semicolons between the elements in $[N]$—maxima does not do this. Often you must doctor the output from such programs a little before it gets really readable.

25.5 Strain Distribution Matrix

As usual, we get the strain distribution matrix $[B(\xi)]$ from $[N(\xi)]$ by proper differentiations. Here, this is very easy, because the axial strain $\varepsilon(\xi)$ is given as the first derivative of the axial displacement $u(\xi)$. Thus,

Strain distribution

$$\varepsilon(\xi) = [B(\xi)]\{v\} \tag{25.9}$$

where

Strain distribution matrix $[B]$

$$B_i(\xi) = \frac{dN_i(\xi)}{d\xi}\ , \quad i = 1, \ldots, N_{No} \tag{25.10}$$

where we mention that the quantities are nondimensional, which is why the length of the element does not enter in (25.10). All of of the programs mentioned above have a function to compute derivatives and for $N_{No} = 3$ maxima gives

Strain distribution matrix $[B]$

$$[B] = \begin{bmatrix} 4\xi - 3\ ;\ -8\xi + 4\ ;\ 4\xi - 1 \end{bmatrix} \tag{25.11}$$

which, again may be checked by hand.

25.6 Stiffness Matrix

The stiffness matrix $[k]$ is computed as

Element stiffness matrix $[k]$

$$[k] = \int_0^1 [B]^T[D][B]d\xi \tag{25.12}$$

where ^{T} denotes the matrix transpose, and where I emphasize that $[D] = [D(\xi)]$. The way the axial stiffness varies is input to the program and may be arbitrary as long as it can be given in a form that can be understood by maxima. As an example, I have taken[25.4]

Simple choice of axial stiffness

$$\widetilde{EA} = 1 - \alpha\xi \tag{25.13}$$

and for $N_{No} = 3$ maxima gives[25.5]

$$\begin{aligned} k_{11} &= \int_0^1 \left(-24\xi - 9\xi\alpha + 16\xi^2 + 24\xi^2\alpha - 16\xi^3\alpha + 9\right) d\xi \\ k_{12} &= \int_0^1 \left(40\xi + 12\xi\alpha - 32\xi^2 - 40\xi^2\alpha + 32\xi^3\alpha - 12\right) d\xi \\ k_{13} &= \int_0^1 \left(-16\xi - 3\xi\alpha + 16\xi^2 + 16\xi^2\alpha - 16\xi^3\alpha + 3\right) d\xi \end{aligned} \tag{25.14}$$

with similar expressions for the remaining elements of $[k]$.

[25.4] Remember to multiply all stiffness matrix elements below by $(EA)_0$ in order to get the proper results.
[25.5] Again, I have manipulated the output a little to make it look nice.

The result is quite simple

$$[k] = \begin{bmatrix} \left(-\frac{\alpha}{2}+\frac{7}{3}\right) ; & \left(\frac{2\alpha}{3}-\frac{8}{3}\right) ; & \left(-\frac{\alpha}{6}+\frac{1}{3}\right) \\ \left(\frac{2\alpha}{3}-\frac{8}{3}\right) ; & \left(-\frac{8\alpha}{3}+\frac{16}{3}\right) ; & \left(2\alpha-\frac{8}{3}\right) \\ \left(-\frac{\alpha}{6}+\frac{1}{3}\right) ; & \left(2\alpha-\frac{8}{3}\right) ; & \left(-\frac{11\alpha}{6}+\frac{7}{3}\right) \end{bmatrix} \qquad (25.15)$$

Element stiffness matrix $[k]$

where we must remember the factor $(EA)_0$, see footnote on page 464.

25.7 Elimination of Internal Nodes

We may note that the internal nodes—in the present case there is only one internal node, namely number 2—are not connected to the rest of the structure. Therefore, it may seem reasonable to try to eliminate them at the element level and in that way get a smaller system of equations.[25.6]

Internal nodes may be eliminated at element level

It is important to note that the following procedure for eliminating the internal degrees of freedom in no way is limited to the present example. On the contrary, it works for solids, beams, plates, shells, etc., but in this example we may follow not only the matrix algebra, but also its interpretation step by step and see the contents of the matrices involved in the manipulations.

The elimination procedure can be applied to many other kinds of elements

It seems fair to mention that we may achieve our objective of elimination in several different ways. Here we choose to base the derivation on the *Potential Energy* $\mathbf{\Pi}_{\mathrm{P}}$ of the finite element, which is

$$\mathbf{\Pi}_{\mathrm{P}}(\{v\}) = \tfrac{1}{2}\{v\}^T[k]\{v\} - \{v\}^T\{\bar{r}\} \qquad (25.16)$$

Here, we use the potential energy $\mathbf{\Pi}_{\mathrm{P}}$ to eliminate internal degrees of freedom

where $\{v\}$ is the vector of nodal displacements and $\{\bar{r}\}$ denotes the vector of nodal loads which may stem from loading distributed over the length of the element. As usual, the overbar indicates that the nodal loads are prescribed. In order to eliminate the internal degrees of freedom we must divide $\{v\}$ into a vector $\{v_{\mathrm{i}}\}$, which contains the displacements of the internal, or free nodes, and another vector $\{v_{\mathrm{e}}\}$ that contains the displacements of the external, or end nodes. The latter vector has 2 elements, while the former has $(N_{No}-2)$ elements. The connection between $\{v_{\mathrm{i}}\}$ and $\{v_{\mathrm{e}}\}$ and $\{v\}$ may be given as follows

$\{v_{\mathrm{i}}\}$ contains internal degrees of freedom
$\{v_{\mathrm{e}}\}$ contains the external ones

$$\{v_{\mathrm{i}}\} = [T_{\mathrm{i}}]\{v\} \text{ and } \{v_{\mathrm{e}}\} = [T_{\mathrm{e}}]\{v\} \qquad (25.17)$$

Connection between $\{v_{\mathrm{i}}\}$ and $\{v\}$ and between $\{v_{\mathrm{e}}\}$ and $\{v\}$

respectively.

[25.6] The following elimination of the internal node(s) can also be done when the stiffness matrices, etc. have been computed numerically—the matrix algebra works on the matrices whether they have been computed analytically or not.

In the present case we get

$[T_i]$ and $[T_e]$ for $N_{No} = 3$

$$[T_{\rm i}] = [0\ 1\ 0] \ \text{ and } \ [T_{\rm e}] = \begin{bmatrix} 1 & 0 & 0 \\ 0 & 0 & 1 \end{bmatrix} \tag{25.18}$$

but see also the maxima program, where the algorithms for computing $\{v_{\rm i}\}$ and $\{v_{\rm e}\}$ are shown—they are not at all difficult to establish, as you will see. Actually, we need to write $\{v\}$ in terms of $\{v_{\rm i}\}$ and $\{v_{\rm e}\}$. You may easily check that the following formula is correct

$\{v\}$ in terms of $\{v_i\}$ and $\{v_e\}$

$$\{v\} = [T_{\rm i}]^T\{v_{\rm i}\} + [T_{\rm e}]^T\{v_{\rm e}\} \tag{25.19}$$

Using (25.19) and in that way having divided $\{v\}$ in $\{v_{\rm i}\}$ and $\{v_{\rm e}\}$ we may write the potential energy of (25.16) as follows

$$\begin{aligned} &\mathbf{\Pi}_{\rm P}(\{v_{\rm i}\},\{v_{\rm e}\}) \\ &= \tfrac{1}{2}\big[\{v_{\rm i}\}^T[T_{\rm i}] + \{v_{\rm e}\}^T[T_{\rm e}]\big][k]\big[[T_{\rm i}]^T\{v_{\rm i}\} + [T_{\rm e}]^T\{v_{\rm e}\}\big] \\ &\quad - \big[\{v_{\rm i}\}^T[T_{\rm i}] + \{v_{\rm e}\}^T[T_{\rm e}]\big]\{\bar{r}\} \end{aligned} \tag{25.20}$$

or

$$\begin{aligned} \mathbf{\Pi}_{\rm P}(\{v_{\rm i}\},\{v_{\rm e}\}) = \tfrac{1}{2}\big[\{v_{\rm i}\}^T \; ; \; \{v_{\rm e}\}^T\big] \begin{bmatrix} [k_{\rm ii}] & [k_{\rm ie}] \\ [k_{\rm ei}] & [k_{\rm ee}] \end{bmatrix} \begin{bmatrix} \{v_{\rm i}\} \\ \{v_{\rm e}\} \end{bmatrix} \\ - \big[\{v_{\rm i}\}^T \; ; \; \{v_{\rm e}\}^T\big] \begin{bmatrix} \{\bar{r}_{\rm i}\} \\ \{\bar{r}_{\rm e}\} \end{bmatrix} \end{aligned} \tag{25.21}$$

where the meaning of the notation of the matrices $[k_{\rm ii}]$, $[k_{\rm ie}]$, $[k_{\rm ei}]$, and $[k_{\rm ee}]$, and of the vectors $\{\bar{r}_{\rm i}\}$ and $\{\bar{r}_{\rm e}\}$ should be obvious. By comparing (25.20) and (25.21) we may find the various stiffness matrices

$$[k_{\rm ii}] = [T_{\rm i}][k][T_{\rm i}]^T = \left[\left(-\frac{8\alpha}{3} + \frac{16}{3}\right)\right] \tag{25.22}$$

$$[k_{\rm ie}] = [T_{\rm i}][k][T_{\rm e}]^T = \left[\left(\frac{2\alpha}{3} - \frac{8}{3}\right) ; \left(2\alpha - \frac{8}{3}\right)\right] \tag{25.23}$$

$$[k_{\rm ei}] = [T_{\rm e}][k][T_{\rm i}]^T = \begin{bmatrix} \left(\dfrac{2\alpha}{3} - \dfrac{8}{3}\right) \\ \left(2\alpha - \dfrac{8}{3}\right) \end{bmatrix} \tag{25.24}$$

$$[k_{\rm ee}] = [T_{\rm e}][k][T_{\rm e}]^T = \begin{bmatrix} \left(-\dfrac{\alpha}{2} + \dfrac{7}{3}\right) ; & \left(-\dfrac{\alpha}{6} + \dfrac{1}{3}\right) \\ \left(-\dfrac{\alpha}{6} + \dfrac{1}{3}\right) ; & \left(-\dfrac{-11\alpha}{6} + \dfrac{7}{3}\right) \end{bmatrix} \tag{25.25}$$

The right-hand side vectors, the loading terms are

$$\{\bar{r}_{\rm i}\} = [T_{\rm i}]\{\bar{r}\} \tag{25.26}$$

$$\{\bar{r}_{\rm e}\} = [T_{\rm e}]\{\bar{r}\} \tag{25.27}$$

where $\{\bar{r}_{\rm i}\}$ and $\{\bar{r}_{\rm e}\}$ depend on the applied loads.

Since $[k]$ is symmetric, as usual, the following relation always holds

$$[k_{\rm ie}] = [k_{\rm ei}]^T \tag{25.28}$$

see also (25.23) and (25.24).

We may now rewrite (25.21)

$$\begin{aligned}\boldsymbol{\Pi}_{\rm P}(\{v_{\rm i}\},\{v_{\rm e}\}) = &+\tfrac{1}{2}\{v_{\rm i}\}^T[k_{\rm ii}]\{v_{\rm i}\} + \tfrac{1}{2}\{v_{\rm i}\}^T[k_{\rm ie}]\{v_{\rm e}\}\\ &+\tfrac{1}{2}\{v_{\rm e}\}^T[k_{\rm ei}]\{v_{\rm i}\} + \tfrac{1}{2}\{v_{\rm e}\}^T[k_{\rm ee}]\{v_{\rm e}\}\\ &-\{v_{\rm i}\}^T\{\bar{r}_{\rm i}\} - \{v_{\rm e}\}^T\{\bar{r}_{\rm e}\}\end{aligned} \tag{25.29}$$

The crux of the matter now is that we may vary $\boldsymbol{\Pi}_{\rm P}$ with respect to $\{v_{\rm i}\}$ independently for each element because the internal nodes are not connected to other elements. It is *not* permitted to vary with respect to $\{v_{\rm e}\}$ without taking the connectedness of the structure into account. Bearing this in mind we may demand

We may vary $\{v_{\rm i}\}$, *not* $\{v_{\rm e}\}$, at the element level

$$\delta\boldsymbol{\Pi}_{\rm P}\Big|_{\{v_{\rm i}\}} = 0 \;\forall\; \{\delta v_{\rm i}\} \tag{25.30}$$

which becomes

$$\begin{aligned} 0 &= \{\delta v_{\rm i}\}^T[k_{\rm ii}]\{v_{\rm i}\} + \{\delta v_{\rm i}\}^T[k_{\rm ie}]\{v_{\rm e}\} - \{\delta v_{\rm i}\}^T\{\bar{r}_{\rm i}\} \;\forall\; \{\delta v_{\rm i}\}\\ \Rightarrow \{0\} &= [k_{\rm ii}]\{v_{\rm i}\} + [k_{\rm ie}]\{v_{\rm e}\} - \{\bar{r}_{\rm i}\}\end{aligned} \tag{25.31}$$

Since $\boldsymbol{\Pi}_{\rm P}$ is positive definite $[k_{\rm ii}]$ is non-singular and $[k_{\rm ii}]^{-1}$ exists.[25.7] Therefore, we may solve for $\{v_{\rm i}\}$

$$\{v_{\rm i}\} = [k_{\rm ii}]^{-1}\Big(\{\bar{r}_{\rm i}\} - [k_{\rm ie}]\{v_{\rm e}\}\Big) \tag{25.32}$$

When we introduce (25.32) in (25.29) and exploit (25.28) we may realize that the second and third terms of (25.29) are equal and arrive at

$$\begin{aligned}\boldsymbol{\Pi}_{\rm P}(\{v_{\rm e}\}) = &+\tfrac{1}{2}\big(\{\bar{r}_{\rm i}\}^T - \{v_{\rm e}\}^T[k_{\rm ie}]^T\big)[k_{\rm ii}]^{-1}[k_{\rm ii}]\\ &\qquad [k_{\rm ii}]^{-1}\big(\{\bar{r}_{\rm i}\} - [k_{\rm ie}]\{v_{\rm e}\}\big)\\ &+\big(\{\bar{r}_{\rm i}\}^T - \{v_{\rm e}\}^T[k_{\rm ie}]^T\big)\,[k_{\rm ii}]^{-1}[k_{\rm ie}]\{v_{\rm e}\}\\ &+\tfrac{1}{2}\{v_{\rm e}\}^T[k_{\rm ee}]\{v_{\rm e}\}\\ &-\big(\{\bar{r}_{\rm i}\}^T - \{v_{\rm e}\}^T[k_{\rm ie}]^T\big)\,[k_{\rm ii}]^{-1}\{\bar{r}_{\rm i}\}\\ &-\{v_{\rm e}\}^T\{\bar{r}_{\rm e}\}\end{aligned} \tag{25.33}$$

[25.7] Also, physical reasoning shows that $[k_{\rm ii}]$ is non-singular.

Written out, this becomes

$$\begin{aligned}\mathbf{\Pi}_{\mathrm{P}}(\{v_{\mathrm{e}}\}) = & -\tfrac{1}{2}\{\bar{r}_{\mathrm{i}}\}^T[k_{\mathrm{ii}}]^{-1}\{\bar{r}_{\mathrm{i}}\} + \{v_{\mathrm{e}}\}^T[k_{\mathrm{ie}}]^T[k_{\mathrm{ii}}]^{-1}\{\bar{r}_{\mathrm{i}}\} \\ & -\tfrac{1}{2}\{v_{\mathrm{e}}\}^T[k_{\mathrm{ie}}]^T[k_{\mathrm{ii}}]^{-1}[k_{\mathrm{ie}}]\{v_{\mathrm{e}}\} + \tfrac{1}{2}\{v_{\mathrm{e}}\}^T[k_{\mathrm{ee}}]\{v_{\mathrm{e}}\} \\ & -\{v_{\mathrm{e}}\}^T\{\bar{r}_{\mathrm{e}}\}\end{aligned} \tag{25.34}$$

where (25.28) has been exploited several times.

Terms that do not contain $\{v_{\mathrm{e}}\}$ as a factor are constant under variation with respect to $\{v_{\mathrm{e}}\}$ and may therefore be excluded. Thus,

$$\begin{aligned}\mathbf{\Pi}_{\mathrm{P}}(\{v_{\mathrm{e}}\}) = & +\tfrac{1}{2}\{v_{\mathrm{e}}\}^T\left([k_{\mathrm{ee}}] - [k_{\mathrm{ie}}]^T[k_{\mathrm{ii}}]^{-1}[k_{\mathrm{ie}}]\right)\{v_{\mathrm{e}}\} \\ & -\{v_{\mathrm{e}}\}^T\left(\{\bar{r}_{\mathrm{e}}\} - [k_{\mathrm{ie}}]^T[k_{\mathrm{ii}}]^{-1}\{\bar{r}_{\mathrm{i}}\}\right)\end{aligned} \tag{25.35}$$

When we introduce

Stiffness matrix and load term after elimination of internal degrees of freedom

$$[k^*_{\mathrm{ee}}] \equiv [k_{\mathrm{ee}}] - [k_{\mathrm{ie}}]^T[k_{\mathrm{ii}}]^{-1}[k_{\mathrm{ie}}] \tag{25.36}$$

and

$$\{\bar{r}^*_{\mathrm{e}}\} \equiv \{\bar{r}_{\mathrm{e}}\} - [k_{\mathrm{ie}}]^T[k_{\mathrm{ii}}]^{-1}\{\bar{r}_{\mathrm{i}}\} \tag{25.37}$$

we may write the potential energy of the finite element as

$$\mathbf{\Pi}_{\mathrm{P}}(\{v_{\mathrm{e}}\}) = \tfrac{1}{2}\{v_{\mathrm{e}}\}^T[k^*_{\mathrm{ee}}]\{v_{\mathrm{e}}\} - \{v_{\mathrm{e}}\}^T\{\bar{r}^*_{\mathrm{e}}\} \tag{25.38}$$

As we may note, (25.38) has the same form as (25.16) which means that the new finite element stiffness matrix $[k^*_{\mathrm{ee}}]$ and the associated new right-hand side vector $\{\bar{r}^*_{\mathrm{e}}\}$ may be used in a conventional finite element program without much difficulty.

In our example[25.8]

$$[k^*_{\mathrm{ee}}] = \begin{bmatrix} \dfrac{6\alpha - \alpha^2 - 6}{3\alpha - 6} & \dfrac{-6\alpha + \alpha^2 + 6}{3\alpha - 6} \\ \dfrac{-6\alpha + \alpha^2 + 6}{3\alpha - 6} & \dfrac{6\alpha - \alpha^2 - 6}{3\alpha - 6} \end{bmatrix} \tag{25.39}$$

No Free Lunches

There are no free lunches

You may see that in order to compute the axial strain ε we need more information than we possess after having eliminated the internal node(s)—as usual, there is no such thing as a free lunch. It is possible to compute the axial strain, but it becomes a little involved because we must determine the

[25.8] Remember the factor $(EA)_0$, see footnote on page 464.

displacements $\{v\}$ of *all* nodes, and we only know the displacements $\{v_e\}$ of the external nodes. Go back to (25.19)

$$\{v\} = [T_i]^T\{v_i\} + [T_e]^T\{v_e\} \tag{25.40}$$

and utilize (25.32)

$$\{v_i\} = [k_{ii}]^{-1}(\{\bar{r}_i\} - [k_{ie}]\{v_e\}) \tag{25.41}$$

to get

$$\{v\} = \Big([T_e]^T - [T_i]^T[k_{ii}]^{-1}[k_{ie}]\Big)\{v_e\} + [T_i]^T[k_{ii}]^{-1}\{\bar{r}_i\} \tag{25.42}$$

Define

$$[T_e^*] \equiv [T_e]^T - [T_i]^T[k_{ii}]^{-1}[k_{ie}] \tag{25.43}$$

and

$$\{\bar{r}^*\} \equiv [T_i]^T[k_{ii}]^{-1}\{\bar{r}_i\} \tag{25.44}$$

where, in our example

$$[T_e^*] = \begin{bmatrix} 1 & 0 \\ \dfrac{\alpha-4}{4\alpha-8} & \dfrac{3\alpha-4}{4\alpha-8} \\ 0 & 1 \end{bmatrix} \tag{25.45}$$

and introduce these quantities in (25.42)

$$\{v\} = [T_e^*]\{v_e\} + \{\bar{r}^*\} \tag{25.46}$$

Finally, we may compute the axial strain from, see (25.9)

$$\varepsilon = [B]\Big([T_e^*]\{v_e\} + \{\bar{r}^*\}\Big) \quad \text{or} \quad \varepsilon = [B_e^*]\{v_e\} + \varepsilon^* \tag{25.47}$$

where

$$[B_e^*] \equiv [B][T_e^*] \text{ and } \varepsilon^* \equiv [B]\{\bar{r}^*\} \tag{25.48}$$

We may see that the last term in the parenthesis of (25.47a) depends on the load applied to the internal node that we wished to eliminate in the first place. There is, however, no mystery hidden here because the full vector $\{v\}$ must depend on these loads. On the other hand, if the applied loads are acting on the external nodes only, then the $\{\bar{r}^*\}$ vanishes, and the computation of ε is almost as usual.

25.8 Program written in maxima

maxima program

Regarding the program listed below I do not claim that it is perfect or particularly efficient. Its main virtue is that it works and, since it is not supposed to be run more than very few times, efficiency is of minor concern compared with readability. As you may see below, the program consists of a main program which loads various sub-programs. I have done this because it makes checking the program much easier.

The contents of the main program `BarMaxima.mac` is

```
/* We wish to be able to choose the number of nodes.                    */
print("Input number of nodes, NNodes");
read("Write, NNodes: <NNodes>; (don't forget the ; at the end) ");

/* We wish to output in TeX format.                                     */
TeXFile: openw("barelm.out");
/* We would also like to be able to output c code.                      */
CFile: openw("barelm.c");

printf(TeXFile,"NNodes = ~%");
tex(NNodes,TeXFile);

printf(CFile,"NNodes = ~d;~%",NNodes);

load("RowMat2CFileProc.mac");
load("ColMat2CFileProc.mac");
load("Matrix2CFileProc.mac");
load("CoorMatProc.mac");
load("NMatProc.mac");
load("CheckNMatProc.mac");
load("BMatProc.mac");
load("DMatProc.mac");
load("KMatProc.mac");
load("TMatExtProc.mac");
load("TMatIntProc.mac");
load("KMatExtExtStarProc.mac");
load("TMatExtStarProc.mac");

/* The coordinates of the nodes will be stored in                       */
/* CoorMat[1..NNodes]. Here, the coordinates are the                    */
/* abscissae of the points along the bar, but in other cases            */
/* we shall need to store the values in matrices, whose size            */
/* depends on the spatial dimension of the structure.                   */
/* Therefore, NMat instead of NVec. We assume that the nodal            */
/* points are distributed evenly along the length of the bar,           */
/* whose length is 1. The values of CoorMat are written to              */
/* the TeX-file and the c-file.                                         */
CoorMat: CoorMatProc(NNodes);

/* Now define the displacement interpolation vector NMat.               */
/* Here, we insist that NMat is a row vector. When there are            */
/* more than one generalized strain information about                   */
/* interpolation must be stored in a matrix. Hence, NMat                */
/* instead of NVec.                                                     */
NMat: NMatProc(NNodes, CoorMat);

/* Let's check that NMat takes the values we expect at the              */
/* nodes, i.e. either a zero or a one. To this end we use the           */
/* function CheckNMatProc.                                              */
CheckNMatProc(NNodes,NMat);
```

```
/* We compute the B-matrix.                                    */
BMat: BMatProc(NNodes, NMat);

/* Now you must input the constitutive matrix, i.e. DMat,     */
/* which here is a 1 by 1 matrix, i.e. DMat[1,1].             */
DMat: DMatProc();

/* Now we are ready to compute the stiffness matrix, i.e.     */
KMat: KMatProc(BMat, DMat, NNodes);



/* If the total number of nodes NNodes is greater than 2 we   */
/* would probably like to eliminate the internal degrees of   */
/* freedom since the bar is only connected to the rest of the */
/*  world through its end nodes.  This is somewhat complicated */
/* in theory but fairly straightforward to do, once you have  */
/* understood the theory.  First we must establish the matrix */
/* TMatExt which determines the end node displacement vector  */
/* VVecExt from the the total displacement vector VVec as well */
/* as the matrix TMatInt which gives the displacement vector  */
/* VVecInt in terms of VVec.                                  */
if (NNodes > 2) then
  (TMatExt: TMatExtProc(NNodes),
   TMatInt: TMatIntProc(NNodes),
   KMatExtExtStar:
     KMatExtExtStarProc(KMat, TMatExt, TMatInt, NNodes),
   TMatExtStar:
     TMatExtStarProc(TMatExt, TMatInt, KMatIntExt, NNodes));

/* Remember to close the files.                               */
close(TeXFile);
close(CFile);

quit;
```

The contents of the functions used in `BarMaxima.mac` are:

```
/* In order for our program to be reasonably flexible we write */
/* a function to print row vectors to the CFile.              */
/*                                                            */
RowMat2CFileProc(MatName,Mat,NNodes)
  := block(local(i),
     for i:1 thru NNodes do
       (printf(CFile,"~a(~d) = ~a;~%", MatName, i, Mat[1,i])));
```

```
/* In order for our program to be reasonably flexible we write */
/* a function to print column vectors to the CFile.           */
/*                                                            */
ColMat2CFileProc(MatName,Mat,NNodes)
  := block(local(i),
     for i:1 thru NNodes do
       (printf(CFile,"~a(~d) = ~a;~%", MatName, i, Mat[i])));
```

```
/* In order for our program to be reasonably flexible we write */
/* a function to print two-dimensional matrices to the CFile. */
/*                                                            */
Matrix2CFileProc(MatName,Mat,ilim, jlim)
  := block(local(i,j),
     for i:1 thru ilim do
       for j:1 thru jlim do
         (printf(CFile,"~a(~d,~d) = ~a;~%", MatName, i, j,
                        Mat[i,j])));
```

```
/* In order for our program to be reasonably flexible we write */
/* a function to compute CoorMat.                               */
/*                                                              */
CoorMatProc(NNodes)
        := block(local(i,CoorAll,xi),
           /* Define CoorMat[1..NNodes] containing abscissae   */
           /*  of the nodes.                                    */
           CoorMat: zeromatrix(NNodes,1),
           /* We assume that the nodal points are distributed */
           /* evenly along the length of the bar, whose        */
           /* length is 1.  Then, their coordinates are        */
           /* computed as follows.                             */
           for i:1 thru NNodes do
             (CoorMat[i,1]: (i-1)/(NNodes-1)),
           /* Write the values of CoorMat to the TeX-file and */
           /* the c-file.                                      */
           printf(TeXFile, "CoorMat: "),
           tex(CoorMat,TeXFile),
           ColMat2CFileProc("CoorMat",CoorMat,NNodes),
           return(CoorMat));
```

```
/* In order for our program to be reasonably flexible we write */
/* a function to compute NMat.                                  */
/*                                                              */
NMatProc(NNodes, CoorMat)
        := block(local(i,CoorAll,xi),
           /* The displacement interpolation vector is NMat,   */
           /* NMat is a row vector: NMat[1,1..NNodes].         */
           /* In maxima a row vector is like a matrix with     */
           /* row.                                             */
           NMat: transpose(zeromatrix(NNodes,1)),
           /* Construct an expression that vanishes at all    */
           /* nodal points.                                    */
           CoorAll: 1,
           for i:1 thru NNodes do
             (CoorAll: CoorAll*(xi -CoorMat[i,1])),
           /* We know that NMat[1,i] should vanish at all     */
           /* other nodal points than point number i and      */
           /* compute:                                         */
           for i:1 thru NNodes do
             (NMat[1,i]: CoorAll/(xi -CoorMat[i,1])),
           /* The value of NMat at node i should be 1, so we  */
           /* do a little more arithmetic.                     */
           for i:1 thru NNodes do
             (NMat[1,i]:
                NMat[1,i]/subst(CoorMat[i,1],xi,NMat[1,i])),
           /* Write the values of NMat to the TeX-file and    */
           /* the c-file.                                      */
           printf(TeXFile, "NMat: "),
           tex(NMat,TeXFile),
           RowMat2CFileProc("NMat",NMat,NNodes),
           return(NMat));
```

```
CheckNMatProc(NNodes,NMat)
          :=block(local(i, xih),
             for i:1 thru NNodes do
               (xih: (i-1)/(NNodes-1),
                printf(TeXFile,"Value of $\\xi$: "),
                printf(TeXFile,"~a~%",xih),
                tex("Value of NMat:",TeXFile),
                NMatH:subst(xih,xi,NMat),
                tex(NMatH, TeXFile)));
```

```
/* In order for our program to be reasonably flexible we write */
/* a function to compute BMat.                                 */
/*                                                             */
BMatProc(NNodes, NMat)
        := block(
          /* The strain distribution vector is BMat.           */
          /* BMat is a row vector: BMat[1,1..NNodes].          */
          /* In maxima a row vector is like a matrix with      */
          /* row.                                              */
          BMat: expand(diff(NMat,xi)),
          /* Write the values of BMat to the TeX-file and      */
          /* the c-file.                                       */
          printf(TeXFile, "BMat: "),
          tex(BMat,TeXFile),
          RowMat2CFileProc("BMat",BMat,NNodes),
          return(BMat));
```

```
/* In order for our program to be reasonably flexible we write */
/* a function to compute DMat.                                 */
/*                                                             */
DMatProc()
        := block(
          /* The constitutive matrix is DMat. Here, DMat is a */
          /* matrix: DMat[1..1,1..1].                          */
          DMat: zeromatrix(1,1),
          print("Input the value of the axial stiffness, ~%"),
          print("i.e. EA (actually DMat[1,1])"),
          print("You may write letters instead of a value "),
          print("for EA.~%"),
          read("Write, EA: (don't forget the ; at the end) "),
          DMat[1,1]: EA,
          /* Write the values of DMat to the TeX-file and      */
          /* the c-file.                                       */
          printf(TeXFile, "DMat: "),
          tex(DMat,TeXFile),
          /* RowMat2CFileProc("DMat",DMat,1), */
          Matrix2CFileProc("DMat",DMat,1,1),
          return(DMat));
```

```
/* This function computes KMat without elimination of the     */
/* the internal degrees of freedom.                           */
/*                                                            */
KMatProc(BMat, DMat, NNodes)
  := block(local(i,j),
     KMat: transpose(BMat) . DMat . BMat,
     KMat: expand(KMat),
     KMat: integrate(KMat,xi,0,1),
     /* Write the values of KMat to the TeX-file and          */
     /* the c-file.                                           */
     printf(TeXFile, "KMat: "),
     tex(KMat,TeXFile),
     Matrix2CFileProc("KMat", KMat, NNodes, NNodes),
     return(KMat));
```

```
/* Here, we establish the matrix TMatExt that determines the  */
/*  end node displacement vector VVecExt from the the total   */
/* displacement vector VVec.                                  */
/*                                                            */
TMatExtProc(NNodes)
  := block(local(NExtDisp,NIntDisp,i,j),
     NExtDisp: 2,
     NIntDisp: NNodes -NExtDisp,
     TMatExt: zeromatrix(NExtDisp,NNodes),
     TMatExt[1,1]: 1,
```

```
      TMatExt[NExtDisp,NNodes]: 1,
      /* Write the values of TMatExt to the TeX-file              */
      /* and the c-file.                                          */
      printf(TeXFile, "TMatExt: "),
      tex(TMatExt,TeXFile),
      Matrix2CFileProc("TMatExt", TMatExt, NExtDisp, NNodes),
      return(TMatExt));
```

```
/* Here, we establish the matrix TMatInt which determines the   */
/* internal node displacement vector VVecInt from the the       */
/*  total displacement vector VVec as well as the matrix.       */
/*                                                              */
TMatIntProc(NNodes)
  := block(local(NExtDisp,NIntDisp,i,j),
      NExtDisp: 2,
      NIntDisp: NNodes -NExtDisp,
      TMatInt: zeromatrix(NIntDisp,NNodes),
      for i:2 thru NNodes -1 do
        (TMatInt[i-1,i]: 1),
      /* Write the values of TMatExt and TMatINt to the          */
      /* TeX-file and the c-file.                                */
      printf(TeXFile, "TMatInt: "),
      tex(TMatInt,TeXFile),
      Matrix2CFileProc("TMatInt", TMatInt, NIntDisp, NNodes),
      return(TMatInt));
```

```
/* This function computes KMatExtExtStar with elimination of    */
/* the the internal degrees of freedom.                         */
/*                                                              */
KMatExtExtStarProc(KMat, TMatExt, TMatInt, NNodes)
  := block(local(NExtDisp,NIntDisp),
      NExtDisp: 2,
      NIntDisp: NNodes -NExtDisp,
      KMatExtExt: TMatExt . KMat . transpose(TMatExt),
      KMatExtInt: TMatExt . KMat . transpose(TMatInt),
      KMatIntExt: transpose(KMatExtInt),
      KMatIntInt: TMatInt . KMat . transpose(TMatInt),
      /* Write the values of the various Kimatrices to the       */
      /* TeX-file and the c-file.                                */
      printf(TeXFile, "KMatExtExt: "),
      tex(KMatExtExt,TeXFile),
      Matrix2CFileProc("KMatExtExt", KMatExtExt, NExtDisp,
                                     NExtDisp),
      printf(TeXFile, "KMatExtInt: "),
      tex(KMatExtInt,TeXFile),
      Matrix2CFileProc("KMatExtInt", KMatExtInt, NExtDisp,
                                     NIntDisp),
      printf(TeXFile, "KMatIntExt: "),
      tex(KMatIntExt,TeXFile),
      Matrix2CFileProc("KMatIntExt", KMatIntExt, NIntDisp,
                                     NExtDisp),
      printf(TeXFile, "KMatIntInt: "),
      tex(KMatIntInt,TeXFile),
      Matrix2CFileProc("KMatIntInt", KMatIntInt, NIntDisp,
                                     NIntDisp),
      if (NIntDisp = 1) then
        (Kh: KMatIntInt^(-1) * transpose(KMatExtInt))
      else
        (Kh: KMatIntInt^^(-1) . transpose(KMatExtInt)),
      KMatExtExtStar: KMatExtExt
                     -KMatExtInt . Kh,
      KMatExtExtStar: factor(KMatExtExtStar),
      printf(TeXFile, "KMatExtExtStar: "),
      tex(KMatExtExtStar,TeXFile),
      Matrix2CFileProc("KMatExtExtStar", KMatExtExtStar,
```

```
                    NExtDisp, NExtDisp),
    return(KMatExtExtStar));
```

```
/* This function computes TMatExtStar which is used in the       */
/* computation of the axial strains when the internal degrees    */
/* of freedom have been eliminated.                              */
/*                                                               */
TMatExtStarProc(TMatExt, TMatInt, KMatIntExt, NNodes)
  := block(local(NExtDisp,NIntDisp,i,j),
    NExtDisp: 2,
    NIntDisp: NNodes -NExtDisp,
    if (NIntDisp = 1) then
      (Th: KMatIntInt^(-1) * KMatIntExt)
    else
      (Th: KMatIntInt^^(-1) . KMatIntExt),
    TMatExtStar:  transpose(TMatExt)
                 -transpose(TMatInt) . Th,
    TMatExtStar: factor(TMatExtStar),
    /* Write the values of TMatExtStar to the TeX-file and       */
    /* the c-file.                                               */
    printf(TeXFile, "TMatExtStar: "),
    tex(TMatExtStar,TeXFile),
    Matrix2CFileProc("TMatExtStar", TMatExtStar, NNodes, NExtDisp),
    return(TMatExtStar));
```

You may note that, especially in the main program, there are many comments and that the number of real code lines is much shorter than the total number. As usual, it is relatively inexpensive and a very good idea to write comments in the program listing. If the syntax of the language is compact it is especially important to write comments.[25.9] Fortunately, maxima's syntax makes for reasonably readable programs. Yet, I urge you to document your programs through comments.

[25.9] Once, a computer programmer with an M.Sc. told me that APL had such a compact notation that he could not read his own programs after a couple of weeks if he had not written a lot of comments.

Chapter 26

Circular Beam Finite Elements, Problems and Solutions

26.1 Introduction

While it is a fairly easy task to develop straight beam finite elements for the case of kinematic linearity, curved beam elements present much more severe problems. There are two causes for the difficulties. One has to do with the strain definitions where certain terms may not match and create internal incompatibilities in the element, see below. This effect is known as *locking*—a term that to my opinion is a misnomer—and should rather be called *internal mismatch* or something similar.[26.1] The other cause must be sought in the problems associated with satisfying the *Rigid-Body Criterion*, which simply states that if a finite element is subjected to a rigid body translation and rotation, then it must not experience strains. This requirement seems evident but is sometimes not easily obeyed. Most shell finite elements, in particular in the early days, exhibited self-straining.

Curved beam elements exhibit locking = Internal incompatibilities = internal mismatch

Rigid-body criterion

Self-straining

In this connection I do not intend to set up procedures for curved beam elements of arbitrary geometry, but limit the task to the case of circular beam elements. The reason for this is that general curved beam elements entail other problems than the ones I wish to focus on here. For example, the geometric description of the element becomes much more involved and the manipulations leading to the stiffness matrix cannot be done analytically. So, in order to bring out the message as clearly as possible we shall only concern ourselves with circular elements.

Stiffness matrices of general curved beam elements cannot be found analytically

Before we begin the derivations leading to two different circular finite elements it may be justified mentioning that in this particular case we could

[26.1] The topic of "locking" and the use of *Lagrange Multipliers* are two pet subjects of mine, so bear with me—this chapter may be a little long, but I think that it is justified. For my treatment of "locking" in a particular connection, see (Byskov 1989), for my advocacy of the use of Lagrange Multipliers you may consult other sections of this book, e.g. Section 33.7 and Example Ex 32-5.

E. Byskov, *Elementary Continuum Mechanics for Everyone*,
Solid Mechanics and Its Applications 194, DOI: 10.1007/978-94-007-5766-0_26,

For circular finite elements it is possible to find the exact expression for their stiffness matrix, but don't do it—the stiffness matrix has poor qualities for short element length

have based the development on exact displacement assumptions which entail trigonometric functions instead of polynomials. There is, however, a problem in this connection in that such a finite element becomes less and less numerically well-conditioned when its length decreases. Look at Fig. 26.2, p. 481, and imagine the element length becoming smaller and smaller. Then, let us notice that the exact displacement field in a circular element consists of trigonometric functions, while the the displacements of a straight element are composed by polynomials. There is, however, no way that the trigonometric functions can turn into polynomials by themselves. For very small lengths one would have to use Taylor expansions of the trigonometric functions in order that the finite element behaves well in such cases. The conclusion is that the exact stiffness matrix must become less and less reliable the shorter the element.

26.2 Strains

As mentioned above, the structure we are dealing with is a beam of circular shape. Its radius is R, and we describe the position on the beam by the the axial coordinate s, which is the length along the beam, see Fig. 26.1, but shift to an angle Φ later, see (26.4). We denote the axial displacement component v and the transverse displacement component w. We shall also need the rotation of the beam, which we call ω.

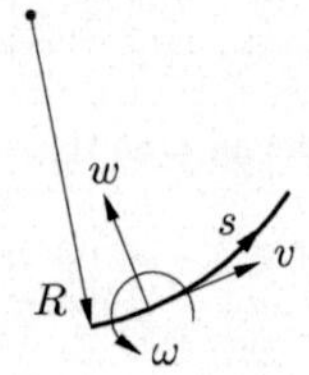

Fig. 26.1: Circular structure

We assume that the Bernoulli-Euler beam theory holds, i.e. the beam does not experience shear deformations. Then, there are only two strain components, the *generalized strains*, namely the axial strain ε and the bending strain and κ. For kinematic linearity the expressions for these strains are, see Section 8.3.1

Generalized strains ε and κ

$$\varepsilon \equiv v' - \frac{w}{R} \quad \text{and} \quad \kappa \equiv w'' + \frac{v'}{R} \tag{26.1}$$

where prime ($'$) designates differentiation with respect to s

$$(\,)' \equiv \frac{d(\,)}{ds} \tag{26.2}$$

and s is the physical coordinate.

The rotation is given by

$$\omega \equiv w' + \frac{v}{R} \tag{26.3}$$

Rotation ω

Later, it proves convenient for us to use the angle Φ instead of s as the independent variable

$$\Phi \equiv \frac{s}{R} \tag{26.4}$$

Nondimensional variable Φ

Let an over-dot denote differentiation with respect to Φ, then

$$\dot{(\ )} \equiv \frac{d(\ )}{d\Phi} = R\frac{d(\ )}{ds} \tag{26.5}$$

Differentiation with respect to Φ

and we may rewrite (26.1) and (26.3)

$$\varepsilon = \frac{1}{R}(\dot{v} - w)\,, \quad \kappa = \frac{1}{R^2}(\ddot{w} + \dot{v}) \quad \text{and} \quad \omega = \frac{1}{R}(\dot{w} + v) \tag{26.6}$$

Strains and rotation

26.3 Stresses

The *generalized stresses* that are the (work) conjugate of the generalized strains ε and κ are the axial force N and the bending moment M, respectively, see Section 8.3.3. There, the equilibrium equations are shown to be, see (8.76)

$$\left.\begin{aligned} N' + \frac{1}{R}M' + \bar{p}_v = 0 \\ M'' - \frac{1}{R}N - \bar{p}_w = 0 \end{aligned}\right\} s \in]0; L[\tag{26.7}$$

Equilibrium equations

where $\bar{p}_v$ and $\bar{p}_w$ designate the distributed axial and transverse loads, respectively. Since these equations express equilibrium they may, of course, be established by considering equilibrium, but this is left to the reader as an exercise.

26.4 Linear Elasticity

For linear elasticity the relations between the generalized stresses and the generalized strains are

$$N(s) = EA\varepsilon(s) \quad \text{and} \quad M(s) = EI\kappa(s) \tag{26.8}$$

Linear elasticity

26.5 Potential Energy

For a circular beam of length L the *Potential Energy* $\mathbf{\Pi}_\mathrm{P}(v, w)$ is

$$\begin{aligned} \mathbf{\Pi}_\mathrm{P}(v, w) &= \tfrac{1}{2}\int_0^L \left(EA\varepsilon(v, w)^2 + EI\kappa(v, w)^2\right) ds + \text{load terms} \\ &= W(v, w) + \text{load terms} \end{aligned} \tag{26.9}$$

Potential energy $\mathbf{\Pi}_\mathrm{P}(v, w)$

where the dependency of ε and κ on v an w is indicated explicitly, the load terms cover end load, $W(v,w)$ denotes the *strain energy*, and

Strain energy $W(v,w)$

$$W(v,w) \equiv \tfrac{1}{2}\int_0^L \big(EA\varepsilon(v,w)^2 + EI\kappa(v,w)^2\big)ds \tag{26.10}$$

In the following sections we pay little attention to the load terms because our main objective is to develop finite element stiffness matrices. Of course, the load terms depend on the way the stiffness matrix is established, but for simple cases where only the finite element nodes are loaded we may center on the strain energy.

26.5.1 Matrix Formulation

As in the previous sections on finite elements, in order to write the formulas below in a compact form we collect the kinematic and static variables in vectors. Thus, the displacements v and w are given as the column vector $\{u\}$

Generalized displacements $\{u(s)\}$

$$\{u(s)\} \equiv \begin{bmatrix} v(s) \\ w(s) \end{bmatrix} \tag{26.11}$$

the strains $\varepsilon(s)$ and $\kappa(s)$ as the column vector $\{\varepsilon\}$

Generalized strains $\{\varepsilon(s)\}$

$$\{\varepsilon(s)\} \equiv \begin{bmatrix} \varepsilon(s) \\ \kappa(s) \end{bmatrix} \tag{26.12}$$

and the stresses $N(s)$ and $M(s)$ as the column vector $\{\sigma\}$

Generalized stresses $\{\sigma(s)\}$

$$\{\sigma(s)\} \equiv \begin{bmatrix} N(s) \\ M(s) \end{bmatrix} \tag{26.13}$$

Because of (26.12) and (26.13) the stiffnesses $EA(s)$ and $EI(s)$ of the beam must be written in terms of a square matrix $[D]$

Constitutive matrix $[D(s)]$

$$[D(s)] \equiv \begin{bmatrix} EA(s) & 0 \\ 0 & EI(s) \end{bmatrix} \tag{26.14}$$

Then, the constitutive relation for linear elasticity is

Constitutive relation

$$\begin{bmatrix} N(s) \\ M(s) \end{bmatrix} = \begin{bmatrix} EA(s) & 0 \\ 0 & EI(s) \end{bmatrix} \begin{bmatrix} \varepsilon(s) \\ \kappa(s) \end{bmatrix} \tag{26.15}$$

or using our matrix notation

Constitutive relation

$$\{\sigma(s)\} = [D(s)]\{\varepsilon(s)\} \tag{26.16}$$

With (26.11)–(26.16) in hand we may express the strain energy W, see (26.10), as

$$W(\{u\}) = \tfrac{1}{2}\int_0^L \{\varepsilon\}^T[D]\{\varepsilon\}ds \tag{26.17}$$

Strain energy W

where the dependence of $\{\varepsilon\}$ on $\{u\}$ is not indicated. In most of the derivations below we abandon the explicit indications of this kind—otherwise the formulas become too lengthy.

26.5.2 Discretization

When we discretize the beam, i.e. subdivide the beam in N_{El} elements, the potential energy of the system becomes

$$\mathbf{\Pi}_{\mathrm{P}}(\{v\}_i) = \tfrac{1}{2}\sum_{j=1}^{N_{El}}\int_0^{L_j} \{\varepsilon\}_j^T[D]_j\{\varepsilon\}_j ds + \text{load terms} \tag{26.18}$$

$\mathbf{\Pi}_{\mathrm{P}}$ of discretized beam

We assume that the displacements vary along the length of the beam according to

$$\{u(s)\} = [N(s)]\{v\} \tag{26.19}$$

Displacement interpolation

where the *Displacement Interpolation Matrix* $[N]$ must not be confused with the axial force N.

For the elements we are dealing with here there is a problem concerning the establishing of $[N]$. In a number of other cases, we may determine the elements of $[N]$ fairly easily, see e.g. Sections 25.4 and 24.2, but here we shall shift gears and write the displacements in terms of functions that do not a

Displacement functions which do not satisfy the conditions for $[N]$

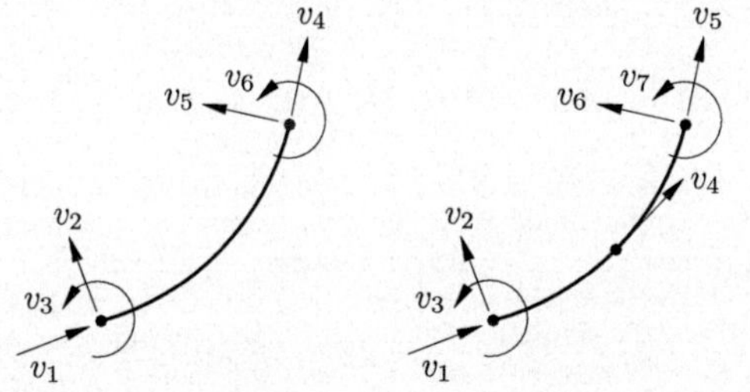

Fig. 26.2: Circular beam finite elements

priori satisfy the conditions that must be fulfilled by $[N]$. Whether we look at the element with 6 or the one with 7 degrees of freedom in Fig. 26.2 we may see that v_1 must contribute not only to the axial displacement v, but also to the transverse displacement w. In order to see the justification of the procedure used below we may consider the fields resulting from $v_1 = 1$ and all other $v_j = 0$. While the axial displacement is quite easy to obtain and may be determined in the same fashion as in Section 25.4, the transverse one seems less obvious because $v_1 \neq 0$ must not result in non-vanishing transverse displacement w and rotation ω at the ends of the element, while both quantities are non-zero between the ends.[26.2] Later, we arrive at an $[N]$ that possesses the relevant properties. First, let us write

$[N^c(s)] \neq [N(s)]$

$$\{u(s)\} = [N^c(s)]\{C\} \tag{26.20}$$

where $\{C\}$ denotes a column vector containing constants that are closely connected to $\{v\}$, and where $[N^c]$ is a two-dimensional matrix whose elements we may choose to our liking. In most cases we choose them to be simple polynomials.

To be specific, let us assume that our finite element has N_v degrees of axial freedom and N_w transverse degrees of freedom, see Fig. 26.2, where the element to the left in the figure has $N_v = 2$ and $N_w = 4$ and the element to the right has the same N_w but $N_v = 3$. It may be worthwhile mentioning that the transverse degrees of freedom cover the rotation ω of the nodes as well as the transverse displacement component w. We choose

Displacement assumption

$$[N^c] = \begin{bmatrix} 1; \left(\frac{\Phi}{\Psi}\right);\ldots;\left(\frac{\Phi}{\Psi}\right)^{N_v-1} & ;0; & 0 & ;\ldots; & 0 \\ 0; \quad 0 \quad ;\ldots; \quad 0 & ;1; & \left(\frac{\Phi}{\Psi}\right) & ;\ldots; & \left(\frac{\Phi}{\Psi}\right)^{N_w-1} \end{bmatrix} \tag{26.21}$$

where

$0 \le \dfrac{\Phi}{\Psi} \le 1$

$$\Psi \equiv \frac{L_i}{R} \tag{26.22}$$

and L_i is the length of the element. The reason that we use (Φ/Ψ) instead of just Φ, or s for that matter, in the expressions for the elements of $[N^c]$ is that (Φ/Ψ) takes the values 0 and 1 rather than 0 and Ψ at the end of the interval. Our task is now to establish $[N]$ from $[N^c]$. First, however, we need one more matrix, namely $[G^c]$, which is given by differentiation of $[N^c]$, because we must express the rotation ω as a function of $\{C\}$, see (26.3)

$$[G^c] = \frac{1}{R\Psi}\begin{bmatrix} 0;1;\ldots;(N_v-1)\left(\frac{\Phi}{\Psi}\right)^{N_v-2} & ;0;0;\ldots; & 0 \\ 0;0;\ldots; \quad 0 & ;0;1;\ldots; & (N_w-1)\left(\frac{\Phi}{\Psi}\right)^{N_w-2} \end{bmatrix} \tag{26.23}$$

where the presence of the factor $1/R$ is due to the fact that $[G^c]$ is derived from $[N^c]$ by differentiation with respect to the *physical* coordinate s.

[26.2] The task is possible to perform here, but in other cases it is even more difficult.

Then,

$$\omega = \sum_{j=1}^{N_{tot}} \left(G_{2j}^c + \frac{1}{R} N_{1j}^c \right) C_j \tag{26.24}$$

where

$$N_{tot} \equiv (N_v + N_w) \tag{26.25}$$

For the moment, we arrange the nodal displacement vector $\{v\}$ such that the axial degrees of freedom occupy the first N_v rows and the transverse degrees of freedom, including rotatory degrees of freedom, the last N_w rows. Then, we may express

$$\begin{aligned} v_i &= \sum_{j=1}^{N_{tot}} N_{1j}^c(\Phi_i) C_j \ , \quad i = 1, \ldots, N_v \\ v_i &= \sum_{j=1}^{N_{tot}} N_{2j}^c(\Phi_i) C_j \ , \quad i = N_v + 1, \ldots, N_{tot} \end{aligned} \tag{26.26}$$

First N_v degrees of freedom: axial Last N_w degrees of freedom: transverse and rotational

where Φ_i is the coordinate of the node associated with the ith component of $\{v\}$. As shown in Fig. 26.2 the axial and the transverse nodal displacements are not necessarily associated with the same points of the beam.[26.3]

Using matrix notation, (26.26) may be written

$$\{v\} = [T^c]\{C\} \tag{26.27}$$

$\{v\}$ in terms of $\{C\}$

This implies

$$\{C\} = [T^c]^{-1}\{v\} \tag{26.28}$$

From (26.19), (26.20), (26.27), and (26.28) we may see that

$$[N] = [N^c][T^c]^{-1} \tag{26.29}$$

$[N]$ in terms of $[N^c]$

and analogously (26.24) provides

$$[G] = [G^c][T^c]^{-1} \tag{26.30}$$

$[G]$ in terms of $[G^c]$

In order to express $W(\{u\})$ in (26.17) we need to establish the relation

$$\{\varepsilon(s)\} = [B^c(s)]\,\{C\} \tag{26.31}$$

ε in terms of $\{C\}$

[26.3] Actually, in many cases, it is best to choose only 4 transverse degrees of freedom, which must be at the ends of the finite element, and 3 axial degrees of freedom, with 2 at the ends and one at the middle of the element, as shown at the right in the figure. This is certainly the true for the case of moderate kinematic nonlinearity, but it would be too much of a diversion to expand on this here.

Obviously, in the end we wish to establish

ε in terms of $\{v\}$

$$\{\varepsilon(s)\} = [B(s)]\{v\} \tag{26.32}$$

Clearly, by analogy with (26.29) and (26.30)

$$[B] = [B^c][T^c]^{-1} \tag{26.33}$$

From (26.12) and (26.6) we see that

$$\{\varepsilon\} = \begin{bmatrix} \left(\frac{\dot{v}}{R} - \frac{w}{R}\right) \\ \left(\frac{\ddot{w}}{R^2} + \frac{\dot{v}}{R^2}\right) \end{bmatrix} \tag{26.34}$$

and

ε in terms of $\{C\}$ and $\{v\}$

$$\{\varepsilon\} = \begin{bmatrix} \left(G^c_{1j} - \frac{1}{R}N^c_{2j}\right) \\ \left(\frac{1}{R}\dot{G}^c_{2j} + \frac{1}{R}G^c_{1j}\right) \end{bmatrix} C_j = \begin{bmatrix} \left(G_{1j} - \frac{1}{R}N_{2j}\right) \\ \left(\frac{1}{R}\dot{G}_{2j} + \frac{1}{R}G_{1j}\right) \end{bmatrix} v_j\,, \tag{26.35}$$

sum over $j = 1, \ldots, N_{tot}$

and thus

$$\begin{bmatrix} B^c_{1j} \\ B^c_{2j} \end{bmatrix} = \begin{bmatrix} \left(G^c_{1j} - \frac{1}{R}N^c_{2j}\right) \\ \left(\frac{1}{R}\dot{G}^c_{2j} + \frac{1}{R}G^c_{1j}\right) \end{bmatrix}, \quad j = 1, \ldots, N_{tot} \tag{26.36}$$

Then,

Strain distribution matrix $[B]$

$$\begin{bmatrix} B_{1j} \\ B_{2j} \end{bmatrix} = \begin{bmatrix} \left(G_{1j} - \frac{1}{R}N_{2j}\right) \\ \left(\frac{1}{R}\dot{G}_{2j} + \frac{1}{R}G_{1j}\right) \end{bmatrix}, \quad j = 1, \ldots, N_{tot} \tag{26.37}$$

26.5.3 Stiffness Matrix

The strain energy W of (26.17) is

$$\begin{aligned} W(\{u\}) &= \tfrac{1}{2}\sum_{j=1}^{N_{El}} \int_0^{L_j} \{\varepsilon\}_j^T [D]_j \{\varepsilon\}_j ds \\ &= \tfrac{1}{2}\sum_{j=1}^{N_{El}} R\{v\}_j^T \int_0^{\Psi_j} [B]_j^T [D]_j [B]_j d\Phi \{v\}_j \end{aligned} \tag{26.38}$$

or

$$W(\{v\}_i) = \frac{1}{2}\sum_{j=1}^{N_{El}} \{v\}_j^T [k]_j \{v\}_j \tag{26.39}$$

where the stiffness matrix $[k]_j$ of element j is

$$[k]_j \equiv \int_0^{L_j} [B]_j^T [D]_j [B]_j ds = R\int_0^{\Psi_j} [B]_j^T [D]_j [B]_j d\Phi \tag{26.40}$$

Element stiffness matrix $[k]$

26.6 Internal Mismatch—Locking

Expect internal mismatch (locking) in strains

If we go back to our assumption (26.21) for $[N^c]$ and the expression (26.23) for $[G^c]$, which is a consequence of the choice (26.21), and consider the formula (26.36) for $[B^c]$, we may see that the terms in the elements of $[B]$ do not match each other in the sense that terms coming from the axial degrees of freedom are of another power of Φ than the terms from the transverse degrees of freedom.

For the elements in the first row of $[B^c]$, i.e. the elements associated with the axial strain ε we get

$$B_{1j}^c = G_{1j}^c - \frac{1}{R}N_{2j}^c \tag{26.41}$$

and thus

$$\varepsilon = \sum_{j=1}^{N_{tot}} \left(G_{1j}^c - \frac{1}{R}N_{2j}^c \right) C_j \tag{26.42}$$

or, more explicitly

$$\varepsilon = \frac{1}{R\Psi}\sum_{j=2}^{N_v} C_j (j-1)\left(\frac{\Phi}{\Psi}\right)^{j-2} - \frac{1}{R\Psi}\sum_{j=1}^{N_w} C_{N_v+j}\left(\frac{\Phi}{\Psi}\right)^{j-1} \tag{26.43}$$

In order to satisfy the rigid-body criterion which states that the element must not experience self-straining when moved as a rigid body we require that the terms from the first sum cancel the terms from the second and may choose

$$N_w = N_v - 1 \tag{26.44}$$

which excludes the choice $N_w = 4$ combined with anything less than $N_v = 5$, where our most obvious choice would be either $N_v = 2$ or $N_v = 3$. However, let us accept (26.44) and investigate the consequences for the bending strain κ, which is given by

$$\kappa = \sum_{j=1}^{N_{tot}} \left(\frac{1}{R}\dot{G}_{2j}^c + \frac{1}{R}N_{2j}^c \right) C_j \tag{26.45}$$

or

$$\kappa = \frac{1}{(R\Psi)^2}\sum_{j=3}^{N_w} C_{N_v+j}(j-1)(j-2)\left(\frac{\Phi}{\Psi}\right)^{j-3} + \frac{1}{R^2}\sum_{j=2}^{N_v}\Phi^{j-2}C_j \quad (26.46)$$

On the assumption of (26.44) the terms of highest power in Φ in these sums are

$$\begin{aligned} &\text{First sum:} \quad N_w - 3 = N_v - 4 \\ &\text{Second sum:} \; N_v - 2 \end{aligned} \quad (26.47)$$

These elements violate the rigid-body criterion

As we may see, in general there is no way that the two types of terms can balance each other in the expressions for both the axial and the bending strain and thus there is no way that we can satisfy the rigid-body criterion for this element. We may choose N_v and N_w such that $\varepsilon \equiv 0$ for rigid-body displacements, or we may choose their values to satisfy $\kappa \equiv 0$, but we cannot have both. Whether one is better than the other is not possible to determine from the above considerations.[26.4] Even if the element suffers from this kind of deficiency, it does not necessarily mean that it is useless—we may just have to use (many) more elements than seems reasonable, see Fig. Ex. 26-1.2–Fig. Ex. 26-1.5.

26.7 Rigid-Body Displacements—Self-Strain

Rigid-body displacement must be seen in context

As mentioned earlier, a finite element is not supposed to experience strains when we subject it to a rigid-body displacement field. As obvious as this requirement may seem we need to modify it by saying that the magnitude of the rigid-body displacements must be in accord with the theory we apply. If, for instance, we assume a kinematically linear theory to be valid, as we do here, it is not reasonable to require that the element does not exhibit self-straining when it is rotated an angle as large as π.

Both elements shown in Fig. 26.2 do self-strain because the displacement field given by (26.21) is not capable of describing a circle, i.e. the circle will be distorted by *any* such displacement field. On the other hand, when the opening angle Ψ is small the element is almost straight, and the displacement field becomes more and more capable of approximating the shape of the undeformed element. Thus, we may expect that the self-straining depends on Ψ to some power that is characteristic of the element. For the two elements mentioned above there is a difference in the way they behave as far as self-straining is concerned. Without any indication of the analysis behind the results, the order of self-straining of these two elements is given

[26.4] Intuitively one may prefer to satisfy $\varepsilon \equiv 0$ rather than $\kappa \equiv 0$ for rigid-body displacements because the resistance to axial deformation of a beam is much larger than that associated with bending and the energy from an error on the axial strains could therefore be larger than that from an error on bending strains.

in Table 26.1, which shows that the element with 7 degrees of freedom is the superior because a high exponent is better than a lower one.

Degrees of freedom	ε	κ
6	$O(\Psi^1)$	$O(\Psi^1)$
7	$O(\Psi^2)$	$O(\Psi^3)$

Table 26.1: Worst order of self-straining of usual circular elements.

Isoparametric Elements

It seems worth mentioning that there is a way of avoiding self-strain, namely when the shapes of the element and the displacement field are given by the same functions. Such elements are called *isoparametric*. The name might as well have been *homoparametric* or something like that, but the term *isoparametric* is the one which is used universally.

Isoparametric elements: no self-straining

Establishing such elements is not always an easy task, and in many cases we must learn to live with some amount of self-straining.

26.8 Modified Potential Energy

As mentioned above there are two reasons why the "usual" finite elements developed do not behave as well as we like. In the following we shall pay attention to the problem of "locking," or "internal mismatch," as I prefer to call it. The culprits here are the strain definitions (26.1) in combination with the displacement assumptions (26.21). There seems to be no reason for choosing another set of displacements, except that we might choose the exact field associated with constant properties, i.e. $EA =$ const. and $EI =$ const. This choice has some inherent problems, which we do not intend to go deeper into here, but again it may be mentioned that this solution results in finite elements that behave very poorly in terms of numeric stability when the opening angle Ψ is small, i.e. when the number of elements is large. Also, this solution entails trigonometric functions that have no natural role in the case of varying cross-sectional properties. This leaves the strain definitions to be considered, but they are in a way built-in in the *principle of stationary value of the potential energy* $\mathbf{\Pi}_\mathrm{P}$ in that the strains must be derived from the (assumed) displacements in compliance with (26.1)—the strain definitions are *auxiliary conditions* or *side conditions* on the potential energy $\mathbf{\Pi}_\mathrm{P}$.

Internal mismatch = Locking

However, there is a way that we may consider, namely giving up (26.1) as side conditions and introduce strain fields $\tilde{\varepsilon}$ and $\tilde{\kappa}$ that are independent fields, i.e. not derived directly from the displacements v and w, and include them in the potential by use of *Lagrange Multipliers* η_ε and η_κ which are

Lagrange multipliers η_ε and η_κ

Modified potential $\mathbf{\Pi}_{\mathrm{PM}}$

associated with the two auxiliary conditions we abandon.[26.5] Obviously, the potential energy must be altered in order to take account of the new situation. Below, we postulate a *Modified potential* $\mathbf{\Pi}_{\mathrm{PM}}$ and see if it makes sense to the extent that after proper variations it yields not only the equilibrium equations as the potential energy $\mathbf{\Pi}_{\mathrm{P}}$ does, but also the strain definitions.

Let it be said that there are no real secrets behind the way we establish $\mathbf{\Pi}_{\mathrm{PM}}$ and that the the below postulate of the $\mathbf{\Pi}_{\mathrm{PM}}$ comes as a direct consequence of abandoning (26.1), see below. In the present case the side conditions are given by (26.1) and may be rewritten

$$0 = \varepsilon - \left(v' - \frac{w}{R}\right) \quad \text{and} \quad 0 = \kappa - \left(w'' + \frac{v'}{R}\right) \tag{26.48}$$

We simply multiply each of these conditions by a Lagrange multiplier (field), η_ε and η_κ, respectively, integrate the products over the structure, and subtract the results from the original functional $\mathbf{\Pi}_{\mathrm{P}}$ to get the modified potential $\mathbf{\Pi}_{\mathrm{PM}}$ for the entire structure. In principle, the only difficulty lies in the the sign of the new terms—if we choose the "wrong" sign we may have to interpret the Lagrange multiplier η_ε as *minus* the axial force instead of the axial force itself. Likewise, η_κ may be interpreted as plus or minus the bending moment depending on the sign of the added term. On the other hand, such "mistakes" are easily corrected a posteriori. Whether this procedure leads to an improvement is not certain in advance, but to my experience it is worth trying before going to other measures.

In our case, the modified potential $\mathbf{\Pi}_{\mathrm{PM}}$ is

Modified potential energy $\mathbf{\Pi}_{\mathrm{PM}}$

$$\begin{aligned}
&\mathbf{\Pi}_{\mathrm{PM}}(v, w, \tilde{\varepsilon}, \tilde{\kappa}, \eta_\varepsilon, \eta_\kappa) \\
&= \tfrac{1}{2}\int_0^L EA\tilde{\varepsilon}^2 ds + \tfrac{1}{2}\int_0^L EI\tilde{\kappa}^2 ds - \int_0^L (\bar{p}_v v + \bar{p}_w w)\, ds \\
&\quad - \int_0^L \eta_\varepsilon \left(\tilde{\varepsilon} - \left(v' - \frac{1}{R}w\right)\right) ds \\
&\quad - \int_0^L \eta_\kappa \left(\tilde{\kappa} - \left(w'' + \frac{1}{R}v'\right)\right) ds
\end{aligned} \tag{26.49}$$

where we have specified the load terms to be the same as in Section 26.3.

Construction of a modified potential

Thus, the way we construct a modified potential consists in abandoning the side conditions that cause the problems and include them in the modified potential by use of Lagrange multipliers.

[26.5] The reason why I use η instead of λ, which would be more natural since the name refers to the French mathematician *Lagrange*, is because I use λ to designate a *load factor*—especially in connection with structural stability problems.

26.8.1 Justification of the Modified Potential

We must justify the modified potential

It is our intention to show that the modified potential is a decent one in that it may furnish the usual field equations by proper variations.

When we take variations with respect to $v, w, \tilde{\varepsilon}, \tilde{\kappa}, \eta_\varepsilon, \eta_\kappa$ we get

$$
\begin{aligned}
&\delta\mathbf{\Pi}_{\text{PM}}(v, w, \tilde{\varepsilon}, \tilde{\kappa}, \eta_\varepsilon, \eta_\kappa) \\
=&\int_0^L \delta\tilde{\varepsilon} EA\tilde{\varepsilon} ds + \int_0^L \delta\tilde{\kappa} EI\tilde{\kappa} ds - \int_0^L (\delta v \bar{p}_v + \delta w \bar{p}_w)\, ds \\
&- \int_0^L \delta\eta_\varepsilon \left(\tilde{\varepsilon} - \left(v' - \frac{1}{R}w\right)\right) ds - \int_0^L \delta\eta_\kappa \left(\tilde{\kappa} - \left(w'' + \frac{1}{R}v'\right)\right) ds \\
&- \int_0^L \eta_\varepsilon \left(\delta\tilde{\varepsilon} - \left(\delta v' - \frac{1}{R}\delta w\right)\right) ds - \int_0^L \eta_\kappa \left(\delta\tilde{\kappa} - \left(\delta w'' + \frac{1}{R}\delta v'\right)\right) ds
\end{aligned}
\tag{26.50}
$$

As with other functionals, we insist that the first variation vanishes, i.e. that

$$
\delta\mathbf{\Pi}_{\text{PM}} = 0\ \forall\, (\delta v, \delta w, \delta\tilde{\varepsilon}, \delta\tilde{\kappa}, \delta\eta_\varepsilon, \delta\eta_\kappa) \tag{26.51}
$$

where $\forall$ must be understood as "all admissible" in the usual sense of continuity and differentiability.

Before getting any further in our attempt of showing that $\mathbf{\Pi}_{\text{PM}}$ is a decent functional we must rearrange terms and get rid of derivatives of the variations. For instance, this is necessary because the variations $\delta w'$ and $\delta w''$ depend directly on δw, and we wish to exploit the fact that the coefficient to δw must vanish in $]0; L[$ in order that $\delta\mathbf{\Pi}_{\text{PM}}$ vanishes. Therefore, we rewrite $\mathbf{\Pi}_{\text{PM}}$

$$
\begin{aligned}
&\delta\mathbf{\Pi}_{\text{PM}}(v, w, \tilde{\varepsilon}, \tilde{\kappa}, \eta_\varepsilon, \eta_\kappa) \\
=&\int_0^L \delta\tilde{\varepsilon}\, (EA\tilde{\varepsilon} - \eta_\varepsilon)\, ds + \int_0^L \delta\tilde{\varepsilon}\, (EI\tilde{\kappa} - \eta_\kappa)\, ds \\
&+ \int_0^L \delta\eta_\varepsilon \left(\tilde{\varepsilon} - \left(v' - \frac{1}{R}w\right)\right) ds + \int_0^L \delta\eta_\kappa \left(\tilde{\kappa} - \left(w'' + \frac{1}{R}v'\right)\right) ds \\
&- \int_0^L \delta v \left(\eta_\varepsilon' + \frac{1}{R}\eta_\kappa' + \bar{p}_v\right) ds + \int_0^L \delta w \left(\eta_\kappa'' - \frac{1}{R}\eta_\varepsilon - \bar{p}_w\right) ds \\
&+ \text{boundary terms} \\
=&0\ \forall\, (\delta v, \delta w, \delta\tilde{\varepsilon}, \delta\tilde{\kappa}, \delta\eta_\varepsilon, \delta\eta_\kappa)
\end{aligned}
\tag{26.52}
$$

When we insist that (26.51) is observed, then comparison of the third and fourth integrals of (26.52) with (26.1a) and (26.1b), respectively, shows that

$\tilde{\varepsilon}$ and $\tilde{\kappa}$ may indeed be interpreted as the axial strain ε and the curvature strain κ, respectively. Having noticed this, we may utilize (26.8) to see that the first and second integrals of (26.52) indicate that η_ε may be interpreted as the axial force N and that η_κ may be interpreted as the bending moment M. Finally, comparison between the fifth and sixth integrals of (26.52) with the equilibrium equations (26.7a) and (26.7b), respectively, indeed prove that η_ε is the axial force N and that η_κ is the bending moment M. Thus, we should have gained some confidence in the validity of the modified potential.

26.8.2 Other Ways to Handle "Locking"

Reduced (Numerical) Integration

In the finite element literature, and other literature too, there are many other suggestions for coping with problems such as locking, most notably the idea of *Reduced (Numerical) Integration*, see e.g. (Noor & Peters 1981), (Stolarski & Belytschko 1982), (Cook et al. 2002) or (Bathe 1996), which may work, but whose application requires much more care than the method described here. Reduced integration may easily lead to degenerate finite elements that are worthless. Other methods whose application requires a fair amount of intuition, such as the *Mode Decomposition Projection Method*, see e.g. (Belytschko, Stolarski, W K. Liu, Carpenter & Ong 1985) or (Mau & El-Mabsout 1989), have been proposed and utilized, but their range of applicability is usually smaller than that of a modified potential.

26.8.3 Matrix Formulation

We shall still utilize the displacement assumption (26.5.2), but need more field assumptions, namely to describe η_ε, η_κ, $\tilde{\varepsilon}$, and $\tilde{\kappa}$. First, however, let us collect the Lagrange multipliers in the vector $\{\eta\}$

Lagrange multiplier vector $\{\eta(s)\}$

$$\{\eta(s)\} \equiv \begin{bmatrix} \eta_\varepsilon(s) \\ \eta_\kappa(s) \end{bmatrix} \tag{26.53}$$

By analogy with (26.12) we define a vector $\{\tilde{\varepsilon}(s)\}$ containing strains that are not derived directly from the displacements

Independent strain vector $\{\tilde{\varepsilon}(s)\}$

$$\{\tilde{\varepsilon}(s)\} \equiv \begin{bmatrix} \tilde{\varepsilon}(s) \\ \tilde{\kappa}(s) \end{bmatrix} \tag{26.54}$$

Of course, in the end $\{\tilde{\varepsilon}(s)\}$ is connected to the displacements, but in a more indirect way than through mere differentiation of the displacements.

The side, or auxiliary, conditions (26.48) may be written in terms of $\{\tilde{\varepsilon}\}$ and $\{\varepsilon\}$

Auxiliary condition

$$\{0\} = \{\tilde{\varepsilon}\} - \{\varepsilon(v, w)\} \tag{26.55}$$

where it is important to note that $\{\varepsilon\}$ depends on the displacements v and w.

Finally, collect the load terms in a vector $\{\bar{p}\}$

$$\{\bar{p}(s)\} \equiv \begin{bmatrix} \bar{p}_v(s) \\ \bar{p}_w(s) \end{bmatrix} \tag{26.56}$$

Load vector $\{\bar{p}\}$

When we recall (26.11) and (26.14), we may write the modified potential $\mathbf{\Pi}_{\text{PM}}$ for a finite element

$$\begin{aligned} &\mathbf{\Pi}_{\text{PM}}(\{v\},\{\eta\},\{\tilde{\varepsilon}\}) \\ &= \tfrac{1}{2}\int_0^L \{\tilde{\varepsilon}\}^T[D]\{\tilde{\varepsilon}\}ds - \int_0^L \{\eta\}^T\Big(\{\tilde{\varepsilon}\} - \{\varepsilon\}\Big)ds \\ &\quad - \int_0^L \{u\}^T\{\bar{p}\}ds \end{aligned} \tag{26.57}$$

Modified potential $\mathbf{\Pi}_{\text{PM}}$

where it is important to note that, because of continuity conditions, we may not vary all the quantities independently in each element, see later.

26.8.4 Discretization

In order to utilize the modified potential $\mathbf{\Pi}_{\text{PM}}$ as a basis for writing finite element equations we need to discretize all fields. Fortunately the discretization of $\{\tilde{\varepsilon}\}$ and $\{\eta\}$ is straightforward, as we shall see, and for the more difficult discretization of $\{\varepsilon\}$ we can exploit the expressions from Section 26.5.2, in particular (26.29) and (26.30) in connection with (26.21) and (26.22). One reason why discretization of $\{\tilde{\varepsilon}\}$ and $\{\eta\}$ presents no problem is that there are no inter-element compatibility conditions on the strains and the Lagrange multipliers. Another reason is that while v and w both enter the expressions for the axial strain ε and the curvature strain κ there is no similar coupling between $\{\tilde{\varepsilon}\}$ and $\{\eta\}$, nor between these and other fields. We may therefore take them to be simple polynomials of Φ.

No inter-element compatibility conditions on $\{\tilde{\varepsilon}\}$ and $\{\eta\}$

No coupling between $\{\tilde{\varepsilon}\}$ and $\{\eta\}$ and other fields

The necessary discretizations for this element are

$$\begin{aligned} \{u(s)\}_j &= [N(s)]_j\{v\}_j\,, \quad \{\varepsilon(s)\}_j = [B(s)]_j\{v\}_j \\ \{\tilde{\varepsilon}(s)\}_j &= [N_\varepsilon(s)]_j\{v_\varepsilon\}_j\,, \quad \{\eta(s)\}_j = [N_\eta(s)]_j\{v_\eta\}_j \end{aligned} \tag{26.58}$$

Discretizations

where subscript $_j$ indicates the element number, and $[N]$ and $[B]$ may be taken to be the same as in Section 26.5.2 and, as in Section 26.5.2, $\{v\}$ is the nodal displacement vector. The new vectors $\{v_\varepsilon\}$ and $\{v_\eta\}$ contain strain parameters and Lagrange Multiplier parameters, respectively.

Because of the above mentioned fact that there are no inter-element continuity requirement as regards $\{\tilde{\varepsilon}\}$ and $\{\eta\}$ the values of $\{v_\varepsilon\}$ and $\{v_\eta\}$ are local to each element, which is a property we intend to exploit later in that we eliminate them at the element level, see Section 26.8.5.

$\{v_\varepsilon\}$ and $\{v_\eta\}$ are local to each element

We may choose $[N_\varepsilon]$ and $[N_\eta]$ almost without restrictions as long as they

Hardly any restrictions on choice of $[N_\varepsilon]$ and $[N_\eta]$—no inter-element continuity required

are smooth within each element. However, since $\tilde{\varepsilon}$ contains the strains and η contains the Lagrange multipliers, which may be interpreted as the axial force and the bending moment, it does not pay to use more terms in $[N_\varepsilon]$ and $[N_\eta]$ than would be justified from the number of degrees of freedom of the element. For instance, for the element with 6 degrees of freedom the "natural" number of terms in the assumption associated with the axial component of $[N_\varepsilon]$ and $[N_\eta]$ is 1, while the number associated with the transverse component is 2. For the element with 7 degrees of freedom the best number of axial component is 2.

Don't use (too) many terms in $[N_\varepsilon]$ and $[N_\eta]$

You may also realize that using a large number of terms in $[N_\varepsilon]$ and $[N_\eta]$ must be counterproductive. If we took an infinite number of terms, then the smoothing effect of $\tilde{\varepsilon}$ and η would vanish because in that case they would be able to accommodate *any* variation, in particular the variations that are the result of the strain definitions (26.1) in connection with the interpolations given by (26.19) and (26.20), which are the ones that we try to avoid. Therefore we must expect that utilization of a large number of terms in $[N_\varepsilon]$ and $[N_\eta]$ may result in inferior finite elements.[26.6]

Using these expressions we may write the modified potential as

Discretized modified potential

$$
\begin{aligned}
&\mathbf{\Pi}_{\mathrm{PM}}(\{v\},\{v_\varepsilon\},\{v_\eta\})\\
&=\frac{1}{2}\sum_{j=1}^{N_{El}}\{v_\varepsilon\}_j^T\left(\int_0^{L_j}[N_\varepsilon]_j^T[D]_j[N_\varepsilon]_j ds\right)\{v_\varepsilon\}_j\\
&\quad-\sum_{j=1}^{N_{El}}\{v_\eta\}_j^T\left(\int_0^{L_j}[N_\eta]_j^T[N_\varepsilon]_j ds\right)\{v_\varepsilon\}_j\\
&\quad+\sum_{j=1}^{N_{El}}\{v_\eta\}_j^T\left(\int_0^{L_j}[N_\eta]_j^T[B]_j ds\right)\{v\}_j+\text{load terms}
\end{aligned}
\tag{26.59}
$$

where we have omitted the expression for the load term because we intend to eliminate $\{v_\varepsilon\}_i$ and $\{v_\eta\}_i$ at the *element level*, and the load term is the usual one, which only depends on $\{v\}_i$ and thus not on $\{v_\varepsilon\}_i$ and $\{v_\eta\}_i$.

Introduce the following definitions

$$
\left.
\begin{aligned}
[k_{\varepsilon\varepsilon}]_j &\equiv \int_0^{L_j}[N_\varepsilon]_j^T[D]_j[N_\varepsilon]_j ds\\
[k_{\eta\varepsilon}]_j &\equiv \int_0^{L_j}[N_\eta]_j^T[N_\varepsilon]_j ds\\
[k_{\eta v}]_j &\equiv \int_0^{L_j}[N_\eta]_j^T[B]_j ds
\end{aligned}
\right\}\ j=1,\ldots,N_{El}
\tag{26.60}
$$

[26.6] My own experience, also from other problems, justifies the recommendation and guidelines outlined above. See for instance Section 27.3.

Note that $[k_{\varepsilon\varepsilon}]$ is symmetric, i.e. $[k_{\varepsilon\varepsilon}] = [k_{\varepsilon\varepsilon}]^T$, and rewrite $\mathbf{\Pi}_{\mathrm{PM}}$

$[k_{\varepsilon\varepsilon}]$ is symmetric

$$\begin{aligned}\mathbf{\Pi}_{\mathrm{PM}}(\{v\},\{v_\varepsilon\},\{v_\eta\}) = \tfrac{1}{2} & \sum_{j=1}^{N_{El}} \{v_\varepsilon\}_j^T [k_{\varepsilon\varepsilon}]_j \{v_\varepsilon\}_j \\ & - \sum_{j=1}^{N_{El}} \{v_\eta\}_j^T [k_{\eta\varepsilon}]_j \{v_\varepsilon\}_j \\ & + \sum_{j=1}^{N_{El}} \{v_\eta\}_j^T [k_{\eta v}]_j \{v\}_j \\ & + \text{load terms}\end{aligned} \tag{26.61}$$

Now, since there are no requirements on inter-element compatibility of $\tilde{\varepsilon}$ and η, and since the load terms are independent of the strain and Lagrange Multiplier field, we may vary $\mathbf{\Pi}_{\mathrm{PM}}$ with respect to $\{v_\varepsilon\}_j$ and $\{v_\eta\}_j$ independently for each element and demand that these variations of $\mathbf{\Pi}_{\mathrm{PM}}$ vanish. Therefore we may get

$$\left.\begin{aligned}\delta\mathbf{\Pi}_{\mathrm{PM}}\left(\{v\},\{v_\varepsilon\},\{v_\eta\}\right)\Big|_{\{v_\varepsilon\}_j} &= 0 \\ \delta\mathbf{\Pi}_{\mathrm{PM}}\left(\{v\},\{v_\varepsilon\},\{v_\eta\}\right)\Big|_{\{v_\eta\}_j} &= 0\end{aligned}\right\} \; j = 1,\ldots,N_{El} \tag{26.62}$$

Variation with respect to $\{v_\varepsilon\}$ and $\{v_\eta\}$

26.8.5 Elimination of Lagrange Multipliers and Strain Parameters

In the expression (26.59) for the modified potential there are three sets of parameters, namely $\{v\}$, $\{v_\varepsilon\}$, and $\{v_\eta\}$, but since only the displacements are subject to continuity conditions, we may eliminate $\{v_\varepsilon\}$ and $\{v_\eta\}$ at element level, as mentioned above. This leaves us with only the nodal displacement vector as the unknown, and therefore our modified element may be implemented in a usual finite element code. If we are able to carry out the manipulations below analytically it is particularly easy.

Remember: no inter-element continuity conditions on $\tilde{\varepsilon}$ and η

For brevity and convenience, omit the explicit indication of the element number below. Then, (26.62) provides

$$\begin{aligned}\{\delta v_\varepsilon\}^T\Big([k_{\varepsilon\varepsilon}]\{v_\varepsilon\} - [k_{\eta\varepsilon}]^T\{v_\eta\}\Big) &= 0 \;\forall\, \{\delta v_\varepsilon\} \\ \{\delta\eta\}^T\Big([k_{\eta\varepsilon}]\{v_\varepsilon\} - [k_{\eta v}]\{v\}\Big) &= 0 \;\forall\, \{\delta\eta\}\end{aligned} \tag{26.63}$$

If we choose $[N_\varepsilon]$ such that its elements are linearly independent—any other choice would be strange, to say the least—then $[k_{\varepsilon\varepsilon}]$ is non-singular,[26.7] and $[k_{\varepsilon\varepsilon}]^{-1}$ exists.

$[k_{\varepsilon\varepsilon}]$ may be inverted

[26.7] This is certain because under this condition the scalar $\{v_\varepsilon\}^T[k_{\varepsilon\varepsilon}]\{v_\varepsilon\}$ is positive for all $\{v_\varepsilon\} \neq \{0\}$.

$\{v_\varepsilon\}$ in terms of $\{v_\eta\}$

Therefore (26.63a) may provide

$$\{v_\varepsilon\} = [k_{\varepsilon\varepsilon}]^{-1}[k_{\eta\varepsilon}]^T\{v_\eta\} \tag{26.64}$$

Now, (26.63b) yields

$$[k_{\eta\varepsilon}][k_{\varepsilon\varepsilon}]^{-1}[k_{\eta\varepsilon}]^T\{v_\eta\} = [k_{\eta v}]\{v\} \tag{26.65}$$

Introduce the matrix $[k^*_{\eta\eta}]$ by

$$[k^*_{\eta\eta}] \equiv [k_{\eta\varepsilon}][k_{\varepsilon\varepsilon}]^{-1}[k_{\eta\varepsilon}]^T \tag{26.66}$$

and note that $[k^*_{\eta\eta}]$ is symmetric.

Assuming that $[k^*_{\eta\eta}]$ is non-singular (26.65) may provide[26.8]

$\{v_\eta\}$ in terms of $\{v\}$

$$\{v_\eta\} = [k^*_{\eta v}]\{v\} \tag{26.67}$$

where the matrix $[k^*_{\eta v}]$ is defined by

$$[k^*_{\eta v}] \equiv [k^*_{\eta\eta}]^{-1}[k_{\eta v}] \tag{26.68}$$

Now that we have the expression (26.67) for $\{v_\eta\}$ we can go back to (26.64) and get

$$\{v_\varepsilon\} = [k_{\varepsilon v}]\{v\} \tag{26.69}$$

where the matrix $[k_{\varepsilon v}]$ is defined by

$$[k_{\varepsilon v}] \equiv [k_{\varepsilon\varepsilon}]^{-1}[k_{\eta\varepsilon}]^T[k^*_{\eta\eta}]^{-1}[k_{\eta v}] \tag{26.70}$$

The modified potential may now be written

New expression for the $\mathbf{\Pi}_{\mathrm{PM}}$
Now $\mathbf{\Pi}_{\mathrm{PM}}$ only depends on $\{v\}$

$$\begin{aligned}
&\mathbf{\Pi}^*_{\mathrm{PM}}(\{v\}) \\
&= \tfrac{1}{2}\sum_{j=1}^{N_{\mathrm{El}}}\{v\}_j^T\Big([k_{\eta v}]^T[k^*_{\eta\eta}]^{-1}[k_{\eta\varepsilon}][k_{\varepsilon\varepsilon}]^{-1}[k_{\varepsilon\varepsilon}][k_{\varepsilon\varepsilon}]^{-1} \\
&\qquad\qquad [k_{\eta\varepsilon}]^T[k^*_{\eta\eta}]^{-1}[k_{\eta v}]\Big)_j\{v\}_j \\
&\quad - \sum_{j=1}^{N_{\mathrm{El}}}\{v\}_j^T\Big([k_{\eta v}]^T[k^*_{\eta\eta}]^{-1}[k_{\eta\varepsilon}][k_{\varepsilon\varepsilon}]^{-1}[k_{\eta\varepsilon}]^T \\
&\qquad\qquad [k^*_{\eta\eta}]^{-1}[k_{\eta v}]\Big)_j\{v\}_j \\
&\quad + \sum_{j=1}^{N_{\mathrm{El}}}\{v\}_j^T\Big([k_{\eta v}]^T[k^*_{\eta\eta}]^{-1}[k_{\eta v}]\Big)_j\{v\}_j \\
&\quad + \text{load terms}
\end{aligned} \tag{26.71}$$

where the asterisk in $\mathbf{\Pi}^*_{\mathrm{PM}}$ has been introduced in order to make clear that the new modified potential $\mathbf{\Pi}^*_{\mathrm{PM}}$ is different from $\mathbf{\Pi}_{\mathrm{PM}}$ in that the only

[26.8] Only if the choice of Lagrange multiplier and strain fields contained linearly dependent terms would $[k^*_{\eta\eta}]$ be singular, so don't worry.

variable now is $\{v\}$. When we recall the definition (26.66) of $[k^*_{\eta\eta}]$ and after some trivial derivations (26.71) we may see that the last two sums cancel each other with the result that $\mathbf{\Pi}_{\text{PM}}$ may be written

$$\mathbf{\Pi}_{\text{PM}}(\{v\}) = \frac{1}{2}\sum_{j=1}^{N_{\text{El}}}\{v\}_j^T[k^*_{vv}]_j\{v\}_j + \text{load terms} \tag{26.72}$$

Final expression for the $\mathbf{\Pi}_{\text{PM}}$

where the modified element stiffness matrix $[k^*_{vv}]$ is defined by

$$[k^*_{vv}] \equiv [k_{\eta v}]^T[k^*_{\eta\eta}]^{-1}[k_{\eta v}] \tag{26.73}$$

New element stiffness matrix $[k^*_{vv}]$

where $[k^*_{vv}]$ clearly is symmetric and index $_j$ has been omitted.

At a first glance, the above derivations maybe a little lengthy, but the resulting set of formulas for the various "stiffness matrices" is quite small.

It is convenient that the structure of the modified potential $\mathbf{\Pi}^*_{\text{PM}}$ is the same as that of the potential energy $\mathbf{\Pi}_{\text{P}}$ because then the structure of the ensuing finite element equations is the same as usual.[26.9] The extra work involved in establishing the modified finite element equations is relatively small because all the new matrices $[k_{\varepsilon\varepsilon}]$, $[k_{\eta\varepsilon}]$, $[k_{\eta v}]$, $[k^*_{\eta\eta}]$, and $[k_{\varepsilon v}]$ are smaller than the original element stiffness matrix $[k]$ (and the modified element stiffness matrix $[k^*_{vv}]$). Of course, when we compute the strains of the beam we must go through a somewhat more complicated procedure than usual because, in order to determine the strains $\tilde{\varepsilon}$, we must determine $\{v_\varepsilon\}$ from (26.69) before we can utilize (26.58b) to determine the strains. Note that this last part is even simpler than the usual since $[N_\varepsilon]$ and $\{v_\varepsilon\}$ contain fewer elements than $[B]$ and $\{v\}$, respectively. The computation of the stresses becomes simpler than usual because the the stresses are identified as the Lagrange multipliers, and $\{v_\eta\}$ is given by (26.67). Thus, we do not have to go through the constitutive matrix $[D]$ in order to compute the stresses, but can use (26.58d) instead. Again, $[N_\eta]$ and $\{v_\eta\}$ contain relatively few elements.

The structure of the finite element equations is as usual.
The extra work involved in getting the new "stiffness matrices" is small

Computation of the stresses is simpler than usual

26.9 Rigid-Body Displacements—Self-Strain

As in Section 26.7 we cite the results for the order of self-straining of the two

Degrees of freedom	ε	κ
6	$O(\Psi^4)$	$O(\Psi^1)$
7	$O(\Psi^3)$	$O(\Psi^3)$

Table 26.2: Worst order of self-straining of modified circular elements. Degrees of freedom refers to displacements.

modified elements without indication of the analysis behind, see Table 26.2, which shows that the element with 7 degrees of freedom is the superior.

[26.9] Recall that the load terms are the usual ones.

26.10 Analytic Results—Matrices

Below, we summarize analytic results for the usual and the modified beam elements with 6 displacement degrees of freedom, see Section 26.10.1 where the analytic expressions for $[N]$, $[G]$, $[B]$, and $[k]$ are given.[26.10]

Expressions for the matrices $[N^c]$ and $[G^c]$ are given in Section 26.5.2 and will not be repeated here. The formulas for $[N]$, $[G]$, and $[B]$ require some work, but only the final results are given below.

Here, as in most similar cases, I have let maxima perform the analytic manipulations. But, as is the rule rather than the exception I have had to rewrite the LaTeX-output from maxima in order to get a reasonably nice structure of the formulas. I do not consider this an error of maxima because there is no unique way to define the "best" appearance of a formula.

26.10.1 Fundamental Matrices for 6 Displacement Degrees of Freedom

The displacement interpolation is here given by

$$[N] = \left[\begin{matrix} \left(-\frac{\Phi}{\Psi}+1\right) & ; & 0 & ; & 0 \\ \left(-\Phi+\frac{2\Phi^2}{\Psi}-\frac{\Phi^3}{\Psi^2}\right) & ; & \left(-\frac{3\Phi^2}{\Psi^2}+\frac{2\Phi^3}{\Psi^3}+1\right) & ; & R\left(\Phi-\frac{2\Phi^2}{\Psi}+\frac{\Phi^3}{\Psi^2}\right) \end{matrix}\right.$$

$$\left.\begin{matrix} \left(\frac{\Phi}{\Psi}\right) & ; & 0 & ; & 0 \\ \left(\frac{\Phi^2}{\Psi}-\frac{\Phi^3}{\Psi^2}\right) & ; & \left(\frac{3\Phi^2}{\Psi^2}-\frac{2\Phi^3}{\Psi^3}\right) & ; & R\left(-\frac{\Phi^2}{\Psi}+\frac{\Phi^3}{\Psi^2}\right) \end{matrix}\right]$$

The other matrices associated with displacements, namely $[G]$ and $[B]$ follow from $[N]$ according to the definition of the strains ε and κ and the rotation ω, see (26.6).

In order to determine the modified stiffness matrix $[k^*_{vv}]$ we need $[N_\varepsilon]$ and $[N_\eta]$. Since we assume linear elasticity there is no reason for choosing them differently. Our choice for the element with six displacement degrees of freedom then is

$[N_\varepsilon]$ and $[N_\eta]$ for element with 6 nodes

$$[N_\varepsilon] = [N_\eta] = \left[\begin{array}{c|cc} 1 & 0 & 0 \\ 0 & 1 & \Phi \end{array}\right] \tag{26.74}$$

where the vertical line divides the terms associated with the axial component from the transverse ones.

[26.10] The expressions for the better element with 7 displacement degrees of freedom take up too much more space and would not fit on the page.

For the following discussion divide the element stiffness matrices $[k]$ and $[k^*_{vv}]$

$$[k] = EA[k]_{EA} + EI[k]_{EI} \text{ and } [k^*_{vv}] = [k^*_{vv}]_{EA} + [k^*_{vv}]_{EI} \tag{26.75}$$

which makes the structure of $[k]$ more clear.

At this point it seems worth mentioning that in order to compute the element stiffness matrix $[k]$ it is necessary that we know the analytic formulas for the matrices $[N]$, $[G]$, and $[B]$, and for $[k^*_{vv}]$ also the analytic contents of $[N_\varepsilon]$ and $[N_\eta]$ is pertinent. On the other hand, analytic integration of the various stiffness matrices does not always lead to the computationally most efficient finite element code. The reason for using analytic manipulations here is that analytic expressions for the elements of $[k]$ and $[k^*_{vv}]$ facilitate the study of the influence of various parameters, such as the opening angle Ψ. Because the expressions for these matrices would take up quite a substantial amount of space they are not given here, so you must trust me regarding the following findings.

The part of the stiffness matrix that depends on EI is not altered by the modifications in this case, i.e. $[k]_{EI} = [k^*_{vv}]_{EI}$, while the modifications make the matrix $[k^*_{vv}]_{EA}$ different from $[k]_{EA}$.

If you expand the elements of $[k]_{EA}$ and of $[k^*_{vv}]_{EA}$ you will find that some of them agree as far as the dominating term is concerned while others differ quite substantially which may explain why they provide results of different quality, as we shall see below.

Ex 26-1 Numerical Example

We shall use the structural example shown in Fig. Ex. 26-1.1 as a basis for our study of the sources of error of the finite elements developed above.

Ex 26-1.1 *Exact Results*

A fairly elementary analysis provides the following formulas[26.11]

$$N(\Phi) = -\bar{P}_A \cos(\Phi)\,, \qquad M(\Phi) = +\bar{P}_A R\big(1 + \cos(\Phi)\big)$$
$$v_A = \frac{3\pi}{2}\left(1 + \frac{I}{3AR^2}\right)\frac{\bar{P}_A R^3}{EI}\,, \; w_A = \frac{2\bar{P}_A R^3}{EI} \tag{Ex. 26-1.1}$$
$$\omega_A = \frac{\pi \bar{P}_A R^2}{EI}$$

Ex 26-1.2 *Specific Structural Parameters*

Table 26.3 contains values of the load and of the geometric and constitutive properties used in the computations below.

[26.11] Do not try to solve the differential equations for v and w—use the *Principle of Virtual Forces* to determine v_A, w_A, and ω_A.

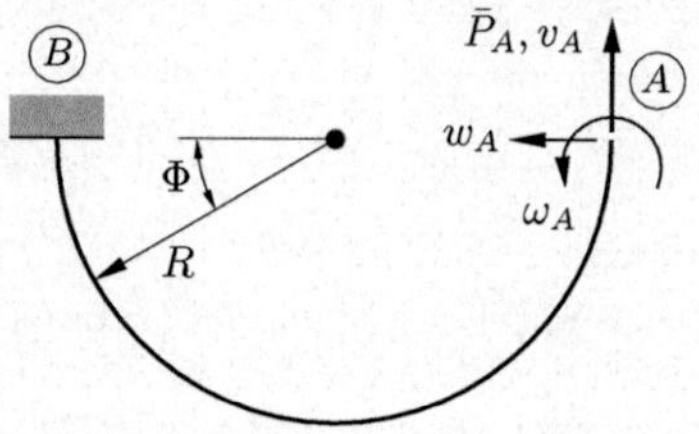

Fig. Ex. 26-1.1: Structural example.

Property	Name	Value
Load	$\bar{P}_A$	$\frac{4}{1000}$
Radius	R	10
Axial stiffness	EA	$\frac{1}{2} \times 10^4$
Bending stiffness	EI	$\frac{1}{96} \times 10^4$

Table 26.3: Values of load and of geometric and constitutive properties.

Be careful with drawing conclusions from just one numerical example

While the value of the load $\bar{P}_A$ is unimportant because the solution depends linearly on the load, the relation between the axial and the bending stiffness, i.e. the slenderness, is essential in that the relative magnitude of the axial and the bending strain strongly depends on the slenderness, and thus influences the relation between the various terms of the strain measures. Therefore, we must be cautious when we draw conclusions based on numerical results. However, more examples where the structural parameters are varied show that the qualitative features of the studies of convergence and accuracy are quite universally valid.

Ex 26-1.3 Errors

Two kinds of errors: Errors on displacement quantities Errors on stress quantities

The errors fall into two categories. The first covers displacement quantities, i.e. the axial displacement component v, the transverse displacement component w, and the rotation ω. The other comprises errors on strain and stress quantities, i.e. the axial strain ε and the bending strain κ, and the axial force N, and the bending moment M, respectively. In both the usual and the modified analysis the displacements are direct results of the finite element solution and may be expected to be of similar quality, but this does not imply that we expect the results obtained by the two methods to be identical. Actually, we hope that the results found by the modified procedure will be superior. As regards determination of the the strain and stress quantities, they are found by differentiation in the usual computation, but are determined much more directly by the modified procedure.

As indicated in the plots below the curves labeled U6 and U7 show results from the usual computation with 6 and 7 displacement degrees of freedom, respectively, while the equivalent curves found by the modified procedure are labeled M6 and M7.

Ex 26-1.3.1 Errors on Displacements

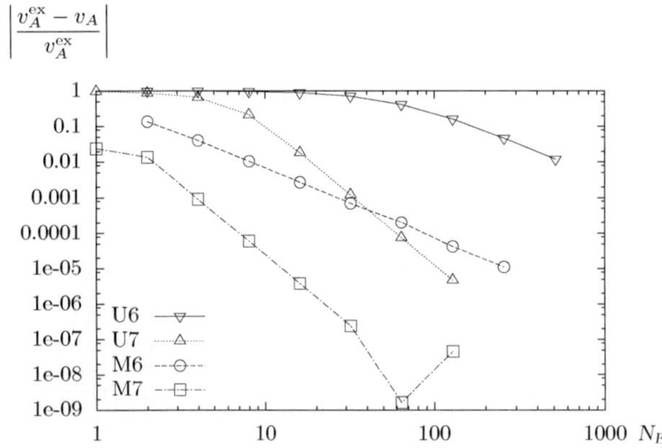

Fig. Ex. 26-1.2: *Relative errors on the displacement v_A of the load.*

Figure Ex. 26-1.2 clearly shows that the modified procedure is better than the usual one as far as the value of the displacement v_A of the load, not because the convergence rate is better—you may see that the rate of the two methods is the same for the same number of element displacement degrees of freedom, but because the modified method provides a much better result for the same number of system degrees of freedom. A decent rate of convergence is in itself a nice quality, but if the starting point is very inaccurate, see U6, it does not help much. In order to obtain a relative error of about 1% the usual method with 6 degrees of freedom needs around 500 elements along the arc which is completely unacceptable. The modified procedure with 6 displacement degrees of freedom only needs 8 finite elements to provide the same accuracy, and the modified procedure with 7 displacement degrees of freedom only requires 2–3 elements. Clearly, the modified analysis with 7 displacement degrees of freedom is by far the best of the 4 methods in this regard, but note that around 30 elements the plot ceases to be a straight line indicating that numerical errors take over due to the limited number of digits ($\approx$ 16) in the computer's representation of floating point numbers. It may be noted that long and slender structures that are only supported at very few places are the most likely to suffer from loss of digits in the solution of the finite element equations. It seems to be a common problem associated with finite elements based

The modified procedure seems superior

Convergence rate is not everything

The modified analysis with 7 displacement degrees of freedom is by far the best regarding accuracy of displacements on modified potentials that the ensuing equations are numerically less stable than the ones resulting from application of displacement based finite elements. However, 8 modified elements with 7 degrees of freedom provide a relative error of less than 10^{-4}, which is more than sufficient for most computations.

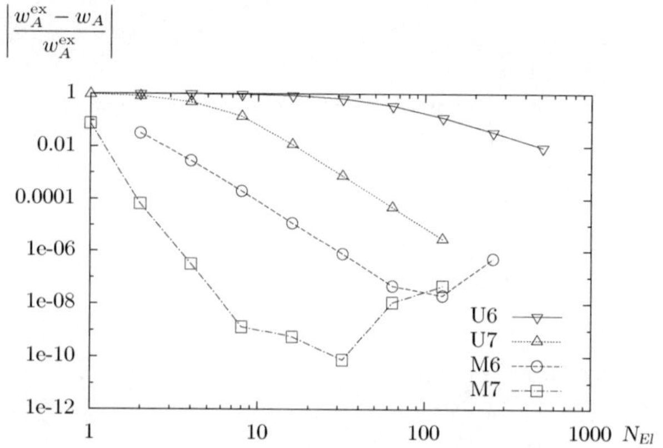

Fig. Ex. 26-1.3: Relative errors on the displacement at right angles to the load w_A.

The the displacement component w_A at right angles to the load is much smaller than v_A, and therefore its relative errors are less indicative.

The picture of the errors on the rotation ω_A at the load resembles that of v_A, but the plot is not shown here.

Ex 26-1.3.2 Errors on Stress Quantities

As indicated above, the errors on the stress quantities, here the axial force N_B and the bending moment M_B at the support, are much greater than the errors on the displacement quantities, see Fig. Ex. 26-1.4 and Fig. Ex. 26-1.5, but at the same time it is clear from the figures that the modified procedure furnishes better results.

N_B from the usual elements are unacceptable M_B is not much better

The values of axial force N_B obtained by the usual elements are clearly not acceptable, and of the usual elements only the one with 7 degrees of freedom provides reasonable results for the bending moment M_B. There is a fundamental theoretical difference between the convergence of v_A and the other quantities in that v_A is directly associated with the applied load—it is the "characteristic" displacement, see also Section 33.4.3.1. Therefore, we may not expect as good a convergence for the other quantities as for v_A.

A Beam-Specific Trick.

Beams are *internally* statically determinate

There is a certain property of beams that we may exploit to get better results for the stress quantities, namely the fact that beams are *internally* statically determinate. Therefore, for beams, the nodal forces,

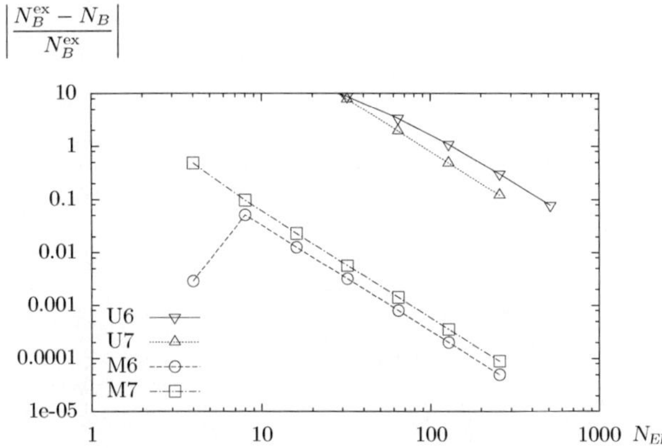

Fig. Ex. 26-1.4: Errors on the axial force N_B at the support.

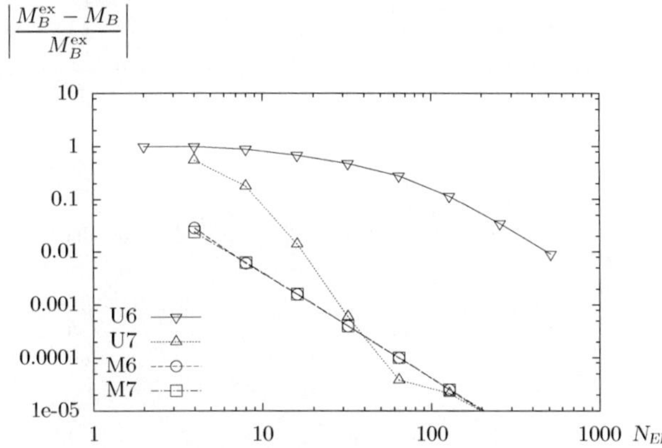

Fig. Ex. 26-1.5: Errors on the bending moment M_B at the support.

which are a direct outcome of the finite element analysis, may be interpreted as the boundary values of the axial force, the shear force, and the bending moment. This trick only works for beams and we shall not go further into this, but mention that the values of the axial forces, which are the most inaccurate ones, improve by using this trick.

Ex 26-1.4 Variation of Stress Quantities Along the Length

As mentioned above, the axial force N_B and the bending moment M_B at the support are not completely indicative of the stress picture. Therefore, we plot the variation of the axial force $N(\Phi)$ and of the bending moment $M(\Phi)$ for various numbers of finite elements.

Neither U6 nor U7 furnishes acceptable values of $N(\Phi)$

Ex 26-1.4.1 Results Obtained with 4 Elements

The first results are obtained by use of 4 elements, see Figs. Ex. 26-1.6 and Ex. 26-1.7. It is clear that the usual elements provide very inaccurate values of both stress quantities in that their values barely

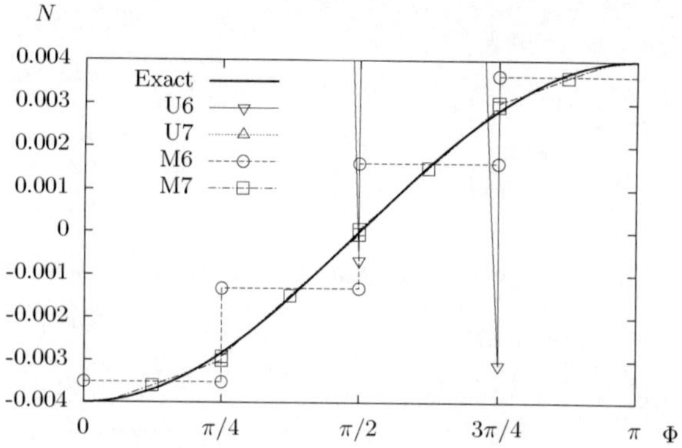

Fig. Ex. 26-1.6: Axial force $N(\Phi)$, 4 elements.

Interpreted in the right way even M6 provides reasonable values of $N(\Phi)$

lie inside the plots, while both modified elements furnish reasonably values. The behavior of the axial force of the modified element with 6

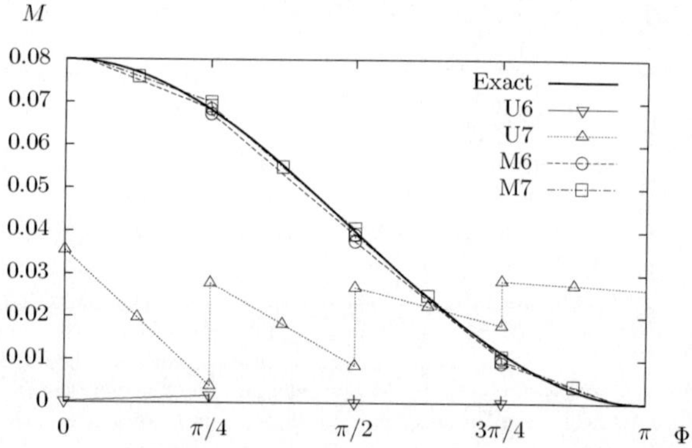

Fig. Ex. 26-1.7: Bending moment $M(\Phi)$, 4 elements.

displacement degrees of freedom deserves a comment because we may see that if we take the value of the axial force to be associated with the midpoint of the element, then the accuracy is remarkably good. This is not a trick, because if a quantity, be it a stress or another quantity, is constant over any type of element, then that quantity should be referred to the center of the element. This, by the way, seems intuitively reasonable—at least to me.

The values of the bending moment M are also unacceptable for the usual element with 6 degrees of freedom, and the results from application of 7 degrees of freedom does not do much better, while both modified elements furnish results of high quality.

Both usual element perform poorly regarding the $M(\Phi)$

Ex 26-1.4.2 Results Obtained with 16 Elements

For the axial force neither 8 nor 16 usual elements provide acceptable results. see Fig. Ex. 26-1.8.

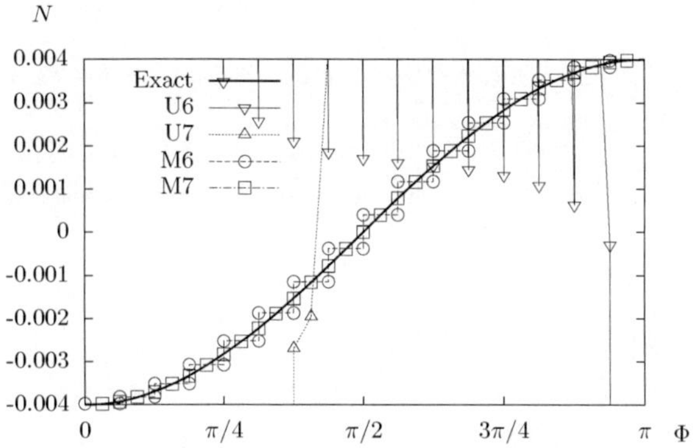

Fig. Ex. 26-1.8: *Axial force $N(\Phi)$, 16 elements.*

As regards the bending moment the usual finite element with 6 degrees of freedom it does not provide adequate results, while the bending moment obtained by the usual element with 7 degrees of freedom $M(\Phi)$ is close to accurate, see Fig. Ex. 26-1.9.

Both modified elements provide almost exact results.

For as simple an example as the present one you would expect that 4, maybe 6 finite elements could do the job. And, as we saw, we need as few as about 4 of the modified finite element with 7 degrees of freedom for that purpose, while the usual finite elements fail even with s many as 16 elements.

In order to see if the usual finite elements at all are capable of providing useful results we shall employ an unreasonably high number of elements, namely 256.

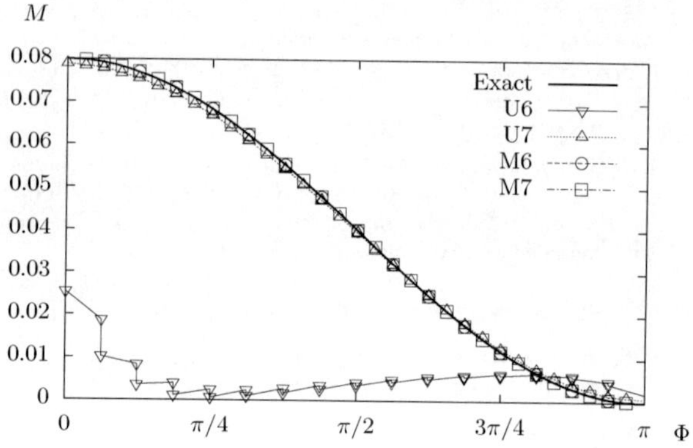

Fig. Ex. 26-1.9: Bending moment $M(\Phi)$, 16 elements.

Ex 26-1.4.3 Results Obtained with 256 Elements

For clarity only the values of $N(\Phi)$ obtained by application of the usual element with 7 degrees of freedom are shown in Fig. Ex. 26-1.10. The

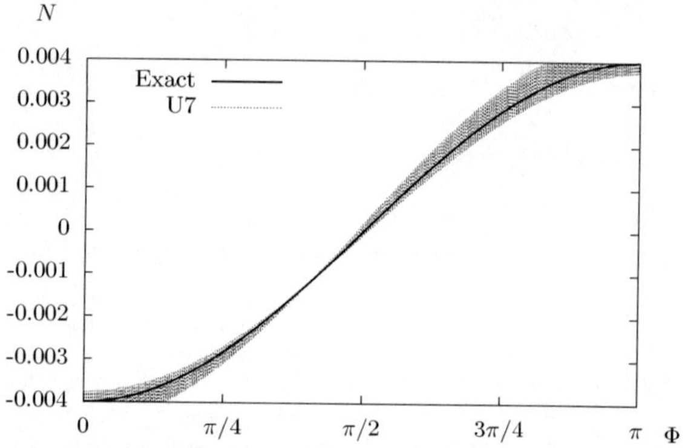

Fig. Ex. 26-1.10: Axial force $N(\Phi)$, 256 elements.

Even 256 elements of U7-type are not enough to determine $N(\Phi)$

basic trend of the results is reasonable, but the values do not lie on a curve but are confined to a fairly broad band around the correct curve. The usual element with 6 degrees of freedom provides results so poor that the curve would blacken a great deal of the plot. This must be ascribed to the fact that this element suffers severely from self-straining as far as the axial strain ε, and therefore also the axial force N, is concerned.

The usual element furnishes so inferior predictions of the bending moment that even 256 elements do not reproduce the correct curve, see Fig. Ex. 26-1.11, although the shape of the curve resembles the correct one.

$M(\Phi)$ with 256 U6-elements is inaccurate

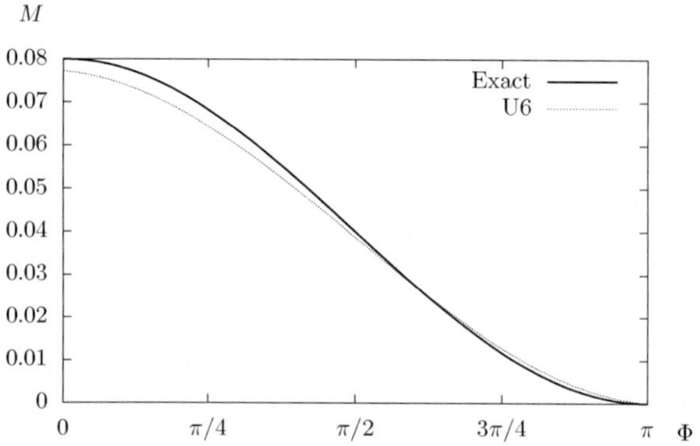

Fig. Ex. 26-1.11: *Bending moment $M(\Phi)$, 256 elements.*

26.11 Concluding Comments

As mentioned at the end of Example Ex 26-1.2, we must not draw too wide conclusions based on only one structural example. However, more examples where the structural parameters are varied show that the qualitative features of the study of convergence and accuracy are quite universally valid resulting in two conclusions regarding the above finite elements. The first has to do with the fact that, because of their inherent problem of "locking," the usual elements perform poorly, while the modified elements provide rather good results. The other conclusion is related to the fact that the modified elements furnish numerically less stable systems of equations, but that this effect does not come into play until the number of elements is much larger than needed to get useful results.[26.12]

The usual elements "lock" causing poor performance.
The modified elements provide less well-conditioned equations

[26.12] In all fairness one should acknowledge that the structure studied above is so slender and has so few supports that almost all other examples will provide healthier systems of equations.

Chapter 27

Modified Complementary Energy and Stress Hybrid Finite Elements

In this chapter we establish the foundation for the so-called *Stress Hybrid Finite Elements* which often present advantages over the ones that are based on displacement assumptions.

27.1 Modified Complementary Energy

It turns out that establishing so-called equilibrium finite elements, which are based on the complementary energy $\mathbf{\Pi}_{\mathrm{C}}$, often presents significant difficul-

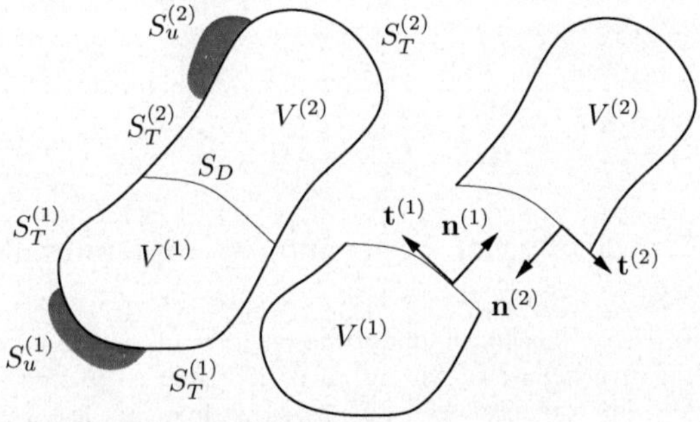

Fig. 27.1: A body containing a discontinuity surface.

ties because the stresses tend to be discontinuous over the element boundaries. We may therefore try a different approach and abandon the requirement (4.133) of stress continuity over boundaries. In Fig. 27.1 the volume V is subdivided into two bodies $V^{(1)}$ and $V^{(2)}$ which are separated by the discontinuity surface S_D. As usual, the outer boundary consists of a static part S_T and a kinematic part S_u. For clarity S_T is subdivided into $S_T^{(1)}$

E. Byskov, *Elementary Continuum Mechanics for Everyone*,
Solid Mechanics and Its Applications 194, DOI: 10.1007/978-94-007-5766-0_27,

and $S_T^{(2)}$, and S_u is subdivided into $S_u^{(1)}$ and $S_u^{(2)}$, as shown in the figure. Note that at the discontinuity surface the tangent $\mathbf{t}^{(2)}$ and the normal $\mathbf{n}^{(2)}$ to $V^{(2)}$ have the opposite directions of the tangent $\mathbf{t}^{(1)}$ and the normal $\mathbf{n}^{(1)}$ to $V^{(1)}$. This is especially clear when you think of a two-dimensional body, where you integrate with the body to the left of the integration path.

Now, the complementary energy of the body is, see (4.131)

Complementary energy $\mathbf{\Pi}_{\mathrm{C}}(\sigma_{ij})$

$$\mathbf{\Pi}_{\mathrm{C}}(\sigma_{ij}) = \frac{1}{2}\int_{V^{(1)}+V^{(2)}} C_{ijkl}\sigma_{ij}\sigma_{kl}dV - \int_{S_u^{(1)}+S_u^{(2)}} T_i\bar{u}_i dS \tag{27.1}$$

where $V^{(1)}$ and $V^{(2)}$ mathematically speaking are open domains, and where the auxiliary conditions on (27.1) are

Auxiliary conditions on $\mathbf{\Pi}_{\mathrm{C}}$

$$\begin{aligned} &\left.\begin{aligned} \sigma_{ij,j} + \bar{q}_i = 0\,, x_i \in V^{(\alpha)} \\ \sigma_{ij}n_j = \overline{T}_i\,, \quad x_i \in S_T^{(\alpha)} \end{aligned}\right\} \alpha = (1,2) \\ &(\sigma_{ij}n_j)^{(2)} = -(\sigma_{ij}n_j)^{(1)}\,, x_i \in S_D \end{aligned} \tag{27.2}$$

We may rewrite (27.2c) in a more convenient form when we introduce the unit normal $\mathbf{n}^{(12)}$

$$\mathbf{n}^{(12)} \equiv \mathbf{n}^{(1)} = -\mathbf{n}^{(2)} \tag{27.3}$$

to get

Auxiliary stress continuity condition on $\mathbf{\Pi}_{\mathrm{C}}$

$$\left(\sigma_{ij}^{(2)} - \sigma_{ij}^{(1)}\right) n_j^{(12)} = 0\,,\ x_i \in S_D = S^{(12)} \tag{27.4}$$

27.1.1 Establishing of a Modified Complementary Energy

The following derivations which, unfortunately, take up quite a lot of space and are rather involved serve two purposes. The first is proving that the new, modified complementary energy $\mathbf{\Pi}_{\mathrm{CM}}$ is legitimate in the sense that it, together with its auxiliary conditions, may furnish the same equations as the original complementary energy $\mathbf{\Pi}_{\mathrm{C}}$ with its auxiliary conditions. The second purpose involves interpretations of the *Lagrange Multipliers* associated with the modified complementary energy.

We insist on fulfilling (27.2a) exactly, but abandon the other auxiliary conditions (27.2b) and (27.4) and include these in a *Modified Complementary Energy* $\mathbf{\Pi}_{\mathrm{CM}}$ by use of the *Lagrange Multipliers* $\eta_i^{(T)}$ and $\eta_i^{(12)}$ where the first is associated with the so-called static boundary and the second with the discontinuity surface.

The Lagrange Multiplier terms appear in the third and fourth integrals below

$$\begin{aligned}\mathbf{\Pi}_{\mathrm{CM}}(\sigma_{ij},\eta_i^{(T)},\eta_i^{(12)}) = \tfrac{1}{2}\int_{V^{(1)}+V^{(2)}} & C_{ijkl}\sigma_{ij}\sigma_{kl}dV \\ -\int_{S_u^{(1)}+S_u^{(2)}} & \sigma_{ij}n_j\bar{u}_i dS \\ -\int_{S_T^{(1)}+S_T^{(2)}} & \eta_i^{(T)}\left(\sigma_{ij}n_j-\overline{T}_i\right)dS \\ +\int_{S^{(12)}} & \eta_i^{(12)}\left(\sigma_{ij}^{(2)}-\sigma_{ij}^{(1)}\right)n_j^{(12)}dS\end{aligned} \tag{27.5}$$

Modified complementary energy $\mathbf{\Pi}_{\mathrm{CM}}$ with Lagrange Multipliers

where the only remaining auxiliary conditions are the static field equations (4.119) or (27.2a). As usual, we demand that the variation of the functional vanishes

$$\begin{aligned}0 = & \int_{V^{(1)}+V^{(2)}} C_{ijkl}\delta\sigma_{ij}\sigma_{kl}dV - \int_{S_u^{(1)}+S_u^{(2)}} \delta\sigma_{ij}n_j\bar{u}_i dS \\ & -\int_{S_T^{(1)}+S_T^{(2)}} \delta\eta_i^{(T)}\left(\sigma_{ij}n_j-\overline{T}_i\right)dS - \int_{S_T^{(1)}+S_T^{(2)}} \eta_i^{(T)}\delta\sigma_{ij}n_j dS \\ & +\int_{S^{(12)}} \delta\eta_i^{(12)}\left(\sigma_{ij}^{(2)}-\sigma_{ij}^{(1)}\right)n_j^{(12)}dS \\ & +\int_{S^{(12)}} \eta_i^{(12)}\left(\delta\sigma_{ij}^{(2)}-\delta\sigma_{ij}^{(1)}\right)n_j^{(12)}dS\end{aligned} \tag{27.6}$$

Variation of $\mathbf{\Pi}_{\mathrm{CM}}=0$

It is important to note that we may vary $\eta_i^{(T)}$ independently for each sub-volume because each static boundary belongs to only one sub-volume

$$\begin{aligned}& 0 = \int_{S_T^{(\alpha)}} \delta\eta_i^{(T)}\left(\sigma_{ij}n_j-\overline{T}_i\right)dS\,,\ \alpha=1,2 \\ & \Rightarrow \sigma_{ij}n_j = \overline{T}_i\,,\ x_i\in S_T^{(\alpha)}\,,\ \alpha=1,2\end{aligned} \tag{27.7}$$

Static boundary of an element independent of all other elements

and that $\eta_i^{(12)}$ is confined to the discontinuity surface $S^{(12)}$, which means that its variation therefore is independent of all other variations with the result that

$$\begin{aligned}& 0 = \int_{S^{(12)}} \delta\eta_i^{(12)}\left(\sigma_{ij}^{(2)}-\sigma_{ij}^{(1)}\right)n_j^{(12)}dS \\ & \Rightarrow \left(\sigma_{ij}^{(2)}-\sigma_{ij}^{(1)}\right)n_j^{(12)} = 0\,,\ x_i\in S^{(12)}\end{aligned} \tag{27.8}$$

Discontinuity surface $S^{(12)}$ belongs only to $V^{(1)}$ and $V^{(2)}$

Thus we have proved that requiring $\delta\mathbf{\Pi}_{\mathrm{CM}}=0$ implies that the the static boundary conditions (27.2b) and the static continuity conditions (27.4) are fulfilled. The remaining terms of (27.6) may be rewritten

Remaining terms

$$
\begin{aligned}
0 = &\sum_{\alpha=1}^{2} \int_{V^{(\alpha)}} C_{ijkl} \delta\sigma_{ij} \sigma_{kl} dV \\
&- \sum_{\alpha=1}^{2} \int_{S_u^{(\alpha)}} \delta\sigma_{ij} n_j \bar{u}_i dS - \sum_{\alpha=1}^{2} \int_{S_T^{(\alpha)}} \eta_i^{(T)} \delta\sigma_{ij} n_j dS \\
&- \sum_{\alpha=1}^{2} \int_{S^{(12)}} \eta_i^{(12)} \delta\sigma_{ij} n_j dS
\end{aligned}
\tag{27.9}
$$

and thus the contributions from the two sub-volumes $V^{(1)}$ and $V^{(2)}$ are now separated.

For any of the two sub-volumes $V^{(\alpha)}$, $\alpha = 1, 2$ we may get

Boundary terms from $\delta\mathbf{\Pi}_{\mathrm{CM}}$

$$
\begin{aligned}
&\int_{S^{(\alpha)}} \delta\sigma_{ij} n_j u_i dS \\
&= \int_{S_u^{(\alpha)}} \delta\sigma_{ij} n_j \bar{u}_i dS + \int_{S_T^{(\alpha)}} \delta\sigma_{ij} n_j u_i dS + \int_{S^{(12)}} \delta\sigma_{ij} n_j u_i dS
\end{aligned}
\tag{27.10}
$$

and

Body terms from $\delta\mathbf{\Pi}_{\mathrm{CM}}$

$$
\begin{aligned}
&\int_{S^{(\alpha)}} \delta\sigma_{ij} n_j u_i dS = \int_{V^{(\alpha)}} \left(\delta\sigma_{ij} u_i\right)_{,j} dV \\
&= \int_{V^{(\alpha)}} \delta\sigma_{ij,j} u_i dV + \int_{V^{(\alpha)}} \delta\sigma_{ij} u_{i,j} dV \\
&= \int_{V^{(\alpha)}} \delta\sigma_{ij} \varepsilon_{ij} dV
\end{aligned}
\tag{27.11}
$$

Recall: $\delta\sigma_{ij,j} = 0$ and $\sigma_{ij} = \sigma_{ji}$

where we have exploited that $\delta\sigma_{ij,j}$ vanishes because σ_{ij} satisfies the static field equations. Furthermore, since the stress tensor is symmetric any antisymmetric part of $u_{i,j}$ does not contribute to the sum $\delta\sigma_{ij} u_{i,j}$ with the result that we may write $\delta\sigma_{ij}\varepsilon_{ij}$ instead. In view of (27.10) and (27.11) the outcome of (27.9) is

Constitutive equation follows from of $\delta\mathbf{\Pi}_{\mathrm{CM}} = 0$

$$
\begin{aligned}
&0 = \sum_{\alpha=1}^{2} \int_{V^{(\alpha)}} \left(C_{ijkl}\sigma_{kl} - \varepsilon_{ij}\right) \delta\sigma_{ij} dV \\
&\Rightarrow \varepsilon_{ij} = C_{ijkl}\sigma_{kl}\,,\ x_i \in V^{(\alpha)}
\end{aligned}
\tag{27.12}
$$

and thus requiring $\delta\mathbf{\Pi}_{\mathrm{CM}} = 0$ results in fulfillment of the constitutive relation (4.129).

Introduce a parameter $\eta_i^{(\alpha)}$, which is associated with the entire boundary of the sub-body $V^{(\alpha)}$, in the following way

$$\eta_i^{(\alpha)} = \begin{cases} \bar{u}_i \,,\ x_i \in S_u^{(\alpha)} \\ \eta_i^{(T)} = u_i \,,\ x_i \in S_T^{(\alpha)} \\ \eta_i^{(12)} = u_i \,,\ x_i \in S^{(12)} \end{cases} \tag{27.13}$$

Lagrange multiplier $\eta_i^{(\alpha)}$ $\sim$ boundary displacements

and we may realize that the Lagrange Multiplier in the present case may be interpreted as the *boundary* displacements of $V^{(1)}$ and $V^{(2)}$. The displacement fields in the *interior* of the two sub-volumes appear nowhere in the expression for the modified complementary energy $\mathbf{\Pi}_{\mathrm{CM}}$, which may now be written

$$\mathbf{\Pi}_{\mathrm{CM}}(\sigma_{ij}, u_i^B) = \tfrac{1}{2}\sum_{\alpha=1}^{2}\int_{V^{(\alpha)}} C_{ijkl}\sigma_{ij}\sigma_{kl}dV - \sum_{\alpha=1}^{2}\int_{S^{(\alpha)}} \sigma_{ij}n_j u_i^B dS \tag{27.14}$$

New expression for $\mathbf{\Pi}_{\mathrm{CM}}$

where u_i^B signifies the displacement on the entire boundary of the sub-volumes.

It may be worthwhile thinking of the boundary terms of (27.14) in a more physical way than was done above. The individual terms of $\mathbf{\Pi}_{\mathrm{CM}}$ all express virtual work. On $S_T^{(\alpha)}$ the term is a weighted average of the virtual work done by the error due to the fact that we may not be able to fulfill the static boundary condition on $S_T^{(\alpha)}$ exactly. Similarly, on $S^{(12)}$ the term expresses a weighted average of work done by the stress discontinuity here. In view of this, it seems reasonable that in both cases the boundary displacement serves as the weight function.

The Lagrange Multiplier field serve as weight functions

27.2 Stress Hybrid Finite Elements

Apparently T.H.H. Pian is the first to apply the idea of using the modified complementary energy derived in Section 27.1 to derive finite elements, see (Pian 1964). Since then a vast number of researchers have applied this idea to almost any type of finite element that you may think of.

Based on the derivations in Section 27.1 we may establish a version of the Finite Element Method when we interpret the discontinuity surface as an inter-element boundary. The boundaries of an element, say number J, may be of the following kinds

Stress hybrid finite element

1. An inter-element boundary $S^{(JK)}$, where K signifies the element number of all neighboring elements,
2. A static boundary $S_T^{(J)}$,
3. A kinematic boundary $S_u^{(J)}$.

For our present purpose the expression (27.5) for the modified complementary energy is more convenient than (27.14). For a body divided into N_{El} finite elements we get

$\mathbf{\Pi}_{\text{CM}}$ for a collection of finite elements

$$\begin{aligned}\mathbf{\Pi}_{\text{CM}} = \sum_{J=1}^{N_{El}} \Bigg(& \frac{1}{2}\int_{V^{(J)}} C_{ijkl}\sigma_{ij}\sigma_{kl} dV - \int_{S_u^{(J)}} \sigma_{ij} n_j \bar{u}_i dS \\ & - \int_{S_T^{(J)}} \eta_i^{(J)} \left(\sigma_{ij} n_j - \overline{T}_i\right) dS \\ & + \sum_{K=1}^{J-1} \int_{S^{(JK)}} \eta_i^{(JK)} \left(\sigma_{ij}^{(K)} - \sigma_{ij}^{(J)}\right) n_j^{(JK)} dS \Bigg)\end{aligned} \tag{27.15}$$

where the upper limit on the internal sum insures that no inter-element boundary is encountered more than once. As it stands, (27.15) is not a convenient basis for matrix formulations. Therefore, we rearrange the stresses and strains, see e.g. Section 5.1, (5.1) and (5.2)

Rearrangement of stresses and strains

$$\begin{bmatrix}\sigma_1\\ \sigma_2\\ \sigma_3\\ \sigma_4\\ \sigma_5\\ \sigma_6\end{bmatrix} \equiv \begin{bmatrix}\sigma_{11}\\ \sigma_{22}\\ \sigma_{33}\\ \sigma_{23}\\ \sigma_{31}\\ \sigma_{12}\end{bmatrix} \quad \text{and} \quad \begin{bmatrix}\varepsilon_1\\ \varepsilon_2\\ \varepsilon_3\\ \varepsilon_4\\ \varepsilon_5\\ \varepsilon_6\end{bmatrix} \equiv \begin{bmatrix}\varepsilon_{11}\\ \varepsilon_{22}\\ \varepsilon_{33}\\ 2\varepsilon_{23}\\ 2\varepsilon_{31}\\ 2\varepsilon_{12}\end{bmatrix} \tag{27.16}$$

The constitutive relation (4.129) is then transformed into

Constitutive relation

$$\varepsilon_j = C_{ij}\sigma_i\,,\ (i,j) = (1,2,\ldots,6) \tag{27.17}$$

or, using matrix notation

Constitutive relation in matrix form

$$\{\varepsilon\} = [C]\{\sigma\} \tag{27.18}$$

Now, introduce

Auxiliary condition on tractions

$$T_i = \sigma_{ij} n_j \tag{27.19}$$

as an auxiliary condition on $\mathbf{\Pi}_{\text{CM}}$.[27.1]

In order to express the tractions T_i in vector form we introduce a matrix $[n]$, which contains the elements of the unit normal vector n_j and which in

[27.1] This is an auxiliary condition which is easily satisfied.

the three-dimensional case has 3 rows and 6 columns

$$[n] \equiv \begin{bmatrix} n_1 & 0 & 0 & 0 & n_3 & n_2 \\ 0 & n_2 & 0 & n_3 & 0 & n_1 \\ 0 & 0 & n_3 & n_2 & n_1 & 0 \end{bmatrix} \tag{27.20}$$

Boundary normal vector in matrix form

and write

$$\{T\} = [n]\{\sigma\} \tag{27.21}$$

Boundary tractions in matrix form

This way of writing the tractions is especially convenient in connection with developing finite element matrices, see (27.24), where the matrices $[F_\beta]$ and the vector $\{T^0\}$ may easily be found from $[N_\beta]$ and $\{\sigma^0\}$ after premultiplication by $[n]$.

The expression (27.15) for the modified complementary energy $\mathbf{\Pi}_{\mathrm{CM}}$ may now be cast in terms of matrices

$$\begin{aligned} \mathbf{\Pi}_{\mathrm{CM}} = \sum_{J=1}^{N_{\mathrm{El}}} \Bigg(& \frac{1}{2}\int_{V^{(J)}} \{\sigma\}^T[C]\{\sigma\}dV - \int_{S_u^{(J)}} \{T\}^T\{\bar{u}\}dS \\ & - \int_{S_T^{(J)}} \{\eta^{(J)}\}^T \Big(\{T\} - \{\overline{T}\}\Big)dS \\ & + \sum_{K=1}^{J-1} \int_{S^{(JK)}} \{\eta^{(JK)}\}^T \Big(\{T^{(KJ)}\} - \{T^{(JK)}\}\Big)dS \Bigg) \end{aligned} \tag{27.22}$$

$\mathbf{\Pi}_{\mathrm{CM}}$ in matrix form

27.2.1 Discretization

As long as the assumed stress distribution satisfies the static field equations (4.120) it is valid. Let

$$\{\sigma(x_i)\} = [N_\beta(x_i)]\{v_\beta\} + \{\sigma^0(x_i)\}\,,\ x_i \in V^{(J)} \tag{27.23}$$

Discretization of stresses

where $[N_\beta(x_i)]$ is the *Stress Distribution Matrix*, $\{v_\beta\}$ is a vector containing the stress field parameters, and $\{\sigma^0(x_i)\}$ is a solution to (4.120). Therefore, $\{\sigma(x_i)\}$ must satisfy the *homogeneous* part of (4.120). On the boundaries of element J (27.23) yields

$$\{T(x_i)\} = [F_\beta(x_i)]\{v_\beta\} + \{T^0(x_i)\}\,,\ x_i \in S^{(J)} \tag{27.24}$$

Discretization of boundary tractions

where $[F_\beta(x_i)]$ and $\{T^0(x_i)\}$ both depend on the unit normal to the boundary of the element in question. In order to find the dependency we will need to revert to the notation σ_{ij} instead of σ_i because the product $\sigma_{ij}n_j$ is not expressible using the latter notation. There are no severe difficulties associated with this, as you may see later in Section 27.3.

On the *entire* boundary of the element we choose

Discretization of Lagrange multipliers

$$\{\eta(x_i)\} = [N_\eta(x_i)]\{v_\eta\}\,,\ x_i \in S^{(J)} \tag{27.25}$$

where we note that the displacement interpolation is confined to the element boundaries and thus the displacements inside the element are not defined by (27.25). In my experience it is, however, prudent to choose the boundary displacements such that they are indeed boundary values of a displacement field defined over the entire element, which probably agrees with one's intuition. The assumption (27.25) entails that we assume that the prescribed boundary displacements $\bar{u}$ with sufficient accuracy can be represented by the same interpolation as the other boundary displacements

Possible small error at $S_u^{(J)}$

$$\{\bar{u}(x_i)\} \approx [N_\eta(x_i)]\{\bar{v}_\eta\}\,,\ x_i \in S_u^{(J)} \tag{27.26}$$

In compliance with (27.13) we extend the definition of $\{v_\eta\}$ to include $\{\bar{v}_\eta\}$ and may discretize the modified complementary energy $\mathbf{\Pi}_{\text{CM}}$, see (27.22), as follows

$$\begin{aligned}
&\mathbf{\Pi}_{\text{CM}}(\{v_\beta\},\{v_\eta\}) \\
&= \sum_{J=1}^{N_{\text{El}}} \Bigg(\frac{1}{2}\int_{V^{(J)}} \Big(\{v_\beta\}^T[N_\beta(x_i)]^T + \{\sigma^0(x_i)\}^T\Big)[C(x_i)] \\
&\qquad \Big([N_\beta(x_i)]\{v_\beta\} + \{\sigma^0(x_i)\}\Big)dV \\
&\quad - \int_{S_u^{(J)}} \Big(\{v_\beta\}^T[F_\beta(x_i)]^T + \{T^0(x_i)\}^T\Big) \\
&\qquad [N_\eta(x_i)]\{v_\eta\}dS \\
&\quad - \int_{S_T^{(J)}} \Big(\{v_\beta\}^T[F_\beta(x_i)]^T + \{T^0(x_i)\}^T - \{\overline{T}(x_i)\}^T\Big) \\
&\qquad [N_\eta(x_i)]\{v_\eta\}dS \\
&\quad - \int_{S^{(J)}-S_u^{(J)}-S_T^{(J)}} \Big(\{v_\beta\}^T[F_\beta(x_i)]^T + \{T^0(x_i)\}^T\Big) \\
&\qquad [N_\eta(x_i)]\{v_\eta\}dS \Bigg)_J
\end{aligned} \tag{27.27}$$

where the indication that all vectors and matrices belong to element J is omitted.

Written out, (27.27) provides

$$
\begin{aligned}
&\mathbf{\Pi}_{\mathrm{CM}}(\{v_\beta\},\{v_\eta\})\\
&=\sum_{J=1}^{N_{\mathrm{El}}}\Bigg(\tfrac{1}{2}\{v_\beta\}^T\Big(\int_{V^{(J)}}[N_\beta]^T[C][N_\beta]dV\Big)\{v_\beta\}\\
&\qquad+\{v_\beta\}^T\Big(\int_{V^{(J)}}[N_\beta]^T[C]\{\sigma^0\}dV\Big)\\
&\qquad+\tfrac{1}{2}\Big(\int_{V^{(J)}}\{\sigma^0\}^T[C]\{\sigma^0\}dV\Big)\\
&\qquad-\{v_\beta\}^T\Big(\int_{S^{(J)}}[F_\beta]^T[N_\eta]dS\Big)\{v_\eta\}\\
&\qquad-\{v_\eta\}^T\Big(\int_{S^{(J)}}[N_\eta]^T\{T^0\}dS\Big)\\
&\qquad+\{v_\eta\}^T\Big(\int_{S_T^{(J)}}[N_\eta]^T\{\overline{T}\}dS\Big)\Bigg)_J
\end{aligned}
\tag{27.28}
$$

Introduce the following vectors and matrices

$$[k_{\beta\beta}]\equiv\int_{V^{(J)}}[N_\beta]^T[C][N_\beta]dV \tag{27.29}$$

Flexibility matrix $[k_{\beta\beta}]$

$$\{r_\beta\}\equiv\int_{V^{(J)}}[N_\beta]^T[C]\{\sigma^0\}dV \tag{27.30}$$

Right-hand side $\{r_\beta\}$

$$[k_{\beta\eta}]\equiv\int_{S^{(J)}}[F_\beta]^T[N_\eta]dS \tag{27.31}$$

Mixed matrix $[k_{\beta\eta}]$

$$\{r_\eta\}\equiv\int_{S^{(J)}}[N_\eta]^T\{T^0\}dS-\int_{S_T^{(J)}}[N_\eta]^T\{\overline{T}\}dS \tag{27.32}$$

Right-hand side $\{r_\eta\}$

and rewrite (27.28) in the much simpler form

$$
\begin{aligned}
\mathbf{\Pi}_{\mathrm{CM}}(\{v_\beta\},\{v_\eta\})=\sum_{J=1}^{N_{\mathrm{El}}}\Big(&\tfrac{1}{2}\{v_\beta\}^T[k_{\beta\beta}]\{v_\beta\}+\{v_\beta\}^T\{r_\beta\}\\
&-\{v_\beta\}^T[k_{\beta\eta}]\{v_\eta\}-\{v_\eta\}^T\{r_\eta\}\\
&+\text{const.}\Big)_J
\end{aligned}
\tag{27.33}
$$

$\mathbf{\Pi}_{\mathrm{CM}}$ in matrix form

where the term "const." means a term that is constant with respect to variations of $\{v_\beta\}$ and $\{v_\eta\}$.

27.2.2 Elimination of Stress Field Parameters

Stress continuity across element boundaries no longer enforced

Since we have abandoned the condition of stress continuity across element boundaries as an auxiliary condition there are no conditions which tie the stress field parameters $\{v_\beta\}^{(J)}$ of element J directly to the stress field parameters $\{v_\beta\}^{(K)}$ of element K. We may therefore expect to be able to eliminate these parameters at element level. First, we take the variation of $\mathbf{\Pi}_{\text{CM}}$ given by (27.33)

$\delta\mathbf{\Pi}_{\text{CM}} = 0$

$$\begin{aligned} 0 = {} & \delta\mathbf{\Pi}_{\text{CM}}(\{v_\beta\},\{v_\eta\}) \\ = {} & \sum_{J=1}^{N_{El}} \Big(\{\delta v_\beta\}^T[k_{\beta\beta}]\{v_\beta\} + \{\delta v_\beta\}^T\{r_\beta\} \\ & -\{\delta v_\beta\}^T[k_{\beta\eta}]\{v_\eta\} - \{\delta v_\eta\}^T[k_{\beta\eta}]^T\{v_\beta\} \\ & -\{\delta v_\eta\}^T\{r_\eta\}\Big)_J \end{aligned} \tag{27.34}$$

which, in view of the comments above, provides

$$0 = \{\delta v_\beta\}^T\Big([k_{\beta\beta}]\{v_\beta\} + \{r_\beta\} - [k_{\beta\eta}]\{v_\eta\}\Big)\ \forall\{\delta v_\beta\} \tag{27.35}$$

Connection between stress and displacement variables

and thus

$$\{v_\beta\} = [k_{\beta\beta}]^{-1}\Big([k_{\beta\eta}]\{v_\eta\} - \{r_\beta\}\Big) \tag{27.36}$$

where it has been assumed that $[k_{\beta\beta}]$ is non-singular. For a compressible material this is always the case, unless you have chosen $[N_\beta(x_i)]$ such that its rows and columns are not linearly independent which would be unwise. A singular $[k_{\beta\beta}]$ would mean that we could subject the element to a stress state without producing stress energy in the element. Now, (27.34) and (27.36) yield

$$\begin{aligned} 0 = {} & \delta\mathbf{\Pi}_{\text{CM}}(\{v_\eta\}) \\ = {} & \sum_{J=1}^{N_{El}} \Big(\{\delta v_\eta\}^T \big([k_{\beta\eta}]^T[k_{\beta\beta}]^{-1}[k_{\beta\eta}]\{v_\eta\} \\ & -[k_{\beta\eta}]^T[k_{\beta\beta}]^{-1}\{r_\beta\} - \{r_\eta\}\big)\Big)_J\ \forall\{\delta v_\eta\} \end{aligned} \tag{27.37}$$

or

$$0 = \sum_{J=1}^{N_{El}} \Big\{ \{\delta v_\eta\}^T\big([k^*_{\eta\eta}]\{v_\eta\} - \{r^*_\eta\}\big)\Big\}_J\ \forall\{\delta v_\eta\} \tag{27.38}$$

where

Stiffness matrix $[k^*_{\eta\eta}]$

$$[k^*_{\eta\eta}] \equiv [k_{\beta\eta}]^T[k_{\beta\beta}]^{-1}[k_{\beta\eta}] \tag{27.39}$$

serves as an element stiffness matrix in the same sense as a stiffness matrix based on assumed displacements.

The vector $\{r^*_\eta\}$ below is the associated right-hand side, the modified load vector

$$\{r^*_\eta\} \equiv [k_{\beta\eta}]^T [k_{\beta\beta}]^{-1} \{r_\beta\} + \{r_\eta\} \tag{27.40}$$

Right-hand side $\{r^*_\eta\}$

When we note that $\{v_\eta\}$ contains the nodal displacements it is obvious that (27.38) furnishes system finite element equations of the same structure as the usual ones that are based on the potential energy $\mathbf{\Pi}_\mathrm{P}$. Thus, at system level, there is nothing fundamentally new. In order to compute the stresses, however, we must compute $\{v_\beta\}$ for each element from (27.36) and finally utilize (27.23) to determine the stresses. This is, we must admit, a more lengthy process than computing the matrix product $[D][B]\{v\}$ of the usual stiffness based finite element method. On the other hand, all matrix operations are performed at element level and are therefore computationally inexpensive.

27.3 A Rectangular Stress Hybrid Finite Element

Quadrilateral Stress Hybrid Finite Element

In this Section we derive the necessary formulas for a rectangular finite element for analysis of in-plane states, i.e. a stress hybrid version of the element treated in Section 24.2 and shown in Fig. 24.2. For our present purpose it proves more convenient placing the coordinate axes differently, as shown below.

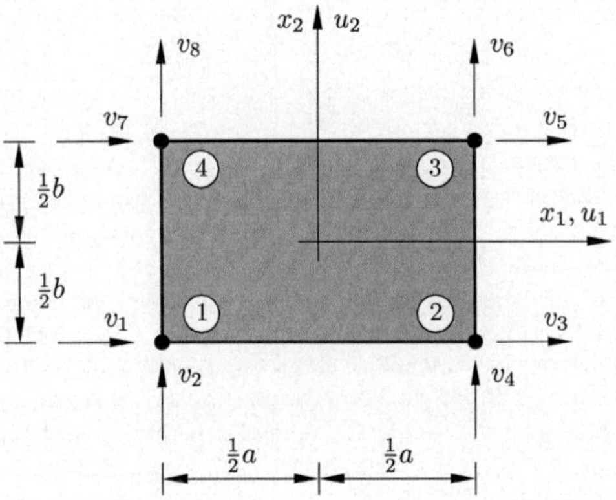

Fig. 27.2: Rectangular plate finite element.

Recall that the condition on the stress distribution matrix $[N_\beta(x_i)]$ is that the stresses it produces must satisfy the homogeneous equilibrium equations (9.19a)

$$N_{\alpha\beta,\beta} = 0\,,\;(\alpha,\beta) = (1,2) \tag{27.41}$$

We take the physical coordinate system (x_1, x_2) to have its origin at the centroid of the rectangle. The length of its sides are a and b, respectively. Introduce the nondimensional coordinates ξ_1 and ξ_2

Nondimensional coordinates

$$\xi_1 \equiv \frac{x_1}{2a} \text{ and } \xi_2 \equiv \frac{x_2}{2b}\,,\;(\xi,\eta) \in [-1,+1] \tag{27.42}$$

in order to simplify the following computations.

Then, the following choices all satisfy the requirement of internal equilibrium[27.2]

$$[N_\beta(x_i)] = \begin{bmatrix} 1 & 0 & 0 \\ 0 & 1 & 0 \\ 0 & 0 & 1 \end{bmatrix} \tag{27.43}$$

Experience proves this to be the best choice

$$[N_\beta(x_i)] = \begin{bmatrix} 1 & 0 & 0 & \xi_2 & 0 \\ 0 & 1 & 0 & 0 & \xi_1 \\ 0 & 0 & 1 & 0 & 0 \end{bmatrix} \tag{27.44}$$

$$[N_\beta(x_i)] = \begin{bmatrix} 1 & 0 & 0 & \xi_2 & 0 & \xi_1 & 0 \\ 0 & 1 & 0 & 0 & \xi_1 & 0 & \xi_2 \\ 0 & 0 & 1 & 0 & 0 & -\frac{b}{a}\xi_2 & -\frac{a}{b}\xi_1 \end{bmatrix} \tag{27.45}$$

The first of these choices entails the assumption that all stress components are constant over the element which may be too simple. If we make the second choice, we must live with a strong asymmetry as regards the variation of the stresses in that the shear stress is assumed constant over the element, while the normal stresses are allowed to vary, but only in one direction. The third possibility let all stresses vary as first degree polynomials in both directions implying a great deal of freedom which may seem appealing. The "correct" choice depends on the element. My experience shows that you should limit the number of stress parameters to the number of strain terms associated with the analogous displacement based element. For the element of Fig. 27.2 we should therefore use (27.44).

"Correct" choice of stress distribution

For this element the boundary displacements are determined in an easy way in that we may employ the boundary values of the displacement interpolation matrix of Section 24.2, which satisfies all the pertinent requirements.

[27.2] You may continue establishing assumptions with more rows, but they prove to furnish less good results than the ones provided by choosing (27.44).

Thus,

$$[N_\eta(x_i)] = \begin{cases} \frac{1}{2}\begin{bmatrix} \xi_1^M & 0 & \xi_1^P & 0 & 0 & 0 & 0 & 0 \\ 0 & \xi_1^M & 0 & \xi_1^P & 0 & 0 & 0 & 0 \end{bmatrix} & \text{on 1–2} \\ \frac{1}{2}\begin{bmatrix} 0 & 0 & \xi_2^M & 0 & \xi_2^P & 0 & 0 & 0 \\ 0 & 0 & 0 & \xi_2^M & 0 & \xi_2^P & 0 & 0 \end{bmatrix} & \text{on 2–3} \\ \frac{1}{2}\begin{bmatrix} 0 & 0 & 0 & 0 & \xi_1^P & 0 & \xi_1^M & 0 \\ 0 & 0 & 0 & 0 & 0 & \xi_1^P & 0 & \xi_1^M \end{bmatrix} & \text{on 3–4} \\ \frac{1}{2}\begin{bmatrix} \xi_2^M & 0 & 0 & 0 & 0 & 0 & \xi_2^P & 0 \\ 0 & \xi_2^M & 0 & 0 & 0 & 0 & 0 & \xi_2^P \end{bmatrix} & \text{on 4–1} \end{cases} \tag{27.46}$$

Boundary displacements

where

$$\xi_1^P \equiv 1+\xi_1\,,\ \xi_1^M \equiv 1-\xi_1\,,\ \xi_2^P \equiv 1+\xi_2 \ \text{ and } \ \xi_2^M \equiv 1-\xi_2 \tag{27.47}$$

Whether you proceed analytically from here or use a purely numerical scheme is a matter of taste, but the analytic expressions become rather horrendous in length. For instance, the full formula for $k^\star_{\eta\eta}[1,2]$, which is one of the shorter ones, is so big that it barely fits the width of the page, even in a very small font

$$\begin{aligned} &k^\star_{\eta\eta}[1,2] \\ &= \frac{abC_{11}C_{22}-abC_{12}C_{33}+abC_{13}C_{23}-abC_{12}{}^2-a^2C_{11}C_{23}+a^2C_{12}C_{13}+b^2C_{12}C_{23}-b^2C_{13}C_{22}}{4ab(C_{11}C_{22}C_{33}+2C_{12}C_{13}C_{23}-C_{11}C_{23}{}^2-C_{13}{}^2C_{22}-C_{12}{}^2C_{33})} \end{aligned} \tag{27.48}$$

where it seems worth mentioning that $[C]$ is inversely proportional to the plate thickness t and for the case of plane stress and isotropy is given by (9.49)–(9.54).

For the case of plane stress and isotropy the expressions for $k^\star_{\eta\eta}[i,j]$ become more manageable, e.g. $k^\star_{\eta\eta}[1,1]$, which is one of the longest expressions, is

$$k^\star_{\eta\eta}[1,1] = \frac{-3a^2-8b^2+3a^2\nu+2b^2\nu^2}{24ab\,(\nu-1)\,(\nu+1)}\,Et \tag{27.49}$$

The fact that the formulas become very long in the general case—and rather complicated even for plane stress and isotropy—may not be important in itself, but the implication is that the number of floating point operations becomes very large, and therefore it seems better to establish $[k_{\beta\beta}]$ and $[k_{\beta\eta}]$ analytically, invert $[k_{\beta\beta}]$ numerically and compute the final expression for $[k^\star_{\eta\eta}]$ numerically.

For convenience, the formulas for $[k_{\beta\beta}]$ and $[k_{\beta\eta}]$ are given below

$[k_{\beta\beta}]$

$$[k_{\beta\beta}] = ab \begin{bmatrix} C_{11} & C_{12} & C_{13} & 0 & 0 \\ C_{12} & C_{22} & C_{23} & 0 & 0 \\ C_{13} & C_{23} & C_{33} & 0 & 0 \\ 0 & 0 & 0 & \frac{1}{3}C_{11} & 0 \\ 0 & 0 & 0 & 0 & \frac{1}{3}C_{22} \end{bmatrix} \tag{27.50}$$

In order to compute $[k_{\beta\eta}]$ we utilize the idea of rewriting the normal vector n_α as a matrix $[n]$, see (27.20). In the two-dimensional case we get the following expression for $[n]$

Normal vector in matrix form $[n]$

$$[n] = \begin{bmatrix} n_1 & 0 & n_2 \\ 0 & n_2 & n_1 \end{bmatrix} \tag{27.51}$$

which, of course, depends on the boundary

Normal vector on the boundary in matrix form $[n]$

$$[n] = \begin{cases} \begin{bmatrix} 0 & 0 & -1 \\ 0 & -1 & 0 \end{bmatrix} & \text{on 1–2} \\[2ex] \begin{bmatrix} 1 & 0 & 0 \\ 0 & 0 & 1 \end{bmatrix} & \text{on 2–3} \\[2ex] \begin{bmatrix} 0 & 0 & 1 \\ 0 & 1 & 0 \end{bmatrix} & \text{on 3–4} \\[2ex] \begin{bmatrix} -1 & 0 & 0 \\ 0 & 0 & -1 \end{bmatrix} & \text{on 4–1} \end{cases} \tag{27.52}$$

At this point we may expect that the derivations become much more complicated if the element is a more general type of quadrangle than rectangles.

By use of (27.31) and (27.24) we may find the following rather simple result for $[k_{\beta\eta}]$

$$[k_{\beta\eta}] = \begin{bmatrix} -\frac{1}{2}b & 0 & \frac{1}{2}b & 0 & \frac{1}{2}b & 0 & -\frac{1}{2}b & 0 \\ 0 & -\frac{1}{2}a & 0 & -\frac{1}{2}a & 0 & \frac{1}{2}a & 0 & \frac{1}{2}a \\ -\frac{1}{2}a & -\frac{1}{2}b & -\frac{1}{2}a & \frac{1}{2}b & \frac{1}{2}a & \frac{1}{2}b & \frac{1}{2}a & -\frac{1}{2}b \\ \frac{1}{6}b & 0 & -\frac{1}{6}b & 0 & \frac{1}{6}b & 0 & -\frac{1}{6}b & 0 \\ 0 & \frac{1}{6}a & 0 & -\frac{1}{6}a & 0 & \frac{1}{6}a & 0 & -\frac{1}{6}a \end{bmatrix} \tag{27.53}$$

With (27.50) and (27.53) in hand it is straightforward to compute $[k^*_{\eta\eta}]$ from (27.39). Once the finite element equations have been solved the stress field parameters of each element may be determined from (27.36), and, finally, the stresses may be found from (27.23) with $[N_\beta]$ given by (27.44).

Ex 27-1 Comparison Between a Stress Hybrid Element and the Melosh Element

In order to test the possible virtues of the stress hybrid element developed above and compare it with the Melosh element from Section 24.2 let us analyze plates such as the one sketched in Fig. Ex. 27-1.1. The

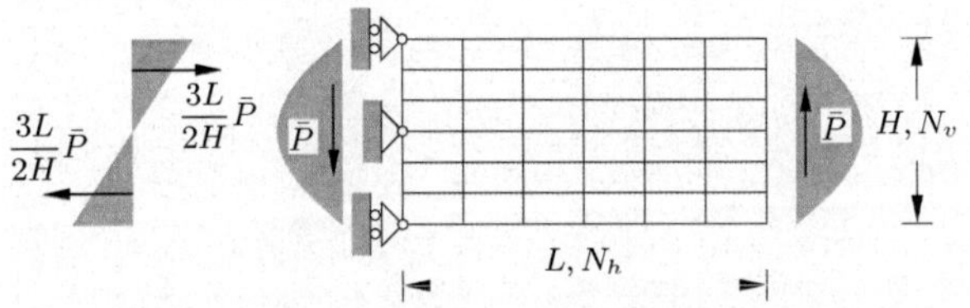

Fig. Ex. 27-1.1: *Test case for comparison between a Stress Hybrid Element and the Melosh element*

plate is supported as shown in the Fig. Ex. 27-1.1, and the loading consists of parabolically distributed shear loads at the vertical ends and two triangular normal stress distributions at the left-hand end. With the values of these loads given as indicated in the figure, all reactions vanish. In this case there exists an exact analytic solution which may be found from a solution in (Timoshenko & Goodier 1970), where the kinematic boundary conditions are different in that the midpoint of the left-hand edge is restricted against translation in the two directions and is not permitted to rotate. This last condition is not compatible with usual plate theories, such as the ones given in (Timoshenko & Goodier 1970) and in the present book. Furthermore, it is not possible to prescribe such a boundary condition for the two types of elements utilized here. Therefore, we choose to support the plate as shown in Fig. Ex. 27-1.1. In the following we concentrate on the "characteristic displacement" which in this case is the vertical component $v_2(L)$ of the displacement of the midpoint at the right-hand edge. The exact value $v_2^{ex}(L)$ is

$$v_2^{ex}(L) = \frac{\bar{P}L^3}{3EI}\left(1 + (2 + \tfrac{5}{2}\nu)\left(\frac{H}{2L}\right)^2\right) \qquad (Ex.\ 27\text{-}1.1)$$

Exact solution

Note that the structure of (*Ex. 27-1.1*) is the same as that of (*Ex. 7-4.11*) which indicates that the Timoshenko beam theory may provide useful results for short beams, see also Example Ex 12-6.

Ex 27-1.1 Long "Beam"

First, we shall investigate a fairly long "beam" with $L = 10H$ where bending dominates over the contribution from shear as is clear from (*Ex. 27-1.1*). Results obtained by use of the so-called Melosh element, see Section 24.2, and by the stress hybrid element developed above are compared in Fig. Ex. 27-1.2. In both computations the finite elements are squares as indicated by "ratio 1:1" in the figure. As expected the

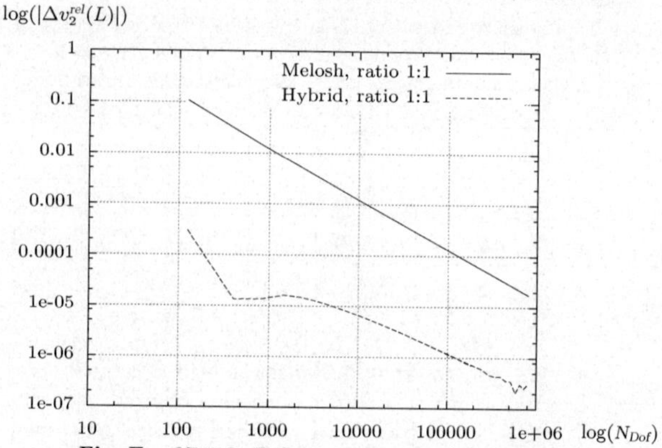

Fig. Ex. 27-1.2: *Relative error on $v_2(L)$. Comparison between the Melosh and the stress hybrid element for long beam.*
The quotient between the horizontal and vertical element length is given by the ratio.

Melosh element does not perform well here, see Section 24.2.5, where the inherent deficiency of the Melosh element is discussed. If we accept a relative error of 1% on the displacement it is necessary to use about 360 Melosh elements, while the stress hybrid element provides a relative error more than 10 times smaller with only 40 elements. In some cases one might insist that the relative error is about 0.1% which requires a little less than 6,000 Melosh elements but still only 40 stress hybrid elements.

Application of the Melosh element will always result in a too stiff behavior, see Section 33.4.3. Furthermore, as long as numerical errors do not enter the picture, the errors from the Melosh element must lie on a straight line in the plot as proved by Strang & Fix (1973). The convergence of the Melosh element is clearly better behaved than that of the hybrid element and, as a matter of fact, up to 160 hybrid elements the resulting prediction is too stiff, while it is too flexible for higher numbers of elements. It may be observed that the Melosh element is the more robust in that the results obtained by use of more than about

300,000 hybrid elements are affected by numerical problems, while the Melosh results stay on the straight line. On the other hand, it does not seem likely that, for the present case, any sane person would insist on a relative error as small as 1×10^{-6}.

Ex 27-1.2 Short "Beam"

A shorter "beam" with $L = 3H$ is so stocky that the (in-plane) bending

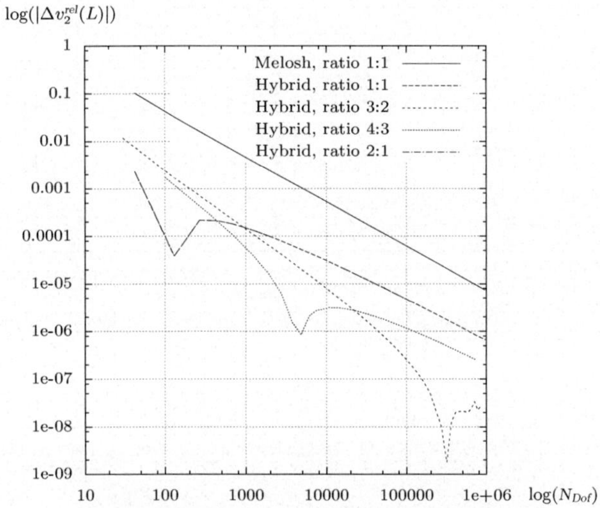

Fig. Ex. 27-1.3: *Relative error on $v_2(L)$. Comparison between the Melosh and the stress hybrid element for short beam.*
The quotient between the horizontal and vertical element length is given by the ratio.

deformation does not dominate completely over the shear deformation, thus making the above mentioned inherent deficiency of the Melosh element less pronounced. Results obtained by use of the Melosh element and by the stress hybrid element are compared in Fig. Ex. 27-1.3. Again, the Melosh element displays a convergence more uniform than that of the stress hybrid element, but also in this example the latter element provides superior accuracy. For some purposes a relative error of about 1% it may be enough, and, independent of the ratio between the sides of the elements, the hybrid elements all provide that accuracy with 8–12 finite elements, while it takes 108 Melosh elements to obtain the same error. If we wish a relative error ten times smaller, i.e. 0.1%, the hybrid elements provide that for 72 or less elements, while the Melosh version needs 448 elements.

Ex 27-1.3 Extremely Short "Beam"

The structure analyzed in our last example can hardly be called a beam because its length and depth are the same as given by $L = H$. This does, however, not imply that (*Ex. 27-1.1*) no longer is valid. In

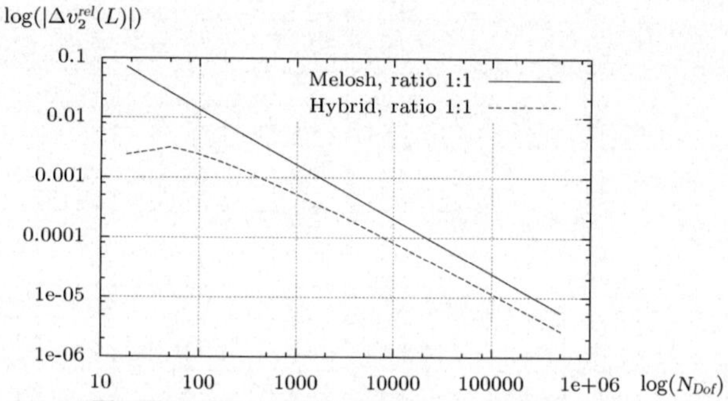

Fig. Ex. 27-1.4: *Relative error on $v_2(L)$. Comparison between the Melosh and the stress hybrid element for "long beam.*
The quotient between the horizontal and vertical element length is given by the ratio.

this case, shear deformation is more important than in the previous examples, which may be seen from (*Ex. 27-1.1*) where the second term is about 0.6 and thus contributes significantly to the displacements. The trend in the plot looks very similar to that of the previous ones, except that the stress hybrid element is not as much better than the Melosh element, which was to be expected because in-plane bending contributes less than in the previous examples.

27.3.1 Why Does the Hybrid Element Perform so Well in Bending?

From Example Ex 27-1 it appears that the stress hybrid finite element copes well with in-plane bending which, at a first glance, might seem strange because its faces deform exactly like the faces of the Melosh element and, as discussed above, see Section 24.2.5, it was concluded that it was the strain field which resulted from the displacement field that caused the deficiency of the Melosh element. It is, however, important noticing that the displacements in the interior of the hybrid element are not given directly by the displacements of the faces—as a matter of fact the displacements in the interior are not defined in the model.[27.3] In the interior only the stresses

[27.3] The only reasonable way to compute the displacements in the interior of the element is, of course, to assume that they follow from the boundary displacements in the same are defined and, to a large extent, it is these stresses that determine the behavior of the element.

We intend to dive a little into the subject of how the hybrid element performs under in-plane bending. Referring to Fig. 24.5 we may see that the displacement vector $\{v\}$ or $\{v_\eta\}$, see Figs. 24.2, and (27.25) for this case is given by

$$\{v\} = \{v_\eta\}^T = [\ \text{-1},\ 0,\ \ 1,\ \ 0,\ \ \text{-1},\ 0,\ \ 1,\ \ 0\ ] \tag{27.54}$$

and the stresses are found from (27.23) with $\sigma^0(x_i) = 0$

Discretization of stresses

$$\{\sigma(x_i)\} = [N_\beta(x_i)]\{v_\beta\} \tag{27.55}$$

where $[N_\beta]$ is given by (27.44)

$$[N_\beta(x_i)] = \begin{bmatrix} 1 & 0 & 0 & \xi_2 & 0 \\ 0 & 1 & 0 & 0 & \xi_1 \\ 0 & 0 & 1 & 0 & 0 \end{bmatrix} \tag{27.56}$$

and $\{v_\beta\}$ is determined by (27.36) with $\{r_\beta\} = 0$

$$\{v_\beta\} = [k_{\beta\beta}]^{-1}[k_{\beta\eta}]\{v_\eta\} \tag{27.57}$$

The product of $[k_{\beta\eta}]$ and $\{v_\eta\}$ may be found from (27.53) and (27.54)

$$\left([k_{\beta\eta}]\{v_\eta\}\right)^T = [\ 0,\ \ 0,\ \ 0,\ -\tfrac{2}{3}b,\ 0\ ] \tag{27.58}$$

It is quite remarkable that, even for cases without orthotropy or isotropy, the structure of $[k_{\beta\beta}]$ and of $[k_{\beta\beta}]^{-1}$ is as shown in Fig. 27.5. From (27.58) we may see that the structure of $[k_{\beta\eta}]\{v_\eta\}$ is as shown in the figure. The only contribution to the stresses then comes from one element of $[k_{\beta\beta}]^{-1}$, namely the one from row 4 and column 4. It is

$$\text{Element (4,4) of } [k_{\beta\beta}]^{-1} = \frac{3}{abC_{11}} \tag{27.59}$$

When we combine these findings with (27.56) we may see that the result for $\{\sigma\}$ is

$$\sigma_1 = -\frac{2\xi_2}{aC_{11}}\,,\ \sigma_2 = 0\,,\ \sigma_3 = 0 \tag{27.60}$$

which shows that the stress hybrid finite element does not produce spurious shear strains or stresses when it is subjected to in-plane bending.

After some, but not much, work this result may be confirmed by use of Fig. 24.5.

---

way as is the case for the Melosh element. But, don't use the displacement field in the interior for anything else that plotting.

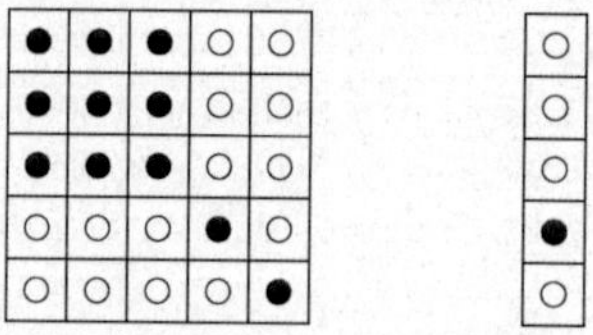

$[k_{\beta\beta}]$ and $[k_{\beta\beta}]^{-1}$ $\qquad$ $[k_{\beta\eta}]\{v_\eta\}$

Fig. 27.5: Structure of $[k_{\beta\beta}]$ (and $[k_{\beta\beta}]^{-1}$) and $[k_{\beta\eta}]\{v_\eta\}$. Filled circles indicate non-zero elements, open circles designate zero.

27.3.2 Isoparametric Version

Isoparametric version of the stress hybrid element may require special attention

If the element is of a more general shape than a rectangle, then you might be tempted to establish an isoparametric version of this element, see also Section 26.7 where the term *isoparametric* is introduced, you will have to resort to numerical quadratures and, in that case, you may just as well abandon the analytic approach as early as possible. In this connection it may be worth mentioning that in an isoparametric version of the element described above you must be careful regarding the coordinate system used for the stress distribution matrix $[N_\beta]$. If the coordinate system used for this purpose in a square element lies at an angle of $\pi/4$ with respect to the sides of the square the rank of the ensuing element matrix $[k^*_{\eta\eta}]$ is less than 5.

Chapter 28

Linear Elastic Finite Element Analysis of Torsion

In the following we shall see how finite elements based on the stress function concept may be derived in a fairly simple fashion.

One way to construct finite elements would be to discretize the potential $\mathbf{\Pi}$, see (*Ex. 13-6.4*), of Navier. It is, however, not the best way because it entails second derivatives of the stress function T. Since the compatibility equation (13.53) is a second order partial differential equation with no first order derivatives we may expect that a functional containing partial derivatives of first order can be established. This is indeed the case, as we shall see.

Navier's potential not the best—it contains second derivatives of T

28.1 A Functional for Torsion

For convenience we consider only a simply connected region but mention that the derivations may be extended to cover multiply connected regions in a manner similar to the one applied in Section 13.6.3.2.

We hope that the following functional $\mathbf{\Pi_T}$

$$\mathbf{\Pi_T}(T) = \tfrac{1}{2}\int_{A_0} T_{,\alpha}T_{,\alpha}dA - 2\int_{A_0} G\theta T dA \tag{28.1}$$

The functional $\mathbf{\Pi_T}$ contains first derivatives of T

is valid. The first variation $\delta\mathbf{\Pi_T}$ of $\mathbf{\Pi_T}$ is

$$\begin{aligned}\delta\mathbf{\Pi_T}(T) &= \int_{A_0} T_{,\alpha}\delta T_{,\alpha}dA - 2\int_{A_0} G\theta\delta T dA \\ &= \int_{A_0} \Big(\big(T_{,\alpha}\delta T\big)_{,\alpha} - T_{,\alpha\alpha}\delta T\Big)dA - 2\int_{A_0} G\theta\delta T dA\end{aligned} \tag{28.2}$$

Variation $\delta\mathbf{\Pi_T}(T)$ of $\mathbf{\Pi_T}(T)$

or, by application of the divergence theorem

$$\delta\mathbf{\Pi_T}(T) = \int_{\Gamma_0} T_{,\alpha}n_\alpha\delta T d\Gamma - \int_{A_0} T_{,\alpha\alpha}\delta T dA - 2\int_{A_0} G\theta\delta T dA \tag{28.3}$$

E. Byskov, *Elementary Continuum Mechanics for Everyone*,
Solid Mechanics and Its Applications 194, DOI: 10.1007/978-94-007-5766-0_28,

$\delta\mathbf{\Pi_T}(T) = 0$ provides boundary condition and differential equation

Require that $\delta\mathbf{\Pi_T}$ vanishes

$$\delta\mathbf{\Pi_T}(T) = 0 \Rightarrow \begin{cases} T_{,\alpha} n_\alpha \delta T = 0\,, & x_\alpha \in \Gamma_0 \\ (T_{,\alpha\alpha} + 2G\theta)\delta T = 0\,, & x_\alpha \in A_0 \end{cases} \tag{28.4}$$

where the first of these conditions is satisfied because T is constant on Γ_0 and, therefore $\delta T = 0$ on Γ_0.[28.1] Since δT is arbitrary for $x_\alpha \in A_0$ the second condition provides the compatibility condition (13.53)

$\mathbf{\Pi_T}$ is a valid functional

$$T_{,\beta\beta} = -2G\theta \tag{28.5}$$

and thus, we have proved that $\mathbf{\Pi_T}$ is a valid functional.

$\mathbf{\Pi_T}$ requires continuous values of T. Navier's functional also requires continuous first derivatives of T

Because $\mathbf{\Pi_T}$ does not contain derivatives of higher order than one, trial functions $\widetilde{T}$ used in $\mathbf{\Pi_T}$ are required to satisfy C_0-continuity, i.e. their value must be continuous, while there are no conditions on their derivatives.This would not have been the case if we had utilized the potential of Navier since it contains second order derivatives.

In the following we shall see how the functional $\mathbf{\Pi_T}$ may be utilized as a basis for finite elements.

28.2 Discretization

In an element let us discretize the stress function T as follows

Interpolation of T

$$T(x_\alpha) = [N_T(x_\alpha)]\,\{v_T\} \tag{28.6}$$

where $\{v_T\}$ denotes the nodal values of the stress function and $[N_T(x_\alpha)]$ is the stress function interpolation matrix—actually a row vector. We may now find the derivatives of the stress function and cast them in matrix form

Derivatives of T

$$\begin{bmatrix} T_{,1} \\ T_{,2} \end{bmatrix} = [B_T]\,\{v_T\} \quad \text{where} \quad [B_T] = \begin{bmatrix} N_{T,1} \\ N_{T,2} \end{bmatrix} \tag{28.7}$$

By use of (28.7) it now a straightforward task to write the finite element version of the functional $\mathbf{\Pi_T}$ given by (28.1)

Discretized $\mathbf{\Pi_T}$ at element level

$$\mathbf{\Pi_T}(\{v_T\}) = \sum_{i=1}^{N_{El}} \Big(\tfrac{1}{2} \{v_T\}^T [k_T] \{v_T\} - \{v_T\} \{\bar{r}_T\} \Big)_i \tag{28.8}$$

where the element "stiffness matrix" $[k_T]$ is defined by

Element "stiffness matrix" $[k_T]$

$$[k_T] \equiv \int_{A_i} [B_T]^T [B_T]\, dA \tag{28.9}$$

[28.1] In the present context, the condition $T = \text{const.}, x_\alpha \in \Gamma_0$ plays the same role as a kinematic boundary condition associated with the displacement field u_j in a potential energy $\mathbf{\Pi_P}(u_j)$ and must therefore be satisfied by all trial functions.

The element "load vector" $\{\bar{r}_T\}$ is given by

$$\{\bar{r}_T\} \equiv 2\theta \int_{A_i} G\,[N_T]^T\,dA \tag{28.10}$$

Element "load vector" $\{\bar{r}_T\}$

where, without loss of generality, we may take $\theta = 1$.

In order to compute the torsional moment M_T we need, see (13.38) and (13.47)

$$\begin{aligned}\int_{A_i} T dA &= \int_{A_i} [N_T]\{v_T\}\,dA = \{v_T\}^T \int_{A_i} [N_T]^T\,dA \\ &= \{v_T\}^T \{p_T\}\end{aligned} \tag{28.11}$$

where

$$\{p_T\} \equiv \int_{A_i} [N_T]^T\,dA \tag{28.12}$$

As usual, we may assemble the element stiffness matrices and the element load vectors and get the functional $\mathbf{\Pi_T}(\{V_T\})$ in terms of the system variables

$$\mathbf{\Pi_T} = \tfrac{1}{2}\{V_T\}^T[K_T]\{V_T\} - \{V_T\}^T\{\bar{R}_T\} \tag{28.13}$$

Discretized $\mathbf{\Pi_T}$ at system level

which is expressed in terms of the "system stiffness matrix" $[K_T]$, the vector of system nodal values $\{V_T\}$ and the "system load vector" $\{\bar{R}_T\}$. After introduction of the boundary conditions we may vary $\mathbf{\Pi_T}(\{V_T\})$ with respect to $\{V_T\}$, require that the variation vanishes and get a conventional set of finite element equations

$$[K_T]\{V_T\} = \{\bar{R}_T\} \tag{28.14}$$

System finite element equations

When a particular finite element has been developed, we may utilize almost any finite element code which allows introduction of user-defined elements to compute approximations to T and M_T.

Below, we shall derive a simple rectangular finite element and present results obtained by this and another, more sophisticated, element.

Ex 28-1 A Simple Rectangular Finite Element for Torsion

Four node finite element for torsion

As an illustrative example of how to develop finite elements based on $\mathbf{\Pi_T}$ we shall consider the simple rectangular element shown in Fig. Ex. 28-1.1. This element is the equivalent of the well-know *Melosh* element for plates with in-plane loading, see Section 24.2. Since we must require that T is continuous across element boundaries the sides

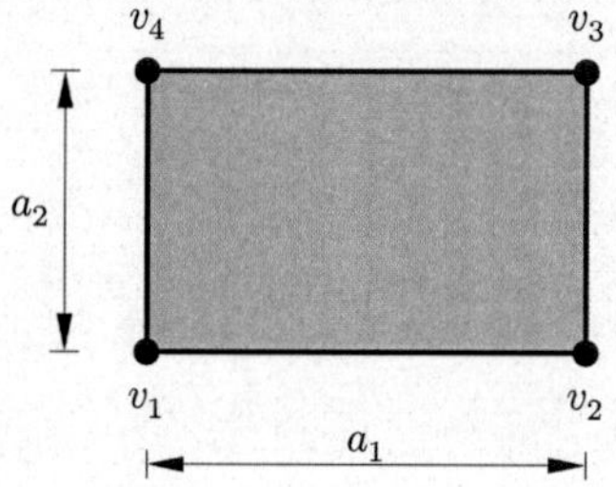

Fig. Ex. 28-1.1: *A simple rectangular finite Element for torsion.*

must remain straight after "deformation." This is the case when we choose

$$[N_T] = \left[\frac{(a_1 - x_1)\,(a_2 - x_2)}{a_1\,a_2} \;;\; \frac{x_1\,(a_2 - x_2)}{a_1\,a_2} \;; \right. \\ \left. \frac{x_1\,x_2}{a_1\,a_2} \;;\; \frac{x_2\,(a_1 - x_1)}{a_1\,a_2} \right] \qquad \text{(Ex. 28-1.1)}$$

As you may see, this is very much alike the rectangular plate finite element for in-plane states developed in Section 24.2.

Then, the element "load vector" $\{\bar{r}_T\}$ is

$$\{\bar{r}_T\}^T = \left[\frac{G\,a_1\,a_2}{2} \;;\; \frac{G\,a_1\,a_2}{2} \;;\; \frac{G\,a_1\,a_2}{2} \;;\; \frac{G\,a_1\,a_2}{2} \right] \qquad \text{(Ex. 28-1.2)}$$

The vector $\{p_T\}$, which is used for computation of the torsional moment M_T, is

$$\{p_T\}^T = \left[\frac{a_1\,a_2}{4} \;;\; \frac{a_1\,a_2}{4} \;;\; \frac{a_1\,a_2}{4} \;;\; \frac{a_1\,a_2}{4} \right] \qquad \text{(Ex. 28-1.3)}$$

Here, $[B_T]$, which is the equivalent of the strain distribution matrix $[B]$ associated with the above mentioned element for in-plane for in-plane states, see in particular (24.8), becomes

$$[B_T] = \begin{bmatrix} -\dfrac{a_2 - x_2}{a_1\,a_2} \;;\; \dfrac{a_2 - x_2}{a_1\,a_2} \;;\; \dfrac{x_2}{a_1\,a_2} \;;\; -\dfrac{x_2}{a_1\,a_2} \\ -\dfrac{a_1 - x_1}{a_1\,a_2} \;;\; -\dfrac{x_1}{a_1\,a_2} \;;\; \dfrac{x_1}{a_1\,a_2} \;;\; \dfrac{a_1 - x_1}{a_1\,a_2} \end{bmatrix} \qquad \text{(Ex. 28-1.4)}$$

The elements of the "stiffness matrix" $[k_T]$ then are

$$\begin{aligned}
k_T[1,1] &= k_T[2,2] \\
&= k_T[3,3] = k_T[4,4] = \frac{(a_1)^2 + (a_2)^2}{3\,a_1\,a_2} \\
k_T[1,2] &= k_T[2,1] \\
&= k_T[3,4] = k_T[4,3] = \frac{(a_1)^2 - 2\,(a_2)^2}{6\,a_1\,a_2} \\
k_T[1,3] &= k_T[3,1] \\
&= k_T[2,4] = k_T[4,2] = \frac{-(a_1)^2 - (a_2)^2}{6\,a_1\,a_2} \\
k_T[1,4] &= k_T[4,1] \\
&= k_T[1,3] = k_T[3,1] = \frac{(a_2)^2 - 2\,(a_1)^2}{6\,a_1\,a_2}
\end{aligned} \qquad \textit{(Ex. 28-1.5)}$$

Ex 28-2 An Eight-Nodes Rectangular Finite Element for Torsion

Eight node finite element for torsion

Our next example is the eight-node rectangular element shown in Fig. Ex. 28-2.1.

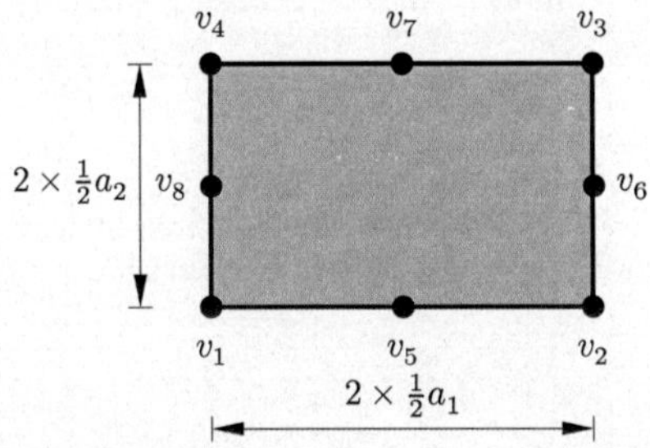

Fig. Ex. 28-2.1: *An eight-node rectangular finite Element for torsion.*

The stress function interpolation matrix $[N_T]$ may be easily found from various sources, such as (Cook et al. 2002), but, due to the length of the formulas involved, we do not show any of the ensuing formulas for $[B_T]$ etc. We shall, however, use it in presentation of results, see Example Ex 28-3.

Ex 28-3 Finite Element Results

As an example of use of the finite elements described above we analyze a square divided into square finite elements. The results are presented in Figs. Ex. 28-3.1 and (Ex. 28-3.2) where *Dof* indicates the total number of degrees of freedom of the finite element discretization.

That the results for the torsional moment M_T are better than those for the maximum shear stress $\max(|\tau_\alpha|)$ should not be surprising since

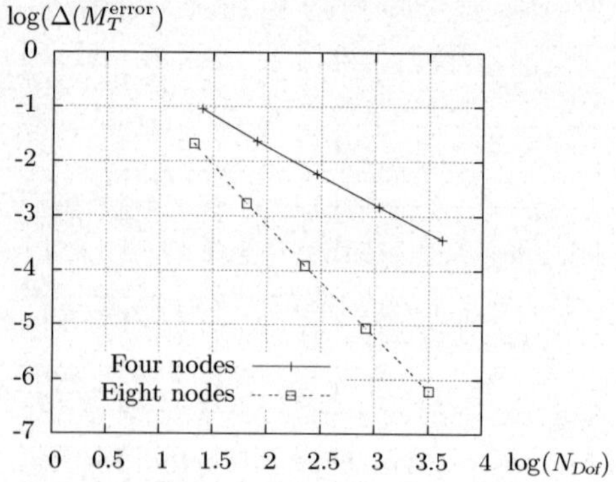

Fig. Ex. 28-3.1: Error on the torsional moment M_T. Finite element results.

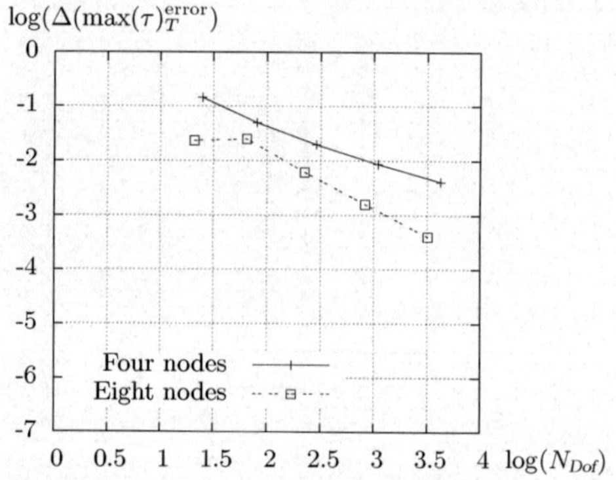

Fig. Ex. 28-3.2: Error on the maximum shear stress max ($|\tau_\alpha|$). Finite element results.

the torsional moment is computed by integration of the value of the primary variable, namely the stress function T, while differentiation provides the shear stress τ_α and, in general, you lose accuracy by differentiation.

It is clear that measured in terms of number of degrees of freedom the eight-node element is the superior. On the other hand, the 4-node element is not treated completely fair as regards the stress computation because the shear stress in that element is associated with the center of the element—not its boundary. Therefore, this element predicts the value of the shear stress at a distance from the boundary of the cross-section, never at its boundary.

Part VI

Mathematical Preliminaries

Chapter 29

Introduction

This part, Part VI, contains a number of subjects that are necessary for understanding the main topics of this book, namely continuum mechanics, including specialized continua, buckling, and the finite element method. I do not pretend to give a rigorous exposition. In spite of this, it is my hope that the basic ideas become clear once the reader has gone through this chapter.[29.1] In my opinion the best way to go about it is not to read any of the material covering the mathematical preliminaries until the reader finds it inevitable when he or she is reading the other parts.

I am sure that the conventional way of structuring a book on continuum mechanics, namely beginning with a long chapter introducing vectors, matrices, tensors, summation convention, etc., etc., has turned many students off because they may have known much of it in advance and because the rest seemed too abstract when it was not presented in connection with the "real" subject. Therefore, I have put this material at the end rather than at the beginning. Yet, I believe that it should be fairly easy to find the relevant mathematical background material.

It has been the experience of myself and others that many students think that subjects such as *the Index Notation*, *Functionals*, *Variational Principles* etc. that are touched on here find no other applications than continuum mechanics. None of these methods were invented in connection with continuum mechanics, although you might say that *Variational Principles* almost were. Thus, except for some of the specific formulas the topics in this part of the book have a much wider range than just continuum mechanics.

[29.1] I do hope that I have avoided the pitfall that one of my younger colleagues once fell into when he told me how he wanted to explain some difficult material to the students. His explanation simply was wrong, but his comment to me when I pointed this out to him was: "I know that, but I think that the students understand it better that way."

E. Byskov, *Elementary Continuum Mechanics for Everyone*,
Solid Mechanics and Its Applications 194, DOI: 10.1007/978-94-007-5766-0_29,

Chapter 30

Notation

The notations shown below are the ones that are most commonly used in this book. Some that appear less often are defined locally.

30.1 Overbar

An overbar $\bar{}$ indicates that the quantity is prescribed.[30.1] As an example $\bar{p}$ denotes a prescribed load, while $\bar{u}_1$ indicates that the displacement in the x_1-direction is prescribed.

Overbar $\bar{}$ indicates prescribed quantity

30.2 Tilde

A tilde $\tilde{}$ or a wider version $\widetilde{}$ is used for several purposes, which, in each particular case, is stated explicitly.

Tilde $\tilde{}$ serves several purposes

In some cases it indicates an approximation, e.g. $\tilde{w}$ may designate an approximation to the transverse displacement component w of a beam or a plate.

Sometimes a quantity is furnished with a tilde to distinguish it from a quantity with known properties. As an example of this, consider the quantities $\widetilde{N}$, $\widetilde{V}$ and $\widetilde{M}$ that are derived directly from equilibrium equations in Section 7.3.3 and are shown to be the same as the generalized quantities N, V and M whose properties were established earlier by application of the *Principle of Virtual Displacements.*

30.3 Indices

This subject is addressed in some detail in Chapter 31, but here it is mentioned that superscripts (upper indices) in this book serve as labels, not as tensor indices. On the other hand, subscripts (lower indices) play both roles depending on the circumstances. For instance, in connection with vectors and matrices used in Part V on finite elements they are used as labels and not as tensor indices, see Sections 31.1–31.3.3, while in connection with continuum mechanics they are tensor indices, see also Section 30.4.

Superscripts are labels

Subscripts may be labels or tensor indices

[30.1] Do *not* confuse this notation with the overbar used to signify a complex conjugate, a mean value, or something else.

E. Byskov, *Elementary Continuum Mechanics for Everyone*,
Solid Mechanics and Its Applications 194, DOI: 10.1007/978-94-007-5766-0_30,

30.4 Vectors and Matrices

Two types of vector

It is important to distinguish between vectors that describe geometric quantities, whose components vary in a specified fashion with changes of coordinate system, and vectors that are just ordered collections of numbers, such as column vectors. The first kind is said to be *first order tensors.*

"Geometric" vector

A "geometric" vector, i.e. a first order tensor, is denoted by a boldface upright Roman or Greek letter such as $\mathbf{t}$, $\mathbf{F}$, $\boldsymbol{\varepsilon}$ or $\boldsymbol{\alpha}$, or in component form by an italicized letter with a subscript, e.g. t_j, F_k, or α_k.

Column vector
Row vector

A column vector is denoted by an italicized letter enclosed in braces, e.g. $\{q\}$, while a row vector is given either as the transpose of a column vector such as $\{n\}^T$, where ^{T} indicates the transpose, or by an italicized letter enclosed in brackets such as $[N]$.

Two-dimensional matrix

A two-dimensional matrix is given by an italicized letter enclosed in brackets such as $[B]$. This means that a row vector may be viewed as a special case of a two-dimensional matrix.

30.5 Fields

Fields

In general, fields are denoted by boldface italicized letters, either Roman or Greek, such as $\boldsymbol{u}$, $\boldsymbol{\alpha}$ or $\boldsymbol{\sigma}$.

30.6 Operators

Operators

Operators are denoted by boldface upright Roman letters, such as $\mathbf{l}_1$, $\mathbf{l}_{11}$ and $\mathbf{H}$.

Chapter 31

Index Notation, the Summation Convention, and a Little About Tensor Analysis

In most cases it proves too cumbersome to write all terms of the equations which represent some physical phenomenon. Therefore, some shorthand notation is preferable. One of these is the so-called *Index Notation*, which is often useful, in particular when it is combined with the *Summation Convention*, see below.

Index Notation
Summation
Convention

In the present chapter we give an introduction to the subject of the index notation and the summation convention, but postpone many of the details to the sections on continuum mechanics, e.g. Section 2.2, which seems to be a more natural place to introduce the definitions of several topics.

31.1 Index Notation

In a three-dimensional Cartesian[31.1] coordinate system the axes are often denoted the x-, y- and z-axes. This is rather inconvenient for many purposes as we shall see shortly. And, by the way, at the same time the *base vectors* of the (x, y, z) coordinate system are often called $\mathbf{i}_1, \mathbf{i}_2, \mathbf{i}_3$, respectively, which does not seem consistent. Furthermore, if we wish to indicate that a scalar valued function f depends on all three coordinates we must write $f(x, y, x)$ which for one thing is not very elegant and for another rather lengthy.

Cartesian
Coordinate System

In terms of its components the *divergence* $\operatorname{div} \mathbf{v}$ of a vector field $\mathbf{v}$ may

[31.1] In Cartesian coordinate systems the axes are straight lines, which are at right angles to each other, and the unit measure along the axes is the same and constantly independent of the position of the point in question. Although the axes of a polar coordinate system are at right angles, it is not a Cartesian coordinate system because one of the families of coordinate lines consists of circles, while the other is formed by straight lines. As a consequence, the length measure varies with the distance from the pole.

E. Byskov, *Elementary Continuum Mechanics for Everyone*,
Solid Mechanics and Its Applications 194, DOI: 10.1007/978-94-007-5766-0_31,

Divergence div v

be written

$$\operatorname{div} \mathbf{v} = \frac{\partial v_x}{\partial x} + \frac{\partial v_y}{\partial y} + \frac{\partial v_z}{\partial z} \tag{31.1}$$

Index Notation

which is longer than necessary in that the structure of all terms is the same. Therefore, we seek a notation which is compact and yet displays the contents of the equations in a reasonably decipherable form. Here, the *Index Notation* proves to be of good help.

Lowercase Roman subscripts take the values $1, 2, 3$

The first step is to replace x by x_1, y by x_2, and z by x_3. The next step is to introduce a counter, the index $i, i \in [1, 2, 3]$. Then[31.2] we may refer to the above-mentioned scalar function f as $f(x_i)$, or $f(x_k)$ for that matter because we let all lowercase Roman subscripts (lower indices) take the values $1, 2, 3$. With this index notation in hand we may rewrite (31.1)

Divergence div v

$$\operatorname{div} \mathbf{v} = \frac{\partial v_1}{\partial x_1} + \frac{\partial v_2}{\partial x_2} + \frac{\partial v_3}{\partial x_3} \tag{31.2}$$

which, however, is even longer and more complicated than (31.1). We shall therefore introduce further simplifications, see Sections 31.3 and 31.2.

31.2 Comma Notation

Another convenient shorthand notation is the convention that comma indicates partial differentiation. Thus,

Comma notation $()_{,j}$

$$()_{,j} \equiv \frac{\partial()}{\partial x_j} \tag{31.3}$$

which means that we may write (31.2) in a somewhat shorter form

Divergence div v

$$\operatorname{div} \mathbf{v} = v_{1,1} + v_{2,2} + v_{3,3} \tag{31.4}$$

31.3 Summation Convention

Roman indices: range $[1, 3]$

First, note that we may write v_i instead of $\mathbf{v}$ to indicate the vector. Then, introduce the *Summation Convention*, which states that a repeated lowercase Roman index indicates a sum from 1 to 3 over the index and must appear *twice*—not three or four times—in each term of an expression. Then, e.g.

Summation Convention: Sum over repeated lower-case index

$$x_i x_i \equiv \sum_{i=1}^{3} (x_i)^2 \quad \text{or} \quad x_j x_j \equiv \sum_{k=1}^{3} (x_k)^2 \tag{31.5}$$

Summation index Dummy index

where repeated indices, here i and j in (31.5a) and (31.5b), respectively, are *summation indices.* Such indices are also called a *dummy indices* because their names are irrelevant.

[31.2] In Section 31.3.1 we introduce lowercase Greek indices to cover the values $1, 2$.

Now we can write (31.1) in a very compact and convenient form

$$\text{div}\,\mathbf{v} = \frac{\partial v_i}{\partial x_i} \tag{31.6}$$

Divergence div v

or, even shorter

$$\text{div}\,\mathbf{v} = v_{i,i} \tag{31.7}$$

Divergence div v

There is a self-evident result, which we shall utilize time and again, namely

$$\frac{\partial x_i}{\partial x_j} = \delta_{ij} \tag{31.8}$$

where δ_{ij} denotes the *Kronecker delta* which is defined by

$$\delta_{ij} \equiv \begin{cases} 1 & \text{for} \quad j = i \\ 0 & \text{for} \quad j \neq i \end{cases} \tag{31.9}$$

Kronecker delta δ_{ij}

The Kronecker delta may be used to change index of some quantity

$$v_i = \delta_{ij} v_j \tag{31.10}$$

Use the Kronecker delta to change index

see also (2.3) and the derivation following that equation.

For the Kronecker delta δ_{ij} the following formula is useful and follows from the summation convention

$$\delta_{jj} = 3 \tag{31.11}$$

$\delta_{jj} = 3$

Another useful symbol is the *Permutation Symbol* e_{ijk}, whose definition in the 3-dimensional case is[31.3]

$$e_{ijk} \equiv \begin{cases} 0 & \text{if any two of the subscripts are equal} \\ +1 & \text{if } (i,j,k) = (1,2,3) \text{ or } (2,3,1) \text{ or } (3,1,2) \\ -1 & \text{if } (i,j,k) = (3,2,1) \text{ or } (2,1,3) \text{ or } (1,3,2) \end{cases} \tag{31.12}$$

Permutation symbol e_{ijk}

The *vector product* of two vectors **a** and **b** is

$$\mathbf{v} = \mathbf{a} \times \mathbf{b} \tag{31.13}$$

Vector product $\mathbf{v} = \mathbf{a} \times \mathbf{b}$

which, by use of the permutation symbol the vector product, may be written in component form.

$$v_i = e_{ijk} a_j b_k \tag{31.14}$$

Vector product $v_i = e_{ijk} a_j b_k$

Here, index i takes the values 1, 2 and 3 resulting in the three expressions

[31.3] Do not confuse the permutation symbol with a linear strain measure. As mentioned before, the many different notations for strain cause problems.

Free index

for v_1, v_2 and v_3, respectively. Thus, while j and k are dummy indices in (31.14) i plays another role and is called a called a *free index.*[31.4]

In some cases we consider the *curl*, also called the *rotation*, of a vector field and in this connection the permutation symbol proves to be useful. The curl of a vector field $\mathbf{v}$ with components v_j is defined by

Curl v

$$\text{curl}\,\mathbf{v} \equiv \begin{vmatrix} \mathbf{i}_1 & \mathbf{i}_2 & \mathbf{i}_3 \\ \dfrac{\partial}{\partial x_1} & \dfrac{\partial}{\partial x_2} & \dfrac{\partial}{\partial x_3} \\ v_1 & v_2 & v_3 \end{vmatrix} \tag{31.15}$$

In components this may be written

Curl v $\sim c_i$

$$\text{curl}(v_j) = -e_{ijk} v_{j,k} \tag{31.16}$$

where c_i denotes the components of the curl of v_k.

31.3.1 Lowercase Greek Indices

Lowercase Greek indices: range $[1, 2]$

Sometimes we shall deal with two-dimensional bodies such as plates, and in that connection it is convenient to be able to distinguish between the summation convention for two- and three-dimensional bodies. For two-dimensional bodies it is customary to let Greek lowercase letters play the same role as the lowercase Roman letters for the three-dimensional bodies.

Summation Convention: Sum over repeated lower-case Greek index

The two-dimensional equivalent of (31.5) clearly is

$$x_\alpha x_\alpha \equiv \sum_{\alpha=1}^{2} (x_\alpha)^2 \quad \text{or} \quad x_\beta x_\beta \equiv \sum_{\gamma=1}^{2} (x_\gamma)^2 \tag{31.17}$$

and the two-dimensional version of (31.11) is

$\delta_{\alpha\alpha} = 2$

$$\delta_{\alpha\alpha} = 2 \tag{31.18}$$

The obvious definition of the *Permutation Symbol* $e_{\alpha\beta}$ for the two-dimensional case is[31.5]

Two-dimensional case: Permutation symbol $e_{\alpha\beta}$

$$e_{\alpha\beta} \equiv \begin{cases} 0 \text{ for } (\alpha,\beta) = (1,1) \text{ or } (\alpha,\beta) = (2,2) \\ +1 \text{ for } (\alpha,\beta) = (1,2) \\ -1 \text{ for } (\alpha,\beta) = (2,1) \end{cases} \tag{31.19}$$

and thus it is seen from (31.12) and (31.19) that the *Permutation Symbol* may be generalized to cases with more indices in that its value is always 0 when two indices are equal, while it is +1 when the indices are $(1, 2, \ldots)$ or an even permutation thereof, and −1 otherwise.

[31.4] In a way the term free index is a little misleading in that i must be the same on both sides of the equation, while j and k could be substituted by n and m without altering the value of v_i.

[31.5] Also in the two-dimensional case there is a problem regarding notation in that $e_{\alpha\beta}$ might be confused with the linear strain in a plate.

31.3.2 Symmetric and Antisymmetric Quantities

31.3.2.1 Product of a Symmetric and an Antisymmetric Matrix

For our immediate purpose we consider a matrix A_{jk}[31.6] which is symmetric in its indices

$$A_{kj} = A_{jk} \tag{31.20}$$

Symmetric matrix

and another matrix B_{jk} which is antisymmetric[31.7]

$$B_{kj} = -B_{jk} \tag{31.21}$$

Antisymmetric matrix

and try to compute the value of their *inner product* $A_{jk}B_{jk}$

$$A_{jk}B_{jk} = \tfrac{1}{2}A_{jk}B_{jk} + \tfrac{1}{2}A_{kj}B_{kj} = \tfrac{1}{2}A_{jk}B_{jk} + \tfrac{1}{2}A_{jk}B_{kj} \tag{31.22}$$

where in the second term we have interchanged dummy indices.

Then,

$$A_{jk}B_{jk} = \tfrac{1}{2}A_{jk}(B_{jk} + B_{kj}) = 0 \tag{31.23}$$

The inner product between a symmetric and an antisymmetric matrix i zero

because of (31.21). We have then shown that the inner product between a symmetric and an antisymmetric matrix vanishes.

31.3.2.2 Product of a Symmetric and a General Matrix

The above finding has an important side effect. Given two matrices, namely A_{jk} which is symmetric and another. B_{mn} which may be of a more general character in that it is neither symmetric, nor antisymmetric. Then we may split B_{mn} in a symmetric part $B^S_{mn} = B^S_{nm}$ and an antisymmetric part $B^A_{mn} = -B^A_{nm}$

$$B_{mn} = B^S_{mn} + B^A_{mn} \tag{31.24}$$

The inner product of A_{jk} and B_{mn} then is

$$A_{mn}B_{mn} = A_{mn}(B^S_{mn} + B^A_{mn}) = A_{mn}B^S_{mn} \tag{31.25}$$

In the inner product between a symmetric and general matrix the antisymmetric part is wiped out

because $A_{mn}B^A_{mn} = 0$ according to the above result. Thus, the inner product between a symmetric and a general matrix does not contain any information about the antisymmetric part of the general matrix.

[31.6] Actually, the following derivation is valid for matrices as well as (Cartesian) tensors, see page 544.

[31.7] The term antisymmetric is somewhat funny in that symmetric means "with the metric" and, therefore, antisymmetric must be "against with the metric." The term antimetric, which is used in some European countries therefore makes more sense.

31.3.3 Summation Convention Results in Brevity

Let us consider the following expression—don't try to interpret it

$$a_\alpha = b_{\alpha\zeta\omega} c_\zeta d_\omega \tag{31.26}$$

and observe that the summation convention is indeed a very handy tool, because written out (31.26) not only signifies two equations, but also a double sum, i.e. 4 terms on the right-hand side of each equation. Had the indices been Roman instead of Greek, the number of equations would have been 3 and the number of terms in each equation 9 instead of 4.

In Section 2.6 and Chapter 5 we encounter formulas like

$$\sigma_{ij} = E_{ijkl} \varepsilon_{kl} \tag{31.27}$$

which is exceedingly lengthy if written out in full, and it becomes obvious that there is a vast saving in terms of writing effort by using the index notation in conjunction with the summation convention—*without sacrificing the possibility to identify each individual term of the equations.* Let us pick the expression for σ_{13} as an example. Using the summation convention

$$\sigma_{13} = E_{13jk} \varepsilon_{jk} \tag{31.28}$$

and in full

$$\begin{aligned}\sigma_{13} = &+ E_{1311}\varepsilon_{11} + E_{1312}\varepsilon_{12} + E_{1313}\varepsilon_{13} \\ &+ E_{1321}\varepsilon_{21} + E_{1322}\varepsilon_{22} + E_{1323}\varepsilon_{23} \\ &+ E_{1331}\varepsilon_{31} + E_{1332}\varepsilon_{32} + E_{1333}\varepsilon_{33}\end{aligned} \tag{31.29}$$

There are 9 of these expressions, so the amount of space taken up by writing (31.27) out in full is so large that you—at least I—very easily lose perspective.

Tensor Analysis

Tensor Analysis is much more than we cover here

It seems fair to mention that the subjects covered in this chapter are very specialized examples of *Tensor Analysis*, which deals with the description of quantities in *any* coordinate system, e.g. curvilinear coordinate systems. In the present context of an introduction to continuum mechanics we do not need more than *Cartesian Tensors*, i.e. quantities in Cartesian coordinate systems, see Section 31.1.

31.4 Generalized Coordinates

The concept of a coordinate is conveniently broadened to cover vectors and functions.

31.4.1 Vectors as Generalized Coordinates

Vectors as Generalized Coordinates

The Cartesian coordinate system may be said to be spanned by its *base vectors* $\mathbf{i}_j$, see Section 2.2.1. In the same spirit we may use any set of linearly independent vectors, such as $\mathbf{j}_k, k \in [1, N]$, as the basis for an N-dimensional space.[31.8] We may simply choose $\mathbf{j}_k$ such that its only non-vanishing component is the kth, which is taken to be 1. Then any vector in the N-dimensional space may be written as a sum of its components, i.e. coordinates, in the direction of the N base vectors just as any vector in the three-dimensional space may be resolved in terms of the three base vectors $\mathbf{i}_j$. Some people find it helpful to think in this way, while others find it confusing. The latter may as well proceed without further speculations along this line.

For the sake of introducing vectors as generalized coordinates, consider the following problem: We may wish to know how the length of a vector $\mathbf{v}$ varies with respect to one of its components, say v_j. For mathematical ease we shall consider the *square* of the length instead of the length itself[31.9]

Functional $\mathbf{\Pi}$

$$\mathbf{\Pi}(\mathbf{v}) = \mathbf{\Pi}(v_k) = v_i v_i \tag{31.30}$$

where $\mathbf{\Pi}$ designates the square of the length.[31.10] Then, differentiation with respect to v_j provides

Partial derivative with respect to a vector

$$\frac{\partial \mathbf{\Pi}(v_i)}{\partial v_j} = 2\frac{\partial v_i}{\partial v_j} v_i \tag{31.31}$$

Obviously, we need to be able to compute the value of $\partial v_i/\partial v_j$. This is very easy when we realize that the only meaningful rule is

$$\frac{\partial v_i}{\partial v_j} \equiv \begin{cases} 1 & \text{for } j = i \\ 0 & \text{for } j \neq i \end{cases} \tag{31.32}$$

The right-hand side of (31.32) is the *Kronecker delta* δ_{ij}, which was defined in (31.9) and in Section 2.2, i.e.

$$\frac{\partial v_i}{\partial v_j} = \delta_{ij} \tag{31.33}$$

Application of the definition of the *Kronecker delta* makes it possible to write (31.31) as

$$\frac{\partial \mathbf{\Pi}(v_i)}{\partial v_j} = 2\delta_{ij} v_i = 2v_j \tag{31.34}$$

[31.8] Here and below lowercase Roman indices may take values from 1 to N, where N may be larger than 3, but the summation convention is still supposed to apply.
[31.9] I admit that I am somewhat sloppy as regards the notation in (31.30) and onwards in that $\mathbf{\Pi}$ is used as a functional of a vector $\mathbf{v}$ as well as a functional of the components v_k of the vector–mathematicians would faint.
[31.10] In this book *Potentials*—see Chapter 32—are denoted $\mathbf{\Pi}$, and the square of the length falls in this class, which is the reason for using the somewhat strange notation.

because in the sum over i all terms except the one for $i = j$ vanish.

Scalar product = Inner product

More generally, we may also be interested in the behavior of *Scalar Products*, also denoted *Inner Products*, of vectors or matrices. As an introduction consider the scalar product of two vectors $\mathbf{u}$ and $\mathbf{v}$ with components u_i and v_j, respectively

Inner product of vectors $\mathbf{u}$ and $\mathbf{v}$

$$\boldsymbol{\Phi}(u_i, v_j) = \mathbf{u} \cdot \mathbf{v} = u_i v_i \tag{31.35}$$

As before we may wish to discover the behavior of the functional $\boldsymbol{\Phi}$ close to its value for u_k and v_m. Therefore, we differentiate $\boldsymbol{\Phi}$ with respect to fixed components of $\mathbf{u}$ and $\mathbf{v}$, namely u_k and v_m

$$\begin{aligned} \frac{\partial \boldsymbol{\Phi}(u_i, v_j)}{\partial u_k} &= \frac{\partial u_i}{\partial u_k} v_i = \delta_{ik} v_i = v_k \\ \frac{\partial \boldsymbol{\Phi}(u_i, v_j)}{\partial v_m} &= \frac{\partial v_i}{\partial v_m} u_i = \delta_{im} u_i = u_m \end{aligned} \tag{31.36}$$

Finally, compute the derivatives of the product of two vectors u_i and v_j with a two-dimensional matrix A_{ij} whose elements do not depend on either of the vectors. The product $\boldsymbol{\Phi}$ is defined as

$$\boldsymbol{\Phi}(u_i, v_j) = A_{ij} u_i v_j \quad \text{or} \quad \boldsymbol{\Phi}(\mathbf{u}, \mathbf{v}) = \mathbf{u}^T \mathbf{A} \mathbf{v} = \mathbf{v}^T \mathbf{A}^T \mathbf{u} \tag{31.37}$$

where ^{T} signifies the matrix transpose. The partial derivatives then are

$$\begin{aligned} \frac{\partial \boldsymbol{\Phi}(u_i, v_j)}{\partial u_k} &= A_{ij} \frac{\partial u_i}{\partial u_k} v_j = A_{ij} \delta_{ik} v_j = A_{kj} v_j = \mathbf{v}^T \mathbf{A}^T \\ \frac{\partial \boldsymbol{\Phi}(u_i, v_j)}{\partial v_m} &= A_{ij} \frac{\partial v_j}{\partial v_m} u_i = A_{ij} \delta_{jm} u_i = A_{im} u_i = \mathbf{A}\mathbf{u} \end{aligned} \tag{31.38}$$

In Part I we repeatedly encounter products like the one in (31.37) except that it contains only one vector

$$\boldsymbol{\Pi}(u_i) = \tfrac{1}{2} A_{ij} u_i u_j \tag{31.39}$$

where the factor $\frac{1}{2}$ is chosen for convenience and consistency with most of our applications, and where the matrix is symmetric

$$A_{ij} = A_{ji} \tag{31.40}$$

Then, the partial derivative is

$$\begin{aligned} \frac{\partial \boldsymbol{\Pi}(u_i)}{\partial u_k} &= \tfrac{1}{2} A_{ij} \frac{\partial (u_i u_j)}{\partial u_k} = \tfrac{1}{2} A_{ij} \frac{\partial u_i}{\partial u_k} u_j + \tfrac{1}{2} A_{ij} \frac{\partial u_j}{\partial u_k} u_i \\ &= \tfrac{1}{2} A_{ij} \delta_{ik} u_j + \tfrac{1}{2} A_{ij} \delta_{jk} u_i = \tfrac{1}{2} A_{kj} u_j + \tfrac{1}{2} A_{ik} u_i \\ &= A_{ik} u_i \end{aligned} \tag{31.41}$$

where the symmetry of A_{ij} has been exploited in the last step of the derivation. In the Part I and II we carry out manipulations like the ones above very frequently and do not continue our efforts here.

31.4.2 Functions as Generalized Coordinates

Functions as Generalized Coordinates

The idea of generalized coordinates may be extended further in that we may choose *functions* instead of *vectors* as the basis. We do not intend to go into any depth here and at this point limit ourselves to a specific example. Consider the following formula[31.11]

$$\mathbf{\Phi}(u,v) = \int_0^1 u(x)^2 v(x) dx \tag{31.42}$$

where the notation $\mathbf{\Phi}(u,v)$ indicates that the value of $\mathbf{\Phi}$ depends on the functions u and v. Obviously, $u(x)$ and $v(x)$ play the same role here as do u_i and v_j in the preceding examples. Again, in some instances it is necessary to investigate the behavior of $\mathbf{\Phi}$ for functions that are close to u and v. Therefore we compute

$$\frac{\partial \mathbf{\Phi}(u,v)}{\partial u} = 2\int_0^1 u(x)v(x)dx \frac{\partial \mathbf{\Phi}(u,v)}{\partial v} = \int_0^1 u(x)^2 dx \tag{31.43}$$

We do not continue along this line here, but refer to Chapter 32, and as regards applications to Parts I and II.

[31.11] Do not try to give an interpretation of (31.42)—I didn't.

Chapter 32

Introduction to Variational Principles

32.1 Introduction

Calculus of variations
Functionals
Potential Energy

In broad terms, *Calculus of Variations* deals with neighboring states. One of the reasons that neighboring states are interesting to us is that often we seek a stationary value of some *Functional*, e.g. the *Potential Energy*. and calculus of variations provides us with very efficient tools for this purpose. But, already at this point let it be mentioned that calculus of variations may be used in cases where no functional exists. In many of our applications we deal with problems with infinitely many degrees of freedom in that we treat continua, but in other cases we need to be able to handle systems with a finite number of degrees of freedom. We begin with an example with infinitely many degrees of freedom, which is not connected with solid or structural mechanics, but may serve as an illustrative introduction to the topic. Since many consider finite degree problems somewhat more instructive we continue by introducing the concepts of *Variations*, *Functionals* and *Potential Energy* via two examples that lie within this realm. On the other hand, others may consider this a departure from the direct course to the continuous systems[32.1] and I suggest that they jump to Section 32.5.

The concepts of functionals and potentials, see below, are very useful for at least two purposes. One is to establish certain equations, such as static equations for specialized continua in a consistent way, see Part II. The other has to do with obtaining approximate solutions that are "good" in some sense, see later.

If the reader wishes to get acquainted with a more strict introduction to variational principles than the one below there are many excellent sources to consult, e.g. (Arfken & Weber 1995), (Mikhlin 1964), (Sokolnikoff 1956), or (Strang & Fix 1973).

This chapter contains introductory examples of two kinds. The first example does not presuppose any knowledge about structural mechanics,

[32.1] I may agree with this.

E. Byskov, *Elementary Continuum Mechanics for Everyone*,
Solid Mechanics and Its Applications 194, DOI: 10.1007/978-94-007-5766-0_32,

while the others treat simple structural mechanics problems. Depending on his or her individual taste the reader may skip the first or second kind of examples, or disregard this chapter completely and go directly to chapter 33. Personally, I recommend the last way only for readers who are well versed in the subject.

32.2 Functionals

Functionals versus functions

To set the stage, first we consider *Functionals* and begin by introducing that concept. But, in order to relate to a known concept let us start with a loose definition of a *function*, which says that fed a number a function provides a (new) number, e.g. when we feed the function $\sin(x)$ the number $\pi/6$ it yields the result $1/2$. As a consequence of this, when we feed a function a function we get a function as the result, e.g. $\sin(\exp(x))$ is a new function. Just as loosely speaking a *functional* is a recipe which takes a function as input and provides a number as the result. As an example of a functional consider the area $A(f)$ under a curve $f(x)$ from 0 to 1

$$A(f) = \int_0^1 f(x)dx$$

Clearly, the value of $A(f)$ depends on the function $f(x)$. As an example, when we supply the function $f(x) = x$ the functional $A(f)$ provides the *number* $1/2$, etc. Thus, we may summarize the difference between functions and functionals by

$$\begin{aligned}
&\text{Function(Number)} && \Rightarrow \text{Number}\\
&\text{Function(Function)} && \Rightarrow \text{Function}\\
&\text{Functional(Function)} && \Rightarrow \text{Number}
\end{aligned}$$

But, instead of going into further detail on this subject we consider the examples below.

Ex 32-1 A Broken Pocket Calculator

While the other examples of variational problems in this chapter are associated with structural problems of some kind, here we consider an example that comes from a completely different world.

Assume that you are out in the sticks of a remote continent and are surveying or maybe doing structural analysis. Then, you discover that your programmable pocket calculator is incapable of computing the value of the functions[32.2] $\sin(x)$ and $\cos(x)$. The easiest remedy is, of course, to write back home and ask them to ship a new calculator to you, but time is a prime concern, so you have no other choice than to write a little program to compute $\sin(x)$ (and one for $\cos(x)$). First, it is a good idea to realize that you can make do with a routine that

[32.2] Once a major company actually sold a pocket calculator which had an faulty chip in that it gave wrong values of $\sin(x)$ for large arguments.

works for x in the interval $[0, \pi/2]$ because for all other angles you may subtract or add the necessary multiples of $\pi/2$.[32.3]

Using a *Taylor Expansion* may work, but it is not a very good idea since such an expansion by its very nature concentrates on the values of x in the neighborhood of 0 which means that the error becomes too large for finite values of x unless you employ a great number of terms. So, you must look for an alternative. If we stop for a moment and analyze the problem we realize that we may formulate it somewhat like this: write a series expansion or some other expansion that computes $\sin(x)$ with a small enough error over the interval $[0, \pi/2]$ *with as few terms as possible* without you having to go through too much advanced analysis before writing the program.[32.4] Therefore, we must search for another, simpler, and yet fast procedure. Ideally, we would like the *maximum* error to be below some limit, but this leads to continued fractions, so we must be satisfied with something less wonderful than that. The idea that comes to mind almost immediately is to minimize the *mean* error.

Let us denote our approximating function $f(x)$ and compute half the mean error,[32.5] which we designate $\boldsymbol{\Pi}(f)$[32.6]

$$\boldsymbol{\Pi}(f) = \tfrac{1}{2}\int_0^{\pi/2} \big(f(x) - \sin(x)\big)^2 dx \qquad (Ex.\ 32\text{-}1.1)$$

Potential $\boldsymbol{\Pi}(f)$

where $\boldsymbol{\Pi}(f)$ is a *Potential* whose value depends on the choice of f. A potential may be considered a special case of a *functional*, see Section 31.4.1 and Ex 32-2–32.5, whose particular characteristic is that it attains a *minimum* or *maximum*—not just a *stationary value*—for the correct solution. In our case, the potential attains the minimum value 0 for $f(x) = \sin(x)$. In order to compute the minimum we differentiate $\boldsymbol{\Pi}(f)$ with respect to f and require that the derivative vanishes. This is in the spirit of Section 31.4.2, but later we shall spend some space on the concept of variations.

The simplest choice of expansion must be a polynomial in x

$$f(x) = a_0 + a_1 x + a_2 x^2 + a_3 x^3 + \cdots \qquad (Ex.\ 32\text{-}1.2)$$

Simplest choice of expansion

where you may object to the presence of the term a_0 because, as everyone knows, the value of $\sin(0)$ is 0 and not some finite value given by a_0.

Why include a_0?

32.3 Actually, we could limit ourselves to the interval, say $[0, \pi/4]$, but the ensuing algebra would become more involved, so we won't do that.

32.4 The reason for the last statement is that computer scientists seem to prefer *Continued Fractions*, but the theory and manipulations behind getting such an expansion lie way above your possibilities in the jungle—and certainly much above my knowledge, anyway.

32.5 There are two reasons for computing *half* the mean error. The first is that in our structural examples there is always the factor 1/2, and the other has to do with convenience in that introduction of the factor makes some of the equations below a little nicer. But, since we are interested in the minimum it does not matter which factor we introduce.

32.6 The reason for calling the mean error $\boldsymbol{\Pi}$ is that in this book this is the notation used for *all* potentials.

We shall look a little into this question, but mention that you might just as well insist that $f(\pi/2) = 1$ instead, or that $f(\pi/4) = 1/\sqrt{2}$ for that matter. Obviously, the condition $f(0) = 0$ is much easier to impose than the other two, but this fact does not seem to be sufficient justification—at least not to me. You may also oppose the idea of using *even* terms in the expansion since $\sin(x)$ is *odd* in x. On the other hand, we need not compute $f(x)$ for negative values of x. Now that we have limited the degrees of freedom of $f(x)$ by the choice (*Ex. 32-1.2*) we may as well write (*Ex. 32-1.1*) as

Why include even terms?

Potential $\mathbf{\Pi}(a_j)$

$$\mathbf{\Pi}(a_j) = \tfrac{1}{2}\int_0^{\pi/2} \left(a_j x^j - \sin(x)\right)^2 dx \qquad (Ex.\ 32\text{-}1.3)$$

where summation over the repeated lower-case index $j \in [0, \infty]$ is implied

Sum over repeated lower-case index $j \in [0, \infty]$

$$a_j x^j \equiv \sum_{K=0}^{\infty} a_K x^K \qquad (Ex.\ 32\text{-}1.4)$$

see also Chapter 31, but bear in mind that in the present context the range of j is not limited to $[1, 3]$.

In order that $\mathbf{\Pi}(a_j)$ is stationary—here minimum

$$\frac{\partial \mathbf{\Pi}(a_j)}{\partial a_k} = 0 \;\forall\; a_k \qquad (Ex.\ 32\text{-}1.5)$$

or

$$0 = \int_0^{\pi/2} \left(a_j x^j - \sin(x)\right)\left(\frac{\partial a_m}{\partial a_k} x^m\right) dx \qquad (Ex.\ 32\text{-}1.6)$$

where the reason for the change of *dummy index* from j to m in the last factor is that a dummy index must appear *exactly* twice in a term, see e.g. Chapter 31. Then,

$$\begin{aligned} 0 &= \int_0^{\pi/2} \left(a_j x^j - \sin(x)\right)\left(x^k\right) dx \\ &= \left(\int_0^{\pi/2} x^{(j+k)} dx\right) a_j - \left(\int_0^{\pi/2} \left(x^k \sin(x)\right) dx\right) \end{aligned} \qquad (Ex.\ 32\text{-}1.7)$$

where we have utilized the fact that

$$\frac{\partial a_j}{\partial a_i} = \delta_{ij} \qquad (Ex.\ 32\text{-}1.8)$$

where the *Kronecker delta* δ_{ij} is defined in (31.9)

Kronecker delta δ_{ij}

$$\delta_{ij} \equiv \begin{cases} 1 & \text{for} \quad j = i \\ 0 & \text{for} \quad j \neq i \end{cases} \qquad (Ex.\ 32\text{-}1.9)$$

From (*Ex. 32-1.6*) we conclude that we need the value of the following kinds of integrals

We need two types of integrals

$$\int_0^{\pi/2} x^{(j+k)} dx \quad \text{and} \quad \int_0^{\pi/2} x^k \sin(x) dx \qquad (Ex.\ 32\text{-}1.10)$$

We know the first kind by heart[32.7]

$$\int_0^{\pi/2} x^{(j+k)} dx = \frac{1}{j+k+1}\left(\frac{\pi}{2}\right)^{(j+k+1)} \qquad (Ex.\ 32\text{-}1.11)$$

while determination of the other kind requires some work. The first six of these are listed here

k	$\int_0^{\pi/2} x^k \sin(x) dx$
0	$+1$
1	$+1$
2	$\pi - 2$
3	$+3\left(\frac{\pi}{2}\right)^2 - 6$
4	$+16\left(\frac{\pi}{2}\right)^3 - 24\frac{\pi}{2} + 24$
5	$+5\left(\frac{\pi}{2}\right)^4 - 60\left(\frac{\pi}{2}\right)^2 + 120$

(*Ex. 32-1.12*)

where all values have been determined through integration by parts without the aid of any kind of table of integrals, which you might have forgotten to bring along.

How many terms?

The first question which comes to mind is: how many—and which—terms are necessary? When we recall the shape of a sine function, we may see that is looks pretty much like a parabola, so in our first attempt we will employ a parabolic approximation with three terms, denoted f_{012}. A linear approximation would be much too crude, of course.

Ex 32-1.1 Parabola with Three Terms

In this example we utilize the first three terms in (*Ex. 32-1.2*)

$$f_{012}(x) = a_0 + a_1 x + a_2 x^2 \qquad (Ex.\ 32\text{-}1.13)$$

where index $_{012}$ indicates the terms that are used.

Here, the equations are

$$\begin{aligned} \tfrac{\pi}{2} a_0 + \tfrac{1}{2}\left(\tfrac{\pi}{2}\right)^2 a_1 + \tfrac{1}{3}\left(\tfrac{\pi}{2}\right)^3 a_2 &= +1 \\ \tfrac{1}{2}\left(\tfrac{\pi}{2}\right)^2 a_0 + \tfrac{1}{3}\left(\tfrac{\pi}{2}\right)^3 a_1 + \tfrac{1}{4}\left(\tfrac{\pi}{2}\right)^4 a_2 &= +1 \\ \tfrac{1}{3}\left(\tfrac{\pi}{2}\right)^3 a_0 + \tfrac{1}{4}\left(\tfrac{\pi}{2}\right)^4 a_1 + \tfrac{1}{5}\left(\tfrac{\pi}{2}\right)^5 a_2 &= \pi - 2 \end{aligned} \qquad (Ex.\ 32\text{-}1.14)$$

[32.7] Note that these integrals are symmetric in j and k which reduces the work involved in the computation of the coefficients.

The solution is

$$\begin{aligned} a_0 &= -60\left(\tfrac{2}{\pi}\right)^3 + 24\left(\tfrac{2}{\pi}\right)^2 + 9\left(\tfrac{2}{\pi}\right) \\ a_1 &= +360\left(\tfrac{2}{\pi}\right)^4 - 168\left(\tfrac{2}{\pi}\right)^3 - 36\left(\tfrac{2}{\pi}\right)^2 \\ a_2 &= -360\left(\tfrac{2}{\pi}\right)^5 + 180\left(\tfrac{2}{\pi}\right)^4 + 30\left(\tfrac{2}{\pi}\right)^3 \end{aligned} \qquad (Ex.\ 32\text{-}1.15)$$

or

$$\begin{aligned} a_0 &= -0.024\,324\,944 \\ a_1 &= +1.195\,745\,064 \\ a_2 &= -0.338\,240\,011 \end{aligned} \qquad (Ex.\ 32\text{-}1.16)$$

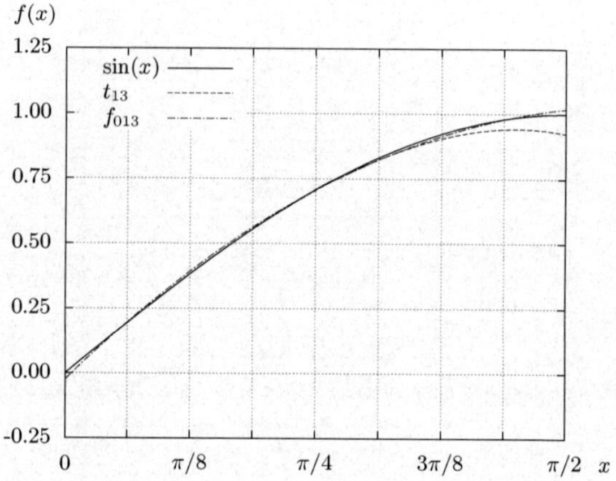

Fig. Ex. 32-1.1: *The sine function* $\sin(x)$ *and two approximations.*

Before investigating the merits of $f_{012}(x)$ we note that the MacLaurin expansion of $\sin(x)$, which is the same as a Taylor expansion about $x = 0$, is

$$t(x) = x - \frac{x^3}{3!} + \frac{x^5}{5!} - \frac{x^7}{7!} \cdots \qquad (Ex.\ 32\text{-}1.17)$$

and denote the approximation that entails the first two terms $t_{13}(x)$ and so forth. As mentioned earlier, this approximation centers on the values for small x. Thus, the approximation is likely to get worse for larger values of x and we may therefore seek an approximation similar to the expansion (*Ex. 32-1.2*), but only retain the terms of degrees 1 and 3 in x because we know that $\sin(x)$ is antisymmetric, as indicated by (*Ex. 32-1.17*).

Ex 32-1.2 First and Third Degree Terms

Here, the only terms are the ones of first and third power of x are, and the equations are

$$\begin{aligned} \tfrac{1}{3}\left(\tfrac{\pi}{2}\right)^3 b_1 + \tfrac{1}{5}\left(\tfrac{\pi}{2}\right)^5 b_3 &= +1 \\ \tfrac{1}{5}\left(\tfrac{\pi}{2}\right)^5 b_1 + \tfrac{1}{7}\left(\tfrac{\pi}{2}\right)^7 b_3 &= +3\left(\tfrac{\pi}{2}\right)^2 - 6 \end{aligned} \qquad (Ex.\ 32\text{-}1.18)$$

where, in order to differentiate from the previous solutions, the coefficients are denoted b_j instead of a_j. with the solution

$$\begin{aligned} b_1 &= +\tfrac{315}{2}\left(\tfrac{2}{\pi}\right)^5 - 60\left(\tfrac{2}{\pi}\right)^3 \\ b_3 &= -\tfrac{525}{2}\left(\tfrac{2}{\pi}\right)^7 + 105\left(\tfrac{2}{\pi}\right)^5 \end{aligned} \qquad (Ex.\ 32\text{-}1.19)$$

or

$$\begin{aligned} b_1 &= +0.988\,792\,233 \quad \text{cf. } +1 \\ b_3 &= -0.145\,061\,813 \quad \text{cf. } -\tfrac{1}{6} \end{aligned} \qquad (Ex.\ 32\text{-}1.20)$$

where the values of the coefficients are compared with the coefficients in the MacLaurin expansion. The behavior of the above three approximations is illustrated in Fig. Ex. 32-1.2, which shows the errors.

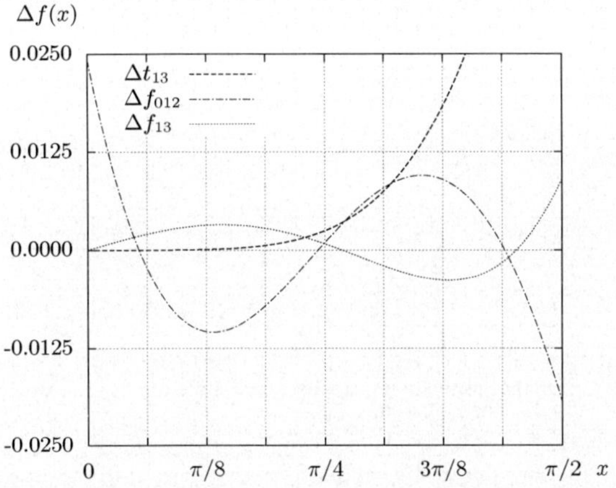

Fig. Ex. 32-1.2: *Error on various approximations to* $\sin(x)$.

Fig. Ex. 32-1.2 clearly shows that the error Δt_{13} grows rapidly with increasing values of x. Both approximations based on the minimization process, i.e. f_{012} and f_{13}, provide errors that are fairly uniform over the interval $[0, \pi/2]$ with f_{13} as the winner. It may be mentioned that if we limited ourselves to third degree polynomials, but allowed the even

The error on t_{13} grows with x. On our approximations it does not terms x^0 and x^2 to get the approximation f_{0123}, the results would be more accurate than those obtained from f_{13}. The work involved, however, is much greater, and we lose the feature that the value of the approximation is zero for $x = 0$.

Ex 32-1.3 Fifth Degree Approximations

The error on t_{135} grows with x. On our approximations it does not

The largest error on f_{13} is around 1 per cent and inclusion of just one more term makes it dramatically smaller, as is obvious from Fig. Ex. 32-1.3, where the errors on only t_{135} and f_{135} are shown. Again, the solution based on minimization is better over the interval $[0, \pi/2]$, while t_{135} is the superior for values of x less than about $\pi/4$.

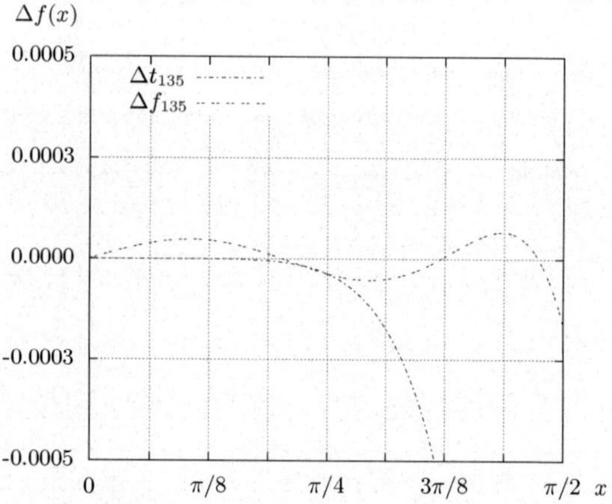

Fig. Ex. 32-1.3: *Error on t_{135} and f_{135}.*

For the sake of completeness the coefficients in f_{135}, which we denote $c_i, i = 1, 3, 5$, are given below

$$\begin{aligned}
c_1 &= +\tfrac{155925}{8}\left(\tfrac{2}{\pi}\right)^7 - 8505\left(\tfrac{2}{\pi}\right)^5 + \tfrac{1995}{8}\left(\tfrac{2}{\pi}\right)^3 \\
c_3 &= -\tfrac{363825}{8}\left(\tfrac{2}{\pi}\right)^9 + \tfrac{80325}{2}\left(\tfrac{2}{\pi}\right)^7 - \tfrac{5355}{4}\left(\tfrac{2}{\pi}\right)^5 \\
c_5 &= +\tfrac{654885}{8}\left(\tfrac{2}{\pi}\right)^{11} - \tfrac{72765}{8}\left(\tfrac{2}{\pi}\right)^9 + \tfrac{10395}{8}\left(\tfrac{2}{\pi}\right)^7
\end{aligned} \qquad \text{(Ex. 32-1.21)}$$

or

$$\begin{aligned}
c_1 &= +0.999\,771\,408 \quad \text{cf. } +1 \\
c_3 &= -0.165\,827\,042 \quad \text{cf. } -\tfrac{1}{6} \\
c_5 &= +0.007\,574\,247 \quad \text{cf. } +\tfrac{1}{120}
\end{aligned} \qquad \text{(Ex. 32-1.22)}$$

Ex 32-1.4 Variations

We may get the same approximations in another way, namely by use of *Variational Principles*, see Chapter 32 and Chapter 33. For convenience we provide the basic formulas for doing that below.

Consider the potential $\mathbf{\Pi}(f)$ for two different choices of the function $f(x)$. The first we denote $g(x)$ and the next $g(x)+\epsilon\delta g(x)$, where $\delta g(x)$ is the (shape of the) *variation* and the "American epsilon" ϵ is the *amplitude* of the variation. Then,

Shape of variation $\delta g(x)$. Amplitude of variation ϵ

$$\mathbf{\Pi}(g) = \tfrac{1}{2}\int_0^{\pi/2} \big(g(x) - \sin(x)\big)^2 dx \qquad (Ex.\ 32\text{-}1.23)$$

as before, except that g appears instead of f, and

$$\begin{aligned}
\mathbf{\Pi}(g+\epsilon\delta g) &= \tfrac{1}{2}\int_0^{\pi/2} \big(g(x) + \epsilon\delta g(x) - \sin(x)\big)^2 dx \\
&= \tfrac{1}{2}\int_0^{\pi/2} \big(g(x) - \sin(x)\big)^2 dx \\
&\quad + \epsilon\int_0^{\pi/2} \big(g(x) - \sin(x)\big)\delta g(x) dx \\
&\quad + \epsilon^2\int_0^{\pi/2} \delta g(x)^2 dx \\
&= \mathbf{\Pi}(g) + \epsilon\delta\mathbf{\Pi}(g) + \epsilon^2\delta^2\mathbf{\Pi}(g)
\end{aligned} \qquad (Ex.\ 32\text{-}1.24)$$

where $\delta\mathbf{\Pi}(g)$ is the (first) variation and $\delta^2\mathbf{\Pi}(g)$ is the second variation of $\mathbf{\Pi}(g)$. If we demand that the first variation vanishes for all $\delta g(x)$

First and second variation

$$\delta\mathbf{\Pi}(g) = 0 \;\forall\; \delta g(x) \qquad (Ex.\ 32\text{-}1.25)$$

we get

$$g(x) - \sin(x) = 0 \qquad (Ex.\ 32\text{-}1.26)$$

and $g(x)$ becomes $\sin(x)$. Since the condition $\delta\mathbf{\Pi}(g) = 0 \;\forall\; \delta g(x)$ leads to the correct solution if $g(x)$ has infinitely many degrees of freedom it seems obvious to impose the same condition when $g(x)$ only has a finite number of degrees of freedom. As in (*Ex. 32-1.2*) with (*Ex. 32-1.4*) assume

$$g(x) = a_j x^j\,,\ j = [1, N] \qquad (Ex.\ 32\text{-}1.27)$$

and recover (*Ex. 32-1.5*) or (*Ex. 32-1.6*).

Purposes of Example Ex 32-1

Example Ex 32-1 is intended to serve two purposes. The first is to introduce the idea of a potential and to show its usefulness. For the above introduction I chose an example which ought to be fairly straightforward and did not require that the reader knows anything about structural or solid mechanics. The other has to do with the issue of approximations and their quality. If

What is a good approximations?

you characterize an approximation as being "good" you must qualify that statement. Do you mean very accurate?[32.8] Or is it an approximation which demands very little work before it is implemented?[32.9] Or is it maybe an approximation that is computationally efficient?[32.10] Or, does it have other virtues?

32.3 Variations

While the example of the broken pocket calculator may be instructive—at least I hope so—it is a little different than most of problems in structural and solid mechanics in that the functional $\mathbf{\Pi}(f)$, see (*Ex. 32-1.1*), or $\mathbf{\Pi}(g)$, see (*Ex. 32-1.23*), does not entail derivatives of the functions involved. In our applications both first and also often second derivatives usually appear. Therefore, as a more relevant basis for the following I utilize the *Strain Energy* $W(\kappa(w))$, where κ is the bending strain and w is the transverse displacement component of a beam and reintroduce the concept of a variation. At the same time, I intend to make the definition of a variation more precise.

In the following, as in Example Ex 32-1, when we prepend a δ to a quantity it signifies the variation of the quantity. To be specific, let us assume that the solution to some beam problem is $w(x)$. Then, we may write a (slightly)[32.11] different field $\widetilde{w}(x)$

$$\widetilde{w}(x) = w(x) + \epsilon\delta w(x) \tag{32.1}$$

"American epsilon" ϵ

where the "American epsilon"[32.12] ϵ is the amplitude of the variation, and $\delta w(x)$ is the shape of the variation, see Fig. 32.4.[32.13]

Now, suppose that another quantity, say the strain energy $W(w)$[32.14] of a linearly elastic beam, contains $(w)^2$ or the square of some derivative of $w(x)$, say the second, which is the correct one in this case, then

$$W(w) = \tfrac{1}{2}\int_0^L EI\big(w''(x)\big)^2 dx \tag{32.2}$$

[32.8] I am fairly sure that this is what most people understand by the term *good* in this connection. But, everything comes at a cost which, in this case, may be many terms or much work before the approximation is achieved.

[32.9] In some cases, such as the one described in Example Ex 32-1, this is a proper definition of the word.

[32.10] Computer scientists must prefer this, even when the algebraic work behind the approximation is very heavy.

[32.11] Variations do not have to be small, but in most—in fact all—cases we intend to limit the variations to being infinitesimal.

[32.12] The "American epsilon" ϵ must not be confused with the other epsilon ε which is used to signify strains.

[32.13] If you suffer from *déjà vu* here it is not surprising because the figure is almost the same as Fig. 2.6, p. 50.

[32.14] For the sake of brevity we write $W(w)$ instead of $W(\kappa(w))$ knowing that a mathematician would faint if he saw the expression $W(w) = W(\kappa(w))$.

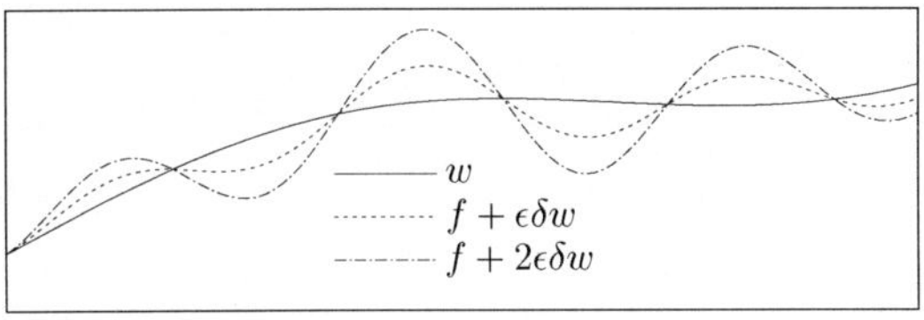

Fig. 32.4: Solution and variations.

where EI signifies the bending stiffness of the beam, L is its length and prime ($'$) denotes differentiation with respect to x. In some derivations we need the variation $\delta W(w)$ of $W(w)$, so let us write

$$\begin{aligned} W(w+\epsilon\delta w) &= \tfrac{1}{2}\int_0^L EI\Big(\big(w(x)+\epsilon\delta w(x)\big)''\Big)^2 dx \\ &= \tfrac{1}{2}\int_0^L EI\big(w''(x)\big)^2 dx + \epsilon\int_0^L EIw''(x)\delta w''(x)dx \\ &\quad + \epsilon^2\tfrac{1}{2}\int_0^L EI\big(\delta w''(x)\big)^2 dx \\ &= W(w) + \epsilon\delta W(w) + O(\epsilon^2) \end{aligned} \tag{32.3}$$

where $O(\epsilon^2)$ denotes a quantity of order ϵ^2. Thus,

$$\delta W(w) = \int_0^L EIw''(x)\delta w''(x)dx \tag{32.4}$$

We may introduce two more formal definitions of the variation of $W(w)$ in the following way

$$\delta W(w) \equiv \lim_{\epsilon\to 0}\frac{W\big(w(x)+\epsilon\delta w(x)\big) - W\big(w(x)\big)}{\epsilon} \tag{32.5}$$

Gateaux Derivative

or

$$\delta W(w) \equiv \left.\frac{\partial W\big(w(x)+\epsilon\delta w(x)\big)}{\partial\epsilon}\right|_{\epsilon=0} \tag{32.6}$$

which are both seen to provide the same expression for $\delta W(w)$ as (32.3)–(32.4).

The above definition (32.5) utilizes the concept of a *Gateaux Derivative*, see e.g. (Budiansky 1974) and Section 33.3.

"Taking variations" $\sim$ differentiating

Once you have tried it a number of times you will find that "taking variations" is as easy as—and much similar to—computing derivatives. For instance, using a quite common notation, the variation $\delta W(w)$ of the example above can be found in the following way

See below for interpretation

$$\delta W(w) = \frac{\partial W\big(w(x)\big)}{\partial w''(x)}\,\delta w''(x) \tag{32.7}$$

with the result

$$\begin{aligned}\delta W(w) &= \frac{\partial}{\partial w''(x)}\left(\tfrac{1}{2}\int_0^L EI\big(w''(x)\big)^2 \delta w''(x)dx\right)\\ &= \int_0^L EIw''(x)\delta w''(x)dx\end{aligned} \tag{32.8}$$

The right-hand side of (32.7) entails integration of the its entirety

which agrees with (32.4). But, as you may observe immediately, there is a problem regarding the interpretation of the term on the right-hand side of (32.7) in that also $\delta w''(x)$ must be understood to be covered by the integration, which is not clear from the notation. So, to quote one of my younger colleagues, "we must misunderstand [the right-hand side of (32.7)] correctly." This notational problem becomes even more complicated when a functional depends on more than, say, the second derivative of $w(x)$, but also on, maybe, $w(x)$ itself. As an example, the *Potential Energy* $\mathbf{\Pi}_{\mathrm{P}}$ of a particular beam, see Fig. Ex. 32-4.1, according to (*Ex. 32-4.2*), is

$$\mathbf{\Pi}_{\mathrm{P}}(w) = \tfrac{1}{2}\int_0^L EI(x)\big(w''(x)\big)^2 dx - \int_0^L \bar{p}(x)w(x)dx \tag{32.9}$$

In this case, the analogy of (32.7) is

The right-hand side entails integration of all terms

$$\delta W(w) = \frac{\partial W\big(w(x)\big)}{\partial w(x)}\,\delta w(x) + \frac{\partial W\big(w(x)\big)}{\partial w''(x)}\,\delta w''(x) \tag{32.10}$$

and, again, the integration inherent in the expression for $W(w)$ must be understood to be extended to the variations.

32.4 Systems with a Finite Number of Degrees of Freedom

The next examples deal with structural problems of a simple kind. Two of them are concerned with structures with finite degrees of freedom, while the last one returns to the topic of infinite number of degrees of freedom.

The first of the structural mechanics examples shows some of the aspects rather clearly, while other aspects are better revealed in the second.

Ex 32-2 A "Structure" with One Degree of Freedom

The linearly elastic spring in Fig. Ex. 32-2.1, whose stiffness is c, supports the load $\bar{P}$—recall that an overbar indicates that the quantity is prescribed. For instructional purposes it is our intention to find the

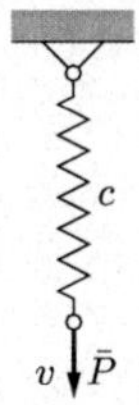

Fig. Ex. 32-2.1: A "Structure" with one degree of freedom.

downward displacement v of the load by use of the potential energy of the structure. First, however, acknowledge that from elementary structural analysis we know that the result is

$$v = \frac{\bar{P}}{c} \qquad \text{(Ex. 32-2.1)}$$

Known result

We postulate that the total *Potential Energy* $\mathbf{\Pi}_\text{P}$ of the structure is[32.15]

$$\mathbf{\Pi}_\text{P}(v) = \tfrac{1}{2}c\,v^2 - \bar{P}v \qquad \text{(Ex. 32-2.2)}$$

Potential energy $\mathbf{\Pi}(v)$

where the first term is the *Strain Energy* stored in the spring, the second term is the *Potential Energy of the load*, and we have emphasized that $\mathbf{\Pi}_\text{P}$ depends on the value of v by writing $\mathbf{\Pi}_\text{P}(v)$. Let us consider a small perturbation $\epsilon\delta v$ of the displacement, where[32.16]

$$|\epsilon| \ll 1 \qquad \text{(Ex. 32-2.3)}$$

Amplitude of variation ϵ
Shape of variation δv

is an *amplitude*,[32.17] and δv is the *shape*[32.18] of the *variation* $\epsilon\delta v$. Compute the value of $\mathbf{\Pi}_\text{P}$ for the perturbed displacement $(v + \epsilon\delta v)$ instead of v

$$\begin{aligned}\mathbf{\Pi}_\text{P}(v+\epsilon\delta v) &= \tfrac{1}{2}c(v+\epsilon\delta v)^2 - \bar{P}\,(v+\epsilon\delta v)\\ &= \tfrac{1}{2}cv^2 - \bar{P}v + \epsilon\left(cv\delta v - \bar{P}\delta v\right)\\ &\quad + \epsilon^2\tfrac{1}{2}c\delta v^2\\ &= \mathbf{\Pi}_\text{P}(v) + \epsilon\delta\mathbf{\Pi}_\text{P} + \epsilon^2\delta^2\mathbf{\Pi}_\text{P}\end{aligned} \qquad \text{(Ex. 32-2.4)}$$

Potential energy of perturbed displacement $\mathbf{\Pi}(v + \epsilon\delta v)$

where $\delta\mathbf{\Pi}_\text{P}$ is the (first) variation of $\mathbf{\Pi}_\text{P}$ and $\delta^2\mathbf{\Pi}_\text{P}$ is the second vari-

[32.15] You probably know this already.
[32.16] In this book the "American epsilon" ϵ is reserved for use as a small parameter, while the other epsilon ε is used as a symbol for strains.
[32.17] In a one-degree-of-freedom system ϵ is not necessary, but later we shall see its usefulness.
[32.18] For a system with only one degree of freedom the term *shape* does not make much sense, but the next examples should justify its existence.

(First) variation of $\boldsymbol{\Pi}$

$\delta\boldsymbol{\Pi}_{\mathrm{P}} = 0 \; \forall \; \delta v$

ation of $\boldsymbol{\Pi}_{\mathrm{P}}(v)$, and

$$\delta\boldsymbol{\Pi}_{\mathrm{P}} = \left(cv - \bar{P}\right)\delta v \qquad (Ex.\ 32\text{-}2.5)$$

From (*Ex. 32-2.1*) it is clear that

$$\delta\boldsymbol{\Pi}_{\mathrm{P}} = 0 \; \forall \; \delta v \qquad (Ex.\ 32\text{-}2.6)$$

which is a statement that is always true for the equilibrium state of an elastic structure because the first variation of $\boldsymbol{\Pi}_{\mathrm{P}}$ simply is the *Principle of Virtual Displacements*, see e.g. Section 2.4, with the constitutive equations exploited to write the *generalized stresses*[32.19] in terms of the *generalized strains*. In the present case, the generalized stress is the spring force N, and the generalized strain is the elongation of the spring v. The *Principle of Virtual Displacements* for the present structure is

$$N\delta\varepsilon - \bar{P}\delta v = 0 \qquad (Ex.\ 32\text{-}2.7)$$

"American epsilon" ϵ versus the other epsilon ε

where N is the spring force, and $\varepsilon = v$ is the (generalized) strain of the spring. When we introduce the constitutive equation

$$N = c\varepsilon \qquad (Ex.\ 32\text{-}2.8)$$

and the kinematic equation

$$\varepsilon = v \qquad (Ex.\ 32\text{-}2.9)$$

we may rewrite the *Principle of Virtual Displacements*

$$cv\delta v - \bar{P}\delta v = 0 \; \forall \; \delta v \qquad (Ex.\ 32\text{-}2.10)$$

which proves the statement in *Ex. 32-2.6* that, in this case, the first variation of the potential energy vanishes for the solution, see also (*Ex. 32-2.1*).

Let us compute the variation $\delta\boldsymbol{\Pi}_{\mathrm{P}}$ according to (32.6)

$$\begin{aligned} \delta\boldsymbol{\Pi}_{\mathrm{P}}(v) &\equiv \left.\frac{\partial\boldsymbol{\Pi}_{\mathrm{P}}\left(v + \epsilon\delta v\right)}{\partial\epsilon}\right|_{\epsilon=0} \\ &= \left.\frac{\partial\left(\frac{1}{2}c(v + \epsilon\delta v)^2 - \bar{P}(v + \epsilon\delta v)\right)}{\partial\epsilon}\right|_{\epsilon=0} \\ &= (cv - \bar{P})\delta v \end{aligned} \qquad (Ex.\ 32\text{-}2.11)$$

If we require that $\delta\boldsymbol{\Pi}_{\mathrm{P}}(v)$ vanishes for all values of δv we get (*Ex. 32-2.1*).[32.20]

As our second structural example, we consider a structure with two degrees of freedom.

[32.19] For the term *generalized*, see e.g. Section 2.4.2.

[32.20] In view of the greater generality of the *Principle of Virtual Displacements* you might ask why I base so many of the derivations in this book on the *Potential Energy*. The reason is that, according to my experience, the potential energy furnishes a foundation that is easier for students to handle in connection with for example the *Finite Element Method*, see Part V.

Ex 32-3 A "Structure" with Two Degrees of Freedom

Both springs in Fig. Ex. 32-3.1 are linearly elastic with the stiffnesses c_1 and c_2, respectively. The rigid crossbar is prevented from displace-

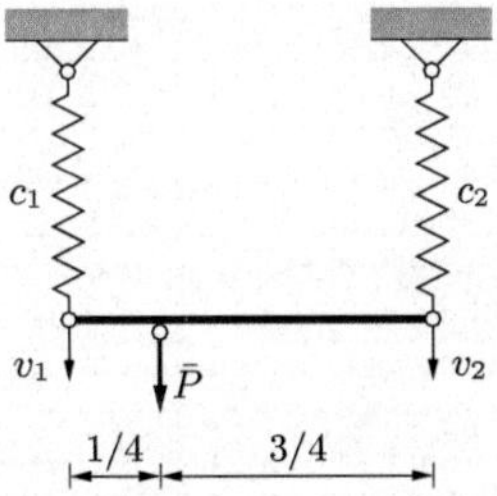

Fig. Ex. 32-3.1: *A "Structure" with two degrees of freedom.*

ments in its own direction. Here, the total *Potential Energy* $\mathbf{\Pi}_P$ of the structure is

Potential energy $\mathbf{\Pi}_P(v_1, v_2)$

$$\mathbf{\Pi}_P(v_1, v_2) = \tfrac{1}{2}\left(c_1 v_1^2 + c_2 v_2^2\right) - \bar{P}v \qquad (Ex.\ 32\text{-}3.1)$$

where, from kinematics

$$v = \tfrac{3}{4}v_1 + \tfrac{1}{4}v_2 \qquad (Ex.\ 32\text{-}3.2)$$

Thus,

$$\mathbf{\Pi}_P(v_1, v_2) = \tfrac{1}{2}\left(c_1 v_1^2 + c_2 v_2^2\right) - \bar{P}\left(\tfrac{3}{4}v_1 + \tfrac{1}{4}v_2\right) \qquad (Ex.\ 32\text{-}3.3)$$

Consider the value of $\mathbf{\Pi}_P$ for v_1 and v_2 perturbed to $(v_1 + \epsilon\delta v_1)$ and $(v_2 + \epsilon\delta v_2)$, respectively

$$\begin{aligned}&\mathbf{\Pi}_P(v_1 + \epsilon\delta v_1, v_2 + \epsilon\delta v_2)\\ =\;& \tfrac{1}{2}\left(c_1(v_1 + \epsilon\delta v_1)^2 + c_2(v_2 + \epsilon\delta v_2)^2\right)\\ &- \bar{P}\left(\tfrac{3}{4}(v_1 + \epsilon\delta v_1) + \tfrac{1}{4}(v_2 + \epsilon\delta v_2)\right)\end{aligned} \qquad (Ex.\ 32\text{-}3.4)$$

which provides

$$\begin{aligned}&\mathbf{\Pi}_P(v_1 + \epsilon\delta v_1, v_2 + \epsilon\delta v_2)\\ =\;& \tfrac{1}{2}c_1 v_1^2 + \tfrac{1}{2}c_2 v_2^2 - \bar{P}\left(\tfrac{3}{4}v_1 + \tfrac{1}{4}v_2\right)\\ &+ \epsilon\left(c_1 v_1 \delta v_1 + c_2 v_2 \delta v_2 - \bar{P}\tfrac{3}{4}\delta v_1 - \bar{P}\tfrac{1}{4}\delta v_2\right)\\ &+ \epsilon^2\left(\tfrac{1}{2}c_1 \delta v_1^2 + \tfrac{1}{2}c_2 \delta v_2^2\right)\end{aligned} \qquad (Ex.\ 32\text{-}3.5)$$

Again, as in Example Ex 32-2, since $\mathbf{\Pi}_P$ is a smooth function of ϵ we may always write

$$\begin{aligned}&\mathbf{\Pi}_P(v_1 + \epsilon\delta v_1, v_2 + \epsilon\delta v_2)\\ =\;& \mathbf{\Pi}_P(v_1, v_2)\\ &+ \epsilon\delta\mathbf{\Pi}_P(v_1, v_2, \delta v_1, \delta v_2)\\ &+ \epsilon^2\delta^2\mathbf{\Pi}_P(v_1, v_2, \delta v_1, \delta v_2)\end{aligned} \qquad (Ex.\ 32\text{-}3.6)$$

In our example

Potential $\mathbf{\Pi}_P$

First variation $\delta\mathbf{\Pi}_P$

Second variation $\delta^2\mathbf{\Pi}_P$

$$\begin{aligned}\mathbf{\Pi}_P(v_1,v_2) &= \tfrac{1}{2}c_1v_1^2+\tfrac{1}{2}c_2v_2^2 \\ &\quad -\bar{P}\left(\tfrac{3}{4}v_1+\tfrac{1}{4}v_2\right) \\ \delta\mathbf{\Pi}_P(v_1,v_2,\delta v_1,\delta v_2) &= c_1v_1\delta v_1+c_2v_2\delta v_2 \\ &\quad -\bar{P}\tfrac{3}{4}\delta v_1-\bar{P}\tfrac{1}{4}\delta v_2 \\ \delta^2\mathbf{\Pi}_P(v_1,v_2,\delta v_1,\delta v_2) &= \tfrac{1}{2}c_1\delta v_1^2+\tfrac{1}{2}c_2\delta v_2^2\end{aligned} \qquad \text{(Ex. 32-3.7)}$$

The reason that $\delta^2\mathbf{\Pi}_P$ does not contain terms entailing v_1 and v_2, but only their variations δv_1 and δv_2, is that the structural problem is linear in v_1 and v_2. This property is displayed by the fact that in $\mathbf{\Pi}_P$ the *strain energy term*, which expresses the energy stored in the springs, is quadratic in the displacements, and that the load term is linear in the displacements.[32.21]

While in Example Ex 32-2 it was difficult to justify the terms shape and amplitude of the perturbation both make sense here, in particular if we introduce the column vectors $\{v\}$ and $\{\delta v\}$

Displacement vector $\{v\}$ and its variation $\{\delta v\}$

$$\{v\}=\begin{bmatrix}v_1\\v_2\end{bmatrix} \quad \text{and similarly} \quad \{\delta v\}=\begin{bmatrix}\delta v_1\\ \delta v_2\end{bmatrix} \qquad \text{(Ex. 32-3.8)}$$

Introduce the square matrix $[K]$, which—in this case rightfully—deserves the name stiffness matrix[32.22]

Stiffness matrix $[K]$

$$[K]=\begin{bmatrix}c_1 & 0\\ 0 & c_2\end{bmatrix} \qquad \text{(Ex. 32-3.9)}$$

Furthermore, a column vector $\{\bar{R}\}$, the load vector, may be given as[32.23]

Load vector $\{\bar{R}\}$

$$\{\bar{R}\}=\begin{bmatrix}\tfrac{3}{4}\bar{P}\\ \tfrac{1}{4}\bar{P}\end{bmatrix} \qquad \text{(Ex. 32-3.10)}$$

With these expressions in hand we may write

$\mathbf{\Pi}_P$ in matrix notation

$$\mathbf{\Pi}_P(\{v\})=\tfrac{1}{2}\{v\}^T[K]\{v\}-\{v\}^T\{\bar{R}\} \qquad \text{(Ex. 32-3.11)}$$

where ^{T} signifies the transpose.

Now, the variation of $\{v\}$ is $\epsilon\{\delta v\}$, where ϵ clearly is a measure of the amplitude and $\{\delta v\}$ is an expression of the shape. When we exploit

32.21 Note that the second variation is positive meaning that this structure is stable. We shall not go further into this subject here, but see Part IV.

32.22 I prefer the notation $[K]$ over the more intuitive $[c]$ because it is common in literature on the *Finite Element Method.*

32.23 The reason for using the name $\{\bar{R}\}$ instead of $\{\bar{P}\}$ is that it is the right-hand side in the ensuing system of equations.

(*Ex. 32-3.8*)–(*Ex. 32-3.11*) we may rewrite (*Ex. 32-3.5*) in the more compact form

$$\begin{aligned}\boldsymbol{\Pi}_{\mathrm{P}}(\{v\}+\epsilon\{\delta v\}) &= \tfrac{1}{2}\{v\}^T[K]\{v\} - \{v\}^T\{\bar{R}\}\\ &\quad+\epsilon\left(\{\delta v\}^T[K]\{v\} - \{\delta v\}^T\{\bar{R}\}\right)\\ &\quad+\epsilon^2\tfrac{1}{2}\{\delta v\}^T[K]\{\delta v\}\\ &= \boldsymbol{\Pi}_{\mathrm{P}}(\{v\})\\ &\quad+\epsilon\delta\boldsymbol{\Pi}_{\mathrm{P}}(\{v\},\{\delta v\})\\ &\quad+\epsilon^2\delta^2\boldsymbol{\Pi}_{\mathrm{P}}(\{v\},\{\delta v\})\end{aligned} \qquad (Ex.\ 32\text{-}3.12)$$

In deriving (*Ex. 32-3.12*) we have exploited the fact that the transpose of a scalar is the scalar itself. The expressions for the first and second variation of $\boldsymbol{\Pi}_{\mathrm{P}}$ are seen to be

$$\begin{aligned}\delta\boldsymbol{\Pi}_{\mathrm{P}} &= \{\delta v\}^T\left([K]\{v\} - \{\bar{R}\}\right)\\ \delta^2\boldsymbol{\Pi}_{\mathrm{P}} &= \tfrac{1}{2}\{\delta v\}^T[K]\{\delta v\}\end{aligned} \qquad (Ex.\ 32\text{-}3.13)$$

If we solve the structural problem in hand we may realize that for the solution the first variation vanishes for all $\{\delta v\}$

$$\begin{aligned}&\delta\boldsymbol{\Pi}_{\mathrm{P}} = \{\delta v\}^T\left([K]\{v\} - \{\bar{R}\}\right) = 0 \quad \forall\,\{\delta v\}\\ &\Rightarrow \{v\} = [K]^{-1}\{\bar{R}\}\end{aligned} \qquad (Ex.\ 32\text{-}3.14)$$

where $^{-1}$ signifies the matrix inverse. The solution is seen to be

$$\{v\} = \begin{bmatrix}\dfrac{3}{4}\dfrac{\bar{P}}{c_1}\\[2ex] \dfrac{1}{4}\dfrac{\bar{P}}{c_2}\end{bmatrix} \qquad (Ex.\ 32\text{-}3.15)$$

Solution $\{v\}$

As mentioned in Example Ex 32-2 the requirement $\delta\boldsymbol{\Pi}_{\mathrm{P}} = 0$ is equivalent to the *Principle of Virtual Displacements* (when the constitutive equations have been introduced), which in turn is another way of writing the equilibrium equations. Here, however, we shall not spend more space on this issue, but refer to Section 2.4.2.

When we study (*Ex. 32-3.12*) we may realize that

$$\delta\boldsymbol{\Pi}_{\mathrm{P}} = \left.\frac{\partial\boldsymbol{\Pi}_{\mathrm{P}}(\{v\}+\epsilon\{\delta v\})}{\partial\epsilon}\right|_{\epsilon=0} \qquad (Ex.\ 32\text{-}3.16)$$

it agrees with (32.6).

32.5 Systems with Infinitely Many Degrees of Freedom

The previous examples with springs are so simple that they are incapable of illustrating enough features of variations of the potential energy because they cover examples with a finite number of degrees of freedom.

Ex 32-4 A Structure with Infinitely Many Degrees of Freedom

An example which is somewhat more complicated is shown in Fig. Ex. 32-4.1. The beam is taken to be linearly elastic with the bending stiffness

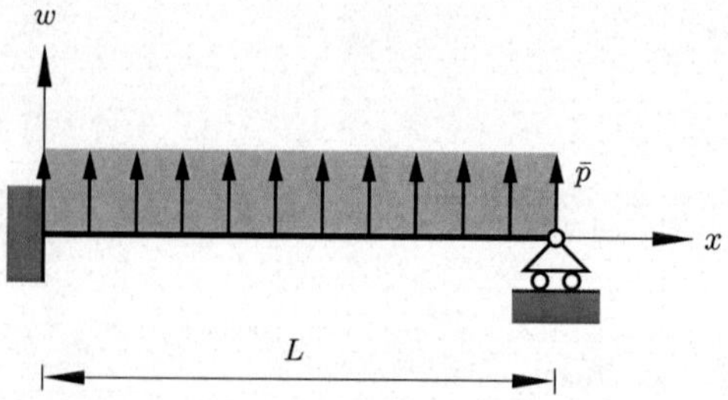

Fig. Ex. 32-4.1: *A structure with infinitely many degrees of freedom—a beam.*

$EI(x)$, and the *Kinematically Linear Bernoulli-Euler Beam Theory* (infinitesimal displacements, infinitesimal strains, and no shear strains) is assumed valid. This beam theory is derived in Section 7.5, but see also Section 7.7, where the constitutive relation $M = EI\kappa$ is introduced. When we consider only transverse displacements the *Generalized Strain* of the beam is the *Bending Strain*, also called the *Curvature Strain*, κ, which on the above assumptions, is defined as

Definition of bending strain κ

$$\kappa(x) = w''(x) \quad \text{(Ex. 32-4.1)}$$

The potential energy of the beam then is:[32.24]

$\mathbf{\Pi}_\mathrm{P}(w)$

$$\mathbf{\Pi}_\mathrm{P}(w) = \tfrac{1}{2}\int_0^L EI(x)\big(w''(x)\big)^2 dx - \int_0^L \bar{p}(x)w(x)dx \quad \text{(Ex. 32-4.2)}$$

Ex 32-4.1 Equilibrium Equations Obtained by Variation of the Potential Energy

Let us see if we can get something useful out of taking the variation of $\mathbf{\Pi}_\mathrm{P}$ and demand that $\delta\mathbf{\Pi}_\mathrm{P}$ vanishes

Variation $\delta\mathbf{\Pi}_\mathrm{P}$ of $\mathbf{\Pi}_\mathrm{P}$

$$\begin{aligned}\delta\mathbf{\Pi}_\mathrm{P}(w) &= \int_0^L EI(x)w''(x)\delta w''(x)dx \\ &\quad - \int_0^L \bar{p}(x)\delta w(x)dx = 0 \;\forall\; \delta w(x)\end{aligned} \quad \text{(Ex. 32-4.3)}$$

"Taking variations" $\sim$ differentiate

You may get (*Ex. 32-4.3*) by use of (32.5) or (32.6) and note that "taking variations" is very similar to differentiation, as mentioned earlier,

[32.24] We postulate this, but realize that the first term is the *Strain Energy* and the second the *Load Potential*—exactly as was the case for the spring.

see Section 32.3, in particular (32.7)–(32.10) and the accompanying text. Another possibility consists in computing $\mathbf{\Pi}_{\mathrm{P}}(w+\epsilon\delta w)$ and proceed along the same line as in Example Ex 32-2 and Example Ex 32-3 and isolate terms of different powers of ϵ to identify $\delta\mathbf{\Pi}_{\mathrm{P}}$—it is a good exercise.

Before we can get any further we must realize that in order to exploit the arbitrariness of $\delta w(x)$ we must get rid of its derivatives in the integrals. Therefore, we rewrite (*Ex. 32-4.3*)

$$\begin{aligned} 0 &= \Big[EI(x)w''(x)\delta w'(x)\Big]_0^L - \Big[\big(EI(x)w''(x)\big)'\delta w(x)\Big]_0^L \\ &+ \int_0^L \Big(\big(EI(x)w''(x)\big)'' - \bar{p}(x)\Big)\delta w(x)dx \;\forall\, \delta w(x) \end{aligned} \qquad (Ex.\ 32\text{-}4.4)$$

Due to the fact that the left-hand end of the beam is clamped the variations $\delta w(0)$ and $\delta w'(0)$ both vanish, and we do not get anything from these terms.[32.25] At the right-hand end $\delta w(L) = 0$, but $\delta w'(L) \not\equiv 0$, and thus

$$\begin{aligned} & EI(L)w''(L)\delta w'(L) = 0 \;\forall\, \delta w'(L) \\ \Rightarrow\; & EI(L)w''(L) = 0 \end{aligned} \qquad (Ex.\ 32\text{-}4.5)$$

The constitutive relation, sometimes called the constitutive "law," for the beam is

Constitutive relation

$$M(x) = EI(x)\kappa(x) \qquad (Ex.\ 32\text{-}4.6)$$

which with (*Ex. 32-4.1*) may be written

$$M(x) = EI(x)w''(x) \qquad (Ex.\ 32\text{-}4.7)$$

By comparing (*Ex. 32-4.7*) and (*Ex. 32-4.5*) we conclude that (*Ex. 32-4.5*) provides us with the *static boundary condition* $M(L) = 0$ for the bending moment at $x = L$, written in terms of the displacement derivative $w''(L)$.

Now that we have taken care of the boundary terms we can concentrate on the field terms

Equilibrium equation

$$\begin{aligned} 0 &= \int_0^L \Big(\big(EI(x)w''(x)\big)'' - \bar{p}(x)\Big)\delta w(x)dx \;\forall\, \delta w(x) \\ \Rightarrow\; & \Big(EI(x)w''(x)\Big)'' = \bar{p}(x) \\ \Rightarrow\; & M''(x) = \bar{p}(x) \end{aligned} \qquad (Ex.\ 32\text{-}4.8)$$

You may recognize the last equation of (*Ex. 32-4.8*) as the equilibrium equation in the field $x \in]0; L[$. The fact that we have arrived at the usual differential equation and static boundary conditions for the beam problem of Fig. Ex. 32-4.1 is a good indication that the idea of requiring the *Potential Energy* $\mathbf{\Pi}_{\mathrm{P}}$ to be stationary may be valuable.

[32.25] It is permissible to let both these variations be non-vanishing, but for our purposes it is better not to allow this.

The present example is, however, so simple that this may not be obvious. Without going further into this subject it may be worthwhile mentioning that the use of the stationarity property of Π_P becomes particularly important in connection with approximate solutions, see e.g. Examples Ex 32-4.2 and Ex 32-5 and Part V.

Ex 32-4.2 Approximate Solution by Use of the Potential Energy

Let us now assume that the moment of inertia $EI(x)$ of the beam of Fig. Ex. 32-4.1 varies linearly from the value EI_0 at the left-hand end to $(1-\alpha)EI_0$ at the right-hand end, where

Linearly varying bending stiffness

$$EI(\xi) = (1-\alpha\xi)EI_0 \, , \quad \xi \equiv \frac{x}{L} \, , \quad (\dot{\ }) \equiv L(\ )' \qquad (Ex.\ 32\text{-}4.9)$$

In this case, the expression for the exact solution becomes horrendous, although the problem looks very simple, and even when we take the fixed value $\alpha = \frac{1}{2}$ the analytic result is

Exact solution is complicated, even for fixed value of α

$$w_{ex}(\xi; \alpha = \tfrac{1}{2}) = \Bigg(-\tfrac{1}{6}\xi^3 + \frac{6\ln(2)-5}{12\ln(2)-6}\xi^2 + \frac{2\ln(2)-2\ln(2-\xi)+2}{6\ln(2)-3}\xi + \frac{4\ln(2-\xi)-4\ln(2)}{6\ln(2)-3}\Bigg)\frac{\bar{p}L^4}{EI_0} \qquad (Ex.\ 32\text{-}4.10)$$

Use analytic manipulation programs to find w_{ex}

which I consider quite complicated, too. In order to find the general solution, even for a fixed value of α, programs such as maxima, MuPAD, Maple or Mathematica, which do analytic manipulations are of great help. The exact solution (*Ex. 32-4.10*) may be interesting in itself, but our use of it is to study the convergence of the approximate method used below. Some of the programs have built-in functions which make solving differential equations easy. With maxima you must do a little more by yourself, see my program below. For the sake of making the program easier to handle by maxima the problem is nondimensionalized. No matter which of the above-mentioned programs you use you may have to rearrange the LaTeX 2_ε-output by hand to suit your particular taste.

```
/* Program to solve the problem of a clamped-simply       */
/* supported beam loaded by a distributed, constant       */
/* load 1.                                                */
/* The length of the beam is 1.                           */
/* The bending stiffness varies as (1-alpha*xi)           */
/* where xi is the nondimensional axial coordinate.       */
TeXFile: openw("Beam.out");   /* (La)TeX file.           */
DatFile: openw("Beam.d");     /* Output file.            */

assume(alpha <= 1);
alpha: 1/2;
```

```
printf(TeXFile,"Value of $\\alpha$: ~f~%~%",alpha);

assume(xi <= 1);
lognegint: true;
odeh: ode2(integrate(integrate('diff((1-alpha*xi)
          *'diff(w,xi,2),xi,2)=1,xi),xi),w,xi);
w: rhs(odeh);
eq1: expand(subst(0,xi,w));
eq2: expand(subst(1,xi,w));
eq3: expand(subst(0,xi,diff(w,xi)));
eq4: expand(subst(1,xi,(1-alpha*xi)*diff(w,xi,2)));
/* In order to solve the equations you must first        */
/* see that the constants entering eq1,...,eq4 are       */
/* %k1, %k2, %c1 and %c2.                                */
/* Therefore, you must run the first part of the         */
/* program in interactive mode.                          */
ks: linsolve([eq1,eq2,eq3,eq4],[%k1,%k2,%c1,%c2]);
K1: rhs(ks[1]);
K2: rhs(ks[2]);
K3: rhs(ks[3]);
K4: rhs(ks[4]);
wh: w;
wh: subst(K1,%k1,wh);
wh: subst(K2,%k2,wh);
wh: subst(K3,%c1,wh);
wh: subst(K4,%c2,wh);
wrat: ratsimp(wh);
numme: num(wrat);
denne: denom(wrat);
/* Check that the solution is real                       */
realdenne: realpart(denne);
realnumme: expand(realpart(numme));
imagdenne: imagpart(denne);
imagnumme: expand(imagpart(numme));
printf(TeXFile,
       "Imaginary part of numerator   of w: ~f\\\\ ",
       imagnumme);
printf(TeXFile,
       "Imaginary part of denominator of w: ~f\\\\ ",
       imagdenne);

wfinal: realnumme/realdenne;
printf(TeXFile,"$w$ = ~a~%",tex(wfinal,false));
wp: diff(wfinal,xi,1);
printf(TeXFile,"$w'$ = ~a~%",tex(wp,false));
wpp: diff(wp,xi,1);
printf(TeXFile,"$w''$ = ~a~%",tex(wpp,false));
M: (1-alpha*xi)*wpp;
M: ratsimp(M);
printf(TeXFile,"$M$ = ~a~%",tex(M,false));

/* Check solution                                        */
CheckDiffEq: diff((1-alpha*xi)*diff(wfinal,xi,2),xi,2);
CheckDiffEq: ratsimp(CheckDiffEq);
printf(TeXFile,
       "Check of differential equation: ~g~%~%",
       CheckDiffEq);
printf(TeXFile,
```

```
       "Check of w(0): ~g~%~%",subst(0,xi,wfinal));
printf(TeXFile,
       "Check of w'(0): ~g~%~%",
       subst(0,xi,diff(wfinal,xi)));
printf(TeXFile,
       "Check of w(1): ~g~%~%",subst(1,xi,wfinal));
M: (1-alpha*xi)*diff(wfinal,xi,2);
M:ratsimp(M);
printf(TeXFile,"Check of M(1): ~g~%~%~%",subst(1,xi,M));

/* Print values of the abscissa, the displacement           */
/* and the bending moment.                                  */
printf(TeXFile,
    "\\par\\vspace*{3.0ex} $\\xi$ and $w$ and $M(xi)$~%");
Np: 25;
wordinates: zeromatrix(Np,1);
wabscissae: zeromatrix(Np,1);
for i:0 thru Np-1 do
  (xival: i/float(Np-1),
   ih: i+1,
   wabscissae[ih,1]: xival,
   wordin: subst(xival,xi,wfinal),
   wordin: float(wordin),
   wordinates[ih,1]: expand(wordin),
   Mval: subst(xival,xi,M),
   printf(TeXFile,
      "~f\\qquad   ~f\\qquad ~f~%~%",xival,wordin,Mval),
   printf(DatFile,"~f\    ~f\ ~f~%~%",xival,wordin,Mval)
   );

close(TeXFile);
close(DatFile);

quit();
```

We shall let the potential energy $\mathbf{\Pi}_{\mathrm{P}}$ given by (*Ex. 32-4.2*) be our foundation for the computations below, which we perform in the spirit of Example Ex 32-1. We must assume a *kinematically admissible* displacement field, see Ex 32-4.1 above. The simplest set of functions—one term is probably not enough—is

Displacement assumption

$$\widetilde{w} = [N(\xi)]\{V\} \qquad (Ex.\ 32\text{-}4.11)$$

where the shape of the so-called *trial* functions is given by

Trial functions vector $[N(\xi)]$

$$[N(\xi)] = [N_1(\xi)\,,\; N_2(\xi)\,,\; \ldots\,,\; N_j(\xi)\,,\; \ldots\,,\; N_Q(\xi)]$$
$$N_j(\xi) = \frac{1}{j}\left(\xi^{j+2} - \xi^2\right) \qquad (Ex.\ 32\text{-}4.12)$$

and the unknown of our problem is the vector $\{V\}$

Unknown displacement vector $\{V\}$

$$\{V\}^T = \left[V_1\;,\; V_2\;,\; \cdots\;,\; V_j\;,\; \cdots\;,\; V_Q\right] \qquad (Ex.\ 32\text{-}4.13)$$

Above, Q is the number of terms we use.

Note that none of the above terms satisfies the *static* boundary condition $M(1) = 0 \Rightarrow \ddot{w}(1) = 0$. We could have utilized a set of terms with this property. In that case, the elements of $[N]$ may be taken to be

$$N_j(\xi) = -\frac{4}{j^2+j}\xi^{j+3} + \frac{j+4}{j}\xi^3 - \frac{j+5}{j+1}\xi^2 \qquad (Ex.\ 32\text{-}4.14)$$

Trial functions satisfying the static boundary condition

Clearly, these functions are more complicated, but there are two important issues to note here. First, for other kinds of structures, such as solids, plates and shells, often we cannot choose fields that satisfy static boundary conditions. Second, the fields utilized in the application of the principle of stationary potential energy do not have to satisfy any other conditions than the kinematic conditions. So, it would be a little like cheating if we based our computations below on (*Ex. 32-4.14*).

Two reasons for not choosing trial functions satisfying static (boundary) conditions

In order to compute the strain energy term of (*Ex. 32-4.2*) we need the second derivative $[B]$, the *Strain Distribution Matrix*,[32.26] of $[N]$

$$[B(\xi)] \equiv [N''(\xi)] = \frac{1}{L^2}[\ddot{N}(\xi)] \qquad (Ex.\ 32\text{-}4.15)$$

Strain distribution matrix $[B]$

With these expressions in hand we may now compute the potential energy

$$\begin{aligned}
&\mathbf{\Pi}_{\mathrm{P}}(\{V\}) \\
&= \tfrac{1}{2}L\int_0^1 \{V\}^T[B(\xi)]^T EI_0(1-\alpha\xi)[B(\xi)]\{V\}d\xi \\
&\quad -L\int_0^1 \{V\}^T\bar{p}[N(\xi)]^T d\xi \\
&= \tfrac{1}{2}\{V\}^T\left(EI_0L\int_0^1 [B(\xi)]^T(1-\alpha\xi)[B(\xi)]d\xi\right)\{V\} \\
&\quad -\{V\}^T\left(\bar{p}L\int_0^1 [N(\xi)]^T d\xi\right) \\
&= \tfrac{1}{2}\{V\}^T[K]\{V\} - \{V\}^T\{\bar{R}\}
\end{aligned} \qquad (Ex.\ 32\text{-}4.16)$$

$\mathbf{\Pi}_{\mathrm{P}}$ for trial functions

where

$$\begin{aligned}
[K] &\equiv EI_0L\int_0^1 [B(\xi)]^T(1-\alpha\xi)[B(\xi)]d\xi \\
\{\bar{R}\} &\equiv \bar{p}L\int_0^1 [N(\xi)]^T d\xi
\end{aligned} \qquad (Ex.\ 32\text{-}4.17)$$

"Stiffness" matrix $[K]$

Load vector $\{\bar{R}\}$

In the present case—and this is quite common—the integrands of both $[K]$ and $\{\bar{R}\}$ only contain polynomials which can be integrated analytically. The variation of $\mathbf{\Pi}_{\mathrm{P}}(\{V\})$ is

$$\begin{aligned}
&\delta\mathbf{\Pi}_{\mathrm{P}}(\{V\}) = \{\delta V\}^T\left([K]\{V\} - \{\bar{R}\}\right) = 0\ \forall\,\{\delta V\} \\
&\Rightarrow \{V\} = [K]^{-1}\{\bar{R}\}
\end{aligned} \qquad (Ex.\ 32\text{-}4.18)$$

where $[K]^{-1}$ signifies the matrix inverse of $[K]$. It is customaryto write the solution in the above way, although only for very small systems of equations should you compute $[K]^{-1}$. Instead, you should solve the equations by *Gauss elimination* or some other elimination scheme, but

Don't compute $[K]^{-1}$, solve the equations

[32.26] In this case the strain distribution matrix contains only one row.

that subject lies beyond the scope of this book, so you will have to use other sources.

When we have solved the linear equations of (*Ex. 32-4.18*) we can compute any kinematic or static quantity of the structure, and in order to compute the rotation $\omega(1)$ at $\xi = 1$, which we shall pay some attention to later, we need the first derivative $[G]$ of $[N]$

$$[G(\xi)] \equiv [N'(\xi)] = \frac{1}{L}[\dot{N}(\xi)] \qquad (Ex.\ 32\text{-}4.19)$$

Rotation at right-hand end $\omega(1)$

and thus

$$\omega(1) = \frac{1}{L}\dot{w}(1) = [G(1)]\{V\} \qquad (Ex.\ 32\text{-}4.20)$$

Without going into any details, here are results for the case of $\alpha = \frac{1}{2}$

Relative errors

Q	$\dfrac{\Delta\mathbf{\Pi}_{\mathrm{P}}}{\mathbf{\Pi}_{\mathrm{P}}^{\mathrm{ex}}}$	$\dfrac{\Delta\omega(1)}{\omega_{\mathrm{ex}}(1)}$	$\dfrac{\Delta M(0)}{M_{\mathrm{ex}}(0)}$
1	3.1×10^{-1}	1.3×10^{-1}	5.1×10^{-1}
2	9.4×10^{-3}	3.8×10^{-3}	-1.0×10^{-1}
3	2.8×10^{-4}	1.1×10^{-4}	2.0×10^{-2}
4	8.4×10^{-6}	3.4×10^{-6}	-3.7×10^{-3}
5	2.5×10^{-7}	1.0×10^{-7}	7.0×10^{-4}
6	7.4×10^{-9}	3.0×10^{-9}	-1.3×10^{-4}
7	2.2×10^{-10}	8.9×10^{-11}	2.3×10^{-5}
8	6.4×10^{-12}	2.7×10^{-12}	-4.3×10^{-6}

(*Ex. 32-4.21*)

As you can see, the number of terms needed to provide a satisfactory answer regarding the value of the potential energy $\mathbf{\Pi}_{\mathrm{P}}$, which in itself is not a very valuable quantity, is around 2–3.

"Characteristic" displacement $\omega(1)$

The same holds for the value of the "characteristic" displacement[32.27] $\omega(1)$. The number of terms to produce a sufficiently accurate value of the bending moment $M(0)$ at the support, on the other hand, lies in the range 4–5. This fact is not surprising since the bending moment follows from the displacement field as the second derivative, and for each time we differentiate the accuracy decreases in any approximative or numerical method. There are ways to circumvent this, see e.g. Section 33.6 and Ex 32-5, but we shall not pursue that issue here.

To sum up: in a fairly simple case as the one treated above the exact solution proves to be complicated,[32.28] but by use of the principle of

[32.27] The term "characteristic" displacement is here taken in a somewhat more loose sense than in Section 33.4.3.1, see in particular (33.42), where the issue is a point load, but, as you probably acknowledge, $\omega(1)$ is indeed a displacement that is characteristic of the displacement field.

[32.28] If the bending stiffness varies as a polynomial of degree 3, for example, the exact solution becomes even worse than (*Ex. 32-4.10*).

stationarity of the potential energy and a simple set of trial functions with just a little work we have found solutions that are sufficiently accurate.

32.6 Lagrange Multipliers

In a number of cases it is necessary to fulfill some *Auxiliary Conditions*, often denoted *Side Conditions*, such as a coupling of some displacement components. Sometimes it is possible to (re)formulate the problem such that these conditions are satisfied *a priori*, but in other cases it is either not feasible or maybe not worth the effort. Then, the method of *Lagrange Multipliers*, see e.g. (Arfken & Weber 1995), comes in handy. Below, we introduce the idea via an example.

Auxiliary condition = Side condition

Lagrange Multipliers

Ex 32-5 A Structure with Auxiliary Conditions —The Euler Column

The Euler Column with auxiliary conditions

We take the *Euler Column*, see Example Ex 7-3, Example Ex 18-1 and Fig. Ex. 32-5.1, as our point of departure. In Example Ex 7-3 it was

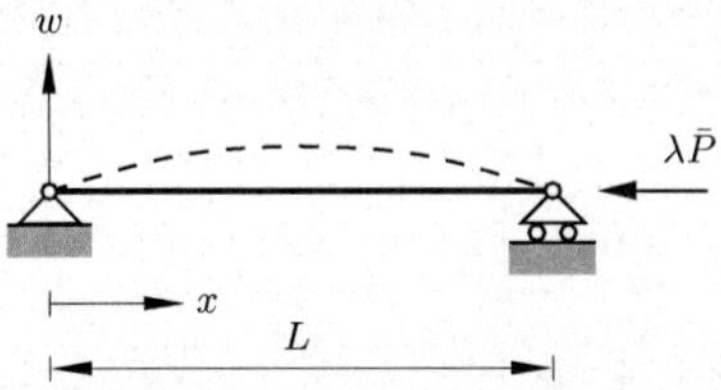

Fig. Ex. 32-5.1: *The Euler Column.*

found that the potential energy $\mathbf{\Pi}_\mathrm{P}$ is, see (*Ex. 7-3.9*)

$$\begin{aligned}\mathbf{\Pi}_\mathrm{P}(w) &= \tfrac{1}{2}EI\int_0^L (w'')^2\,dx - \tfrac{1}{2}\int_0^L N_0\,(w')^2\,dx\\ &= \tfrac{1}{2}EI\int_0^L (w'')^2\,dx - \tfrac{1}{2}\lambda\bar{P}\int_0^L (w')^2\,dx\end{aligned} \qquad (Ex.\ 32\text{-}5.1)$$

Note, that the values of N_0 on the right-hand side result from the solution of the prebuckling problem, see (*Ex. 7-3.8*). Only in simple cases, such as the present one, can you express this term as the work done by the applied load. The value of $-\frac{1}{2}\int_0^L (w')^2 dx$ in the last term of (*Ex. 32-5.1*) signifies the so-called "geometric shortening" Δ of the column due to the displacement w, which may be seen from the following computation.

$$\begin{aligned}\Delta &= \int_0^L \cos(\omega)dx - L = \int_0^L \big(\cos(w') - 1\big)dx\\ &\approx \int_0^L \Big(\big(1 - \tfrac{1}{2}(w')^2\big) - 1\Big)dx\end{aligned} \qquad (Ex.\ 32\text{-}5.2)$$

In particular you will get erroneous results from approximate analyses of buckling of plates with concentrated loads if you make this assumption.[32.29] which provides the solution (*Ex. 7-3.7*)[32.30]

Solution

$$\left.\begin{aligned} \lambda_{(n)} &= n^2 \\ w_{(n)} &= \xi_{(n)} \sin\left(n\pi\frac{x}{L}\right) \end{aligned}\right\} n \in [1, 2, \ldots, \infty] \qquad (Ex.\ 32\text{-}5.3)$$

where we shall be concerned only with the value $\lambda_{(1)}$ and its associated displacement field $w_{(1)}$, i.e. with the buckling load λ_c and the buckling mode $w_{(1)}$.

Ex 32-5.1 Potential Energy

Our objective is to find an approximate value $\tilde{\lambda}_c$ of λ_c based on an assumed approximation $\widetilde{w}_1$—a *trial* function—to w_1. We shall attempt to get a "good" value of $\tilde{\lambda}_c$ with only one term in $\widetilde{w}_1$. First, we must realize that $\widetilde{w}_1$ has to satisfy the *kinematic*[32.31] conditions

- The kinematic boundary conditions $\widetilde{w}_1(0) = 0$ and $\widetilde{w}_1(L) = 0$
- Differentiability with a continuous second derivative

Simplest polynomial trial function

Except for the exact solution the simplest field we may think of is

$$\widetilde{w}_1(x) = \xi_1 x(L - x) = \xi_1(Lx - x^2) \qquad (Ex.\ 32\text{-}5.4)$$

where ξ_1 denotes the amplitude of the approximate buckling mode. The relevant derivatives of $\widetilde{w}_1(x)$ are

$$\widetilde{w}_1'(x) = \xi_1(L - 2x) \text{ and } \widetilde{w}_1''(x) = -2\xi_1 \qquad (Ex.\ 32\text{-}5.5)$$

$\widetilde{w}_1(x)$ has inherent problems

It seems obvious that this cannot be a very good approximation to the real buckling mode because the second derivative, which is proportional to the bending moment, is constant, while the real bending moment must satisfy the condition that it vanishes at the supports. However, we shall not let such minor issues bother us, but employ $\widetilde{w}_1(x)$ given by (*Ex. 32-5.4*) as our trial function in (*Ex. 32-5.1*).

[32.29] Even one of the greatest figures in the theory of stability, namely Timoshenko made the mistake, see (Timoshenko & Gere 1961), p. 388, (d). In some cases, Timoshenko underestimates the buckling load by as much as 40% as shown by Damkilde & Byskov (1979).

[32.30] In Example Ex 7-3 we proved that $\mathbf{\Pi}_\mathrm{P}$ given by (*Ex. 32-5.1*) may yield the differential equation and associated boundary conditions, which result in the solution (*Ex. 32-5.3*).

[32.31] The *static* conditions may or may not be satisfied by the *trial function*—it may be a good idea, but it is not necessary. The reason for this is that the static boundary conditions may be derived from the first variation of $\mathbf{\Pi}_\mathrm{P}(w)$, while w, which is the so-called *trial function*, must satisfy the kinematic conditions because of the assumptions inherent in the derivation of the potential energy.

Observe that now we use the notation $\mathbf{\Pi}_\mathrm{P}(\xi_1)$ to indicate that the only degree of freedom is given by the value of ξ_1

$$\begin{aligned}\mathbf{\Pi}_\mathrm{P}(\xi_1) &= (\xi_1)^2\,\tfrac{1}{2}EI\int_0^L 4dx \\ &\quad - (\xi_1)^2\,\tfrac{1}{2}\tilde{\lambda}_c\bar{P}\int_0^L \left(L^2 - 4Lx + 4x^2\right)dx \\ &= (\xi_1)^2\,\tfrac{1}{2}EI\left(4L - \tilde{\lambda}_c\left(\frac{\pi}{L}\right)^2\frac{L^3}{3}\right)\end{aligned} \qquad (Ex.\ 32\text{-}5.6)$$

Potential energy with one degree of freedom

The first variation of $\mathbf{\Pi}_\mathrm{P}(\xi_1)$ is

$$\delta\mathbf{\Pi}_\mathrm{P}(\xi_1) = \xi_1\delta\xi_1 EI\left(4L - \tilde{\lambda}_c\left(\frac{\pi}{L}\right)^2\frac{L^3}{3}\right) \qquad (Ex.\ 32\text{-}5.7)$$

Variation of $\mathbf{\Pi}_\mathrm{P}(\xi_1)$

The requirement $\delta\mathbf{\Pi}_\mathrm{P}(\xi_1) = 0\,\forall\,\delta\xi_1$ can only be satisfied either if $\xi_1 = 0$, which is a trivial solution that is uninteresting to us, or if

$$\tilde{\lambda}_c^{(1)} = \frac{12}{\pi^2} \approx 1.22 \text{ cf. } \lambda_c = 1 \qquad (Ex.\ 32\text{-}5.8)$$

$\tilde{\lambda}_c^{(1)} \approx 1.22$ too inaccurate

The relative error then is 22%, which is much more than is acceptable. There are now several ways to improve on this result

More terms

- We may add more terms to the trial function and write

  $$\widetilde{w}_1^{(2)} = \xi_1(Lx - x^2) + \xi_2(Lx - x^2)^2 \qquad (Ex.\ 32\text{-}5.9)$$

  etc., where ξ_j are coefficients and $^{(2)}$ indicates that this is our second attempt. This, however, requires solution of a matrix eigenvalue problem, which is somewhat more involved than solving one linear equation with one unknown. The lowest of the two eigenvalues is

  $$\tilde{\lambda}_c^{(2)} = \frac{2\sqrt{1605} - 90}{\pi^2} \approx 1.00056 \qquad (Ex.\ 32\text{-}5.10)$$

  $\tilde{\lambda}_c^{(2)} \approx 1.00056$ very accurate

  which is a very good approximation.

- We may try to get a better trial function by letting it satisfy the static boundary conditions also. Such a function is

  Satisfy static boundary conditions

  $$\widetilde{w}_1^{(3)} = \xi_1\left(xL^3 - 2x^3L + x^4\right) \qquad (Ex.\ 32\text{-}5.11)$$

  The result of this assumption is

  $$\tilde{\lambda}^{(3)} = 168/(17\pi^2) \approx 1.0013 \qquad (Ex.\ 32\text{-}5.12)$$

  $\tilde{\lambda}^{(3)} \approx 1.0013$ accurate enough

  which also is a very decent result.

- We may reconsider our original choice $\widetilde{w}_1^{(1)}$, but try to mend the deficiencies as regards the second derivative. We shall do this and employ the method of a *Lagrange Multiplier.*

  Lagrange Multiplier

Recall that in Section 18.3.7 we proved that any approximate value of λ_c which is obtained from stationary values of the potential energy is

$\mathbf{\Pi}_{\mathrm{PM}}$ may provide too small or too large approximations of λ_c.

larger than or equal to the correct value. This is not always the case for values obtained by use of a modified potential energy $\mathbf{\Pi}_{\mathrm{PM}}$.

Ex 32-5.2 Modified Potential Energy

Cause of error behind $\tilde{\lambda}_c^{(1)}$: $\widetilde{w}_1'' =$ const.

First, let us once more realize that the main cause of error inherent in the choice $\widetilde{w}_1(x) = \xi_1 x(L-x)$ comes from the poor second derivative. Having done that we may try to introduce an additional independent field that approximates the second derivative—or the bending moment—in a better way. This cannot be done directly in the potential energy because one of the *auxiliary conditions* on $\mathbf{\Pi}_{\mathrm{P}}$ is

Auxiliary condition

$$\kappa = w'' \qquad (Ex.\ 32\text{-}5.13)$$

Lagrange Multiplier Field $\tilde{\eta}$

We may, abandon this condition and include it in a *Modified Potential Energy* $\mathbf{\Pi}_{\mathrm{PM}}$ by use of a *Lagrange Multiplier Field* $\tilde{\eta}$, see also Chapter 33, in particular Section 33.7.

Ex 32-5.2.1 Is $\mathbf{\Pi}_{\mathrm{PM}}$ a Legitimate Potential?

Instead of going through theoretical derivations, in the present connection we postulate that the modified potential energy $\mathbf{\Pi}_{\mathrm{PM}}$ is

Three independent fields: $\widetilde{w}$, $\tilde{\kappa}$ and $\tilde{\eta}$

$$\mathbf{\Pi}_{\mathrm{PM}}(\widetilde{w}, \tilde{\kappa}, \tilde{\eta}) = \tfrac{1}{2}\int_0^L EI\tilde{\kappa}^2 dx - \tfrac{1}{2}\lambda\bar{P}\int_0^L (\widetilde{w}')^2\, dx - \int_0^L \tilde{\eta}(\tilde{\kappa} - \widetilde{w}'')dx \qquad (Ex.\ 32\text{-}5.14)$$

$\tilde{\kappa} \not\equiv \widetilde{w}''$

where $\widetilde{w}(x)$, $\tilde{\kappa}(x)$ and $\tilde{\eta}(x)$ are independent fields. Still, $\widetilde{w}$ must satisfy the kinematic conditions, except $\kappa = w''$. The new fields $\tilde{\kappa}$ and $\tilde{\eta}$ must be continuous and "sufficiently" smooth,[32.32] but are not subject to any other conditions à priori.

We demand that the first variation of $\mathbf{\Pi}_{\mathrm{PM}}$ vanishes

$$0 = \delta\mathbf{\Pi}_{\mathrm{PM}}(\widetilde{w}, \tilde{\kappa}, \tilde{\eta})\ \forall\, (\delta\widetilde{w}, \delta\tilde{\kappa}, \delta\tilde{\eta}) \qquad (Ex.\ 32\text{-}5.15)$$

i.e.

$$0 = \int_0^L EI\tilde{\kappa}\delta\tilde{\kappa}dx - \lambda\bar{P}\int_0^L \widetilde{w}'\delta\widetilde{w}'dx - \int_0^L \delta\tilde{\eta}(\tilde{\kappa} - \widetilde{w}'')dx - \int_0^L \tilde{\eta}(\delta\tilde{\kappa} - \delta\widetilde{w}'')dx \quad \forall\, (\delta\widetilde{w}, \delta\tilde{\kappa}, \delta\tilde{\eta}) \qquad (Ex.\ 32\text{-}5.16)$$

This statement proves to be very useful to us later in connection with the approximate solution, but first we need to see if it provides the correct differential equations. In order to get these, we must get rid of $\delta\widetilde{w}'$ and $\delta\widetilde{w}''$, which follow directly from $\delta\widetilde{w}$ and therefore are not independent.

[32.32] By this I imply that the functions should be so smooth that the second derivative of $\tilde{\eta}$, see (*Ex. 32-5.17*), as well as $\tilde{\kappa}$ itself is continuous.

We integrate by parts and rearrange in (*Ex. 32-5.16*) to get

$$\begin{aligned}0 = &\int_0^L (EI\tilde{\kappa} - \tilde{\eta})\,\delta\tilde{\kappa}dx - \int_0^L (\tilde{\kappa} - \tilde{w}'')\delta\tilde{\eta}dx \\ &+ \int_0^L (\tilde{\eta}'' + \lambda\bar{P}\tilde{w}'')\delta\tilde{w}dx \\ &+ [\tilde{\eta}\delta\tilde{w}']_0^L - [\tilde{\eta}'\delta\tilde{w}]_0^L - \lambda\bar{P}\,[\tilde{w}'\delta\tilde{w}]_0^L \\ &\forall\,(\delta\tilde{w},\delta\tilde{\kappa},\delta\tilde{\eta})\end{aligned} \qquad (Ex.\ 32\text{-}5.17)$$

Vanishing of the integrals implies

$$\left.\begin{array}{l}\tilde{\kappa} = \tilde{w}'' \\ \tilde{\eta} = EI\tilde{\kappa} \\ \tilde{\eta}'' + \lambda\bar{P}\tilde{w}'' = 0\end{array}\right\} x \in\,]\,0; L\,[\qquad (Ex.\ 32\text{-}5.18)$$

and thus

$$EI\tilde{w}^{\text{iv}} + \lambda\bar{P}\tilde{w}'' = 0\,,\ x \in]\,0; L\,[\qquad (Ex.\ 32\text{-}5.19)$$

as required, while the contributions for $x = 0$ and $x = L$ provide

$$\left.\begin{array}{l}\delta\tilde{w}'(0) \\ \delta\tilde{w}'(L)\end{array}\right\} \text{arbitrary} \Rightarrow \left\{\begin{array}{l}\tilde{\eta}(0) = 0 \\ \tilde{\eta}(L) = 0\end{array}\right. \qquad (Ex.\ 32\text{-}5.20)$$

and

$$\left.\begin{array}{l}\delta\tilde{w}(0) \\ \delta\tilde{w}(L)\end{array}\right\} = 0 \Rightarrow \text{ no condition on } \left\{\begin{array}{l}\tilde{\eta}'(0) \\ \tilde{\eta}'(L)\end{array}\right. \qquad (Ex.\ 32\text{-}5.21)$$

Since we demand that the fields are continuous in $[0; L]$ (*Ex. 32-5.20*) together with (*Ex. 32-5.18a–b*) means that

$$\tilde{\kappa}(0) = 0\,,\ \tilde{\kappa}(L) = 0\,,\ \tilde{w}''(0) = 0 \text{ and } \tilde{w}''(L) = 0 \qquad (Ex.\ 32\text{-}5.22)$$

This concludes the proof that (*Ex. 32-5.15*) or, equivalently, (*Ex. 32-5.16*) results in the correct differential equations and boundary conditions. We may therefore, with some confidence, utilize (*Ex. 32-5.14*) as a basis for approximate solutions.

Π_{PM} provides the correct equations. It is legitimate

We may easily see from the above equations that the *Lagrange Multiplier Field* $\tilde{\eta}$ may be interpreted as the bending moment associated with buckling. Already (*Ex. 32-5.14*) gives a strong indication that this is the case because there $\tilde{\eta}$ produces energy together with the (approximate) bending strains $\tilde{\kappa}$. On the other hand, if we had chosen the sign on the last term in (*Ex. 32-5.14*) to be positive, then $\tilde{\eta}$ would have to be interpreted as the *negative* of the bending moment.[32.33]

Here, the Lagrange Multiplier Field $\tilde{\eta}$ may be thought of as the bending moment M, but beware of its sign

[32.33] Sometimes we write a modified potential energy and later discover that the Lagrange term should change sign in order for the *Lagrange Multiplier* to have a meaningful physical interpretation.

Simplest assumptions

Ex 32-5.2.2 Parabolic Approximation

As the approximating fields we choose essentially the same function as in (*Ex. 32-5.4*)

$$\left.\begin{aligned}\widetilde{w}^{(4)}(\zeta) &= a_1\zeta(1-\zeta)\\ \widetilde{\kappa}^{(4)}(\zeta) &= a_2\zeta(1-\zeta)\\ \widetilde{\eta}^{(4)}(\zeta) &= a_3\zeta(1-\zeta)\end{aligned}\right\}\text{ where } \zeta \equiv \frac{x}{L} \tag{Ex. 32-5.23}$$

Then,

$$\begin{aligned}\mathbf{\Pi}_{\mathrm{PM}}(a_j) = &\ (a_2)^2\,\tfrac{1}{2}EIL\int_0^1 (\zeta-\zeta^2)^2\,d\zeta\\ &-\tfrac{1}{2}(a_1)^2\,\lambda\frac{\bar{P}}{L}\int_0^1 (1-2\zeta)^2\,d\zeta\\ &-a_3a_2L\int_0^1(\zeta-\zeta^2)^2\,d\zeta\\ &+a_3a_1\frac{1}{L}\int_0^1(\zeta-\zeta^2)(-2)d\zeta\end{aligned} \tag{Ex. 32-5.24}$$

or

$$\begin{aligned}\mathbf{\Pi}_{\mathrm{PM}}(a_j) = &\ \frac{EIL}{60}(a_2)^2 - \lambda\frac{\bar{P}}{6L}(a_1)^2\\ &-\frac{L}{30}a_3a_2 - \frac{1}{3L}a_3a_1\end{aligned} \tag{Ex. 32-5.25}$$

Therefore,

$$\begin{aligned}\delta\mathbf{\Pi}_{\mathrm{PM}}(a_j) = &\ \left(-\lambda\frac{\bar{P}}{3L}a_1 - \frac{1}{3L}a_3\right)\delta a_1\\ &+\left(\frac{EIL}{30}a_2 - \frac{L}{30}a_3\right)\delta a_2\\ &+\left(-\frac{1}{3L}a_1 - \frac{L}{30}a_2\right)\delta a_3\end{aligned} \tag{Ex. 32-5.26}$$

Recall: the amplitude of the buckling mode is arbitrary

Quite arbitrarily [32.34] we may choose $a_1 = L$ and by requiring that $\delta\mathbf{\Pi}_{\mathrm{PM}}$ vanishes we get the solution

$$a_1 = L \Rightarrow a_2 = -10\frac{1}{L} \Rightarrow a_3 = -10\frac{EI}{L} \tag{Ex. 32-5.27}$$

Buckling problem

and the eigenvalue problem becomes

$$\frac{10EI}{3L^2} - \lambda\frac{\bar{P}}{3} = 0 \tag{Ex. 32-5.28}$$

$\tilde{\lambda}^{(4)} \approx 1.013$ is a fair approximation

The approximate $\tilde{\lambda}^{(4)}$ value of λ_c is then

$$\tilde{\lambda}^{(4)} = 10\frac{EI}{\bar{P}L^2} = \frac{10}{\pi^2} \approx 1.013 \tag{Ex. 32-5.29}$$

[32.34] The reason why this is permissible is that (*Ex. 32-5.26*) is an eigenvalue problem with the eigenfunction given by (*Ex. 32-5.23*) whose amplitude is arbitrary.

This is a fair approximation, but not as good as $\tilde{\lambda}^{(2)}$ and $\tilde{\lambda}^{(3)}$, which were based on fields of higher degrees in x or ζ.

It may be in order to make a comment on the dimension of a_1, a_2 and a_3. We chose $a_1 = L$, which insured that it has the right dimension of length. The dimension of a_2 is force×length, i.e. the dimension of a bending moment, as desired. Finally, the dimension of a_3 is length^{-1}, which is the dimension of a curvature strain and therefore in accord with (*Ex. 32-5.18b*).

Choose reasonably dimensions

Ex 32-5.2.3 Quartic Approximation

If we employ a set of better trial functions, namely a set which is essentially the same as the one leading to $\tilde{\lambda}^{(3)}$, the result of applying the modified potential energy is

$$\tilde{\lambda}^{(5)} = \frac{306}{(31\pi^2)} \approx 1.000138 \qquad (Ex.\ 32\text{-}5.30)$$

Quartic approximation provides excellent $\tilde{\lambda}^{(5)} \approx 1.013$ a fair approximation

which is even better than $\tilde{\lambda}^{(2)}$, $\tilde{\lambda}^{(3)}$ and $\tilde{\lambda}^{(4)}$ and much better than engineering practice requires, but it takes solution of a matrix eigenvalue problem, albeit a simple one.

In the examples above, application of the method of Lagrange Multipliers resulted in substantial improvements of the accuracy. This is not necessarily always the case, but experience—my experience—shows that if you utilize the method judiciously you will get better solutions. The crux of the matter is to determine the probable cause of inaccuracies. In Example Ex 32-5 the problem lies in the inadequate moment distribution that follows from the assumed parabolic displacement field given by (*Ex. 32-5.4*). Obviously, the idea was to abandon the relation $\kappa = w''$ and introduce the independent curvature strain field $\tilde{\kappa}$ at the expense of introducing yet another field, namely the Lagrange multiplier field $\tilde{\eta}$. Although it is true that there is no such thing as a free lunch the additional work associated with the application of the Lagrange Multiplier method proved to be rather limited in our case.

Often, but not always, do we get better results from modified potentials

32.7 General Treatment

For a more general treatment of variational principles associated with continuum mechanics the reader is referred to Chapter 33, where the *Principle of Virtual Displacements* and the *Principle of Minimum Potential Energy* are presented in a compact notation.

Chapter 33

Budiansky-Hutchinson Notation

In many instances it proves convenient to utilize a more compact notation than one which entails the use of indices. There are several such notations available, and here we choose the so-called *Budiansky-Hutchinson Notation* (Budiansky & Hutchinson 1964) which is introduced below. Another, quite popular, notation utilizes *dyads* and *polyads*, see e.g. (Malvern 1969).[33.1]

33.1 Linear, Quadratic and Bilinear Operators

Note that the Lagrange Strain as given by (2.19) is a the sum of a term which is linear in the displacements (and their gradients) and a term which is quadratic in the displacements (and their gradients)

$$\gamma_{mn} = \tfrac{1}{2}\left(u_{m,n} + u_{n,m}\right) + \tfrac{1}{2}u_{k,m}u_{k,n} \tag{33.1}$$

Lagrange strain γ_{mn}

or, with a change of notation[33.2]

$$\varepsilon_{mn} = \tfrac{1}{2}\left(u_{m,n} + u_{n,m}\right) + \tfrac{1}{2}u_{k,m}u_{k,n} \tag{33.2}$$

If our concern is not so much the individual components of the strain, or for the purpose of deriving general statements,[33.3] we may write (33.2) in the following form

Generalized strain-displacement relation

$$\varepsilon = \mathrm{l}_1(\boldsymbol{u}) + \tfrac{1}{2}\mathrm{l}_2(\boldsymbol{u}) \tag{33.3}$$

which may, at a first glance, seem to contain almost no real information, but, as we shall see later, it is part of a convenient basis for general statements. We shall also see how it can be interpreted for various kinds of structures.

In (33.3), $\boldsymbol{u}$ denotes all components of the generalized displacements,[33.4]

Generalized displacement $\boldsymbol{u}$
Generalized strain ε

[33.1] For our present purposes, I have judged that it offers too little in compensation for a more involved set of rules, but acknowledge that under other circumstances it certainly has its important virtues.

[33.2] In most of our applications the strain tensor is called ε_{mn} and not γ_{mn}, whether it is a linear or nonlinear measure.

[33.3] In this book focus is on the general statements.

E. Byskov, *Elementary Continuum Mechanics for Everyone*,
Solid Mechanics and Its Applications 194, DOI: 10.1007/978-94-007-5766-0_33,

$\mathbf{l}_1$ is a *Linear Operator* working on $\boldsymbol{u}$, $\mathbf{l}_2$ is a *Quadratic Operator* on $\boldsymbol{u}$, and ε signifies all components of the generalized strains.[33.5]

The linearity of $\mathbf{l}_1$ is given by the following rules

Linear operator $\mathbf{l}_1$

$$\mathbf{l}_1(k\boldsymbol{u}) = k\mathbf{l}_1(\boldsymbol{u}) \text{ and } \mathbf{l}_1(\boldsymbol{u}_a + \boldsymbol{u}_b) = \mathbf{l}_1(\boldsymbol{u}_a) + \mathbf{l}_1(\boldsymbol{u}_b) \tag{33.4}$$

where k is a constant, and $\boldsymbol{u}_a$ and $\boldsymbol{u}_b$ are two different displacement fields.

The operator $\mathbf{l}_2$ shows its quadratic property in the following rule, which is the analogy of (33.4)

Quadratic operator $\mathbf{l}_2$

$$\mathbf{l}_2(k\boldsymbol{u}) = k^2\mathbf{l}_2(\boldsymbol{u}) \tag{33.5}$$

In order to establish a rule for the quadratic operator $\mathbf{l}_2$, which corresponds to (33.4b), it proves necessary to introduce a *Bilinear Operator* $\mathbf{l}_{11}$ by

Bilinear operator $\mathbf{l}_{11}$

$$\mathbf{l}_2(\boldsymbol{u}_a + \boldsymbol{u}_b) = \mathbf{l}_2(\boldsymbol{u}_a) + 2\mathbf{l}_{11}(\boldsymbol{u}_a, \boldsymbol{u}_b) + \mathbf{l}_2(\boldsymbol{u}_b) \tag{33.6}$$

where it is obvious that $\mathbf{l}_{11}$ is symmetric

$\mathbf{l}_{11}$ is symmetric

$$\mathbf{l}_{11}(\boldsymbol{u}_a, \boldsymbol{u}_b) = \mathbf{l}_{11}(\boldsymbol{u}_b, \boldsymbol{u}_a) \tag{33.7}$$

and that

$\mathbf{l}_{11}(\boldsymbol{u}) = \mathbf{l}_2(\boldsymbol{u})$

$$\mathbf{l}_{11}(\boldsymbol{u}, \boldsymbol{u}) = \mathbf{l}_2(\boldsymbol{u}) \tag{33.8}$$

The bi-linearity of $\mathbf{l}_{11}$ shows itself in the relation

$$\mathbf{l}_{11}(k_a\boldsymbol{u}_a, k_b\boldsymbol{u}_b) = k_a k_b \mathbf{l}_{11}(\boldsymbol{u}_b, \boldsymbol{u}_a) \tag{33.9}$$

where k_a and k_b are two constants.

We may easily see that the original formula (33.2) behaves in accordance with the rules given above for the operators $\mathbf{l}_1$, $\mathbf{l}_{11}$ and $\mathbf{l}_2$. In order to do so, we write (33.2) in the more explicit form

$$\begin{aligned} \varepsilon_{mn}(u_j^a + u_j^b) = &+ \tfrac{1}{2}(u_{m,n}^a + u_{m,n}^b + u_{n,m}^a + u_{n,m}^b) \\ &+ \tfrac{1}{2}(u_{k,m}^a + u_{k,m}^b)(u_{k,n}^a + u_{k,n}^b) \end{aligned} \tag{33.10}$$

where the labeling of the two displacement fields now is given by upper indices ^{a} and ^{b}.

[33.4] Sometimes the generalized displacements include rotations which is the case for Timoshenko beam theories, see Sections 7.4 and 7.6 as displacements.

[33.5] For instance, the generalized strains may include axial, shear (in Timoshenko beams) and curvature strains in beams, and in plates membrane, shear (in Mindlin plates), bending and torsional strains in plates, see Part II.

Elementary manipulations on (33.10) provide

$$\begin{aligned}\varepsilon_{mn}(u_j^a + u_j^b) = &+ \tfrac{1}{2}\left(u_{m,n}^a + u_{n,m}^a\right) + \tfrac{1}{2}u_{k,m}^a u_{k,n}^a \\ &+ \tfrac{1}{2}\left(u_{m,n}^b + u_{n,m}^b\right) + \tfrac{1}{2}u_{k,m}^b u_{k,n}^b \\ &+ \left(u_{k,m}^a u_{k,n}^b + u_{k,m}^b u_{k,n}^a\right)\end{aligned} \qquad (33.11)$$

and, clearly, this is equivalent to

$$\begin{aligned}\varepsilon(\boldsymbol{u}_a + \boldsymbol{u}_b) = &+ \mathbf{l}_1(\boldsymbol{u}_a) + \tfrac{1}{2}\mathbf{l}_2(\boldsymbol{u}_a) \\ &+ \mathbf{l}_1(\boldsymbol{u}_b) + \tfrac{1}{2}\mathbf{l}_2(\boldsymbol{u}_b) \\ &+ \mathbf{l}_{11}(\boldsymbol{u}_a, \boldsymbol{u}_b)\end{aligned} \qquad (33.12)$$

For more interpretations of $\mathbf{l}_1$, $\mathbf{l}_2$ and $\mathbf{l}_{11}$, see Part II, where expressions for these operators are given for beams and plates.

33.2 Principle of Virtual Displacements

The Budiansky-Hutchinson notation proves particularly convenient in connection with principles such as the *Principle of Virtual Displacements*, see Section (2.4), where the principle is written in (2.97)

Principle of Virtual Displacements for $\delta u_i = 0,\ x_j \in S_u^0$

$$\int_{V^0} t_{ij}\delta\gamma_{ij}dV^0 = \int_{S_T^0} \bar{\tau}_i \delta u_i dS^0 + \int_{V^0} \bar{q}_j \delta u_j dV^0 \qquad (33.13)$$
$$\forall\ \delta u_j = 0,\ x_j \in S_u^0$$

see Section 2.4.1 where the Budiansky-Hutchinson Notation is also mentioned.

In order to write this equation in a compact fashion we introduce two more symbols in addition to $\boldsymbol{u}$ and ε. The first is $\boldsymbol{\sigma}$ and denotes the generalized stresses.[33.6] The second is $\overline{\boldsymbol{T}}$ and signifies the generalized prescribed loads, which may include point loads, line loads, surface tractions and body forces.

Generalized stress $\boldsymbol{\sigma}$
Generalized load $\overline{\boldsymbol{T}}$

Further, let a dot ($\cdot$) between two fields indicate a (generalized) *Inner Product* of the fields. In the present, three-dimensional case the dot implies volume integrals over the body of the internal virtual work and of the work of the body forces, while it signifies surface integrals over the relevant parts of the surface of the work of the surface loads (and in the case of (2.95), which does not assume $\delta u_j = 0$ on S_u^0, also the work of the reactions on S_u^0).

A dot ($\cdot$) implies inner product

With the Budiansky-Hutchinson Notation in hand we may write either of the principles of virtual displacements (2.95) or (2.97) in the very short

[33.6] For instance, the generalized strains may include normal and shear (in Timoshenko beams) force and bending moments in beams, and in plates membrane and shear (in Mindlin plates) stresses as well as bending and torsional moments, see Part II.

Principle of Virtual Displacements

form, see also (2.98)

$$\boldsymbol{\sigma} \cdot \delta\boldsymbol{\varepsilon} = \overline{\boldsymbol{T}} \cdot \delta\boldsymbol{u} \tag{33.14}$$

The interpretation of the dot may require some further explanation, so here is a list which is not complete:[33.7]

- In some cases the dot just indicates the product of two quantities. For instance, this is the case of the internal work done by a spring force and (the variation of)[33.8] its associated elongation. This interpretation is also valid for a point load and its displacement.
- The dot may imply a sum as well as a product, for instance over a number of springs or point loads and their displacements.
- The internal virtual work in bars and beams are formed as integrals over the length of the body, so here the dot implies integration.
- In a plate or a shell the internal virtual work is given as a surface integral over the body meaning that here, the dot signifies integration. The external virtual work in these structures may be the result of "body forces"[33.9] working with the displacements over the body and, again it is a surface integral. Along the boundaries and at line loads the dot implies line integrals.
- For three-dimensional bodies the internal virtual work is given by a volume integral, and some external work may be associated with body forces whose virtual work is also computed as a volume integral. For other contributions to the external virtual work in this case you may quite easily apply some of the above definitions.

Before we can derive or apply the principle of virtual displacements we need to know the expression giving $\delta\boldsymbol{\varepsilon}$ in terms of $\boldsymbol{u}$ and $\delta\boldsymbol{u}$. In the spirit of Section 2.4 let the total displacement field be $\boldsymbol{u}_{\text{tot}}$, where

$$\boldsymbol{u}_{\text{tot}} = \boldsymbol{u} + \epsilon\delta\boldsymbol{u} \tag{33.15}$$

and the total strain field $\boldsymbol{\varepsilon}_{\text{tot}}$

$$\boldsymbol{\varepsilon}_{\text{tot}} = \boldsymbol{\varepsilon} + \epsilon\delta\boldsymbol{\varepsilon} + O(\delta\boldsymbol{\varepsilon}^2) \quad \text{where} \quad |\epsilon| \ll 1 \tag{33.16}$$

Here, $|\epsilon| \ll 1$ is the amplitude of the variations whose shapes are given as $\delta\boldsymbol{u}$ and $\delta\boldsymbol{\varepsilon}$.

Then, (33.3)–(33.9) make it possible to write the following expression for the total strain provided that it is composed by a linear and a quadratic term in the displacements, which certainly is the case for the Lagrange Strain, the strains of the kinematically moderately nonlinear beam theory

[33.7] For cases which are not in the list you should be able to extend it yourself.
[33.8] In the interest of brevity the following I omit "the variation of."
[33.9] For the meaning of that term see Section 9.1.4.

developed in Sections 7.3 and 8.2 and of the plate theory of Chapter 9. We may get

$$\begin{aligned}\varepsilon_{\text{tot}} &= \mathbf{l}_1(\boldsymbol{u}_{\text{tot}}) + \tfrac{1}{2}\mathbf{l}_2(\boldsymbol{u}_{\text{tot}})\\ &= \mathbf{l}_1(\boldsymbol{u}+\epsilon\delta\boldsymbol{u}) + \tfrac{1}{2}\mathbf{l}_2(\boldsymbol{u}+\epsilon\delta\boldsymbol{u})\\ &= \mathbf{l}_1(\boldsymbol{u}) + \tfrac{1}{2}\mathbf{l}_2(\boldsymbol{u}) + \epsilon(\mathbf{l}_1(\delta\boldsymbol{u}) + \mathbf{l}_{11}(\boldsymbol{u},\delta\boldsymbol{u})) + \epsilon^2\left(\tfrac{1}{2}\mathbf{l}_2(\delta\boldsymbol{u})\right)\\ &= \varepsilon + \epsilon(\mathbf{l}_1(\delta\boldsymbol{u}) + \mathbf{l}_{11}(\boldsymbol{u},\delta\boldsymbol{u})) + O(\epsilon^2)\end{aligned} \tag{33.17}$$

with the consequence that

$$\delta\varepsilon = \mathbf{l}_1(\delta\boldsymbol{u}) + \mathbf{l}_{11}(\boldsymbol{u},\delta\boldsymbol{u}) \tag{33.18}$$

Strain variation $\delta\varepsilon$

which means that we may express the principle of virtual displacements (33.14) in the alternative way

$$\boldsymbol{\sigma}\cdot\left(\mathbf{l}_1(\delta\boldsymbol{u}) + \mathbf{l}_{11}(\boldsymbol{u},\delta\boldsymbol{u})\right) = \overline{\boldsymbol{T}}\cdot\delta\boldsymbol{u} \tag{33.19}$$

At this point we may discuss the implications and use of the principle of virtual displacements.

- We may try to establish it separately for each kind of structure the way we did it for the three-dimensional body in Section 2.4, which lead to (2.97) which also appears above as (33.13). To do this we need to be able to derive the equilibrium equations for the structure we consider and this is not always an easy task.
- The other way is to postulate that the principle, given by (33.14) or (33.19), is valid and derive the static equations, the equilibrium equations, under this assumption. This is how we exploit the principle in Part II.

Stresses and strains must be *generalized* = Work conjugate

In the present context the most important implication of the principle of virtual work is that we insist that all strain and stress quantities of theories valid for specialized continua, such as beams and plates, are *Generalized* in that they are work conjugate in the sense of (33.14). Therefore, in almost all cases, we shall choose the second of the above possibilities.

33.3 Variation of a Potential

Before we proceed with the more continuum mechanics oriented material we may take a break and realize that we have talked about variations a number of times, but admit that each time it has been in an *ad hoc* fashion. In order that the following derivations do not get cluttered in the same way we interject this section.

Potential $\boldsymbol{\Pi}(\boldsymbol{\alpha})$ and its variation $\delta\boldsymbol{\Pi}(\boldsymbol{\alpha})$

Let us consider the functional $\boldsymbol{\Pi}(\boldsymbol{\alpha})$, which—as indicated—depends on the field[33.10] $\boldsymbol{\alpha}$ and define the variation $\delta\boldsymbol{\Pi}(\boldsymbol{\alpha})$ of $\boldsymbol{\Pi}(\boldsymbol{\alpha})$ by a *Gateaux Deriva-*

[33.10] You may think of $\boldsymbol{\alpha}$ as a displacement field, if you like, but it is probably better not to do so.

tive, see e.g. (Budiansky 1974)

$\delta\mathbf{\Pi}(\boldsymbol{\alpha})$ by Gateaux derivative

$$\delta\mathbf{\Pi}(\boldsymbol{\alpha}) = \lim_{\epsilon\to 0}\frac{\mathbf{\Pi}(\boldsymbol{\alpha}+\epsilon\delta\boldsymbol{\alpha}) - \mathbf{\Pi}(\boldsymbol{\alpha})}{\epsilon} \tag{33.20}$$

This may also be written

$\delta\mathbf{\Pi}(\boldsymbol{\alpha})$

$$\delta\mathbf{\Pi}(\boldsymbol{\alpha}) = \left.\frac{\partial\mathbf{\Pi}(\boldsymbol{\alpha}+\epsilon\delta\boldsymbol{\alpha})}{\partial\epsilon}\right|_{\epsilon=0} \tag{33.21}$$

These are the two most common and simple notations, or definitions, of the variation of a potential. There exist a number of other, more sophisticated and general definitions, see e.g. (Budiansky 1974), but the ones given here should suffice for our purposes.

33.4 Potential Energy for Linear Elasticity

The potential energy, see Section 2.7, may also be expressed in a convenient form by use of the operator and dot notation. This is especially true when the constitutive model is linear, i.e. expresses linear (hyper)elasticity, see (2.101)

Linear constitutive model

$$W(\gamma_{ij}) = \tfrac{1}{2}E_{ijkl}\gamma_{ij}\gamma_{kl} \tag{33.22}$$

and (2.102)

Linear constitutive model

$$t_{ij} = E_{ijkl}\gamma_{kl} \tag{33.23}$$

When we introduce the compact notation

Linear constitutive operator $\mathbf{H}$

$$\boldsymbol{\sigma} = \mathbf{H}(\boldsymbol{\varepsilon}) \tag{33.24}$$

where $\mathbf{H}$ is a linear constitutive operator, then the potential energy (2.106)

Potential energy $\mathbf{\Pi}_{\mathrm{P}}(u_i)$

$$\mathbf{\Pi}_{\mathrm{P}}(u_i) = \int_{V^0} W(\gamma_{ij})dV^0 - \int_{V^0}\bar{q}_i u_i dV^0 - \int_{S_T^0}\bar{T}_i u_i dS^0 \tag{33.25}$$

may be written

Potential energy $\mathbf{\Pi}_{\mathrm{P}}(\boldsymbol{u})$

$$\mathbf{\Pi}_{\mathrm{P}}(\boldsymbol{u}) = \tfrac{1}{2}\mathbf{H}(\boldsymbol{\varepsilon})\cdot\boldsymbol{\varepsilon} - \overline{\boldsymbol{T}}\cdot\boldsymbol{u} \tag{33.26}$$

Note that $\boldsymbol{\varepsilon}$ depends on $\boldsymbol{u}$, which justifies the notation $\mathbf{\Pi}_{\mathrm{P}}(\boldsymbol{u})$.

The following reciprocity relation is a consequence of the nature of the constitutive "law," namely that it is linearly elastic and, therefore, the stresses depend directly on the strains and not on the loading history

Reciprocity relation

$$\mathbf{H}(\boldsymbol{\varepsilon}_a)\boldsymbol{\varepsilon}_b = \mathbf{H}(\boldsymbol{\varepsilon}_b)\boldsymbol{\varepsilon}_a \tag{33.27}$$

where subscripts $_a$ and $_b$ indicate two different fields. Note that the statement is valid without a dot, i.e. it is valid for all points of the structure, individually.

A result of this is that

Integrated reciprocity relation

$$\mathbf{H}(\varepsilon_a) \cdot \varepsilon_b = \mathbf{H}(\varepsilon_b) \cdot \varepsilon_a \tag{33.28}$$

It is worth noticing that in (33.27) and (33.28) one of the fields may be the variation $\delta\varepsilon$.

33.4.1 Stationarity of the Potential Energy for Linear Elasticity

By application of (33.28) and of the operator rules given above in Section 33.3 the variation of $\mathbf{\Pi}_{\mathrm{P}}$ is

Variation of Potential Energy

$$\delta\mathbf{\Pi}_{\mathrm{P}}(\boldsymbol{u}) = \mathbf{H}(\varepsilon) \cdot \delta\varepsilon - \overline{\boldsymbol{T}} \cdot \delta\boldsymbol{u} \tag{33.29}$$

where the conditions on the variations are given below.

By comparison of (33.29) and (33.14) and using (33.24) we may realize that the first variation $\delta\mathbf{\Pi}_{\mathrm{P}}$ of the potential energy $\mathbf{\Pi}_{\mathrm{P}}$ vanishes for the exact solution in that

Principle of Stationary Potential Energy

$$\delta\mathbf{\Pi}_{\mathrm{P}}(\boldsymbol{u}) = 0 \tag{33.30}$$

because (33.30) in connection with (33.29) shows that the requirement $\delta\mathbf{\Pi}_{\mathrm{P}} = 0$ is equivalent to the principle of virtual displacements (33.14), but with the stresses expressed in terms of the constitutive relation (33.24). Realizing this, we may conclude that the variations must be kinematically admissible. Thus, the variations must be "sufficiently" smooth and satisfy the homogeneous kinematic boundary conditions.

Variations must be kinematically admissible

We shall refer to (33.30) as an expression of the *Principle of Stationary Potential Energy*.

33.4.2 Minimum of the Potential Energy for Linear Elasticity and Kinematic Linearity

For linear elasticity and kinematic linearity the potential energy is not only *stationary* at the exact solution, it is also a *minimum*, which we prove below. Let the exact solution be denoted $\boldsymbol{u}$ and an approximate solution (which satisfies the kinematic conditions) be $\tilde{\boldsymbol{u}} = \boldsymbol{u} + \epsilon\delta\boldsymbol{u}$, where ϵ is the amplitude of the difference between $\tilde{\boldsymbol{u}}$ and $\boldsymbol{u}$. Note that $\delta\boldsymbol{u}$ satisfies the *homogeneous* boundary conditions.

Assuming kinematic linearity

Kinematically linear strain

$$\varepsilon = \mathrm{l}_1(\boldsymbol{u}) \text{ and } \tilde{\varepsilon} = \mathrm{l}_1(\boldsymbol{u}) + \epsilon\mathrm{l}_1(\delta\boldsymbol{u}) \tag{33.31}$$

the potential energy of (33.26) for the exact field $\boldsymbol{u}$ becomes

$\mathbf{\Pi}_{\mathrm{P}}(\boldsymbol{u})$ for kinematic linearity

$$\mathbf{\Pi}_{\mathrm{P}}(\boldsymbol{u}) = \tfrac{1}{2}\mathbf{H}(\mathrm{l}_1(\boldsymbol{u})) \cdot \mathrm{l}_1(\boldsymbol{u}) - \overline{\boldsymbol{T}} \cdot \boldsymbol{u} \tag{33.32}$$

and for the approximate field $\tilde{\boldsymbol{u}}$ the expression for the potential energy looks

$\Pi_P(\tilde{u})$ for kinematic linearity

just as simple

$$\Pi_P(\tilde{\boldsymbol{u}}) = \tfrac{1}{2}\mathbf{H}(\mathbf{l}_1(\tilde{\boldsymbol{u}})) \cdot \mathbf{l}_1(\tilde{\boldsymbol{u}}) - \overline{\boldsymbol{T}} \cdot \tilde{\boldsymbol{u}} \tag{33.33}$$

but, when we note that $\mathbf{l}_1$ is a linear operator, see also (33.31b), we may see that it is more complicated

$$\begin{aligned}\Pi_P(\tilde{\boldsymbol{u}}) &= \tfrac{1}{2}\mathbf{H}(\mathbf{l}_1(\boldsymbol{u}+\epsilon\delta\boldsymbol{u})) \cdot \mathbf{l}_1(\boldsymbol{u}+\epsilon\delta\boldsymbol{u}) - \overline{\boldsymbol{T}} \cdot (\boldsymbol{u}+\epsilon\delta\boldsymbol{u}) \\ &= \tfrac{1}{2}\mathbf{H}(\mathbf{l}_1(\boldsymbol{u})) \cdot \mathbf{l}_1(\boldsymbol{u}) - \overline{\boldsymbol{T}} \cdot \boldsymbol{u} \\ &\quad + \epsilon\big(\mathbf{H}(\mathbf{l}_1(\boldsymbol{u})) \cdot \mathbf{l}_1(\delta\boldsymbol{u}) - \overline{\boldsymbol{T}} \cdot \delta\boldsymbol{u}\big) \\ &\quad + \tfrac{1}{2}\epsilon^2\mathbf{H}(\mathbf{l}_1(\delta\boldsymbol{u})) \cdot \mathbf{l}_1(\delta\boldsymbol{u})\end{aligned} \tag{33.34}$$

The term of order ϵ is the *Principle of Virtual Displacements* with $\boldsymbol{\sigma} = \mathbf{H}(\mathbf{l}_1(\boldsymbol{u}))$ and therefore vanishes. For $\epsilon\delta\boldsymbol{u} \not\equiv \mathbf{0}$ the term of order ϵ^2 is greater than zero because the strain energy is positive definite and is quadratic in $\mathbf{l}_1(\delta\boldsymbol{u})$. Since the term of order ϵ^0 is the potential energy for the exact field we may conclude that

Π_P is minimum for the exact solution

$$\Pi_P(\tilde{\boldsymbol{u}}) > \Pi_P(\boldsymbol{u}) \,, \tilde{\boldsymbol{u}} \not\equiv \boldsymbol{u} \tag{33.35}$$

which is a much stronger statement than that of stationarity, but we can get an even more useful result, see the next section.

33.4.3 Minimization of the Potential Energy for Linear Elasticity and Kinematic Linearity Results in Too Stiff Structures

$\Pi_P(\tilde{u}) \geq \Pi_P(u)$ is useful, but we want more

Below, as in Section 33.4.2, we assume kinematic and constitutive linearity. In most cases, the fact that the potential energy for an assumed displacement field is greater than for the exact one does not provide us with very valuable information, except that it can be used to choose the best of a set of approximate solutions. Rather, we would like to know the quality of the displacements or the stresses, and in particular their maximum values. As regards the displacements we show that use of an assumed displacement field, which is different from the exact one, in the principle of stationarity of the potential energy results in a behavior which is too stiff.[33.11] The stresses, however, present more severe difficulties, and we shall not address the question of their properties except to mention that the stress field usually follows from the displacement field by a process which entails differentiation, and, therefore, the stresses usually are of poorer quality than the displacements.

From (33.35) it follows that

$$\Pi_P(\boldsymbol{u}) = \tfrac{1}{2}\mathbf{H}(\boldsymbol{\varepsilon}) \cdot \boldsymbol{\varepsilon} - \overline{\boldsymbol{T}} \cdot \boldsymbol{u} < \Pi_P(\tilde{\boldsymbol{u}}) = \tfrac{1}{2}\mathbf{H}(\tilde{\boldsymbol{\varepsilon}}) \cdot \tilde{\boldsymbol{\varepsilon}} - \overline{\boldsymbol{T}} \cdot \tilde{\boldsymbol{u}} \tag{33.36}$$

where the meaning of $\tilde{\boldsymbol{\varepsilon}}$ should be obvious. According to (33.30), we know

[33.11] As you will see, the proof is quite simple, but—much to my surprise–many books on finite element theory do not touch it.

that for the exact displacement

$$\delta\mathbf{\Pi}_{\mathrm{P}}(\boldsymbol{u}) = 0 \;\Rightarrow\; \mathbf{H}(\varepsilon)\cdot\delta\varepsilon - \overline{\boldsymbol{T}}\cdot\delta\boldsymbol{u} = 0 \tag{33.37}$$

and similarly we demand that

$$\delta\mathbf{\Pi}_{\mathrm{P}}(\tilde{\boldsymbol{u}}) = 0 \;\Rightarrow\; \mathbf{H}(\tilde{\varepsilon})\cdot\delta\tilde{\varepsilon} - \overline{\boldsymbol{T}}\cdot\delta\tilde{\boldsymbol{u}} = 0 \tag{33.38}$$

Provided that all prescribed displacements are zero, i.e. provided that all kinematic boundary conditions are homogeneous, we may introduce $\delta\boldsymbol{u} = \boldsymbol{u}$ in (33.37) and $\delta\tilde{\boldsymbol{u}} = \tilde{\boldsymbol{u}}$ in (33.38) to get

$$\mathbf{H}(\varepsilon)\cdot\varepsilon = \overline{\boldsymbol{T}}\cdot\boldsymbol{u} \;\text{ and }\; \mathbf{H}(\tilde{\varepsilon})\cdot\tilde{\varepsilon} = \overline{\boldsymbol{T}}\cdot\tilde{\boldsymbol{u}} \tag{33.39}$$

respectively. Note that in (33.39a) the field $\boldsymbol{u}$ and in (33.39b) the field $\tilde{\boldsymbol{u}}$ are the ones determined by (33.37) and (33.38), respectively. Thus, both fields are fixed in the present context and it would therefore not make sense to vary with respect to these fields in (33.39a) and (33.39b).[33.12] Insert (33.39a) and (33.39b) in (33.36)

min($\mathbf{\Pi}_{\mathrm{P}}(\tilde{\boldsymbol{u}})$) for ($\tilde{\boldsymbol{u}} \neq \boldsymbol{u}$) $\Rightarrow$ too stiff structure

$$-\tfrac{1}{2}\overline{\boldsymbol{T}}\cdot\boldsymbol{u} < -\tfrac{1}{2}\overline{\boldsymbol{T}}\cdot\tilde{\boldsymbol{u}} \;\Rightarrow\; \overline{\boldsymbol{T}}\cdot\boldsymbol{u} > \overline{\boldsymbol{T}}\cdot\tilde{\boldsymbol{u}} \tag{33.40}$$

which means that the work done by the loading $\overline{\boldsymbol{T}}$ is greater for the exact solution. This statement is very useful because it tells us that among a set of approximate solutions we should always choose the one that entails the largest work of the loading.

33.4.3.1 Single Point Force

When the loading consists of only a single point force, say $\bar{P}$ at point P we can derive a very strong statement regarding its work conjugate displacement u_P. In this case, (33.40) yields

$$\bar{P}u_P > \bar{P}\tilde{u}_P \tag{33.41}$$

and, since $\bar{P}$ is the same in both cases, we may conclude that

Displacement of load is largest for exact solution

$$u_P > \tilde{u}_P \tag{33.42}$$

implying that the value of the "characteristic" displacement u_P is larger for the exact solution than for any other. Therefore, an assumed displacement field $\tilde{\boldsymbol{u}} \neq \boldsymbol{u}$, which is used in the principle of stationarity of the potential energy results in predicting too stiff a structural behavior. The result (33.42) is even more convenient than (33.40) since, for the case of a single point force $\bar{P}$, among approximate solutions we should choose the solution for which the displacement $\tilde{u}_P$ of the force is the largest.[33.13]

[33.12] We may also see this from the fact that, otherwise, we would get results that disagree with (33.37) and (33.38).

[33.13] In the above derivations it is tacitly assumed that the characteristic displacement is finite. This is not the case for a plate loaded in-plane with a point force, see e.g. (Muskhelishvili 1963).

33.5 Complementary Energy for Linear Elasticity

As in the case of the potential energy it proves convenient to formulate the complementary energy in terms of the *Budiansky-Hutchinson Notation*, see Section 33.4. Below, as in Section 33.4.2, we assume kinematic and constitutive linearity. Noting (4.134) we may easily see that the equivalent of (33.26) is

Complementary energy $\mathbf{\Pi}_\mathrm{C}$

$$\mathbf{\Pi}_\mathrm{C}(\boldsymbol{\sigma}) = \tfrac{1}{2}\mathbf{C}(\boldsymbol{\sigma}) \cdot \boldsymbol{\sigma} - \boldsymbol{T} \cdot \bar{\boldsymbol{u}} \tag{33.43}$$

where I emphasize that $\mathbf{\Pi}_\mathrm{C}(\boldsymbol{\sigma})$ is valid only for kinematic linearity, where $\bar{\boldsymbol{u}}$ denotes the prescribed displacements, $\boldsymbol{T}$ are the reactions to $\bar{\boldsymbol{u}}$, and the linear constitutive operator $\mathbf{C}$ is the inverse of $\mathbf{H}$ and signifies *Hooke's "Law,"* and, formally, we may write

Linear constitutive relation $\mathbf{C} = \mathbf{H}^{-1}$

$$\mathbf{C} = \mathbf{H}^{-1} \;\Rightarrow\; \boldsymbol{\varepsilon} = \mathbf{C}(\boldsymbol{\sigma}) \tag{33.44}$$

see (4.129).

Variation of $\mathbf{\Pi}_\mathrm{C}(\boldsymbol{\sigma})$ provides

Variation of complementary energy $\mathbf{\Pi}_\mathrm{C}$

$$\delta\mathbf{\Pi}_\mathrm{C}(\boldsymbol{\sigma}) = \mathbf{C}(\boldsymbol{\sigma}) \cdot \delta\boldsymbol{\sigma} - \delta\boldsymbol{T} \cdot \bar{\boldsymbol{u}} \tag{33.45}$$

and requiring that $\delta\mathbf{\Pi}_\mathrm{C}(\boldsymbol{\sigma})$ vanishes for all statically admissible stress variations, i.e. stress variations satisfying the equilibrium equations, results in

$$\mathbf{C}(\boldsymbol{\sigma}) \cdot \delta\boldsymbol{\sigma} = \delta\boldsymbol{T} \cdot \bar{\boldsymbol{u}} \tag{33.46}$$

The *Principle of Virtual Forces*, see (4.124), is expressed as

Principle of Virtual Forces

$$\boldsymbol{\varepsilon} \cdot \delta\boldsymbol{\sigma} = \delta\boldsymbol{T} \cdot \bar{\boldsymbol{u}} \tag{33.47}$$

and when the constitutive relation (33.44) is introduced, we recover (33.46). This is comforting because it means that requiring $\delta\mathbf{\Pi}_\mathrm{P}(\boldsymbol{\sigma}) = 0$ is equivalent to to the principle of virtual forces with the strains expressed in terms of the stresses.

33.5.1 Minimum Complementary Energy

Let $\boldsymbol{\sigma}$ denote the exact solution to some problem and let $\tilde{\boldsymbol{\sigma}} = \boldsymbol{\sigma} + \epsilon\delta\boldsymbol{\sigma}$, where, as usual $|\epsilon \ll 1|$, be another field, an approximate solution, which satisfies the equilibrium equations. Then, the complementary energy of the approximate solution is

Complementary energy $\mathbf{\Pi}_\mathrm{C}(\tilde{\boldsymbol{\sigma}})$ is minimum for the exact solution

$$\begin{aligned}\mathbf{\Pi}_\mathrm{C}(\tilde{\boldsymbol{\sigma}}) &= \tfrac{1}{2}\mathbf{C}(\boldsymbol{\sigma} + \epsilon\delta\boldsymbol{\sigma}) \cdot (\boldsymbol{\sigma} + \epsilon\delta\boldsymbol{\sigma}) - (\boldsymbol{T} + \delta\boldsymbol{T}) \cdot \bar{\boldsymbol{u}} \\ &= \tfrac{1}{2}\mathbf{C}(\boldsymbol{\sigma}) \cdot \boldsymbol{\sigma} + \epsilon\mathbf{C}(\boldsymbol{\sigma}) \cdot \delta\boldsymbol{\sigma} + \tfrac{1}{2}\epsilon^2\mathbf{C}(\delta\boldsymbol{\sigma}) \cdot \delta\boldsymbol{\sigma} \\ &\qquad -\boldsymbol{T} \cdot \bar{\boldsymbol{u}} - \epsilon\delta\boldsymbol{T} \cdot \bar{\boldsymbol{u}} \\ &= \mathbf{\Pi}_\mathrm{C}(\boldsymbol{\sigma}) + \tfrac{1}{2}\epsilon^2\mathbf{C}(\delta\boldsymbol{\sigma}) \cdot \delta\boldsymbol{\sigma} \\ &\geq \mathbf{\Pi}_\mathrm{C}(\boldsymbol{\sigma})\end{aligned} \tag{33.48}$$

The terms of order ε^1 vanish because of (33.46), and thus the *Complementary Energy* attains a minimum for the correct solution.

33.5.2 Minimization of the Complementary Energy for Linear Elasticity and Kinematic Linearity Results in Too Flexible Structures

As a consequence of (33.48)

$$\boldsymbol{\Pi}_{\mathrm{C}}(\boldsymbol{\sigma}) = \tfrac{1}{2}\mathbf{C}(\boldsymbol{\sigma}) \cdot \boldsymbol{\sigma} - \boldsymbol{T} \cdot \bar{\boldsymbol{u}} \le \boldsymbol{\Pi}_{\mathrm{C}}(\tilde{\boldsymbol{\sigma}}) = \tfrac{1}{2}\mathbf{C}(\tilde{\boldsymbol{\sigma}}) \cdot \tilde{\boldsymbol{\sigma}} - \widetilde{\boldsymbol{T}} \cdot \bar{\boldsymbol{u}} \quad (33.49)$$

Here, $\tilde{\boldsymbol{\sigma}}$ and $\widetilde{\boldsymbol{T}}$ denote fields that are associated with an approximation which satisfies all equilibrium conditions. Variation of the complementary energy for the exact and for the approximate field provides

$$\delta\boldsymbol{\Pi}_{\mathrm{C}}(\boldsymbol{\sigma}) = 0 \;\Rightarrow\; \mathbf{C}(\boldsymbol{\sigma}) \cdot \delta\boldsymbol{\sigma} - \delta\boldsymbol{T} \cdot \bar{\boldsymbol{u}} = 0 \quad (33.50)$$

and

$$\delta\boldsymbol{\Pi}_{\mathrm{C}}(\tilde{\boldsymbol{\sigma}}) = 0 \;\Rightarrow\; \mathbf{C}(\tilde{\boldsymbol{\sigma}}) \cdot \delta\tilde{\boldsymbol{\sigma}} - \delta\widetilde{\boldsymbol{T}} \cdot \bar{\boldsymbol{u}} = 0 \quad (33.51)$$

respectively. Since the stress fields $(\boldsymbol{\sigma}, \boldsymbol{T})$ and $(\tilde{\boldsymbol{\sigma}}, \widetilde{\boldsymbol{T}})$ satisfy the same conditions as $(\delta\boldsymbol{\sigma}, \delta\boldsymbol{T})$ and $(\delta\tilde{\boldsymbol{\sigma}}, \delta\widetilde{\boldsymbol{T}})$, respectively, we may get

$$\mathbf{C}(\boldsymbol{\sigma}) \cdot \boldsymbol{\sigma} = \boldsymbol{T} \cdot \bar{\boldsymbol{u}} \text{ and } \mathbf{C}(\tilde{\boldsymbol{\sigma}}) \cdot \tilde{\boldsymbol{\sigma}} = \widetilde{\boldsymbol{T}} \cdot \bar{\boldsymbol{u}} \quad (33.52)$$

and thus

$$\boldsymbol{\Pi}_{\mathrm{C}}(\boldsymbol{\sigma}) = -\tfrac{1}{2}\boldsymbol{T} \cdot \bar{\boldsymbol{u}} \text{ and } \boldsymbol{\Pi}_{\mathrm{C}}(\tilde{\boldsymbol{\sigma}}) = -\tfrac{1}{2}\widetilde{\boldsymbol{T}} \cdot \bar{\boldsymbol{u}} \quad (33.53)$$

with the comment that in (33.53a) and (33.53b) variations are not permitted because the stress fields occurring here are those given by (33.50) and (33.51), and therefore are fixed fields.

Because of the result (33.48) we get[33.14]

$$-\tfrac{1}{2}\widetilde{\boldsymbol{T}} \cdot \bar{\boldsymbol{u}} \ge -\tfrac{1}{2}\boldsymbol{T} \cdot \bar{\boldsymbol{u}} \;\Rightarrow\; \widetilde{\boldsymbol{T}} \cdot \bar{\boldsymbol{u}} \le \boldsymbol{T} \cdot \bar{\boldsymbol{u}} \quad (33.54)$$

min($\boldsymbol{\Pi}_{\mathrm{C}}(\tilde{\boldsymbol{\sigma}})$) for ($\tilde{\boldsymbol{\sigma}} \neq \boldsymbol{\sigma}$) $\Rightarrow$ too flexible structure

This result is in itself helpful because it says that the work done by the reactions of the approximate field is less than (or equal to, if we have guessed right) that of the exact field, i.e. the structure is predicted to be more flexible than the real one.

[33.14] I hope that you can see that this derivation looks very much like the one which we performed in Section 33.4.3. This is one of the many examples of the duality between statics and kinematics.

33.5.2.1 Single Point Force

Now, the question is whether the structure responds by a too flexible behavior when it is subjected to a load instead of prescribed displacements. Let the prescribed displacement be confined to a particular point and denote it $\bar{u}$, then for the reactions, which we shall denote P and $\widetilde{P}$, the relation is[33.15]

$$\widetilde{P} \leq P \tag{33.55}$$

and thus the value of the reaction to a prescribed displacement is underestimated if the approximate stress field differs from the exact one, meaning that the structure is predicted to be too flexible. Based on this, it seems a reasonable conjecture that the approximate solution, based on the minimum of the complementary energy, also for an applied *load* predicts a larger displacement than the correct one. The effect of distributed loads may be found by use of integration of point loads.

Ex 33-1 A Clamped-Clamped Beam

As an example of use of the *Principle of Minimum Complementary Energy* consider the clamped-clamped beam shown in Fig. Ex. 33-1.1. The elastic properties, given by the bending stiffness EI, and the load $\bar{p}$ are independent of x.

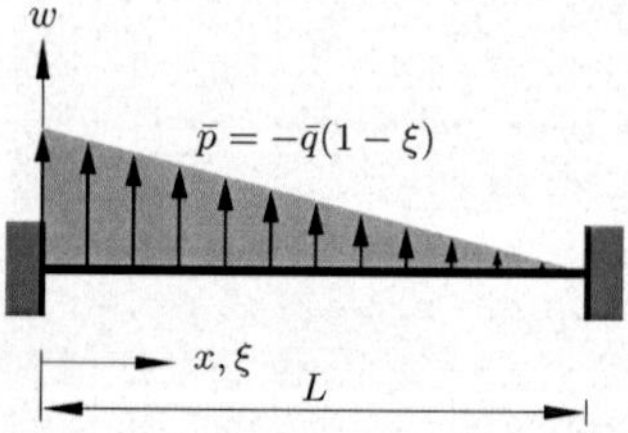

Fig. Ex. 33-1.1: A clamped-clamped Beam.

We shall only deal with the transverse displacements of the beam, and, therefore, the complementary energy is

Complementary energy $\Pi_C(M)$

$$\Pi_C(M) = \tfrac{1}{2}\int_0^L \frac{1}{EI} M^2 dx \tag{Ex. 33-1.1}$$

where the bending moment M must satisfy the auxiliary condition[33.16]

Load $\bar{p} = \bar{q}(1-\xi)$

$$M'' = \bar{p} = -\bar{q}(1-\xi)\ ,\ \bar{q} = \text{const.} > 0\ ,\ \xi \equiv x/L \tag{Ex. 33-1.2}$$

[33.15] The remarks in the footnote p. 589 regarding point forces are still valid but must be understood in the sense that here the reaction to a prescribed displacement in a plate, which is subjected to in-plane loading, is zero.

[33.16] Note that because $\bar{q}$ is negative the load actually points downwards.

We may get an admissible moment field $\widetilde{M}$ by use of the field for the equivalent simply supported beam, see later, and add an arbitrary *linear* field, but it is even easier instead of using the beam result just to integrate twice in (*Ex. 33-1.2*) to get

$$\widetilde{M}(\xi) = \bar{r}(\xi) + v_\beta(1) + v_\beta(2)\xi \qquad \text{(Ex. 33-1.3)}$$

Assumed bending moment $\widetilde{M}(\xi)$

where $v_\beta(j)$ are stress field parameters and

$$\bar{r}(\xi) = -\tfrac{1}{6}\bar{q}L^2\xi^2(3-\xi) \qquad \text{(Ex. 33-1.4)}$$

Load term $\bar{r}(\xi)$

By this choice the only pertinent equilibrium equation, namely (*Ex. 33-1.2*) is satisfied—and there are no other possible moment fields which satisfy this equation. Changing the notation slightly (*Ex. 33-1.3*) may be written

$$\widetilde{M}(\xi) = [N_\beta]\{v_\beta\} + \bar{r} \qquad \text{(Ex. 33-1.5)}$$

Assumed bending moment $\widetilde{M}(\xi)$

where

$$\begin{aligned} [N_\beta] &= [1\ ;\ \xi] \\ \{v_\beta\}^T &= [v_\beta(1)\ ;\ v_\beta(2)] \end{aligned} \qquad \text{(Ex. 33-1.6)}$$

Stress matrix $[N_\beta]$
Stress field parameters $\{v_\beta\}$

Introduce the constitutive matrix $[C]$, which is the material flexibility matrix, by

$$[C] = \left[\frac{1}{EI}\right] \qquad \text{(Ex. 33-1.7)}$$

Constitutive matrix $[C]$

Rewrite (*Ex. 33-1.1*)

$$\mathbf{\Pi}_C(M) = \tfrac{1}{2}L\int_0^1 ([N_\beta]\{v_\beta\}+\bar{r})^T[C]([N_\beta]\{v_\beta\}+\bar{r})d\xi \qquad \text{(Ex. 33-1.8)}$$

Complementary energy $\mathbf{\Pi}_C(M)$

or

$$\mathbf{\Pi}_C(M) = \tfrac{1}{2}\{v_\beta\}^T[k_{\beta\beta}]\{v_\beta\} + \{v_\beta\}^T\{\bar{r}_\beta\} + \text{const.} \qquad \text{(Ex. 33-1.9)}$$

Complementary energy $\mathbf{\Pi}_C(M)$

where the constant term is independent of $\{v_\beta\}$ and, therefore, contributes nothing to the variation of $\mathbf{\Pi}_C$, and

$$\begin{aligned} [k_{\beta\beta}] &\equiv L\int_0^1 [N_\beta]^T[C][N_\beta]d\xi \\ \{\bar{r}_\beta\} &\equiv L\int_0^1 [N_\beta]^T[C]\bar{r}d\xi \end{aligned} \qquad \text{(Ex. 33-1.10)}$$

Coefficient matrix $[k_{\beta\beta}]$
Right-hand side vector $\{\bar{r}_\beta\}$

where $[k_{\beta\beta}]$ is a flexibility matrix and $\{\bar{r}_\beta\}$ is the right-hand side vector of the ensuing matrix equation.

Variation of $\mathbf{\Pi}_C(M)$

Variation of $\mathbf{\Pi}_C$ provides

$$\delta\mathbf{\Pi}_C(M) = \delta\{v_\beta\}^T\Big([k_{\beta\beta}]\{v_\beta\} + \{\bar{r}_\beta\}\Big) \tag{Ex. 33-1.11}$$

System of equations

When we require that the variation of $\mathbf{\Pi}_C$ vanishes

$$[k_{\beta\beta}]\{v_\beta\} = -\{\bar{r}_\beta\} \tag{Ex. 33-1.12}$$

With $\{v_\beta\}$ in hand we may determine the bending moments from (*Ex. 33-1.5*).

Coefficient matrix $[k_{\beta\beta}]$ Right-hand side vector $\{\bar{r}_\beta\}$

In the present case

$$[k_{\beta\beta}] = \frac{L}{EI}\begin{bmatrix} 1 & \frac{1}{2} \\ \frac{1}{2} & \frac{1}{3} \end{bmatrix} \quad \text{and} \quad \{\bar{r}_\beta\} = -\frac{\bar{q}L^3}{8EI}\begin{bmatrix} 1 \\ \frac{11}{15} \end{bmatrix} \tag{Ex. 33-1.13}$$

Solution vector $\{v_\beta\}$

with the solution

$$\{v_\beta\} = -\frac{\bar{q}L^2}{20}\begin{bmatrix} 1 \\ -7 \end{bmatrix} \tag{Ex. 33-1.14}$$

Resulting bending moment $\widetilde{M}(\xi)$

and thus

$$\widetilde{M} = -\tfrac{1}{60}\bar{q}L^2(3 - 21\xi + 30\xi^2 - 10\xi^3) \tag{Ex. 33-1.15}$$

Supporting moments

The support moments are

$$\widetilde{M}(0) = -\tfrac{1}{20} \text{ and } \widetilde{M}(1) = -\tfrac{1}{30}\bar{q}L^2 \tag{Ex. 33-1.16}$$

where it makes sense that the absolute value of the support moment is larger at the end where the load intensity is the higher.

The maximum value $\widetilde{M}_{\max}$ of the bending moment is

Maximum bending moment $\widetilde{M}_{\max}$

$$\begin{aligned} \widetilde{M}_{\max} &= -\frac{3\sqrt{30}-30}{300}\bar{q}L^2 = -0.0214389qL^2 \\ &\text{for } \xi = 1 - \frac{\sqrt{30}}{10} = 0.4522774 \end{aligned} \tag{Ex. 33-1.17}$$

where it is reasonable to expect the maximum bending moment to occur closest to the highest load intensity.

Ex 33-1.1 Is Our Solution the Exact One?

We may see that $\widetilde{M}$ is the exact solution

You may check against the exact solution and see that indeed $\widetilde{M} = M$ in this case. When you think about it, this is not surprising because we started out with a field comprising the only possible components of the exact field, namely the first term of (*Ex. 33-1.3*), which is a particular solution of the governing differential equation (*Ex. 33-1.3*), and the only two terms, the second and third term of (*Ex. 33-1.3*), which together constitute the full solution to the corresponding homogeneous differential equation. At a first glance it seems remarkable that we may find a solution to a statically indeterminate problem without invoking the kinematic boundary conditions. However, since (*Ex. 33-1.1*) does not contain any boundary terms, cf. (33.43), this means that the boundary displacements are all prescribed to be zero. But, because our solution is formulated in terms of the bending moment, which may be expressed by the constitutive relation and the strain-displacement relation, only the second derivative of the transverse displacement is given. Thus, the transverse displacement is determined to within a linear function, and in order to find the displacement we must integrate twice and enforce the kinematic boundary condition explicitly. This will not be done here.

We need to do more to get the displacements

Ex 33-1.2 Another, Better(?) Assumption

The choice (*Ex. 33-1.3*) is, of course, not the only possible one. We may, for instance, choose

A choice which makes interpretation of $\{v_\beta\}$ transparent

$$\begin{aligned}\widetilde{M}(\xi) &= \tfrac{1}{6}\bar{q}L^2\xi(1-\xi)(2-\xi) + [N_\beta]\{v_\beta\} \\ \text{with } & [N_\beta] = [(1-\xi)\ ;\ \xi]\end{aligned} \qquad \text{(Ex. 33-1.18)}$$

The first term on the right-hand side of the expression for $\widetilde{M}$ is the solution for the equivalent simply supported beam, and the second term has the convenient property that here we may then identify $v_\beta(1)$ and $v_\beta(2)$ as the bending moments at the supports.

33.6 Auxiliary Conditions

It is worthwhile noticing that there are a number of conditions that must be satisfied by the fields in the principle of minimum of the potential energy and in the principle of minimum of the complementary energy. Here, we shall focus on the potential energy and, as far as the complementary energy is concerned, refer to Chapter 27 where the issue of auxiliary conditions on the complementary energy is treated in some detail.

For the potential energy as given by (33.26) the most fundamental of these conditions, i.e. those which must always be satisfied, are *kinematic* conditions

1. The displacements $\boldsymbol{u}$ must be sufficiently smooth.
2. The displacements $\boldsymbol{u}$ must satisfy the *kinematic* boundary conditions.
3. The strains $\boldsymbol{\varepsilon}$ must be derived from the displacements $\boldsymbol{u}$ according to (33.3).
4. The displacement variations $\delta\boldsymbol{u}$ must be sufficiently smooth.
5. The displacement variations $\delta\boldsymbol{u}$ must satisfy the *homogeneous kinematic* boundary conditions, i.e. $\delta\boldsymbol{u}$ must vanish on the kinematic boundary.
6. The strain variations $\delta\boldsymbol{\varepsilon}$ must be derived from the displacements $\boldsymbol{u}$ and displacement variations $\delta\boldsymbol{u}$ according to (33.18).[33.17]

In addition to these conditions, there may be others, which are specific to the particular problem at hand, e.g. inextensibility of a beam or a membrane; incompressibility of a solid; a condition, which expresses a coupling between a number of displacement components; etc.

[33.17] Obviously, when we utilize the alternative formulation of the *Principle of Virtual Displacements* (33.19) we satisfy (33.18) right from the outset.

Auxiliary conditions = side conditions

Such conditions are called *Auxiliary Conditions* or *Side Conditions.* Sometimes it proves to be impossible to satisfy a number of these conditions, and then the method of *Lagrange Multipliers*, see Section 33.7, comes in handy.

33.7 Lagrange Multipliers

Lagrange multiplier

It is not always possible or feasible in advance to satisfy all auxiliary conditions that apply to the *Principle of Virtual Displacements* or the *Principle of Stationarity Potential Energy*, see Sections 33.2 and 33.4. If this is the case, the method of *Lagrange Multipliers*, see e.g. (Arfken & Weber 1995), proves to be valuable. This is particularly true in connection with approximate solutions, but here we concentrate on the derivation of the method itself.

33.7.1 Principle of Virtual Displacements

Suppose that we must fulfill an auxiliary condition on the *Principle of Virtual Displacements*, and that the condition may be written

$$\boldsymbol{\Psi}(\boldsymbol{u}) = 0 \tag{33.56}$$

where it is indicated explicitly that the condition depends on the displacements.[33.18] Then, we may augment the *Principle of Virtual Displacements* (33.14) with a term $\delta(\boldsymbol{\Psi}(\boldsymbol{u}) \cdot \boldsymbol{\eta})$ and write

Lagrange multiplier (field) $\boldsymbol{\eta}$

$$\boldsymbol{\sigma} \cdot \delta\boldsymbol{\varepsilon} - \overline{\boldsymbol{T}} \cdot \delta\boldsymbol{u} + \delta(\boldsymbol{\Psi}(\boldsymbol{u}) \cdot \boldsymbol{\eta}) = 0 \tag{33.57}$$

where $\boldsymbol{\eta}$ denotes the *Lagrange Multiplier* (Field).[33.19] The last term of (33.57) is, of course, the most interesting in the current connection. Written out (33.57) becomes

$$\boldsymbol{\sigma} \cdot \delta\boldsymbol{\varepsilon} - \overline{\boldsymbol{T}} \cdot \delta\boldsymbol{u} + \delta\boldsymbol{\Psi}(\boldsymbol{u}) \cdot \boldsymbol{\eta} + \boldsymbol{\Psi}(\boldsymbol{u}) \cdot \delta\boldsymbol{\eta} = 0 \tag{33.58}$$

Now, because there are no conditions on $\delta\boldsymbol{\eta}$, except maybe continuity conditions, (33.58) implies that

$$\boldsymbol{\Psi}(\boldsymbol{u}) = 0 \tag{33.59}$$

as required.

[33.18] You may think of this condition as one of incompressibility or some other kinematic constraint.

[33.19] Since the term Lagrange Multiplier is used in honor of the French mathematician J.L. Lagrange, the symbol $\boldsymbol{\eta}$ is somewhat unnatural and, indeed, in most literature symbols such as $\boldsymbol{\lambda}$ or $\boldsymbol{\Lambda}$ are used. The reason why I have chosen $\boldsymbol{\eta}$ instead is that the Greek letter *lambda*, frequently λ, in literature associated with structural stability is the most commonly used symbol for load parameter. In the present context this justification is, admittedly, not a very valid one, but all authors have their quirks, and I hope that you—and Lagrange—will forgive me.

Further, (33.59) means that

$$\delta\boldsymbol{\Psi}(\boldsymbol{u}) = 0 \tag{33.60}$$

and therefore (33.14) is recovered. Just from reading the above derivation the idea of applying *Lagrange Multipliers* may seem like an awkward way of writing the original equations, but in some cases—in particular in connection with approximate solutions—the method proves to be extremely efficient, see for instance Example Ex 32-5.2.

Are Lagrange Multipliers at all useful?

33.8 Interpretation of the Budiansky-Hutchinson Notation for Selected Examples

The previous derivations may be rather straightforward to follow, but my experience is that application of the formalism may be less obvious. Therefore, we interpret the Budiansky-Hutchinson Notation for some of the examples given in Part II and in earlier sections of Part VI and begin with the latter.

Interpretation of Budiansky-Hutchinson Notation

33.8.1 Interpretations Related to Example Ex 32-2

For the one-degree "structure"

One-degree "structure"

$$\begin{aligned} &\boldsymbol{u} \sim v\,, \quad \boldsymbol{\varepsilon} \sim v\,, \quad \mathbf{H}(\boldsymbol{\varepsilon}) \sim cv \\ &\boldsymbol{\sigma} \cdot \delta\boldsymbol{\varepsilon} = \overline{\boldsymbol{T}} \cdot \delta\boldsymbol{u} \sim F\delta v = \bar{P}\delta v \end{aligned} \tag{33.61}$$

where $F = cv$ is the force in the spring.

33.8.2 Interpretations Related to Example Ex 32-3

For the "structure" with two degrees of freedom

Two-degree "structure"

$$\boldsymbol{u} \sim \{v\} = \begin{bmatrix} v_1 \\ v_2 \end{bmatrix}\,, \quad \boldsymbol{\varepsilon} \sim \{v\} = \begin{bmatrix} v_1 \\ v_2 \end{bmatrix} \tag{33.62}$$

$$\mathbf{H}(\boldsymbol{\varepsilon}) \sim [c]\{v\} \sim \begin{bmatrix} c_1 & 0 \\ 0 & c_2 \end{bmatrix} \begin{bmatrix} v_1 \\ v_2 \end{bmatrix} \tag{33.63}$$

$$\boldsymbol{\sigma} \cdot \delta\boldsymbol{\varepsilon} = \overline{\boldsymbol{T}} \cdot \delta\boldsymbol{u} \sim c \begin{bmatrix} F_1 & F_2 \end{bmatrix} \begin{bmatrix} \delta v_1 \\ \delta v_2 \end{bmatrix} = \bar{P}\delta v \tag{33.64}$$

33.8.3 Interpretations Related to Example Ex 32-5

For the structure with auxiliary conditions

Structure with auxiliary conditions, the Euler column

$$\begin{aligned} &\boldsymbol{u} \sim w(x)\,, \quad \boldsymbol{\varepsilon} \sim \kappa(x) = \varepsilon\,, \quad \mathbf{H}(\boldsymbol{\varepsilon}) \sim EI\kappa(x) \\ &\boldsymbol{\Psi}(\boldsymbol{u}) = 0 \sim (\kappa - w'') = 0\,, \quad \boldsymbol{\eta} \sim \eta \\ &\boldsymbol{\sigma} \cdot \delta\boldsymbol{\varepsilon} = \overline{\boldsymbol{T}} \cdot \delta\boldsymbol{u} \sim \int_0^a (N\delta\varepsilon + M\delta\kappa)dx = -\bar{P}\delta u(L) \end{aligned} \tag{33.65}$$

33.8.4 Interpretations Related to Sections 7.3 and 7.7

Moderately kinematically nonlinear Bernoulli-Euler beam

For the moderately kinematically nonlinear Bernoulli-Euler beams we cite the following relations, which can also be found in Section 7.3

$$\boldsymbol{u} \sim \begin{bmatrix} u \\ w \end{bmatrix} \,,\ \boldsymbol{\varepsilon} \sim \begin{bmatrix} \varepsilon \\ \kappa \end{bmatrix} = \mathbf{l}_1(\boldsymbol{u}) + \tfrac{1}{2}\mathbf{l}_2(\boldsymbol{u}) \tag{33.66}$$

$$\mathbf{l}_1(\boldsymbol{u}) \sim \begin{bmatrix} u' \\ w'' \end{bmatrix} \,,\ \mathbf{l}_2(\boldsymbol{u}) \sim \begin{bmatrix} (w')^2 \\ 0 \end{bmatrix} \,,\ \mathbf{l}_{11}(\boldsymbol{u}^a, \boldsymbol{u}^b) \sim \begin{bmatrix} w'_a w'_b \\ 0 \end{bmatrix} \tag{33.67}$$

$$\boldsymbol{\sigma} \sim \begin{bmatrix} N \\ M \end{bmatrix} \,,\ \mathbf{H}(\varepsilon) \sim \begin{bmatrix} EA & 0 \\ 0 & EI \end{bmatrix} \begin{bmatrix} \varepsilon \\ \kappa \end{bmatrix} \tag{33.68}$$

$$\begin{aligned} & \boldsymbol{\sigma} \cdot \delta\varepsilon = \overline{\boldsymbol{T}} \cdot \delta\boldsymbol{u} \\ \sim\ & \int_a^b (N\delta\varepsilon + M\delta\kappa)dx \\ =\ & \int_a^b (\bar{p}_u \delta u + \bar{p}_w \delta w)dx \\ & -P_u(a)\delta u(a) - P_w(a)\delta w(a) - C(a)\delta w'(a) \\ & +P_u(b)\delta u(b) + P_w(b)\delta w(b) + C(b)\delta w'(b) \end{aligned} \tag{33.69}$$

33.8.5 Interpretations Related to Section 9.1

Kinematically moderately nonlinear plane plates

For the kinematically moderately nonlinear plane plates we cite the following relations, which can also be found in Chapter 9

$$\boldsymbol{u} \sim \begin{bmatrix} u_\alpha \\ w \end{bmatrix} \,,\ \boldsymbol{\varepsilon} \sim \begin{bmatrix} \varepsilon_{\alpha\beta} \\ \kappa_{\alpha\beta} \end{bmatrix} = \mathbf{l}_1(\boldsymbol{u}) + \tfrac{1}{2}\mathbf{l}_2(\boldsymbol{u}) \tag{33.70}$$

$$\mathbf{l}_1(\boldsymbol{u}) \sim \begin{bmatrix} \frac{1}{2}(u_{\alpha,\beta} + u_{\beta,\alpha}) \\ w_{,\alpha\beta} \end{bmatrix} \,,\ \mathbf{l}_2(\boldsymbol{u}) \sim \begin{bmatrix} w_{,\alpha} w_{,\beta} \\ 0 \end{bmatrix} \tag{33.71}$$

$$\mathbf{l}_{11}(\boldsymbol{u}^a, \boldsymbol{u}^b) \sim \begin{bmatrix} \frac{1}{2}\left(w^a_{,\alpha} w^b_{,\beta} + w^b_{,\alpha} w^a_{,\beta} \right) \\ 0 \end{bmatrix} \tag{33.72}$$

$$\mathbf{H}(\varepsilon) \sim \begin{bmatrix} EA & 0 \\ 0 & EI \end{bmatrix} \begin{bmatrix} \varepsilon \\ \kappa \end{bmatrix} \quad \boldsymbol{\sigma} \sim \begin{bmatrix} N_{\alpha\beta} \\ M_{\alpha\beta} \end{bmatrix} \tag{33.73}$$

$$\boldsymbol{\sigma} \cdot \delta\varepsilon = \int_{A^0} (N_{\alpha\beta}\delta\varepsilon_{\alpha\beta} + M_{\alpha\beta}\delta\kappa_{\alpha\beta})\, dA^0 \tag{33.74}$$

Bibliography

Arfken, G. B. & Weber, H. J. (1995). *Mathematical Methods for Physicists, Fourth Edition*, Academic Press, San Diego, New York, Boston, London, Sydney, Tokyo, Toronto.

Bathe, K. J. (1996). *Finite Element Procedures*, Prentice-Hall, Upper Saddle River, New Jersey. 1037 pages.

Belytschko, T., Stolarski, H., W K. Liu, W., Carpenter, N. & Ong, J. S. J. (1985). Stress projection for membrane and shear locking in shell finite elements, *Computer Methods in Applied Mechanics and Engineering* **51**: 221–258.

Brazier, L. G. (1927). On the flexure of thin cylindrical shells and other "thin" sections, *Technical report*, Proc. Royal Society Series A.

Brush, D. O. & Almroth, B. O. (1975). *Buckling of Bars, Plates, and Shells*, McGraw-Hill. xiii + 379 pages.

Budiansky, B. (1966). Dynamic buckling of elastic structures: Criteria and estimates, *Dynamic Stability of Structures*, Pergamon Press, Oxford, New York, pp. 83–106.

Budiansky, B. (1974). Theory of buckling and post-buckling behavior of elastic structures, *Advances in Applied Mechanics* **14**: 1–65.

Budiansky, B. & Hutchinson, J. W. (1964). Dynamic buckling of imperfection-sensitive structures, *in* H. Görtler (ed.), *Proceedings of the 11th International Congress on Applied Mechanics*, Springer–Verlag, Berlin, pp. 636–651.

Budiansky, B. & Hutchinson, J. W. (1979). Buckling: Progress and challenge, *Trends in Solid Mechanics 1979*, Delft University Press, Sijthoff and Noordhoff International Publishers, pp. 93–116.

Byskov, E. (1979). Applicability of an asymptotic expansion for elastic buckling problems with mode interaction, *AIAA Journal* **17**(6): 630–633.

Byskov, E. (1982). Imperfection sensitivity of elastic-plastic truss columns, *AIAA Journal* **20**(2): 263–267.

Byskov, E. (1982–83). Plastic symmetry of Roorda's frame, *J. Struct. Mech.* **10**(3): 311–328.

Byskov, E. (1983). An asymptotic expansion applied to van der Neut's column, *in* J. M. T. Thompson & G. W. Hunt (eds), *Collapse—the buckling of structures in theory and practice*, Cambridge University

E. Byskov, *Elementary Continuum Mechanics for Everyone*,
Solid Mechanics and Its Applications 194, DOI: 10.1007/978-94-007-5766-0,

Press.

Byskov, E. (1987–88). Elastic buckling with infinitely many local modes, *Mechanics of Structures and Machines* **15**(4): 413–435.

Byskov, E. (1989). Smooth postbuckling stresses by a modified finite element method, *International Journal for Numerical Methods in Engineering* **28**: 2877–2888.

Byskov, E. (2004). Mode interaction–an overview, *in* Z. H. Yao, M. W. Yuan & W. X. Zhong (eds), *Computational Mechanics WCCM VI in conjunction with APCOM'04*, Tsinghua University Press & Springer-Verlag, pp. 1–11. WCCMVI_CDROM_1/MS_PDF/M-63.htm.

Byskov, E., Christensen, C. D. & Jørgensen, K. (1996). Elastic postbuckling with nonlinear constraints, *Int. J. Solids. Structures* **33**(17): 2417–2436.

Byskov, E., Damkilde, L. & Jensen, K. J. (1988). Multimode interaction in axially stiffened cylindrical shells, *Mechanics of Structures and Machines* **16**(3): 387–405.

Byskov, E. & Hansen, J. C. (1980). Postbuckling and imperfection sensitivity analysis of axially stiffened cylindrical shells with mode interaction,', *J. Struct. Mech.* **8**(2): 205–224.

Byskov, E. & Hutchinson, J. W. (1977). Mode interaction in axially stiffened cylindrical shells, *AIAA Journal* **15**(7): 941–948.

Cauchy, A. (1829). *Exercises de mathematiques*, Paris.

Christensen, C. D. & Byskov, E. (2008). Advanced postbuckling and imperfection sensitivity of the elastic-plastic shanley–hutchinson model column, *JoMMS* **3**(3): 459–492.

Christensen, D. D. & Byskov, E. (2010). An enhanced asymptotic expansion for the stability of nonlinear elastic structures, *JoMMS* **5**(6): 925–961.

Cook, R. D., Malkus, D. S. & Plesha, M. E. (1989). *Concepts and Applications of Finite Element Analysis*, third edn, John Wiley & Sons, New York.

Cook, R. D., Malkus, D. S. & Plesha, M. E. (2002). *Concepts and Applications of Finite Element Analysis*, fourth edn, John Wiley & Sons, New York. ISBN 0-471-35605-0.

Coulomb, C. A. (1787). *Histoire de l'Academie*, Paris, pp. 229–269.

Crawford, R. F. & Hedgepeth, J. M. (1975). Effects of initial waviness on the strength and design built-up structures, *AIAA Journal* pp. 672–675.

Damkilde, L. & Byskov, E. (1979). Buckling of elastic plates under concentrated loads, *Proceedings of the Seventh Canadian Congress on Applied Mechanics*, Sherbrooke, pp. 241–242.

Duberg, J. E. & Wilder, T. W. (1952). Inelastic column behavior, *Technical Report 1072*, National Advisory Committee for Aeronautics.

Euler, L. (1744). *Methodus inveniendi lineas curvas maximi minimive pro-*

prieta gaudentes, Bousquet, Lausanne and Geneva.

Fitch, J. R. (1968). The buckling and post-buckling behavior of spherical caps under concentrated load, *Int. J. Solids. Structures* **4**: 421–446.

Flügge, W. (1972). *Tensor Analysis and Continuum Mechanics*, Springer-Verlag, Berlin, Heidelberg, New York.

Frandsen, P. M. (1946). *Elasticitetsteori*, Jul. Gjellerups Forlag, København.

Fung, Y. C. (1965). *Foundations of Solid Mechanics*, Prentice-Hall, Inc., Englewood Cliffs, New Jersey.

Fung, Y. C. & Tong, P. (2001). *Classical and Computational Solid Mechanics*, World Scientific Publishing Co. Pte. Ltd., Suite 1B, 1060 Main Street, River Edge, NJ 07661.

Galassi, M., Davies, J., Theiler, J., Gough, B., Jungman, G., Alken, P., Booth, M. & Rossi, F. (2009). *GNU Scientific Library — Reference Manual*, 1.13 edn.

Green, A. E. & Zerna, W. (1960). *Theoretical Elasticity*, Oxford University Press, Amen House, London E.C.4.

Hutchinson, J. W. (1972). On the postbuckling behavior of imperfection-sensitive structures in the plastic range, *Journal of Applied Mechanics* **39**: 155–162.

Hutchinson, J. W. (1973a). Imperfection-sensitivity in the plastic range, *Journal of Mechanics and Physics in Solids* **21**: 191–204.

Hutchinson, J. W. (1973b). Post-bifurcation behavior in the plastic range, *Journal of Mechanics and Physics in Solids* **21**: 163–190.

Hutchinson, J. W. (1974a). Notes on postbuckling and imperfection sensitivity. These notes were hand-outs at a course in Udine, Italy.

Hutchinson, J. W. (1974b). Plastic buckling, *Advances in Applied Mechanics* **14**: 67–144.

Hutchinson, J. W. & Amazigo, J. C. (1967). Imperfection-sensitivity of eccentrically stiffened cylindrical shells, *AIAA Journal* **5**(3): 392–401.

Hutchinson, J. W. & Koiter, W. T. (1970). Postbuckling theory, *Applied Mechanics Reviews* **23**: 1353–1366.

Johansen, K. W. (1963). *Yield Line Theory (in Danish)*, Dr.Techn. thesis, Copenhagen. Danish title: Brudlinieteorier. First edition 1943.

Johansen, K. W. (1972). *Yield Line Formulas for Slabs*, Cement and Concrete Association.

Kachanov, L. M. (1974). *Fundamentals of the Theory of Plasticity*, MIR Publishers, Moscow. Revised translation of the 1969 Russian edition.

Keillor, G. (1987). *Leaving Home*, Penguin Books, London, England; New York, New York; Markham, Ontario; Auckland, New Zealand.

Koiter, W. T. (1945). *On the Stability of Elastic Equilibrium*, H.J. Paris,

Amsterdam. In Dutch, also *NASA TT-F10,833*, (1967); also Air Force Flight Dynamics Laboratory, *AFFDL-TR-70-25*, (1970).

Koiter, W. T. (1966). Post-buckling analysis of a simple two-bar frame, *Recent Progress in Applied Mechanics*, Almquist and Wiksell.

Koiter, W. T. (1976). General theory of mode interaction in stiffened plate and shell structures, *Technical Report Rept. WTHD-91*, Delft University of Technology, Delft, The Netherlands.

Koiter, W. T. (2008). *W. T. Koiter's Elastic Stability of Solids and Structures*, Cambridge University Press. Also published online 2009.

Koiter, W. T. & Kuiken, G. D. C. (1971). The interaction between local buckling and overall buckling on the behaviour of built-up columns, *Technical Report Rept. WTHD-23*, Delft University of Technology, Delft, The Netherlands.

Koiter, W. T. & van der Neut, A. (1979). Interaction between local and overall buckling of stiffened compression panels, *TWS International Conference On Thin-Walled Structures*, University of Strathclyde, Scotland.

Kreyszig, E. (1993). *Advanced Engineering Mathematics, Seventh Edition*, John Wiley & Sons, Inc., New York, Chichester, Brisbane, Toronto, Singapore.

Kuznetsov, V. V. & Levyakov, S. V. (2002). Complete solution of the stability problem for elastica of Euler's column, *Int. J. Non-Linear Mechanics* **37**(6): 1003–1009.

Kyriakides, S. & Corona, E. (2007). *Mechanics of offshore pipelines: Buckling and collapse*, Elsevier.

Kyriakides, S. & Shaw, P. K. (1982). Response and stability of elastoplastic circular pipes under combined bending and external pressure, *Int. J. Solids Structures* **18**(11): 957–973.

Lakes, R. S. (1987). Foam structures with a negative poisson's ratio, *Science* **235**: 1038–1040.

Lévy, M. (1899). *Comptes rendus 9'Octbr.*, pp. 535–539.

Levy, M. & Salvadori, M. (1992). *Why Buildings Fall Down*, W.W. Norton & Company, New York, London.

Malvern, L. E. (1969). *Introduction to the Mechanics of a Continuous Medium*, Prentice-Hall, Inc., Englewood Cliffs, New Jersey.

Mau, S. T. & El-Mabsout, M. (1989). Inelastic buckling of reinforcing bars, *Journal of Engineering Mechanics* **115**: 1–17.

Melosh, R. J. (1963). Basis of derivation of matrices for the direct stiffness method, *AIAA Journal* **1**: 1631–1637.

Mikhlin, S. G. (1964). *Variational Methods in Mathematical Physics*, Pergamon Press, Oxford, Edinburgh, London, New York, Paris, Frankfurt.

Translated by T. Boddington.

Møllmann, H. & Goltermann, P. (1989a). Interactive buckling in thin-walled beams—i. theory, *Int. J. Solids Structures* **25**(7): 715–728.

Møllmann, H. & Goltermann, P. (1989b). Interactive buckling in thin-walled beams—ii. applications, *Int. J. Solids Structures* **25**(7): 729–749.

Muskhelishvili, N. I. (1963). *Some Basic Problems of the Mathematical Theory of Elasticity*, P. Noordhoff Ltd., Groningen, The Netherlands. Translated from the Russian by J.R.M. Radok.

Navier, C. M. L. H. (1827). *Memoires de l'Academie des Sciences* **7**.

Needleman, A. & Tvergaard, V. (1976). An analysis of the imperfection sensitivity of square elastic-plastic plates under compression,', *International Journal of Solids and Structures* **12**: 185–201.

Needleman, A. & Tvergaard, V. (1982). Aspects of plastic postbuckling behavior, *in* H. Hopkins & M. Sewell (eds), *Mechanics of Solids. The Rodney Hill 60th Anniversary Volume*, pp. 453–498.

Noor, A. K. & Peters, J. M. (1981). Mixed models and reduced/selective integration displacement models for nonlinear analysis of curved beams, *International Journal for Numerical Methods in Engineering* **17**: 615–631.

Ottosen, N. S. & Petersson, H. (1992). *Introduction to the Finite Element Method*, Prentice Hall, New York, London, Toronto, Tokyo, Sydney, Singapore.

Pearson, C. E. (1959). *Theoretical Elasticity*, Harvard University Press, Cambridge, Massachusetts.

Peek, R. & Kheyrkhahan, M. (1993). Postbuckling behavior and imperfection sensitivity of elastic structures by the Lyapunov-Schmidt-Koiter approach, *Computer Methods in Applied Mechanics and Engineering* **108**: 261–279.

Pian, T. H. H. (1964). Derivation of element stiffness matrices by assumed stress distribution, *AIAA Journal* **2**: 1333–1336.

Prager, W. & Hodge, Jr., P. G. (1968). *Theory of Perfectly Plastic Solids*, Dover Publications, Inc., New York. Paperback edition, first published in 1951 by John Wiley and Sons.

Prandtl, L. (1904). Eine neue Darstellung der Torsionsspannungen, *Jahresbericht d. deutsch. Math. Vereinigung* .

Reismann, H. & Pawlik, P. S. (1980). *Elasticity, Theory and Applications*, Wiley, New York.

Renton, J. D. (1991). Generalized beam theory applied to shear stiffness, *International Journal of Solids and Structures* **27**: 1955–1967.

Roorda, J. (1965). Stability of structures with small imperfections, *Journal of the Engineering Mechanics Division, ASCE (EMI)* **91**: 87–106.

Saint-Venant, B. (1855). *Mémoire sur la torsion des prismes*, Paris.

Shanley, F. R. (1947). Inelastic column theory, *Journal of the Aeronautical Sciences* **14**: 261–267.

Sokolnikoff, I. S. (1956). *Mathematical Theory of Elasticity*, second edn, McGraw-Hill Book Company, Inc., New York, Toronto, London.

Stephens, W. B. (1971). Imperfection sensitivity of axially compressed stringer reinforced cylindrical panels under internal pressure, *AIAA J.* **9**: 1713–1719.

Stolarski, H. & Belytschko, T. (1982). Membrane locking and reduced integration for curved elements, *Journal of Applied Mechanics* **49**: 172–176.

Strang, G. & Fix, G. J. (1973). *An Analysis of the Finite Element Method*, Prentice-Hall, Inc., Englewood Cliffs, N.J.

Sundström, B. (ed.) (2010). *Handbook of Solid Mechanics*, Department of Solid Mechanics, KTH, Stockholm, Sweden.

Thompson, J. M. T. & Hunt, G. W. (1973). *A General Theory of Elastic Stability*, Wiley, New York.

Timoshenko, S. P. & Gere, J. M. (1961). *Theory of Elastic Stability*, second edn, McGraw-Hill, New York, Toronto, London.

Timoshenko, S. P. & Goodier, J. N. (1970). *Theory of Elasticity*, third edn, McGraw-Hill, New York, Toronto, London.

Timoshenko, S. P. & Woinowsky-Krieger, S. (1959). *Theory of Plates and Shells*, second edn, McGraw-Hill, New York, Toronto, London.

Truesdell, C. (1965). *The Elements of Continuum Mechanics*, Springer-Verlag New York, Inc., New York.

Tvergaard, V. (1973a). Imperfection-sensitivity of a wide integrally stiffened panel under compression, *International Journal of Solids and Structures* **9**: 177–192.

Tvergaard, V. (1973b). Influence of post-buckling behaviour on optimum design of stiffened panels, *International Journal of Solids and Structures* **9**: 1519–1534.

Tvergaard, V. (1976). Buckling behaviour of plate and shell structures, *Proceedings of the 14th IUTAM Congress*, pp. 233–247.

Tvergaard, V. & Needleman, A. (1975). Buckling of eccentrically stiffened elastic-plastic panels on two simple supports or multiply supported, *International Journal of Solids and Structures* **11**: 647–663.

van der Neut, A. (1969). The interaction of local buckling and column failure of thin-walled compression members, *in* M. Hetényi & W. Vincent (eds), *Proceedings of the* 12th *International Congress on Applied Mechanics*, Springer-Verlag, Berlin, pp. 389–399.

Washizu, K. (1982). *Variational Methods in Elasticity and Plasticity*, Pergamon, Oxford.

Wunderlich, W., Obrecht, H. & Schrödter, V. (1986). Nonlinear analysis and elastic-plastic load-carrying behaviour of thin-walled spatial beam structures with warping constraints, *International Journal for Numerical Methods in Engineering* **22**: 671–695.

Yu, T. & Johnson, W. (1982). The Plastica: The large elastic-plastic deflections of a strut, *International Journal of Non-Linear Mechanics* **17**: 195–209. Online 2002.

Zienkiewicz, O. C. & Taylor, R. L. (2000a). *Finite Element Method—Volume 1: The Basis, Fifth Edition*, Butterworth-Heinemann, Oxford. xvi+669 pages.

Zienkiewicz, O. C. & Taylor, R. L. (2000b). *Finite Element Method—Volume 2: Solid Mechanics, Fifth Edition*, Butterworth-Heinemann, Oxford. xii+459 pages.

Index

E. Byskov, *Elementary Continuum Mechanics for Everyone*,
Solid Mechanics and Its Applications 194, DOI: 10.1007/978-94-007-5766-0,

C

D

F

G

H

I

J

K

L

M

N

Q

R

S

T

U

V

W

Y

Z